Glossary of Notation

Numerical
Analysis

SECOND EDITION

Richard L. Burden
Youngstown State University

J. Douglas Faires
Youngstown State University

Albert C. Reynolds
University of Tulsa

Prindle, Weber & Schmidt
Boston, Massachusetts

Prindle, Weber & Schmidt is a division of Wadsworth, Inc.

Library of Congress Cataloging in Publication Data
Burden, Richard L
    Numerical analysis.

    Bibliography: p.
    Includes index.
    1. Numerical analysis.   I. Faires, J. Douglas,
joint author.   II. Reynolds, Albert C., joint
author.   III. Title.
QA297.B84   1981   519.4   80-29558
ISBN 0-87150-314-X
ISBN 0-534-98020-1 (International Edition)

Printed in the United States of America.
Second printing: July, 1981

Text design by Eileen Katin and the staff of Prindle, Weber & Schmidt. Text composition in Monophoto Times Roman and Univers by Composition House Ltd. Printed and bound by Alpine Press. Cover design by Julie Gecha.

Preface

About the Text

The material in this text was developed over a period of years to meet the need for a sequence of courses in the theory and methods of numerical analysis. The students in this sequence are primarily junior-level science and engineering majors who have completed the basic college calculus sequence and have some knowledge of a high-level programming language, generally Fortran. Additional knowledge of the fundamentals of matrix theory and differential equations is helpful in the latter portions of the text, but adequate introductory material is presented to make these topics unnecessary as prerequisites.

Although sufficient material is included to serve as a basis for a full year of study on the undergraduate level, the text can also be used for an introductory single-term course in numerical methods. In such a course, the student learns the types of problems that can be solved by numerical techniques and the difficulties associated with these methods; the remainder of the text can be used as a reference for problems that may arise in future work. Either treatment is consistent with the philosophy of the text: to give the student an understanding of numerical methods, how and why they work, and their limitations—an understanding that will serve as a firm basis for future study or for adaptation of newer techniques that will be devised in the future.

Virtually every concept in the text is illustrated by an example, and this edition includes over 1000 class-tested exercises; these range from simple applications of the methods and algorithms to sophisticated generalizations and extensions of theory. In addition, a large number of applied problems are presented from the diverse areas of engineering and the physical, biological, and social sciences. Most of these problems are taken from the literature in areas where an approximation technique is required. The applications chosen are concise and demonstrate how the methods can and are applied in "real life" situations. We feel that students should be aware that the details which are often simplified in the classroom are in practice quite complicated.

Changes in the Second Edition

This edition contains an improved treatment of some topics in the first edition, a significant number of new topics, and a rearrangement of topics that more closely reflects the order of coverage by many of our colleagues. The introductory material concerning the solution to a single nonlinear equation (Chaper 2) has been condensed and simplified. Newton's backward difference formula has been added to Chapter 3 and subsequently used for the derivation of multistep methods in Chapter 5. The topics of numerical differentiation and integration (Chapter 4), numerical solutions of differential equations (Chapter 5), and direct methods for solving linear systems (Chapter 6) have been moved forward in the book since these topics are generally discussed in a one-term course. The treatment of approximation theory (Chapter 7) now follows these topics, and the discussion of discrete least squares has been rewritten to be more accessible to the beginning student.

The treatment of numerical integration has been expanded with the inclusion of new sections on adaptive quadrature and multiple integrals. A new section has been added (Chapter 5) discussing stiff differential equations, the extrapolation method for solving differential equations has been rewritten to be both more efficient and easier to employ, and the discussion on stability and error control has been expanded. The presentation of direct methods for solving linear systems (Chapter 5) has been rearranged to allow the reader with a background in introductory linear algebra to omit Section 6.3 with no loss of continuity.

The treatment of approximation theory has had significant reorganization. The material on orthogonal polynomials has been completely rewritten and condensed, without the omission of any important topics. New sections have been added discussing approximation by rational functions and trigonometric polynomials—including an algorithm for the Fast Fourier Transform. The iterative techniques for solving linear systems have been introduced earlier in Chapter 8, and the treatment of eigenvalue approximation techniques has been improved by the addition of the inverse power method and the replacement of the QR method by the more highly recommended QL method. A new section has been added to Chapter 9 to discuss quasi-Newton methods for solving nonlinear systems, including a secant update technique known as Broyden's method.

The final two chapters are structurally unchanged from the first edition, with the exception that the method of characteristics in Chapter 11 has been moved to a separate section following the finite element method.

Use of Algorithms

As in the previous edition, a detailed algorithm without program listing is given for each significant method presented in the text. The algorithms are presented in a form that can be coded by a student with even limited programming experience. The algorithms have been completely rewritten for this edition to reflect the trend toward the use of structured programs, particularly in view of the increased use of FORTRAN 77. Actual programs are not included in the text because it has been our experience that, in conflict with our aims, listing programs permits students to generate results without understanding the method involved. A Fortran listing of the programs is, however, included in a manual for instructors using the text, which can be obtained by contacting the publisher. Most of the results listed in the answer section, as well as the example listings, were generated using these programs, run in Structured WATFIV on an AMDAHL 470 V/5 computer at Youngstown State University. When the primary concern was to illustrate truncation error, the programs were run using double precision; when the stability of a technique was being illustrated, the example and exercise results were generated using single precision.

It should be emphasized that while the algorithms as listed lead to correct logical programs, they are not always listed in the most efficient manner in terms of time or storage requirements for computer implementation. When the listing of an algorithm in a form that leads to an extremely efficient program conflicts with the ability to understand how the method operates, the most understandable form has been chosen. If a general-purpose program library is needed, a number of high-quality software packages are commercially available and have been adapted to run on all common computers. The most complete source of reference on this software is IMSL (International Mathematical and Statistical Software Library, Inc.,), Sixth Floor, NBC Building, 7500 Bellaire Blvd., Houston, Texas 77036. Their basic library program is continually expanding, and at the time this text was prepared, contained over 400 supported mathematical and statistical subroutines written in ANSI FORTRAN. In addition, they serve as a distribution center for high-quality software provided by other sources.

Suggested Course Outlines

The text has been designed to allow instructors flexibility in choice of topics and in the levels of rigor and application. In line with these aims, detailed references are listed for all results not specifically demonstrated in the text, as well as for extensions of the theory and application. The references have been chosen, when possible, on the basis of being the most generally available sources in college libraries.

Listed below is a flow chart indicating approximate chapter prerequisites. Most of the possible sequences that can be generated from this flow chart have been successfully taught to undergraduate students at Youngstown State University.

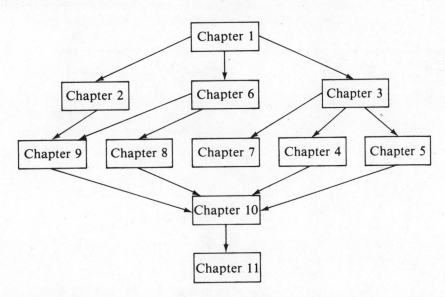

Acknowledgments

A number of people deserve our thanks for making the preparation of this edition possible. First, we would like to thank the students at Youngstown State University for their diligence in locating errors in the first edition and their candid remarks regarding improvements. We would like to express our appreciation to our assistants and typists Kathy Bosak, Pam Spon, and Rosa Bianco, who helped put this book together. We also wish to extend thanks to our colleagues and reviewers for making valuable suggestions for improvement: William Briggs, Richard Falk, Myron Henry, Joseph Hintz, Thomas Price, David Rosen, Sherwood Silliman. We single Joe Hintz from this list to express our deep appreciation for his detailed review of the text and algorithms, a review he was forced to prepare at very short notice.

In addition, we would like to express our thanks to Ann Burden, who has worked diligently on the algorithms and programs, and to Barbara Faires, who proofread the entire manuscript and worked around the clock with us on the last few days of preparation.

Finally, our thanks goes to our editor, Theron Shreve, who extended himself far beyond what is required of his position. He is a quality individual in a quality organization.

Richard L. Burden
J. Douglas Faires

Contents

Introduction

The subject of numerical analysis is concerned with devising methods for approximating, in an efficient manner, the solutions to mathematically expressed problems. The efficiency of the method depends both upon the accuracy required of the method and the ease with which it can be implemented. In a practical situation, the mathematical problem is derived from a physical phenomenon where some simplifying assumptions have been made to allow the mathematical representation to develop. Generally a relaxation on the physical assumptions leads to a more appropriate mathematical model, but at the same time one which is more difficult or impossible to solve explicitly. Since the mathematical problem ordinarily does not solve the physical problem exactly in any case, it is often more appropriate to find an approximate solution to a more complicated mathematical model of a physical problem than to find an exact solution of a simplified model. To obtain such an approximation a method called an algorithm is devised. The algorithm consists of a sequence of algebraic and logical operations that produces the approximation to the mathematical problem, and, it is hoped, to the physical problem as well, within a prescribed tolerance or accuracy.

Since the efficiency of a method depends upon its ease of implementation, the choice of the appropriate method for approximating the solution to a problem is influenced significantly by changes in calculator and computer technology. Twenty-five years ago, before the widespread use of digital computing equipment, methods requiring a large amount of computational effort could not be reasonably applied. Since that time, however, the advances in computing equipment have made some of these methods increasingly attractive. At present, the limiting factor generally involves the amount of computer storage requirements of the method, although the cost factor associated with a large amount of computation time is, of course, also important. The recent availability of relatively low-cost calculators with programmable characteristics is also an influencing factor in the choice of an approximation method, since these can be used to solve many relatively simple problems and are readily accessible.

The basic ideas which underlie most current numerical techniques, however, have been known for some time, as have the methods used in predicting bounds for the maximum error that can be produced in an application of the methods. It is of primary interest, then, to determine the way in which these methods have developed and how their error can be estimated, since variations of these techniques will undoubtedly be used to develop and apply numerical procedures in the future, irrespective of the technology.

The methods we will discuss in this text include those which are commonly used at the present time and those on which improvements will most likely be based in the near future.

1

1

Mathematical Preliminaries

In beginning chemistry courses, students are confronted with a relationship known as the *ideal gas law*,

$$PV = NRT,$$

which relates the pressure P, volume V, temperature T, and number of moles N of an "ideal" gas. The R in this equation is a constant depending only on the measurement system being used.

Suppose two experiments are conducted to test this law, using the same gas in each case. In the first experiment,

$$P = 1.0 \text{ atmosphere}, \quad V = .10 \text{ cubic meter},$$
$$N = .0042 \text{ mole}, \quad R = .082.$$

Using the ideal gas law, we predict the temperature of the gas to be

$$T = \frac{PV}{NR} = \frac{(1.0)(.10)}{(.082)(.0042)} = 290° \text{ Kelvin or } 17° \text{ Celsius}.$$

When we measure the temperature of the gas, we find that the true temperature is 15° Celsius.

The experiment is then repeated, using the same values of R and N, but increasing the pressure by a factor of four while reducing the volume by the same factor. Since the product PV remains the same, the predicted temperature would still be 290° Kelvin (or 17° Celsius), but now we find that the actual temperature of the gas is 32° Celsius.

Clearly the ideal gas law is suspect when an error of this magnitude is obtained. Before concluding that the law is invalid in this situation, however, we should examine our data to determine whether the error could be attributed to the experimental results. If so, it would be of interest to determine how much more accurate our experimental results would need to be to ensure that an error of this magnitude could not occur.

This type of analysis of the error involved in calculations is an important topic in numerical analysis and will be introduced in the second section of this chapter.

This chapter contains a short review of those topics from elementary single-variable calculus that will repeatedly be needed in later chapters, together with an introduction to the terminology used in discussing convergence, error analysis, and the machine representation of numbers.

1.1   Review of Calculus

Fundamental to the study of calculus are the concepts of **limit** and **continuity** of a function.

Definition 1.1   Let f be a function defined on a set X of real numbers; f is said to have the **limit** L at x_0, written $\lim_{x \to x_0} f(x) = L$, if, given any real number $\varepsilon > 0$, there exists a real number $\delta > 0$ such that $|f(x) - L| < \varepsilon$, whenever $x \in X$ and $0 < |x - x_0| < \delta$. (See Fig. 1.1.)

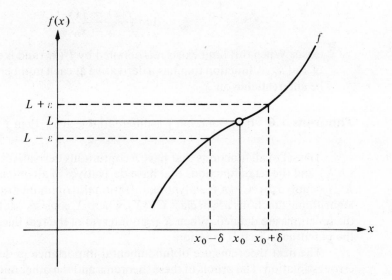

FIGURE 1.1

Definition 1.2   Let f be a function defined on a set X of real numbers and $x_0 \in X$; f is said to be **continuous** at x_0 if $\lim_{x \to x_0} f(x) = f(x_0)$. The function f is said to be continuous on X if it is continuous at each number in X; $C(X)$ denotes the set of all functions continuous on X. When X is an interval of the real line, the parentheses in this notation will be omitted. For example, the set of all functions continuous on the closed interval $[a, b]$ will be denoted $C[a, b]$.

In a similar manner, the **limit of a sequence** of real or complex numbers can be defined.

Definition 1.3   Let $\{x_n\}_{n=1}^{\infty}$ be an infinite sequence of real or complex numbers. The sequence is said to **converge** to a number x (called the limit) if, for any $\varepsilon > 0$, there exists a positive integer $N(\varepsilon)$ such that $n > N(\varepsilon)$ implies

$|x_n - x| < \varepsilon$. The notation $\lim_{n \to \infty} x_n = x$, or $x_n \to x$ as $n \to \infty$, means that the sequence $\{x_n\}_{n=1}^{\infty}$ converges to x.

The following theorem relates the concepts of convergence and continuity.

Theorem 1.4   If f is a function defined on a set X of real numbers and $x_0 \in X$, then the following are equivalent:

a) f is continuous at x_0;

b)   if $\{x_n\}_{n=1}^{\infty}$ is any sequence in X converging to x_0, then

$$\lim_{n \to \infty} f(x_n) = f(x_0).$$

Definition 1.5   If f is a function defined in an open interval containing x_0, f is said to be **differentiable** at x_0 if

$$\lim_{x \to x_0} \frac{f(x) - f(x_0)}{x - x_0}$$

exists. When this limit exists it is denoted by $f'(x_0)$ and is called the **derivative** of f at x_0. A function that has a derivative at each number in a set X is said to be **differentiable** on X.

Theorem 1.6   If the function f is differentiable at x_0, then f is continuous at x_0.

The set of all functions that have n continuous derivatives on X is denoted by $C^n(X)$, and the set of functions that have derivatives of all orders at each number in X is denoted by $C^{\infty}(X)$. Polynomial, rational, trigonometric, exponential, and logarithmic functions are in class $C^{\infty}(X)$, where X consists of all numbers at which the functions are defined. When X is an interval of the real line, we will again omit the parentheses in this notation.

The next theorems are of fundamental importance in deriving methods for error estimation. The proofs of these theorems and the other unreferenced results in this section can be found in any elementary calculus text.

Theorem 1.7 (Rolle's Theorem)   Suppose $f \in C[a, b]$ and f is differentiable on (a, b). If $f(a) = f(b) = 0$, then a number c, $a < c < b$, exists with $f'(c) = 0$. (See Fig. 1.2.)

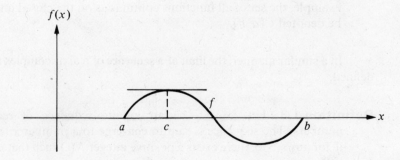

FIGURE 1.2

Theorem 1.8 (Mean Value Theorem)   If $f \in C[a, b]$ and f is differentiable on (a, b), then a number c, $a < c < b$, exists such that

$$f'(c) = \frac{f(b) - f(a)}{b - a}. \qquad \text{(See Fig. 1.3.)}$$

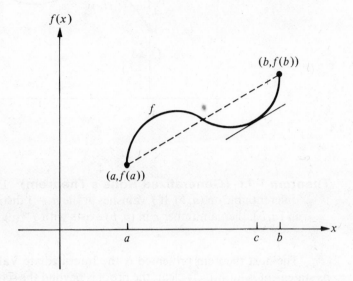

FIGURE 1.3

Theorem 1.9 (Extreme Value Theorem)   If $f \in C[a, b]$, then $c_1, c_2 \in [a, b]$ exist with $f(c_1) \leq f(x) \leq f(c_2)$ for each $x \in [a, b]$. If, in addition, f is differentiable on (a, b), then either $c_i = a$, $c_i = b$, or $f'(c_i) = 0$ for each $i = 1, 2$.

Two other results will be needed in our study of numerical methods. The first is a generalization of the usual Mean Value Theorem for Integrals.

Theorem 1.10 (Weighted Mean Value Theorem for Integrals) If $f \in C[a, b]$, g is integrable on $[a, b]$, and $g(x) \geq 0$, then there exists a number c, $a < c < b$, such that:

$$\int_a^b f(x)g(x) \, dx = f(c) \int_a^b g(x) \, dx.$$

When $g(x) \equiv 1$, this theorem gives what is called the **average value** of the function over the interval $[a, b]$ (see Fig. 1.4). The proof of Theorem 1.10 is not generally given in a basic calculus course, but can be found in any standard advanced calculus text (see, for example, Fulks [40], page 125).

The other theorem we will need that is not generally presented in a basic calculus course is derived by applying Rolle's Theorem (Theorem 1.7) successively to $f, f', \ldots$, and finally to $f^{(n-1)}$.

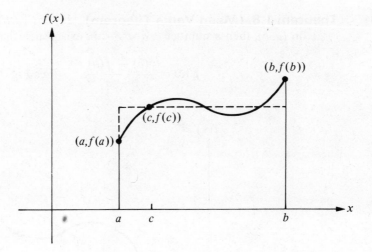

FIGURE 1.4

Theorem 1.11 (Generalized Rolle's Theorem)   Let $f \in C[a,b]$ be n times differentiable on (a, b). If f vanishes at the $n + 1$ distinct numbers $x_0, \ldots, x_n$ in $[a, \underline{b}]$, then a number c in (a, b) exists with $f^{(n)}(c) = 0$.

The next theorem presented is the Intermediate Value Theorem. Although its statement is intuitively clear, the proof is beyond the scope of the usual calculus course. The proof can be found in most advanced calculus texts (see, for example, Fulks [40], page 59).

Theorem 1.12 (Intermediate Value Theorem)   If $f \in C[a, b]$ and K is any number between $f(a)$ and $f(b)$, then there exists c in (a, b) for which $f(c) = K$ (see Fig. 1.5).

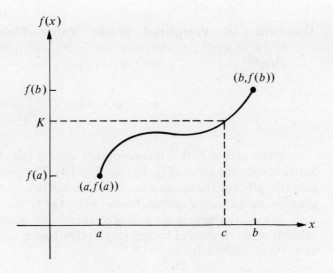

FIGURE 1.5

EXAMPLE 1 To show that $x^5 - 2x^3 + 3x^2 - 1 = 0$ has a solution on the interval $[0, 1]$, consider the function $f(x) = x^5 - 2x^3 + 3x^2 - 1$. Clearly f is continuous on $[0, 1]$ and $f(0) = -1$ while $f(1) = 1$. Since $f(0) < 0 < f(1)$, the Intermediate Value Theorem implies that there is a number x, with $0 < x < 1$, for which $x^5 - 2x^3 + 3x^2 - 1 = 0$. □

As seen in Example 1, the Intermediate Value Theorem is important as an aid to determine when solutions to certain problems exist. It does not, however, give a means for finding these solutions. This topic will be discussed more thoroughly in Chapter 2.

The final theorem in this review from calculus describes the development of the Taylor polynomials. The importance of the Taylor polynomials to the study of numerical analysis cannot be overemphasized, and the following result will be used repeatedly. An indication of the proof of this theorem will be given in Section 3.2, Exercise 13.

Theorem 1.13 (Taylor's Theorem) Suppose $f \in C^n[a, b]$ and $f^{(n+1)}$ exists on $[a, b]$. Let $x_0 \in [a, b]$. For every $x \in [a, b]$, there exists $\xi(x)$ between x_0 and x with

$$f(x) = P_n(x) + R_n(x),$$

where

$$P_n(x) = f(x_0) + f'(x_0)(x - x_0) + \frac{f''(x_0)}{2!}(x - x_0)^2 + \cdots$$

$$+ \frac{f^{(n)}(x_0)}{n!}(x - x_0)^n$$

$$= \sum_{k=0}^{n} \frac{f^{(k)}(x_0)}{k!}(x - x_0)^k,$$

and

$$R_n(x) = \frac{f^{(n+1)}(\xi(x))}{(n+1)!}(x - x_0)^{n+1}.$$

Here $P_n(x)$ is called the **nth-degree Taylor polynomial** for f about x_0 and $R_n(x)$ is called the **remainder term** (or **truncation error**) associated with $P_n(x)$. The infinite series obtained by taking the limit of $P_n(x)$ as $n \to \infty$ is called the **Taylor Series** for f about x_0.

The term **truncation error** generally refers to the error involved in using a truncated or finite summation to approximate the sum of an infinite series. This terminology will be reintroduced in subsequent chapters.

EXAMPLE 2 Let $f(x) = \cos x$. Since $f \in C^\infty(R)$, Theorem 1.13 can be applied for any $n > 0$. For $n = 2$ and $x_0 = 0$, Theorem 1.13 gives

$$\cos x = 1 - \tfrac{1}{2}x^2 + \tfrac{1}{6}x^3 \sin \xi(x),$$

where $\xi(x)$ is a number between 0 and x.

With $x = .001$, the Taylor polynomial and remainder term is

$$\cos .001 = 1 - \tfrac{1}{2}(.001)^2 + \tfrac{1}{6}(.001)^3 \sin \xi(x)$$
$$= .9999995 + (.16\bar{6}) \cdot 10^{-9} \sin \xi(x),$$

where $0 < \xi(x) < .001$. (The bar over the last digit in .166 is used to indicate that this digit repeats indefinitely.)

Since $|\sin \xi(x)| < 1$, .9999995 can be used as an approximation to $\cos .001$ with assurance of at least nine decimal-place accuracy. Using standard tables, it can be found that

$$\cos .001 = .999999500000042;$$

so there is actually 13-decimal-place accuracy.

If, in this example, the third-degree Taylor polynomial had been used with $x_0 = 0$, then

$$\cos x = 1 - \tfrac{1}{2}x^2 + \tfrac{1}{24}x^4 \cos \xi(x),$$

where $0 < \xi(x) < .001$, since $f'''(0) = 0$. The approximating polynomial remains the same, and the approximation would still be .9999995, but 13-decimal-place accuracy would be expected since

$$\left| \tfrac{1}{24}x^4 \cos \xi(x) \right| \le \tfrac{1}{24}(.001)^4 (1) \approx 4.2 \times 10^{-14}.$$

This corresponds more closely to the actual accuracy obtained. ◻

Exercise Set 1.1

1. Let $f(x) = 1 - e^x + (e - 1)\sin((\pi/2)x)$. Show that $f'(x)$ is zero at least once in $[0, 1]$.

2. Show that the equation $x = 3^{-x}$ has a solution in $[0, 1]$.

3. Find the Taylor polynomial of degree four for f expanded about $x_0 = 0$ if $f(x) = e^x \cos x$. Use this polynomial to approximate $f(\pi/16)$, and find a bound for the error in this approximation.

4. Find the Taylor polynomial of degree four for f expanded about $x_0 = 1$ if $f(x) = e^{x^2}$. Use this polynomial to approximate $f(1.1)$, and find a bound for the error in this approximation.

5. Let $f(x) = e^{-x}$. Find the third-degree Taylor polynomial for f expanded about $x_0 = 1$, and approximate $e^{-.99}$ using the Taylor polynomial. How many decimal places of accuracy are expected?

6. Let $f(x) = |x|$ for $x \in R$. Show that f is differentiable at any $x \ne 0$ but that f is not differentiable at $x = 0$.

7. Determine whether the Mean Value Theorem, Theorem 1.8, applies in the following situations. If so, find a number c satisfying the conclusion of this theorem; if not, show that no such number exists.

    a) $f(x) = x^{2/3}$, $[a, b] = [-1, 8]$ b) $f(x) = x^{2/3}$, $[a, b] = [0, 8]$
    c) $f(x) = |x|$, $[a, b] = [-1, 1]$ d) $f(x) = |x|$, $[a, b] = [0, 1]$.

8. Using Taylor's Theorem, Theorem 1.13, with $n = 2$ and $x_0 = 0$, find an approximation for $\sin .001$. Do you expect the accuracy in this problem to be that obtained for $\cos .001$ in Example 2?

9. A function $f : [a, b] \to R$ is said to satisfy a Lipschitz condition with Lipschitz constant L on $[a, b]$ if, for every $x, y \in [a, b]$,

$$|f(x) - f(y)| \le L|x - y|.$$

a) Show that if f satisfies a Lipschitz condition with Lipschitz constant L on an interval $[a, b]$, then $f \in C[a, b]$.

b) Show that if f has a derivative that is bounded on $[a, b]$ by L, then f satisfies a Lipschitz condition with Lipschitz constant L on $[a, b]$.

c) Give an example of a function that is continuous on a closed interval but does not satisfy a Lipschitz condition on the interval.

1.2 Round-Off Errors and Computer Arithmetic

When a calculator or digital computer is used to perform numerical calculations, an unavoidable error, called **round-off error**, must be considered. This error arises because the arithmetic performed in a machine involves numbers with only a finite number of digits, with the result that many calculations are performed with approximate representations of the actual numbers. In a typical computer, only a relatively small subset of the real number system is used for the representation of all real numbers. This subset contains only rational numbers, both positive and negative, and stores a fractional part, called the **mantissa**, together with an exponential part, called the **characteristic**. For example, a single-precision floating-point number used in the IBM 370 or 3000 series consists of a 1-binary-digit sign indicator, a 7-binary-digit exponent with a base of 16, and a 24-binary-digit mantissa. Since 24 binary digits correspond to between 6 and 7 decimal digits, we can assume that this number has at least six decimal digits of precision for the floating-point number system. The exponent of seven binary digits gives a range of 0 to 127, but because of an exponential bias the range is actually -64 to $+63$, that is, 64 is automatically subtracted from the listed exponent.

The **machine number**

0	1000010	10110011000001000000000

precisely represents the decimal number

$$+ \left(\left(\frac{1}{2}\right)^1 + \left(\frac{1}{2}\right)^3 + \left(\frac{1}{2}\right)^4 + \left(\frac{1}{2}\right)^7 + \left(\frac{1}{2}\right)^8 + \left(\frac{1}{2}\right)^{14} \right) \times 16^{66-64} = 179.015625$$

since the first binary digit represents the sign, 0 for plus and 1 for minus, the next seven binary digits represent the exponent, and the last twenty-four binary digits represent the mantissa. This machine number is actually used to represent any real number in the interval

$$[179.01561737060546875, \ 179.01563262939453125]$$

since the next smallest machine number is

0	1000010	10110011000000111111111	$= 179.0156097412109375$

and the next largest machine number is

0	1000010	10110011000001000000001	= 179.0156402587890625.

With this representation, the smallest positive number that can be expressed is $16^{-64} \approx 10^{-77}$, and the largest is $16^{63} \approx 10^{76}$. At least one of the four leftmost binary digits for any nonzero number greater than 16^{-64} is required to be one. Consequently, there are 15×2^{28} numbers of the form

$$\pm .d_1 d_2 \cdots d_{24} \times 16^{e_1 e_2 \cdots e_7},$$

which are used by this system to represent all real numbers. Numbers occurring in calculations that have a magnitude of less than 16^{-64} result in what is called **underflow**, and are often set to zero, while numbers greater than 16^{63} result in an **overflow** condition and cause the computations to halt.

The number representation system described above is not standard for all computing machines, but gives an indication of the possible difficulties that can occur. For the remainder of this discussion, we will, for simplicity, assume that machine numbers are represented in the normalized decimal form

$$(1.1) \qquad \pm .d_1 d_2 \cdots d_k \times 10^n, \qquad 1 \le d_1 \le 9, 0 \le d_i \le 9,$$

for each $i = 2, \ldots, k$, where, from what we have just discussed, the **IBM** machines have approximately $k = 6$ and $-77 \le n \le 76$.

It is useful to consider the representation of an arbitrary real number in the **floating-point form** (1.1). Any positive real number y can be normalized to achieve the form

$$y = .d_1 d_2 \cdots d_k d_{k+1} d_{k+2} \cdots \times 10^n,$$

if we assume y is within the numerical range of the machine. The floating-point form (1.1), denoted by $fl(y)$, is obtained by terminating the mantissa of y at k decimal digits. There are two ways of performing this termination. One method is to simply chop off the digits $d_{k+1} d_{k+2} \cdots$ to obtain

$$fl(y) = .d_1 d_2 \cdots d_k \times 10^n.$$

This method is quite accurately called **chopping** the number. The other method is to add $5 \times 10^{n-(k+1)}$ to y and then chop to obtain

$$fl(y) = .\delta_1 \delta_2 \cdots \delta_k \times 10^n.$$

The latter method is often referred to as **rounding** the number. In this method if $d_{k+1} \ge 5$, we add one to d_k to obtain $fl(y)$; that is, we round up. If $d_{k+1} < 5$, we merely chop off all but the first k digits; so we round down. For example, if $k = 5$ and rounding is used, we represent π and e as $.31416 \times 10^1$ and $.27183 \times 10^1$, respectively. If $k = 5$ and chopping is used, we represent π and e as $.31415 \times 10^1$ and $.27182 \times 10^1$, respectively.

Since the real numbers with which we are familiar cannot always be represented exactly inside a machine, it is necessary to consider the error due to this finite-digit approximation. The following definition specifies two methods for measuring approximation errors. These methods will be used throughout the text.

Definition 1.14   If p^* is an approximation to p, **absolute error** is given by $|p - p^*|$, and the **relative error** is given by $|p - p^*|/|p|$, provided that $p \neq 0$.

Consider the absolute and relative errors in representing p by p^* in the following example.

EXAMPLE 1

a)   If $p = .3000 \times 10$ and $p^* = .3100 \times 10$, the absolute error is .1 and the relative error is $.333\overline{3} \times 10^{-1}$.

b)   If $p = .3000 \times 10^{-3}$ and $p^* = .3100 \times 10^{-3}$, the absolute error is $.1 \times 10^{-4}$ and the relative error is $.333\overline{3} \times 10^{-1}$.

c)   If $p = .3000 \times 10^4$ and $p^* = .3100 \times 10^4$, the absolute error is $.1 \times 10^3$ and the relative error is $.333\overline{3} \times 10^{-1}$.

This example shows that the same relative error, $.333\overline{3} \times 10^{-1}$, occurs for widely varying absolute errors. Consequently, as a measure of accuracy, the absolute error may be misleading and the relative error more meaningful. As the following definition indicates, the relative error can be used to tell something about the number of correct digits of an approximation or representation.   □

Returning to the machine representation of numbers we see that the floating-point representation $fl(y)$ for the number y has the relative error

$$\left| \frac{y - fl(y)}{y} \right|.$$

Using k decimal digits in the representation produces an error bound of 10^{-k+1} for chopping and 5×10^{-k} for rounding.

Definition 1.15   The number p^* is said to approximate p to t **significant digits** (or figures) if t is the largest nonnegative integer for which

$$\frac{|p - p^*|}{|p|} < 5 \times 10^{-t}.$$

The reason for using relative error in the definition is to obtain a continuous concept. Consider the numbers 1000, 5000, 9990, and 10,000. In order for p^* to approximate 1000 to four significant figures by this definition, p^* has to satisfy

$$\left| \frac{p^* - 1000}{1000} \right| < 5 \times 10^{-4}.$$

This requires that $999.5 < p^* < 1000.5$ and agrees with the intuitive definition of significant digits. Considering this same problem for the numbers 5000 and 9990, p^* would need to satisfy $4997.5 < p^* < 5002.5$ and $9985.005 < p^* < 9994.995$, respectively, in order to be accurate to four significant digits. This may not agree with the intuitive idea of significant digits. Note, however, that for p^* to be a four-significant-digit approximation to 10,000, p^* must satisfy $9995 < p^* < 10005$, which again agrees with intuition. The following table illustrates the continuous nature of this concept by listing, for the various values of p, the least upper bound of $|p - p^*|$, denoted $\max|p - p^*|$, when p^* agrees with p to four significant digits.

p	.1	.5	100	1000	5000	9990	10000
$\max|p - p^*|$	.00005	.00025	.05	.5	2.5	4.995	5.

In addition to inaccurate representation of numbers, the arithmetic performed in a computer is not exact. The arithmetic generally involves manipulating binary digits, or **bits**, by various shifting or logical operations. Since the actual mechanics of these operations are not pertinent to this presentation, we shall devise our own approximation to computer arithmetic. Although our arithmetic will not give the exact picture, it should suffice to explain the problems that occur. (For an explanation of the manipulations actually involved, the reader is urged to consult more technically oriented computer-science texts, such as Chu [27], *Introduction to Computer Organization*.)

Assume the floating-point representations $fl(x)$ and $fl(y)$ are given for the real numbers x and y and that the symbols $\oplus, \ominus, \otimes, \oslash$ represent the machine addition, subtraction, multiplication, and division operations, respectively.

We will assume a k-digit arithmetic given by

$$x \oplus y = fl(fl(x) + fl(y)),$$
$$x \ominus y = fl(fl(x) - fl(y)),$$
$$x \otimes y = fl(fl(x) \times fl(y)),$$
$$x \oslash y = fl(fl(x) \div fl(y)).$$

This idealized arithmetic corresponds to performing exact arithmetic on the floating-point representations of x and y and then converting the exact result to its floating-point representation.

EXAMPLE 2 Suppose that $x = \frac{1}{3}$, $y = \frac{5}{7}$ and that 5-digit chopping is used for arithmetic calculations involving x and y. Table 1.1 lists the values of these computer-type operations on $fl(x) = .33333$ and $fl(y) = .71428$.

TABLE 1.1

Operation	Result	Actual value	Absolute error	Relative error
$x \oplus y$	$.10476 \times 10^1$	22/21	$.190 \times 10^{-4}$	$.182 \times 10^{-4}$
$y \ominus x$	$.38095$	8/21	$.238 \times 10^{-5}$	$.625 \times 10^{-5}$
$x \otimes y$	$.23809$	5/21	$.524 \times 10^{-5}$	$.220 \times 10^{-4}$
$y \oslash x$	$.21428 \times 10^1$	15/7	$.571 \times 10^{-4}$	$.267 \times 10^{-4}$

Since the maximum relative error for these operations is $.267 \times 10^{-4}$, the arithmetic produces satisfactory five-digit results. Suppose, however, that we also have $u = .714251$, $v = 98765.9$, and $w = .111111 \times 10^{-4}$ so that $fl(u) = .71425$, $fl(v) = .98765 \times 10^5$, and $fl(w) = .11111 \times 10^{-4}$. These numbers were chosen to illustrate some problems that can arise with finite-digit arithmetic. In Table 1.2, $y \ominus u$ results in a small absolute error but a large relative error. The subsequent

TABLE 1.2

Operation	Result	Actual value	Absolute error	Relative error
$y \ominus u$	$.30000 + 10^{-4}$	$.34714 \times 10^{-4}$	$.471 \times 10^{-5}$	$.136$
$(y \ominus u) \oplus w$	$.27000 \times 10^{1}$	$.31243 \times 10^{1}$	$.424$	$.136$
$(y \ominus u) \otimes v$	$.29629 \times 10^{1}$	$.34285 \times 10^{1}$	$.465$	$.136$
$u \oplus v$	$.98765 \times 10^{5}$	$.98766 \times 10^{5}$	$.161 \times 10^{1}$	$.163 \times 10^{-4}$

division by the small number w or multiplication by the large number v magnifies the absolute error without modifying the relative error. The addition of the large and small numbers u and v produces large absolute error but not large relative error.

The results in this table serve to illustrate that the ordering of arithmetic operations should be arranged to avoid subtracting nearly equal numbers, dividing by very small numbers, or multiplying by very large numbers, whenever possible. □

The loss of significant digits can often be avoided by a reformulation of the problem.

EXAMPLE 3   Evaluate $f(x) = x^3 - 6x^2 + 3x - .149$ at $x = 4.71$ using three-digit arithmetic.

	x	x^2	x^3	$6x^2$	$3x$
Exact	4.71	22.1841	104.487111	133.1046	14.13
3-digit (chopping)	4.71	22.1	104.	132.	14.1
3-digit (rounding)	4.71	22.2	105.	133.	14.1

Note that the three-digit chopping values simply retain the leading three digits, with no rounding involved, and differ significantly from the three-digit rounding values.

Exact:
$$f(4.71) = 104.487111 - 133.1046 + 14.13 - .149$$
$$= -14.636489;$$

3-digit: (chopping)
$$f(4.71) = 104. - 132. + 14.1 - .149$$
$$= -14.0;$$

3-digit: (rounding)
$$f(4.71) = 105. - 133. + 14.1 - .149$$
$$= -14.0.$$

The relative error for both the three-digit methods is

$$\left| \frac{-14.636489 + 14.0}{-14.636489} \right| \approx .04.$$

As an alternative approach, $f(x)$ could be written:

$$f(x) = x^3 - 6x^2 + 3x - .149 = ((x - 6)x + 3)x - .149,$$

which gives

3-digit
(chopping) $f(4.71) = ((4.71 - 6)4.71 + 3)4.71 - .149 = -14.5,$

and a 3-digit rounding answer of -14.6. The new relative errors are

3-digit
(chopping) $\left| \dfrac{-14.636489 + 14.5}{-14.636489} \right| \approx .0093$

and

3-digit
(rounding) $\left| \dfrac{-14.636489 + 14.6}{-14.636489} \right| \approx .0025.$

Although both three-digit chopping approximations are correct to two significant figures, the relative error has been reduced to about one-fourth that of the original procedure. In the case of the three-digit rounding procedure, an additional significant figure has been obtained.

The decreased error is due to the fact that the number of computations has been reduced from four multiplications and three additions to two multiplications and three additions. Evidently, one way to reduce this error is to reduce the number of error-producing computations. □

The preceding examples demonstrate ways that machine calculations involving approximations can result in the growth of rounding errors. Throughout this text we will be examining approximation procedures involving sequences of calculations called **algorithms**. We are primarily interested in choosing methods that will produce dependably accurate results. One criteria we will impose on an algorithm whenever possible is that small changes in the initial data produce correspondingly small changes in the final results. An algorithm that satisfies this property is called **stable**; it is **unstable** when this criterion is not fulfilled. Some algorithms will be stable for certain choices of initial data but not for all choices. We will attempt to characterize the stability properties of algorithms whenever possible.

To consider further the subject of rounding error growth and its connection to algorithm stability, suppose that an error ε is introduced at some stage in the calculations and that the error after n subsequent operations is denoted by E_n. Two cases arise often in practice and are defined below.

Definition 1.16   If $|E_n| \approx Cn\varepsilon$, where C is a constant independent of n, the growth of error is said to be **linear**. If $|E_n| \approx k^n\varepsilon$, for some $k > 1$, the growth of error is **exponential**.

Linear growth of error is usually unavoidable, and when C and ε are small, the results are generally acceptable. Exponential growth of error should be avoided, since the term k^n becomes large for even relatively small values of n. This leads to unacceptable inaccuracies, regardless of the size of ε. As a consequence, an algorithm

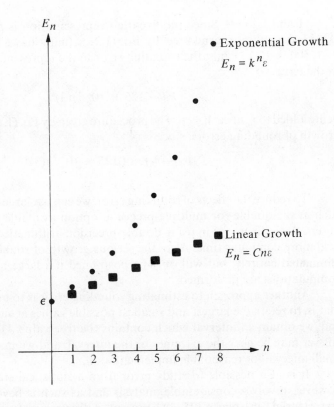

E_n

• Exponential Growth

$E_n = k^n \varepsilon$

■ Linear Growth

$E_n = Cn\varepsilon$

ϵ

n

1   2   3   4   5   6   7   8

FIGURE 1.6

that exhibits linear growth of error is stable, while an algorithm exhibiting exponential-error growth is unstable. (See Fig. 1.6.)

EXAMPLE 4   Consider the sequence of numbers $1, \frac{1}{3}, \frac{1}{9}, \frac{1}{27}, \frac{1}{81}, \ldots$. This sequence can be generated by defining $p_0 = 1$, $p_1 = \frac{1}{3}$, and

(1.2) $$p_n = \tfrac{10}{3} p_{n-1} - p_{n-2}.$$

If $\{p_n\}$ is generated using five-digit arithmetic with rounding, we obtain the values shown in Table 1.3.

Equation (1.2) is satisfied by $p_n = C_1(\frac{1}{3})^n + C_2(3)^n$ for any real numbers C_1 and C_2. In order to have $p_0 = 1$ and $p_1 = \frac{1}{3}$, we must choose these constants as

TABLE 1.3

n	Computed p_n	Correct value p_n
0	1.0000	1.0000
1	.33333	.33333
2	.11110	.11111
3	$.37000 \times 10^{-1}$	$.37037 \times 10^{-1}$
4	$.12230 \times 10^{-1}$	$.12346 \times 10^{-1}$
5	$.37660 \times 10^{-2}$	$.41152 \times 10^{-2}$
6	$.32300 \times 10^{-3}$	$.13717 \times 10^{-2}$
7	$-.26893 \times 10^{-2}$	$.45725 \times 10^{-3}$
8	$-.92872 \times 10^{-2}$	$.15242 \times 10^{-3}$

$C_1 = 1$ and $C_2 = 0$. Since the five-digit representation is $\hat{p}_0 = 1.0000$ and $\hat{p}_1 = .33333$, the solution generated by Eq. (1.2) actually has $C_1 = 1.0000$ and $C_2 = -.12500 \times 10^{-5}$. Thus, the rounding error in the representation of $\hat{p}_1 = \frac{1}{3}$ results in the term

$$(-.125 \times 10^{-5})(3)^n$$

being added to p_n at each step. The procedure given by Eq. (1.2) exhibits exponential growth of rounding errors, since

$$|p_n - \hat{p}_n| = (.125 \times 10^{-5})(3)^n. \qquad \square$$

To reduce the effects of rounding error, we can use large-order digit arithmetic such as the double- or multiple-precision option available on most digital computers. A disadvantage in using double-precision arithmetic is that it takes a great deal more computer time. Also, the serious growth of rounding error will not be eliminated entirely, but will only be postponed if a large number of subsequent computations are performed.

Another approach to estimating rounding error is to use interval arithmetic, that is, to retain the largest and smallest possible values at each step, so that, in the end, we obtain an interval which contains the true value. Unfortunately, the true answer may be near the extremes of the interval, and we may have to find a very small interval for reasonable implementation.

It is also possible to study error from a statistical standpoint. This study, however, involves considerable analysis and as such is beyond the scope of this text. Henrici [46], pages 305–309, presents a discussion of a statistical approach to estimate accumulated round-off error.

Since iterative techniques involving sequences are often used, this section will conclude with a brief discussion of the concept of **order of convergence**.

Definition 1.17   Suppose $\{\alpha_n\}_{n=1}^{\infty}$ is a sequence that converges to a number α. We say that $\{\alpha_n\}_{n=1}^{\infty}$ converges to α with $O(\beta_n)$ **rate of convergence**, where $\{\beta_n\}_{n=1}^{\infty}$ is another sequence with $\beta_n \neq 0$ for each n, if

$$\frac{|\alpha - \alpha_n|}{|\beta_n|} \leq K \qquad \text{for sufficiently large } n,$$

where K is a constant independent of n. This case is often indicated by writing $\alpha_n = \alpha + O(\beta_n)$ or $\alpha_n \to \alpha$ with rate of convergence $O(\beta_n)$.

EXAMPLE 5   Let

$$\alpha_n = \frac{\sin n}{n} \qquad \text{and} \qquad \beta_n = \frac{1}{n} \qquad \text{for } n \geq 1.$$

Then $\{\alpha_n\}_{n=1}^{\infty} \to 0$ with rate of convergence $O(1/n)$, since

$$\left| \frac{(\sin n)/n - 0}{1/n} \right| = |\sin n| \leq 1 \qquad \text{for all } n. \qquad \square$$

This concept generalizes to functions as follows:

Definition 1.18   If $\lim_{h \to 0} F(h) = L$, the convergence is said to be $O(G(h))$ if there exists a number $K > 0$, independent of h, for which

$$\frac{|F(h) - L|}{|G(h)|} \leq K \qquad \text{for sufficiently small } h > 0.$$

This situation is often indicated by writing $F(h) = L + O(G(h))$ or $F(h) \to L$ with rate of convergence $O(G(h))$.

EXAMPLE 6   By L'Hôpital's rule,

$$\lim_{h \to 0} \frac{\ln(1 + h)}{h} = 1. \qquad \text{(See Fig. 1.7.)}$$

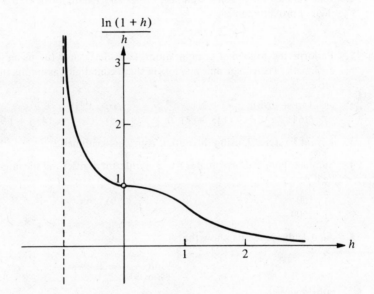

FIGURE 1.7

To determine the rate of convergence of $\ln(1 + h)/h$ as $h \to 0$, consider

$$\lim_{h \to 0} \frac{[(\ln(1 + h)/h) - 1]}{h} = \lim_{h \to 0} \frac{\ln(1 + h) - h}{h^2}.$$

Using L'Hôpital's rule,

$$\lim_{h \to 0} \frac{\ln(1 + h) - h}{h^2} = \lim_{h \to 0} \frac{(1/(1 + h)) - 1}{2h} = \lim_{h \to 0} \frac{-h}{2h(1 + h)} = -\frac{1}{2}.$$

Thus,

$$\lim_{h \to 0} \frac{|(\ln(1 + h)/h) - 1|}{|h|} = \frac{1}{2}.$$

This implies that

$$\frac{\ln(1 + h)}{h} = 1 + O(h);$$

so

$$\lim_{h \to 0} \frac{\ln(1 + h)}{h} = 1 \qquad \text{with convergence } O(h). \qquad \square$$

Theorem 1.19 If $\alpha_n = \alpha + O(\beta_n)$ and $\alpha'_n = \alpha' + O(\beta'_n)$, where $\beta'_n = O(\beta_n)$, then the following equations hold:

a) $\alpha_n \pm \alpha'_n = \alpha \pm \alpha' + O(\beta_n)$;

b) $\alpha_n \alpha'_n = \alpha\alpha' + O(\beta_n)$;

c) $\dfrac{\alpha_n}{\alpha'_n} = \dfrac{\alpha}{\alpha'} + O(\beta_n)$, provided $\alpha' \neq 0$.

Proof   See Exercise 7.   □

Exercise Set 1.2

1. Find bounds for x, using standard tables and Definition 1.15, if x is a four-significant-digit approximation to

   a) π b) e

2. Perform the following computations (i) exactly and (ii) using three-digit chopping arithmetic. Then determine any loss in significant digits, assuming that the given numbers are exact.

   a) $14.1 + .0981$ b) $.0218 \times 179$
   c) $(164. + .913) - (143. + 21.0)$ d) $(164. - 143.) + (.913 - 21.0)$

3. Repeat Exercise 2, using three-digit rounding arithmetic.

4. Suppose that p^* approximates p to 3 significant digits. Find the interval in which p^* must lie if p is

   a) 150   b) 900
   c) 1500

5. All calculus students know that

$$\lim_{h \to 0} \frac{\sin h}{h} = 1,$$

   but show that

$$\frac{\sin h}{h} = 1 + O(h^2).$$

   and that the convergence is $O(h^2)$.

6. $\lim\limits_{h \to 0} \dfrac{1 - \cos h}{h} = 0$ at what rate of convergence?

7. Prove Theorem 1.19.

8. The quadratic formula tells us that the roots of $ax^2 + bx + c = 0$, where $a \neq 0$, are given by

$$x_1 = \frac{-b + \sqrt{b^2 - 4ac}}{2a} \quad \text{and} \quad x_2 = \frac{-b - \sqrt{b^2 - 4ac}}{2a}.$$

   a) Show that $x_1 + x_2 = -b/a$ and $x_1 \cdot x_2 = c/a$.
   b) Use part (a) to show that an alternate form for the roots of $ax^2 + bx + c = 0$ is

$$x_1 = \frac{-2c}{b + \sqrt{b^2 - 4ac}} \quad \text{and} \quad x_2 = \frac{-2c}{b - \sqrt{b^2 - 4ac}}.$$

c) Use both formulas for x_1 to compute the largest root of $x^2 + 62.10x + 1 = 0$, using four-digit rounding arithmetic. The correct value for x_1 to seven significant digits is $-.01610723$. How do you explain the difference in your two computations for x_1?

9. The Taylor polynomial of degree n for $f(x) = e^x$ is $\sum_{i=0}^n (x^i/i!)$. Use the Taylor polynomial of degree 9 to find an approximation to e^{-5} by:

a) $e^{-5} \approx \sum_{i=0}^9 \dfrac{(-5)^i}{i!} = \sum_{i=0}^9 \dfrac{(-1)^i 5^i}{i!}$ b) $e^{-5} \approx \dfrac{1}{e^5} \approx \dfrac{1}{\sum_{i=0}^9 (5^i/i!)}$

An approximate value of e^{-5} correct to three digits is 6.74×10^{-3}. Which formula, (a) or (b), gives the most accuracy, and why?

10. Consider the equations:

(1) $31.69x + 14.31y = 45.00,$

(2) $13.11x + 5.89y = 19.00.$

The unique solution to this set of equations is $x = 7.2$ and $y = -12.8$. A method often presented in elementary algebra courses for solving problems of this type is to multiply Eq. (1) by the coefficient of x in Eq. (2), multiply Eq. (2) by the coefficient of x in Eq. (1), and then subtract the resulting equations. For this problem, we would obtain

(3) $[(13.11)(14.31) - (31.69)(5.89)]y = (13.11)(45.00) - (31.69)(19.00).$

a) Perform this operation using four-digit chopping arithmetic, and use your result to find four-digit values of y and x.

b) Explain why the answers obtained in (a) differ significantly from the actual values for x and y.

11. The sequence of numbers $1, \frac{1}{3}, \frac{1}{9}, \frac{1}{27}, \ldots$ considered in Example 4 can be generated by defining $P_0 = 1$ and

$$P_n = \tfrac{1}{3}P_{n-1}, \qquad n = 1, 2, \ldots$$

a) Compute P_n for $n = 1, \ldots, 8$, using five-digit arithmetic and compare the results to those obtained in Example 4.

b) Show that this algorithm exhibits linear growth of rounding error.

12. Use 3-digit chopping arithmetic to compute the sum $\sum_{i=1}^{10} (1/i^2)$ first by $\frac{1}{1} + \frac{1}{4} + \cdots + \frac{1}{100}$ and then by $\frac{1}{100} + \frac{1}{81} + \cdots + \frac{1}{1}$. Which method is most accurate and why?

13. A rectangular parallelepiped has sides 3 cm (centimeters), 4 cm, and 5 cm, measured only to the nearest centimeter. What are the best upper and lower bounds for the volume of this parallelpiped? What are the best upper and lower bounds for the surface area?

Solutions of Equations in One Variable

The growth of large populations can be modeled over short periods of time by assuming that the population grows continuously with time at a rate proportional to the number of individuals present at that time. If we let $N(t)$ denote the number of individuals at time t and λ denote the constant birth rate of the population, the population satisfies the differential equation

$$\frac{dN(t)}{dt} = \lambda N(t).$$

The solution to this equation is $N(t) = N_0 e^{\lambda t}$, where N_0 denotes the initial population.

This model is valid only when the population is isolated with no immigration from outside the community. If immigration is permitted at a constant rate v, the differential equation governing the situation becomes

$$\frac{dN(t)}{dt} = \lambda N(t) + v,$$

whose solution is

$$N(t) = N_0 e^{\lambda t} + \frac{v}{\lambda}(e^{\lambda t} - 1).$$

Suppose a certain population contains one million individuals initially, that 435,000 individuals immigrate into the community in the first year, and that 1,564,000 individuals are present at the end of one year. To determine the birth rate of this population necessitates solving for λ in the equation

$$1,564,000 = 1,000,000 e^{\lambda} + \frac{435,000}{\lambda}(e^{\lambda} - 1).$$

The numerical methods discussed in this chapter are used to find approximations to solutions of equations of this type, when the exact solutions cannot be obtained by algebraic methods. The solution to this particular problem is considered in Exercise 15 of Section 2.3.

In this chapter we will discuss one of the most basic problems in numerical a-nalysis: finding values of the variable x that satisfy $f(x) = 0$ for a given f. A solution to this problem is called a **zero** of f or a **root** of $f(x) = 0$. This is one of the oldest numerical approximation problems and yet research is continuing actively in this area at the present time. The procedures we will discuss range from the classical Newton–Raphson method, basically developed by Isaac Newton over 300 years ago, to the Quotient–Difference method for polynomial functions, first published quite recently.

2.1   The Bisection Algorithm

The first technique, based upon the Intermediate Value Theorem (Theorem 1.12, p. 6, is called the **bisection algorithm**, or **binary-search method**. Suppose a continuous function f, defined on the interval $[a, b]$, is given, with $f(a)$ and $f(b)$ of opposite sign. Then by Theorem 1.12, there exists p, $a < p < b$, for which $f(p) = 0$. Although the procedure will work for the case when there is more than one root in the interval $[a, b]$, for simplicity, it will be assumed that the root in this interval is unique. The method calls for a repeated halving of subintervals of $[a, b]$ and, at each step, locating the "half" containing p. To begin, set $a_1 = a$ and $b_1 = b$, and let p_1 be the midpoint of $[a, b]$; that is,

$$p_1 = \tfrac{1}{2}(a_1 + b_1).$$

If $f(p_1) = 0$, then $p = p_1$; if not, then $f(p_1)$ has the same sign as either $f(a_1)$ or $f(b_1)$. If $f(p_1)$ and $f(a_1)$ have the same sign, then $p \in (p_1, b_1)$, and we set $a_2 = p_1$ and $b_2 = b_1$. If $f(p_1)$ and $f(b_1)$ are of the same sign, then $p \in (a_1, p_1)$, and we set $a_2 = a_1$ and $b_2 = p_1$. Now we reapply the process to the interval $[a_2, b_2]$. This produces the following algorithm. (See Fig. 2.1.)

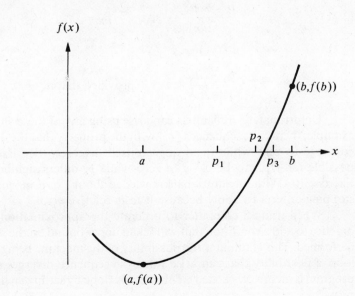

FIGURE 2.1

Bisection Algorithm 2.1

To find a solution to $f(x) = 0$ given the continuous function f on the interval $[a, b]$ where $f(a)$ and $f(b)$ have opposite signs:

INPUT endpoints a, b; tolerance TOL; maximum number of iterations N_0.

OUTPUT approximate solution p or message of failure.

Step 1 Set $i = 1$.

Step 2 While $i < N_0$ do Steps 3–6.

Step 3 Set $p = a + (b - a)/2$.   (*Compute p_i.*)

Step 4 If $f(p) = 0$ or $(b - a)/2 < TOL$ then
　　　　OUTPUT (p);   (*Procedure completed successfully.*)
　　　　STOP.

Step 5 Set $i = i + 1$.

Step 6 If $f(a)f(p) > 0$ then set $a = p$ (*Compute a_i, b_i.*)
　　　　　　　　else set $b = p$.

Step 7 OUTPUT ('Method failed after N_0 iterations, $N_0 = $ ', N_0);
(*Procedure completed unsuccessfully.*)
STOP.

We briefly mention some other stopping procedures which could also be applied in Step 4 of Algorithm 2.1, the first three of which apply to any iterative technique considered in this chapter. Select a tolerance $\varepsilon > 0$ and generate $p_1, \ldots, p_N$ until one of the following conditions is met:

(2.1) $\qquad\qquad |p_N - p_{N-1}| < \varepsilon,$

(2.2) $\qquad\qquad \dfrac{|p_N - p_{N-1}|}{|p_N|} < \varepsilon, \qquad p_N \neq 0,$

(2.3) $\qquad\qquad |f(p_N)| < \varepsilon$

or $\qquad\qquad \dfrac{b_N - a_N}{|a_N|} < \varepsilon, \qquad$ provided that $a_N \neq 0$.

Unfortunately, difficulties can arise using any of these stopping criteria. For example, there exist sequences $\{p_n\}$ with the property that the differences $p_n - p_{n-1}$ converge to zero while the sequence itself diverges. (See Exercise 9.) It is also possible for $f(p_n)$ to be close to zero while p_n differs significantly from p. (See Exercise 10.) Without additional knowledge about f or p, inequality (2.2) is the best stopping criteria to apply because it tests relative error.

When using a computer to generate the approximations, it is also a good practice to add a condition that will set an upper bound on the number of iterations performed. This eliminates the possibility of the machine being put into an infinite loop, a possibility that can arise when the sequence diverges (and also when the program is incorrectly coded). This is easily done by setting an initial bound N_0 and requiring the procedure to terminate if $i > N_0$, This was done in Step 2 of the Algorithm 2.1.

Note that in order to start the bisection algorithm, an interval $[a, b]$ must be found with $f(a) \cdot f(b) < 0$. At each step of the bisection algorithm, the length of the interval known to contain a zero of f is reduced by a factor of two; hence it is advantageous to choose the initial interval $[a, b]$ as small as possible. For example, if $f(x) = 2x^3 - x^2 + x - 1$,

$$f(-4) \cdot f(4) < 0 \quad \text{and} \quad f(0) \cdot f(1) < 0,$$

so the bisection algorithm could be used on either of the intervals $[-4, 4]$ or $[0, 1]$. Starting the bisection algorithm on $[0, 1]$ instead of $[-4, 4]$ will reduce the number of iterations required to achieve any specified accuracy by three.

To illustrate the bisection algorithm, consider the following example. The iteration in this example is terminated when $(b_n - a_n)/|a_n| < 5 \times 10^{-4}$ to insure 5 significant digits of accuracy.

EXAMPLE 1   The function $f(x) = x^3 + 4x^2 - 10$ has a root in $[1, 2]$ since $f(1) = -5$ and $f(2) = 14$. It is easily seen that there is only one root in $[1, 2]$. The bisection algorithm gives the values in Table 2.1.

TABLE 2.1

n	a_n	b_n	p_n	$f(p_n)$
1	1.0	2.0	1.5	2.375
2	1.0	1.5	1.25	-1.79687
3	1.25	1.5	1.375	.16211
4	1.25	1.375	1.3125	$-.84839$
5	1.3125	1.375	1.34375	$-.35098$
6	1.34375	1.375	1.359375	$-.09641$
7	1.359375	1.375	1.3671875	.03236
8	1.359375	1.3671875	1.36328125	$-.03215$
9	1.36328125	1.3671875	1.365234375	.000072
10	1.36328125	1.365234375	1.364257813	$-.01605$
11	1.364257813	1.365234375	1.364746094	$-.00799$
12	1.364746094	1.365234375	1.364990235	$-.00396$

After 12 iterations, we can see that $p_{12} = 1.364990235$ approximates the root p with an error

$$|p - p_{12}| \le |b_{12} - a_{12}| = .000488281,$$

and since

$$\frac{|b_{12} - a_{12}|}{|a_{12}|} \approx 3.6 \times 10^{-4},$$

the approximation is correct to at least four significant digits. The correct value of p, to eight decimal places, is $p = 1.36523001$. It is interesting to note that p_9 is closer to p than is the final approximation p_{12}, but there is no way of determining this unless the true answer is known. $\square$

The bisection algorithm, though conceptually clear, has significant draw-backs. It is very slow in converging (that is, N may become quite large before $|p - p_N|$ is sufficiently small) and, moreover, a good intermediate approximation may be inadvertently discarded. However, the method has the important property that it will always converge to a solution and for that reason is often used as a "starter" for the more efficient methods presented later in this chapter.

Theorem 2.1   Let $f \in C[a, b]$ and suppose $f(a) \cdot f(b) < 0$. The bisection pro-cedure, (Algorithm 2.1) generates a sequence $\{p_n\}$ approximating p with the property

$$(2.4) \qquad\qquad |p_n - p| \le \frac{b - a}{2^n}, \qquad n \ge 1.$$

Proof   It is clear that, for each $n \ge 1$, we have

$$b_n - a_n = \frac{1}{2^{n-1}}(b - a) \qquad \text{and} \qquad p \in (a_n, b_n).$$

Since $p_n = \frac{1}{2}(a_n + b_n)$, for all $n \ge 1$, it follows that

$$|p_n - p| \le \tfrac{1}{2}(b_n - a_n) = 2^{-n}(b - a). \qquad\qquad \square$$

According to Definition 1.17 (p. 16), inequality (2.4) implies that $\{p_n\}_{n=1}^{\infty}$ con-verges to p with $O(2^{-n})$ rate of convergence. With a given tolerance $\varepsilon > 0$, we could iteratively determine p_n until

$$2^{-n}(b - a) < \varepsilon,$$

and then $|p_n - p| < \varepsilon$. Alternatively, the number of iterations N necessary to achieve a certain tolerance ε could be estimated; that is, we could require that:

$$2^{-N}(b - a) < \varepsilon \qquad \text{or} \qquad 2^{-N} < \frac{\varepsilon}{(b - a)}.$$

Using logarithms, this implies

$$-N \ln 2 < \ln \varepsilon - \ln(b - a) \qquad \text{or} \qquad N > \frac{\ln(b - a) - \ln \varepsilon}{\ln 2}.$$

Thus $[(\ln(b - a) - \ln \varepsilon)/\ln 2]$ iterations are needed if application of Theorem 2.1 is to ensure that $|p_n - p| < \varepsilon$.

Note that although it is indicated here that the natural logarithm is used in calculating this bound, at times, it is computationally more convenient to use base-10 logarithms. This is due to the fact that error tolerances are often given in powers of 10. The formula for the number of iterations needed to ensure accuracy within a tolerance of ε, using base-10 logarithms, would correspondingly be:

$$N > \frac{\log_{10}(b - a) - \log_{10} \varepsilon}{\log_{10} 2}.$$

EXAMPLE 2   Determine approximately how many iterations are necessary to solve $f(x) = x^3 + 4x^2 - 10 = 0$ with an accuracy of $\varepsilon = 10^{-5}$ for $a_1 = 1$ and $b_1 = 2$. This requires finding an integer N that will satisfy:

$$|p_N - p| \le 2^{-N}(b - a) = 2^{-N} < 10^{-5}.$$

Using logarithms with base 10,

$$-N \log_{10} 2 < -5 \quad \text{or} \quad N > \frac{5}{\log_{10} 2} \approx 16.6.$$

It would require 17 iterations to obtain an approximation accurate to 10^{-5}. With $\varepsilon = 10^{-3}$, $N \ge 10$ iterations are required and the value of $p_9 = 1.36523475$ is accurate to within 10^{-4}. It is important to note that these techniques give only a *bound* for the number of iterations necessary, and in many cases this bound is much larger than the actual number required.   □

Exercise Set 2.1

1. Show that $f(x) = x^3 - x - 1$ has exactly one zero in the interval $[1, 2]$.

2. Use the Bisection Algorithm to find solutions accurate to 10^{-5} for the following problems:

   a) $x - 2^{-x} = 0$ for $0 \le x \le 1$,
   b) $e^x + 2^{-x} + 2 \cos x - 6 = 0$ for $1 \le x \le 2$,
   c) $e^x - x^2 + 3x - 2 = 0$ for $0 \le x \le 1$.

3. Find an approximation to $\sqrt[3]{25}$ correct to two significant digits, using the Bisection Algorithm. [*Hint*: Consider $f(x) = x^3 - 25$.]

4. Find an approximation to $\sqrt{3}$ correct to four significant digits using the Bisection Algorithm.

5. Use Theorem 2.1 to find a bound for the number of iterations needed to achieve an approximation with accuracy 10^{-4} to the solution of $x^3 - x - 1 = 0$, lying in the interval $[1, 2]$. Find an approximation to the root with this degree of accuracy.

6. Use Theorem 2.1 to find a bound for the number of iterations needed to achieve an approximation with accuracy 10^{-3} to the solution of $x^3 + x - 4 = 0$, lying in the interval $[1, 4]$. Find an approximation to the root with this degree of accuracy.

7. Apply the Bisection Algorithm to the equation

$$\frac{4x - 7}{(x - 2)^2} = 0,$$

using the intervals $[1.2, 2.2]$ and $[1.5, 2.5]$. Explain the results graphically.

8. Find all the zeros of $f(x) = x^2 + 10 \cos x$ to four significant digits. [*Hint*: Consider the graph of f.]

9. Let $\{p_n\}$ be the sequence defined by $p_n = \sum_{k=1}^{n} (1/k)$. Show that $\lim_{n \to \infty} (p_n - p_{n-1}) = 0$, but that $\{p_n\}$ diverges.

10. Let $f(x) = (x - 1)^{10}$, $p = 1$, and $p_n = 1 + 1/n$. Show that $|f(p_n)| < 10^{-3}$ whenever $n > 1$, but that $|p - p_n| < 10^{-3}$ requires that $n > 1000$.

2.2   Fixed-Point Iteration

The methods in this section and most of those in the remainder of this chapter are based upon changing the equation $f(x) = 0$ into the form $g(x) = x$. There are various ways of accomplishing this task. One procedure is to define $g(x)$ to be $x - f(x)$. In this case, $\bar{x}$ satisfies $f(\bar{x}) = 0$ if and only if $g(\bar{x}) = \bar{x}$. Consequently if a general method for solving a problem of the form $g(x) = x$ can be found, a solution to $f(x) = 0$ can be obtained by defining the function $g(x) = x - f(x)$. As will be seen later in the chapter, other choices of g will also permit finding a solution of $f(x) = 0$, and many are more appropriate than the choice above.

Definition 2.2   If g is defined on $[a, b]$ and $g(p) = p$ for some $p \in [a, b]$, then the function g is said to have the **fixed point** p in $[a, b]$.

EXAMPLE 1

   a)   The function $g(x) = x, 0 \le x \le 1$, has a fixed point at each x in $[0, 1]$.
   b)   The function $g(x) = x - \sin \pi x$ has exactly two fixed points in $[0, 1]$, $x = 0$ and $x = 1$. (See Fig. 2.2.)   □

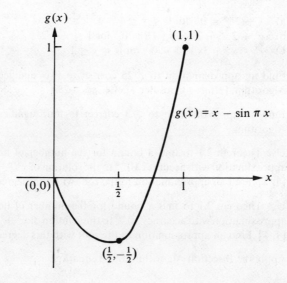

FIGURE 2.2

The following theorem gives sufficient conditions for the existence and uniqueness of a fixed point.

Theorem 2.3   If $g \in C[a, b]$ and $g(x) \in [a, b]$ for all $x \in [a, b]$, then g has a fixed point in $[a, b]$. Further, suppose $g'(x)$ exists on (a, b) and

(2.5) $|g'(x)| \le k < 1$ for all $x \in [a, b]$.

Then g has a unique fixed point p in $[a, b]$. (See Fig. 2.3.)

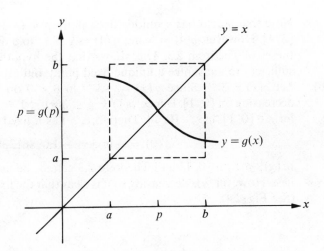

FIGURE 2.3

Proof   If $g(a) = a$ or $g(b) = b$, the existence of a fixed point is obvious. Suppose not, then it must be true that $g(a) > a$ and $g(b) < b$. Define $h(x) = g(x) - x$; h is continuous on $[a, b]$, and, moreover,

$$h(a) = g(a) - a > 0, \qquad h(b) = g(b) - b < 0.$$

The Intermediate Value Theorem implies that there exists $p \in (a, b)$ for which $h(p) = 0$. Thus, $g(p) - p = 0$ and p is a fixed point of g.

Suppose in addition that inequality (2.5) holds and that p and q are both fixed points in $[a, b]$ with $p \neq q$. By the Mean Value Theorem (Theorem 1.8 p. 5), ξ in $[a, b]$ exists with

$$|p - q| = |g(p) - g(q)| = |g'(\xi)||p - q| \le k|p - q| < |p - q|,$$

which is a contradiction. Hence, $p = q$ and the fixed point in $[a, b]$ is unique. $\square$

EXAMPLE 2

a)   Let $g(x) = (x^2 - 1)/3$ on $[-1, 1]$. Using the Extreme Value Theorem (Theorem 1.9, p. 5), it is easy to show that the absolute minimum of g occurs at $x = 0$ and is $g(0) = -\frac{1}{3}$. Similarly, the absolute maximum of g occurs at $x = \pm 1$ and has the value $g(\pm 1) = 0$. Moreover, g is continuous and

$$|g'(x)| = \left|\frac{2x}{3}\right| \le \frac{2}{3} \qquad \text{for all } x \in [-1, 1],$$

so g satisfies the hypotheses of Theorem 2.3 and has a unique fixed point in $[-1, 1]$.

Let p be the unique fixed point in the interval $[-1, 1]$. Then

$$p = g(p) = \frac{p^2 - 1}{3} \qquad \text{and} \qquad p^2 - 3p - 1 = 0,$$

which, by the quadratic formula, implies that

$$p = \frac{3 - \sqrt{13}}{2}.$$

Note that g also has a unique fixed point $p = (3 + \sqrt{13})/2$ for the interval $[3, 4]$. However, $g(4) = 5$ and $g'(4) = \frac{8}{3} > 1$; so g does not satisfy the hypotheses of Theorem 2.3. This shows that the hypotheses of Theorem 2.3 are sufficient to guarantee a unique fixed point, but are not necessary.

b)   Let $g(x) = 3^{-x}$. Since $g'(x) = -3^{-x} \ln 3 < 0$ on $[0, 1]$, the function g is decreasing on $[0, 1]$. Hence, $g(1) = \frac{1}{3} \le g(x) \le 1 = g(0)$ for $0 \le x \le 1$. Thus for $x \in [0, 1]$, $g(x) \in [0, 1]$. Therefore, g has a fixed point in $[0, 1]$. Since,

$$g'(0) = -\ln 3 = -1.098612289,$$

$|g'(x)| \nleq 1$ on $[0, 1]$ and Theorem 2.3 cannot be used to determine uniqueness. However, g is decreasing so it is clear that the fixed point must be unique. (See Fig. 2.4). ☐

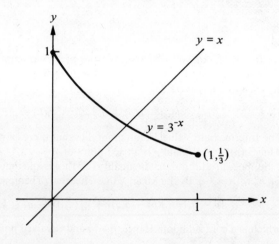

FIGURE 2.4

In order to approximate the fixed point of a function g, we choose an initial approximation p_0 and generate the sequence $\{p_n\}_{n=0}^{\infty}$ by letting $p_n = g(p_{n-1})$ for each $n \ge 1$. If the sequence converges to p and g is continuous, then, by Theorem 1.4

$$p = \lim_{n \to \infty} p_n = \lim_{n \to \infty} g(p_{n-1}) = g\left(\lim_{n \to \infty} p_{n-1}\right) = g(p),$$

and a solution to $x = g(x)$ is obtained. This technique is called the **fixed-point iterative technique**, or **functional iteration**, and is formally given by the following algorithm. The procedure is detailed in Fig. 2.5.

Fixed-Point Algorithm 2.2

To find a solution to $p = g(p)$ given an initial approximation p_0:

INPUT initial approximation p_0; tolerance TOL; maximum number of iterations N_0.

OUTPUT approximate solution p or message of failure.

Step 1 Set $i = 1$.

Step 2 While $i < N_0$ do Steps 3–6.

Step 3 Set $p = g(p_0)$. *(Compute p_i.)*

Step 4 If $|p - p_0| < TOL$ then
    OUTPUT (p); *(Procedure completed successfully.)*
    STOP.

Step 5 Set $i = i + 1$.

Step 6 Set $p_0 = p$. *(Update p_0.)*

Step 7 OUTPUT ('Method failed after N_0 iterations, $N_0 = $ ', N_0);
    (Procedure completed unsuccessfully.)
    STOP.

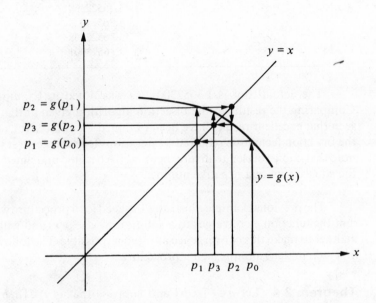

FIGURE 2.5

Factors involving the determination, in Step 4, of whether to continue could also involve the value of $|p_i - p_{i-1}|/|p_i|$.

To illustrate this algorithm, consider the following example.

EXAMPLE 3 The equation $x^3 + 4x^2 - 10 = 0$ has a unique root in $[1, 2]$. There are many ways to change the equation to the form $x = g(x)$ by simple algebraic manipulation. It can be verified easily that some of these are:

a) $x = g_1(x) = x - x^3 - 4x^2 + 10,$ b) $x = g_2(x) = \left(\dfrac{10}{x} - 4x\right)^{1/2},$

c) $x = g_3(x) = \frac{1}{2}(10 - x^3)^{1/2},$ d) $x = g_4(x) = \left(\dfrac{10}{4 + x}\right)^{1/2},$

e) $x = g_5(x) = x - \dfrac{x^3 + 4x^2 - 10}{3x^2 + 8x}.$

With $p_0 = 1.5$, Table 2.2 lists the results of the fixed-point iteration method for all five choices of g.

TABLE 2.2	n	(a)	(b)	(c)	(d)	(e)
	0	1.5	1.5	1.5	1.5	1.5
	1	$-.875$	.8165	1.28695377	1.34839973	1.37333333
	2	6.732	2.9969	1.40254080	1.36737637	1.36526201
	3	-469.7	$(-8.65)^{1/2}$	1.34545838	1.36495701	1.36523001
	4	1.03×10^8		1.37517025	1.36526475	
	5			1.36009419	1.36522559	
	6			1.36784697	1.36523058	
	7			1.36388700	1.36522994	
	8			1.36591673	1.36523002	
	9			1.36487822	1.36523001	
	10			1.36541006		
	15			1.36522368		
	20			1.36523024		
	23			1.36522998		
	25			1.36523001		

The actual root is 1.36523001, as was noted in Example 1 of Section 2.1. Comparing the results to the bisection algorithm given in that example, it can be seen that excellent results have been obtained for choices (c), (d), and (e), whereas the bisection technique required 27 iterations for such accuracy. It is interesting to note that choice (a) led to divergence and (b) became undefined because it involved the square root of a negative number.   □

The previous example illustrates the need for a procedure which will guarantee that the function g converges to a solution of $x = g(x)$ and will also choose g in a manner to make this convergence as rapid as possible. The following theorem is the first step in determining such a procedure.

Theorem 2.4   Let $g \in C[a, b]$ and suppose that $g(x) \in [a, b]$ for all $x \in [a, b]$. Further, let g' exist on (a, b) with

(2.6) $$|g'(x)| \leq k < 1 \qquad \text{for all } x \in (a, b).$$

If p_0 is any number in $[a, b]$, then the sequence defined by

$$p_n = g(p_{n-1}), \qquad n \geq 1,$$

will converge to the unique fixed point p in $[a, b]$.

Proof   By Theorem 2.3, a unique fixed point exists in $[a, b]$. Since g maps $[a, b]$ into itself, the sequence $\{p_n\}_{n=0}^{\infty}$ is defined for all $n \geq 0$ and $p_n \in [a, b]$ for all n. Using inequality (2.6) and the Mean Value Theorem,

(2.7) $$|p_n - p| = |g(p_{n-1}) - g(p)| \leq |g'(\xi)||p_{n-1} - p| \leq k|p_{n-1} - p|,$$

where $\xi \in (a, b)$. Applying inequality (2.7) inductively gives:

(2.8) $$|p_n - p| \leq k|p_{n-1} - p| \leq k^2|p_{n-2} - p| \leq \cdots \leq k^n|p_0 - p|.$$

Since $k < 1$,

$$\lim_{n \to \infty} |p_n - p| \leq \lim_{n \to \infty} k^n|p_0 - p| = 0$$

and $\{p_n\}_{n=0}^{\infty}$ converges to p.   □

Reconsidering Example 3 in light of the previous theorem:

EXAMPLE 4

a)   When $g_1(x) = x - x^3 - 4x^2 + 10$, $g_1'(x) = 1 - 3x^2 - 8x$. There is no interval $[a, b]$, containing p, for which $|g_1'(x)| < 1$. Although Theorem 2.4 does not guarantee that the method must fail for this choice of g, there is no reason to suspect convergence.

b)   With $g_2(x) = [(10/x) - 4x]^{1/2}$, we can see that g_2 does not map $[1, 2]$ into $[1, 2]$ and the sequence $\{p_n\}_{n=0}^{\infty}$ is not defined with $p_0 = 1.5$. Moreover, there is no interval containing p such that

$$|g_2'(x)| < 1, \qquad \text{since} \qquad |g_2'(p_0)| \approx 3.4.$$

c)   For the function $g_3(x) = \frac{1}{2}(10 - x^3)^{1/2}$,

$$g_3'(x) = -\tfrac{3}{4}x^2(10 - x^3)^{-1/2} < 0 \qquad \text{on } [1, 2],$$

so g is strictly decreasing on $[1, 2]$. However, $|g_3'(2)| \approx 2.12$, so inequality (2.6) does not hold on $[1, 2]$. A closer examination of the sequence $\{p_n\}_{n=0}^{\infty}$ starting with $p_0 = 1.5$ will show that it suffices to consider the interval $[1, 1.5]$ instead of $[1, 2]$. On this interval it is still true that $g_3'(x) < 0$ and g is strictly decreasing, but, additionally, $1 < 1.28 \approx g_3(1.5) \leq g_3(x) \leq g_3(1) = 1.5$ for all $x \in [1, 1.5]$. This shows that g_3 maps the interval $[1, 1.5]$ into itself. Since it is also true that $|g_3'(x)| \leq |g_3'(1.5)| \approx .66$ on this interval, Theorem 2.4 confirms the convergence of which we were already aware.

The other parts of Example 3 could be handled in a similar manner. $\square$

Using results obtained in Theorem 2.4, precise error estimates can be obtained, as indicated by the following corollaries.

Corollary 2.5   If g satisfies the hypotheses of Theorem 2.4, a bound for the error involved in using p_n to approximate p is given by

(2.9) $$|p_n - p| \leq k^n \max\{p_0 - a, b - p_0\} \qquad \text{for all } n \geq 1.$$

Proof   From inequality (2.8),

$$|p_n - p| \leq k^n|p_0 - p| \leq k^n \max\{p_0 - a, b - p_0\},$$

since $p \in [a, b]$. $\square$

Corollary 2.6   If g satisfies the hypotheses of Theorem 2.4, then

(2.10) $$|p_n - p| \leq \frac{k^n}{1 - k}|p_0 - p_1| \qquad \text{for all } n \geq 1.$$

Proof   For $n \geq 1$, the procedure used in the proof of Theorem 2.4 implies that

$$\begin{aligned}
|p_{n+1} - p_n| &= |g(p_n) - g(p_{n-1})| \\
&\leq k|p_n - p_{n-1}| \\
&\;\;\vdots \\
&\leq k^n|p_1 - p_0|.
\end{aligned}$$

Thus, for $m > n \geq 1$,

$$\begin{aligned}
|p_m - p_n| &= |p_m - p_{m-1} + p_{m-1} - \cdots + p_{n+1} - p_n| \\
&\leq |p_m - p_{m-1}| + |p_{m-1} - p_{m-2}| + \cdots + |p_{n+1} - p_n| \\
&\leq k^{m-1}|p_1 - p_0| + k^{m-2}|p_1 - p_0| + \cdots + k^n|p_1 - p_0| \\
&= k^n(1 + k + k^2 + \cdots + k^{m-n-1})|p_1 - p_0|.
\end{aligned}$$

By Theorem 2.4, $\lim_{m \to \infty} p_m = p$, so

$$|p - p_n| = \lim_{m \to \infty} |p_m - p_n| \leq k^n|p_1 - p_0| \sum_{i=0}^{\infty} k^i = \frac{k^n}{1 - k}|p_1 - p_0|. \qquad \square$$

Both corollaries relate the rate of convergence to the bound k on the first derivative. It is clear that the rate of convergence depends upon the factor $k^n/(1 - k)$, and that the smaller k can be made, the faster the convergence. The convergence may be very slow if k is close to 1.

To illustrate the use of the corollaries in estimating the number of iterations required to achieve a specified accuracy, consider Example 5.

EXAMPLE 5   Consider again the functions in Examples 3 and 4. It was found that for $g_3(x) = \frac{1}{2}(10 - x^3)^{1/2}$, $|g_3'(x)| < .66$, whenever $x \in [1, 1.5]$. If the same analysis is done on the function

$$g_4(x) = \left(\frac{10}{4 + x}\right)^{1/2},$$

given in Example 3(d), it can be seen that

$$|g_4'(x)| = \left| \frac{-5}{\sqrt{10}(4 + x)^{3/2}} \right| \leq \frac{5}{\sqrt{10}(5)^{3/2}} < .15 \qquad \text{for all } x \in [1, 2].$$

The bound on the magnitude of $g_4'(x)$ is much smaller than the bound on the magnitude of $g_3'(x)$, which explains the reason for the more rapid convergence using $g_4(x)$.

Suppose that we need to determine a bound N for the number of iterations required to obtain

$$|p_N - p| < 10^{-5}.$$

To accomplish this, we will use Corollary 2.6 with $k = .15$ and the values $p_0 = 1.5$ and $p_1 = 1.3439973$ taken from Table 2.2. With these values, inequality (2.10) becomes

$$|p_N - p| \leq \frac{(.15)^N}{1 - .15}|1.5 - 1.3439973| \leq .19\,(.15)^N.$$

To ensure that $|p_N - p| < 10^{-5}$, it remains to calculate N so that

$$.19(.15)^N < 10^{-5}.$$

Thus,
$$N \geq \frac{\log_{10} 10^{-5} - \log_{10} .19}{\log_{10} .15} \approx 6.94,$$

and 7 is a bound for the number of required iterations.

Note that the entries in Table 2.2 imply that, in actuality, five iterations are necessary to obtain accuracy to within 10^{-5}. As a result, the bound on the number of iterations is somewhat conservative. $\square$

Exercise Set 2.2

1. Use Theorem 2.3 to show that $g(x) = 2^{-x}$ has a unique fixed point on $[\frac{1}{3}, 1]$. Use fixed-point iteration to find an approximation to the fixed point accurate to within 10^{-4}. Use Corollary 2.5 or 2.6 to estimate the number of iterations required to achieve 10^{-4} accuracy, and compare this theoretical estimate to the number actually needed.

2. Use Theorem 2.3 to show that $g(x) = \pi + .5 \sin x$ has a unique fixed point on $[0, 2\pi]$. Use fixed-point iteration to find an approximation to the fixed point that is accurate to within 10^{-2}. Use Corollary 2.5 or 2.6 to estimate the number of iterations required to achieve 10^{-2} accuracy, and compare this theoretical estimate to the number actually needed.

3. Solve $x^3 - x - 1 = 0$ for the root in $[1, 2]$, using fixed-point iteration. Obtain an approximation to the root accurate to within 10^{-5}.

4. Use a fixed-point iteration procedure to find an approximation to $\sqrt{3}$ that is accurate to four significant digits. Compare your result and the number of iterations required with the answer obtained in Exercise 4 of Section 2.1.

5. Use a fixed-point iteration procedure to find an approximation to $\sqrt[3]{25}$ that is accurate to four significant digits. Compare your result and the number of iterations required with the answer obtained in Exercise 3 of Section 2.1.

6. For each of the following equations, determine an interval $[a, b]$ such that fixed-point iteration will converge. Estimate the number of iterations necessary to obtain approximations accurate to within 10^{-5}, and perform the calculations.

   a) $x = \dfrac{2 - e^x + x^2}{3}$ \qquad b) $x = 4^{-x}$

   c) $x = 5^{-x}$ \qquad d) $x = 6^{-x}$

   e) $x = 1.75 + \dfrac{4x - 7}{x - 2}$

7. For each of the following equations, determine a function g and an interval $[a, b]$ on which fixed-point iteration will converge to a positive solution of the equation.

   a) $3x^2 - e^x = 0$ \qquad b) $x - \cos x = 0$.

   Find the solutions to within 10^{-5}.

8. Find all the zeros of $f(x) = x^2 + 10 \cos x$ by using the fixed-point iteration method for an appropriate iteration function g. Find the zeros to four significant digits and compare the number of required iterations to the number required in Exercise 8, Section 2.1.

9. Prove that the sequence defined by

$$x_n = \frac{1}{2}\left(x_{n-1} + \frac{2}{x_{n-1}}\right), \qquad \text{for } n \geq 1.$$

   converges to $\sqrt{2}$ for any $x_0 > 0$.

10. Replace the assumption in inequality (2.6) of Theorem 2.4 with "g satisfies a Lipschitz condition on the interval $[a, b]$ with Lipschitz constant $L < 1$." (See Exercise 9, Section 1.1.) Show that the conclusions of this theorem are still valid.

11. Suppose that g is continuously differentiable on some interval (c, d) which contains the fixed point p of g. Show that if $|g'(p)| < 1$, then there exists a $\delta > 0$ such that the fixed-point iteration converges for any initial approximation p_0, whenever $|p_0 - p| \leq \delta$.

2.3 The Newton–Raphson Method

The **Newton–Raphson** (or simply **Newton's**) method is one of the most powerful and well-known numerical methods for finding a root of $f(x) = 0$. There are at least three common ways of introducing Newton's method. The most common approach is by considering the technique graphically (see Exercise 1). Another possibility is to derive Newton's method as a simple technique to obtain faster convergence than offered by many other types of functional iteration (see Section 2.4). The third means of introducing Newton's method, which will be discussed below, is an intuitive approach based upon the Taylor polynomial defined in Theorem 1.13.

Suppose that the function f is twice continuously differentiable on the interval $[a, b]$; that is, $f \in C^2[a, b]$. Let $\bar{x} \in [a, b]$ be an approximation to p such that $f'(\bar{x}) \neq 0$ and $|\bar{x} - p|$ is "small." Consider the second-degree Taylor polynomial for $f(x)$, expanded about $\bar{x}$,

$$(2.11) \qquad f(x) = f(\bar{x}) + (x - \bar{x})f'(\bar{x}) + \frac{(x - \bar{x})^2}{2} f''(\xi(x)),$$

where $\xi(x)$ lies between x and $\bar{x}$. Since $f(p) = 0$, Eq. (2.11), with $x = p$, gives

$$(2.12) \qquad 0 = f(\bar{x}) + (p - \bar{x})f'(\bar{x}) + \frac{(p - \bar{x})^2}{2} f''(\xi(p)).$$

Since $|\bar{x} - p|$ was assumed to be small, $|\bar{x} - p|^2$ is even smaller, and if this quantity is assumed to be negligible, then

$$(2.13) \qquad 0 \approx f(\bar{x}) + (p - \bar{x})f'(\bar{x}).$$

Solving for p in this equation gives:

$$(2.14) \qquad p \approx \bar{x} - \frac{f(\bar{x})}{f'(\bar{x})},$$

which should be a better approximation to p than is $\bar{x}$. This sets the stage for the Newton–Raphson method, which involves generating the sequence $\{p_n\}$ defined by

$$(2.15) \qquad p_n = p_{n-1} - \frac{f(p_{n-1})}{f'(p_{n-1})}, \qquad n \geq 1.$$

Figure 2.6 illustrates graphically how the approximations are obtained using successive tangents. (See also Exercise 1.)

Newton–Raphson Algorithm 2.3

To find a solution to $f(x) = 0$ given an initial approximation p_0:

INPUT initial approximation p_0: tolerance TOL; maximum number of iterations N_0.

OUTPUT approximate solution p or message of failure.

Step 1 Set $i = 1$.

Step 2 While $i < N_0$ do Steps 3–6.

Step 3 Set $p = p_0 - f(p_0)/f'(p_0)$.   (*Compute p_i.*)

Step 4 If $|p - p_0| < TOL$ then
      OUTPUT (p);   (*Procedure completed successfully.*)
      STOP.

Step 5 Set $i = i + 1$.

Step 6 Set $p_0 = p$.   (*Update p_0.*)

Step 7 OUTPUT ('Method failed after N_0 iterations, $N_0 = $ ', N_0);
      (*Procedure completed unsuccessfully.*)
      STOP.

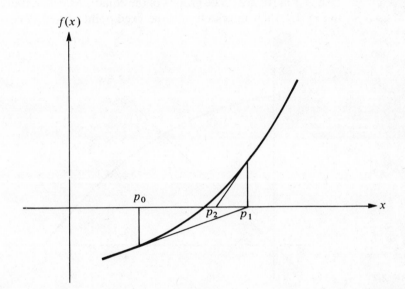

FIGURE 2.6

    The stopping-technique inequalities (2.1), (2.2), and (2.3) are applicable to Newton's method; that is: select a tolerance $\varepsilon > 0$ and construct $p_1, \ldots, p_N$ until

(2.16) $$|p_N - p_{N-1}| < \varepsilon,$$

(2.17) $$\frac{|p_N - p_{N-1}|}{|p_N|} < \varepsilon, \qquad p_N \neq 0,$$

or

(2.18) $$|f(p_N)| < \varepsilon.$$

A form of inequality (2.16) is used in Step 4 of Algorithm 2.3. Note that inequality (2.18) may not give much information about the actual error $|p_N - p|$. (See Exercise 10, Section 2.1.)

    Newton's method is a functional iteration technique $p_n = g(p_{n-1})$, $n \geq 1$ for which

$$g(p_{n-1}) = p_{n-1} - \frac{f(p_{n-1})}{f'(p_{n-1})}, \qquad n \geq 1.$$

It is clear from this equation that Newton's method cannot be continued if $f'(p_{n-1})$ = 0 for some n. We will see that the method is most effective when f' is bounded away from zero near the fixed point p.

EXAMPLE 1

a)   Suppose a solution to the equation $x = \cos x$ is desired. Let $f(x) = \cos x - x$. Then

$$f\left(\frac{\pi}{2}\right) = -\frac{\pi}{2} < 0 < 1 = f(0),$$

and, by the Intermediate Value Theorem (Theorem 1.12, p. 6), there exists a zero of f in $[0, \pi/2]$. The graphs of the equations $y = x$ and $y = \cos x$ appear in Fig. 2.7; their intersection is the fixed point of $g(x) = \cos x$.

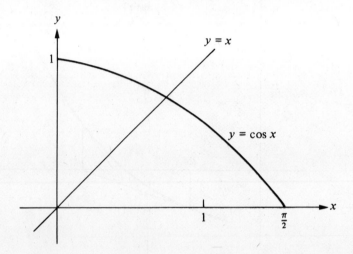

FIGURE 2.7

From the graph it is clear that $f(x) = 0$ has a unique solution in $[0, \pi/2]$. Since $f'(x) = -\sin x - 1$, Newton's method has the form

$$p_n = p_{n-1} - \frac{(\cos p_{n-1} - p_{n-1})}{(-\sin p_{n-1} - 1)}, \qquad n \geq 1,$$

where p_0 is yet to be selected. For some problems it suffices to let p_0 be arbitrary, while for others it is important to select a good initial approximation. For the problem under consideration, the graph in Fig. 2.7 suggests $p_0 = \pi/4$ as an initial approximation. With $p_0 = \pi/4$, the approximations in Table 2.3 are generated. An excellent approximation is obtained with $n = 3$.

TABLE 2.3

n	p_n	n	p_n
0	.7853981635	3	.7390851332
1	.7395361337	4	.7390851332
2	.7390851781	5	.7390851332

b) To obtain the unique solution to $x^3 + 4x^2 - 10 = 0$ on the interval $[1, 2]$ by Newton's method, generate the sequence $\{p_n\}_{n=1}^{\infty}$ by

$$p_n = p_{n-1} - \frac{p_{n-1}^3 + 4p_{n-1}^2 - 10}{3p_{n-1}^2 + 8p_{n-1}}, \qquad n \geq 1.$$

Selecting $p_0 = 1.5$ produces the results of Example 3(e) of Section 2.2, in which $p_3 = 1.36523001$ is correct to the eighth decimal place.

To illustrate the importance of a good initial approximation, we use the above procedure with $p_0 = -100$. The results of such a poor initial approximation are presented in Table 2.4. The convergence began slowly; in fact, at $n = 18$ it appeared that there would be no convergence at all. Finally, when $p_{22} \in [1, 2]$, the convergence to the solution became very rapid. It would appear from this example that rapid convergence would be expected from Newton's method when the initial approximation is close to the actual root, but that the method would converge slowly when the initial approximation differed substantially from the solution. In actuality, many problems will give divergence for Newton's method if the initial approximation is not sufficiently close to the actual root. □

TABLE 2.4

n	p_n	n	p_n
0	-100	13	-2.0757
1	-67.1229	14	-2.5401
2	-45.2107	15	-3.1421
3	-30.6110	16	-2.8007
4	-20.8903	17	-2.2742
5	-14.4276	18	-2.6754
6	-10.1440	19	4.7255
7	-7.3217	20	2.9616
8	-5.4823	21	1.9405
9	-4.3043	22	1.4793
10	-3.5648	23	1.3711
11	-3.0995	24	1.36525
12	-2.7643	25	1.36523001

The following convergence theorem for Newton's method illustrates the importance of the choice of p_0.

Theorem 2.7 Let $f \in C^2[a, b]$. If $p \in [a, b]$ is such that $f(p) = 0$ and $f'(p) \neq 0$, then there exists $\delta > 0$ such that Newton's method generates a sequence $\{p_n\}_{n=1}^{\infty}$ converging to p for any initial approximation $p_0 \in [p - \delta, p + \delta]$.

Proof The proof will be based upon analyzing Newton's method as a functional iteration scheme $p_n = g(p_{n-1})$, for $n \geq 1$, with

$$g(x) = x - \frac{f(x)}{f'(x)}.$$

The object is to find an interval $[p - \delta, p + \delta]$ such that g maps the interval $[p - \delta, p + \delta]$ into itself and $|g'(x)| \le k < 1$ for $x \in [p - \delta, p + \delta]$, where k is a fixed constant in $(0, 1)$.

Since $f'(p) \ne 0$ and f' is continuous, there exists $\delta_1 > 0$ such that $f'(x) \ne 0$ for $x \in [p - \delta_1, p + \delta_1] \subset [a, b]$. Thus, g is defined and continuous on $[p - \delta_1, p + \delta_1]$. Also,

$$g'(x) = 1 - \frac{f'(x)f'(x) - f(x)f''(x)}{[f'(x)]^2} = \frac{f(x)f''(x)}{[f'(x)]^2}$$

for $x \in [p - \delta_1, p + \delta_1]$; and since $f \in C^2[a, b], g \in C^1[p - \delta_1, p + \delta_1]$. By assumption, $f(p) = 0$, so

(2.19)
$$g'(p) = \frac{f(p)f''(p)}{[f'(p)]^2} = 0.$$

Since g' is continuous, Eq. (2.19) implies that there exists a δ with $0 < \delta < \delta_1$ and

$$|g'(x)| \le k < 1 \qquad \text{for } x \in [p - \delta, p + \delta].$$

It remains to show that $g: [p - \delta, p + \delta] \to [p - \delta, p + \delta]$. If $x \in [p - \delta, p + \delta]$, then

$$|g(x) - p| = |g(x) - g(p)| = |g'(\xi)||x - p| \le k|x - p| < |x - p|$$

for some ξ between x and p. Since $x \in [p - \delta, p + \delta]$, it follows that $|x - p| < \delta$ and that $|g(x) - p| < \delta$. This implies $g: [p - \delta, p + \delta] \to [p - \delta, p + \delta]$.

All the hypotheses of Theorem 2.4 (p. 30) are now satisfied for $g(x) = x - f(x)/f'(x)$, so the sequence $\{p_n\}_{n=1}^{\infty}$ defined by

$$p_n = g(p_{n-1}) \qquad \text{for } n = 1, 2, 3, \ldots$$

converges to p for any $p_0 \in [p - \delta, p + \delta]$.   □

Note that Theorem 2.7 states that, under reasonable assumptions, Newton's method will always converge provided a sufficiently accurate initial approximation is chosen. Sometimes, as in Example 1(b), Newton's method will converge with a very poor initial approximation, but in many cases it is imperative that a good initial approximation be chosen, as illustrated by Exercise 9. In Theorem 2.7 it is assumed that $f(p) = 0$ and $f'(p) \ne 0$. If $f'(p) = 0$, Newton's method may still converge, although the convergence will generally be slower. The following example considers such a case.

EXAMPLE 2   Let $f(x) = e^x - x - 1$. (See Fig. 2.8.) Then $f'(x) = e^x - 1$. Note that $f(x) = 0$ has the unique solution $p = 0$ and that $f'(0) = 0$. With $p_0 = 1$, the sequence $\{p_n\}_{n=1}^{\infty}$ with

$$p_n = p_{n-1} - \frac{e^{p_{n-1}} - p_{n-1} - 1}{e^{p_{n-1}} - 1} \qquad \text{for } n \ge 1,$$

given by Newton's method is listed in Table 2.5.   □

A major difficulty often encountered in Newton's method is that while $f(x)$ may be easy to evaluate, $f'(x)$ is not. For example, if $f(x) = x^2 3^x \cos 2x$, then $f'(x) = 2x3^x \cos 2x + x^2 3^x(\cos 2x)\ln 3 - 2x^2 3^x \sin 2x$, which is extremely

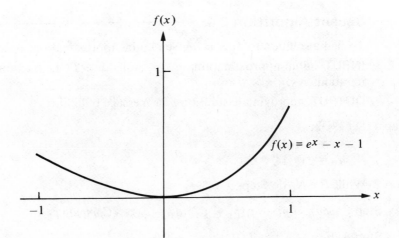

FIGURE 2.8

TABLE 2.5

n	p_n	n	p_n
0	1.0	9	2.7750×10^{-3}
1	.58198	10	1.3881×10^{-3}
2	.31906	11	6.9411×10^{-4}
3	.16800	12	3.4703×10^{-4}
4	.08635	13	1.7416×10^{-4}
5	.04380	14	8.8041×10^{-5}
6	.02206	15	4.2610×10^{-5}
7	.01107	16	1.9142×10^{-5}
8	.005545		

tiresome to evaluate, particularly if the method is being performed on a non-programmable calculator. To circumvent this problem, we can derive a slight variation of Newton's method. By definition,

$$f'(p_{n-1}) = \lim_{x \to p_{n-1}} \frac{f(x) - f(p_{n-1})}{x - p_{n-1}}.$$

If $x = p_{n-2}$, then

$$f'(p_{n-1}) \approx \frac{f(p_{n-2}) - f(p_{n-1})}{p_{n-2} - p_{n-1}} = \frac{f(p_{n-1}) - f(p_{n-2})}{p_{n-1} - p_{n-2}}.$$

Using this approximation for $f'(p_{n-1})$ in Newton's formula gives

$$p_n = p_{n-1} - \frac{f(p_{n-1})(p_{n-1} - p_{n-2})}{f(p_{n-1}) - f(p_{n-2})}.$$

The technique using this formula is called the **Secant method** and is presented in Algorithm 2.4.

┌─Secant Algorithm 2.4─

To find a solution to $f(x) = 0$ given initial approximations p_0 and p_1:

INPUT initial approximations p_0, p_1; tolerance TOL; maximum number of iterations N_0.

OUTPUT approximate solution p or message of failure.

Step 1 Set $i = 2$;
$$q_0 = f(p_0);$$
$$q_1 = f(p_1).$$

Step 2 While $i < N_0$ do Steps 3–6.

    Step 3 Set $p = p_1 - q_1(p_1 - p_0)/(q_1 - q_0)$.   (*Compute p_i.*)

    Step 4 If $|p - p_1| < TOL$ then
        OUTPUT (p);   (*Procedure completed successfully.*)
        STOP.

    Step 5 Set $i = i + 1$.

    Step 6 Set $p_0 = p_1$;   (*Update p_0, q_0, p_1, q_1.*)
        $q_0 = q_1$;
        $p_1 = p$;
        $q_1 = f(p)$.

Step 7 OUTPUT ('Method failed after N_0 iterations, $N_0 = $ ', N_0);
(*Procedure completed unsuccessfully.*)
STOP.

The next example involves a problem that we considered in Example 1(a), where we used Newton's method with $p_0 = \pi/4$.

EXAMPLE 3   Find a zero of $f(x) = \cos x - x$, using the Secant method.

    In Example 1 we used an initial approximation of $p_0 = \pi/4$. Here we need two initial approximations. Table 2.6 lists the calculations with $p_0 = .5$, $p_1 = \pi/4$, and the formula

$$p_n = p_{n-1} - \frac{(p_{n-1} - p_{n-2})(\cos p_{n-1} - p_{n-1})}{(\cos p_{n-1} - p_{n-1}) - (\cos p_{n-2} - p_{n-2})} \qquad \text{for } n \geq 2,$$

from Algorithm 2.4.   □

TABLE 2.6

n	p_n
0	.5
1	.7853981635
2	.7363841390
3	.7390581394
4	.7390851492
5	.7390851334

By comparing the results here with those in Example 1, we can see that p_5 is accurate to the tenth decimal place. It is interesting to note that the convergence of the Secant method is slightly slower in this example than that of Newton's method, which obtained this degree of accuracy with p_3. This result is generally true, as will be shown in Section 2.4. (See Exercise 14 of Section 2.4.)

Newton's method (or the Secant method) is often used to refine an answer obtained by another technique, such as the Bisection method. Since these methods require a good first approximation, but generally give rapid convergence, they serve this purpose well.

Exercise Set 2.3

1. The following describes Newton's method graphically: Suppose that $f'(x)$ exists on $[a, b]$ and that $f'(x) \neq 0$ on $[a, b]$. Further, suppose there exists one $p \in [a, b]$ such that $f(p) = 0$. Let $p_0 \in [a, b]$ be arbitrary. Let p_1 be the point at which the tangent line to f at $(p_0, f(p_0))$ crosses the x-axis. For each $n \geq 1$, let p_n be the x-intercept of the line tangent to f at $(p_{n-1}, f(p_{n-1}))$. Derive the formula describing this method.

2. Find an approximate root of $x^3 - x - 1 = 0$ in $[1, 2]$ with 10^{-5} accuracy, first by Newton's method and then by the Secant method.

3. Describe the Secant method geometrically. [*Hint*:   Refer to Exercise 1 and the definition of the secant line.]

4. Use Newton's method to approximate the solutions of the following equations to within 10^{-5}.

   a) $x = \dfrac{2 - e^x + x^2}{3}$

   b) $3x^2 - e^x = 0$

   c) $e^x + 2^{-x} + 2 \cos x - 6 = 0$

   d) $x^2 + 10 \cos x = 0$

5. Repeat Exercise 4 using the Secant method.

6. Solve $4 \cos x = e^x$ with accuracy 10^{-4}, by using:

   a)   Newton's method with $p_0 = 1$.
   b)   The Secant method with $p_0 = \pi/4$ and $p_1 = \pi/2$.

7. Use Newton's method to solve the equation

$$0 = \left(\sin x - \frac{x}{2}\right)^2 \quad \text{with} \quad p_0 = \frac{\pi}{2}.$$

   Iterate using Newton's method until an accuracy of 10^{-5} is obtained for the approximate root with $f(x) = (\sin x - x/2)^2$. Do the results seem unusual for Newton's method? Explain.

8. a)   Compute an approximation to $\sqrt{3}$ accurate to four significant digits, using Newton's method with $p_0 = 2$.
   b)   Repeat part (a), using the Secant method, and then compare your results with those obtained in Exercise 4 of Section 2.2 and Exercise 4 of Section 2.1.

9. The function $f(x) = (4x - 7)/(x - 2)$ has a zero at $p = 1.75$. Use Newton's method with the following initial approximations.

   a) $p_0 = 1.625$
   c) $p_0 = 1.5$

   b) $p_0 = 1.875$
   d) $p_0 = 1.95$

   Explain the results graphically.

10.   For what values of p_0 and p_1 can the Secant method be used to solve the following equation?

$$f(x) = \frac{4x - 7}{x - 2} = 0.$$

11.   The iteration equation for the Secant method can also be written as:

$$p_n = \frac{f(p_{n-1})p_{n-2} - f(p_{n-2})p_{n-1}}{f(p_{n-1}) - f(p_{n-2})}.$$

Can you explain why, in general, this iteration equation is inferior to the one given in Algorithm 2.4?

12.   **Method of False Position (or *Regular Falsi*)** This is another method for finding a root of the equation $f(x) = 0$ lying in the interval $[a, b]$. The method is similar to the Bisection technique in that intervals $[a_i, b_i]$ are generated bracketing a root, and the method is similar to the Secant method in the manner of obtaining new approximate intervals. Assuming that the interval $[a_i, b_i]$ contains a root of $f(x) = 0$, compute the value of the x-intercept of the line joining the points $(a_i, f(a_i))$ and $(b_i, f(b_i))$ and label this point p_i. If $f(p_i)f(a_i) < 0$, define $a_{i+1} = a_i$ and $b_{i+1} = p_i$ otherwise, define $a_{i+1} = p_i$ and $b_{i+1} = b_i$. Construct an algorithm, similar to Algorithms 2.1 and 2.4, which describes the Method of False Position.

13.   Repeat Exercise 4 using the Method of False Position described in Exercise 12.

14.   Use the algorithm constructed in Exercise 12 to find an approximate root of $\cos x - x = 0$ lying on the interval $[.5, \pi/4]$.

15.   Use Newton's method to find an approximation for λ, accurate to within 10^{-4}, for the population equation

$$1{,}564{,}000 = 1{,}000{,}000e^{\lambda} + \frac{435{,}000}{\lambda}(e^{\lambda} - 1),$$

discussed in the introduction to this chapter. Use this value to predict the population at the end of the second year, assuming that the immigration rate during this year remains at 435,000 individuals per year.

16.   Problems relating the amount of money required to pay off a mortgage over a fixed period of time involve a formula,

$$A = \frac{P}{i}[1 - (1 + i)^{-n}],$$

known as an *ordinary annuity equation*. In this equation A is the amount of the mortgage, P is the amount of each payment, and i is the interest rate per period for the n payment periods.

   Suppose that a 30-year home mortgage in the amount of $50,000 is needed and that the borrower can afford house payments of at most $450 per month. What is the maximal interest rate the borrower can afford to pay?

17.   The accumulated value of a savings account based on regular periodic payments can be determined from the *annuity due equation*,

$$A = \frac{P}{i}[(1 + i)^n - 1].$$

In this equation A is the amount in the account, P is the amount regularly deposited, and i is the rate of interest per period for the n deposit periods.

   An engineer would like to have a savings account valued at $75,000 upon retirement in 20 years and can afford to put $150 per month toward this goal. What is the minimal interest rate at which this amount can be deposited, assuming that the interest is compounded quarterly? What is the minimal interest rate if the interest is compounded daily? (Assume a 360-day year, the common banking practice.)

18. The logistic population growth model is described by an equation of the form

$$P(t) = \frac{P_L}{1 - ce^{-kt}},$$

where P_L, c, and k are constants and $P(t)$ is the population at time t. P_L represents the limiting value of the population since $\lim_{t \to \infty} P(t) = P_L$, provided that $k > 0$. Use the census data for the years 1950, 1960, and 1970 listed in the table on page 73 to determine the constants P_L, c, and k for a logistic growth model. Use the logistic model to predict the population of the United States in 1980 and in 2000. Compare the 1980 prediction to the actual value.

19. The Gompertz population growth model is described by

$$P(t) = P_L e^{-ce^{-kt}}$$

where P_L, c, and k are constants and $P(t)$ is the population at time t. Repeat Exercise 18 using the Gompertz growth model in place of the logistic model.

20. In the design of terrain vehicles, it is necessary to consider the failure of the vehicle when attempting to negotiate two types of obstacles. One type of failure is called Hang-Up Failure (HUF) and typically occurs when the vehicle attempts to cross an obstacle that causes the bottom of the vehicle to touch the ground (or the obstacle). The other type of failure is called Nose-In Failure (NIF) and commonly occurs when the vehicle descends into a ditch and its nose touches the ground.

The following figure, adapted from Bekker [9], shows the components associated with the NIF of a vehicle. In that reference it is shown that the maximum angle α, which can be negotiated by a vehicle when β is the maximum angle at which HUF does *not* occur, satisfies the equation

$$A \sin \alpha \cos \alpha + B \sin^2 \alpha - C \cos \alpha - E \sin \alpha = 0,$$

where

$$A = l \sin \beta_1,$$
$$B = l \cos \beta_1,$$
$$C = (h + .5D)\sin \beta_1 - .5D \tan \beta_1,$$
$$E = (h + .5D)\cos \beta_1 - .5D$$

(See the following figure.)

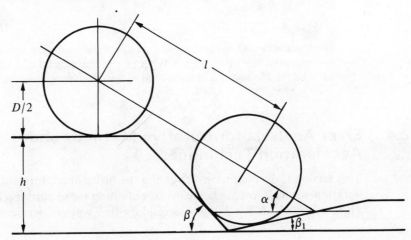

a) It is stated that when $l = 89$ in., $h = 49$ in., $D = 55$ in., and $\beta_1 = 11.5°$, angle α is approximately 33°. Verify this result.

b) Find α for the situation when l, h, and β_1 are the same as in part (a) but $D = 30$ in.

21. In a paper entitled "Holdup and Axial Mixing in Bubble Columns Containing Screen Cylinders" [25], B. H. Chen computes the gas holdup in a gas–liquid bubble column by first approximating the quantity

(1)
$$2 \sum_{n=1}^{\infty} \frac{S_n \sin S_n}{S_n^2 + M^2 + 2M} \exp\left[M - \frac{S_n^2 + M^2}{2M} \cdot \frac{t}{\theta} \right],$$

where t and M are physical parameters and the S_n's are the smallest values (in magnitude) satisfying

(2)
$$S_n \tan\left(\frac{S_n}{2}\right) = M, \qquad \text{when } n \text{ is odd, and}$$

(3)
$$S_n \cot\left(\frac{S_n}{2}\right) = -M, \qquad \text{when } n \text{ is even.}$$

a) Assuming $M = 3.7$, find S_1, S_2, S_3, and S_4.

b) Use the results in part (a) to approximate the sum in Eq. (1), when $t = 0$.

22. In order to find approximations for the tension, T, and angle of inclination from the horizontal, ϕ, at a particular point on a cable or pipeline run underwater, a number of equations of the form

(1)
$$\phi = \tan^{-1}\left[\frac{T_0 \sin \phi_0 - F}{T_0 \cos \phi_0 - G} \right], \quad \text{and}$$

(2)
$$T = [(T_0 \sin \phi_0 - F)^2 + (T_0 \cos \phi_0 - G)^2]^{1/2}$$

must be solved for ϕ and T (see reference [88]). The functions F and G in (1) and (2) both involve ϕ and have the form

$$F = \frac{s}{2}[-f(\phi_0)\sin \phi_0 - f(\phi)\sin \phi + g(\phi_0)\cos \phi_0 + g(\phi)\cos \phi] - ws$$

and $$G = -\frac{s}{2}[f(\phi_0)\cos \phi_0 + f(\phi)\cos \phi + g(\phi_0)\sin \phi_0 + g(\phi)\sin \phi],$$

for given tangential and normal hydrodynamic loading functions f and g.

Suppose that $\phi_0 = \pi/2$, $s = .1$, $T_0 = 2$, $w = 1$, and the loading functions f and g are given by

$$f(\phi) = .02 \cos \phi \qquad \text{and} \qquad g(\phi) = .98 \sin^2 \phi + .02 \sin \phi.$$

a) Use the Secant method to find an approximation to ϕ starting with both initial approximations close to $\pi/2$. What type of accuracy for ϕ is sufficient?

b) Use the Method of False Position to find this approximation. (See Exercise 12.)

2.4 Error Analysis for Iterative Methods and Acceleration Techniques

This section is devoted to investigating the order of convergence of the functional iteration schemes and, as a means of obtaining rapid convergence, rediscovering Newton's method. We will also consider other ways of accelerating convergence.

Before methods for accelerating convergence can be presented, it is necessary to define a procedure for measuring the speed of convergence.

Definition 2.8 Suppose that $\{p_n\}_{n=0}^{\infty}$ is a sequence that converges to p and that $e_n = p_n - p$ for each $n \geq 0$. If positive constants λ and α exist, with

$$\lim_{n \to \infty} \frac{|e_{n+1}|}{|e_n|^{\alpha}} = \lambda,$$

then $\{p_n\}_{n=0}^{\infty}$ is said to **converge to p of order α**, with asymptotic error constant λ.

An iterative technique for solving a problem of the form $x = g(x)$ is said to be of order α if, whenever the method gives convergence for a sequence $\{p_n\}_{n=0}^{\infty}$, where $p_n = g(p_{n-1})$ for $n \geq 1$, then the sequence $\{p_n\}_{n=0}^{\infty}$ converges to the solution of order α.

Two cases will be given special attention. If $\alpha = 1$, then the method is called **linear**. For example, the Bisection Algorithm 2.1 gives linear convergence (see Exercise 12). A second-order method, $\alpha = 2$, is called **quadratic**, and the convergence in such a case is called quadratic convergence. The following example shows that fixed-point iteration is generally linear, and also compares linear to quadratic convergence.

EXAMPLE 1

a)   Suppose we want to find an approximate solution to $g(x) = x$, using the fixed-point iteration scheme $p_n = g(p_{n-1})$, for all $n \geq 1$.

Assume that g maps the interval $[a, b]$ into itself and that a positive number k exists with $|g'(x)| \leq k < 1$ for all $x \in [a, b]$. Theorem 2.4 (p. 30) implies that g has a unique fixed point $p \in [a, b]$ and if $p_0 \in [a, b]$, then the fixed-point sequence $\{p_n\}_{n=0}^{\infty}$ converges to p. The convergence will be shown to be linear, provided that $g'(p) \neq 0$. If n is any positive integer, then

$$e_{n+1} = p_{n+1} - p = g(p_n) - g(p) = g'(\xi_n)(p_n - p) = g'(\xi_n)e_n,$$

where ξ_n is between p_n and p. Since $\{p_n\}_{n=0}^{\infty}$ converges to p, $\{\xi_n\}_{n=0}^{\infty}$ also converges to p. Assuming that g' is continuous on $[a, b]$, we have

$$\lim_{n \to \infty} g'(\xi_n) = g'(p).$$

Thus,

$$\lim_{n \to \infty} \frac{e_{n+1}}{e_n} = \lim_{n \to \infty} g'(\xi_n) = g'(p) \quad \text{and} \quad \lim_{n \to \infty} \frac{|e_{n+1}|}{|e_n|} = |g'(p)|.$$

Hence, fixed-point iteration exhibits linear convergence if $g'(p) \neq 0$. Higher-order convergence can occur only when $g'(p) = 0$.

b)   To compare linear and quadratic convergence, assume that we have two iterative schemes described by :

(2.20) $\displaystyle\lim_{n\to\infty} \frac{|e_{n+1}|}{|e_n|} = \lambda$ where $0 < \lambda < 1$.

(2.21) $\displaystyle\lim_{n\to\infty} \frac{|\tilde{e}_{n+1}|}{|\tilde{e}_n|^2} = \tilde{\lambda}$ where $\tilde{\lambda} > 0$ and $|\tilde{\lambda}\tilde{e}_0| < 1$.

Also assume that

$$|e_{n+1}| \leq \lambda|e_n|$$

and $$|\tilde{e}_{n+1}| \leq \tilde{\lambda}|\tilde{e}_n|^2 \qquad \text{for all } n \geq 0.$$

Then, $$|e_{n+1}| \leq \lambda^{n+1}|e_0|$$

and $$|\tilde{e}_{n+1}| \leq \tilde{\lambda}|\tilde{e}_n|^2 \leq \tilde{\lambda}\tilde{\lambda}^2|\tilde{e}_{n-1}|^4 \leq \cdots$$
$$\leq \tilde{\lambda}^{(2^{n+1}-1)}|\tilde{e}_0|^{2^{n+1}} = (\tilde{\lambda}|\tilde{e}_0|)^{(2^{n+1}-1)}|\tilde{e}_0| \qquad \text{for all } n \geq 0.$$

Suppose that $|e_0| = |\tilde{e}_0| = 1$, $\lambda = \tilde{\lambda} = .75$ and that we need to know the number of iterations necessary to obtain an error less than or equal to 10^{-8}. For the linear method N, the number of necessary iterations, must have the property that

$$(.75)^{N+1} \leq 10^{-8} \qquad \text{or} \qquad N + 1 \geq \frac{-8}{\log_{10}.75} \approx 64.$$

Thus, the linear method (2.20) requires about 63 iterations.

Analogously, the quadratic scheme requires that

$$(.75)^{(2^{N+1}-1)} \leq 10^{-8} \qquad \text{or} \qquad 2^{N+1} \geq \frac{-8}{\log_{10}.75} + 1 \approx 65,$$

and the quadratic scheme requires no more than five or six iterations. In this circumstance, the quadratically convergent scheme is vastly superior.

Another striking property of quadratic schemes can be illustrated as follows: Suppose that the iterates (p_n's) are approximately 1, and that $|e_n| = 10^{-b_n}$, $b_n > 0$; that is, b_n is essentially the number of correct digits in the nth iterate. Since $|e_{n+1}| \leq \lambda|e_n|^2$,

$$10^{-(b_{n+1})} \leq \lambda 10^{-2b_n}.$$

Solving for b_{n+1},

$$-b_{n+1} \leq \log_{10}\lambda - 2b_n;$$

so $$b_{n+1} \geq 2b_n - \log_{10}\lambda,$$

and the number of correct digits is approximately doubled at each step in a quadratic scheme. $\square$

Having discussed the desirability of quadratically convergent schemes, the next step is to determine and characterize quadratic functional iteration schemes.

Theorem 2.9   Let p be a solution of $x = g(x)$. Suppose that $g'(p) = 0$ and g'' is continuous in an open interval containing p. Then there exists a $\delta > 0$ such that, for $p_0 \in [p - \delta, p + \delta]$, fixed-point iteration becomes a second-order method. Furthermore, if $|g''(x)|/2 \leq M$ for $p - \delta \leq x \leq p + \delta$ and $M|e_0| < 1$, then fixed-point iteration will converge and the convergence will be quadratic.

Proof   Choose $\delta > 0$ such that on the interval $[p - \delta, p + \delta]$, $|g'(x)| \leq k < 1$ and g'' is continuous. Since $|g'(x)| \leq k < 1$, it follows that the terms of the sequence $\{p_n\}_{n=0}^{\infty}$ are contained in $[p - \delta, p + \delta]$. Expanding $g(x)$ in a linear Taylor polynomial for $x \in [p - \delta, p + \delta]$ gives

$$g(x) = g(p) + g'(p)(x - p) + \frac{g''(\xi)}{2}(x - p)^2,$$

where ξ lies between x and p. Using the hypotheses $g(p) = p$ and $g'(p) = 0$, we infer that:

$$g(x) = p + \frac{g''(\xi)}{2}(x - p)^2.$$

In particular, when $x = p_n$ for some n,

$$p_{n+1} = g(p_n) = p + \frac{g''(\xi_n)}{2}(p_n - p)^2$$

with ξ_n between p_n and p. Thus,

(2.22) $$e_{n+1} = \frac{g''(\xi_n)}{2}e_n^2.$$

Since $|g'(x)| \leq k < 1$ on $[p - \delta, p + \delta]$, and g maps $[p - \delta, p + \delta]$ into itself, it follows from Theorem 2.4 (p. 30) that $\{p_n\}_{n=0}^{\infty}$ converges to p. Since ξ_n is between p and p_n for each n, $\{\xi_n\}_{n=0}^{\infty}$ converges to p also, and

$$\lim_{n \to \infty} \frac{|e_{n+1}|}{|e_n|^2} = \frac{|g''(p)|}{2}.$$

This shows that the scheme is second-order. If $|g''(x)|/2 \leq M$ for all $x \in [p - \delta, p + \delta]$, Eq. (2.22) implies that

$$|e_{n+1}| \leq M|e_n|^2 \leq M^3|e_{n-1}|^4 \leq \cdots \leq (M|e_0|)^{(2^{n+1}-1)}|e_0|,$$

so the assumption $M|e_0| < 1$ is sufficient to imply that the scheme is quadratically convergent. $\square$

To use Theorem 2.9 to solve an equation of the form $f(x) = 0$, suppose the equation $f(x) = 0$ has a solution p for which $f'(p) \neq 0$. Consider a fixed-point scheme

$$p_n = g(p_{n-1}), \qquad n \geq 1,$$

with g of the form

$$g(x) = x - \phi(x)f(x),$$

where $\phi(x)$ is an arbitrary function to be chosen later.

If $\phi(x)$ is bounded, then $g(p) = p$, and, in order for the iterative procedure derived from g to be quadratically convergent, it suffices to have $g'(p) = 0$. But

$$g'(x) = 1 - \phi'(x)f(x) - f'(x)\phi(x) \qquad \text{and} \qquad g'(p) = 1 - f'(p)\phi(p).$$

Consequently, $g'(p) = 0$ if and only if $\phi(p) = 1/f'(p)$.

In particular, quadratic convergence holds for the scheme

$$(2.23) \qquad\qquad p_n = g(p_{n-1}) = p_{n-1} - \frac{f(p_{n-1})}{f'(p)}$$

under suitable conditions on f. However, since p, and consequently $f'(p)$, are generally unknown, a reasonable approach is to let $\phi(x) = 1/f'(x)$, so that $\phi(p) = 1/f'(p)$. The procedure then becomes

$$(2.24) \qquad\qquad p_n = g(p_{n-1}) = p_{n-1} - \frac{f(p_{n-1})}{f'(p_{n-1})},$$

which can be recognized as Newton's method.

In the discussion above it was necessary that at the solution p, $f'(p) \neq 0$. Whenever p is a solution of the equation $f(x) = 0$, a unique function $q(x)$ and positive integer m can be found with

$$(2.25) \qquad\qquad f(x) = (x - p)^m q(x),$$

where p is not a root of $q(x) = 0$. The integer m is called the **multiplicity** of the root p. It is easy to show (see Exercise 9) that $f'(p) \neq 0$ at a solution of $f(x) = 0$ precisely when $m = 1$. Consequently, in the previous discussion it has actually been assumed that p is a root of multiplicity one. In the case when p is a root of multiplicity $m > 1$, we can often show that Newton's method leads to convergence but not in a quadratic manner.

A slight modification of Newton's method can be made to handle the case when the multiplicity is known. Instead of the usual procedure, let

$$(2.26) \qquad\qquad g(x) = x - \frac{mf(x)}{f'(x)}.$$

From the result in Exercise 11, $g'(p) = 1 - (m/m) = 0$ and quadratic convergence is obtained.

While this modification of the Newton–Raphson method may seem to handle the difficulty presented by multiple roots, in actual practice the multiplicity of the root is unknown at the outset, so the difficulty remains. Another, more successful, method of handling the problem is to define a function $\mu(x)$ by

$$(2.27) \qquad\qquad \mu(x) = \frac{f(x)}{f'(x)}.$$

If p is a root of multiplicity $m \geq 1$ and $f(x) = (x - p)^m q(x)$, then

$$\mu(x) = \frac{(x - p)^m q(x)}{m(x - p)^{m-1} q(x) + (x - p)^m q'(x)} = \frac{(x - p)q(x)}{mq(x) + (x - p)q'(x)}$$

also has a root at p, but of multiplicity one. Newton's method can then be applied to the function μ to give

$$g(x) = x - \frac{\mu(x)}{\mu'(x)} = x - \frac{f(x)/f'(x)}{\{[f'(x)]^2 - [f(x)][f''(x)]\}/[f'(x)]^2}$$

or

$$\text{(2.28)} \qquad g(x) = x - \frac{f(x)f'(x)}{[f'(x)]^2 - [f(x)][f''(x)]}.$$

If g has the required continuity conditions, the functional iteration applied to g will be quadratically convergent regardless of the multiplicity of the root. Theoretically, the only drawback to this method is the additional calculation of $f''(x)$ and the more laborious procedure of calculating the iterates. In practice, however, the presence of a multiple root can cause severe round-off problems.

EXAMPLE 2

a) In Example 3 of Section 2.2 we solved $f(x) = x^3 + 4x^2 - 10 = 0$ for the root $p = 1.36523001$. To compare convergence for a root of multiplicity one by Newton's method and the modified Newton's method listed in Eq. (2.28), let

i) $\qquad p_n = p_{n-1} - \dfrac{p_{n-1}^3 + 4p_{n-1}^2 - 10}{3p_{n-1}^2 + 8p_{n-1}},$ from Newton's method

and, from Eq. (2.28),

ii) $\qquad p_n = p_{n-1} - \dfrac{(p_{n-1}^3 + 4p_{n-1}^2 - 10)(3p_{n-1}^2 + 8p_{n-1})}{(3p_{n-1}^2 + 8p_{n-1})^2 - (p_{n-1}^3 + 4p_{n-1}^2 - 10)(6p_{n-1} + 8)}.$

With $p_0 = 1.5$, the first three iterates for (i) and (ii) are as follows.

	(i)	(ii)
p_1	1.37333333	1.35689898
p_2	1.36526201	1.36519585
p_3	1.36523001	1.36523001

b) For an illustration of the situation that occurs at a multiple root, consider the equation $f(x) = x^4 - 4x^2 + 4 = 0$, which has a root of multiplicity two at $x = \sqrt{2}$. Since $f'(\sqrt{2}) = 0$, the method described by Eq. (2.23) cannot be applied even though the actual root is known beforehand. Instead we will compare the results from three methods:

i) $\qquad p_n = p_{n-1} - \dfrac{(p_{n-1}^2 - 2)}{2p_{n-1}},$ from Eq. (2.26) with $m = 2$,

ii) $\qquad p_n = p_{n-1} - \dfrac{(p_{n-1}^2 - 2)}{4p_{n-1}},$ from Newton's method,

and

iii) $\qquad p_n = p_{n-1} - \dfrac{(p_{n-1}^2 - 2)p_{n-1}}{(p_{n-1}^2 + 2)}$ from Eq. (2.28).

With $p_0 = 1.5$, the first three iterates for (i), (ii), and (iii) are as follows.

	(i)	(ii)	(iii)
p_1	1.416666667	1.458333333	1.411764706
p_2	1.414215686	1.436607143	1.414211438
p_3	1.414213562	1.425497619	1.414213562

The actual answer correct to 10^{-9} is the value listed for p_3 in either (i) or (iii). To obtain this accuracy by the standard Newton–Raphson method requires 20 iterations. □

We now present a second-order method that does not involve f'. It is an accelerating technique called **Aitken's Δ^2 process**, and can be used to accelerate the convergence of any sequence that is linearly convergent, regardless of its origin.

Assume that $\{p_n\}_{n=0}^{\infty}$ is a linearly convergent sequence with limit p; that is, for $e_n = p_n - p$,

$$\lim_{n \to \infty} \frac{|e_{n+1}|}{|e_n|} = \lambda \quad \text{and} \quad 0 < \lambda < 1.$$

To investigate the construction of a sequence $\{\hat{p}_n\}_{n=0}^{\infty}$, which converges more rapidly to p, suppose that the limiting case actually occurs for all positive integers n, and that all of the e_n's have the same sign. In this case,

$$e_{n+1} = \lambda e_n \quad \text{for each } n \geq 0.$$

So
$$p_{n+2} = e_{n+2} + p = \lambda e_{n+1} + p, \quad \text{or}$$

(2.29)
$$p_{n+2} = \lambda(p_{n+1} - p) + p \quad \text{for each } n \geq 0.$$

Replacing $(n + 1)$ by n in Eq. (2.29) gives:

(2.30)
$$p_{n+1} = \lambda(p_n - p) + p;$$

and solving equations (2.29) and (2.30) for p while eliminating λ leads to

$$
\begin{aligned}
p &= \frac{p_{n+2} p_n - p_{n+1}^2}{p_{n+2} - 2p_{n+1} + p_n} \\
&= \frac{p_n^2 + p_n p_{n+2} + 2p_n p_{n+1} - 2p_n p_{n+1} - p_n^2 - p_{n+1}^2}{p_{n+2} - 2p_{n+1} + p_n} \\
&= \frac{(p_n^2 + p_n p_{n+2} - 2p_n p_{n+1}) - (p_n^2 - 2p_n p_{n+1} + p_{n+1}^2)}{p_{n+2} - 2p_{n+1} + p_n} \\
&= p_n - \frac{(p_{n+1} - p_n)^2}{p_{n+2} - 2p_{n+1} + p_n}.
\end{aligned}
$$

In general, the original assumption $e_{n+1} = \lambda e_n$ will not be true; nevertheless, it is expected that the sequence $\{\hat{p}_n\}_{n=0}^{\infty}$, defined by

(2.31)
$$\hat{p}_n = p_n - \frac{(p_{n+1} - p_n)^2}{p_{n+2} - 2p_{n+1} + p_n},$$

converges more rapidly to p than the original sequence $\{p_n\}_{n=0}^{\infty}$.

EXAMPLE 3   The sequence $\{p_n\}_{n=1}^{\infty}$, where $p_n = n \ln(1 + (1/n))$, converges linearly to $p = 1$. The first few terms of the sequences $\{p_n\}_{n=1}^{\infty}$ and $\{\hat{p}_n\}_{n=1}^{\infty}$ are given in Table 2.7.

TABLE 2.7

n	p_n	$\hat{p}_n$
1	.6931471805	.9044075720
2	.8109302162	.9311744212
3	.8630462166	.9461287492
4	.8925742052	.9557185842
5	.9116077840	.9624007828
6	.9249040806	
7	.9347197496	

It appears that $\{\hat{p}_n\}_{n=1}^{\infty}$ converges more rapidly to $p = 1$ than $\{p_n\}_{n=1}^{\infty}$.   □

This technique of constructing a more rapidly convergent sequence is called *Aitken's Δ^2 method* because of the following definition:

Definition 2.10   Given the sequence $\{p_n\}_{n=0}^{\infty}$, define the **forward difference** Δp_n by

$$\Delta p_n = p_{n+1} - p_n \qquad \text{for } n \geq 0.$$

Higher powers $\Delta^k p_n$ are defined recursively by

$$\Delta^k p_n = \Delta^{k-1}(\Delta p_n) \qquad \text{for } k \geq 2.$$

Because of the definition,

$$\begin{aligned}
\Delta^2 p_n &= \Delta(p_{n+1} - p_n) \\
&= \Delta p_{n+1} - \Delta p_n \\
&= (p_{n+2} - p_{n+1}) - (p_{n+1} - p_n) \\
&= p_{n+2} - 2p_{n+1} + p_n.
\end{aligned}$$

Thus, the formula for $\hat{p}_n$ can be written as

$$\hat{p}_n = p_n - \frac{(\Delta p_n)^2}{\Delta^2 p_n} \qquad \text{for all } n \geq 0.$$

To this point in our discussion of Aitken's Δ^2 method, we have stated that the sequence $\{\hat{p}_n\}_{n=0}^{\infty}$ converges to p more rapidly than does the original sequence $\{p_n\}_{n=0}^{\infty}$, but we have not said exactly what is meant by the term "more rapid" convergence. The theorem presented below explains and justifies this terminology.

Theorem 2.11   Let p_n be any sequence converging to the limit p and $e_n = p_n - p$, with $e_n \neq 0$ for all $n \geq 0$. Suppose

$$\lim_{n \to \infty} \frac{e_{n+1}}{e_n} = \lambda \qquad \text{where } 0 < \lambda < 1;$$

then the sequence $\{\hat{p}_n\}_{n=0}^{\infty}$ converges to p faster than $\{p_n\}_{n=0}^{\infty}$ in the sense that

$$\lim_{n\to\infty} \frac{\hat{p}_n - p}{p_n - p} = 0.$$

Proof   See Exercise 15.   □

Aitken's Δ^2 method accelerates the convergence of any linearly convergent sequence regardless of the origin of the sequence. Applying Aitken's Δ^2 method to fixed-point iteration gives rise to a procedure known as **Steffensen's method**.

Steffensen's Algorithm 2.5

To find a solution to $p = g(p)$ given an initial approximation p_0:

INPUT initial approximation p_0; tolerance TOL; maximum number of iterations N_0.

OUTPUT approximate solution p or message of failure.

Step 1 Set $i = 1$.

Step 2 While $i < N_0$ do Steps 3–6.

Step 3 Set $p_1 = g(p_0)$;   (*Compute $p_1^{(i-1)}$.*)
$\qquad\quad$ $p_2 = g(p_1)$;   (*Compute $p_2^{(i-1)}$.*)
$\qquad\quad$ $p = p_0 - (p_1 - p_0)^2/(p_2 - 2p_1 + p_0)$.   (*Compute $p_0^{(i)}$.*)

Step 4 If $|p - p_0| < TOL$ then
$\qquad\quad$ OUTPUT (p);   (*Procedure completed successfully.*)
$\qquad\quad$ STOP.

Step 5 Set $i = i + 1$.

Step 6 Set $p_0 = p$.   (*Update p_0.*)

Step 7 OUTPUT ('Method failed after N_0 iterations, $N_0 = $ ', N_0);
$\qquad\quad$ (*Procedure completed unsuccessfully.*)
$\qquad\quad$ STOP.

Note that $\Delta^2 p_n$ may be zero. If that should happen, we terminate the sequence and select $p_2^{(n-1)}$ as the approximate answer, since otherwise this would introduce a zero in the denominator of the next iterate.

EXAMPLE 4   To solve $x^3 + 4x^2 - 10 = 0$ using Steffensen's method, let $x^3 + 4x^2 = 10$ and solve for x by dividing by $x + 4$. Thus, if

$$g(x) = \left(\frac{10}{x + 4}\right)^{1/2},$$

then $x = g(x)$ implies $x^3 + 4x^2 - 10 = 0$.

With $p_0 = 1.5$, Steffensen's procedure gives

k	$p_0^{(k)}$	$p_1^{(k)}$	$p_2^{(k)}$
0	1.5	1.348399725	1.367376372
1	1.365265224	1.365225534	1.365230583
2	1.365230013		

The iterate $p_0^{(2)} = 1.365230013$ is accurate to the ninth decimal place. It is interesting to note that, for this example, Steffensen's method gave about the same rate of convergence as Newton's method (see Example 2). □

From Example 4, it appears that Steffensen's method gives quadratic convergence without evaluating a derivative. Theorem 2.12 verifies that this is the case. The proof of this theorem is not presented here but can be found in Henrici [46], pages 90–92, or Isaacson and Keller [52], pages 103–107.

Theorem 2.12   Suppose that $x = g(x)$ has the solution p with $g'(p) \neq 1$. If there exists a $\delta > 0$ such that $g \in C^3[p - \delta, p + \delta]$, then Steffensen's method gives quadratic convergence for any $p_0 \in [p - \delta, p + \delta]$.

Exercise Set 2.4

1. Solve $x^3 - x - 1 = 0$ for the root in $[1, 2]$ to an accuracy of 10^{-4} using Steffensen's method and the results of Exercise 3 of Section 2.2.

2. Solve $x - 2^{-x} = 0$ for the root in $[0, 1]$ to an accuracy of 10^{-4} using Steffensen's method, and compare to the results of Exercise 1 of Section 2.2.

3. Use Steffensen's method with $p_0 = 2$ to compute an approximation to $\sqrt{3}$ accurate to four significant digits. Compare this result with those obtained in Exercises 8 of Section 2.3, 4 of Section 2.2, and 4 of Section 2.1.

4. Approximate the solutions to within 10^{-5} of the following equations, using Steffensen's method.

   a) $x = \dfrac{2 - e^x + x^2}{3}$, where g is the function in Exercise 6(a) of Section 2.2.

   b) $3x^2 - e^x = 0$, where g is the function in Exercise 7(a) of Section 2.2.

   c) $x - \cos x = 0$, where g is the function in Exercise 7(b) of Section 2.2.

5. Use (2.26) with $m = 2$ and functional iteration to find a root, accurate within 10^{-5} of $f(x) = (\sin x - (x/2))^2 = 0$, starting with $p_0 = \pi/2$. Compare your results with those of Exercise 7 in Section 2.3.

6. Use Eq. (2.28) and functional iteration to find a root, accurate within 10^{-5}, of $f(x) = (\sin x - (x/2))^2 = 0$, starting with $p_0 = \pi/2$. Compare your results with those obtained in Exercise 5 above and in Exercise 7 in Section 2.3.

7. Use the procedure described in Eq. (2.26) with $m = 1, 2,$ and 3 to find an approximation to a root of the equation

$$f(x) = x^2 + 2xe^x + e^{2x} = 0$$

starting with $p_0 = 0$ and performing 10 iterations in each case. Can you determine the multiplicity of the root found? If you are doing the work on a computer, redo your calculations using extended precision and observe the effects of round-off error on your single-precision calculations. Can you determine why the round-off error is so severe?

8. Use the Modified Newton–Raphson method described in Eq. (2.28) to find an approximation to a root of

$$f(x) = x^2 + 2xe^x + e^{2x} = 0,$$

starting with $p_0 = 0$ and performing 10 iterations. If you are using a computer, redo the calculations using extended precision and compare your results with those obtained in Exercise 7.

9. Show that p is a zero of multiplicity one of the differentiable function f if and only if $f(p) = 0$, but $f'(p) \neq 0$.

10. Suppose that f has $m + 1$ continuous derivatives. Show that f has a zero of multiplicity m at p if and only if

$$0 = f(p) = f'(p) = \cdots = f^{(m-1)}(p), \qquad \text{but } f^{(m)}(p) \neq 0.$$

11. Suppose that p is a root of $f(x) = 0$ of multiplicity m and that $g(x) = x - mf(x)/f'(x)$. Show that $g'(p) = 0$.

12. Show that the Bisection Algorithm 2.1 gives linear convergence.

13. Show that the iterative method to solve $f(x) = 0$, given by

$$p_n = p_{n-1} - \frac{f(p_{n-1})}{f'(p_{n-1})} - \frac{f''(p_{n-1})}{2f'(p_{n-1})} \left[\frac{f(p_{n-1})}{f'(p_{n-1})} \right]^2, \qquad n = 1, 2, 3, \ldots,$$

will generally yield cubic ($\alpha = 3$) convergence. Using the analysis of Example 1(b), compare quadratic and cubic convergence.

14. Assume that the Secant method converges to the root p of $f(x) = 0$, that f is twice continuously differentiable, and that $f'(p) \neq 0$. Show that

$$\lim_{n \to \infty} \frac{e_{n+1}}{e_n e_{n-1}} = \frac{f''(p)}{2f'(p)}.$$

(This implies that the convergence of the Secant method is more than linear yet not quite quadratic.)

15. Prove Theorem 2.11.

16. The concept stated in Definition 2.8 could have been defined to include sequences that do not converge. A sequence $\{p_n\}_{n=0}^{\infty}$ is sometimes said to be of order α, with respect to a number p, if

$$\lim_{n \to \infty} \frac{|p_{n+1} - p|}{|p_n - p|^{\alpha}} = \lambda > 0 \qquad \text{for some } \lambda.$$

a) Suppose that $\{p_n\}_{n=0}^{\infty}$ is of order α, with respect to p, for $\alpha = 1$.

    i)   Show that if $\lambda < 1$, then $\{p_n\}_{n=0}^{\infty}$ converges to p.
    ii)   Show that if $\lambda > 1$, then $\{p_n\}_{n=0}^{\infty}$ does not converge to p.
    iii)   What can be concluded when $\lambda = 1$?

b)   Suppose that $\{p_n\}_{n=0}^{\infty}$ is of order α, with respect to p, for $\alpha > 1$. Show that if $|\lambda(p - p_0)| < 1$, then $\{p_n\}_{n=0}^{\infty}$ converges to p.

17.   A sequence $\{p_n\}$ is said to be **superlinearly convergent** to p if

$$\lim_{n \to \infty} \frac{p_{n+1} - p}{p_n - p} = 0.$$

a)   Show that if $p_n \to p$ of order α for $\alpha > 1$, then $\{p_n\}$ is superlinearly convergent to p.
b)   Find a sequence $\{p_n\}$ that is superlinearly convergent to zero, but does not converge to zero of order α for any $\alpha > 1$.

18.   Suppose that $\{p_n\}$ is superlinearly convergent to p. Show that

$$\lim_{n \to \infty} \frac{|p_{n+1} - p_n|}{|p_n - p|} = 1.$$

2.5   Zeros of Real Polynomials

A function of the form

$$P(x) = a_n x^n + a_{n-1} x^{n-1} + \cdots + a_1 x + a_0,$$

where the a_i's are constants and $a_n \neq 0$, is called a **polynomial of degree n**. All polynomials considered in this section are assumed to be real; that is, all the a_i's are real numbers. The zero function, $P(x) = 0$ for all values of x, is considered a polynomial but is assigned no degree.

Theorem 2.13 (Fundamental Theorem of Algebra)   If P is a polynomial of degree $n \geq 1$, then $P(x) = 0$ has at least one (possibly complex) root.

Although Theorem 2.13 is basic to any study of elementary functions, the usual proof requires techniques from the study of complex-function theory. The reader is referred to Dettman [31], page 116, for the culmination of a systematic development of the topics needed to prove Theorem 2.13.

An important consequence of Theorem 2.13 is:

Corollary 2.14   If $P(x) = a_n x^n + a_{n-1} x^{n-1} + \cdots + a_1 x + a_0$ is a polynomial of degree $n \geq 1$, then there exist unique constants $x_1, x_2, \ldots, x_k$, possibly complex, and positive integers, $m_1, m_2, \ldots, m_k$, such that $\sum_{i=1}^{k} m_i = n$ and

$$P(x) = a_n(x - x_1)^{m_1}(x - x_2)^{m_2} \cdots (x - x_k)^{m_k}.$$

Proof   See Exercise 3(a).   □

If $(x - x_i)^{m_i}$ is a factor of $P(x)$, x_i is said to be a root of $P(x) = 0$ (or zero of P) of multiplicity m_i. Corollary 2.14 states that the zeros of a polynomial are unique and that if each zero x_i is counted as many times as its multiplicity m_i, then a polynomial of degree n has exactly n zeros.

The following corollary of the Fundamental Theorem of Algebra will be used often in this section and in later chapters. The proof of this result is considered in Exercise 3(b).

Corollary 2.15 Let P and Q be polynomials of degree at most n. If $x_1, x_2, \ldots, x_k$, $k > n$, are distinct numbers with $P(x_i) = Q(x_i)$ for $i = 1, 2, \ldots, k$, then $P(x) = Q(x)$ for all values of x.

Theorem 2.16 If $z = a + bi$ is a complex zero of multiplicity m of the polynomial P, then $\bar{z} = a - bi$ is also a zero of multiplicity m of the polynomial P and $(x^2 - 2ax + a^2 + b^2)^m$ is a factor of P.

Proof   See Exercise 4.   □

In order to use the Newton–Raphson procedure to locate approximate zeros of a polynomial P, it is necessary to evaluate P and its derivative at specified values. If the nth-degree polynomial $P(x) = a_n x^n + \cdots + a_1 x + a_0$ is to be evaluated at $x = x_0$, a "natural" procedure is to first calculate $x_0^2, x_0^3, \ldots, x_0^n$, a procedure which requires $(n - 1)$ multiplications, then calculate $a_i x_0^i$ for $i = 1, 2, \ldots, n$, which requires n more multiplications, and finally do n additions to obtain the value $P(x_0) = a_0 + a_1 x_0 + a_2 x_0^2 + \cdots + a_n x_0^n$. Therefore, to evaluate $P(x_0)$ in this manner requires $(2n - 1)$ multiplications and n additions. The next theorem presents an algorithm known as **Horner's method** or, in its condensed form, **synthetic division**. Horner's method requires only n multiplications and n additions to evaluate a polynomial and consequently is more efficient in terms of time required for the evaluation of a polynomial than is the "natural" procedure described above. Since Horner's method requires fewer mathematical operations than the "natural" procedure, it is also likely that Horner's method is better than that procedure in terms of reducing the effect of round-off error.

Theorem 2.17 (Horner's Method)   Let

$$P(x) = a_n x^n + a_{n-1} x^{n-1} + \cdots + a_1 x + a_0 \qquad \text{and} \qquad b_n = a_n.$$

If $b_k = a_k + b_{k+1} x_0$ for $k = n - 1, n - 2, \ldots, 1, 0$,

then $b_0 = P(x_0)$. Moreover, if

$$Q(x) = b_n x^{n-1} + b_{n-1} x^{n-2} + \cdots + b_2 x + b_1,$$

then

(2.32) $$P(x) = (x - x_0)Q(x) + b_0.$$

Proof   By the definition of $Q(x)$,

$$
\begin{aligned}
(x - x_0)Q(x) + b_0 &= (x - x_0)(b_n x^{n-1} + \cdots + b_2 x + b_1) + b_0 \\
&= (b_n x^n + b_{n-1} x^{n-1} + \cdots + b_2 x^2 + b_1 x) \\
&\quad - (b_n x_0 x^{n-1} + \cdots + b_2 x_0 x + b_1 x_0) + b_0 \\
&= b_n x^n + (b_{n-1} - b_n x_0)x^{n-1} + \cdots + (b_1 - b_2 x_0)x + (b_0 - b_1 x_0).
\end{aligned}
$$

By the hypotheses of the theorem, $b_n = a_n$ and $b_k - b_{k+1} x_0 = a_k$, so

$$(x - x_0)Q(x) + b_0 = P(x) \qquad \text{and} \qquad b_0 = P(x_0). \qquad □$$

EXAMPLE 1   Evaluate $P(x) = 2x^4 - 3x^2 + 3x - 4$ at $x_0 = -2$ using Horner's method. Using Theorem 2.17.

$$b_4 = 2, \qquad b_3 = 2(-2) + 0 = -4,$$

$$b_2 = (-4)(-2) - 3 = 5, \qquad b_1 = 5(-2) + 3 = -7$$

and finally,

$$P(-2) = b_0 = (-7)(-2) - 4 = 10.$$

Moreover, Theorem 2.17 gives

$$P(x) = (x + 2)(2x^3 - 4x^2 + 5x - 7) + 10. \qquad \square$$

When we use hand calculation in Horner's method, we first construct a table, which suggests the "synthetic division" name often applied to Horner's method. For the problem in the preceding example, the table would appear as:

	$\begin{pmatrix} \text{Coefficient} \\ \text{of } x^4 \end{pmatrix}$	$\begin{pmatrix} \text{Coefficient} \\ \text{of } x^3 \end{pmatrix}$	$\begin{pmatrix} \text{Coefficient} \\ \text{of } x^2 \end{pmatrix}$	$\begin{pmatrix} \text{Coefficient} \\ \text{of } x \end{pmatrix}$	$\begin{pmatrix} \text{Constant} \\ \text{term} \end{pmatrix}$
$x_0 = -2$	$a_4 = 2$	$a_3 = 0$	$a_2 = -3$	$a_1 = 3$	$a_0 = -4$
		$b_4 x_0 = -4$	$b_3 x_0 = 8$	$b_2 x_0 = -10$	$b_1 x_0 = 14$
	$b_4 = 2$	$b_3 = -4$	$b_2 = 5$	$b_1 = -7$	$b_0 = 10$

An additional advantage of using the Horner's (or synthetic-division) procedure is that, since

$$P(x) = (x - x_0)Q(x) + b_0,$$

where $\qquad Q(x) = b_n x^{n-1} + b_{n-1} x^{n-2} + \cdots + b_2 x + b_1,$

differentiating with respect to x gives

(2.33) $\qquad P'(x) = Q(x) + (x - x_0)Q'(x) \qquad$ and

(2.34) $\qquad P'(x_0) = Q(x_0).$

Therefore, when the Newton–Raphson method is being used to find an approximate zero of a polynomial P, both P and P' can be evaluated in the same manner. The following algorithm computes $P(x_0)$ and $P'(x_0)$ using Horner's method.

Horner's Algorithm 2.6

To evaluate the polynomial

$$P(x) = a_n x^n + a_{n-1} x^{n-1} + \cdots + a_1 x + a_0$$

and its derivative at x_0:

INPUT degree n; coefficients $a_0, a_1, \ldots, a_n$; x_0.

OUTPUT $y = P(x_0); z = P'(x_0)$.

Step 1 Set $y = a_n;$ (*Compute b_n for P.*)
  $z = a_n.$ (*Compute b_{n-1} for Q.*)

Step 2 For $j = n - 1, n - 2, \ldots, 1$
    set $y = x_0 y + a_j$;   (*Compute b_j for P.*)
    $z = x_0 z + y$.   (*Compute b_{j-1} for Q.*)

Step 3 Set $y = x_0 y + a_0$.   (*Compute b_0 for P.*)

Step 4 OUTPUT (y, z);
    STOP.

EXAMPLE 2   Find an approximation to one of the zeros of

$$P(x) = 2x^4 - 3x^2 + 3x - 4$$

by using the Newton–Raphson procedure and using synthetic division to evaluate $P(x_n)$ and $P'(x_n)$ for each iterate x_n. Using $x_0 = -2$ as an initial approximation, we obtained $P(-2)$ in Example 1 by:

$$
\begin{array}{r|rrrrr}
x_0 = -2 & 2 & 0 & -3 & 3 & -4 \\
         &   & -4 & 8 & -10 & 14 \\
\hline
         & 2 & -4 & 5 & -7 & 10 = P(-2)
\end{array}
$$

Using Theorem 2.17 and Eq. (2.34), we obtain:

$$Q(x) = 2x^3 - 4x^2 + 5x - 7 \qquad \text{and} \qquad P'(-2) = Q(-2);$$

so $P'(-2)$ can be found by evaluating $Q(-2)$ in a similar manner:

$$
\begin{array}{r|rrrr}
x_0 = -2 & 2 & -4 & 5 & -7 \\
         &   & -4 & 16 & -42 \\
\hline
         & 2 & -8 & 21 & -49 = Q(-2) = P'(-2)
\end{array}
$$

and $x_1 = x_0 - \dfrac{P(x_0)}{P'(x_0)} = -2 - \dfrac{10}{-49} \approx -1.796.$

Repeating the procedure to find x_2, we have:

$$
\begin{array}{r|rrrr}
-1.796 & 2 & 0 & -3 & 3 & -4 \\
       &   & -3.592 & 6.451 & -6.197 & 5.742 \\
\hline
       & 2 & -3.592 & 3.451 & -3.197 & 1.742 = P(x_1) \\
       &   & -3.592 & 12.902 & -29.368 & \\
\hline
       & 2 & -7.184 & 16.353 & -32.565 = Q(x_1) = P'(x_1)
\end{array}
$$

So $P(-1.796) = 1.742$, $P'(-1.796) = -32.565$, and

$$x_2 = -1.796 - \frac{1.742}{-32.565} \approx -1.7425.$$

An actual zero to five decimal places is -1.74259.   □

Note that the polynomial denoted by Q depends on the approximation being used and changes from iterate to iterate.

If the Nth iterate, x_N, in the Newton–Raphson procedure is an approximate zero of P, then

$$P(x) = (x - x_N)Q(x) + b_0 = (x - x_N)Q(x) + P(x_N) \approx (x - x_N)Q(x);$$

so $x - x_N$ is an approximate factor of $P(x)$. Letting $\hat{x}_1 = x_N$ be the approximate zero of P and $Q_1(x)$ the approximate factor,

(2.35) $$P(x) \approx (x - \hat{x}_1)Q_1(x),$$

we can find a second approximate zero of P by applying the Newton–Raphson procedure to $Q_1(x)$. If P is an nth-degree polynomial with n real zeros, this procedure applied repeatedly will eventually result in $(n - 2)$ approximate zeros of P and an approximate quadratic factor $Q_{n-2}(x)$. At this stage, $Q_{n-2}(x) = 0$ can be solved by the quadratic formula to find the last two approximate zeros of P. Although this method can be used to find approximate zeros of many polynomials, it depends on repeated use of approximations and on occasion can lead to very inaccurate approximations.

Reviewing the procedure just described, we can see that the difficulty with accuracy is due to the fact that, when we obtained the approximate zeros of P, the Newton–Raphson procedure was used on the reduced polynomial Q_k, i.e., the polynomial having the property that

$$P(x) \approx (x - \hat{x}_1)(x - \hat{x}_2) \cdots (x - \hat{x}_k)Q_k(x).$$

An approximate zero $\hat{x}_{k+1}$ of Q_k was found and the new reduced polynomial Q_{k+1} was used to find the next approximate zero where

$$Q_k(x) \approx (x - \hat{x}_{k+1})Q_{k+1}(x).$$

One way to eliminate this difficulty is to use the reduced equations, that is, the approximate factors of the original polynomial P, to find approximations, $\hat{x}_2, \hat{x}_3, \ldots, \hat{x}_k$, to the zeros of P and then improve these approximations by applying the Newton–Raphson procedure to the original polynomial P. This procedure will be illustrated after the following discussion, which details a method for obtaining initial approximations to the zeros of polynomials.

It has been noted previously that the success of Newton's method often depends on obtaining a good initial approximation. Since polynomials are continuous on the real line, a basic method for finding an approximation to a zero of a polynomial is given by the Intermediate Value Theorem (p. 6).

The basic idea for finding approximate zeros of P is as follows: evaluate P at points x_i for $i = 1, 2, \ldots, k$. If $P(x_i)P(x_j) < 0$, then P has a zero between x_i and x_j. The problem becomes a matter of choosing the x_i's so that the chance of missing a change of sign is minimized, while keeping the number of x_i's reasonably small.

To illustrate the possible difficulty of this problem, consider the polynomial

$$P(x) = 16x^4 - 40x^3 + 5x^2 + 20x + 6.$$

If x_i is any integer, it can be shown that $P(x_i) > 0$. In fact, if $x_i = i/4$ for any integer i, it still remains true that $P(x_i) > 0$; hence, evaluating $P(x)$ at this infinite number of x_i's would not locate any intervals (x_i, x_j) containing zeros of P. However, the reader will see in Example 3 that P does have real zeros. It happened that the above choice of the x_i's is inappropriate for this polynomial because of the closeness of the roots.

The basic question is: how does one choose x_i's that are appropriate? Generally, if P has real zeros and one chooses enough x_i's, then eventually intervals will be located that contain the zeros of P; however, it is clear that this could become an extremely time-consuming procedure.

A suggested method for finding intervals containing real zeros of $P(x)$ is as follows: to find an initial approximation to a real root of

$$0 = P(x) = a_n x^n + a_{n-1} x^{n-1} + \cdots + a_1 x + a_0,$$

first note that $y \in (0, 1]$ precisely when $1/y \in [1, \infty)$, and $y \in [-1, 0)$ precisely when $1/y \in (-\infty, -1]$. Also,

$$P\left(\frac{1}{y}\right) = a_n\left(\frac{1}{y}\right)^n + a_{n-1}\left(\frac{1}{y}\right)^{n-1} + \cdots + a_1\left(\frac{1}{y}\right) + a_0$$

$$= \frac{1}{y^n}(a_n + a_{n-1}y + \cdots + a_1 y^{n-1} + a_0 y^n).$$

Hence, if $\hat{P}$ is defined by

$$\hat{P}(y) = a_0 y^n + a_1 y^{n-1} + \cdots + a_{n-1}y + a_n,$$

then $\hat{y} \neq 0$ is a zero of $\hat{P}$ if and only if $1/\hat{y}$ is a zero of P.

To find approximations to all real zeros of P, it suffices to search the intervals $[-1, 0]$ and $[0, 1]$ evaluating both P and $\hat{P}$ to determine sign changes. By the Intermediate Value Theorem, if $P(x_i)P(x_j) < 0$, then P has a zero between x_i and x_j. (If some $P(x_i) = 0$, then a zero of P has been located.) If $\hat{P}(x_i)\hat{P}(x_j) < 0$, then $\hat{P}$ has a zero between x_i and x_j and P has a zero between $1/x_i$ and $1/x_j$. We consider the intervals $[-1, 0]$ and $[0, 1]$ separately in order to eliminate confusion in the case where $\hat{P}(x_i)\hat{P}(x_j) < 0$ for x_i in $[-1, 0)$ and x_j in $(0, 1]$.

There are many ways to choose the x_i's; however, for most practical problems the following choice of the x_i's should be sufficient: choose an even integer N that will give a bound for the number of points to be evaluated. Let $x_0 = -1$, and $x_i = x_{i-1} + 2/N$ for each $i = 1, \ldots, N$. This choice results in using $N + 1$ points with $x_0 = -1$, $x_{N/2} = 0$, and $x_N = 1$. The following example shows a practical application of this procedure.

EXAMPLE 3   To find the roots of

$$0 = P(x) = 16x^4 - 40x^3 + 5x^2 + 20x + 6,$$

let $y = 1/x$; then

$$P\left(\frac{1}{y}\right) = 16\left(\frac{1}{y}\right)^4 - 40\left(\frac{1}{y}\right)^3 + 5\left(\frac{1}{y}\right)^2 + 20\left(\frac{1}{y}\right) + 6$$

$$= \frac{1}{y^4}(16 - 40y + 5y^2 + 20y^3 + 6y^4);$$

so

$$\hat{P}(y) = 6y^4 + 20y^3 + 5y^2 - 40y + 16,$$

or

$$\hat{P}(x) = 6x^4 + 20x^3 + 5x^2 - 40x + 16,$$

and the zeros of $\hat{P}$ are reciprocals of the zeros of P. It is now necessary to search $[-1, 1]$ to find intervals (x_i, x_j) which contain zeros of P or zeros of $\hat{P}$. Synthetic division (Horner's method) will be used to evaluate $P(x_i)$ and $\hat{P}(x_i)$. For example, evaluating $P(-1)$ and $\hat{P}(-1)$ involves the following scheme:

$$
\begin{array}{r|rrrrr}
& & P & & & \\
x = -1 & 16 & -40 & 5 & 20 & 6 \\
& & -16 & 56 & -61 & 41 \\
\hline
& 16 & -56 & 61 & -41 & \underline{47} = P(-1)
\end{array}
$$

$$
\begin{array}{r|rrrrr}
& & \hat{P} & & & \\
x = -1 & 6 & 20 & 5 & -40 & 16 \\
& & -6 & -14 & 9 & 31 \\
\hline
& 6 & 14 & -9 & -31 & \underline{47} = \hat{P}(-1)
\end{array}
$$

Table 2.8 lists the points used in the systematic search of $[-1, 1]$ and the values of P and $\hat{P}$ at these points. In this example, the number of points to be evaluated is nine.

TABLE 2.8

x_i	$P(x_i)$	$\hat{P}(x_i)$	x_i	$P(x_i)$	$\hat{P}(x_i)$
-1.00	47.000	47.000	.25	10.75	6.6484
$-.75$	15.750	42.273	.50	13.25	.1250
$-.50$	3.250	35.125	.75	12.00	$-.8516$
$-.25$	2.000	26.023	1.00	7.00	7.0000
.00	6.000	16.000			

Since $\hat{P}(.5) = .125$, $\hat{P}(.75) = -.8516$, and $\hat{P}(1) = 7$, $\hat{P}$ has one zero in $(.5, .75)$ and another zero in $(.75, 1)$. Consequently, P has one zero in $(1/.75, 1/.5) = (\frac{4}{3}, 2)$ and another zero in $(1, \frac{4}{3})$. Since $|\hat{P}(.5)|$ is smaller than either $|\hat{P}(1)|$ or $|\hat{P}(.75)|$, it is likely that .5 is closer to a zero of $\hat{P}$ than either 1 or .75. Consequently, it is reasonable to suspect that 2 is closer to a zero of P than either 1 or $\frac{4}{3}$. We will use Newton's method with an initial approximation of $x_0 = 2$ to locate the first approximate zero of P. With $x_0 = 2$, using the procedure in Example 2 to find $x_1 = x_0 - P(x_0)/P'(x_0)$ gives:

$$
\begin{array}{r|rrrrr}
2 & 16 & -40 & 5 & 20 & 6 \\
& & 32 & -16 & -22 & -4 \\
\hline
& 16 & -8 & -11 & -2 & \underline{2} = P(2) \\
& & 32 & 48 & 74 & \\
\hline
& 16 & 24 & 37 & \underline{72} = P'(2)
\end{array}
$$

and $x_1 = 2 - \frac{2}{72} \approx 1.97$. Repeating this procedure,

$$
\begin{array}{r|rrrrr}
1.97 & 16 & -40 & 5 & 20 & 6 \\
& & 31.52 & -16.7056 & -23.0600 & -6.0283 \\
\hline
& 16 & -8.48 & -11.7056 & -3.0600 & \underline{-.0283} = P(1.97) \\
& & 31.52 & 45.3888 & 66.3559 & \\
\hline
& 16 & 23.04 & 33.6832 & \underline{63.2959} = P'(1.97)
\end{array}
$$

and

$$
x_2 = x_1 - \frac{P(x_1)}{P'(x_1)} \approx 1.9705.
$$

Recall that the synthetic-division procedure also yields an approximate factorization of P. For this example, the approximate factorization is

$$P(x) \approx (x - 1.97)Q(x),$$

where $$Q(x) = 16x^3 - 8.48x^2 - 11.7056x - 3.0600.$$

The polynomial Q can now be used to find additional approximate zeros of P. Since Q is only an approximate factor of P, the exact zeros of Q are only approximate zeros of P; however, since Q is a polynomial of lower degree than P, there are fewer computations involved in finding approximate zeros of Q. As previously discussed, the recommended procedure is to first use the Newton–Raphson method to find approximate zeros of Q and then iterate with the original polynomial P to improve the accuracy.

Carrying Example 3 further, let $x_0 = 1.33 \approx \frac{4}{3}$ and apply the synthetic-division procedure to Q to obtain:

$$x_1 = 1.33 - \frac{Q(1.33)}{Q'(1.33)} = 1.33 - \frac{4.0135}{50.6448} \approx 1.251,$$

an approximate root of Q and, consequently, an approximate root of P. To improve the approximation, do one iteration of Newton's method with P. This yields

$$x_2 = 1.251 - \frac{P(1.251)}{P'(1.251)} = 1.251 - \frac{-.280}{-29.990} \approx 1.2417.$$

Now that approximate zeros $\hat{x}_2 = 1.2417$ and $\hat{x}_1 = 1.9705$ are known, we can use synthetic division one more time to factor $P(x)$:

$$P(x) \approx (x - \hat{x}_1)(x - \hat{x}_2)\hat{Q}(x).$$

The procedure is as follows:

1.2417	16	-40	5	20	6
		19.8672	-24.9989	-24.8326	-6.0007
1.9705	16	-20.1328	-19.9989	-4.8326	$-.0007$
		31.528	22.4542	4.8382	
	16	11.3952	2.4553	.0056	

Thus

(2.36) $P(x) \approx (x - 1.2417)(x - 1.9705)(16x^2 + 11.3952x + 2.4553),$

where the remainders $-.0007$ and $.0056$ are omitted. Approximations, $\hat{x}_3$ and $\hat{x}_4$, for the remaining two zeros of P can now be obtained by solving

$$16x^2 + 11.3952x + 2.4553 = 0$$

to obtain $\hat{x}_3 = -.3561 + .16325i$ and $\hat{x}_4 = -.3561 - .16325i$. $\square$

To improve the approximation for $\hat{x}_3$ and $\hat{x}_4$, the methods previously discussed would require an additional application of Newton's method, starting with $\hat{x}_3 = -.3561 + .16325i$ and using the original polynomial

$$P(x) = 16x^4 - 40x^3 + 5x^2 + 20x + 16.$$

Since this procedure would call for a considerable amount of arithmetic with complex numbers, we will instead develop a general method for finding complex zeros of a polynomial. The basic idea in the procedure is to find quadratic factors corresponding to complex zeros and then find the roots of these quadratic polynomials by the quadratic formula. If a polynomial equation

$$0 = P(x) = a_n x^n + a_{n-1} x^{n-1} + \cdots + a_1 x + a_0$$

has a complex root $x = \alpha + \beta i$, then Theorem 2.16 implies that $\bar{x} = \alpha - \beta i$ is also a zero of P and that $R(x) = x^2 - 2\alpha x + \alpha^2 + \beta^2$ is a factor of P. It is more convenient to work with the quadratic factor $R(x)$ than with the zeros themselves, since $R(x)$ involves only real numbers.

In general, if $R(x)$ is a quadratic polynomial, then dividing $P(x)$ by $R(x)$ will give

$$P(x) = R(x)Q(x) + Ax + B,$$

where A and B are constants and Q is a polynomial whose degree is two less than the degree of P. We shall explain below a synthetic-division procedure for dividing a polynomial P by a quadratic polynomial of the form $x^2 + b_1 x + b_0$. Although the procedure is somewhat more complicated than ordinary synthetic division, it can be performed quite easily and has, like Horner's method, the advantage of using significantly less computations than the ordinary evaluation procedure.

If R and P are given by

$$R(x) = x^2 + b_1 x + b_0,$$
$$P(x) = a_n x^n + a_{n-1} x^{n-1} + \cdots + a_1 x + a_0,$$

then the scheme for dividing P by R by synthetic division is as follows:

$-b_1, -b_0$	a_n a_{n-1}	a_{n-2}	a_{n-3}	$\cdots$ a_1 a_0
	0 $\quad 0$	$-b_0 a_n$	$-b_0(a_{n-1} - b_1 a_n)$ $\cdots$	
	0 $\quad -b_1 a_n$	$-b_1(a_{n-1} - b_1 a_n)$ $\cdots$		0
	a_n $\quad a_{n-1} - b_1 a_n$ $\cdots$	$\cdots$		A $\quad B$

Zeros are first placed in positions that are not to be filled by calculated values. Specifically, the entries in column one of rows two and three, in column two of row two, and the last column of row three are zero. The entries in the first row are the coefficients of P. The entries in the second row are $-b_0$ multiplied by the sum of the entries in the column two places to the left, or more generally, the entry in column j of the second row is $-b_0$ times the sum of the entries in column $j - 2$, where the columns are numbered 1 to $n + 1$ from left to right. For $j = 1, 2, \ldots, n$, the entry in column j of the third row of the synthetic division table is $-b_1$ multiplied by the sum of the entries in column $j - 1$. The entry in the last column of row three is always zero. The sum of the next-to-last column gives the value of A in the remainder term $Ax + B$, and the sum of the last column gives B.

EXAMPLE 4 Use the quadratic synthetic-division procedure to divide $x^6 - 2x^5 + 7x^4 - 4x^3 + 11x^2 - 2x + 5$ by $x^2 - 2x + 1$.

$2, -1$	1	-2	7	-4	11	-2	5
	0	0	-1	0	-6	-8	-21
	0	2	0	12	16	42	0
	1	0	6	8	21	$\underline{32}$	$\underline{-16}$

The table gives

$$x^6 - 2x^5 + 7x^4 - 4x^3 + 11x^2 - 2x + 5$$
$$= (x^2 - 2x + 1)(x^4 + 6x^2 + 8x + 21) + 32x - 16. \qquad \square$$

For ease in the implementation of the quadratic synthetic-division procedure, the following algorithm is given.

Quadratic Synthetic-Division Algorithm 2.7

To divide the polynomial

$$P(x) = a_n x^n + a_{n-1} x^{n-1} + \cdots + a_1 x + a_0$$

by the quadratic polynomial

$$R(x) = x^2 + b_1 x + b_0:$$

INPUT degree n; coefficients of P, $a_0, a_1, \ldots, a_n$; coefficients of R, $B0$, $B1$.

OUTPUT Z, C (*Remainder upon dividing P by R is $Zx + C$.*)

Step 1 Set $Z = a_n$; (*Compute z_n.*)
$\qquad\qquad C = a_{n-1}$. (*Compute c_{n-1}.*)

Step 2 For $k = n - 1, n - 2, \ldots, 1$
$\qquad\qquad$ set $ZZ = Z$; (*Save old Z.*)
$\qquad\qquad\quad Z = -B1 \cdot ZZ + C$; (*Compute z_k.*)
$\qquad\qquad\quad C = -B0 \cdot ZZ + a_{k-1}$. (*Compute c_k.*)

Step 3 OUTPUT (Z, C); (*Output z_1, c_1.*)
$\qquad\qquad$ STOP.

Note that $z_1 x + c_1$ is the remainder term in the division process, so if $z_1 = c_1 = 0$, then $x^2 + b_1 x + b_0$ is a factor of P. Also note that the coefficient of x^2 must be one in order to use Algorithm 2.7.

EXAMPLE 5   Use Algorithm 2.7 to divide $x^6 - 2x^5 + 7x^4 - 4x^3 + 11x^2 - 2x + 5$ by $x^2 - 2x + 1$.

In this example, $b_1 = -2$ and $b_0 = 1$, and the steps in the algorithm give

Step 1 $z_6 = 1$ $\qquad\qquad\qquad\qquad$ $c_6 = -2$

Step 2 $z_5 = (2)(1) + (-2) = 0$ $\qquad$ $c_5 = (-1)(1) + 7 = 6$

$\qquad\qquad z_4 = (2)(0) + 6 = 6$ $\qquad\qquad$ $c_4 = (-1)(0) + (-4) = -4$

$\qquad\qquad z_3 = (2)(6) + (-4) = 8$ $\qquad\quad$ $c_3 = (-1)(6) + 11 = 5$

$\qquad\qquad z_2 = (2)(8) + 5 = 21$ $\qquad\qquad$ $c_2 = (-1)(8) - 2 = -10$

$\qquad\qquad z_1 = (2)(21) + (-10) = 32$ $\quad$ $c_1 = (-1)(21) + 5 = -16$

Step 3 $x^6 - 2x^5 + 7x^4 - 4x^3 + 11x^2 - 2x + 5$
$$= (x^4 + 6x^2 + 8x + 21)(x^2 + 2x - 1) + 32x - 16.$$

Note that this is the same factorization as obtained in Example 4. $\qquad \square$

Once we have found an approximate quadratic factor corresponding to complex zeros of a polynomial P, we can use the quadratic formula to obtain the approximate complex zeros. Recall that approximate complex zeros $\alpha \pm \beta i$ correspond to the quadratic factor $x^2 - 2\alpha x + (\alpha^2 + \beta^2)$. Once approximate quadratic zeros are found, we can apply Newton's method to improve the accuracy of these zeros, using complex arithmetic. In applying Newton's method in this situation, we will find the results of Algorithm 2.7 useful. If $x_0 = \alpha + \beta i$ is an approximate zero of P and we wish to improve this approximation, we calculate

$$x_1 = x_0 - \frac{P(x_0)}{P'(x_0)}.$$

Since $x_0 = \alpha + \beta i$ corresponds to a quadratic factor,

$$R(x) = x^2 - 2\alpha x + (\alpha^2 + \beta^2) \qquad \text{and} \qquad R(x_0) = 0.$$

If Algorithm 2.7 is used to divide $P(x)$ by $R(x)$, the result is

$$P(x) = R(x)Q(x) + Ax + B \qquad \text{and} \qquad P(x_0) = Ax_0 + B.$$

Moreover,

$$P'(x) = R(x)Q'(x) + R'(x)Q(x) + A;$$

hence, $P'(x_0) = R(x_0)Q'(x_0) + (2x_0 - 2\alpha)Q(x_0) + A = 2\beta i Q(x_0) + A,$

and $P'(x_0)$ can be obtained by first evaluating $Q(x_0)$ by Algorithm 2.7 and then calculating

$$P'(x_0) = 2\beta i Q(x_0) + A.$$

EXAMPLE 6   In Example 3, $\hat{x}_3 = -.3561 + .16325i$ was obtained as an approximate root of

$$0 = P(x) = 16x^4 - 40x^3 + 5x^2 + 20x + 6.$$

We will do one iteration of Newton's method to improve the accuracy of this root. The approximation $\hat{x}_3$ corresponds to the quadratic factor $R(x) = x^2 + .7122x + .1535$. The synthetic-division procedure corresponding to Algorithm 2.7 is now used to obtain:

$-.7122, -.1535$	16	-40	5	20	6
	0	0	-2.456	7.8892	-6.0092
	0	-11.3952	36.6037	27.8810	0
	16	-51.3952	39.1477	.0082	$-.0092$
	0	0	-2.456		
	0	-11.3952	0		
	16	-62.7904	36.6917		

Hence, $\begin{aligned} P(\hat{x}_3) &= P(-.3561 + .16325i) \\ &= (.0082)(-.3561 + .16325i) - .0092 \\ &= -.01212 + .00134i, \end{aligned}$

and $\begin{aligned} P'(\hat{x}_3) &= 2(.16325i)Q(\hat{x}_3) + .0082 \\ &= 2(.16325i)[(-62.7904)(-.3561 + .16325i) + 36.6917] \\ &\quad + .0082 \\ &= 3.35500 + 19.2803i. \end{aligned}$

With
$$x_0 = \hat{x}_3 = -.3561 + .16325i,$$

$$x_1 = x_0 - \frac{P(x_0)}{P'(x_0)} = -.356061 + .162628i$$

is accurate to four decimal places.   □

The weakness of the complex root-finding procedure utilizing Algorithm 2.7 is that an initial approximation to the quadratic factor is required. Recalling that the method outlined for obtaining an initial approximation to a real root was rather complicated, we should not be surprised that this weakness is substantial. A method for finding zeros of polynomials, which has gained recent popularity because it requires no initial approximation, is based on the Quotient–Difference (QD) Algorithm developed by H. Rutishauser. The theory underlying this procedure and the analysis of the situations that are necessary in order for it to be successful are quite involved, and the reader is referred to Henrici [46] or the article by Henrici in Ralston and Wilf [68] for elaboration.

The convergence of the QD method is rather slow and highly susceptible to round-off error, but it can be very useful in obtaining the initial approximations, both to the real roots and to the quadratic factors associated with the complex roots, that are required by the more rapidly convergent Newton's method.

The QD method for approximating the roots of the polynomial equation

$$0 = P(x) = a_n x^n + a_{n-1} x^{n-1} + \cdots + a_1 x + a_0$$

consists of constructing sequences of real numbers:

$$\{e_i^{(k)}\}_{i=1}^{\infty} \quad \text{for each } k = 1, 2, \ldots, (n+1)$$

and
$$\{q_i^{(k)}\}_{i=1}^{\infty} \quad \text{for each } k = 1, 2, \ldots, n,$$

where

$$e_i^{(1)} = 0 \quad \text{for each } i = 1, 2, \ldots,$$

$$e_i^{(n+1)} = 0 \quad \text{for each } i = 1, 2, \ldots,$$

$$e_1^{(k)} = \frac{a_{n-k}}{a_{n-k+1}}, \quad \text{for each } k = 2, 3, \ldots, n,$$

$$q_1^{(1)} = -\frac{a_{n-1}}{a_n},$$

$$q_1^{(k)} = 0, \quad \text{for each } k = 2, 3, \ldots, n,$$

$$q_{i+1}^{(k)} = e_i^{(k+1)} + q_i^{(k)} - e_i^{(k)} \quad \text{for each } k = 1, 2, 3, \ldots, n,$$
$$\text{and } i = 1, 2, \ldots,$$

and finally

$$e_{i+1}^{(k)} = (q_{i+1}^{(k)} e_i^{(k)})/q_{i+1}^{(k-1)} \quad \text{for each } k = 2, 3, \ldots, n, \text{ and } i = 1, 2, \ldots.$$

Although the construction of the sequence appears to be quite complicated, in actuality it can be performed quite simply by the use of a table, which we construct by first entering all the values for $q_1^{(k)}$, $e_1^{(k)}$, $e_i^{(1)}$, and $e_i^{(n+1)}$ (Table 2.9).

TABLE 2.9

i	$e_i^{(1)}$	$q_i^{(1)}$	$e_i^{(2)}$	$q_i^{(2)}$	$e_i^{(3)}$	$q_i^{(3)}$	$\cdots$	$e_i^{(n)}$	$q_i^{(n)}$	$e_i^{(n+1)}$
1	0	$\left(\dfrac{-a_{n-1}}{a_n}\right)$	$\left(\dfrac{a_{n-2}}{a_{n-1}}\right)$	0	$\left(\dfrac{a_{n-3}}{a_{n-2}}\right)$	0		$\left(\dfrac{a_0}{a_1}\right)$	0	0
2	0									0
3	0									0
$\vdots$	$\vdots$									$\vdots$

The next step is to construct the $q_2^{(k)}$ entries in the second row by taking the element to the immediate upper right, $e_1^{(k+1)}$, adding the element directly above, $q_1^{(k)}$, and subtracting the element to the immediate upper left, $e_1^{(k)}$, to obtain Table 2.10.

TABLE 2.10

i	$e_i^{(1)}$	$q_i^{(1)}$	$e_i^{(2)}$	$q_i^{(2)}$	$e_i^{(3)}$	$q_i^{(3)}$	$\cdots$	$e_i^{(n)}$	$q_i^{(n)}$	$e_i^{(n+1)}$
1	0	$\left(\dfrac{-a_{n-1}}{a_n}\right)$	$\left(\dfrac{a_{n-2}}{a_{n-1}}\right) \leftarrow 0 \xleftarrow{+} \left(\dfrac{a_{n-3}}{a_{n-1}}\right)$			0		$\left(\dfrac{a_0}{a_1}\right)$	0	0
2	0	$q_2^{(1)}$		$q_2^{(2)}$		$q_2^{(3)}$	$\cdots$		$q_2^{(n)}$	0
3	0									0
$\vdots$	$\vdots$									$\vdots$

The $e_2^{(k)}$ entries can now be obtained by taking the element to the right, $q_2^{(k)}$, multiplying by the entry above, $e_1^{(k)}$, and dividing by the entry to the left, $q_2^{(k-1)}$ (Table 2.11).

TABLE 2.11

i	$e_i^{(1)}$	$q_i^{(1)}$	$e_i^{(2)}$	$q_i^{(2)}$	$e_i^{(3)}$	$q_i^{(3)}$	$\cdots$	$e_i^{(n)}$	$q_i^{(n)}$	$e_i^{(n+1)}$
1	0	$\left(\dfrac{-a_{n-1}}{a_n}\right)$	$\left(\dfrac{a_{n-2}}{a_{n-1}}\right)$	0	$\left(\dfrac{a_{n-3}}{a_{n-2}}\right)$	0	$\cdots$	$\left(\dfrac{a_0}{a_1}\right)$	0	0
2	0	$q_2^{(1)}$	$e_2^{(2)}$	$q_2^{(2)} \xrightarrow{=} e_2^{(2)}$	$q_2^{(3)}$		$\cdots$	$e_2^{(n)}$	$q_2^{(n)}$	0
3	0									0
$\vdots$	$\vdots$									$\vdots$

If $\lim_{i\to\infty} e_i^{(k)} = \lim_{i\to\infty} e_i^{(k+1)} = 0$ for any $k = 1, 2, \ldots, n$, it can be easily seen that $\lim_{i\to\infty} q_i^{(k)}$ exists (see Exercise 13). The important fact, however, is that this limit is a zero of the polynomial P. Additionally, if $\{e_i^{(k)}\}_{i=1}^{\infty}$ does not converge to zero for some k, then the sequences $\{r_i\}_{i=1}^{\infty}$ and $\{s_i\}_{i=1}^{\infty}$, where

$$r_i^{(k)} = q_i^{(k-1)} + q_i^{(k)} \qquad \text{for each } i = 1, 2, \ldots,$$

and

$$s_i^{(k)} = q_{i-1}^{(k-1)} q_i^{(k)} \qquad \text{for each } i = 2, 3, \ldots,$$

converge to numbers $r^{(k)}$ and $s^{(k)}$ where $x^2 - r^{(k)}x + s^{(k)}$ is a quadratic factor of $P(x)$ corresponding to a pair of complex conjugate roots.

As previously mentioned, this procedure is generally used not to find the actual zeros of the polynomial, but to obtain satisfactory initial approximations to use in a higher order method such as Newton's method.

Quotient-Difference (QD) Algorithm 2.8

To approximate all the zeros and quadratic factors of the real polynomial

$$P(x) = a_n x^n + a_{n-1} x^{n-1} + \cdots + a_1 x + a_0:$$

INPUT degree n; coefficients $a_0, \ldots, a_n$; maximum number of iterations M.

OUTPUT approximations $q_i^{(k)}$ to zeros of P and $x^2 - r_i^{(k)} x + s_i^{(k)}$ to quadratic factors of P or a message that the maximum number of iterations was exceeded.

Step 1 Set $e_1^{(1)} = 0$;

$$e_1^{(n+1)} = 0;$$

$$q_1^{(1)} = -\frac{a_{n-1}}{a_n};$$

$$IND_1 = 1;$$

$IND_{n+1} = 1$.   (*IND is used to note convergence of sequences $e_i^{(k)}$.*)

Step 2 For $k = 2, \ldots, n$ set $q_1^{(k)} = 0$;

$$e_1^{(k)} = \frac{a_{n-k}}{a_{n-k+1}};$$

$$IND_k = 0;$$

$$r_1^{(k)} = 0;$$

$$s_1^{(k)} = 0.$$

Step 3 Set $i = 2$.

Step 4 While $i \leq M$ do Steps 5–13.

Step 5 Set $e_i^{(1)} = 0$;

$$e_i^{(n+1)} = 0;$$

$$q_i^{(1)} = e_{i-1}^{(2)} + q_{i-1}^{(1)} - e_{i-1}^{(1)}.$$

Step 6 For $k = 2, \ldots, n$ set $q_i^{(k)} = e_{i-1}^{(k+1)} + q_{i-1}^{(k)} - e_{i-1}^{(k)}$;

$$e_i^{(k)} = (q_i^{(k)} e_{i-1}^{(k)})/q_i^{(k-1)};$$

$$r_i^{(k)} = q_i^{(k-1)} + q_i^{(k)};$$

$$s_i^{(k)} = q_{i-1}^{(k-1)} q_i^{(k)}.$$

Step 7 For $k = 2, \ldots, n$
if $\{e_j^{(k)}\}$ is converging to zero, then set $IND_k = 1$;
if $\{e_j^{(k)}\}$ is not converging to zero, then set $IND_k = -1$.
(*Note: if either choice is not clear, then IND_k remains zero.*)

Step 8 Set $NS = 0$.   (*NS is used to note that all sequences $\{e_j^{(k)}\}$ are converging.*)

Step 9 For $k = 2, \ldots, n$ if $IND_k = 0$ then set $NS = 1$.

Step 10 If $NS = 0$ then do Steps 11 and 12.

Step 11 For $k = 2, \ldots, n + 1$
if $IND_k = 1$ and $IND_{k-1} = 1$ then
OUTPUT ('APPROX. ZERO', $q_i^{(k)}$);
else
OUTPUT ('APPROX. QUADRATIC
FACTOR', $x^2 - r_i^{(k)}x + s_i^{(k)}$).

Step 12 STOP.

Step 13 Set $i = i + 1$.

Step 14 OUTPUT ('Maximum number of iterations exceeded');
STOP.

EXAMPLE 7   Consider the polynomial $P(x) = 16x^4 - 40x^3 + 5x^2 + 20x + 6$, which we studied in Examples 3 and 6. The QD Algorithm produces the values shown in Table 2.12.

TABLE 2.12

i	$e_i^{(1)}$	$q_i^{(1)}$	$e_i^{(2)}$	$q_i^{(2)}$	$e_i^{(3)}$	$q_i^{(3)}$	$e_i^{(4)}$	$q_i^{(4)}$	$e_i^{(5)}$
1	0	2.5	$-.125$	0	4	0	.3	0	0
2	0	2.375	$-.217105$	4.125	-3.587879	-3.7	.024324	$-.3$	0
3	0	2.157895	$-.075883$	.754226	.417653	$-.087797$	.089855	$-.324324$	0
4	0	2.08201	$-.045477$	1.247762	$-.139108$	$-.415595$	.089549	$-.414179$	0
5	0	2.036535	$-.025772$	1.154130	.022532	$-.186937$	.241301	$-.503728$	0
6	0	2.010763	$-.015412$	1.202434	.000596	.031832	-5.647705	$-.745029$	0
7	0	1.995351	$-.009411$	1.218442	$-.002749$	-5.616470	4.929942	4.902676	0
8	0	1.985940	$-.005806$	1.225104	.001535	$-.683778$	.196581	$-.027266$	0
9	0	1.980135	$-.003613$	1.232444	$-.000609$	$-.488732$	.090037	$-.223846$	0
10	0	1.976521	$-.002259$	1.235449	.000196	$-.398087$	.070992	$-.313883$	0
11	0	1.974263	$-.001416$	1.237904	$-.000052$	$-.327391$	.083483	$-.384875$	0
12	0	1.972846	$-.000890$	1.239268	.000010	$-.243757$	.160405	$-.468358$	0
13	0	1.971957	$-.000560$	1.240168	$-.000001$	$-.083362$	1.209859	$-.628762$	0
14	0	1.971397	$-.000352$	1.240727	$-.000001$	1.126498	-1.974681	-1.838622	0

From the table it seems clear that

$$\lim_{i \to \infty} e_i^{(2)} = 0 \quad \text{and} \quad \lim_{i \to \infty} e_i^{(3)} = 0,$$

which implies that

$$q_{14}^{(1)} = 1.971397 \quad \text{and} \quad q_{14}^{(2)} = 1.240727$$

are approximate roots.

Although the entries $e_9^{(4)}$, $e_{10}^{(4)}$, and $e_{11}^{(4)}$ are small, constructing the later entries $e_{12}^{(4)}$, $e_{13}^{(4)}$, and $e_{14}^{(4)}$ seems to imply that $\{e_i^{(4)}\}_{i=1}^{\infty}$ does not converge to zero. If the values of $(q_i^{(3)} + q_i^{(4)})$ and $q_{i-1}^{(3)}q_i^{(4)}$ are calculated as shown in Table 2.13, an

TABLE 2.13

i	$r_i^{(4)}$	$s_i^{(4)}$
2	-4.000000	.000000
3	$-.412121$	1.200000
4	$-.829774$	.036364
5	$-.690665$	.209347
6	$-.713197$	.139273
7	$-.713794$	.156062
8	$-.711044$	.153139
9	$-.712578$	.153061
10	$-.711970$	.153405
11	$-.712266$	.153214
12	$-.712115$	.153336
13	$-.712124$	.153265
14	$-.712124$	.153271

approximation for the quadratic factor can be seen to be $x^2 + .712124x + .153271$, which has roots $-.356062 \pm .162760i$.

The actual values for the roots of the equation are 1.241677, 1.970446, $-.356062 \pm .162758i$ with all values correct to six decimal places. □

To find better approximations we would use the procedures of Algorithms 2.6 and 2.7, together with Newton's method.

In many cases the QD Algorithm gives useful starting values for Newton's method, but this algorithm can be quite sensitive to rounding errors. In addition, the QD method may not work well in the situation when the polynomial has unequal roots with nearly the same absolute values, since the e_i-columns partition the QD table into sections corresponding to zeros with equal magnitudes. Consequently, other root finding methods often need to be considered. In particular, some of the more popular methods that have not been discussed in this book are Laguerre's method, which can give cubic convergence and also find complex roots (see Householder [49] pages 176–179 for a complete discussion), the Jenkins–Traub method (see Jenkins [55]), and Brent's method (see Brent [15]), which is based upon the bisection and regula-falsi methods. Another popular method, called Muller's method, is considered in Exercise 12 of Section 3.3.

Exercise Set 2.5

1. Use Theorem 2.17 to evaluate

$$P(x) = 5x^4 - 2x^2 + 3x + 4$$

at $x_0 = 3$. Also evaluate P at $x_0 = 3$ by the synthetic division form of Horner's method.

2. Find approximations to all of the zeros of each of the following polynomials by first finding the real zeros and then reducing the polynomials to ones of lower degree to determine those zeros that are complex.

a) $P(x) = x^4 + 5x^3 - 9x^2 - 85x - 136$
b) $P(x) = x^4 - 2x^3 - 12x^2 + 16x - 40$
c) $P(x) = x^4 + x^3 + 3x^2 + 2x + 2$
d) $P(x) = x^5 + 11x^4 - 21x^3 - 10x^2 - 21x - 5$

3. a) Prove Corollary 2.14. b) Prove Corollary 2.15.

4. Prove Theorem 2.16. (*Hint*: First consider the case $m = 1$ and notice what happens to the constants if $a - bi$ is not a zero of $P(x)$.)

5. Prove the following theorem.

Theorem Let $P(x) = a_n x^n + a_{n-1} x^{n-1} + \cdots + a_1 x + a_0$ be a polynomial of degree n and let x_0 be a positive real number with $P(x_0) > 0$. If $Q(x)$ is the polynomial constructed in Algorithm 2.6—that is,

$$P(x) = (x - x_0)Q(x) + P(x_0)$$
$$= (x - x_0)(b_n x^{n-1} + \cdots + b_2 x + b_1) + P(x_0),$$

and $b_i > 0$ for $1 = 1, 2, \ldots, n$, then all zeros of P are less than or equal to x_0.

(Note that this theorem could sometimes be used to reduce the amount of work involved in searching $[-1, 1]$ for zeros of P.)

6. Let $P(x) = a_n x^n + a_{n-1} x^{n-1} + \cdots + a_1 x + a_0$. Show that if $a_i > 0$ for $i = 1, 2, \ldots, n$, then P has no nonnegative zeros. (If P is of this form, search only $[-1, 0]$ for zeros of P.)

7. a) Show that if x_0 is a double root of $0 = P(x) = a_n x^n + \cdots + a_1 x + a_0$, then x_0 is a root of $0 = P'(x)$. What does the graph of P look like near $x = x_0$?
   b) Consider the polynomial

$$P(x) = x^3 - 4.7x^2 + 2.9225x - .49.$$

Search $[-1, 1]$ for zeros of P by evaluating $P(x)$ and $\hat{P}(x) = -.49x^3 + 2.9225x^2 - 4.7x + 1$ at $x_j = -1 + .1j$ for $j = 0, 1, \ldots, 20$. Since P has a double root, you will locate only one real root. Look at the results obtained and see if there is any indication that P might have a double root.

8. $P(x) = 10x^3 - 8.3x^2 + 2.295x - .21141 = 0$ has a root at $x = .29$. Use Newton's method with an initial approximation $x_0 = .26$ to attempt to find this root. What happens? Assume that the only root you desire is $x = .29$; how might you obtain a good enough initial approximation so that Newton's method will converge to $x = .29$?

9. Repeat Exercise 2, first finding initial approximations to the zeros using the QD Algorithm 2.8.

10. Find all roots of

$$16x^4 + 88x^3 + 159x^2 + 76x - 240 = 0$$

to six decimal places by first approximating the roots by the use of Algorithm 2.8 and then using Newton's method to refine the approximations.

11. The QD Algorithm cannot be used to solve for the roots of a polynomial equation

$$P(x) = a_n x^n + a_{n-1} x^{n-1} + \cdots + a_1 x + a_0 = 0,$$

when one of the coefficients $a_0, a_1, \ldots, a_{n-1}$ is zero.

   a) Explain why the statement above is true.
   b) Show that there is always a change of variables of the form $z = x - a$, which can be performed to change the polynomial P into one of the same degree which will have all nonzero coefficients.
   c) Determine the relation between the zeros of P and those of a polynomial Q where Q is obtained from P using a variable substitution of the form $z = x - a$.

12. Use the procedure outlined in Exercise 11 and the QD Algorithm 2.8 to obtain initial approximations to the roots of

$$x^4 - 4x^2 - 3x + 5 = 0$$

and then find approximations to these roots that are accurate to six decimal places.

13. Using the notation of Algorithm 2.8, show that if $\lim_{i \to \infty} e_i^{(k)} = 0$ and $\lim_{i \to \infty} e_i^{(k+1)} = 0$, then $\lim_{i \to \infty} q_i^{(k)}$ exists.

14. In describing a mathematical model for machine-tool chatter [66], the authors need the roots of a polynomial equation of the form

$$P_{2n}(B) = 1 - \phi_1 B - \phi_2 B^2 - \cdots - \phi_{2n} B^{2n} = 0,$$

for various values of n and collections of constants $\phi_1, \phi_2, \ldots, \phi_{2n}$. The value of n depends upon the amount of experimental data being used, and approximations for the constants ϕ_i, $i = 1, 2, \ldots, 2n$ are obtained using this data and a linear least squares technique, a topic discussed in Section 7.1.

a) In an experimental situation involving a Gisholt Turret Lathe, it was found that, for $n = 2$, reasonable approximations for the ϕ_i's are:

$$\phi_1 = 1.8310, \qquad \phi_2 = -.5218$$

$$\phi_3 = -.4754, \qquad \phi_4 = .1595.$$

Use these values to find the zeros of P_4.

b) In the same experiment, with $n = 3$, values of ϕ_i were found to be:

$$\phi_1 = 1.742, \qquad \phi_2 = -.0385, \qquad \phi_3 = -.8133$$

$$\phi_4 = -.1061, \qquad \phi_5 = .2019, \qquad \phi_6 = .0383$$

Use these values to find the zeros of P_6.

Interpolation and
Polynomial Approximation

A census of the population of the United States is taken every 10 years. Below is a table listing, in thousands of people, the population from 1920 to 1970.

Year	1920	1930	1940	1950	1960	1970
Population (In Thousands)	105,711	123,203	131,669	150,697	179,323	203,212

In reviewing this data, we might ask whether it could be used to reasonably estimate the population, say in 1965 or even in the year 2000. Some predictions of this type can be obtained by using a function which fits the given data. This is a topic called *interpolation* and is the subject of this chapter.

One of the most useful and well-known classes of functions mapping the real line into itself is the class of **algebraic polynomials**, i.e., the set of functions of the form

$$P_n(x) = a_0 + a_1 x + \cdots + a_n x^n,$$

where n is a nonnegative integer and $a_0, \ldots, a_n$ are real constants. One major reason for their importance is that they uniformly approximate continuous functions; that is, given any function, defined and continuous on a closed interval, there exists a polynomial that is as "close" to the given function as desired. This result is expressed more precisely in the following theorem. (See Fig. 3.1.)

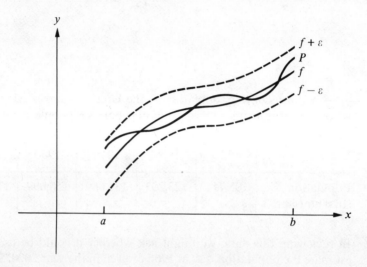

FIGURE 3.1

Theorem 3.1   (Weierstrass Approximation Theorem)   If f is defined and continuous on $[a, b]$, and $\varepsilon > 0$ is given, then there exists a polynomial P, defined on $[a, b]$, with the property that

$$|f(x) - P(x)| < \varepsilon \qquad \text{for all } x \in [a, b].$$

The proof of this theorem is beyond the scope of this book but can be found in any elementary text on real analysis (see, for example, Bartle [9], pages 165–172.)

An additional important reason for considering the class of polynomials in the approximation of functions is that it is easy to determine the derivative and indefinite integral of any polynomial and the result is again a polynomial. For these reasons, the class of polynomials is often used for approximating other functions that are known or assumed to be continuous.

3.1   The Taylor Polynomials

The Weierstrass Theorem is very important from a theoretical standpoint, but it cannot be used effectively for computational purposes. Instead of requiring that a polynomial be found to approximate a function uniformly over an entire interval, it is often of more interest to find a polynomial which satisfies some specific conditions that are useful for the problem under consideration, while still being "close" in some sense.

EXAMPLE 1   Find a polynomial of degree 3 or less to approximate $f(x) = \sin x$ near $x_0 = 0$ and use this polynomial to approximate $\sin .1$. (See Fig. 3.2.)

The example may seem vaguely phrased since no explanation of how to find an approximating polynomial has been given, nor has the term "near x_0" been clearly defined. A logical requirement for an approximating polynomial may be that it agree with the given function and as many of its derivatives as possible at the point x_0. This would have the effect of requiring that the "shape" of the polynomial be as close as possible to that of the given function near x_0.

To proceed in this manner, set

$$P(x) = a_0 + a_1 x + a_2 x^2 + a_3 x^3$$

and determine a_0, a_1, a_2, a_3, so that $P(0) = f(0)$, $P'(0) = f'(0)$, and so on. Upon differentiating P and evaluating at $x = 0$, one has

$$P(x) = a_0 + a_1 x + a_2 x^2 + a_3 x^3, \quad \text{so } P(0) = a_0,$$
$$P'(x) = a_1 + 2a_2 x + 3a_3 x^2, \quad \text{so } P'(0) = a_1,$$
$$P''(x) = 2a_2 + 6a_3 x, \quad \text{so } P''(0) = 2a_2,$$
and $\quad P'''(x) = 6a_3, \quad \text{so } P'''(0) = 6a_3.$

Since $f(0) = \sin 0 = 0$, $f'(0) = \cos 0 = 1$, $f''(0) = -\sin 0 = 0$, and $f'''(0) = -\cos 0 = -1$, it follows that

$$a_0 = 0, \quad a_1 = 1, \quad 2a_2 = 0, \quad \text{and} \quad 6a_3 = -1.$$

The approximating polynomial of degree 3 or less is given by

$$P(x) = x - \tfrac{1}{6}x^3 \quad \text{and} \quad \sin .1 = f(.1) \approx P(.1) = .1 - \tfrac{1}{6}(.001) = .09983333.$$

This approximation agrees with $\sin .1$ to within 10^{-7}. $\square$

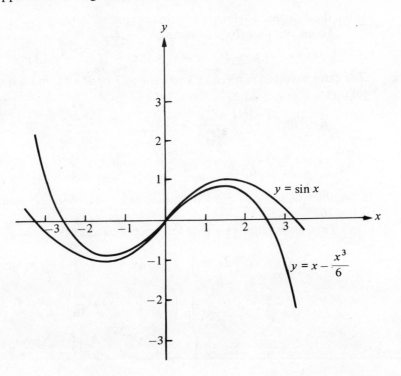

FIGURE 3.2

It should be noted that the procedure followed in Example 1 is precisely that of finding the third-degree Taylor polynomial for the function f, expanded about the point x_0. Consequently, the Taylor polynomials can be used as approximating polynomials "near" a given point.

EXAMPLE 2

a)   Calculate the third-degree Taylor polynomial about $x_0 = 0$ for $f(x) = (1 + x)^{1/2}$.

b)   Use the polynomial in part (a) to approximate $\sqrt{1.1}$, and find a bound for the error involved.

c)   Use the polynomial in part (a) to approximate $\int_0^{.1} (1 + x)^{1/2}\, dx$, and find a bound for the error of this approximation.

Performing the necessary differentiation yields

$$
\begin{aligned}
f(x) &= (1 + x)^{1/2}, & \text{so } f(0) &= 1, \\
f'(x) &= \tfrac{1}{2}(1 + x)^{-1/2}, & \text{so } f'(0) &= \tfrac{1}{2}, \\
f''(x) &= -\tfrac{1}{4}(1 + x)^{-3/2}, & \text{so } f''(0) &= -\tfrac{1}{4}, \\
f'''(x) &= \tfrac{3}{8}(1 + x)^{-5/2}, & \text{so } f'''(0) &= \tfrac{3}{8}, \\
f^{(iv)}(x) &= -\tfrac{15}{16}(1 + x)^{-7/2}, & \text{so } f^{(iv)}(\xi) &= -\tfrac{15}{16}(1 + \xi)^{-7/2},
\end{aligned}
$$

where $0 < \xi < x$. From Taylor's theorem,

$$(3.1) \quad P_3(x) = f(0) + f'(0)x + \frac{f''(0)}{2!}x^2 + \frac{f'''(0)}{3!}x^3$$

$$= 1 + \frac{1}{2}x - \frac{1}{4}\cdot\frac{1}{2!}x^2 + \frac{3}{8}\cdot\frac{1}{3!}x^3 = 1 + \frac{1}{2}x - \frac{1}{8}x^2 + \frac{1}{16}x^3,$$

is the third-degree Taylor polynomial requested in (a).

To answer part (b), we obtain

$$\sqrt{1.1} = f(.1) \approx P_3(.1) = 1 + \tfrac{1}{2}(.1) - \tfrac{1}{8}(.1)^2 + \tfrac{1}{16}(.1)^3 = 1.0488125.$$

The error involved is given in Theorem 1.13 by $R_3(.1)$, and a bound is derived as follows:

$$|R_3(.1)| = \frac{|-\tfrac{15}{16}(1 + \xi)^{-7/2}|}{4!}(.1)^4$$

$$\leq \frac{15}{(16)(24)}(.1)^4 \max_{\xi\in[0,.1]}(1 + \xi)^{-7/2} = \frac{.0005}{128}(1) \leq 3.91 \times 10^{-6}.$$

Since the true value of $\sqrt{1.1}$ (to 8 places) is 1.0488088, the actual error is about 3.7×10^{-6}.

The computations for part (c) proceed as follows:

$$\int_0^{.1} (1 + x)^{1/2}\, dx \approx \int_0^{.1} P_3(x)\, dx$$

$$= \int_0^{.1}\left(1 + \frac{x}{2} - \frac{x^2}{8} + \frac{x^3}{16}\right) dx$$

$$= \left(x + \frac{x^2}{4} - \frac{x^3}{24} + \frac{x^4}{64}\right)\Bigg|_0^{.1} = .1024598958,$$

with a remainder given by $\int_0^{.1} R_3(x)\,dx$. Using techniques similar to those used in part (b), the following error bound is obtained:

$$\left| \int_0^{.1} R_3(x)\,dx \right| = \frac{15}{(16)4!} \int_0^{.1} (1 + \xi)^{-7/2} x^4\,dx$$

$$\leq \frac{5}{128} \int_0^{.1} x^4\,dx = \frac{5}{128} \cdot \frac{x^5}{5} \bigg|_0^{.1} \leq 7.82 \times 10^{-8}.$$

Note that the actual remainder term $\int_0^{.1} R_3(x)\,dx$ is negative, so the integral desired cannot exceed the approximation. Hence,

$$\int_0^{.1} P_3(x)\,dx - \int_0^{.1} |R_3(x)|\,dx \leq \int_0^{.1} (1 + x)^{1/2}\,dx \leq \int_0^{.1} P_3(x)\,dx$$

or

$$.1024598176 \leq \int_0^{.1} (1 + x)^{1/2} \leq .1024598958.$$

Since the actual value of $\int_0^{.1} (1 + x)^{1/2}\,dx$ is .102459822, the true error is about 7.4×10^{-8}. $\square$

At this point the reader may feel that the Taylor polynomial never fails. The next example illustrates a need for other approximating techniques.

EXAMPLE 3   Figure 3.2, page 75, suggests that the Taylor polynomial of degree three for the sine function, expanded about $x_0 = 0$, gives decreasingly accurate results as x moves away from zero. The following table lists the values of the third-degree Taylor polynomial (3.1) for the function $f(x) = \sqrt{1 + x}$ considered in the previous example, and the error associated with using this polynomial for various values of x.

x	.1	.5	1	2	10
$P_3(x)$	1.048813	1.2266	1.438	2.00	56.00
$f(x)$	1.048809	1.2247	1.414	1.73	3.32
$\lvert P_3(x) - f(x) \rvert$	4×10^{-6}	2×10^{-3}	2×10^{-2}	3×10^{-1}	53

Although in some instances, better approximations can be obtained if slightly higher-degree Taylor polynomials are used, this is not always the case. Consider, as an extreme example, the problem of using Taylor polynomials of various degrees for $f(x) = 1/x$ expanded about $x_0 = 1$, to approximate $f(3) = \frac{1}{3}$. Since $f(x) = 1/x$, it is easily verified that the Taylor polynomial for $n \geq 1$ is given by

$$P_n(x) = \sum_{k=0}^{n} \frac{f^{(k)}(1)}{k!} (x - 1)^k = \sum_{k=0}^{n} (-1)^k (x - 1)^k.$$

To approximate $f(3) = \frac{1}{3}$ by $P_n(3)$ for increasing values of n, we obtain the following:

n	0	1	2	3	4	5	6	7
$P_n(3)$	1	-1	3	-5	11	-21	43	-85

The reason this approximating technique fails is that the error term,

$$R_n(x) = \frac{(-1)^{n+1}(x-1)^{n+1}}{\xi^{n+2}} \qquad \text{where } 1 < \xi < x,$$

grows in absolute value as n increases (see Fig. 3.3). The growth in error results because $x = 3$ is not "near enough" to $x_0 = 1$. For a more complete discussion of the difficulties associated with power series, consult a text on analytic function theory, for example, Ahlfors [3].   □

Since the Taylor polynomials have the property that all the information used in the approximation is concentrated at one point, that is, at x_0, the type of difficulty that occurs in Example 3 is quite common. This generally limits the use of approximation by Taylor polynomials to the situation where approximations are needed at points very close to x_0. For ordinary computational purposes, it is more efficient to use methods that include information at various points, and it is the construction of this type of polynomial which will be considered in the remainder of this chapter. The primary use of Taylor polynomials in numerical analysis is not for approximation purposes, but for use in the derivation of numerical techniques.

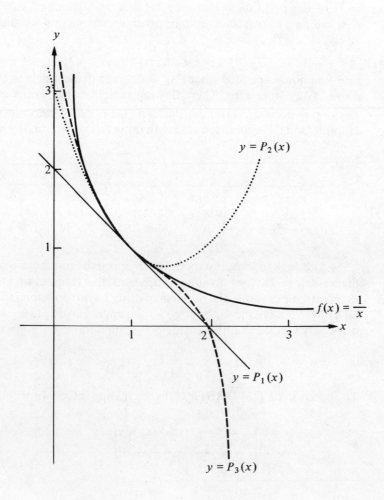

FIGURE 3.3

Exercise Set 3.1

1. Find the Taylor polynomial of degree 2 for $f(x) = x^2 - 3$ expanded about:

   a) $x_0 = 1$.   b) $x_0 = 0$.

2. Obtain the third-degree Taylor polynomial for $f(x) = (1 + x)^{-2}$ about $x_0 = 0$, and use this polynomial to approximate $f(.05)$. Find an error bound for this approximation, and compare your result to the exact value of $f(.05)$.

3. Use the polynomial in Exercise 2 to approximate $\int_0^{.05} (1 + x)^{-2}\, dx$. Find an error bound for this approximation, and compare your result with the actual value for this integral.

4. Find the smallest integer n necessary to approximate $f(x) = 1/x$ at $x = 1.25$ with accuracy 10^{-8}, using the Taylor polynomial of degree n about $x_0 = 1$.

5. Use the error term of a Taylor polynomial to estimate the error involved in using $\sin x \approx x$ to approximate $\sin 1°$?

6. Let $f(x) = \ln(1 + x)$. Find the fourth degree Taylor polynomial for f expanded about $x_0 = 0$, and use it to approximate $\ln(1.1)$. Find a bound for the error in this approximation.

7. Let $F(x) = \int_0^x (1 + t)^{-1}\, dt$. Using the third-degree Taylor polynomial of $f(x) = (1 + x)^{-1}$, expanded about $x_0 = 0$, approximate $F(.1)$. Compare your results to those obtained in Exercise 6.

Bernstein Polynomials   Given a function f defined on $[0, 1]$, the Bernstein polynomial of degree n for f is given by

$$B_n(x) = \sum_{k=0}^{n} \binom{n}{k} f\left(\frac{k}{n}\right) x^k (1 - x)^{n-k}$$

where $\binom{n}{k}$ denotes $\dfrac{n!}{k!(n - k)!}$.

It can be shown that, if f is continuous on $[0, 1]$ and $x_0 \in [0, 1]$, then

$$\lim_{n \to \infty} B_n(x_0) = f(x_0).$$

The polynomials can be used in a constructive proof of the Weierstrass Theorem (see Bartle [9]). Exercises 8–10 refer to the Bernstein polynomials.

8. Find $B_3(x)$ for the functions

   a) $f(x) = x, \quad x \in [0, 1]$,   b) $f(x) = 1, \quad x \in [0, 1]$.

9. Show that for each $k \le n$,

$$\binom{n-1}{k-1} = \left(\frac{k}{n}\right)\binom{n}{k}.$$

10. Use Exercise 9 and the fact that

$$1 = \sum_{k=0}^{n} \binom{n}{k} x^k (1 - x)^{n-k} \qquad \text{for each } n,$$

    to show that, for $f(x) = x^2$,

$$B_n(x) = \left(\frac{n-1}{n}\right) x^2 + \frac{1}{n} x.$$

11. Using Exercise 10, estimate the size of n necessary in order for $|B_n(x) - x^2| \le 10^{-6}$ to hold for all x in $[0, 1]$. Do you think the Bernstein polynomials give a practical means of approximating continuous functions?

3.2 Interpolation and the Lagrange Polynomial

The previous section discussed approximating polynomials that agree with a given function and some of its derivatives at a single point. These polynomials are useful over small intervals for functions whose derivatives exist and are easily evaluated, but this is clearly not always the case. Consequently, the Taylor polynomial is often of little use, and alternative methods of approximation must be sought. The material in this section is concerned with finding approximating polynomials that can be determined simply by specifying certain points on the plane through which they must pass.

Consider the problem of determining a polynomial of degree 1, which passes through the distinct points (x_0, y_0) and (x_1, y_1). This problem is the same as approximating a function f, for which $f(x_0) = y_0$ and $f(x_1) = y_1$, by means of a first-degree polynomial interpolating, or agreeing with, the values of f at the given points. (See Fig. 3.4.)

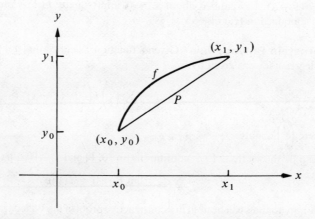

FIGURE 3.4

Consider the first-degree polynomial

$$P(x) = \frac{(x - x_1)}{(x_0 - x_1)} y_0 + \frac{(x - x_0)}{(x_1 - x_0)} y_1.$$

When $x = x_0$,

$$P(x_0) = 1 \cdot y_0 + 0 \cdot y_1 = y_0 = f(x_0)$$

and when $x = x_1$,

$$P(x_1) = 0 \cdot y_0 + 1 \cdot y_1 = y_1 = f(x_1),$$

so P has the required properties.

The technique used to construct P is the method of "interpolation" often used in trigonometric or logarithmic tables. What may not be obvious is that P is the only polynomial of degree 1 or less with the interpolating property. This result, however, follows immediately from Corollary 2.15, page 56.

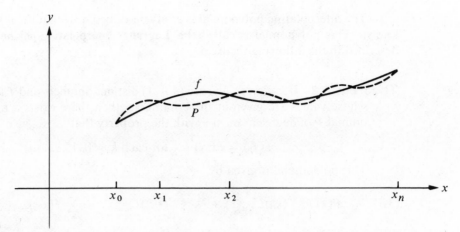

FIGURE 3.5

To generalize the concept of linear interpolation, consider the construction of a polynomial of degree at most n that passes through the $n + 1$ points $(x_0, f(x_0))$, $(x_1, f(x_1)), \ldots, (x_n, f(x_n))$. (See Fig. 3.5.) The linear polynomial passing through $(x_0, f(x_0))$ and $(x_1, f(x_1))$ is constructed by using the quotients

$$L_0(x) = \frac{(x - x_1)}{(x_0 - x_1)} \quad \text{and} \quad L_1(x) = \frac{(x - x_0)}{(x_1 - x_0)}.$$

When $x = x_0$, $L_0(x_0) = 1$ while $L_1(x_0) = 0$. When $x = x_1$, $L_0(x_1) = 0$ while $L_1(x_1) = 1$. For the general case we need to construct, for each $k = 0, 1, \ldots, n$, a quotient $L_{n,k}(x)$ with the property that $L_{n,k}(x_i) = 0$ when $i \neq k$ and $L_{n,k}(x_k) = 1$. To satisfy $L_{n,k}(x_i) = 0$ for each $i \neq k$ requires that the numerator of $L_{n,k}$ contain the term

(3.2) $(x - x_0)(x - x_1) \cdots (x - x_{k-1})(x - x_{k+1}) \cdots (x - x_n).$

To satisfy $L_{n,k}(x_k) = 1$, the denominator of L_k must be equal to (3.2) when $x = x_k$. Thus,

$$L_{n,k}(x) = \frac{(x - x_0) \cdots (x - x_{k-1})(x - x_{k+1}) \cdots (x - x_n)}{(x_k - x_0) \cdots (x_k - x_{k-1})(x_k - x_{k+1}) \cdots (x_k - x_n)}$$

$$= \prod_{\substack{i=0 \\ i \neq k}}^{n} \frac{(x - x_i)}{(x_k - x_i)}.$$

A sketch of the graph of $L_{n,k}$ is shown in Fig. 3.6.

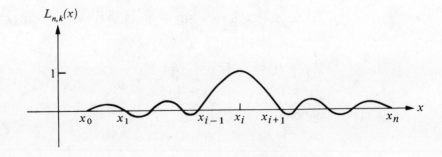

FIGURE 3.6

The interpolating polynomial is easily described now that the form of $L_{n,k}$ is known. This polynomial is called the **Lagrange interpolating polynomial** and is described in the following theorem.

Theorem 3.2 If $x_0, x_1, \ldots, x_n$ are $(n + 1)$ distinct numbers and f is a function whose values are given at these numbers, then there exists a unique polynomial P of degree at most n with the property that

$$f(x_k) = P(x_k) \qquad \text{for each } k = 0, 1, \ldots, n.$$

This polynomial is given by

$$(3.3) \qquad P(x) = f(x_0)L_{n,0}(x) + \cdots + f(x_n)L_{n,n}(x) = \sum_{k=0}^{n} f(x_k)L_{n,k}(x),$$

where

$$(3.4) \quad L_{n,k}(x) = \frac{(x - x_0)(x - x_1) \cdots (x - x_{k-1})(x - x_{k+1}) \cdots (x - x_n)}{(x_k - x_0)(x_k - x_1) \cdots (x_k - x_{k-1})(x_k - x_{k+1}) \cdots (x_k - x_n)}$$

$$= \prod_{\substack{i=0 \\ i \neq k}}^{n} \frac{(x - x_i)}{(x_k - x_i)} \qquad \text{for each } k = 0, 1, \ldots, n.$$

We will write $L_{n,k}(x)$ simply as $L_k(x)$ when there can be no confusion as to its degree.

EXAMPLE 1 Using the numbers, or nodes, $x_0 = 2$, $x_1 = 2.5$, and $x_2 = 4$ to find the second-degree interpolating polynomial for $f(x) = 1/x$ requires that we first determine the coefficient polynomials L_0, L_1, and L_2:

$$L_0(x) = \frac{(x - 2.5)(x - 4)}{(2 - 2.5)(2 - 4)} = x^2 - 6.5x + 10,$$

$$L_1(x) = \frac{(x - 2)(x - 4)}{(2.5 - 2)(2.5 - 4)} \approx -1.333x^2 + 8x - 10.667,$$

and

$$L_2(x) = \frac{(x - 2)(x - 2.5)}{(4 - 2)(4 - 2.5)} \approx .333x^2 - 1.5x + 1.667.$$

Since $f(x_0) = f(2) = .5$, $f(x_1) = f(2.5) = .4$ and $f(x_2) = f(4) = .25$,

$$P(x) = \sum_{k=0}^{2} f(x_k)L_k(x)$$

$$\approx .5(x^2 - 6.5x + 10) + .4(-1.333x^2 + 8x - 10.667)$$

$$+ .25(.333x^2 - 1.5x + 1.667)$$

$$\approx .05x^2 - .425x + 1.15$$

$$\approx (.05x - .425)x + 1.15.$$

An approximation to $f(3) = \frac{1}{3}$ is

$$f(3) \approx P(3) \approx .325.$$

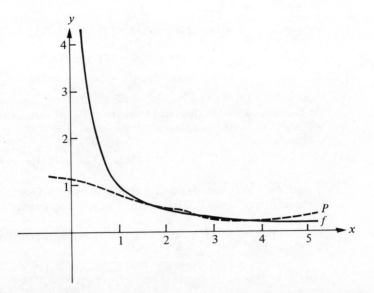

FIGURE 3.7

Compare this example to Example 3 of Section 3.1 where no Taylor polynomial (expanded about $x_0 = 1$) could be used to reasonably approximate $f(3) = \frac{1}{3}$ (see Fig. 3.7). ∎

The next step is to calculate a remainder term or bound for the error involved in approximating a function by an interpolating polynomial. This is done in the following theorem.

Theorem 3.3   If $x_0, x_1, \ldots, x_n$ are distinct points in the interval $[a, b]$ and if $f \in C^{n+1}[a, b]$, then, for each x in $[a, b]$, $\xi(x)$ in (a, b) exists with

$$(3.5) \qquad f(x) = P(x) + \frac{f^{(n+1)}(\xi(x))}{(n+1)!}(x - x_0)(x - x_1)\cdots(x - x_n),$$

where P is the interpolating polynomial given in Eq. (3.3).

Proof   Note first that, if $x = x_k$ for $k = 0, 1, \ldots, n$, then $f(x_k) = P(x_k)$ and choosing $\xi(x_k)$ arbitrarily in (a, b) yields Eq. (3.5). If $x \neq x_k$, for any $k = 0, 1, \ldots, n$, define the function g for any t in $[a, b]$ by

$$g(t) = f(t) - P(t) - [f(x) - P(x)]\frac{(t - x_0)(t - x_1)\cdots(t - x_n)}{(x - x_0)(x - x_1)\cdots(x - x_n)}$$

$$= f(t) - P(t) - [f(x) - P(x)]\prod_{i=0}^{n}\frac{(t - x_i)}{(x - x_i)}.$$

Since $f \in C^{n+1}[a, b]$, $P \in C^{\infty}[a, b]$, and $x \neq x_k$ for any k, it follows that $g \in C^{n+1}[a, b]$. For $t = x_k$

$$g(x_k) = f(x_k) - P(x_k) - [f(x) - P(x)]\prod_{i=0}^{n}\frac{(x_k - x_i)}{(x - x_i)}$$

$$= 0 - [f(x) - P(x)] \cdot 0$$

$$= 0.$$

Moreover, $g(x) = f(x) - P(x) - [f(x) - P(x)]\prod\limits_{i=0}^{n}\dfrac{(x - x_i)}{(x - x_i)}$

$$= f(x) - P(x) - [f(x) - P(x)]$$

$$= 0.$$

Thus, $g \in C^{n+1}[a, b]$ and g vanishes at the $n + 2$ distinct numbers $x, x_0, x_1, \ldots, x_n$. By the Generalized Rolle's Theorem (Theorem 1.11, p. 6), there exists $\xi \equiv \xi(x)$ in (a, b) for which $g^{(n+1)}(\xi) = 0$. Evaluating $g^{(n+1)}$ at ξ gives

(3.6) $0 = g^{(n+1)}(\xi) = f^{(n+1)}(\xi) - P^{(n+1)}(\xi) - [f(x) - P(x)]\dfrac{d^{n+1}}{dt^{n+1}}\left(\prod\limits_{i=0}^{n}\dfrac{(t - x_i)}{(x - x_i)}\right)\Bigg|_{t=\xi}$

Since P is a polynomial of degree at most n, the $(n + 1)$st derivative, $P^{(n+1)}$, must be identically zero. Also, $\prod_{i=0}^{n}[(t - x_i)/(x - x_i)]$ is a polynomial of degree $(n + 1)$, so

$$\prod\limits_{i=0}^{n}\dfrac{(t - x_i)}{(x - x_i)} = \left(\dfrac{1}{\prod_{i=0}^{n}(x - x_i)}\right)t^{n+1} + \text{(lower-degree terms in } t),$$

and $\dfrac{d^{n+1}}{dt^{n+1}}\prod\limits_{i=0}^{n}\dfrac{(t - x_i)}{(x - x_i)} = \dfrac{(n + 1)!}{\prod_{i=0}^{n}(x - x_i)}.$

Equation (3.6) now becomes

$$0 = f^{(n+1)}(\xi) - 0 - [f(x) - P(x)]\dfrac{(n + 1)!}{\prod_{i=0}^{n}(x - x_i)}$$

and, upon solving for $f(x)$, we obtain:

(3.7) $f(x) = P(x) + \dfrac{f^{(n+1)}(\xi)}{(n + 1)!}\prod\limits_{i=0}^{n}(x - x_i).$ □

The error formula (3.5) is of theoretical importance but its practical use is restricted to those functions whose derivatives have known bounds. This is often the case for tabulated functions such as the trigonometric or logarithmic functions.

EXAMPLE 2   Suppose a table is to be prepared for the function $f(x) = e^x$, $0 \le x \le 1$. Assume the number of decimal places to be given per entry will be d; for example, if $d = 5$, then $f(1) = 2.71828$, and that the difference between adjacent x-values, the step size, is h.

a)   Assuming $d \ge 6$, what should h be in order for linear interpolation (i.e., the Lagrange polynomial of degree 1) to give an absolute error of at most 10^{-6}?

b)   If $d < 6$, what should h be in order for linear interpolation to give an absolute error of at most 10^{-6}?

Let $x \in [0, 1]$ and suppose j is such that $x_j \le x \le x_{j+1}$. From Eq. (3.5), the error in linear interpolation is

$$f(x) - P(x) = \dfrac{f^{(2)}(\xi)}{2!}(x - x_j)(x - x_{j+1}).$$

Since the step size is h, it follows that $x_j = jh$, $x_{j+1} = (j + 1)h$, and

$$|f(x) - P(x)| \le \dfrac{|f^{(2)}(\xi)|}{2!}|(x - jh)(x - (j + 1)h)|.$$

Hence

$$|f(x) - P(x)| \leq \tfrac{1}{2} \max_{\xi \in [0,1]} |f^{(2)}(\xi)| \max_{x_j \leq x \leq x_{j+1}} |(x - jh)(x - (j+1)h)|$$

$$= \tfrac{1}{2} \max_{\xi \in [0,1]} e^\xi \max_{x_j \leq x \leq x_{j+1}} |(x - jh)(x - (j+1)h)|$$

$$\leq \tfrac{1}{2} e \max_{x_j \leq x \leq x_{j+1}} |(x - jh)(x - (j+1)h)|.$$

By considering $g(x) = (x - jh)(x - (j+1)h)$ for $jh \leq x \leq (j+1)h$ and using techniques of calculus (see Exercise 14), it can be shown that

$$(3.8) \qquad \max_{x_j \leq x \leq x_{j+1}} |g(x)| = |g((j + \tfrac{1}{2})h)| = \frac{h^2}{4}.$$

Consequently, the error in linear interpolation is bounded by

$$|f(x) - P(x)| \leq \frac{eh^2}{8}.$$

To answer part (a), it is sufficient for h to be chosen so that

$$\frac{eh^2}{8} \leq 10^{-6}, \quad h^2 \leq \frac{8}{e} \cdot 10^{-6}, \quad h^2 < 2.944 \times 10^{-6}, \quad \text{or} \quad h < 1.72 \times 10^{-3}.$$

Letting $h = .001$ would be one logical choice for the step size.

The reason we distinguish between parts (a) and (b) is to emphasize that, if the tables are accurate only to the fifth decimal place, it is impossible to obtain accurate values to the sixth place via interpolation. The errors in rounding to the fifth decimal place will remain, so there is no answer for part (b). □

The last example of this section illustrates interpolation techniques for a situation when the error portion of (3.5) cannot be used. This example also serves to illustrate that we should look for a more efficient way to obtain an approximation via interpolation.

EXAMPLE 3   Table 3.1 lists values of a function (the Bessel function of the first kind of order zero) at various points. The approximations to $f(1.5)$ obtained by various Lagrange polynomials will be compared.

TABLE 3.1

x	$f(x)$
1.0	.7651977
1.3	.6200860
1.6	.4554022
1.9	.2818186
2.2	.1103623

Using $x_0 = 1.3$ and $x_1 = 1.6$, the value of the interpolating polynomial at 1.5 is given by

$$P_1(1.5) = \frac{(1.5 - 1.6)}{(1.3 - 1.6)}(.6200860) + \frac{(1.5 - 1.3)}{(1.6 - 1.3)}(.4554022) = .5102968.$$

Two polynomials of degree two could reasonably be used, one by letting $x_0 = 1.3$, $x_1 = 1.6$, and $x_2 = 1.9$, which gives

$$P_2(1.5) = \frac{(1.5 - 1.6)(1.5 - 1.9)}{(1.3 - 1.6)(1.3 - 1.9)}(.6200860) + \frac{(1.5 - 1.3)(1.5 - 1.9)}{(1.6 - 1.3)(1.6 - 1.9)}(.4554022)$$

$$+ \frac{(1.5 - 1.3)(1.5 - 1.6)}{(1.9 - 1.3)(1.9 - 1.6)}(.2818186)$$

$$= .5112857$$

and the other by letting $x_0 = 1.0$, $x_1 = 1.3$, and $x_2 = 1.6$, in which case

$$\hat{P}_2(1.5) = .5124715.$$

In the third-degree case there are also two choices for the polynomial. One is with $x_0 = 1.3$, $x_1 = 1.6$, $x_2 = 1.9$, and $x_3 = 2.2$, which gives

$$P_3(1.5) = .5118302.$$

The other is obtained by letting $x_0 = 1.0$, $x_1 = 1.3$, $x_2 = 1.6$, and $x_3 = 1.9$, giving

$$\hat{P}_3(1.5) = .5118127.$$

The fourth-degree Lagrange polynomial uses all the entires in the table and is expected to be the most accurate. With $x_0 = 1.0$, $x_1 = 1.3$, $x_2 = 1.6$, $x_3 = 1.9$, and $x_4 = 2.2$, it can be shown that

$$P_4(1.5) = .5118200.$$

Since $P_3(1.5)$, $\hat{P}_3(1.5)$, and $P_4(1.5)$ all agree to within 2×10^{-5} units, we expect $P_4(1.5)$ to be the most accurate approximation and to be correct to within 2×10^{-5} units.

The actual value of $f(1.5)$ is known to be .5118277, so the accuracies of the approximations are as follows:

$$|P_1(1.5) - f(1.5)| \approx 1.53 \times 10^{-3},$$
$$|P_2(1.5) - f(1.5)| \approx 5.42 \times 10^{-4},$$
$$|\hat{P}_2(1.5) - f(1.5)| \approx 6.44 \times 10^{-4},$$
$$|P_3(1.5) - f(1.5)| \approx 2.5 \times 10^{-6},$$
$$|\hat{P}_3(1.5) - f(1.5)| \approx 1.50 \times 10^{-5},$$
$$|P_4(1.5) - f(1.5)| \approx 7.7 \times 10^{-6}.$$

In fact, P_3 is the most accurate approximation, however, with no knowledge of the actual value of $f(1.5)$, P_4 would be accepted as the best approximation. It should be noted that the error or remainder term derived in Theorem 3.3 cannot be applied here, since no knowledge of the fourth derivative of f is available. Unfortunately, this is generally the case. ☐

Exercise Set 3.2

1. Show in two different ways that $f(x) = 3(x - 1)(x - 2)(x + 1)$ and $g(x) = 3x^3 - 6x^2 - 3x + 6$ represent the same polynomial.

2. Use appropriate Lagrange interpolating polynomials of degree one, two, three, and four to approximate $f(2.5)$ if

$$f(2.0) = .5103757 \qquad f(2.4) = .5104147 \qquad f(2.8) = .4359160$$

$$f(2.2) = .5207843 \qquad f(2.6) = .4813306$$

3. Calculate the polynomial of degree 3 or less that agrees with $f(x) = x^2$ at $x_0 = 1$, $x_1 = 3$, $x_2 = 6$, and $x_3 = 7$.

4. Use the values below to construct a third-degree Lagrange polynomial approximation to $f(1.09)$. The function being approximated is $f(x) = \log_{10} \tan x$. Use this knowledge to find a bound for the error in the approximation.

$$f(1.10) = .2933 \qquad f(1.05) = .2414$$

$$f(1.00) = .1924 \qquad f(1.15) = .3492$$

5. Use the values below to construct a fourth-degree Lagrange polynomial approximation to $f(1.25)$. The function being approximated is $f(x) = e^{x^2 - 1}$. Use this knowledge to find a bound for the error in the approximation.

$$f(1) = 1.00000 \qquad f(1.2) = 1.55271 \qquad f(1.4) = 2.61170$$

$$f(1.1) = 1.23368 \qquad f(1.3) = 1.99372$$

6. Use the Lagrange interpolating polynomial of degree three or less to approximate $\cos .750$ using the values below. Find an error bound using Eq. (3.5).

$$\cos .698 = .7661 \qquad \cos .768 = .7193$$

$$\cos .733 = .7432 \qquad \cos .803 = .6946$$

The actual value of $\cos .750$ is .7317 (to four decimal places). If there is a discrepancy between the actual error and your error bound, explain why this occurred.

7. Let $f(x) = 3xe^x - 2e^x$. Approximate $f(1.03)$ using the interpolating polynomial of degree less than or equal to two, using $x_0 = 1$, $x_1 = 1.05$, and $x_2 = 1.07$. Compare the actual error to the error bound obtained from Eq. (3.5).

8. Let $f(x) = (4x - 7)/(x - 2)$ and $x_0 = 1.7$, $x_1 = 1.8$, $x_2 = 1.9$, and $x_3 = 2.1$.

   a) Approximate $f(1.75)$ using the interpolating polynomial of degree at most two on the nodes x_0, x_1, and x_2.
   b) Approximate $f(1.75)$ and $f(2.00)$ using the interpolating polynomial on x_0, x_1, x_2, and x_3.
   c) Can an error bound from Eq. (3.5) be applied to (a) or (b)? What does the bound give as an error estimate?

9. Let $f(x) = e^x$, $0 \le x \le 2$. Using the values given below, perform the following computation:

   a) Approximate $f(.25)$ using linear interpolation with $x_0 = 0$ and $x_1 = .5$.
   b) Approximate $f(.75)$ using linear interpolation with $x_0 = .5$ and $x_1 = 1$.
   c) Approximate $f(.25)$ and $f(.75)$ by using the second-degree interpolating polynomial with $x_0 = 0$, $x_1 = 1$, and $x_2 = 2$.

d)   Which approximations are better? Why?

x	0	.5	1.0	2.0
$f(x)$	1.00000	1.64872	2.71828	7.38906

10.   Suppose it is desired to construct six-place tables for the common or base-10 logarithm function from $x = 1$ to $x = 10$ in such a way that linear interpolation is accurate to the sixth decimal place. Determine the largest possible step size for this table.

11.   Show that

$$\sum_{k=0}^{n} L_k(x) = 1 \qquad \text{for all } x.$$

12.   Let $\omega(x) = \prod_{k=0}^{n} (x - x_k)$. Show that the interpolating polynomial of degree n on $x_0, \ldots, x_n$ can be written as

$$P(x) = \omega(x) \sum_{k=0}^{n} \frac{f(x_k)}{(x - x_k)\omega'(x_k)}.$$

13.   Prove Theorem 1.13, page 7. [*Hint*:   Let

$$g(t) = f(t) - P(t) - [f(x) - P(x)] \cdot \frac{(t - x_0)^n}{(x - x_0)^n},$$

where P is the nth-degree Taylor polynomial, and use Theorem 1.11, page 6.]

14.   Show that

$$\max_{x_j \leq x \leq x_{j+1}} |g(x)| = \frac{h^2}{4}$$

where $g(x) = (x - jh)(x - (j + 1)h)$. This establishes the result in Eq. (3.8).

15.   In the introduction to this chapter, the following table was given, listing the population of the United States from 1920 to 1970.

Year	1920	1930	1940	1950	1960	1970
Population (In Thousands)	105,711	123,203	131,669	150,697	179,323	203,212

Find the Lagrange polynomial of degree 5 fitting this data, and use this polynomial to estimate the population in the years 1910, 1965, and 2000. The population in 1910 was approximately 91,972,000. How accurate do you think your 1965 and 2000 figures are?

3.3   Iterated Interpolation

One difficulty that arises in using the method of Section 3.2 is that since the error term given by Theorem 3.3 is difficult to work with, the degree of the polynomial needed for the desired accuracy is generally not known before the work has been completed. The usual practice is to compute the results given from various polynomials until appropriate agreement is obtained. This was done, for instance, in Example 3. In examining this example, it is seen that the work done in calculating the approximation by the second-degree polynomial does not lessen the work needed to calculate the third-degree approximation; nor is the fourth-degree approximation easier to obtain once the third-degree approximation is known. It is the purpose of this section to derive these approximating polynomials in a manner that utilizes the previous calculations to greatest advantage.

Definition 3.4 Let f be a function defined at $x_0, x_1, x_2, \ldots, x_n$, and suppose that $m_1, m_2, \ldots, m_k$ are k distinct integers with $0 \leq m_i \leq n$ for each i. The Lagrange polynomial of degree $\leq k$ that agrees with f at $x_{m_1}, x_{m_2}, \ldots, x_{m_k}$ is denoted by $P_{m_1, m_2, \ldots, m_k}$.

EXAMPLE 1 If $x_0 = 1$, $x_1 = 2$, $x_2 = 3$, $x_3 = 4$, $x_4 = 6$, and $f(x) = x^3$, then $P_{1,2,4}$ is the polynomial that agrees with f at $x_1 = 2$, $x_2 = 3$, and $x_4 = 6$; that is,

$$P_{1,2,4}(x) = \frac{(x-3)(x-6)}{(2-3)(2-6)}(8) + \frac{(x-2)(x-6)}{(3-2)(3-6)}(27) + \frac{(x-2)(x-3)}{(6-2)(6-3)}(216). \quad \Box$$

Theorem 3.5 Let f be defined at $x_0, x_1, \ldots, x_k$ and let x_j, x_i be two distinct numbers in this set. If

$$(3.9) \quad P(x) = \frac{(x - x_j)P_{0,1,\ldots,j-1,j+1,\ldots,k}(x) - (x - x_i)P_{0,1,\ldots,i-1,i+1,\ldots,k}(x)}{(x_i - x_j)},$$

then P is the Lagrange polynomial of degree less than or equal to k, which interpolates f at $x_0, x_1, \ldots, x_k$.

Proof For ease of notation, let $Q \equiv P_{0,1,\ldots,i-1,i+1,\ldots,k}$ and $\hat{Q} \equiv P_{0,1,\ldots,j-1,j+1,\ldots,k}$. Q and $\hat{Q}$ are polynomials of degree $k - 1$ or less; hence, P must be of degree $\leq k$. If $0 \leq r \leq k$ and $r \neq i, j$, then

$$P(x_r) = \frac{(x_r - x_j)\hat{Q}(x_r) - (x_r - x_i)Q(x_r)}{x_i - x_j} = \frac{(x_i - x_j)}{(x_i - x_j)} f(x_r) = f(x_r).$$

Moreover, $\qquad P(x_i) = \dfrac{(x_i - x_j)\hat{Q}(x_i) - (x_i - x_i)Q(x_i)}{x_i - x_j} = \dfrac{(x_i - x_j)}{(x_i - x_j)} f(x_i) = f(x_i),$

and similarly $P(x_j) = f(x_j)$. But, by definition, $P_{0,1,\ldots,k}$ is the unique polynomial of degree $\leq k$ which agrees with f at $x_0, x_1, \ldots, x_k$. Thus, $P = P_{0,1,\ldots,k}$. $\qquad \Box$

EXAMPLE 2 In Example 3 of Section 3.2, values of various Lagrange polynomials at $x = 1.5$ were obtained, using Table 3.2.

TABLE 3.2

x	$f(x)$
1.0	.7651977
1.3	.6200860
1.6	.4554022
1.9	.2818186
2.2	.1103623

In this example we calculate the approximation of $f(1.5)$ using Theorem 3.5. If $x_0 = 1.0$, $x_1 = 1.3$, $x_2 = 1.6$, $x_3 = 1.9$, $x_4 = 2.2$, the notation of Definition 3.4 implies that $f(1.0) = P_0$, $f(1.3) = P_1$, $f(1.6) = P_2$, $f(1.9) = P_3$, and $f(2.2) = P_4$; so these are the five polynomials of degree zero (constants) that approximate $f(1.5)$. Calculating $P_{0,1}(1.5)$ yields

$$P_{0,1}(1.5) = \frac{(1.5 - 1.0)P_1 - (1.5 - 1.3)P_0}{(1.3 - 1.0)}$$

$$= \frac{.5(.6200860) - .2(.7651977)}{.3}$$

$$= .5233449.$$

Similarly,

$$P_{1,2}(1.5) = \frac{(1.5 - 1.3)(.4554022) - (1.5 - 1.6)(.6200860)}{(1.6 - 1.3)} = .5102968,$$

$$P_{2,3}(1.5) = .5132634,$$

and $P_{3,4}(1.5) = .5104270.$

These give the approximations using first-degree polynomials. $P_{1,2}$ is expected to be the best approximation since 1.5 is between $x_1 = 1.3$ and $x_2 = 1.6$. The approximations using second-degree polynomials are given by

$$P_{0,1,2}(1.5) = \frac{(1.5 - 1.0)(.5102968) - (1.5 - 1.6)(.5233449)}{(1.6 - 1.0)} = .5124715,$$

$$P_{1,2,3}(1.5) = .5112857,$$

and $P_{2,3,4}(1.5) = .5137361.$

The third-degree approximations are determined to be

$$P_{0,1,2,3}(1.5) = \frac{(1.5 - 1.0)(.5112857) - (1.5 - 1.9)(.5124175)}{(1.9 - 1.0)} = .5118127,$$

and $P_{1,2,3,4}(1.5) = .5118302.$

Finally, the fourth-degree approximation is

$$P_{0,1,2,3,4}(1.5) = \frac{(1.5 - 1.0)(.5118302) - (1.5 - 2.2)(.5118127)}{(2.2 - 1.0)} = .5118200.$$

As before, this is expected to be accurate within 2×10^{-5}.

Having calculated the results above, we construct a table (Table 3.3) as follows:

TABLE 3.3

x_0	P_0				
x_1	P_1	$P_{0,1}$			
x_2	P_2	$P_{1,2}$	$P_{0,1,2}$		
x_3	P_3	$P_{2,3}$	$P_{1,2,3}$	$P_{0,1,2,3}$	
x_4	P_4	$P_{3,4}$	$P_{2,3,4}$	$P_{1,2,3,4}$	$P_{0,1,2,3,4}$

which, for this example, appears as in Table 3.4:

TABLE 3.4

1.0	.7651977			
1.3	.6200860	.5233449		
1.6	.4554022	.5102968	.5124715	
1.9	.2818186	.5132634	.5112857	.5118127
2.2	.1103623	.5104270	.5137361	.5118302	.5118200

Suppose that at this point it is decided that the latest approximation $P_{0,1,2,3,4}$ is not as accurate as desired. Another node x_5 could be selected, another row added to the table;

$$x_5, \quad P_5, \quad P_{4,5}, \quad P_{3,4,5}, \quad P_{2,3,4,5}, \quad P_{1,2,3,4,5}, \quad P_{0,1,2,3,4,5},$$

and $P_{0,1,2,3,4}, P_{1,2,3,4,5}$, and $P_{0,1,2,3,4,5}$ could be compared to determine further accuracy. It is easy to see from the formula in Theorem 3.5 that only the fifth row is needed in order to compute the sixth row. In general, only the previous row need be used to add a new row to the table; for example,

$$P_{4,5}(x) = \frac{(x - x_4)P_5 - (x - x_5)P_4}{x_5 - x_4},$$

which involves only x, x_4, x_5, P_4, and P_5.

In the example we have been considering, the value of the Bessel function of the first kind of order zero at 2.5 is $-.0483838$. Using this to construct the new row gives us:

2.5, $-.0483838$, .4807699, .5301984, .5119070, .5118430, .5118277.

The new entry is correct to six decimal places.   □

The procedure outlined above is called Neville's method. Our presentation uses cumbersome notation, and a more useful form of Neville's method will now be considered. As an array is being constructed, only two subscripts are actually needed. Note that proceeding down the table corresponds to using consecutive points x_i with larger i, and proceeding to the right corresponds to increasing the degree of the interpolating polynomial.

Let $Q_{i,j}, i \geq j$, denote the interpolating polynomial of degree j on the $(j + 1)$ numbers $x_{i-j}, x_{i-j+1}, \ldots, x_{i-1}, x_i$. To compute

$$Q_{i,j} = P_{i-j,i-j+1,\ldots,i-1,i}$$

by Neville's method use

$$Q_{i,j-1} = P_{i-j+1,\ldots,i-1,i} \quad \text{and} \quad Q_{i-1,j-1} = P_{i-j,i-j+1,\ldots,i-1}$$

in Eq. (3.9) to obtain:

$$Q_{i,j}(x) = \frac{(x - x_i)Q_{i-1,j-1}(x) - (x - x_{i-j})Q_{i,j-1}(x)}{x_{i-j} - x_i}$$

for each $j = 1, 2, 3, \ldots$, and $i = j, j + 1, \ldots$. Additionally, let $Q_{i,0} = f(x_i)$ for each i. This notation for Neville's method provides the array in Table 3.5. This table

TABLE 3.5

x_0	$Q_{0,0}$			
x_1	$Q_{1,0}$	$Q_{1,1}$		
x_2	$Q_{2,0}$	$Q_{2,1}$	$Q_{2,2}$	
x_3	$Q_{3,0}$	$Q_{3,1}$	$Q_{3,2}$	$Q_{3,3}$
x_4	$Q_{4,0}$	$Q_{4,1}$	$Q_{4,2}$	$Q_{4,3}$	$Q_{4,4}$

and Table 3.3 involving the P's are the same, except for the notation, but Table 3.5 is much easier to set up for computer utilization.

The following algorithm constructs the table by rows.

Neville's Iterated Interpolation Algorithm 3.1

To evaluate the interpolating polynomial P on the $(n + 1)$ distinct numbers $x_0, \ldots, x_n$ at the number x for the function f:

INPUT numbers $x_0, x_1, \ldots, x_n$; values $f(x_0), f(x_1), \ldots, f(x_n)$ as the first column $Q_{0,0}, Q_{1,0}, \ldots, Q_{n,0}$ of Q.

OUTPUT the table Q with $P(x) = Q_{n,n}$.

Step 1 For $i = 1, 2, \ldots, n$
  for $j = 1, 2, \ldots, i$

$$\text{set } Q_{i,j} = \frac{(x - x_i)Q_{i-1,j-1} - (x - x_{i-j})Q_{i,j-1}}{x_{i-j} - x_i}.$$

Step 2 OUTPUT (Q);
STOP.

The algorithm can be modified to allow for the addition of new interpolating nodes. For example, the inequality

$$|Q_{i,i} - Q_{i-1,i-1}| < \varepsilon,$$

could be used as a stopping criterion, where ε is a prescribed error tolerance. If the inequality is true, then $Q_{i,i}$ is a reasonable approximation to $f(x)$. If the inequality is false, then a new interpolation point x_i is added.

Exercise Set 3.3

1. Obtain the approximations for Exercise 2 of Section 3.2 using Theorem 3.5.

2. Approximate $\sqrt{3}$ using Neville's method on the function $f(x) = 3^x$ for the values $x_0 = -2, x_1 = -1, x_2 = 0, x_3 = 1$, and $x_4 = 2$.

3. Use Neville's method to approximate $f(-.78)$ for the function $f(x) = x^2 e^x \cos x$ using $x_0 = -1.0, x_1 = -.9, x_2 = -.8, x_3 = -.7$, and $x_4 = -.6$.

4. Use Neville's method to approximate $f(1.09)$ for the data given in Exercise 4 of Section 3.2.

5. Use Neville's method to approximate $f(1.25)$ for the data given in Exercise 5 of Section 3.2.

6. a) Use Neville's method to approximate $f(1.03)$ with $P_{0,1,2}$ for the function $f(x) = 3xe^x - e^{2x}$ using $x_0 = 1$, $x_1 = 1.05$, and $x_2 = 1.07$.

b) Suppose the approximation of (a) is not sufficiently accurate. Compute $P_{0,1,2,3}$ where $x_3 = 1.04$.

(c) Continue adding nodes $x_4, \ldots, x_n, x_{n+1}$ until $|P_{0,1,\ldots,n+1} - P_{0,1,\ldots,n}| < 10^{-5}$.

7. Repeat Exercise 6 using four-digit arithmetic. Do you think Neville's method is sensitive to rounding errors?

8. Aitken's method, which is similar to Neville's method, constructs the following table of interpolating values:

x_0	P_0				
x_1	P_1	$P_{0,1}$			
x_2	P_2	$P_{0,2}$	$P_{0,1,2}$		
x_3	P_3	$P_{0,3}$	$P_{0,1,3}$	$P_{0,1,2,3}$	
x_4	P_4	$P_{0,4}$	$P_{0,1,4}$	$P_{0,1,2,4}$	$P_{0,1,2,3,4}$
$\vdots$	$\vdots$	$\vdots$	$\vdots$	$\vdots$	$\vdots$

To compute each new value, we use the value at the top of the preceding column with the value in the same row, preceding column; for example,

$$P_{0,1,3}(x) = \frac{(x - x_3)P_{0,1}(x) - (x - x_1)P_{0,3}(x)}{x_1 - x_3}.$$

Using the same notation $Q_{i,j}$ as in Neville's method, construct the algorithm to compute $Q_{i,j}$ for Aitken's method.

9. Approximate $\sqrt{3}$ using Aitken's method on the function $f(x) = 3^x$ for the nodes $x_0 = -2$, $x_1 = -1$, $x_2 = 0$, $x_3 = 1$, $x_4 = 2$. Compare your result with the result of Exercise 2. Which method do you prefer?

10. Repeat Exercise 4 using Aitken's method instead of Neville's method. Compare your approximations to the exact value $f(1.09) = \log_{10} \tan(1.09)$.

11. Repeat Exercise 5 using Aitken's method instead of Neville's method. Compare your approximations to the exact value $f(1.25) = e^{(1.25)^2 - 1}$.

12. Construct a sequence of interpolating values, y_n, to $f(1 + \sqrt{10})$, where $f(x) = (1 + x^2)^{-1}$ for $-5 \le x \le 5$, as follows: For each $n = 1, 2, \ldots, 10$, let $h = 10/n$ and $y_n = P_n(1 + \sqrt{10})$, where $P_n(x)$ is the interpolating polynomial for $f(x)$ at the nodes $x_0^{(n)}, x_1^{(n)}, \ldots, x_n^{(n)}$ and $x_j^{(n)} = -5 + jh$ for each $j = 0, 1, 2, \ldots, n$. Does the sequence $\{y_n\}$ seem to converge to $f(1 + \sqrt{10})$?

Inverse Interpolation   Suppose $f \in C^1[a, b]$, $f'(x) \neq 0$ on $[a, b]$ and f has one zero p in $[a, b]$. Let $x_0, \ldots, x_n$ be $n + 1$ distinct numbers in $[a, b]$ with $f(x_k) = y_k$ for each $k = 0, 1, \ldots, n$. To approximate p, construct the interpolating polynomial of degree n on the nodes $y_0, \ldots, y_n$ for f^{-1}. Since $y_k = f(x_k)$ and $0 = f(p)$, it follows that $f^{-1}(y_k) = x_k$ and $p = f^{-1}(0)$. Using iterated interpolation to approximate $f^{-1}(0)$ is called *iterated inverse interpolation*.

13. Use iterated inverse interpolation to find an approximation to the solution of $x - e^{-x} = 0$, using the data

x	.3	.4	.5	.6
e^{-x}	.740818	.670320	.606531	.548812

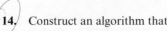

14. Construct an algorithm that can be used for inverse interpolation.

15. Show that iterated inverse interpolation is equivalent to the Secant method. Which method is preferable for the evaluation of p_n and why?

16. Müller's method for approximating a root p of an equation $f(x) = 0$ can be described as follows: given three approximations $p_{n-3}, p_{n-2}, p_{n-1}$ to p, construct the quadratic interpolatory polynomial P_n to f at $p_{n-3}, p_{n-2}, p_{n-1}$. Let p_n denote the root of $P_n(x) = 0$ that is closest to p_{n-1}. Repeat the procedure using p_{n-2}, p_{n-1}, and p_n to find the next approximation p_{n+1}. Construct an algorithm for Müller's method. (Note that the roots of P_n may be complex.)

17. Approximate the real roots of the following polynomials by using Müller's method with $p_0 = -1$, $p_1 = 1$, and $p_2 = 0$. Compare your results to those obtained in Exercise 2 of Section 2.5.

    a) $P(x) = x^4 + 5x^3 - 9x^2 - 85x - 136$
    b) $P(x) = x^4 - 2x^3 - 12x^2 + 16x - 40$
    c) $P(x) = x^4 + x^3 + 3x^2 + 2x + 2$
    d) $P(x) = x^5 + 11x^4 - 21x^3 - 10x^2 - 21x - 5$

18. Use Müller's method to approximate the complex roots of the polynomials in Exercise 17. Compare your results with those obtained in Exercise 2 of Section 2.5.

3.4 Divided Differences

The iterated interpolation techniques discussed in the previous section are useful for determining the values of successively higher-degree interpolating polynomials at a particular point. Each of the entries in the interpolation table, however, depends upon the point being evaluated, so the table cannot be used to give an explicit representation for the interpolating polynomial.

Methods for determining the explicit representation of an interpolating polynomial from tabulated data are known as **divided-difference methods**. These methods were more widely used for computational purposes before digital computing equipment became readily available, but they can also be used to derive methods for approximating the derivatives and integrals of functions, as well as in approximating the solutions to differential equations.

Our treatment of divided-difference methods will be brief since the results in this section will not be used extensively in subsequent material. Most classical texts on the subject of numerical analysis have extensive treatments of divided-difference methods. If a more comprehensive treatment is needed the text by Hildebrand [48] is a particularly good reference.

Suppose that P_n is the polynomial of degree at most n that agrees with the function f at the distinct numbers $x_0, x_1, \ldots, x_n$. The divided differences of f with respect to $x_0, x_1, \ldots, x_n$ can be derived by showing that P_n has the representation

$$(3.10) \qquad P_n(x) = a_0 + a_1(x - x_0) + a_2(x - x_0)(x - x_1) + \cdots$$
$$+ a_n(x - x_0)(x - x_1) \cdots (x - x_{n-1})$$

for appropriate constants $a_0, a_1, \ldots, a_n$.

To determine the first of these constants, a_0, note that if $P_n(x)$ can be written in the form of Eq. (3.10), then evaluating P_n at x_0 leaves only the constant term a_0; that is, $a_0 = P_n(x_0) = f(x_0)$.

Similarly, when P_n is evaluated at x_1, the only nonzero terms in the evaluation of $P_n(x_1)$ are the constant and linear terms,

$$f(x_0) + a_1(x_1 - x_0) = P_n(x_1) = f(x_1);$$

so

(3.11)
$$a_1 = \frac{f(x_1) - f(x_0)}{x_1 - x_0}.$$

At this stage we introduce what is known as the **divided-difference notation**. The zeroth divided difference of the function f, with respect to x_i, is denoted $f[x_i]$ and is simply the evaluation of f at x_i,

$$f[x_i] = f(x_i).$$

The remaining divided differences are defined inductively; the first divided difference of f with respect to x_i and x_j, is denoted $f[x_i, x_j]$ and defined as

$$f[x_i, x_j] = \frac{f[x_j] - f[x_i]}{x_j - x_i}.$$

When the $(k - 1)$st divided differences

$$f[x_i, x_{i+1}, x_{i+2}, \ldots, x_{i+k-1}] \quad \text{and} \quad f[x_{i+1}, x_{i+2}, \ldots, x_{i+k-1}, x_{i+k}]$$

have both been determined, the kth divided difference of f relative to x_i, x_{i+1}, $x_{i+2}, \ldots, x_{i+k}$ is given by

$$f[x_i, x_{i+1}, \ldots, x_{i+k-1}, x_{i+k}] = \frac{f[x_{i+1}, x_{i+2}, \ldots, x_{i+k}] - f[x_i, x_{i+1}, \ldots, x_{i+k-1}]}{x_{i+k} - x_i}.$$

With this notation, Eq. (3.11) can be re-expressed as $a_1 = f[x_0, x_1]$ and the interpolating polynomial in Eq. (3.10) is:

$$P_n(x) = f[x_0] + f[x_0, x_1](x - x_0) + a_2(x - x_0)(x - x_1)$$
$$+ \cdots + a_n(x - x_0)(x - x_1) \cdots (x - x_{n-1}).$$

The other constants in P_n, $a_2, a_3, \ldots, a_n$, can be consecutively obtained in a manner similar to the evaluation of a_0 and a_1, but the algebraic manipulation becomes tedious. (The evaluation of a_2 is considered in Exercise 6. For a general procedure of evaluating these constants, the reader should refer to the previously mentioned book by Hildebrand [48].) As might be expected from the evaluation of a_0 and a_1, the required constants are:

$$a_k = f[x_0, x_1, x_2, \ldots, x_k],$$

for each $k = 0, 1, \ldots, n$; so P_n can be rewritten as

(3.12)
$$P_n(x) = f[x_0] + f[x_0, x_1](x - x_0)$$
$$+ f[x_0, x_1, x_2](x - x_0)(x - x_1) + \cdots$$
$$+ f[x_0, x_1, \ldots, x_n](x - x_0)(x - x_1) \cdots (x - x_{n-1}).$$

The determination of the divided differences from tabulated data points is outlined in Table 3.6. Two fourth and one fifth difference could also be determined from this data.

TABLE 3.6

x	$f(x)$	First divided differences	Second divided differences	Third divided differences
x_0	$f[x_0]$			
		$f[x_0, x_1] = \dfrac{f[x_1] - f[x_0]}{x_1 - x_0}$		
x_1	$f[x_1]$		$f[x_0, x_1, x_2] = \dfrac{f[x_1, x_2] - f[x_0, x_1]}{x_2 - x_0}$	
		$f[x_1, x_2] = \dfrac{f[x_2] - f[x_1]}{x_2 - x_1}$		$f[x_0, x_1, x_2, x_3] = \dfrac{f[x_1, x_2, x_3] - f[x_0, x_1, x_2]}{x_3 - x_0}$
x_2	$f[x_2]$		$f[x_1, x_2, x_3] = \dfrac{f[x_2, x_3] - f[x_1, x_2]}{x_3 - x_1}$	
		$f[x_2, x_3] = \dfrac{f[x_3] - f[x_2]}{x_3 - x_2}$		$f[x_1, x_2, x_3, x_4] = \dfrac{f[x_2, x_3, x_4] - f[x_1, x_2, x_3]}{x_4 - x_1}$
x_3	$f[x_3]$		$f[x_2, x_3, x_4] = \dfrac{f[x_3, x_4] - f[x_2, x_3]}{x_4 - x_2}$	
		$f[x_3, x_4] = \dfrac{f[x_4] - f[x_3]}{x_4 - x_3}$		$f[x_2, x_3, x_4, x_5] = \dfrac{f[x_3, x_4, x_5] - f[x_2, x_3, x_4]}{x_5 - x_2}$
x_4	$f[x_4]$		$f[x_3, x_4, x_5] = \dfrac{f[x_4, x_5] - f[x_3, x_4]}{x_5 - x_3}$	
		$f[x_4, x_5] = \dfrac{f[x_5] - f[x_4]}{x_5 - x_4}$		
x_5	$f[x_5]$			

When $x_0, x_1, \ldots, x_n$ are arranged consecutively with equal spacing, Eq. (3.12) can be expressed in a form which is quite useful for computational purposes. Introducing the notation $h = x_{i+1} - x_i$ for each $i = 0, 1, \ldots, n - 1$ and $x = x_0 + sh$, the difference $x - x_i$ can be written as $x - x_i = (s - i)h$; so Eq. (3.12) becomes

$$P_n(x) = P_n(x_0 + sh) = f[x_0] + shf[x_0, x_1] + s(s - 1)h^2 f[x_0, x_1, x_2]$$

$$+ \cdots + s(s - 1) \cdots (s - n + 1)h^n f[x_0, x_1, \ldots, x_n]$$

$$= \sum_{k=0}^{n} s(s - 1) \cdots (s - k + 1)h^k f[x_0, x_1, \ldots, x_k].$$

Utilizing the binomial-coefficient notation

$$\binom{s}{k} = \frac{s!}{k!(s - k)!} = \frac{s(s - 1) \cdots (s - k + 1)}{k!},$$

we can express $P_n(x)$ compactly as

$$(3.13) \qquad P_n(x) = P_n(x_0 + sh) = \sum_{k=0}^{n} \binom{s}{k} k! h^k f[x_0, x_1, \ldots, x_k].$$

This formula is called the **Newton forward divided-difference formula**. Another form, called the **Newton forward-difference formula**, is constructed by making use of the forward difference notation Δ introduced in Definition 2.10, page 51. With this notation

$$f[x_0, x_1] = \frac{1}{h} \Delta f(x_0),$$

$$f[x_0, x_1, x_2] = \frac{1}{2h} [\Delta f(x_1) - \Delta f(x_0)] = \frac{1}{2h^2} \Delta^2 f(x_0)$$

and, in general,

$$f[x_0, x_1, \ldots, x_k] = \frac{1}{k! h^k} \Delta^k f(x_0).$$

Consequently, Eq. (3.13) becomes

$$P_n(x) = \sum_{k=0}^{n} \binom{s}{k} \Delta^k f(x_0).$$

If the interpolating nodes are reordered as $x_n, x_{n-1}, \ldots, x_0$, a formula similar to Eq. (3.12) results.

$$P_n(x) = f[x_n] + f[x_{n-1}, x_n](x - x_n) + f[x_{n-2}, x_{n-1}, x_n](x - x_n)(x - x_{n-1})$$
$$+ \cdots + f[x_0, \ldots, x_n](x - x_n)(x - x_{n-1}) \cdots (x - x_1).$$

Using equal spacing with $x = x_n + sh$ and $x = x_i + (s + n - i)h$ produces

$$P_n(x) = P_n(x_n - sh)$$
$$= f[x_n] + shf[x_{n-1}, x_n] + s(s + 1)h^2 f[x_{n-2}, x_{n-1}, x_n] + \cdots$$
$$+ s(s + 1) \cdots (s + n - 1)h^n f[x_0, x_1, \ldots, x_n].$$

This form is called the **Newton backward divided-difference formula**. It is used to derive a more commonly applied formula known as the **Newton backward difference formula**. To discuss this formula, we need the following definition.

Definition 3.6 Given the sequence $\{P_n\}_{n=0}^{\infty}$, define the backward difference ∇P_n by

$$\nabla P_n \equiv P_n - P_{n-1} \qquad \text{for } n \geq 1.$$

Higher powers are defined recursively by

$$\nabla P_n = \nabla^{k-1}(\nabla P_n) \qquad \text{for } k \geq 2.$$

Definition 3.6 implies that

$$f[x_{n-1}, x_n] = \frac{1}{h} \nabla f(x_n), \qquad f[x_{n-2}, x_{n-1}, x_n] = \frac{1}{2h^2} \nabla^2 f(x_n),$$

and in general

$$f[x_{n-k}, \ldots, x_{n-1}, x_n] = \frac{1}{k! h^k} \nabla^k f(x_n).$$

Consequently,

$$P_n(x) = f[x_n] + s\nabla f(x_n) + \frac{s(s+1)}{2}\nabla^2 f(x_n) + \cdots$$

$$+ \frac{s(s+1)\cdots(s+n-1)}{n!}\nabla^n f(x_n).$$

Extending the binomial coefficient notation to include negative numbers, we let

$$\binom{-s}{k} = \frac{-s(-s-1)\cdots(-s-k+1)}{k!}$$

so that

$$\binom{-s}{k} = (-1)^k \frac{s(s+1)\cdots(s+k-1)}{k!}$$

and

$$P_n(x) = f(x_n) + (-1)^1\binom{-s}{k}\nabla f(x_n) + (-1)^2\binom{-s}{2}\nabla^2 f(x_n) + \cdots$$

$$+ (-1)^n\binom{-s}{n}\nabla^n f(x_n)$$

to obtain

(3.14)
$$P_n(x) = \sum_{k=0}^{n}(-1)^k\binom{-s}{k}\nabla^k f(x_n).$$

Equation (3.14) is called the **Newton backward difference formula**.

EXAMPLE 1   Consider the table of data given in Examples 3 of Section 3.2 and 2 of Section 3.3. The divided-difference table corresponding to this data is shown in Table 3.7.

TABLE 3.7

		First divided differences	Second divided differences	Third divided differences	Fourth divided differences
1.0	.7651977				
		−.4837057			
1.3	.6200860		−.1087339		
		−.5489460		.0658784	
1.6	.4554022		−.0494433		.0018297
		−.5786120		.0680740	
1.9	.2818186		.0118233		
		−.5715180			
2.2	.1103632				

If an approximation to $f(1.1)$ is required, the reasonable choice for x_0, $x_1, \ldots, x_n$ would be $x_0 = 1.0$, $x_1 = 1.3$, $x_2 = 1.6$, $x_3 = 1.9$, and $x_4 = 2.2$, since this choice makes the greatest possible use of the data points closest to $x = 1.1$ and also makes use of the fourth divided difference. This implies that $h = .3$ and $s = \frac{1}{3}$, so the

Newton forward divided-difference formula is used with the divided differences that have a *solid* underline in the table.

$$f(1.1) = P_4(1.0 + \tfrac{1}{3}(.3)) = .7651997 + \tfrac{1}{3}(.3)(-.4837057)$$
$$+ \tfrac{1}{3}(-\tfrac{2}{3})(.3)^2(-.1087339)$$
$$+ \tfrac{1}{3}(-\tfrac{2}{3})(-\tfrac{5}{3})(.3)^3(.0658784)$$
$$+ \tfrac{1}{3}(-\tfrac{2}{3})(-\tfrac{5}{3})(-\tfrac{8}{3})(.3)^4(.0018297)$$
$$= .7196480.$$

To approximate a value when x is close to the end of the tabulated values, say, $x = 2.0$, we again would like to make maximum use of the data points closest to x. This requires using the Newton backward divided-difference formula with $s = -\tfrac{2}{3}$ and with the divided differences in the table that have a *dashed* underline:

$$f(2.0) = P_4(2.2 - \tfrac{2}{3}(.3)) = .1103632 - \tfrac{2}{3}(.3)(-.5715180)$$
$$- \tfrac{2}{3}(\tfrac{1}{3})(.3)^2(.0118233)$$
$$- \tfrac{2}{3}(\tfrac{1}{3})(\tfrac{4}{3})(.3)^3(.0680740)$$
$$- \tfrac{2}{3}(\tfrac{1}{3})(\tfrac{4}{3})(\tfrac{7}{3})(.3)^4(.0018297)$$
$$= .2238755. \qquad \square$$

The Newton formulas are not appropriate for approximating a value, x, that lies near the center of the table, since employing either the backward or forward method in such a way that the highest-order difference is involved will not allow x_0 to be close to x. A number of divided-difference formulas are available in this instance, however, each of which has situations when it can be used to maximum advantage. These methods are known as **centered-difference formulas**. Because of the number of such methods, we will present only one, Stirling's method, and again refer the interested reader to Hildebrand [48] for a complete presentation.

For the centered-difference formulas we choose x_0 near the point being approximated, label the points directly below x_0 as $x_1, x_2, \ldots$ and those directly above as $x_{-1}, x_{-2}, \ldots$. Using this convention, Stirling's formula is given by:

$$P_n(x) = P_{2m+1}(x) = f[x_0] + \frac{sh}{2}(f[x_{-1}, x_0] + f[x_0, x_1]) + s^2 h^2 f[x_{-1}, x_0, x_1]$$

$$+ \frac{s(s^2 - 1)h^3}{2}(f[x_{-1}, x_0, x_1, x_2] + f[x_{-2}, x_{-1}, x_0, x_1])$$

$$+ \cdots + s^2(s^2 - 1)(s^2 - 4) \cdots (s^2 - (m-1)^2)h^{2m}f[x_{-m}, \ldots, x_m]$$

$$+ \frac{s(s^2 - 1) \cdots (s^2 - m^2)h^{2m+1}}{2}(f[x_{-m}, \ldots, x_{m+1}] + f[x_{-m-1}, \ldots, x_m])$$

if $n = 2m + 1$ is odd, and if $n = 2m$ is even, by the same formula with the deletion of the last term. The entries used for this formula are *circled* in Table 3.8.

EXAMPLE 2 Consider the table of data that was given in the previous example. To use Stirling's formula to approximate $f(1.5)$ with $x_0 = 1.6$, we use the *underlined* entries in the difference Table 3.9.

TABLE 3.8

x	$f(x)$	First divided differences	Second divided differences	Third divided differences	Fourth divided differences
x_{-2}	$f[x_{-2}]$				
		$f[x_{-2}, x_{-1}]$			
x_{-1}	$f[x_{-1}]$		$f[x_{-2}, x_{-1}, x_0]$		
		$\boxed{f[x_{-1}, x_0]}$		$\boxed{f[x_{-2}, x_{-1}, x_0, x_1]}$	
x_0	$\boxed{f[x_0]}$		$\boxed{f[x_{-1}, x_0, x_1]}$		$\boxed{f[x_{-2}, x_{-1}, x_0, x_1, x_2]}$
		$\boxed{f[x_0, x_1]}$		$\boxed{f[x_{-1}, x_0, x_1, x_2]}$	
x_1	$f[x_1]$		$f[x_0, x_1, x_2]$		
		$f[x_1, x_2]$			
x_2	$f[x_2]$				

TABLE 3.9

x	$f(x)$	First divided differences	Second divided differences	Third divided differences	Fourth divided differences
1.0	.7651977				
		−.4837057			
1.3	.6200860		−.1087339		
		−.5489460		.0658784	
1.6	.4554022		−.0494433		.0018297
		−.5786120		.0680740	
1.9	.2818186		−.0118233		
		−.5715180			
2.2	.1103632				

The formula with $h = .3$, $x_0 = 1.6$, and $s = -\frac{1}{3}$ becomes

$$
\begin{aligned}
f(1.5) \approx \hat{P}_4(1.6 + (-\tfrac{1}{3})(.3)) &= .4554022 + (-\tfrac{1}{3})(\tfrac{3}{2})(-.5489460 - .5786120) \\
&+ (-\tfrac{1}{3})^2(.3)^2(-.0494433) \\
&+ \tfrac{1}{2}(-\tfrac{1}{3})((-\tfrac{1}{3})^2 - 1)(.3)^3(.0658784 + .0680740) \\
&+ (-\tfrac{1}{3})^2((-\tfrac{1}{3})^2 - 1)(.3)^4(.0018297) \\
&= .5118200 \qquad \square
\end{aligned}
$$

Exercise Set 3.4

1. Use Eq. (3.12) to construct the interpolating polynomial of degree four for the unequally spaced points given in the following table:

x	$f(x)$
0.0	−7.00000
.1	−5.89483
.3	−5.65014
.6	−5.17788
1.0	−4.28172

2. Suppose the data $f(1.1) = -3.99583$ is added to Exercise 1. Construct the interpolating polynomial of degree five.

3. Approximate $f(.05)$ using the following data and the Newton forward divided-difference formula:

x	0.0	.2	.4	.6	.8
$f(x)$	1.00000	1.22140	1.49182	1.82212	2.22554

4. Using the data of Exercise 3 and the Newton backward divided-difference formula, approximate $f(.05)$.

5. Using the data of Exercise 3 and Stirling's formula, approximate $f(.43)$.

6. Given

   $$P_n(x) = f[x_0] + f[x_0, x_1](x - x_0) + a_2(x - x_0)(x - x_1)$$
   $$+ a_3(x - x_0)(x - x_1)(x - x_2) + \cdots$$
   $$+ a_n(x - x_0)(x - x_1) \cdots (x - x_{n-1}).$$

   Use $P_n(x_2)$ to show that $a_2 = f[x_0, x_1, x_2]$.

7. Show that

   $$f[x_0, x_1, \ldots, x_n, x] = \frac{f^{(n+1)}(\xi(x))}{(n+1)!}$$

   for some $\xi(x)$. [*Hint*: From Eq. (3.5),

   $$f(x) = P_n(x) + \frac{f^{(n+1)}(\xi(x))}{(n+1)!}(x - x_0) \cdots (x - x_n).$$

   Considering the interpolation polynomial of degree $n + 1$ on $x_0, x_1, \ldots, x_n, x$, we have

   $$f(x) = P_{n+1}(x) = P_n(x) + f[x_0, x_1, \ldots, x_n, x](x - x_0) \cdots (x - x_n).]$$

8. The population table listed below was given in the introduction to this chapter and studied in Exercise 15 of Section 3.2.

Year	1920	1930	1940	1950	1960	1970
Population (In Thousands)	105,711	123,203	131,669	150,697	179,323	203,212

   Use an appropriate divided difference method to approximate:

   a) The population in the year 1965. b) The population in the year 2000.

3.5 Hermite Interpolation

The set of **osculating polynomials** is a generalization of both the Taylor polynomials and the Lagrange polynomials. These polynomials have the property that, given $n + 1$ distinct numbers $x_0, x_1, \ldots, x_n$ and nonnegative integers $m_0, m_1, \ldots, m_n$, the osculating polynomial approximating a function $f \in C^m[a, b]$, where $m = \max\{m_0, m_1, \ldots, m_n\}$ and $x_i \in [a, b]$ for each $i = 0, \ldots, n$, is the polynomial of least degree with the property that it agrees with the function f and all of its derivatives of order less than or equal to m_i at x_i for each $i = 0, 1, \ldots, n$. The degree of this osculating polynomial will be at most

$$M = \sum_{i=0}^{n} m_i + n,$$

since the number of conditions to be satisfied is $\sum_{i=0}^{n} m_i + (n + 1)$, and a polynomial of degree M has $M + 1$ coefficients that can be used to satisfy these conditions. For the sake of clarity, the formal definition of an osculating polynomial follows.

Definition 3.7   Let $x_0, x_1, \ldots, x_n$ be $n + 1$ distinct numbers in $[a, b]$ and m_i be a nonnegative integer associated with x_i for $i = 0, 1, \ldots, n$. Let

$$m = \max_{0 \le i \le n} m_i \quad \text{and} \quad f \in C^m[a, b].$$

The osculating polynomial approximating f is the polynomial P of least degree such that

$$\frac{d^k P(x_i)}{dx^k} = \frac{d^k f(x_i)}{dx^k} \quad \text{for each } i = 0, 1, \ldots, n \text{ and } k = 0, 1, \ldots, m_i.$$

Note that when $n = 0$, the osculating polynomial approximating f is simply the Taylor polynomial of degree m_0 for f at x_0. When $m_i = 0$ for $i = 0, 1, \ldots, n$, the osculating polynomial is the polynomial interpolating f on $x_0, x_1, \ldots, x_n$, that is, the Lagrange polynomial.

The situation which occurs when $m_i = 1$ for each $i = 0, 1, \ldots, n$ gives a class of polynomials called **Hermite polynomials**. For a given function f, these polynomials not only agree with f at $x_0, x_1, \ldots, x_n$, but, since their first derivatives agree with those of f, they have the same "shape" as the function at $(x_i, f(x_i))$ in the sense that the *tangent lines* to the polynomial and the function agree. We will restrict our study of osculating polynomials to this situation, and consider first a theorem that describes precisely the procedure for obtaining the polynomials of Hermite type.

Theorem 3.8   If $f \in C^1[a, b]$ and $x_0, \ldots, x_n \in [a, b]$ are distinct, the unique polynomial of least degree agreeing with f and f' at $x_0, \ldots, x_n$ is a polynomial of degree at most $2n + 1$ given by

(3.15) $$H_{2n+1}(x) = \sum_{j=0}^{n} f(x_j) H_{n,j}(x) + \sum_{j=0}^{n} f'(x_j) \hat{H}_{n,j}(x),$$

where

(3.16) $$H_{n,j}(x) = [1 - 2(x - x_j) L'_{n,j}(x_j)] L^2_{n,j}(x)$$

and

(3.17) $$\hat{H}_{n,j}(x) = (x - x_j) L^2_{n,j}(x).$$

In this context, $L_{n,j}$ denotes the jth Lagrange coefficient polynomial of degree n defined in Eq. (3.4).

Moreover, if $f \in C^{(2n+2)}[a, b]$, then

(3.18)
$$f(x) - H_{2n+1}(x) = \frac{(x - x_0)^2 \cdots (x - x_n)^2}{(2n + 2)!} f^{(2n+2)}(\xi)$$

for some ξ with $a < \xi < b$.

Proof   To show that

(3.19)
$$\frac{d^j}{dx^j} H_{2n+1}(x_k) = \frac{d^j f(x_k)}{dx^j}$$

for each $j = 0, 1$, and $k = 0, 1, \ldots, n$, it suffices to show that $H_{n,j}$ and $\hat{H}_{n,j}$, as defined by (3.16) and (3.17), respectively, satisfy the conditions (a)–(d) given below:

a) $H_{n,j}(x_k) = \begin{cases} 0, & \text{if } j \neq k, \\ 1, & \text{if } j = k; \end{cases}$ b) $\dfrac{d}{dx} H_{n,j}(x_k) = 0$,   for all k;

c) $\hat{H}_{n,j}(x_k) = 0$ for all k;   d) $\dfrac{d}{dx} \hat{H}_{n,j}(x_k) = \begin{cases} 0, & \text{if } j \neq k, \\ 1, & \text{if } j = k. \end{cases}$

(See Figs. 3.8 and 3.9.)

These conditions ensure that when either H_{2n+1} or H'_{2n+1} is evaluated at one of the given numbers the appropriate value is obtained since

$$H'_{2n+1}(x_i) = \sum_{j=0}^{n} f(x_j) H_{n,j}(x_i) + \sum_{j=0}^{n} f'(x_j) \hat{H}_{n,j}(x_i)$$

$$= f(x_i) \cdot 1 + \sum_{\substack{j=0 \\ j \neq i}}^{n} f(x_j) \cdot 0 + \sum_{j=0}^{n} f'(x_j) \cdot 0$$

$$= f(x_i),$$

and

$$H'_{2n+1}(x_i) = \sum_{j=0}^{n} f(x_j) H'_{n,j}(x_i) + \sum_{j=0}^{n} f'(x_j) \hat{H}'_{n,j}(x_i)$$

$$= \sum_{j=0}^{n} f(x_j) \cdot 0 + f'(x_i) \cdot 1 + \sum_{\substack{j=0 \\ j \neq i}}^{n} f'(x_j) \cdot 0$$

$$= f'(x_i),$$

for each $i = 0, 1, \ldots, n$.

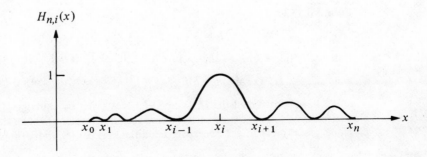

FIGURE 3.8

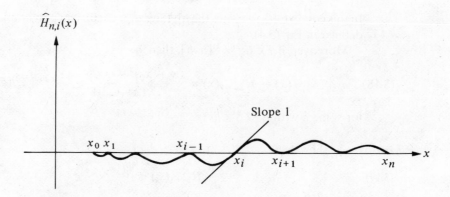

FIGURE 3.9

Considering first the polynomial $\hat{H}_{n,j}$, requirements (c) and (d) imply that $\hat{H}_{n,j}$ must have a double root at x_k for $k \neq j$ and a single root at x_j. A polynomial of degree at most $(2n + 1)$ satisfying these conditions is given by:

$$\hat{H}_{n,j}(x) = \frac{(x - x_0)^2 \cdots (x - x_{j-1})^2 (x - x_j)(x - x_{j+1})^2 \cdots (x - x_n)^2}{(x_j - x_0)^2 \cdots (x_j - x_{j-1})^2 (1)(x_j - x_{j+1})^2 \cdots (x_j - x_n)^2} = L_{n,j}^2(x)(x - x_j).$$

Conditions (a) and (b) imply that x_k, for each $k \neq j$, must be a double root of $H_{n,j}(x)$ and any polynomial of degree at most $(2n + 1)$ satisfying (a) and (b) is given by:

$$H_{n,j}(x) = (x - x_0)^2 \cdots (x - x_{j-1})^2 (x - x_{j+1})^2 \cdots (x - x_n)^2 (\hat{a}x + \hat{b})$$

for some constants $\hat{a}$ and $\hat{b}$. Adjusting the constants $\hat{a}$ and $\hat{b}$ by letting

$$a = \hat{a} \prod_{\substack{i=0, \\ i \neq j}}^{n} (x_i - x_j)^2 \qquad \text{and} \qquad b = \hat{b} \prod_{\substack{i=0, \\ i \neq j}}^{n} (x_i - x_j)^2$$

gives
$$H_{n,j}(x) = L_{n,j}^2(x)(ax + b).$$

Condition (a) implies that

(3.20) $$1 = H_{n,j}(x_j) = L_{n,j}^2(x_j)(ax_j + b) = ax_j + b$$

and using (b) and Eq. (3.20) gives

$$0 = \frac{dH_{n,j}(x_j)}{dx} = 2L_{n,j}(x_j)L'_{n,j}(x_j)(ax_j + b) + L_{n,j}^2(x_j)(a)$$

$$= 2L'_{n,j}(x_j)(ax_j + b) + a$$

$$= 2L'_{n,j}(x_j)(1) + a$$

or $$a = -2L'_{n,j}(x_j).$$
From Eq. (3.20),

$$b = 1 - ax_j = 1 + 2L'_{n,j}(x_j) \cdot (x_j);$$

hence, $$(ax + b) = -2L'_{n,j}(x_j)x + 1 + 2L'_{n,j}(x_j)(x_j) = 1 - 2(x - x_j)L'_{n,j}(x_j)$$

and $$H_{n,j}(x) = (ax + b)L_{n,j}^2(x) = [1 - 2(x - x_j)L'_{n,j}(x)]L_{n,j}^2(x).$$

We have constructed the $H_{n,j}$'s and $\hat{H}_{n,j}$'s as given by equations (3.16) and (3.17) respectively. Note that, since each $H_{n,j}$ and each $\hat{H}_{n,j}$ is of degree at most $(2n + 1)$, H_{2n+1} is of degree at most $2n + 1$. Moreover, conditions (a), (b), (c), and (d) imply that Eq. (3.19) is satisfied.

We will now show uniqueness, that is, that H_{2n+1} is the only polynomial of degree less than or equal to $2n + 1$ such that Eq. (3.19) is satisfied. Let P be any polynomial of degree less than or equal to $2n + 1$ with the property that

$$P(x_k) = f(x_k) \quad \text{and} \quad P'(x_k) = f'(x_k) \qquad \text{for each } k = 0, 1, \ldots, n.$$

Since $D = H_{2n+1} - P$ is also a polynomial of degree at most $(2n + 1)$ and

$$D(x_k) = 0 \quad \text{and} \quad D'(x_k) = 0 \qquad \text{for each } k = 0, 1, \ldots, n,$$

D must have the form

$$D(x) = (x - x_0)^2 (x - x_1)^2 \cdots (x - x_n)^2 g(x),$$

for some polynomial $g(x)$. But D is a polynomial of degree at most $(2n + 1)$ with $(2n + 2)$ roots, so $D \equiv 0$. This implies that $P(x) = H_{2n+1}(x)$ for each value of x and the uniqueness of H_{2n+1} is established.

The calculation of the error term in Eq. (3.18) will be left as Exercise 6.   ☐

EXAMPLE 1   Use the polynomial of least degree that agrees with the data listed in Table 3.10 for the Bessel function of the first kind of order zero to find an approximation of $f(1.5)$.

TABLE 3.10

k	x_k	$f(x_k)$	$f'(x_k)$
0	1.3	.6200860	$-.5220232$
1	1.6	.4554022	$-.5698959$
2	1.9	.2818186	$-.5811571$

First compute the Lagrange polynomials and their derivatives:

$$L_{2,0}(x) = \frac{(x - x_1)(x - x_2)}{(x_0 - x_1)(x_0 - x_2)} = \frac{50}{9} x^2 - \frac{175}{9} x + \frac{152}{9},$$

$$L'_{2,0}(x) = \frac{100}{9} x - \frac{175}{9};$$

$$L_{2,1}(x) = \frac{(x - x_0)(x - x_2)}{(x_1 - x_0)(x_1 - x_2)} = \frac{-100}{9} x^2 + \frac{320}{9} x - \frac{247}{9},$$

$$L'_{2,1}(x) = \frac{-200}{9} x + \frac{320}{9};$$

$$L_{2,2}(x) = \frac{(x - x_0)(x - x_1)}{(x_2 - x_0)(x_2 - x_1)} = \frac{50}{9} x^2 - \frac{145}{9} x + \frac{104}{9},$$

and $$L'_{2,2}(x) = \frac{100}{9} x - \frac{145}{9}.$$

The polynomials $H_{2,j}$ and $\hat{H}_{2,j}$ are then:

$$H_{2,0}(x) = [1 - 2(x - 1.3)(-5)]\left(\frac{50}{9}x^2 - \frac{175}{9}x + \frac{152}{9}\right)^2$$

$$= (10x - 12)\left(\frac{50}{9}x^2 - \frac{175}{9}x + \frac{152}{9}\right)^2,$$

$$H_{2,1}(x) = 1 \cdot \left(\frac{-100}{9}x^2 + \frac{320}{9}x - \frac{247}{9}\right)^2,$$

and
$$H_{2,2}(x) = 10(2 - x)\left(\frac{50}{9}x^2 - \frac{145}{9}x + \frac{104}{9}\right)^2;$$

and
$$\hat{H}_{2,0}(x) = (x - 1.3)\left(\frac{50}{9}x^2 - \frac{175}{9}x + \frac{152}{9}\right)^2,$$

$$\hat{H}_{2,1}(x) = (x - 1.6)\left(\frac{-100}{9}x^2 + \frac{320}{9}x - \frac{247}{9}\right)^2,$$

and
$$\hat{H}_{2,2}(x) = (x - 1.9)\left(\frac{50}{9}x^2 - \frac{145}{9}x + \frac{104}{9}\right)^2.$$

Finally,

$$H_5(x) = .6200860H_{2,0}(x) + .4554022H_{2,1}(x) + .2818186H_{2,2}(x)$$
$$- .5220232\hat{H}_{2,0}(x) - .5698959\hat{H}_{2,1}(x) - .5811571\hat{H}_{2,2}(x)$$

and $$H_5(1.5) = .6200860\left(\frac{4}{27}\right) + .4554022\left(\frac{64}{81}\right) + .2818186\left(\frac{5}{81}\right)$$

$$- .5220232\left(\frac{4}{405}\right) - .5698959\left(\frac{-32}{405}\right) - .5811571\left(\frac{-2}{405}\right)$$

$$= .5118277,$$

a result that is accurate to the places listed.   □

Exercise Set 3.5

1. Use Hermite interpolation to find an approximation to $f(2.5)$, given the following data:

x	$f(x)$	$f'(x)$
2.2	.5207843	−.0014878
2.4	.5104147	−.1004889
2.6	.4813306	−.1883635

2. With $f(x) = 3xe^x - e^{2x}$, approximate $f(1.03)$ by the Hermite interpolating polynomial of degree at most three, using $x_0 = 1$ and $x_1 = 1.05$. Compare the actual error to the error bound obtained from Eq. (3.18).

3. Repeat Exercise 2, with the Hermite interpolating polynomial of degree at most five, using $x_0 = 1$, $x_1 = 1.05$, and $x_2 = 1.07$.

4. Let $f(x) = 3xe^x - e^{2x}$. Does the interpolating polynomial approximation of degree two, $P_2(1.03)$, computed in Exercise 7 of Section 3.2 give a better approximation of $f(1.03)$ than the Hermite polynomial approximation of degree three, $H_3(1.03)$, computed in Exercise 2 of this section? Compare the theoretical error estimates for

$$|P_2(1.03) - f(1.03)| \quad \text{and} \quad |H_3(1.03) - f(1.03)|.$$

5. a) Let $f(x) = e^x$. Find $H_5(x)$, the Hermite polynomial approximating f, using the data points $x_0 = 0$, $x_1 = 1$, and $x_2 = 2$. Compare $H_5(.25)$ to $f(.25)$ and to $P(.25)$ as given in Exercise 9 of Section 3.2.
   b) Use Theorem 3.8 to estimate $|H_5(.25) - f(.25)|$.

6. Derive the error term in Theorem 3.8. [*Hint*: Use the same method as in the Lagrange error derivation, Theorem 3.3 defining

$$g(t) = f(t) - H_{2n+1}(t) + \frac{(t - x_0)^2 \cdots (t - x_n)^2}{(x - x_0)^2 \cdots (x - x_n)^2} [f(x) - H_{2n+1}(x)]$$

and using the fact that $g'(t)$ has $(2n + 2)$ *distinct* zeros in $[a, b]$.]

7. The following table lists data for the function described by $f(x) = e^{.1x^2}$. Approximate $f(1.25)$ by using $H_5(1.25)$ and $H_3(1.25)$ where H_5 uses the nodes $x_0 = 1$, $x_2 = 2$, $x_3 = 3$ and where H_3 uses the nodes $x_0 = 1$, $x_1 = 1.5$. Find error bounds for these approximations.

x	$f(x) = e^{.1x^2}$	$f'(x) = .2xe^{.1x^2}$
$x_0 = \bar{x}_0 = 1$	1.105170918	.2210341836
$\bar{x}_1 = 1.5$	1.252322716	.3756968148
$x_1 = 2$	1.491824698	.5967298792
$x_2 = 3$	2.459603111	1.475761867

8. A car traveling along a straight road is clocked at a number of points. The data from the observations is given in the following table where the time is in seconds, the distance in feet, and the speed in feet per seconds. Use a Hermite polynomial to predict the position of the car and its speed when $t = 10$ sec.

Time	0	3	5	8	13
Distance	0	225	383	623	993
Speed	75	77	80	74	72

3.6 Cubic Spline Interpolation

The previous sections of this chapter were concerned with the approximation of arbitrary functions on closed intervals by the use of polynomials. While this method of approximation is appropriate in many circumstances, the oscillatory nature of high-degree polynomials and the property that a fluctuation over a small portion of the interval can induce large fluctuations over the entire range, restricts their use when approximating many of the functions that arise in actual physical situations.

An alternative approach that can be used to obtain interpolatory functions is to divide the interval into a collection of subintervals and construct a (generally) different approximating polynomial on each subinterval. Approximating by functions of this type is called **piecewise polynomial approximation**.

The simplest type of piecewise polynomial approximation is called piecewise linear interpolation and consists of joining a set of data points

$$\{(x_0, f(x_0)), (x_1, f(x_1)), \ldots, (x_n, f(x_n))\}$$

by a series of straight lines as shown in Fig. 3.10. This is the method of interpolation used in elementary courses involving the study of trigonometric or logarithmic functions when intermediate values are required from a collection of tabulated values.

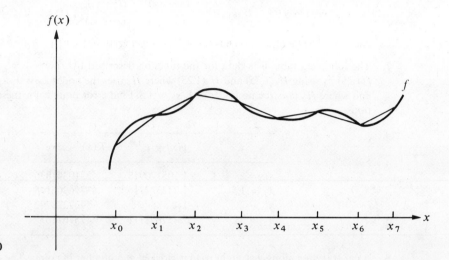

FIGURE 3.10

The disadvantage of approaching an approximation problem using functions of this type, or, say, quadratic polynomials between three consecutive points, is that at each of the endpoints of the subintervals, there is no assurance of differentiability, which, in a geometrical context, means that the interpolating function is not "smooth" at these points. Often it is clear from physical conditions that such a smoothness condition is required and in these cases the approximating function must be continuously differentiable. A procedure that one could follow is to use a piecewise polynomial of Hermite type. For example, if the values of the function f and of f' are known at each of the points $x_0 < x_1 < \cdots < x_n$, a Hermite polynomial of degree 3 could be used on each of the subintervals $[x_0, x_1]$, $[x_1, x_2]$, ..., $[x_{n-1}, x_n]$ to obtain a function that is continuously differentiable on the interval $[x_0, x_n]$. The difficulty with this type of procedure is that the data for the function to be approximated is often known, while the data for its derivative is not.

Perhaps the simplest type of differentiable piecewise polynomial function on an entire interval $[x_0, x_n]$ is the function obtained by fitting a quadratic polynomial between each successive pair of nodes. This is done by constructing a quadratic on $[x_0, x_1]$ agreeing with the function at x_0 and x_1, another quadratic on $[x_1, x_2]$

agreeing with the function at x_1 and x_2, and so on. Since a general quadratic polynomial has three arbitrary constants—the constant term, the coefficient of x, and the coefficient of x^2—and only two conditions are required to fit the data at the endpoints of each subinterval, flexibility exists that allows the quadratic to be chosen so that, in addition, the interpolant has a continuous derivative on $[x_0, x_n]$. The difficulty with this procedure arises when it is desired to specify that the derivative of the interpolant agrees with that of the function at the endpoints x_0 and x_n. In this case, it can be shown that there are not a sufficient number of constants to ensure that the conditions will be satisfied. More elaboration on this difficulty and ways to circumvent it can be found in Exercises 13 and 14.

The type of piecewise polynomial approximation using cubic polynomials between each successive pair of nodes is called **cubic spline interpolation** and is one of the most popular techniques presently in use. A general cubic polynomial involves four constants; so there is sufficient flexibility in the cubic spline procedure to ensure not only that the interpolant is continuously differentiable on the interval, but that it has a continuous second derivative on the interval as well. The construction of the cubic spline does not, however, assume that the derivatives of the interpolant agree with those of the function, even at the nodes.

Definition 3.9 Given a function f defined on $[a, b]$ and a set of numbers, called nodes, $a = x_0 < x_1 < \cdots < x_n = b$, a cubic spline interpolant, S, for f is a function that satisfies the following conditions:

a) S is a cubic polynomial, denoted S_j, on the subinterval $[x_j, x_{j+1}]$ for each $j = 0, 1, \ldots, n - 1$;

b) $S(x_j) = f(x_j)$ for each $j = 0, 1, \ldots, n$;

c) $S_{j+1}(x_{j+1}) = S_j(x_{j+1})$ for each $j = 0, 1, \ldots, n - 2$;

d) $S'_{j+1}(x_{j+1}) = S'_j(x_{j+1})$ for each $j = 0, 1, \ldots, n - 2$;

e) $S''_{j+1}(x_{j+1}) = S''_j(x_{j+1})$ for each $j = 0, 1, \ldots, n - 2$;

f)   one of the following set of boundary conditions is satisfied:

(3.21)   (i) $S''(x_0) = S''(x_n) = 0$ (Free boundary)

(3.22)   (ii) $S'(x_0) = f'(x_0)$ and $S'(x_n) = f'(x_n)$.   (Clamped boundary)

Although cubic splines can be defined with other boundary conditions, the conditions given above are sufficient for our purposes. When the free boundary conditions occur, the spline is called a **natural spline** and its graph approximates the shape that a long flexible rod would assume if forced to go through each of the data points $\{(x_0, f(x_0)), (x_1, f(x_1)), \ldots, (x_n, f(x_n))\}$.

In general, the clamped boundary conditions will lead to more accurate approximations since they include more information about the function; however, in order for this type of boundary condition to hold, it is necessary to have either the values of the derivative at the endpoints or an accurate approximation to those values.

To construct the cubic-spline interpolant for a given function f, the conditions in the definition can be applied to the cubic polynomials.

$$S_j(x) = a_j + b_j(x - x_j) + c_j(x - x_j)^2 + d_j(x - x_j)^3$$

for each $j = 0, 1, \ldots, n - 1$.

Clearly,

$$S_j(x_j) = a_j = f(x_j)$$

and if condition (c) is applied,

$$a_{j+1} = S_{j+1}(x_{j+1}) = S_j(x_{j+1})$$
$$= a_j + b_j(x_{j+1} - x_j) + c_j(x_{j+1} - x_j)^2 + d_j(x_{j+1} - x_j)^3$$

for each $j = 0, 1, \ldots, n - 2$.

Since the term $(x_{j+1} - x_j)$ will be used repeatedly in this development, it is convenient to introduce the simpler notation,

$$h_j = x_{j+1} - x_j,$$

for each $j = 0, 1, \ldots, n - 1$. If, in addition, we define $a_n = f(x_n)$, it can be seen that this implies that the equation

$$(3.23) \qquad a_{j+1} = a_j + b_j h_j + c_j h_j^2 + d_j h_j^3$$

holds for each $j = 0, 1, \ldots, n - 1$.

In a similar manner, define $b_n = S'(x_n)$ and observe that

$$S_j'(x) = b_j + 2c_j(x - x_j) + 3d_j(x - x_j)^2$$

implies $S_j'(x_j) = b_j$ for each $j = 0, 1, \ldots, n - 1$. Applying condition (d),

$$(3.24) \qquad b_{j+1} = b_j + 2c_j h_j + 3d_j h_j^2$$

for each $j = 0, 1, \ldots, n - 1$.

Another relation between the coefficients of S_j can be obtained by defining $c_n = S''(x_n)/2$ and applying condition (e). In this case

$$(3.25) \qquad c_{j+1} = c_j + 3d_j h_j$$

for each $j = 0, 1, \ldots, n - 1$.

Solving for d_j in Eq. (3.25) and substituting this value into equations (3.23) and (3.24) gives the new equations

$$(3.26) \qquad a_{j+1} = a_j + b_j h_j + \frac{h_j^2}{3}(2c_j + c_{j+1})$$

and

$$(3.27) \qquad b_{j+1} = b_j + h_j(c_j + c_{j+1})$$

for each $j = 0, 1, \ldots, n - 1$.

The final relationship involving the coefficients can be obtained by solving the appropriate equation in the form of equation (3.26), first for b_j,

$$(3.28) \qquad b_j = \frac{1}{h_j}(a_{j+1} - a_j) - \frac{h_j}{3}(2c_j + c_{j+1}),$$

and then, with a reduction of the index, for b_{j-1},

$$b_{j-1} = \frac{1}{h_{j-1}}(a_j - a_{j-1}) - \frac{h_{j-1}}{3}(2c_{j-1} + c_j).$$

Substituting these values into the equation derived from Eq. (3.27), when the index is reduced by one, gives the linear system of equations

$$(3.29) \quad h_{j-1}c_{j-1} + 2(h_{j-1} + h_j)c_j + h_j c_{j+1}$$

$$= \frac{3}{h_j}(a_{j+1} - a_j) - \frac{3}{h_{j-1}}(a_j - a_{j-1}), \quad \text{for each } j = 1, 2, \dots, n-1.$$

This system involves, as unknowns, only $\{c_j\}_{j=0}^n$ since the values of $\{h_j\}_{j=0}^{n-1}$ and $\{a_j\}_{j=0}^n$ are given by the spacing of the nodes $\{x_j\}_{j=0}^n$ and the values of f at the nodes.

Note that once the values of $\{c_j\}_{j=0}^n$ are known it is a simple matter to find the remainder of the constants $\{b_j\}_{j=0}^{n-1}$ from Eq. (3.28), and $\{d_j\}_{j=0}^{n-1}$ from Eq. (3.25) and to construct the cubic polynomials $\{S_j\}_{j=0}^{n-1}$.

The major question that arises in connection with this construction is whether the values of $\{c_j\}_{j=0}^n$ can be found using the system of equations given in (3.29) and if so, whether these values are unique. The following theorems indicate that, when either of the boundary conditions given in part (f) of the definition are imposed, the answer to both questions is affirmative. The proofs of these theorems require material from linear algebra. This material is discussed in Sections 6.2, 6.3, 6.5, and 6.6.

Theorem 3.10   If f is a function that is defined on $[a, b]$, then f has a unique natural spline interpolant, i.e., a unique spline interpolant satisfying the free boundary conditions $S''(a) = S''(b) = 0$.

Proof   With the usual notation, $a = x_0 < x_1 < \dots < x_n = b$, the boundary conditions in this case imply that $c_n = S''(x_n)/2 = 0$ and that

$$0 = S''(x_0) = 2c_0 + 6d_0(x_0 - x_0);$$

so $c_0 = 0$.

The two equations $c_0 = 0$ and $c_n = 0$ together with the equations in (3.29) produce a linear system described by the vector equation $A\mathbf{x} = \mathbf{b}$, where A is the matrix

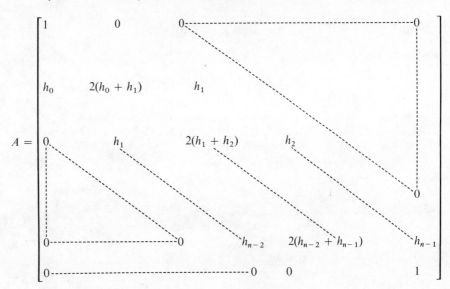

and $\mathbf{b}$ and $\mathbf{x}$ are the vectors

$$
\mathbf{b} = \begin{bmatrix} 0 \\ \dfrac{3}{h_1}(a_2 - a_1) - \dfrac{3}{h_0}(a_1 - a_0) \\ \vdots \\ \dfrac{3}{h_{n-1}}(a_n - a_{n-1}) - \dfrac{3}{h_{n-2}}(a_{n-1} - a_{n-2}) \\ 0 \end{bmatrix}
$$

and

$$
\mathbf{x} = \begin{bmatrix} c_0 \\ c_1 \\ \vdots \\ c_n \end{bmatrix}
$$

The matrix A satisfies the hypotheses of Theorem 6.21, page 299. Therefore, the linear system has a unique solution for $c_0, c_1, \ldots, c_n$. $\quad\square$

The actual solution to the cubic spline problem with the boundary conditions $S''(x_0) = S''(x_n) = 0$ can be obtained by applying the following algorithm.

Natural Cubic Spline Algorithm 3.2

To construct the cubic spline interpolant S for the function f, defined at the numbers $x_0 < x_1 < \cdots < x_n$, satisfying $S''(x_0) = S''(x_n) = 0$:

INPUT n: $x_0, x_1, \ldots, x_n$; either generate $a_i = f(x_i)$ for $i = 0, 1, \ldots, n$ or input a_i for $i = 0, 1, \ldots, n$.

OUTPUT a_j, b_j, c_j, d_j for $j = 0, 1, \ldots, n - 1$.

(*Note*: $S(x) = S_j(x) = a_j + b_j(x - x_j) + c_j(x - x_j)^2 + d_j(x - x_j)^3$ *for* $x_j \le x < x_{j+1}$.)

Step 1 For $i = 0, 1, \ldots, n - 1$ set $h_i = x_{i+1} - x_i$.

Step 2 For $i = 1, 2, \ldots, n - 1$ set

$$
\alpha_i = \frac{3[a_{i+1}h_{i-1} - a_i(x_{i+1} - x_{i-1}) + a_{i-1}h_i]}{h_{i-1}h_i}.
$$

Step 3 Set $l_0 = 1$;   (*Steps* 3, 4, 5, *and part of Step* 6 *solve a tridiagonal linear system using Algorithm* 6.7.)

$\mu_0 = 0$;
$z_0 = 0$.

Step 4 For $i = 1, 2, \ldots, n$
set $l_i = 2(x_{i+1} - x_{i-1}) - h_{i-1}\mu_{i-1}$;
$\mu_i = h_i/l_i$;
$z_i = (\alpha_i - h_{i-1}z_{i-1})/l_i$.

Step 5   Set $l_n = 1$;
$\qquad z_n = 0$;
$\qquad c_n = z_n$.

Step 6   For $j = n - 1, n - 2, \ldots, 0$
$\qquad$ set $c_j = z_j - \mu_j c_{j+1}$;
$\qquad\qquad b_j = (a_{j+1} - a_j)/h_j - h_j(c_{j+1} + 2c_j)/3$;
$\qquad\qquad d_j = (c_{j+1} - c_j)/(3h_j)$.

Step 7   OUTPUT $(a_j, b_j, c_j, d_j \text{ for } j = 0, 1, \ldots, n - 1.)$;
$\qquad$ STOP.

Theorem 3.11   If f is a function that is defined on $[a, b]$, then f has a unique spline interpolant satisfying the clamped boundary conditions $S'(a) = f'(a)$ and $S'(b) = f'(b)$.

Proof   It can be seen, using the fact that $S'(a) = S'(x_0) = b_0$, that Eq. (3.28) with $j = 0$, implies

$$f'(a) = \frac{a_1 - a_0}{h_0} - \frac{h_0}{3}(2c_0 + c_1)$$

Consequently,

$$2h_0 c_0 + h_0 c_1 = \frac{3}{h_0}(a_1 - a_0) - 3f'(a).$$

Similarly,

$$f'(b) = b_n = b_{n-1} + h_{n-1}(c_{n-1} + c_n),$$

so Eq. (3.28) with $j = n - 1$, implies that

$$f'(b) = \frac{a_n - a_{n-1}}{h_{n-1}} - \frac{h_{n-1}}{3}(2c_{n-1} + c_n) + h_{n-1}(c_{n-1} + c_n)$$

$$= \frac{a_n - a_{n-1}}{h_{n-1}} + \frac{h_{n-1}}{3}(c_{n-1} + 2c_n)$$

and

$$h_{n-1}c_{n-1} + 2h_{n-1}c_n = 3f'(b) - \frac{3}{h_{n-1}}(a_n - a_{n-1}).$$

Equations (3.29), together with the equations

$$2h_0 c_0 + h_0 c_1 = \frac{3}{h_0}(a_1 - a_0) - 3f'(a)$$

and

$$h_{n-1}c_{n-1} + 2h_{n-1}c_n = 3f'(b) - \frac{3}{h_{n-1}}(a_n - a_{n-1}),$$

determine the linear system $A\mathbf{x} = \mathbf{b}$ where

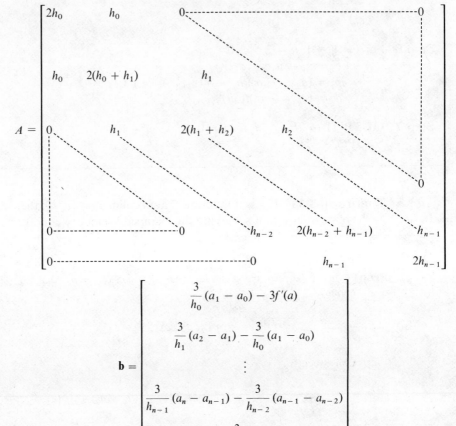

$$A = \begin{bmatrix} 2h_0 & h_0 & 0 & \cdots & & & & 0 \\ h_0 & 2(h_0 + h_1) & h_1 & & & & & \\ 0 & h_1 & 2(h_1 + h_2) & h_2 & & & & \\ \vdots & & & & & & & 0 \\ 0 & \cdots & 0 & & h_{n-2} & 2(h_{n-2} + h_{n-1}) & h_{n-1} \\ 0 & \cdots & & 0 & & h_{n-1} & 2h_{n-1} \end{bmatrix}$$

$$\mathbf{b} = \begin{bmatrix} \dfrac{3}{h_0}(a_1 - a_0) - 3f'(a) \\[2mm] \dfrac{3}{h_1}(a_2 - a_1) - \dfrac{3}{h_0}(a_1 - a_0) \\[2mm] \vdots \\[2mm] \dfrac{3}{h_{n-1}}(a_n - a_{n-1}) - \dfrac{3}{h_{n-2}}(a_{n-1} - a_{n-2}) \\[2mm] 3f'(b) - \dfrac{3}{h_{n-1}}(a_n - a_{n-1}) \end{bmatrix}$$

and

$$\mathbf{x} = \begin{bmatrix} c_0 \\ c_1 \\ \vdots \\ c_n \end{bmatrix}.$$

The matrix A satisfies the conditions of Theorem 6.21, page 299. Therefore, the linear system has a unique solution for $c_0, c_1, \ldots, c_n$.   □

Clamped Cubic Spline Algorithm 3.3

To construct the cubic spline interpolant S for the function f, defined at the numbers $x_0 < x_1 < \cdots < x_n$, satisfying $S'(x_0) = f'(x_0)$ and $S'(x_n) = f'(x_n)$:

INPUT n; $x_0, x_1, \ldots, x_n$; either generate $a_i = f(x_i)$ for $i = 0, 1, \ldots, n$ or input a_i for $i = 0, 1, \ldots, n$; $FPO = f'(x_0)$; $FPN = f'(x_n)$.

OUTPUT a_j, b_j, c_j, d_j for $j = 0, 1, \ldots, n - 1$.

(*Note:* $S(x) = a_j + b_j(x - x_j) + c_j(x - x_j)^2 + d_j(x - x_j)^3$ *for* $x_j \le x < x_{j+1}$.)

Step 1   For $i = 0, 1, \ldots, n - 1$ set $h_i = x_{i+1} - x_i$.

Step 2 Set $\alpha_0 = 3(a_1 - a_0)/h_0 - 3FPO$;
$\alpha_n = 3FPN - 3(a_n - a_{n-1})/h_{n-1}$.

Step 3 For $i = 1, 2, \ldots, n - 1$

$$\text{set } \alpha_i = \frac{3[a_{i+1}h_{i-1} - a_i(x_{i+1} - x_{i-1}) + a_{i-1}h_i]}{h_{i-1}h_i}.$$

Step 4 Set $l_0 = 2h_0$;   (*Steps 4, 5, 6, and part of Step 7 solve a triagonal linear system using Algorithm 6.7.*)

$\mu_0 = .5$;
$z_0 = \alpha_0/l_0$.

Step 5 For $i = 1, 2, \ldots, n$
set $l_i = 2(x_{i+1} - x_{i-1}) - h_{i-1}\mu_{i-1}$;
$\mu_i = h_i/l_i$;
$z_i = (\alpha_i - h_{i-1}z_{i-1})/l_i$.

Step 6 Set $l_n = h_{n-1}(2 - \mu_{n-1})$;
$z_n = (\alpha_n - h_{n-1}z_{n-1})/l_n$;
$c_n = z_n$.

Step 7 For $j = n - 1, n - 2, \ldots, 0$
set $c_j = z_j - \mu_j c_{j+1}$;
$b_j = (a_{j+1} - a_j)/h_j - h_j(c_{j+1} + 2c_j)/3$;
$d_j = (c_{j+1} - c_j)/(3h_j)$.

Step 8 OUTPUT $(a_j, b_j, c_j, d_j \text{ for } j = 0, 1, \ldots, n - 1)$;
STOP.

EXAMPLE 1   We will determine a cubic-spline interpolant to represent the curve constituting the upper portion of the well-known figure shown in Fig. 3.11.

FIGURE 3.11

It is first necessary to set up a grid to describe the points on the curve, as shown in Fig. 3.12. Because the curve does not have a continuous derivative at the points where the paw meets the head and where the head meets the back, it will be necessary to use separate splines for each of these regions. Using the scales in Fig. 3.12 gives the data in Table 3.11, where the values of x are chosen to be more concentrated where abrupt changes occur. (See Fig. 3.13.)

Using Algorithm 3.2 to construct the natural splines corresponding to this data gives the results in Tables 3.12, 3.13, and 3.14.

The graph of this piecewise spline function is shown in Fig. 3.14.

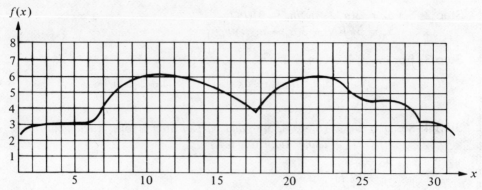

FIGURE 3.12

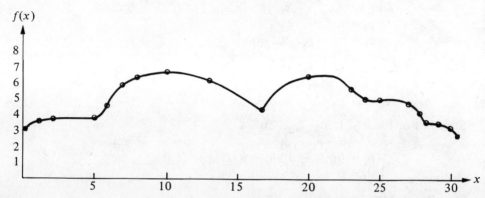

FIGURE 3.13

FIGURE 3.14

TABLE 3.11

	Curve #1			Curve #2			Curve #3	
i	x_i	$f(x_i)$	i	x_i	$f(x_i)$	i	x_i	$f(x_i)$
0	1	3.0	0	17	4.5	0	27.7	4.1
1	2	3.7	1	20	7.0	1	28	4.3
2	5	3.9	2	23	6.1	2	29	4.1
3	6	4.2	3	24	5.6	3	30	3.0
4	7	5.7	4	25	5.8			
5	8	6.6	5	27	5.2			
6	10	7.1	6	27.7	4.1			
7	13	6.7						
8	17	4.5						

TABLE 3.12

| | | | Spline #1 | | |
j	x_j	$a_j = f(x_j)$	b_j	c_j	d_j
0	1	3.0	.786	0	−.086
1	2	3.7	.529	−.257	.034
2	5	3.9	−.086	.052	.334
3	6	4.2	1.019	1.053	−.572
4	7	5.7	1.408	−.664	.156
5	8	6.6	.547	−.197	.024
6	10	7.1	.049	−.052	−.003
7	13	6.7	−.342	−.078	.007
8	17	4.5			

TABLE 3.13

| | | | Spline #2 | | |
j	x_j	$a_j = f(x_j)$	b_j	c_j	d_j
0	17	4.5	1.106	0	−.030
1	20	7.0	.289	−.272	.025
2	23	6.1	−.660	−.044	.204
3	24	5.6	−.137	.567	−.230
4	25	5.8	.306	−.124	−.089
5	27	5.2	−1.263	−.660	.314
6	27.7	4.1			

TABLE 3.14

| | | | Spline #3 | | |
j	x_j	$a_j = f(x_j)$	b_j	c_j	d_j
0	27.7	4.1	.749	0	−.910
1	28	4.3	.503	−.819	.116
2	29	4.1	−.787	−.470	.157
3	30	3.0			

If we approximate the derivative at each of the endpoints of the three smooth pieces of the curve, then three clamped boundary-condition splines can be used to approximate the curve. See Fig. 3.15.

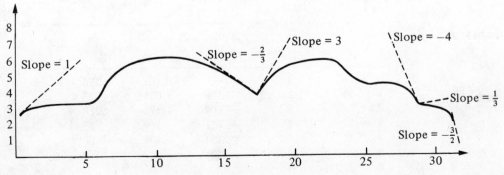

FIGURE 3.15

The data for this situation is the same as that involving the natural splines, with the additional assumptions on the derivatives at the endpoints, and is given in Table 3.15.

Using Algorithm 3.3 to construct the clamped boundary-condition splines corresponding to this data gives the results shown in Tables 3.16, 3.17, and 3.18. The graph of this piecewise spline function is shown in Fig. 3.16.

The reason the natural splines approximate the original curve better than the splines with the clamped boundary conditions is that the derivatives at the endpoints

TABLE 3.15

	Curve #1				Curve #2				Curve #3		
i	x_i	$f(x_i)$	$f'(x_i)$	i	x_i	$f(x_i)$	$f'(x_i)$	i	x_i	$f(x_i)$	$f'(x_i)$
0	1	3.0	1	0	17	4.5	3	0	27.7	4.1	.33
1	2	3.7		1	20	7.0		1	28	4.3	
2	5	3.9		2	23	6.1		2	29	4.1	
3	6	4.2		3	24	5.6		3	30	3.0	−1.5
4	7	5.7		4	25	5.8					
5	8	6.6		5	27	5.2					
6	10	7.1		6	27.7	4.1	−4				
7	13	6.7									
8	17	4.5	−.67								

TABLE 3.16

			Spline #1			
j	x_j	$a_j = f(x_j)$	b_j	c_j	d_j	$f'(x_j)$
0	1	3.0	1	−.347	.049	1
1	2	3.7	.447	−.206	.027	
2	5	3.9	−.074	.033	.342	
3	6	4.2	1.016	1.058	−.575	
4	7	5.7	1.409	−.665	.156	
5	8	6.6	.547	−.196	.024	
6	10	7.1	.048	−.053	−.003	
7	13	6.7	−.339	−.076	.006	
8	17	4.5				−.67

TABLE 3.17

			Spline #2			
j	x_j	$a_j = f(x_j)$	b_j	c_j	d_j	$f'(x_j)$
0	17	4.5	3	−1.101	.126	3
1	20	7.0	−.198	.035	−.023	
2	23	6.1	−.609	−.172	.280	
3	24	5.6	−.111	.669	−.357	
4	25	5.8	.154	−.403	.088	
5	27	5.2	−.401	.126	−2.568	
6	27.7	4.1				−4

TABLE 3.18

			Spline #3			
j	x_j	$a_j = f(x_j)$	b_j	c_j	d_j	$f'(x_j)$
0	27.7	4.1	.330	2.262	-3.800	.33
1	28	4.3	.661	-1.157	.296	
2	29	4.1	$-.765$	$-.269$	$-.065$	
3	30	3.0				-1.5

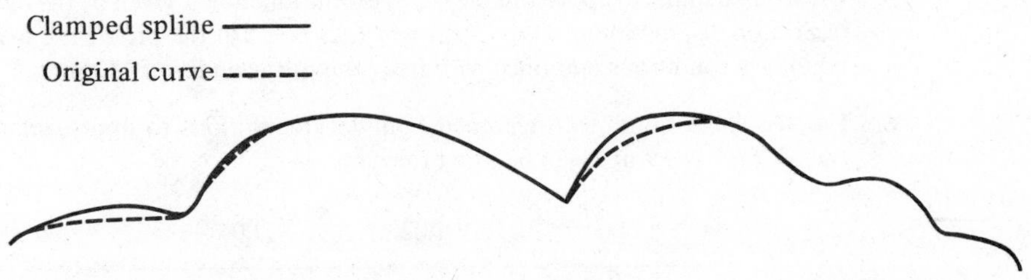

Clamped spline ———
Original curve – – – – –

FIGURE 3.16

of the sections of the curve have been inaccurately estimated. In this example, the natural spline has assumed a better estimate for these derivatives, by using the assumption that the second derivative at each end is zero, than were given for the clamped boundary-condition splines. ☐

It is generally preferable to use the clamped boundary conditions when approximating functions by the use of cubic splines, so it is often necessary to estimate the derivative of the function at the endpoints of the interval. In the case where the nodes are equally spaced near both endpoints, approximations can be obtained by using Eq. (4.23) or any of the other appropriate formulas given in Section 4.1. In the case of unequally spaced nodes, the problem is considerably more difficult.

To conclude this section, we list an error-bound formula for the cubic spline with clamped boundary conditions. The proof of this result can be found in Schultz [77], pages 57–58. A fourth-order error-bound result also holds in the case of free boundary conditions but is more difficult to express. (See Birkhoff and de Boor [12], pages 827–835.)

Theorem 3.12 Let $f \in C^4[a, b]$ with $\max_{a \le x \le b} |f^{(4)}(x)| \le M$. If S is the unique cubic spline interpolant to f with respect to the nodes $a = x_0 < x_1 < \cdots < x_n = b$, which satisfies $S'(a) = f'(a)$ and $S'(b) = f'(b)$, then

(3.30)
$$\max_{a \le x \le b} |f(x) - S(x)| \le \frac{5M}{384} \max_{0 \le j \le n-1} (x_{j+1} - x_j)^4.$$

Exercise Set 3.6

1. Construct a free cubic spline to approximate $f(x) = \cos \pi x$ by using the values given by $f(x)$ at $x = 0, .25, .5, .75, 1.0$. Integrate the spline over $[0, 1]$, and compare the result to $\int_0^1 \cos \pi x \, dx = 0$. Use the spline to approximate $f'(.5)$ and $f''(.5)$. Compare these approximations to the actual values.

2. Construct a free cubic spline to approximate $f(x) = e^{-x}$ by using the values given by $f(x)$ at $x = 0, .25, .75, 1.0$. Integrate the spline over $[0, 1]$, and compare the result to $\int_0^1 e^{-x}\,dx = (1/e)(e - 1)$. Use the spline to approximate $f'(.5)$ and $f''(.5)$. Compare the approximations to the actual values.

3. Repeat Exercise 1, constructing instead the clamped cubic spline with $f'(0) = f'(1) = 0$.

4. Repeat Exercise 2, constructing instead the clamped cubic spline with $f'(0) = -1$, $f'(1) = -e^{-1}$.

5. Use a cubic spline to approximate $f(2.5)$ for the function f given by the data in Exercise 1 of Section 3.5, assuming that $f'(x_0) = f'(x_n) = 0$. Do the problem once with clamped boundary conditions and once with free boundary conditions. Compare your results.

6. Use the cubic spline with clamped boundary conditions to approximate the function $f(x) = 3xe^x - e^{2x}$ at $x = 1.03$ using the data

x	1.0	1.02	1.04	1.06
$f(x)$	.76578939	.79536678	.82268817	.84752226

Estimate the error, using Eq. (3.30) and compare to the actual error.

7. Given the partition $x_0 = 0$, $x_1 = .05$, $x_2 = .1$ of $[0, .1]$ and $f(x) = e^{2x}$:

   a) Find the cubic spline s with clamped boundary conditions that interpolates f.
   b) Find an approximation for $\int_0^1 e^{2x}\,dx$ by evaluating $\int_0^1 s(x)\,dx$.
   c) Use Theorem 3.12 to estimate $\max_{0 \le x \le .1} |f(x) - s(x)|$ and
   $$\left| \int_0^1 f(x)\,dx - \int_0^1 s(x)\,dx \right|.$$

   d) Determine the cubic spline S with free boundary conditions and compare $S(.02)$, $s(.02)$, and $e^{.04} = 1.04081077$.

8. Given the partition $x_0 = 0$, $x_1 = .05$, $x_2 = .1$ of $[0, .1]$, find the piecewise linear interpolating function F for $f(x) = e^{2x}$. Approximate $\int_0^1 e^{2x}\,dx$ with $\int_0^1 F(x)\,dx$. Compare the results to those of Exercise 7.

9. Suppose the following data is given for a function f: $f(0) = 0$, $f(.1) = -.1$, $f(.2) = -.3$, $f(.3) = -.5$, $f(.4) = -.4$, $f(.5) = -.2$, $f(.6) = 0$, $f(.7) = .2$, $f(.8) = .3$, $f(.9) = .7$, $f(1) = .9$. Find the spline that interpolates f with free boundary conditions with respect to the partition $x_j = .1j$ for $j = 0, 1, \ldots, 10$.

10. Extend Algorithms 3.2 and 3.3 to include as output the first and second derivatives of the spline at the nodes.

11. Extend Algorithms 3.2 and 3.3 to include as output the integral of the spline over the interval $[x_0, x_n]$.

12. Let $f \in C^2[a, b]$, and let the nodes $a = x_0 < x_1 < \cdots < x_n = b$ be given. Derive an error estimate similar to that in Theorem 3.12 for the piecewise linear interpolating function F. Use this estimate to derive error bounds for Exercise 8.

13. Let f be defined on $[a, b]$, and let the nodes $a = x_0 < x_1 < x_2 = b$ be given. A quadratic-spline interpolating function S consists of the quadratic polynomial
   $$S_0(x) = a_0 + b_0(x - x_0) + c_0(x - x_0)^2 \qquad \text{on } [x_0, x_1]$$

and the quadratic polynomial

$$S_1(x) = a_1 + b_1(x - x_1) + c_1(x - x_1)^2 \quad \text{on } [x_1, x_2]$$

such that

i) $S(x_0) = f(x_0)$, $S(x_1) = f(x_1)$, and $S(x_2) = f(x_2)$,
ii) $S \in C^1[x_0, x_2]$.

Show that conditions (i) and (ii) lead to five equations in the six unknowns a_0, b_0, c_0, a_1, b_1, and c_1. The problem is to decide what additional condition to impose to make the solution unique, for example,

$$f'(x_0) = S'(x_0) \quad \text{or} \quad f'(x_2) = S'(x_2).$$

Does the condition

$$S \in C^2[x_0, x_2]$$

lead to a meaningful solution?

14. As suggested in a paper by Kammerer, Reddien, and Varga [57], useful quadratic interpolatory splines can be constructed. Given a function f on $[a, b]$ and the points $a = x_0 < x_1 < \cdots < x_n = b$, the quadratic spline satisfies the following:

i)
$$s(x_0) = f(x_0),$$

$$s\left(\frac{x_i + x_{i+1}}{2}\right) = f\left(\frac{x_i + x_{i+1}}{2}\right) \quad \text{for each } i = 0, 1, \ldots, n - 1,$$

$$s(x_n) = f(x_n);$$

ii) $s \in C^1[a, b]$;
iii) For each $i = 0, 1, \ldots, n - 1$,

$$s(x) = a_i + b_i(x - x_i) + c_i(x - x_i)^2, \quad x_i \le x \le x_{i+1}.$$

a) Show that conditions (i), (ii) and (iii) lead to the equations:

$$a_0 = f(x_0);$$
$$a_i + \tfrac{1}{2}h_i b_i + \tfrac{1}{4}h_i^2 c_i = f(x_i + \tfrac{1}{2}h_i), \quad \text{for each } i = 0, 1, \ldots, n - 1,$$
$$a_{n-1} + h_{n-1}b_{n-1} + h_{n-1}^2 c_{n-1} = f(x_n);$$
$$a_i = a_{i-1} + b_{i-1}h_{i-1} + c_{i-1}h_{i-1}^2, \quad \text{for each } i = 1, 2, \ldots, n - 1;$$
$$b_i = b_{i-1} + 2c_{i-1}h_{i-1}, \quad \text{for each } i = 1, 2, \ldots, n - 1.$$

b) Show that the results of part (a) lead to the equations:

$$\tfrac{3}{8}h_0 b_0 + \tfrac{1}{8}h_0 b_1 = f\left(x_0 + \frac{h_0}{2}\right) - f(x_0),$$

$$\tfrac{1}{8}h_{i-1}b_{i-1} + \tfrac{3}{8}(h_i + h_{i-1})b_i + \tfrac{1}{8}h_i b_{i+1} = f\left(x_i + \frac{h_i}{2}\right) - f\left(x_{i-1} + \frac{h_{i-1}}{2}\right),$$

for each $i = 1, 2, \ldots, n - 2$,

$$\tfrac{1}{8}h_{n-2}b_{n-2} + (\tfrac{3}{8}h_{n-2} + \tfrac{1}{3}h_{n-1})b_{n-1}$$

$$= (\tfrac{4}{3})f\left(x_{n-1} + \frac{h_{n-1}}{2}\right) - f\left(x_{n-2} + \frac{h_{n-2}}{2}\right) - \tfrac{1}{3}f(x_n),$$

where $h_i = x_{i+1} - x_i$ for each $i = 0, 1, \ldots, n - 1$.

c) Devise an algorithm similar to Algorithms 3.2 and 3.3 to find $b_0, \ldots, b_{n-1}$.

d) Devise formulas for $a_0, \ldots, a_{n-1}, c_0, \ldots, c_{n-1}$, assuming that $b_0, \ldots, b_{n-1}$ are known.

e) Let $f(x) = e^x, x_0 = 0, x_1 = .2, x_2 = .6,$ and $x_3 = .9$. Find the quadratic interpolating spline s for f, and compute $s(.5)$. Is $s(.5)$ a good approximation to $f(.5)$?

f) Repeat Exercise 6, using the quadratic interpolating spline, and compare your results to those obtained in Exercise 6. Is it reasonable to believe that

$$|f(x) - s(x)| = O(h^k), \qquad \text{for } k = 2, \text{ or } k = 4?$$

15. Carry out in detail the following derivation of an alternate formula for cubic splines. Given a partition

$$a = x_0 < x_1 < \cdots < x_n = b \qquad \text{of } [a, b],$$

the cubic spline s, which interpolates f, is a cubic polynomial $s_j(x)$ on each $[x_j, x_{j+1}]$. Thus, we may write

$$s''(x) = s_j''(x) = \frac{a_j(x_{j+1} - x)}{h_j} + \frac{a_{j+1}(x - x_j)}{h_j}$$

on $[x_j, x_{j+1}]$ for $j = 0, 1, \ldots, n - 1$; that is, $s_j''(x)$ is linear on each subinterval. Show s'' is continuous on $[a, b]$. Using the notation $s_j = s(x_j)$ and $f_j = f(x_j)$ for $j = 0, 1, \ldots, n - 1$, integrate $s_j''(x)$ twice and evaluate the constants of integration by using $s_j = f_j$ and $s_{j+1} = f_{j+1}$ to obtain

$$s_j(x) = a_j \frac{(x_{j+1} - x)^3}{6h_j} + \frac{a_{j+1}}{6h_j} (x - x_j)^3$$

$$+ \left(f_{j+1} - \frac{a_{j+1} h_j^2}{6} \right) \frac{(x - x_j)}{h_j} + \left(f_j - \frac{a_j h_j^2}{6} \right) \frac{(x_{j+1} - x)}{h_j}.$$

Now obtain a system of equations for the a_j's from the requirement that s' is continuous on $[a, b]$.

16. Define

$$(x - \xi)_+^3 = \begin{cases} (x - \xi)^3, & x > \xi, \\ 0, & x \le \xi, \end{cases}$$

Let $a = x_0 < x_1 < x_2 < x_3 = b$ and define

$$S(x) = c_1(x - x_1)_+^3 + c_2(x - x_2)_+^3, \qquad \text{where } x \in [a, b].$$

Show that $S \in C^2[a, b]$.

17. Let $P(x) = a_0 + a_1 x + a_2 x^2 + a_3 x^3$ be any cubic polynomial on $[a, b]$ and $a = x_0 < x_1 < \cdots < x_n = b$ be given nodes. Show that

$$S(x) = P(x) + \sum_{j=1}^{n-1} c_j(x - x_j)_+^3$$

is a cubic spline, in that S is a cubic polynomial on $[x_i, x_{i+1}]$ for each $i = 0, 1, \ldots, n - 1$ and $S \in C^2[a, b]$.

18. Let P be a polynomial of degree at most three, $\{x_j\}_{j=0}^n$ be a partition of $[a, b]$ and

$$s(x) = P(x) + \sum_{j=1}^{n-1} c_j(x - x_j)_+^3.$$

Show that the third derivative of s exists and is continuous on $[a, b]$ if and only if $c_j = 0$ for $j = 1, 2, \ldots, n$.

19. It is suspected that the high amounts of tannin in mature oak leaves inhibits the growth of the winter moth (Operophtera bromata L., Geometridae) larvae that extensively damage these trees in certain years. The following table lists the average weight of two samples of larvae at times in the first 28 days after birth. The first sample was reared on young oak leaves, while the second sample was reared on mature leaves from the same tree.

    a) Use a free cubic spline to approximate the average weight curve for each sample.
    b) Find an approximate maximum average weight for each sample by determining the maximum of the spline.

Day	0	6	10	13	17	20	28
Sample 1 Average Weight (mg.)	6.67	17.33	42.67	37.33	30.10	29.31	28.74
Sample 2 Average Weight (mg.)	6.67	16.11	18.89	15.00	10.56	9.44	8.89

20. Exercise 8 of Section 3.5 lists observed data concerning the speed and distance of a car traveling along a straight road. Use a clamped cubic spline to predict the position of the car and its speed at $t = 10$ sec. Use the spline to determine whether the car ever exceeds a 55 mph speed limit on the road, and if so, what is the first time the car exceeds this speed. What is the predicted maximum speed for the car?

21. The data in the following table lists the population of the United States for the years 1920 to 1970 and was discussed in the introduction to this chapter, as well as in Exercise 15 of Section 3.2 and Exercise 8 of Section 3.4.

Year	1920	1930	1940	1950	1960	1970
Population	105,711	123,203	131,669	150,697	179,323	203,212

    Find a free cubic spline agreeing with this data, and use the spline to predict the population in the years 1910, 1965, and 2000. Compare your approximations with those previously obtained. If you had to make a choice, which interpolation procedure would you choose?

22. The 1979 Kentucky Derby was won by a horse named Spectacular Bid in a time of $2:02\frac{2}{5}$ (2 min, $2\frac{2}{5}$ sec) for the $1\frac{1}{4}$ mile race. Times at the quarter mile, half mile, and mile poles were $25\frac{2}{5}$, $49\frac{2}{5}$, and $1:37\frac{3}{5}$.

    a) Use these values together with the starting time to construct a free cubic spline for Spectacular Bid's race.
    b) Use the spline to predict the time at the three-quarter mile pole, and compare this to the actual time of $1:12\frac{2}{5}$.
    c) Use the spline to approximate Spectacular Bid's starting speed and speed at the finish line.

4

Numerical Differentiation and Integration

A sheet of corrugated roofing is to be constructed using a machine that presses a flat sheet of aluminum into one whose cross section has the form of a sine wave.

Suppose a corrugated sheet of length four feet is needed, the height of each wave is one inch from the center line, and each wave has a period of approximately 2π inches. The problem of finding the length of the initial flat sheet is one of finding the arc length of the curve given by $f(x) = \sin x$ from $x = 0$ inches to $x = 48$ inches. From calculus we know that this length can be expressed by:

$$L = \int_0^{48} \sqrt{1 + \left(\frac{df(x)}{dx}\right)^2}\, dx = \int_0^{48} \sqrt{1 + (\cos x)^2}\, dx,$$

so the problem reduces to evaluating this integral. Although the sine function is one of the most common mathematical functions, the calculation of its arc length gives rise to what is called an elliptic integral of the second kind, which cannot be evaluated by ordinary methods. Approximation methods will be developed in this chapter which will reduce problems of this type to elementary exercises. This particular problem is considered in Exercise 13 of Section 4.4 and in Exercise 4 of Section 4.5.

Because of the many applications that involve the derivatives and integrals of functions, it should be expected that approximation involving these concepts is of interest.

In the introduction to Chapter 3, we mentioned that one reason for using the class of algebraic polynomials to approximate an arbitrary set of data is the property of algebraic polynomials that, given any continuous function defined on a closed interval, there exists a polynomial which is arbitrarily close to the function at every point in the interval. An additional property possessed by this class is that their derivatives and integrals are quite easily obtained and evaluated. It should not be surprising, therefore, to find that the majority of the procedures for approximating integrals and derivatives commence with algebraic polynomials approximating the function.

4.1 Numerical Differentiation

To introduce the subject of numerical differentiation, suppose we are given a function $f \in C^2[a, b]$ and an arbitrary point x_0 in $[a, b]$. A method is needed to approximate $f'(x_0)$. With $x_1 = x_0 + h$ for some $h \neq 0$, small enough to ensure that $x_1 \in [a, b]$, compute $P_{0,1}(x)$. Using the notation in Section 3.3 and the results of Theorem 3.3 (p. 83),

$$f(x) = P_{0,1}(x) + \frac{(x - x_0)(x - x_1)}{2!} f''(\xi)$$

$$= \frac{f(x_0)(x - x_0 - h)}{-h} + \frac{f(x_0 + h)(x - x_0)}{h}$$

$$+ \frac{(x - x_0)(x - x_0 - h)}{2} f''(\xi),$$

for some ξ, dependent upon x, in $[a, b]$. Differentiating gives

$$f'(x) = \frac{f(x_0 + h) - f(x_0)}{h} + \frac{d}{dx}\left[\frac{(x - x_0)(x - x_0 - h)}{2} f''(\xi)\right]$$

$$= \frac{f(x_0 + h) - f(x_0)}{h} + \frac{2(x - x_0) - h}{2} f''(\xi)$$

$$+ \frac{(x - x_0)(x - x_0 - h)}{2} \frac{d}{dx}(f''(\xi)).$$

Although this differentiation requires that $(d/dx)[f''(\xi)]$ exists, our approximation formula can be derived alternately without this assumption by using Taylor's Theorem (see Exercise 11). For $x = x_0$, this equation simplifies to

(4.1) $$f'(x_0) = \frac{f(x_0 + h) - f(x_0)}{h} - \frac{h}{2} f''(\xi).$$

For small values of h, the difference quotient $[f(x_0 + h) - f(x_0)]/h$ can be used to approximate $f'(x_0)$ with an error bounded by $(h/2)M$, where M is a bound on

$f''(x)$ for $x \in [a, b]$. This formula is known as the **forward-difference formula** if $h > 0$ and the **backward-difference formula** if $h < 0$.

EXAMPLE 1 Let $f(x) = \ln x$ and $x_0 = 1.8$. The quotient

$$\frac{f(1.8 + h) - f(1.8)}{h}, \qquad h > 0,$$

will be used to approximate $f'(1.8)$ with error

$$\frac{|hf''(\xi)|}{2} \leq \frac{|h|}{2(1.8)^2} \qquad \text{where } 1.8 < \xi < 1.8 + h.$$

The results in Table 4.1 are for the case when $h = .1, .01,$ and $.001$.

TABLE 4.1

h	$f(1.8 + h)$	$\dfrac{f(1.8 + h) - f(1.8)}{h}$	$\dfrac{|h|}{2(1.8)^2}$
.1	.64185389	.5406722	.0154321
.01	.59332685	.5540180	.0015432
.001	.58834207	.5554000	.0001543

Since $f'(x) = 1/x$, the exact value of $f'(1.8)$ is $.55\bar{5}$ and the error bounds above are appropriate. □

To consider the differentiation procedure in general, let $\{x_0, \ldots, x_n\}$ be $(n + 1)$ distinct points in some interval I and $f \in C^{n+1}(I)$. From Theorem 3.3,

$$f(x) = \sum_{k=0}^{n} f(x_k)L_k(x) + \frac{(x - x_0) \cdots (x - x_n)}{(n + 1)!} f^{(n+1)}(\xi)$$

for some ξ, dependent upon x, in I. Differentiating gives

$$(4.2) \qquad f'(x) = \sum_{k=0}^{n} f(x_k)L_k'(x) + \frac{d}{dx}\left[\frac{(x - x_0) \cdots (x - x_n)}{(n + 1)!}\right] f^{(n+1)}(\xi)$$

$$+ \frac{(x - x_0) \cdots (x - x_n)}{(n + 1)!} \frac{d}{dx}[f^{(n+1)}(\xi)].$$

This formula could be used to approximate $f'(x)$ for any $x \in I$, but the truncation error term,

$$(4.3) \qquad t_n(x) = \frac{d}{dx}\left[\frac{(x - x_0) \cdots (x - x_n)}{(n + 1)!}\right] f^{(n+1)}(\xi)$$

$$+ \frac{(x - x_0) \cdots (x - x_n)}{(n + 1)!} \frac{d}{dx}[f^{(n+1)}(\xi)],$$

is difficult to analyze because little is known about $(d/dx)f^{(n+1)}(\xi)$. Equation (4.2) can, however, be used in the situation when x is an interpolating node. In this case, assuming the quantity $(d/dx)[f^{(n+1)}(\xi)]$ exists and is bounded, the error term simplifies to

$$t_n(x_k) = \left\{ \frac{d}{dx} \left[\frac{(x - x_0) \cdots (x - x_n)}{(n + 1)!} \right] f^{(n+1)}(\xi) \right\} \Bigg|_{x = x_k}$$

$$= \frac{(x_k - x_0) \cdots (x_k - x_{k-1})(x_k - x_{k+1}) \cdots (x_k - x_n)}{(n + 1)!} f^{(n+1)}(\xi).$$

Hence, (4.2) becomes

(4.4) $$f'(x_k) = \sum_{j=0}^{n} f(x_j)L_j'(x_k) + t_n(x_k).$$

Equation (4.4) is called an $(n + 1)$-point formula to approximate $f'(x_k)$, since a linear combination of the $(n + 1)$ values $f(x_j)$ is used for $j = 0, 1, \ldots, n$.

It would appear that using more points would lead to greater accuracy because t_n would decrease, but from a practical point of view this leads to evaluating the function f at more values; and, even if the function is easy to evaluate, the rounding errors tend to discourage the use of a large number of interpolating nodes. To illustrate this point, we will derive some useful three-point formulas and consider aspects of their rounding errors. Since

$$L_0(x) = \frac{(x - x_1)(x - x_2)}{(x_0 - x_1)(x_0 - x_2)},$$

$$L_0'(x) = \frac{2x - x_1 - x_2}{(x_0 - x_1)(x_0 - x_2)}.$$

Similarly,

$$L_1'(x) = \frac{2x - x_0 - x_2}{(x_1 - x_0)(x_1 - x_2)}$$

and

$$L_2'(x) = \frac{2x - x_0 - x_1}{(x_2 - x_0)(x_2 - x_1)}.$$

Hence,

(4.5) $$f'(x_j) = f(x_0) \left[\frac{2x_j - x_1 - x_2}{(x_0 - x_1)(x_0 - x_2)} \right] + f(x_1) \left[\frac{2x_j - x_0 - x_2}{(x_1 - x_0)(x_1 - x_2)} \right]$$

$$+ f(x_2) \left[\frac{2x_j - x_0 - x_1}{(x_2 - x_0)(x_2 - x_1)} \right] + \tfrac{1}{6} f^{(3)}(\xi_j) \prod_{\substack{i=0, \\ i \neq j}}^{2} (x_j - x_i),$$

for each $j = 0, 1, 2$, where the notation ξ_j indicates that this point depends upon x_j.

The three formulas from Eq. (4.5) become especially useful if the nodes are equally spaced, that is, when

$$x_1 = x_0 + h \quad \text{and} \quad x_2 = x_0 + 2h \quad \text{for some } h \neq 0.$$

Using Eq. (4.5) with $x_j = x_0, x_1 = x_0 + h$, and $x_2 = x_0 + 2h$ gives

$$f'(x_0) = \frac{1}{h} \left[-\tfrac{3}{2} f(x_0) + 2f(x_1) - \tfrac{1}{2} f(x_2) \right] + \frac{h^2}{3} f^{(3)}(\xi_0).$$

Doing the same for $x_j = x_1$ and $x_j = x_2$ gives

$$f'(x_1) = \frac{1}{h}\left[-\tfrac{1}{2}f(x_0) + \tfrac{1}{2}f(x_2)\right] - \frac{h^2}{6}f^{(3)}(\xi_1),$$

and

$$f'(x_2) = \frac{1}{h}\left[\tfrac{1}{2}f(x_0) - 2f(x_1) + \tfrac{3}{2}f(x_2)\right] + \frac{h^2}{3}f^{(3)}(\xi_2).$$

Writing the equations in terms of x_0, we obtain:

(4.6) $$f'(x_0) = \frac{1}{h}\left[-\tfrac{3}{2}f(x_0) + 2f(x_0 + h) - \tfrac{1}{2}f(x_0 + 2h)\right] + \frac{h^2}{3}f^{(3)}(\xi_0),$$

(4.7) $$f'(x_0 + h) = \frac{1}{h}\left[-\tfrac{1}{2}f(x_0) + \tfrac{1}{2}f(x_0 + 2h)\right] - \frac{h^2}{6}f^{(3)}(\xi_1),$$

and

(4.8) $$f'(x_0 + 2h) = \frac{1}{h}\left[\tfrac{1}{2}f(x_0) - 2f(x_0 + h) + \tfrac{3}{2}f(x_0 + 2h)\right] + \frac{h^2}{3}f^{(3)}(\xi_2).$$

Replace x_0 by $(x_0 - h)$ in Eq. (4.7) and x_0 by $(x_0 - 2h)$ in Eq. (4.8) to obtain the three formulas to approximate $f'(x_0)$:

(4.9) $$f'(x_0) = \frac{1}{2h}\left[-3f(x_0) + 4f(x_0 + h) - f(x_0 + 2h)\right] + \frac{h^2}{3}f^{(3)}(\xi_0),$$

(4.10) $$f'(x_0) = \frac{1}{2h}\left[-f(x_0 - h) + f(x_0 + h)\right] - \frac{h^2}{6}f^{(3)}(\xi_1),$$

and

(4.11) $$f'(x_0) = \frac{1}{2h}\left[f(x_0 - 2h) - 4f(x_0 - h) + 3f(x_0)\right] + \frac{h^2}{3}f^{(3)}(\xi_2).$$

Finally, note that since Eq. (4.11) can be obtained from Eq. (4.9) by simply replacing h by $-h$, there are actually only two formulas:

(4.12) $$f'(x_0) = \frac{1}{2h}\left[-3f(x_0) + 4f(x_0 + h) - f(x_0 + 2h)\right]$$

$$+ \frac{h^2}{3}f^{(3)}(\xi_0), \qquad h \neq 0,$$

where ξ_0 lies between x_0 and $x_0 + 2h$, and

(4.13) $$f'(x_0) = \frac{1}{2h}\left[-f(x_0 - h) + f(x_0 + h)\right] - \frac{h^2}{6}f^{(3)}(\xi_1), \qquad h \neq 0,$$

where ξ_1 lies between $(x_0 - h)$ and $(x_0 + h)$.

The error in Eq. (4.13) is approximately half the error in Eq. (4.12). This is reasonable since in Eq. (4.13) data is being examined on both sides of x_0 and on only one side in Eq. (4.12). The approximation in Eq. (4.12) is useful near the ends of the interval I since information about f outside the interval is not necessarily available. Note also that in Eq. (4.13), f needs to be evaluated at only two points whereas in Eq. (4.12) three points are needed.

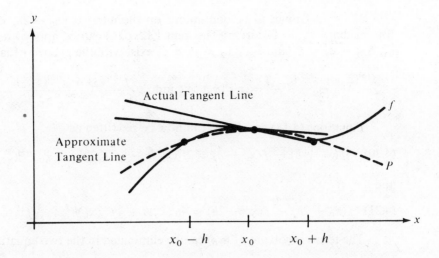

FIGURE 4.1

The methods presented in equations (4.12) and (4.13) are known as three-point formulas. (See Fig. 4.1.) Similarly, there are methods, known as five-point formulas, which involve evaluating the function at more points but whose error term is of the form $O(h^4)$. To derive a typical formula of this type, expand the function f in a fourth-degree Taylor polynomial and evaluate the polynomial at $x_0 - 2h$, $x_0 - h$, $x_0 + h$, and $x_0 + 2h$ to obtain

$$(4.14) \quad f(x_0 + h) = f(x_0) + f'(x_0)h + \tfrac{1}{2}f''(x_0)h^2 + \tfrac{1}{6}f'''(x_0)h^3 \\ + \tfrac{1}{24}f^{(4)}(x_0)h^4 + \tfrac{1}{120}f^{(5)}(\xi_1)h^5$$

and

$$(4.15) \quad f(x_0 - h) = f(x_0) - f'(x_0)h + \tfrac{1}{2}f''(x_0)h^2 - \tfrac{1}{6}f'''(x_0)h^3 \\ + \tfrac{1}{24}f^{(4)}(x_0)h^4 - \tfrac{1}{120}f^{(5)}(\xi_{-1})h^5$$

where $x_0 - h < \xi_{-1} < x_0 < \xi_1 < x_0 + h$. Likewise,

$$(4.16) \quad f(x_0 + 2h) = f(x_0) + 2f'(x_0)h + 2f''(x_0)h^2 + \tfrac{4}{3}f'''(x_0)h^3 \\ + \tfrac{2}{3}f^{(4)}(x_0)h^4 + \tfrac{4}{15}f^{(5)}(\xi_2)h^5,$$

and

$$(4.17) \quad f(x_0 - 2h) = f(x_0) - 2f'(x_0)h + 2f''(x_0)h^2 - \tfrac{4}{3}f'''(x_0)h^3 \\ + \tfrac{2}{3}f^{(4)}(x_0)h^4 - \tfrac{4}{15}f^{(5)}(\xi_{-2})h^5$$

where $x_0 - 2h < \xi_{-2} < x_0 < \xi_2 < x_0 + 2h$.

Subtracting Eq. (4.15) from (4.14), and Eq. (4.17) from (4.16) gives

$$(4.18) \quad f(x_0 + h) - f(x_0 - h) = 2f'(x_0)h + \tfrac{1}{3}f'''(x_0)h^3 \\ + \tfrac{1}{120}[f^{(5)}(\xi_1) + f^{(5)}(\xi_{-1})]h^5$$

and

$$(4.19) \quad f(x_0 + 2h) - f(x_0 - 2h) = 4f'(x_0)h + \tfrac{8}{3}f'''(x_0)h^3 \\ + \tfrac{4}{15}[f^{(5)}(\xi_2) + f^{(5)}(\xi_{-2})]h^5.$$

If $f^{(5)}$ is known to be continuous on the interval $[x_0 - 2h, x_0 + 2h]$, the Intermediate-Value Theorem (Theorem 1.12, p. 6) can be applied to ensure that $\hat{\xi}_1, \xi_{-1} < \hat{\xi}_1 < \xi_1$ and $\hat{\xi}_2, \xi_{-2} < \hat{\xi}_2 < \xi_2$ exist with the property that

$$\frac{f^{(5)}(\xi_1) + f^{(5)}(\xi_{-1})}{2} = f^{(5)}(\hat{\xi}_1) \qquad \text{and} \qquad \frac{f^{(5)}(\xi_2) + f^{(5)}(\xi_{-2})}{2} = f^{(5)}(\hat{\xi}_2).$$

Equations (4.18) and (4.19) can now be rewritten as

(4.20) $f(x_0 + h) - f(x_0 - h) = 2f'(x_0)h + \frac{1}{3}f'''(x_0)h^3 + \frac{1}{60}f^{(5)}(\hat{\xi}_1)h^5$

and

(4.21) $f(x_0 + 2h) - f(x_0 - 2h) = 4f'(x_0)h + \frac{8}{3}f'''(x_0)h^3 + \frac{8}{15}f^{(5)}(\hat{\xi}_2)h^5.$

The terms involving $f'''(x_0)$ can be eliminated in the two equations to give

$$8f(x_0 + h) - 8f(x_0 - h) - f(x_0 + 2h) + f(x_0 - 2h)$$
$$= 12hf'(x_0) + \frac{2}{15}[f^{(5)}(\hat{\xi}_1) - 4f^{(5)}(\hat{\xi}_2)]h^5$$

or

$$f'(x_0) = \frac{1}{12h}[f(x_0 - 2h) - 8f(x_0 - h) + 8f(x_0 + h) - f(x_0 + 2h)]$$

$$+ \frac{h^4}{90}[4f^{(5)}(\hat{\xi}_2) - f^{(5)}(\hat{\xi}_1)].$$

The error term derived in this manner is quite complicated but can be shown to be $(h^4/30)f^{(5)}(\xi)$ for some ξ, $(x_0 - 2h) < \xi < (x_0 + 2h)$. Therefore the formula becomes

(4.22) $f'(x_0) = \frac{1}{12h}[f(x_0 - 2h) - 8f(x_0 - h) + 8f(x_0 + h)$

$$- f(x_0 + 2h)] + \frac{h^4}{30}f^{(5)}(\xi).$$

The usual method of obtaining the error term in this form is to differentiate the Lagrange polynomial and error term given in Theorem 3.3, which is a much more laborious procedure.

Another five-point formula that is useful, particularly with regard to clamped cubic spline interpolation, is

(4.23) $f'(x_0) = \frac{1}{12h}[-25f(x_0) + 48f(x_0 + h) - 36f(x_0 + 2h)$

$$+ 16f(x_0 + 3h) - 3f(x_0 + 4h)] + \frac{h^4}{5}f^{(5)}(\xi)$$

where ξ lies between x_0 and $x_0 + 4h$. Left endpoint approximations can be found using this formula with $h > 0$ and right endpoint approximations with $h < 0$.

EXAMPLE 2   With $h = \pm .1$, use formulas (4.12) and (4.13) to approximate $f'(3.15)$, given the data in Table 4.2.

TABLE 4.2

x	$f(x)$
2.95	13.364875
3.05	14.665125
3.15	16.088375
3.25	17.640625
3.35	19.327875

Using Eq. (4.12) with $h = .1$,

$$f'(3.15) \approx \frac{1}{.2}[-3f(3.15) + 4f(3.25) - f(3.35)] = 14.8475.$$

Using Eq. (4.12) with $h = -.1$,

$$f'(3.15) \approx -\frac{1}{.2}[-3f(3.15) + 4f(3.05) - f(2.95)] = 14.8475.$$

Using Eq. (4.13) with $h = .1$,

$$f'(3.15) \approx \frac{1}{.2}[-f(3.05) + f(3.25)] = 14.8775.$$

In this example, $f(x) = x^3 - 3x^2 + 4x + 2$; so $f'(3.15) = 14.8675$. The error using Eq. (4.12) with $h \pm .1$ is .02, and the error using Eq. (4.13) is .01. Thus, in this example using (4.13) is precisely twice as accurate as using (4.12). It should be noted that all errors in this approximation can be easily determined from the error formulas since $f^{(3)}(x) = 6$, and hence, the particular values of ξ_0 and ξ_1 are irrelevant.

The only five-point formula for which adequate data is given is Eq. (4.22). Using this method leads to

$$f'(3.15) \approx \frac{1}{1.2}[f(2.95) - 8f(3.05) + 8f(3.25) - f(3.35)] = 14.8675,$$

the correct value, as would be expected since the error term in this case involves $f^{(5)}(\xi)$ and $f^{(5)} \equiv 0$.   □

A particularly important subject in the study of numerical differentiation is the effect of round-off error. Let us examine the formula

$$f'(x_0) = \frac{f(x_0 + h) - f(x_0 - h)}{2h} - \frac{h^2}{6} f^{(3)}(\xi_1)$$

more closely. Suppose that, in evaluating $f(x_0 + h)$ and $f(x_0 - h)$, we have encountered rounding errors $e(x_0 + h)$ and $e(x_0 - h)$; that is, our computed values

$\tilde{f}(x_0 + h)$ and $\tilde{f}(x_0 - h)$ are related to the true values $f(x_0 + h)$ and $f(x_0 - h)$ by the formulas

$$f(x_0 + h) = \tilde{f}(x_0 + h) + e(x_0 + h)$$

and

$$f(x_0 - h) = \tilde{f}(x_0 - h) + e(x_0 - h).$$

In this case, the error in approximation,

$$f'(x_0) - \frac{\tilde{f}(x_0 + h) - \tilde{f}(x_0 - h)}{2h} = \frac{e(x_0 + h) - e(x_0 - h)}{2h} - \frac{h^2}{6} f^{(3)}(\xi_1),$$

will have a part due to rounding and a part due to truncating. If it is assumed that the rounding errors $e(x_0 \pm h)$ are bounded by some $\varepsilon > 0$, and that the third derivative of f is bounded by a number $M > 0$, then

$$\left| f'(x_0) - \frac{\tilde{f}(x_0 + h) - \tilde{f}(x_0 - h)}{2h} \right| \leq \frac{\varepsilon}{h} + \frac{h^2}{6} M.$$

If h is small, this implies that the error due to rounding, ε/h, may be large. In practice, then, it is seldom advantageous to let h be too small since the rounding errors become significant.

EXAMPLE 3   Consider approximating $f'(.900)$ for $f(x) = \sin x$, using the values in Table 4.3.

TABLE 4.3

x	$\sin x$	x	$\sin x$
.800	.71736	.901	.78395
.850	.75128	.902	.78457
.880	.77074	.905	.78643
.890	.77707	.910	.78950
.895	.78021	.920	.79560
.898	.78208	.950	.81342
.899	.78270	1.000	.84147

Using the formula

$$f'(.900) \approx \frac{f(.900 + h) - f(.900 - h)}{2h}$$

with different values of h gives the values in Table 4.4.

TABLE 4.4

h	Approximation to $f'(.900)$
.001	.62500
.002	.62250
.005	.62200
.010	.62150
.020	.62150
.050	.62140
.100	.62055

Since $f'(.900) = \cos(.900) = .62161$, it appears that an optimal choice for h lies between .005 and .05. Suppose we perform some analysis on the error term,

$$\frac{\varepsilon}{h} + \frac{h^2}{6} M.$$

Letting

$$e(h) = \frac{\varepsilon}{h} + \frac{h^2}{6} M,$$

we can verify (see Exercise 12) that a minimum for e occurs at $h = \sqrt[3]{3\varepsilon/M}$, where

$$M = \max_{x \in [.800, 1.00]} |f'''(x)| = \max_{x \in [.800, 1.00]} |\cos x| = .69671.$$

Since values of f are given to five decimal places, it is reasonable to assume that $\varepsilon = .000005$. Therefore, the optimal choice of h would be approximately

$$h = \sqrt[3]{\frac{3(.000005)}{.69671}} \approx .028,$$

which is consistent with our results.

In practice, however, we would not be able to compute an actual optimal h to use in approximating the derivative since we would have no knowledge about the third derivative of the function. $\square$

In this section we have examined in varying detail the more commonly used formulas for approximating $f'(x)$. It should be remembered that, as an approximation method, numerical differentiation is not stable, since small values of h can lead to large errors. This is the first method we have studied with this property, and in actual practice this operation is avoided whenever possible. The formulas derived, however, are necessary and have application in approximating the solutions of ordinary and partial-differential equations.

Exercise Set 4.1

1. For $f(x) = e^x$, compute approximations to $f'(1.8)$, using equations (4.12) and (4.13) for $h = .01$. Compare your results to the correct value $e^{1.8}$.

2. Let $f(x) = x^3 e^{x^2} - \sin x$. For $h = .1$ and $h = .01$, approximate $f'(2.19)$, using equations (4.12) and (4.13).

3. Let $f(x) = \cos \pi x$. Use Eq. (4.22) and the values of $f(x)$ at $x = 0, .25, .75$, and 1.0 to approximate $f'(.5)$. Compare this result to the exact value and to the approximation found in Exercise 1 of Section 3.6. Find a bound for the error.

4. Let $f(x) = e^{-x}$. Use Eq. (4.22) and the values of $f(x)$ at $x = 0, .25, .75$, and 1.0 to approximate $f'(.5)$. Compare this result to the exact value and to the approximation found in Exercise 2 of Section 3.6. Find a bound for the error.

5. Let $f(x) = 2^x \sin x$. Approximate $f'(1.05)$ using $h = .05$ and $h = .01$ in the formula (4.13) with the following data:

x	1.0	1.04	1.06	1.10
$f(x)$	1.6829420	1.7732994	1.8188014	1.9103448

6. Repeat Exercise 5 using four-digit arithmetic, and compare the results to those previously obtained.

7. Suppose the following data has been experimentally collected.

x	1.00	1.01	1.02
$f(x)$	1.27	1.32	1.38

    a) Approximate $f'(1.005)$ and $f'(1.015)$ using Eq. (4.13).
    b) Approximate $f''(1.01)$, using Eq. (4.13) and the results of (a).
    c) Suppose the data is accurate to within $\pm.005$. Find the maximum error due to the inaccurate data in parts (a) and (b).

8. In a circuit with impressed voltage $\mathscr{E}(t)$ and inductance L, Kirchhoff's first law gives the relationship

    $$\mathscr{E} = L\frac{di}{dt} + Ri$$

    where R is the resistance in the circuit and i the current. Suppose we measure the current for several values of t and obtain:

t	1.00	1.01	1.02	1.03	1.04
i	3.10	3.12	3.14	3.18	3.24

    where t is measured in seconds, i in amperes, the inductance L is a constant .98 henries, and the resistance is .142 ohms. Approximate the voltage $\mathscr{E}$ at the values $t = 1, 1.01, 1.02, 1.03$, and 1.04, using the appropriate three-point formulas.

9. With $x_1 = x_0 + h$ and $x_2 = x_0 + 3h$, use Eq. (4.4) to derive an approximation to $f'(x_0)$. Using this formula, approximate $f'(1.8)$ for $f(x) = e^x$, and compare your results to those in Exercise 1.

10. For the formula

    $$f'(x_0) = \frac{f(x_0 + h) - f(x_0)}{h} - \frac{h}{2}f''(\xi_0),$$

    analyze the rounding errors as in Example 3. Find an optimal $h > 0$ for the function given in Example 2.

11. Use the Taylor polynomial of degree one expanded about $x = x_0$ to obtain the result given in Eq. (4.1).

12. Consider the function

    $$e(h) = \frac{\varepsilon}{h} + \frac{h^2}{6}M,$$

    where M is a bound for

    $$\left|\frac{d^3(\sin x)}{dx^3}\right| \quad \text{on } [.800, 1.00].$$

    Show that $e(h)$ has a minimum at $\sqrt[3]{3\varepsilon/M}$.

13. To construct a cubic spline with clamped boundary conditions, we need values for f' at both ends of the interval.

   a) Repeat Exercise 6 of Section 3.6, using the clamped boundary conditions. Approximate $f'(x_0)$ using Eq. (4.9) and $f'(x_3)$ using (4.11). Compare your results to Exercise 6, Section 3.6.

   b) Repeat Exercise 9 of Section 3.6, using the clamped boundary conditions. Approximate $f'(x_0)$ and $f'(x_{10})$ using Eq. (4.23).

14. a) Using the derivative of the cubic spline constructed in Exercise 6 of Section 3.6, approximate $f'(1.03)$.

   b) Using the derivative of the cubic constructed in Exercise 13(a), approximate $f'(1.03)$ and compare the answer to the answer in part (a).

4.2 Higher Derivatives and Extrapolation

The methods presented in Section 4.1 can be extended to find approximations to higher derivatives of a function using only tabulated values of the function at various points. The methods involved become algebraically tedious, so only a representative procedure will be presented here.

Expanding a function f in a third-degree Taylor polynomial about a point x_0 and evaluating at $x_0 + h$ and $x_0 - h$, we obtain

$$(4.24) \quad f(x_0 + h) = f(x_0) + f'(x_0)h + \tfrac{1}{2}f''(x_0)h^2 + \tfrac{1}{6}f'''(x_0)h^3 + \tfrac{1}{24}f^{(4)}(\xi_1)h^4$$

and

$$(4.25) \quad f(x_0 - h) = f(x_0) - f'(x_0)h + \tfrac{1}{2}f''(x_0)h^2 - \tfrac{1}{6}f'''(x_0)h^3 + \tfrac{1}{24}f^{(4)}(\xi_{-1})h^4,$$

where $x_0 - h < \xi_{-1} < x_0 < \xi_1 < x_0 + h$. Instead of subtracting to eliminate the higher derivative terms, as was done in Section 4.1, we add Eq. (4.24) and (4.25) to obtain

$$(4.26) \quad f(x_0 + h) + f(x_0 - h) = 2f(x_0) + f''(x_0)h^2 \\ + \tfrac{1}{24}[f^{(4)}(\xi_1) + f^{(4)}(\xi_{-1})]h^4$$

or

$$(4.27) \quad f''(x_0) = \frac{1}{h^2}\left[f(x_0 - h) - 2f(x_0) + f(x_0 + h)\right] \\ - \frac{h^2}{24}\left[f^{(4)}(\xi_1) + f^{(4)}(\xi_{-1})\right].$$

If $f^{(4)}$ is continuous on $[x_0 - h, x_0 + h]$, the Intermediate-Value Theorem (Theorem 1.12, p. 6) permits this to be rewritten as

$$(4.28) \quad f''(x_0) = \frac{1}{h^2}\left[f(x_0 - h) - 2f(x_0) + f(x_0 + h)\right] - \frac{h^2}{12}f^{(4)}(\xi)$$

for some $x_0 - h < \xi < x_0 + h$.

This formula has an error term involving h^2; so in actual practice it would likely be necessary to find a procedure involving a higher-order term for improved accuracy of the approximation. For this purpose, we will introduce a new and important procedure. The technique is known as **Richardson's extrapolation procedure** and is useful in many other situations as well.

Suppose that f had been expanded in a fifth-degree Taylor polynomial, instead of one of degree three. It can be easily shown (see Exercise 10) that an equation similar to (4.26) is obtained with

(4.29)

$$\frac{1}{h^2}[f(x_0 + h) - 2f(x_0) + f(x_0 - h)] = f''(x_0) + \frac{h^2}{12} f^{(4)}(x_0) + \frac{h^4}{360} f^{(6)}(\xi)$$

where, as before, $x_0 - h < \xi < x_0 + h$.

If h is replaced in Eq. (4.29) by qh, where q is any value other than ± 1, then the two equations (4.29) and

(4.30)
$$\frac{1}{q^2 h^2}[f(x_0 + qh) - 2f(x_0) + f(x_0 - qh)]$$

$$= f''(x_0) + \frac{q^2 h^2}{12} f^{(4)}(x_0) + \frac{q^4 h^4}{360} f^{(6)}(\hat{\xi})$$

can be used to eliminate the term involving h^2:

$$\frac{q^2}{h^2(q^2 - 1)}[f(x_0 + h) - 2f(x_0) + f(x_0 - h)]$$

$$- \frac{1}{(q^2 - 1)q^2 h^2}[f(x_0 + qh) - 2f(x_0) + f(x_0 - qh)]$$

$$= f''(x_0) - \frac{q^2 h^4}{(1 - q^2)360}[q^2 f^{(6)}(\hat{\xi}) - f^{(6)}(\xi)].$$

Solving this equation for $f''(x_0)$ produces

(4.31) $f''(x_0) = \dfrac{1}{h^2 q^2(q^2 - 1)}[-f(x_0 + qh) + q^4 f(x_0 + h)$

$$- (2q^4 - 2)f(x_0) + q^4 f(x_0 - h) - f(x_0 - qh)] + O(h^4).$$

In this way an approximation to $f''(x_0)$ is obtained with truncation error $O(h^4)$.

EXAMPLE 1   To find an approximation for the second derivative at 3.15 of the function whose values are given in the table accompanying Example 2 of Section 4.1, it would be necessary to take $q = \pm 2$. These are the only values of q for which the given data suffices. With $q = 2$, Eq. (4.31) assumes the form

$$f''(3.15) = \frac{1}{(.1)^2 2^2(2^2 - 1)}[-f(3.35) + 2^4 f(3.25) - (2 \cdot 2^4 - 2)f(3.15)$$

$$+ 2^4 f(3.05) - f(2.95)] + \frac{2^2(.1)^4}{360} f^{(6)}(\xi).$$

This implies

$$f''(3.15) \approx 12.9.$$

In actuality, this value is exact since, $f^{(6)}(\xi) = 0$ for each value of ξ. $\qquad\square$

An example of another procedure using Richardson's extrapolation technique is given in Exercise 11. In that exercise, two three-point formulas for approximating the first derivative are combined to produce a five-point formula.

Exercise Set 4.2

1. Let $f(x) = \cos \pi x$. Use Eq. (4.28) and the values of $f(x)$ at $x = .25, .5$, and $.75$ to approximate $f''(.5)$. Compare this result to the exact value and to the approximation found in Exercise 1 of Section 3.6. Explain why this method is particularly accurate for this problem. Find a bound for the error.

2. Let $f(x) = e^{-x}$. Use Eq. (4.28) and the values of $f(x)$ at $x = .25, .5$, and $.75$ to approximate $f''(.5)$. Compare this result to the exact value and to the approximation found in Exercise 2 of Section 3.6. Find a bound for the error.

3. For $f(x) = e^x$, compute an approximation to $f''(1.8)$, using Eq. (4.28) with $h = .01$. Compare this result with the value of $e^{1.8}$.

4. Let $f(x) = 3xe^x - \cos x$. Using the data below and Eq. (4.28), approximate $f''(1.3)$ with $h = .1$ and $h = .01$.

x	1.20	1.29	1.30	1.31	1.40
$f(x)$	11.59006	13.78176	14.04276	14.30741	16.86187

Compare your results to $f''(1.3)$

5. Repeat Exercise 3, using instead the Richardson extrapolation formula given in Eq. (4.31) with $q = 2$.

6. By expanding the function f in a fourth-degree Taylor polynomial about x_0 and evaluating at $x_0 \pm h$ and $x_0 \pm 2h$, derive a method for approximating $f'''(x_0)$, whose error term is of order h^2.

7. For $f(x) = e^x$ compute an approximation to $f'''(1.8)$, using the result obtained in Exercise 6 and $h = .01$. Compare this result with the value of $e^{1.8}$.

8. Use the Richardson extrapolation procedure on the formula obtained in Exercise 6 to develop a formula of order h^4 for approximating $f'''(x_0)$.

9. Repeat Exercise 7, using instead the Richardson extrapolation formula obtained in Exercise 8.

10. Derive formula (4.29), using the fifth-degree Taylor polynomial.

11. Using formula (4.13), the Taylor polynomial of degree five, and Richardson's extrapolation, derive formula (4.22).

4.3 Elements of Numerical Integration

The need often arises for evaluating the definite integral of a function that has no explicit antiderivative or whose antiderivative has values that are not easily obtained. The basic method involved in approximating $\int_a^b f(x)\,dx$ is called **numerical quadrature** and uses a sum of the type

$$\sum_{i=0}^{n} a_i f(x_i)$$

to approximate $\int_a^b f(x)\,dx$.

The methods of quadrature we will discuss in this section are based on the interpolation polynomials given in Chapter 3. To proceed, we first select a set of distinct nodes $\{x_0, \ldots, x_n\}$ from an interval containing $[a, b]$. If P_n is the Lagrange interpolating polynomial

$$P_n(x) = \sum_{i=0}^{n} f(x_i)L_i(x),$$

we integrate P_n over $[a, b]$ to obtain the quadrature formula

$$I_{n+1}(f) = \int_a^b \sum_{i=0}^{n} f(x_i)L_i(x)\,dx = \sum_{i=0}^{n} a_i f(x_i),$$

where $\qquad a_i = \int_a^b L_i(x)\,dx, \qquad$ for each $i = 0, 1, \ldots, n.$

Before discussing the general situation of quadrature formulas, we will consider formulas produced by using first and second degree Lagrange polynomials with equally spaced nodes. The formulas are the **trapezoidal rule** and **Simpson's rule**, formulas often discussed in calculus courses.

To derive the trapezoidal rule for approximating $\int_a^b f(x)\,dx$, let $x_0 = a$, $x_1 = b, h = b - a$ and use the Lagrange polynomial:

$$P_1(x) = \frac{(x - x_1)}{(x_0 - x_1)}\,f(x_0) + \frac{(x - x_0)}{(x_1 - x_0)}\,f(x_1).$$

Thus, $\qquad \int_a^b f(x)\,dx \approx \int_{x_0}^{x_1} \left[\frac{(x - x_1)}{(x_0 - x_1)}\,f(x_0) + \frac{(x - x_0)}{(x_1 - x_0)}f(x_1) \right] dx$

$$= \frac{x_1 - x_0}{2}\,[f(x_0) + f(x_1)]$$

or:

(Trapezoidal Rule)

$$\int_a^b f(x)\,dx \approx \frac{h}{2}\,[f(x_0) + f(x_1)].$$

The reason for calling this formula the trapezoidal rule is evident from Fig. 4.2.

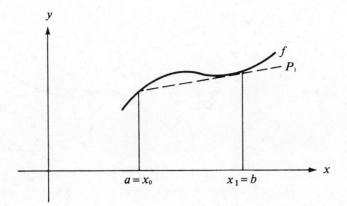

FIGURE 4.2

To derive Simpson's rule, let $x_0 = a$, $x_2 = b$, and $x_1 = a + h$, where $h = (b - a)/2$. The second degree Lagrange polynomial with nodes x_0, x_1, and x_2 is

$$P_2(x) = \frac{(x - x_1)(x - x_2)}{(x_0 - x_1)(x_0 - x_2)} f(x_0) + \frac{(x - x_0)(x - x_2)}{(x_1 - x_0)(x_1 - x_2)} f(x_1)$$

$$+ \frac{(x - x_0)(x - x_1)}{(x_2 - x_0)(x_2 - x_1)} f(x_2).$$

Therefore

$$\int_a^b f(x)\, dx \approx \int_{x_0}^{x_2} \left[\frac{(x - x_1)(x - x_2)}{(x_0 - x_1)(x_0 - x_2)} f(x_0) + \frac{(x - x_0)(x - x_2)}{(x_1 - x_0)(x_1 - x_2)} f(x_1) \right.$$

$$\left. + \frac{(x - x_0)(x - x_1)}{(x_2 - x_0)(x_2 - x_1)} f(x_2) \right] dx$$

$$= \frac{f(x_0)}{2h^2} \int_{x_0}^{x_2} (x - x_1)(x - x_2)\, dx - \frac{f(x_1)}{h^2} \int_{x_0}^{x_2} (x - x_0)(x - x_2)\, dx$$

$$+ \frac{f(x_2)}{2h^2} \int_{x_0}^{x_2} (x - x_0)(x - x_1)\, dx.$$

Evaluating these integrals produces the following:

(Simpson's Rule)

$$\int_a^b f(x)\, dx \approx \frac{h}{3} \left[f(x_0) + 4f(x_1) + f(x_2) \right] \qquad \text{(See Fig. 4.3)}.$$

EXAMPLE 1   The trapezoidal rule for a function f on the interval $[0, 2]$ is

$$\int_0^2 f(x)\, dx \approx f(0) + f(2),$$

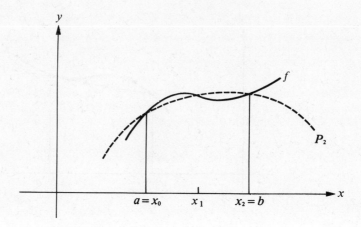

FIGURE 4.3

while Simpson's rule for f on $[0, 2]$ is

$$\int_0^2 f(x)\,dx \approx \tfrac{1}{3}[f(0) + 4f(1) + f(2)].$$

The results for some elementary functions are summarized in Table 4.5.

TABLE 4.5	$f(x) = 1$	$f(x) = x$	$f(x) = x^2$	$f(x) = x^3$	$f(x) = x^4$	$f(x) = e^x$
Exact value	2	2	2.67	4	6.40	6.389
Trapezoidal	2	2	4	8	16	8.389
Simpson's	2	2	2.67	4	6.67	6.421

□

Error formulas for the trapezoidal rule and Simpson's rule can be derived by using the error formulas for the associated Lagrange polynomials. For the first degree Lagrange polynomial P_1, Theorem 3.3, p. 83 implies that

$$f(x) = P_1(x) + \frac{f''(\xi(x))}{2}(x - x_0)(x - x_1)$$

for some $\xi(x) \in (x_0, x_1)$. Thus,

$$\int_a^b f(x)\,dx = \int_{x_0}^{x_1} P_1(x)\,dx + \int_{x_0}^{x_1} \frac{f''(\xi(x))}{2}(x - x_0)(x - x_1)\,dx.$$

The term $(x - x_0)(x - x_1)$ does not change sign in the interval $[x_0, x_1]$, so the Mean Value Theorem for Integrals (Theorem 1.10, p. 5) can be applied to deduce that the approximation error associated with the trapezoidal rule is

$$(4.32)\quad E(\text{Trapezoidal rule}) = \frac{f''(\xi)}{2}\int_{x_0}^{x_1}(x - x_0)(x - x_1)\,dx = -\frac{h^3}{12}f''(\xi),$$

for some $\xi \in (x_0, x_1) = (a, b)$.

While an error formula for Simpson's rule can be derived in a similar manner, a higher order error formula can be derived by using an alternative procedure (See Exercise 15). This procedure is based on the fact that Simpson's rule produces exact results when used as an approximation method for any polynomial of degree three or less. In fact, the standard derivation of quadrature error formulas is based on determining the class of polynomials for which these formulas produce exact results. The following definition is used to facilitate the discussion of this derivation.

Definition 4.1 The **degree of accuracy**, or **precision**, of a quadrature formula is the positive integer n such that $E(P_k) = 0$ for all polynomials P_k of degree less than or equal to n, but for which $E(P_{n+1}) \neq 0$ for some polynomial of degree $(n + 1)$.

Integration and summation are linear operations; that is,

$$\int_a^b (\alpha f(x) + \beta g(x)) \, dx = \alpha \int_a^b f(x) \, dx + \beta \int_a^b g(x) \, dx$$

and

$$\sum_{i=0}^n (\alpha f(x_i) + \beta g(x_i)) = \alpha \sum_{i=0}^n f(x_i) + \beta \sum_{i=0}^n g(x_i),$$

for each pair of integrable functions f and g and each pair of real constants α and β. This implies (see Exercise 12) that the degree of precision of a quadrature formula is n if and only if $E(x^k) = 0$, for all k, $k = 0, 1, \ldots, n$ but $E(x^{n+1}) \neq 0$.

Considering the results in Table 4.5, we see that the trapezoidal rule has degree of precision 1 and Simpson's rule has degree of precision 3.

There are two types of quadrature formulas that are commonly applied, both are known as Newton–Cotes formulas and both employ equally spaced nodes.

The $n + 1$ point **closed Newton–Cotes formula** uses nodes $x_i = x_0 + ih$, for $i = 0, 1, \ldots, n$ where $x_0 = a$, $x_n = b$ and $h = (b - a)/n$. The formula assumes the form

$$\int_a^b f(x) \, dx \approx \sum_{i=0}^n a_i f(x_i)$$

where

$$a_i = \int_{x_0}^{x_n} L_i(x) \, dx = \int_{x_0}^{x_n} \prod_{\substack{j=0 \\ j \neq i}}^n \frac{(x - x_j)}{(x_i - x_j)} \, dx.$$

To evaluate this integral we use the variable substitution $t = (x - x_0)/h$. Then

$$a_i = h \int_0^n \prod_{\substack{j=0 \\ j \neq i}}^n \frac{(t - j)}{(i - j)} \, dt$$

$$= h \int_0^n \frac{(t)(t - 1) \cdots (t - i + 1)(t - i - 1) \cdots (t - n)}{i(i - 1) \cdots 1 \cdot (-1) \cdots (i - n)} \, dt$$

$$= \frac{h}{i!} \int_0^n \frac{t(t - 1) \cdots (t - i + 1)(t - i - 1) \cdots (t - n)}{-1(-2) \cdots (-(n - i))} \, dt;$$

so

$$a_i = \frac{(-1)^{n-i} h}{i!(n - i)!} \int_0^n t(t - 1) \cdots (t - i + 1)(t - i - 1) \cdots (t - n) \, dt.$$

The following theorem details the error analysis associated with the closed Newton–Cotes formulas. For a proof of this theorem, see Isaacson and Keller [52], page 313.

Theorem 4.2   Suppose that $\sum_{i=0}^{n} a_i f(x_i)$ denotes the $n + 1$ point closed Newton–Cotes formula with $x_0 = a$, $x_n = b$, and $h = (b - a)/n$. There exists $\xi \in [a, b]$ for which:

$$(4.33) \quad \int_a^b f(x)\, dx = \sum_{i=0}^{n} a_i f(x_i) + \frac{h^{n+3} f^{(n+2)}(\xi)}{(n+2)!} \int_0^n t^2(t-1)\cdots(t-n)\, dt$$

if n is even and $f \in C^{n+2}[a, b]$, and

$$(4.34) \quad \int_a^b f(x)\, dx = \sum_{i=0}^{n} a_i f(x_i) + \frac{h^{n+2} f^{(n+1)}(\xi)}{(n+1)!} \int_0^n t(t-1)\cdots(t-n)\, dt$$

if n is odd and $f \in C^{n+1}[a, b]$.

Note that when n is an even integer the degree of precision is $n + 1$ although the interpolation polynomial is of degree at most n. In case n is odd, the second part of the theorem shows that the degree of precision is only n. Consequently, if n is even and more nodes are to be added to increase precision, no accuracy is gained by adding only one node; so *nodes should be added in multiples of two*.

Some of the common closed Newton–Cotes formulas with their error terms are listed below:

$n = 1$ **(Trapezoidal Rule):**

$$(4.35) \quad \int_{x_0}^{x_1} f(x)\, dx = \frac{h}{2} [f(x_0) + f(x_1)] - \frac{h^3}{12} f^{(2)}(\xi) \qquad \text{where } x_0 < \xi < x_1;$$

$n = 2$ **(Simpson's Rule):**

$$(4.36) \quad \int_{x_0}^{x_2} f(x)\, dx = \frac{h}{3} [f(x_0) + 4f(x_1) + f(x_2)] - \frac{h^5}{90} f^{(4)}(\xi)$$

$$\text{where } x_0 < \xi < x_2;$$

$n = 3$ **(Simpson's Three-Eighths Rule):**

$$(4.37) \quad \int_{x_0}^{x_3} f(x)\, dx = \frac{3h}{8} [f(x_0) + 3f(x_1) + 3f(x_2) + f(x_3)] - \frac{3h^5}{80} f^{(4)}(\xi)$$

$$\text{where } x_0 < \xi < x_3;$$

$n = 4$:

$$(4.38) \quad \int_{x_0}^{x_4} f(x)\, dx = \frac{2h}{45} [7f(x_0) + 32f(x_1) + 12f(x_2) + 32f(x_3)$$

$$+ 7f(x_4)] - \frac{8h^7}{945} f^{(6)}(\xi) \qquad \text{where } x_0 < \xi < x_4.$$

In the **open Newton–Cotes formulas**, the nodes $x_i = x_0 + ih$ are used for each $i = 0, 1, \ldots, n$, where $h = (b - a)/(n + 2)$ and $x_0 = a + h$. This implies that $x_n = b - h$; so if we label the endpoints by setting $x_{-1} = a$ and $x_{n+1} = b$, the formulas become

$$\int_a^b f(x)\, dx = \int_{x_{-1}}^{x_{n+1}} f(x)\, dx \approx \sum_{i=0}^n a_i f(x_i)$$

where again

$$a_i = \int_a^b L_i(x)\, dx.$$

The following theorem is analogous to Theorem 4.2 and its proof is contained in Isaacson and Keller [52], page 314.

Theorem 4.3 Suppose that $\sum_{i=0}^n a_i f(x_i)$ denotes the $n + 1$ point open Newton–Cotes formula with $x_{-1} = a$, $x_{n+1} = b$, and $h = (b - a)/(n + 2)$. There exists $\xi \in (a, b)$ for which;

$$(4.39) \quad \int_a^b f(x)\, dx = \sum_{i=0}^n a_i f(x_i) + \frac{h^{n+3} f^{(n+2)}(\xi)}{(n + 2)!} \int_{-1}^{n+1} t^2(t - 1) \cdots (t - n)\, dt$$

if n is even and $f \in C^{n+2}[a, b]$, and

$$(4.40) \quad \int_a^b f(x)\, dx = \sum_{i=0}^n a_i f(x_i) + \frac{h^{n+2} f^{(n+1)}(\xi)}{(n + 1)!} \int_{-1}^{n+1} t(t - 1) \cdots (t - n)\, dt$$

if n is odd and $f \in C^{n+1}[a, b]$.

Some of the common open Newton–Cotes formulas with their error terms are listed below:

$n = 0$ **(Midpoint Rule):**

$$(4.41) \quad \int_{x_{-1}}^{x_1} f(x)\, dx = 2hf(x_0) + \frac{h^3}{3} f''(\xi) \qquad \text{where } x_{-1} < \xi < x_1;$$

$n = 1$:

$$(4.42) \quad \int_{x_{-1}}^{x_2} f(x)\, dx = \frac{3h}{2} [f(x_0) + f(x_1)] + \frac{3h^3}{4} f''(\xi) \qquad \text{where } x_{-1} < \xi < x_2;$$

$n = 2$:

$$(4.43) \quad \int_{x_{-1}}^{x_3} f(x)\, dx = \frac{4h}{3} [2f(x_0) - f(x_1) + 2f(x_2)]$$

$$+ \frac{14h^5}{45} f^{(4)}(\xi) \qquad \text{where } x_{-1} < \xi < x_3;$$

$n = 3$:

$$(4.44) \quad \int_{x_{-1}}^{x_4} f(x)\, dx = \frac{5h}{24}\left[11f(x_0) + f(x_1) + f(x_2) + 11f(x_3)\right] + \frac{95}{144} h^5 f^{(4)}(\xi),$$

$$\text{where } x_{-1} < \xi < x_4.$$

EXAMPLE 2   Using the closed and open Newton–Cotes formulas listed as (4.35)–(4.38) and (4.41)–(4.44) to approximate $\int_0^{\pi/4} \sin x\, dx = 1 - (\sqrt{2}/2)$ gives the results in Table 4.6.

TABLE 4.6

n	0	1	2	3	4
Closed formulas		.27768018	.29293264	.29291070	.29289318
Error		.01521303	.00003942	.00001748	.00000004
Open formulas	.30055887	.29798754	.29351798	.29286923	
Error	.00766565	.00509432	.00062477	.00002399	

As illustrated in the preceding example, the closed formulas generally produce results superior to the open formulas of the same order. Consequently, the closed formulas are more frequently used in practice. The open formulas are primarily used for the numerical solution of ordinary differential equations.

To analyze the rounding errors associated with numerical quadrature formulas, suppose that

$$(4.45) \qquad\qquad \int_a^b f(x)\, dx \approx \sum_{i=0}^{n} a_i f(x_i),$$

is one of the Newton–Cotes formulas either of open or closed type. If $a = x_0 + qh$ and $b = x_0 + ph$, where $q = 0$, $p = n$ for closed formulas and $q = -1$, $p = n + 1$ for open formulas, a_i can be expressed as

$$a_i = \frac{(-1)^{n-i}h}{i!(n-i)!} \int_q^p t(t-1)\cdots(t-i+1)(t-i-1)\cdots(t-n)\, dt,$$

or as $a_i = hm_i$,

where $m_i = \dfrac{(-1)^{n-i}}{i!(n-i)!} \displaystyle\int_q^p t(t-1)\cdots(t-i+1)(t-i-1)\cdots(t-n)\, dt$

does not depend on the step size h. Hence, quadrature formula (4.45) becomes

$$\int_a^b f(x)\, dx \approx h\left(\sum_{i=0}^{n} m_i f(x_i)\right).$$

Suppose that $f(x_i) = \tilde{f}(x_i) + e_i$, for each $i = 0, 1, \ldots, n$, where e_i denotes the round-off error associated with using $\tilde{f}(x_i)$ to approximate $f(x_i)$. Then

$$\int_a^b f(x)\, dx \approx h\left(\sum_{i=0}^{n} m_i \tilde{f}(x_i)\right) + h\left(\sum_{i=0}^{n} m_i e_i\right),$$

where the term

$$r_{n+1} = h\left(\sum_{i=0}^{n} m_i e_i\right)$$

is the accumulated round-off error. If $|e_i| < \varepsilon$ for each $i = 0, 1, \ldots, n$, then

(4.46) $$|r_{n+1}| \le h\varepsilon\left(\sum_{i=0}^{n} |m_i|\right).$$

In the special case that all $m_j \ge 0$, which occurs for all the low-order closed formulas and some of the open formulas,

$$|r_{n+1}| \le h\varepsilon\left(\sum_{i=0}^{n} m_i\right).$$

Since all formulas are exact for $f(x) = 1$, it follows that

$$\int_a^b dx = h\left(\sum_{i=0}^{n} m_i\right) \quad \text{and} \quad h\left(\sum_{i=0}^{n} m_i\right) = b - a;$$

so $$|r_{n+1}| \le \varepsilon(b - a).$$

Note that this bound is independent of h and, consequently, the numerical procedure is stable as $h \to 0$. Recall that this was not true in the case of the numerical differentiation procedures studied in Section 4.1.

When some of the m_i are negative, the bound

$$|r_{n+1}| \le h\sum_{i=0}^{n} |m_i|$$

is not as good, but it can again be shown that the procedure is stable.

Exercise Set 4.3

1. Approximate $\int_0^1 x^{1/3}\,dx$, using the trapezoidal and Simpson's rules. Compare the approximations to the actual value. Find a maximum bound for the error in each case, if possible.

2. Use the trapezoidal and Simpson's rules to approximate $\int_1^2 \ln x\,dx$. Compare the approximations to the actual value. Find a maximum bound for the error in each case, if possible.

3. Use the trapezoidal and Simpson's rules to approximate $\int_0^{\pi/2} \sin^2 x\,dx$. Compare the approximations with the actual result. Find a maximum bound for the error in each case, if possible.

4. Use the table below to find an approximation to $\int_{1.1}^{1.5} e^x\,dx$, using:

   a) the trapezoidal rule with $x_0 = 1.1$ and $x_1 = 1.5$;
   b) Simpson's rule with $x_0 = 1.1$, $x_1 = 1.3$, and $x_2 = 1.5$.

x	e^x
1.1	3.0042
1.3	3.6693
1.5	4.4817

5. Use the Newton–Cotes closed formula for $n = 3$ and the open formula for $n = 2$ to approximate $\int_1^3 e^{-x/2}\,dx$. Find a bound for the error in each case and compare the approximations obtained with the actual value .7668010.

6. Repeat Exercise 5, using the closed formula for $n = 4$ and open formula for $n = 3$.

7. Approximate the following integrals using formulas (4.36), (4.37), (4.38), (4.42), (4.43), and (4.44). Are the accuracies of the approximations consistent with the error formulas? Which of parts (d) and (e) gives the better approximation?

    a) $\displaystyle\int_0^{.1} \sqrt{1 + x}\,dx$

    b) $\displaystyle\int_0^{\pi/2} \sin^2 x\,dx$

    c) $\displaystyle\int_{1.1}^{1.5} e^x\,dx$

    d) $\displaystyle\int_1^{10} \frac{1}{x}\,dx$

    e) $\displaystyle\int_1^{5.5} \frac{1}{x}\,dx + \int_{5.5}^{10} \frac{1}{x}\,dx$

    f) $\displaystyle\int_0^1 x^{1/3}\,dx$

8. Given the function f at the values tabulated below:

x	1.8	2.0	2.2	2.4	2.6
$f(x)$	3.12014	4.42569	6.04241	8.03014	10.46675

    Approximate $\int_{1.8}^{2.6} f(x)\,dx$ using all the quadrature formulas of this section that can be applied.

9. Suppose the data of Exercise 8 has rounding errors given by the following table:

x	1.8	2.0	2.2	2.4	2.6
Error in $f(x)$	2×10^{-6}	-2×10^{-6}	$-.9 \times 10^{-6}$	$-.9 \times 10^{-6}$	2×10^{-6}

    Calculate the errors due to rounding in Exercise 8.

10. Use each of the methods given by Eq. (4.35)–(4.38) and (4.41)–(4.44) to obtain an approximation for $\int_0^{1.5} (1 + x)^{-1}\,dx$, and compare the results obtained to the exact value .9162907.

11. Perform the integration involved in Eq. (4.32) to show that the error in the trapezoidal rule is $-f''(\mu)h^3/12$.

12. Prove the statement following Definition 4.1; that is, show that a quadrature formula has degree of precision n precisely when $E(x^k) = 0$ for all $k = 0, 1, \ldots, n$ and $E(x^{n+1}) \neq 0$.

13. Derive Simpson's three-eighths rule, Eq. (4.37), with error term, by the use of Theorem 4.2.

14. Derive Eq. (4.42) with error term by the use of Theorem 4.3.

15. Derive Simpson's rule with error term by using

$$\int_{x_0}^{x_2} f(x)\,dx = a_0\,f(x_0) + a_1\,f(x_1) + a_2\,f(x_2) + k f^{(4)}(\xi).$$

    Find a_0, a_1, and a_2 from the fact that Simpson's rule is exact for $f(x) = x^n$ when $n = 1, 2, 3$. Find k by applying the integration formula with $f(x) = x^4$.

4.4) Composite Numerical Integration

The Newton–Cotes formulas are generally unsuitable for use over large integration intervals. High-degree formulas would be required for use over such intervals and the values of the coefficients in these formulas are difficult to obtain. More importantly, the Newton–Cotes formulas are based on interpolatory polynomials that use equally spaced nodes, a procedure we found, in Section 3.6, to be inaccurate over large intervals because of the oscillatory nature of high-degree polynomials. In this section we will discuss a piecewise approach to numerical integration that utilizes the low order Newton–Cotes formulas. These procedures are among the techniques most often applied in practice.

Consider finding an approximation to $\int_0^4 e^x \, dx$. The interval $[0, 4]$ would be considered a fairly large interval; for the reasons listed above, a polynomial of high degree would likely give an inaccurate approximation to this integral and be difficult to evaluate. Using a polynomial of low degree will not solve the problem, however. Suppose Simpson's rule is used with $h = 2$; then

$$\int_0^4 e^x \, dx \approx \tfrac{2}{3}(e^0 + 4e^2 + e^4) = 56.76958.$$

Since the exact answer in this case is $e^4 - e^0 = 53.59815$, the error of -3.17143 is far larger than would generally be regarded as acceptable.

To apply a piecewise technique to this problem, divide $[0, 4]$ into $[0, 2]$ and $[2, 4]$ and use Simpson's rule, Eq. (4.36), twice with $h = 1$:

$$\int_0^4 e^x \, dx = \int_0^2 e^x \, dx + \int_2^4 e^x \, dx$$
$$\approx \tfrac{1}{3}[e^0 + 4e + e^2] + \tfrac{1}{3}[e^2 + 4e^3 + e^4]$$
$$= \tfrac{1}{3}[e^0 + 4e + 2e^2 + 4e^3 + e^4] = 53.86385.$$

The actual error has been reduced to $-.26570$. Encouraged by our results, we subdivide the intervals $[0, 2]$ and $[2, 4]$ and use Simpson's rule with $h = \tfrac{1}{2}$, giving

$$\int_0^4 e^x \, dx = \int_0^1 e^x \, dx + \int_1^2 e^x \, dx + \int_2^3 e^x \, dx + \int_3^4 e^x \, dx$$
$$\approx \tfrac{1}{6}[e^0 + 4e^{1/2} + e] + \tfrac{1}{6}[e + 4e^{3/2} + e^2]$$
$$+ \tfrac{1}{6}[e^2 + 4e^{5/2} + e^3] + \tfrac{1}{6}[e^3 + 4e^{7/2} + e^4]$$
$$= \tfrac{1}{6}[e^0 + 4e^{1/2} + 2e + 4e^{3/2} + 2e^2 + 4e^{5/2} + 2e^3 + 4e^{7/2} + e^4]$$
$$= 53.61622.$$

The error for this approximation is $-.01807$.

A generalization of this procedure is as follows: Subdivide the interval $[a, b]$ into $2m$ subintervals and use Simpson's rule on each pair of consecutive subintervals. (See Fig. 4.4.)

With $h = (b - a)/2m$ and $a = x_0 < x_1 < \cdots < x_{2m} = b$, where $x_j = x_0 + jh$ for each $j = 0, 1, \ldots, 2m$,

$$\int_a^b f(x) \, dx = \sum_{j=1}^m \int_{x_{2j-2}}^{x_{2j}} f(x) \, dx$$
$$= \sum_{j=1}^m \left\{ \frac{h}{3} [f(x_{2j-2}) + 4f(x_{2j-1}) + f(x_{2j})] - \frac{h^5}{90} f^{(4)}(\xi_j) \right\}$$

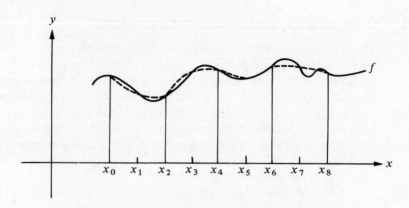

FIGURE 4.4

for some ξ_j with $x_{2j-2} < \xi_j < x_{2j}$. Using the fact that for each $j = 1, 2, \ldots, m - 1$, $f(x_{2j})$ appears in the term corresponding to the interval $[x_{2j-2}, x_{2j}]$ and also, in the term corresponding to the interval $[x_{2j}, x_{2j+2}]$, this reduces to

$$(4.47) \quad \int_a^b f(x)\, dx = \frac{h}{3}\left[f(x_0) + 2\sum_{j=1}^{m-1} f(x_{2j}) + 4\sum_{j=1}^{m} f(x_{2j-1}) + f(x_{2m}) \right]$$
$$- \frac{h^5}{90} \sum_{j=1}^{m} f^{(4)}(\xi_j).$$

Theorem 4.4 If $f \in C^4[a, b]$, there exists a $\mu \in [a, b]$ for which **Simpson's composite rule over 2m subintervals** of $[a, b]$ can be expressed with error term as

$$(4.48) \quad \int_a^b f(x)\, dx = \frac{h}{3}\left[f(a) + 2\cdot\sum_{j=1}^{m-1} f(x_{2j}) + 4\sum_{j=1}^{m} f(x_{2j-1}) + f(b) \right]$$
$$- \frac{(b - a)h^4}{180} f^{(4)}(\mu),$$

where $a = x_0 < x_1 < \cdots < x_{2m} = b$, $h = (b - a)/2m$, and $x_j = x_0 + jh$ for each $j = 0, 1, \ldots, 2m$.

Proof From Eq. (4.47) the error associated with this approximation is

$$E(f) = \frac{-h^5}{90} \sum_{j=1}^{m} f^{(4)}(\xi_j)$$

where $x_{2j-2} < \xi_j < x_{2j}$ for each $j = 1, 2, \ldots, m$. Since $f \in C^4[a, b]$,

$$\min_{x \in [a, b]} f^{(4)}(x) \le f^{(4)}(\xi_j) \le \max_{x \in [a, b]} f^{(4)}(x);$$

so

$$m \min_{x \in [a, b]} f^{(4)}(x) \le \sum_{j=1}^{m} f^{(4)}(\xi_j) \le m \max_{x \in [a, b]} f^{(4)}(x),$$

or

$$\min_{x \in [a, b]} f^{(4)}(x) \le \frac{1}{m} \sum_{j=1}^{m} f^{(4)}(\xi_j) \le \max_{x \in [a, b]} f^{(4)}(x).$$

By the Intermediate Value Theorem (Theorem 1.12, p. 6), there is a $\mu \in [a, b]$ such that

$$f^{(4)}(\mu) = \frac{1}{m} \sum_{j=1}^{m} f^{(4)}(\xi_j).$$

Thus, $$E(f) = \frac{-h^5}{90} m f^{(4)}(\mu).$$

Since $h = (b - a)/2m$,

$$E(f) = \frac{-h^4(b - a)}{180} f^{(4)}(\mu). \qquad \square$$

The following algorithm employs Simpson's composite rule on $2m$ subintervals. This is a quadrature algorithm often used in practice.

Simpson's Composite Algorithm 4.1

To approximate the integral $I = \int_a^b f(x)\,dx$:

INPUT endpoints a, b; positive integer m.

OUTPUT approximation XI to I.

Step 1 Set $h = (b - a)/(2m)$.

Step 2 Set $XI0 = f(a) + f(b)$;
$XI1 = 0$; (*Summation of $f(x_{2i-1})$.*)
$XI2 = 0$. (*Summation of $f(x_{2i})$.*)

Step 3 For $i = 1, \ldots, 2m - 1$ do Steps 4 and 5.

Step 4 Set $X = a + ih$.

Step 5 If i is even then set $XI2 = XI2 + f(X)$
else set $XI1 = XI1 + f(X)$.

Step 6 Set $XI = h(XI0 + 2 \cdot XI2 + 4 \cdot XI1)/3$.

Step 7 OUTPUT (XI);
STOP.

This approach can similarly be applied to any of the lower-order formulas. The extensions of the trapezoidal (see Fig. 4.5) and midpoint rules are given without proof in the following two theorems.

Theorem 4.5 Let $f \in C^2[a, b]$. With $h = (b - a)/m$ and $x_j = a + jh$ for each $j = 0, 1, \ldots, m$, the **trapezoidal rule for m subintervals** is

$$(4.49) \qquad \int_a^b f(x)\,dx = \frac{h}{2} \left[f(a) + f(b) + 2 \sum_{j=1}^{m-1} f(x_j) \right] - \frac{(b - a)h^2}{12} f''(\mu)$$

for some $\mu \in [a, b]$.

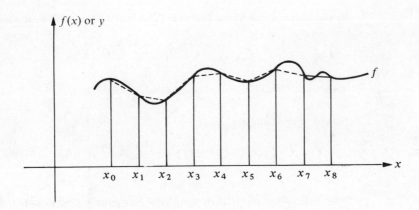

FIGURE 4.5

Theorem 4.6   Let $f \in C^2[a, b]$. With $h = (b - a)/(2m + 2)$ and $x_j = a + (j + 1)h$ for each $j = -1, 0, \ldots, 2m + 1$, **the midpoint rule for** $2m$ **subintervals is**

$$(4.50) \qquad \int_a^b f(x)\, dx = 2h \sum_{j=0}^m f(x_{2j}) + \frac{b - a}{6} h^2 f''(\mu)$$

for some $\mu \in [a, b]$.

EXAMPLE 1   Consider approximating $\int_0^\pi \sin x\, dx$ with an error of at most .00002, using Simpson's composite rule. Applying Eq. (4.47) to the integral $\int_0^\pi \sin x\, dx$:

$$\int_0^\pi \sin x\, dx = \frac{h}{3} \left[2 \sum_{j=1}^{m-1} \sin x_{2j} + 4 \sum_{j=1}^m \sin x_{2j-1} \right] - \frac{\pi h^4}{180} \sin \mu.$$

Since the truncation error is required to be less than .00002, the inequality

$$\left| \frac{\pi h^4}{180} \sin \mu \right| \leq \frac{\pi h^4}{180} = \frac{\pi^5}{2880 m^4} \leq .00002$$

is used to determine m and h. Completing these calculations gives $m \geq 9$. For $m = 10$, $h = \pi/20$, the formula obtained from (4.48) becomes

$$\int_0^\pi \sin x\, dx \approx \frac{\pi}{60} \left[2 \sum_{j=1}^9 \sin\left(\frac{j\pi}{10} \right) + 4 \sum_{j=1}^{10} \sin\left(\frac{(2j - 1)\pi}{20} \right) \right].$$

To be assured of this degree of accuracy using the composite trapezoidal rule requires that

$$\left| \frac{\pi h^2}{12} \sin \mu \right| \leq \frac{\pi h^2}{12} = \frac{\pi^3}{12 m^2} < .00002$$

or that $m \geq 360$. Since this is many more calculations than are needed for Simpson's rule, it is clear that it would be undesirable to use the trapezoidal rule on this problem. For comparison purposes, the trapezoidal rule with $m = 20$ and $h = \pi/20$ gives

$$\int_0^\pi \sin x\, dx \approx \frac{\pi}{40} \left[2 \sum_{j=1}^{19} \sin\left(\frac{j\pi}{20} \right) + \sin 0 + \sin \pi \right] = \frac{\pi}{40} \left[2 \sum_{j=1}^{19} \sin\left(\frac{j\pi}{20} \right) \right].$$

Simpson's rule gives an answer 2.00000679 and the trapezoidal rule gives an answer 1.9958860. The exact answer is 2; so Simpson's rule gave an answer well within the required error bound while the trapezoidal rule with $m = 20$ clearly did not.   □

Exercise Set 4.4

1. Use the extended trapezoidal rule with the indicated values of n to approximate the following definite integrals. Compare the approximations to the exact result.

   a) $\int_1^3 \frac{dx}{x}; n = 4$

   b) $\int_0^2 x^3 \, dx; n = 4$

   c) $\int_0^3 x\sqrt{1 + x^2} \, dx; n = 6$

   d) $\int_0^1 \sin \pi x \, dx; n = 6$

   e) $\int_0^{2\pi} x \sin x \, dx; n = 8$

   f) $\int_0^1 x^2 e^x \, dx; n = 8$

2. Use Simpson's composite rule to approximate the definite integrals given in Exercise 1.

3. Determine the values of m and h needed to approximate $\int_1^3 e^x \sin x \, dx$, with error less than 10^{-6}, using Simpson's composite rule. Determine the approximation.

4. Repeat Exercise 3, using the extended trapezoidal rule.

5. Repeat Exercise 3, using the extended midpoint rule.

6. Exercise 10 of Section 4.3 required that all the methods of that section be used to calculate an approximation to $\int_0^{1.5} (1 + x)^{-1} \, dx$. Use Algorithm 4.1 to find another approximation for this integral with $m = 5$. Compare the result obtained here with the values obtained in that exercise.

7. Repeat Exercise 6 using Theorem 4.5 with $m = 2$ instead of Algorithm 4.1.

8. Approximate $\int_1^{10} \ln x \, dx$ to within 10^{-4}, using Simpson's composite rule.

9. Let f be defined by

$$f(x) = \begin{cases} x^3 + 1, & 0 \le x \le .1, \\ 1.001 + .03(x - .1) + .3(x - .1)^2 + 2(x - .1)^3, & .1 \le x \le .2, \\ 1.009 + .15(x - .2) + .9(x - .2)^2 + 2(x - .2)^3, & .2 \le x \le .3. \end{cases}$$

   a) Investigate the continuity of the derivatives of f.
   b) Approximate $\int_0^3 f(x) \, dx$, using the extended trapezoidal rule with $m = 6$, and estimate the error using Eq. (4.49).
   c) Approximate $\int_0^3 f(x) \, dx$, using Simpson's composite rule with $m = 6$. Are the results more accurate than in part (b)?

10. In an electrical circuit containing an impressed voltage $\mathscr{E}$ and a capacitor with capacitance C, the relationship

$$\mathscr{E} = Ri + \frac{1}{C} \int_0^t i \, dt$$

holds where R is the resistance and i the current in the circuit. Suppose for a given circuit we have $R = .1$ ohms, $C = 1$ farad, and $i = \ln(2t^2 + 1)$ amps. Find the voltage after 2 seconds, using Simpson's composite rule with $h = .1$.

11. A particle of mass m moving through a fluid is subjected to a viscous resistance R, which is a function of the velocity v. The relationship between the resistance R, velocity v, and time t is given by the equation

$$t = \int_{v(t_0)}^{v(t)} \frac{m}{R(u)}\, du.$$

Suppose that $R(v) = -v\sqrt{v}$ for a particular fluid, where R is in newtons and v is in meters/second. If $m = 10$ kg and $v(0) = 10$ m/sec, approximate the time required for the particle to slow to $v = 5$ m/sec, using:

a) Simpson's composite rule with $h = .2$.
b) the extended trapezoidal rule with $h = .2$.
c) Compare these approximations to the actual value.

12. In order to simulate the thermal characteristics of disk brakes (see following figure), D. A. Secrist and R. W. Hornbeck [79] needed to approximate numerically the "area averaged lining temperature," T, of the brake pad from the equation

$$T = \frac{\int_{r_e}^{r_0} T(r) r \theta_p\, dr}{\int_{r_e}^{r_0} r \theta_p\, dr}$$

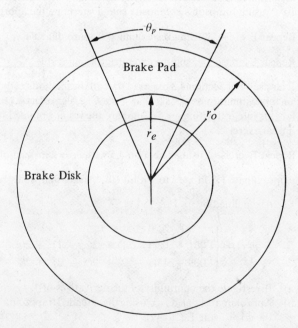

where r_e represents the radius at which the pad–disk contact begins, r_0 represents the outside radius of the pad–disk contact, θ_p represents the angle subtended by sector brake pads, and $T(r)$ is the temperature at each point of the pad, obtained numerically from analyzing the heat equation (see Section 11.3). If $r_e = .308$ ft, $r_0 = .478$ ft, $\theta_p = .7051$ radians, and the temperatures given in the following table have been calculated at the various points on the disk, find an approximation for T using Simpson's composite rule.

r (feet)	$T(r)(°F)$
.308	640
.325	794
.342	885
.359	943
.376	1034
.393	1064
.410	1114
.427	1152
.444	1204
.461	1222
.478	1239

13. Use Simpson's composite rule to find an approximation to within 10^{-4} to the value of the integral

$$\int_0^{48} \sqrt{1 + (\cos x)^2}\, dx.$$

14. Finite-Jump Discontinuity   Let f be defined by

$$f(x) = \begin{cases} x^2 - 1, & 0 \le x \le .5, \\ e^x \sin x, & .5 < x \le 1. \end{cases}$$

a) Is f continuous on $[0, 1]$?
b) Evaluate

$$\int_0^1 f(x)\, dx.$$

c) Approximate the integral in (b) using Simpson's composite rule with $h = .05$.
d) Approximate the integral using Eq. (4.49) with $h = .04$.
e) Approximate

$$\int_0^{.5} f(x)\, dx + \int_{.5}^1 f(x)\, dx$$

using Eq. (4.48) on $[0, .5]$ (with $f(.5) = -.75$) and on $[.5, 1]$ with $h = .05$, and compare your results to those in part (c).
f) Which of (c), (d), and (e) gives the best approximation? Can an error formula be applied in any of the three parts?

15. Infinite Discontinuities   The integral

$$\int_a^b \frac{dx}{(x - a)^p}$$

converges (or exists) if and only if $p < 1$. If the function f can be written in the form

$$f(x) = \frac{g(x)}{(x - a)^p},$$

where $g \in C[a, b]$, the integral

$$\int_a^b f(x)\, dx$$

also exists. If $g \in C^{n+1}[a, b]$ for some $n \geq 0$, the Taylor polynomial P of degree n given by

$$P(x) = \sum_{k=0}^{n} \frac{g^{(k)}(a)}{k!} (x - a)^k$$

can be constructed. The integral of f can then be approximated by

$$\int_a^b f(x)\, dx = \int_a^b \frac{g(x) - P(x)}{(x - a)^p}\, dx + \int_a^b \frac{P(x)}{(x - a)^p}\, dx,$$

where the function $(g(x) - P(x))/(x - a)^p$ has n derivatives at $x = a$. The integral

$$\int_a^b \frac{g(x) - P(x)}{(x - a)^p}\, dx$$

can be approximated by any of the quadrature formulas considered in Sections 4.3 and 4.4, while the integral

$$\int_a^b \frac{P(x)}{(x - a)^p}\, dx$$

can be explicitly evaluated.

a) Show that

$$\int_a^b \frac{P(x)}{(x - a)^p}\, dx = \sum_{k=0}^{n} \frac{g^{(k)}(a)}{k!(k + 1 - p)} (b - a)^{k+1-p}.$$

b) Approximate the integral

$$\int_0^1 \frac{e^x}{\sqrt{x}}\, dx,$$

using Simpson's composite rule with $h = .05$. Can you apply the error formula in (4.48)?

c) Approximate the integral

$$\int_0^1 \frac{e^{-x}}{\sqrt{1 - x}}\, dx$$

using the extended trapezoidal rule with $h = .05$.

16. **Infinite Limits of Integration** The integral $\int_a^{\infty} f(x)\, dx$, $a > 0$, if it exists, can often be approximated using a quadrature formula after the change in variable $t = x^{-1}$.

a) Show that

$$\int_a^{\infty} f(x)\, dx = \int_0^{1/a} t^{-2} f\left(\frac{1}{t}\right) dt.$$

b) Apply part (a) and the extended midpoint rule (4.50) with $h = .05$, to approximate

$$\int_1^{\infty} x^{-2} \sin x\, dx.$$

c) Apply part (a) to approximate

$$\int_0^{\infty} \sqrt{x} e^{-x}\, dx,$$

using the extended midpoint rule (4.50) with $h = .05$.

d)   Apply part (a) to approximate

$$\int_{-\infty}^{\infty} \frac{dx}{1 + x^2},$$

using the extended midpoint rule (4.50) with $h = .05$.

17.   Suppose a body of mass m is traveling vertically upward starting at the surface $x = R$ of the earth. If all resistance except gravity is neglected, then the escape velocity v is given by

$$v^2 = 2gR \int_{1}^{\infty} f(z) \, dz, \qquad \text{where } f(z) = z^{-2} \text{ and } z = \frac{x}{R},$$

and g is the gravitational field strength at the earth's surface. If $g = .00609$ miles/sec^2 and $R = 3960$ miles, approximate the escape velocity v using the extended midpoint rule with $h = .1$.

18.   Verify that the Intermediate Value Theorem (Theorem 1.12) can be applied to obtain the equation

$$f^{(4)}(\mu) = \frac{1}{m} \sum_{j=1}^{m} f^{(4)}(\xi_j),$$

needed in the proof of Theorem 4.4.

4.5   Adaptive Quadrature Methods

The composite formulas require the use of equally spaced nodes. For many problems this is no restriction, but it is not appropriate when integrating a function on an interval that contains both regions with large functional variation and regions with small functional variation. In this situation a smaller step size is needed for the large variation regions than for those with less variation if the approximation error is to be uniformly distributed. An efficient technique for this type of problem is one that can distinquish the amount of functional variation and adapt the step size to the varying requirements of the problem. Such methods are appropriately named **Adaptive Quadrature Methods**. The method we will discuss is based on Simpson's composite rule, but the technique is easily modified to the other composite procedures.

Suppose that we wish to approximate $\int_{a}^{b} f(x) \, dx$ to within a specified tolerance $\varepsilon > 0$. The first step in the procedure is to apply Simpson's rule with step size $h = (b - a)/2$. This results in

(4.51) $$\int_{a}^{b} f(x) \, dx = S(a, b) - \frac{h^5}{90} f^{(4)}(\mu), \quad \text{for some } \mu \text{ in } [a, b]$$

where $$S(a, b) = \frac{h}{3} [f(a) + 4f(a + h) + f(b)].$$

The next step is to determine a way to estimate the accuracy of our approximation, in particular, a way that does not require determining $f^{(4)}(\mu)$. To accomplish this, we first apply Simpson's composite rule to the problem with $m = 2$ and step size $(b - a)/4 = h/2$. Thus,

$$(4.52) \quad \int_a^b f(x)\, dx = \frac{h}{6}\left[f(a) + 4f\left(a + \frac{h}{2}\right) + 2f(a + h) + 4f\left(a + \frac{3h}{2}\right) + f(b) \right]$$

$$- \left(\frac{h}{2}\right)^4 \frac{(b - a)}{180} f^4(\tilde{\mu}),$$

for some $\tilde{\mu}$ in (a, b). To simplify notation, let

$$S\left(a, \frac{a + b}{2}\right) = \frac{h}{6}\left[f(a) + 4f\left(a + \frac{h}{2}\right) + f(a + h) \right]$$

and

$$S\left(\frac{a + b}{2}, b\right) = \frac{h}{6}\left[f(a + h) + 4f\left(a + \frac{3h}{2}\right) + f(b) \right].$$

Then Eq. (4.52) can be rewritten as

$$(4.53) \quad \int_a^b f(x)\, dx = S\left(a, \frac{a + b}{2}\right) + S\left(\frac{a + b}{2}, b\right) - \frac{1}{16}\left(\frac{h^5}{90}\right) f^{(4)}(\tilde{\mu}).$$

The error estimation is derived by assuming that $\mu = \tilde{\mu}$ or, more precisely, that $f^{(4)}(\mu) = f^{(4)}(\tilde{\mu})$. The success of the technique depends upon the accuracy of this assumption. If this is an accurate assumption, then equations (4.51) and (4.52) imply that

$$S\left(a, \frac{a + b}{2}\right) + S\left(\frac{a + b}{2}, b\right) - \frac{1}{16}\left(\frac{h^5}{90}\right) f^{(4)}(\mu) \approx S(a, b) - \frac{h^5}{90} f^{(4)}(\mu),$$

so

$$\frac{h^5}{90} f^{(4)}(\mu) \approx \frac{16}{15}\left[S(a, b) - S\left(a, \frac{a + b}{2}\right) - S\left(\frac{a + b}{2}, b\right) \right].$$

Using this estimate in Eq. (4.53) produces the error estimation

$$(4.54) \quad \left| \int_a^b f(x)\, dx - S\left(a, \frac{a + b}{2}\right) - S\left(\frac{a + b}{2}, b\right) \right|$$

$$\approx \frac{1}{15}\left| S(a, b) - S\left(a, \frac{a + b}{2}\right) - S\left(\frac{a + b}{2}, b\right) \right|.$$

If

$$(4.55) \quad \left| S(a, b) - S\left(a, \frac{a + b}{2}\right) - S\left(\frac{a + b}{2}, b\right) \right| < 15\varepsilon,$$

then

$$(4.56) \quad \left| \int_a^b f(x)\, dx - S\left(a, \frac{a + b}{2}\right) - S\left(\frac{a + b}{2}, b\right) \right| < \varepsilon.$$

In this case

$$S\left(a, \frac{a+b}{2}\right) + S\left(\frac{a+b}{2}, b\right)$$

is assumed to be a sufficiently accurate approximation to $\int_a^b f(x)\,dx$. When inequality (4.55) does not hold, apply the error estimation procedure individually to the subintervals $[a, (a+b)/2]$ and $[(a+b)/2, b]$ to determine if the approximation to the integral on each subinterval is within a tolerance of $\varepsilon/2$. If so, sum the approximations to produce an approximation to $\int_a^b f(x)\,dx$ within the tolerance ε. If the approximation on one of the subintervals fails to be within the tolerance $\varepsilon/2$, that subinterval is itself subdivided and each of its subintervals analyzed to determine if the approximation on that subinterval is accurate to within $\varepsilon/4$. Continue this halving procedure until each portion is within the required tolerance.

The following algorithm details this adaptive quadrature procedure for Simpson's rule. Some technical difficulties arise, which require the implementation of the method to differ slightly from the discussion above. Notice in Step 1 that the tolerance has been set at 10ε rather than the 15ε figure shown in inequality (4.55). This bound was chosen conservatively to compensate for the error in the assumption $f^{(4)}(\mu) = f^{(4)}(\tilde{\mu})$. In problems when $f^{(4)}$ is known to be widely varying, it would be reasonable to lower this bound even further.

The procedure listed in the algorithm always approximates first the integral on the leftmost subinterval in a subdivision. This requires introducing a procedure for efficiently storing and recalling previously computed functional evaluations for the nodes in the right half subintervals. Steps 3, 4, and 5 contain a stacking procedure with an indicator to keep track of the data that will be required for calculating the approximation on the subinterval immediately adjacent and to the right of the subinterval on which the approximation is being generated.

Adaptive Quadrature Algorithm 4.2

To approximate the integral $I = \int_a^b f(x)\,dx$ to within a given tolerance $\varepsilon > 0$:

INPUT endpoints a, b; tolerance ε; limit N to number of levels.

OUTPUT approximation APP or message that N is exceeded.

Step 1 Set $APP = 0$;
$\quad\quad i = 1$;
$\quad\quad TOL_i = 10\varepsilon$;
$\quad\quad a_i = a$;
$\quad\quad h_i = (b - a)/2$;
$\quad\quad FA_i = f(a)$;
$\quad\quad FC_i = f(a + h_i)$;
$\quad\quad FB_i = f(b)$;
$\quad\quad S_i = h_i(FA_i + 4FC_i + FB_i)/3$;   (*Approximation from Simpson's method for entire interval.*)

$\quad\quad L_i = 1$.

Step 2 While $i > 0$ do Steps 3–5.

Step 3 Set $FD = f(a_i + h_i/2)$;
$FE = f(a_i + 3h_i/2)$;
$S1 = h_i(FA_i + 4FD + FC_i)/6$; (*Approximations from Simpson's method for halves of subintervals.*)

$S2 = h_i(FC_i + 4FE + FB_i)/6$;
$v_1 = a_i$; (*Save data at this level.*)
$v_2 = FA_i$;
$v_3 = FC_i$;
$v_4 = FB_i$;
$v_5 = h_i$;
$v_6 = TOL_i$;
$v_7 = S_i$;
$v_8 = L_i$.

Step 4 Set $i = i - 1$. (*Delete the level.*)

Step 5 If $|S1 + S2 - v_7| < v_6$
then set $APP = APP + (S1 + S2)$
else
if $(v_8 \geq N)$
then
OUTPUT ('LEVEL EXCEEDED'); (*Procedure fails.*)
STOP.
Else (*Add one level.*)
set $i = i + 1$; (*Data for right half subinterval.*)
$a_i = v_1 + v_5$;
$FA_i = v_3$;
$FC_i = FE$;
$FB_i = v_4$;
$h_i = v_5/2$;
$TOL_i = v_6/2$;
$S_i = S2$;
$L_i = v_8 + 1$;
set $i = i + 1$; (*Data for left half subinterval.*)

$a_i = v_1$;
$FA_i = v_2$;
$FC_i = FD$;
$FB_i = v_3$;
$h_i = h_{i-1}$;
$TOL_i = TOL_{i-1}$;
$S_i = S1$;
$L_i = L_{i-1}$.

Step 6 OUTPUT (APP); (*APP approximates I to within ε.*)
STOP.

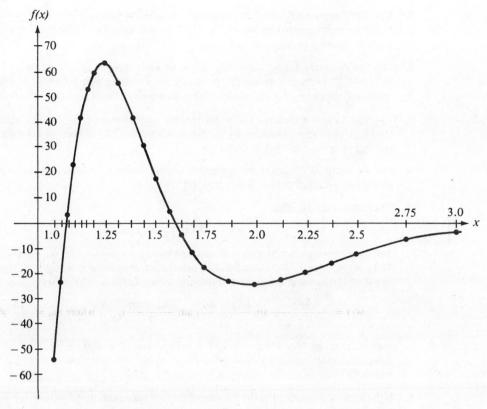

FIGURE 4.6

EXAMPLE 1   The graph of the function $f(x) = (100/x^2)\sin(10/x)$ for x in $[1, 3]$ is shown in Fig. 4.6. Using the Adaptive Quadrature Algorithm 4.2 with tolerance 10^{-4} to approximate $\int_1^3 f(x)\,dx$ produces -1.426014, a result that is accurate to within 1.4×10^{-6}. The approximation required that Simpson's rule with $m = 2$ be performed on the 23 subintervals whose endpoints are shown on the horizontal axis in Fig. 4.6. The total number of functional evaluations required for this approximation is 93.

Using Simpson's composite rule to approximate this integral with $h = \frac{1}{64}$ requires 255 functional evaluations and gives the approximation -1.426059, a result that differs from the actual value by 2.4×10^{-6}.   □

Exercise Set 4.5

1. Use the Adaptive Quadrature procedure to find approximations to within 10^{-2} for the definite integrals in Exercise 1 of Section 4.4. Do not use a computer program to generate these results.

2. Use the Adaptive Quadrature Algorithm 4.2 to approximate $\int_1^3 e^x \sin x\,dx$ to within 10^{-6}. Compare your approximation to the result in Exercise 3 of Section 4.4. Compare the number of nodes required in each case.

3. Use the Adaptive Quadrature Algorithm to approximate $\int_1^{10} \ln x \, dx$ to within 10^{-4}. Compare your approximation to the result in Exercise 8 of Section 4.4. Compare the number of nodes required in each case.

4. Use the Adaptive Quadrature Algorithm to approximate $\int_0^{48} \{1 + (\cos x)^2\}^{1/2} \, dx$ to within 10^{-4}. Does this technique have any advantage in this instance? (Compare the result and the number of functional evaluations with those of Exercise 13 in Section 4.4).

5. Use the Adaptive Quadrature Algorithm to approximate $\int_0^1 x^{1/3} \, dx$ to within 10^{-4}. Compare your approximation to the result in Exercise 7(f) of Section 4.3. Compare the number of nodes required in each case.

6. Use the Adaptive Quadrature Algorithm to approximate $\int_{.1}^2 \sin(1/x) \, dx$ to within 10^{-3}. Sketch the graph of $f(x) = \sin(1/x)$ on $[.1, 1]$.

7. The differential equation

    $$mu''(t) + ku(t) = F_0 \cos \omega t$$

    describes a spring-mass system with mass m, spring constant k, and no applied damping. The term $F_0 \cos \omega t$ describes a periodic external force applied to the system. The solution to the equation when the system is initially at rest ($u'(0) = u(0) = 0$) is

    $$u(t) = \frac{2F_0}{m(\omega_0^2 - \omega^2)} \sin \frac{(\omega_0 - \omega)}{2} t \sin \frac{(\omega_0 + \omega)}{2} t, \quad \text{where } \omega_0 = \sqrt{\frac{k}{m}} \neq \omega.$$

    Sketch the graph of u when $m = 1$, $k = 9$, $F_0 = 1$, $\omega = 2$, and $t \in [0, 2\pi]$. Use both the Adaptive Quadrative method and Simpson's composite rule to determine $\int_0^{2\pi} u(t) \, dt$ to within 10^{-4}.

8. If the term $cu'(t)$ is added to the left side of the motion equation in Exercise 7, the resulting differential equation describes a spring-mass system that is damped with damping constant c. The solution to this equation, when the solution is initially at rest is

    $$u(t) = c_1 e^{r_1 t} + c_2 e^{r_2 t} + \frac{F_0}{\sqrt{m^2(\omega_0^2 - \omega^2) + c^2 \omega^2}} \cos(\omega t - \delta)$$

    where

    $$\delta = \text{Arctan}\left(\frac{c\omega}{m(\omega_0^2 - \omega^2)}\right), \qquad r_1 = \frac{-c + \sqrt{c^2 - 4\omega_0^2 m^2}}{2m},$$

    and

    $$r_2 = \frac{-c - \sqrt{c^2 - 4\omega_0^2 m^2}}{2m}.$$

    Sketch the graph of u when $m = 1$, $k = 9$, $F_0 = 1$, $c = 1$, $\omega = 2$ and $t \in [0, 2\pi]$. Use both the Adaptive Quadrature method and Simpson's composite rule to determine $\int_0^{2\pi} u(t) \, dt$ to within 10^{-4}.

4.6 Romberg Integration

Although the trapezoidal rule is the easiest Newton–Cotes formula to apply, we have shown repeatedly in the previous sections that it lacks the degree of accuracy generally required. Romberg integration is a method that has wide application because it uses the trapezoidal rule to give preliminary approximations, and then applies the Richardson extrapolation process (discussed in Section 4.2) to obtain improvements of the approximations.

To begin the presentation of the Romberg integration scheme, recall (Theorem 4.5) that the extended trapezoidal rule for approximating the integral of a function f on an interval $[a, b]$ using m subintervals is

$$\int_a^b f(x) \, dx = \frac{h}{2}\left[f(a) + f(b) + 2\sum_{j=1}^{m-1} f(x_j) \right] - \frac{(b-a)}{12} h^2 f''(\mu)$$

where $a < \mu < b$, $h = (b-a)/m$ and $x_j = a + jh$ for each $j = 0, 1, \ldots, m$.

The first step in the Romberg process involves obtaining the trapezoidal rule approximations with $m_1 = 1$, $m_2 = 2$, $m_3 = 4$, $\ldots$, $m_n = 2^{n-1}$ where n is some positive integer. The values of the step size h_k corresponding to m_k will be $h_k = (b-a)/m_k = (b-a)/2^{k-1}$, and with this notation the trapezoidal rule becomes

$$(4.57) \qquad \int_a^b f(x) \, dx = \frac{h_k}{2}\left[f(a) + f(b) + 2\left(\sum_{i=1}^{2^{k-1}-1} f(a + ih_k) \right) \right]$$
$$- \frac{(b-a)}{12} h_k^2 f''(\mu_k)$$

where μ_k is a point in $[a, b]$.

If the notation $R_{k,1}$ is introduced to denote that portion of (4.57), which is used for the trapezoidal approximation, then

$$R_{1,1} = \frac{h_1}{2}[f(a) + f(b)] = \frac{(b-a)}{2}[f(a) + f(b)];$$

$$R_{2,1} = \frac{h_2}{2}[f(a) + f(b) + 2f(a + h_2)]$$
$$= \frac{(b-a)}{4}\left[f(a) + f(b) + 2f\left(a + \frac{(b-a)}{2}\right) \right]$$
$$= \tfrac{1}{2}[R_{1,1} + h_1 f(a + \tfrac{1}{2}h_1)];$$

$$R_{3,1} = \frac{h_3}{2}\left\{ f(a) + f(b) + 2\left[f\left(a + \frac{(b-a)}{4}\right) \right.\right.$$
$$\left.\left. + f\left(a + \frac{(b-a)}{2}\right) + f\left(a + \frac{3(b-a)}{4}\right) \right] \right\}$$
$$= \frac{(b-a)}{8}\left\{ f(a) + f(b) + 2\left[f\left(a + \frac{(b-a)}{4}\right) \right.\right.$$
$$\left.\left. + f\left(a + \frac{(b-a)}{2}\right) + f\left(a + \frac{3(b-a)}{4}\right) \right] \right\}$$
$$= \frac{1}{2}\left\{ R_{2,1} + h_2\left[f\left(a + \frac{h_2}{2}\right) + f\left(a + \frac{3h_2}{2}\right) \right] \right\};$$

and, in general,

$$(4.58) \qquad R_{k,1} = \frac{1}{2}\left[R_{k-1,1} + h_{k-1} \sum_{i=1}^{2^{k-2}} f\left(a + \left(i - \frac{1}{2}\right)h_{k-1}\right) \right]$$

for each $k = 2, 3, \ldots, n$.

For the derivation of Eq. (4.58), the reader is referred to Exercises 7 and 8.

EXAMPLE 1   Using Eq. (4.58) to perform the first step of the Romberg integration scheme for approximating $\int_0^\pi \sin x \, dx$ with $n = 6$ leads to:

$$R_{1,1} = \frac{\pi}{2} \left[\sin 0 + \sin \pi \right] = 0;$$

$$R_{2,1} = \frac{1}{2} \left[R_{1,1} + \pi \sin \frac{\pi}{2} \right] = 1.57079633;$$

$$R_{3,1} = \frac{1}{2} \left[R_{2,1} + \frac{\pi}{2} \left(\sin \frac{\pi}{4} + \sin \frac{3\pi}{4} \right) \right] = 1.89611890;$$

$$R_{4,1} = \frac{1}{2} \left[R_{3,1} + \frac{\pi}{4} \left(\sin \frac{\pi}{8} + \sin \frac{3\pi}{8} + \sin \frac{5\pi}{8} + \sin \frac{7\pi}{8} \right) \right] = 1.97423160;$$

$$R_{5,1} = 1.99357034,$$

and $R_{6,1} = 1.99839336.$ ☐

Since the correct value for this integral is 2, it is clear that, although the calculations involved are not difficult, the convergence is very slow.

To speed the convergence the Richardson extrapolation procedure will now be performed. It can be shown, although not easily (see Ralston and Rabinowitz [67], pages 123–126 and accompanying exercises for a presentation), that the extended trapezoidal rule given in Eq. (4.57) can be written with an alternate error term in the form:

$$(4.59) \qquad \int_a^b f(x) \, dx = \frac{h_k}{2} \left[f(a) + f(b) + 2 \left(\sum_{i=1}^{2^{k-1}-1} f(a + ih_k) \right) \right]$$

$$- \frac{h_k^2}{12} \left[f'(b) - f'(a) \right] + \frac{(b-a)h_k^4}{720} f^{(4)}(\mu_k),$$

for each $k = 1, 2, \ldots, n$ and some $a < \mu_k < b$.

With the extended trapezoidal rule in this form, we can eliminate the term involving h_k^2 by combining the equations

$$(4.60) \qquad \int_a^b f(x) \, dx = R_{k-1,1} - \frac{h_{k-1}^2}{12} \left[f'(b) - f'(a) \right]$$

$$+ \frac{(b-a)h_{k-1}^4}{720} f^{(4)}(\mu_{k-1})$$

and

$$(4.61) \qquad \int_a^b f(x) \, dx = R_{k,1} - \frac{h_k^2}{12} \left[f'(b) - f'(a) \right] + \frac{(b-a)h_k^4}{720} f^{(4)}(\mu_k)$$

$$= R_{k,1} - \frac{h_{k-1}^2}{48} \left[f'(b) - f'(a) \right] + \frac{(b-a)h_k^4}{720} f^{(4)}(\mu_k)$$

to obtain

$$(4.62) \quad \int_a^b f(x)\, dx = \frac{4R_{k,1} - R_{k-1,1}}{3} + \frac{(b-a)}{2160} \left[4h_k^4 f^{(4)}(\mu_k) - h_{k-1}^4 f^{(4)}(\mu_{k-1}) \right]$$

$$= \frac{4R_{k,1} - R_{k-1,1}}{3} + O(h_k^4).$$

It is an easy matter to show (see Exercise 6) that the approximation obtained by this technique is actually the approximation given by the Simpson's composite rule with $h = h_k$, so the error of order h_k^4 is expected.

To continue the Romberg scheme, define

$$R_{k,2} = \frac{4R_{k,1} - R_{k-1,1}}{3}$$

for each $k = 2, 3, \ldots, n$, and apply the Richardson extrapolation procedure to these values. It can be shown (see Ralston and Rabinowitz [67]) that the process gives

$$R_{i,j} = \frac{4^{j-1} R_{i,j-1} - R_{i-1,j-1}}{4^{j-1} - 1}$$

for each $i = 2, 3, 4, \ldots, n$, and $j = 2, \ldots, i$, where the values with larger j index correspond to successively higher-order Newton–Cotes formulas. The approximations are often presented in a table of the form of Table 4.7.

TABLE 4.7

$R_{1,1}$				
$R_{2,1}$	$R_{2,2}$			
$R_{3,1}$	$R_{3,2}$	$R_{3,3}$		
$R_{4,1}$	$R_{4,2}$	$R_{4,3}$	$R_{4,4}$	
$\vdots$	$\vdots$	$\vdots$		$\ddots$
$R_{n,1}$	$R_{n,2}$	$R_{n,3}$	$\cdots$	$R_{n,n}$

An elegant summability theorem of Silverman and Toeplitz can be used to show that the terms along the diagonal will converge to the integral provided the values of $R_{n,1}$ converge to this number. A proof of this result, together with necessary conditions for this convergence, can be found in Ralston and Rabinowitz [67], pages 123–126. It is generally expected that the diagonal sequence $\{R_{m,m}\}_{m=1}^{\infty}$ converges much more rapidly than $\{R_{n,1}\}_{n=1}^{\infty}$.

The Romberg technique has the additional desirable feature that it allows an entire new row in the table to be calculated by simply doing one application of the trapezoidal rule and then using the previously calculated values to obtain the succeeding entries in the row. The method generally used to construct a table of this type incorporates this feature by calculating the entries row by row, that is, in the order $R_{1,1}, R_{2,1}, R_{2,2}, R_{3,1}, R_{3,2}, R_{3,3}$, etc. The following algorithm describes this technique in detail.

Romberg Algorithm 4.3

To approximate the integral $I = \int_a^b f(x)\,dx$, select an integer $n > 0$.

INPUT   endpoints a, b; integer n.

OUTPUT   an array R.   ($R_{n,n}$ is the approximation to I. Computed by rows; only 2 rows saved in storage.)

Step 1   Set $h = b - a$;
$$R_{1,1} = h(f(a) + f(b))/2.$$

Step 2   OUTPUT $(R_{1,1})$.

Step 3   For $i = 2, \ldots, n$ do Steps 4–8.

Step 4   Set $R_{2,1} = \dfrac{1}{2}\left[R_{1,1} + h \sum_{k=1}^{2^{i-2}} f(a + (k - .5)h) \right]$.   (*Approximation from trapezoidal method.*)

Step 5   For $j = 2, \ldots, i$

set $R_{2,j} = \dfrac{4^{j-1} R_{2,j-1} - R_{1,j-1}}{4^{j-1} - 1}$.   (*Extrapolation.*)

Step 6   OUTPUT $(R_{2,j}$ for $j = 1, 2, \ldots, i)$.

Step 7   Set $h = h/2$.

Step 8   For $j = 1, 2, \ldots, i$ set $R_{1,j} = R_{2,j}$.   (*Update row 1 of R.*)

Step 9   STOP.

It is often useful not to have predetermined a specific value for n and instead to modify the algorithm slightly to allow the procedure to continue until a value of n is found that satisfies $|R_{n,n} - R_{n-1,n-1}| < \varepsilon$ for a given tolerance ε.

EXAMPLE 2   In Example 1, the values for $R_{1,1}$ through $R_{n,1}$ were obtained for approximating $\int_0^\pi \sin x\,dx$ with $n = 6$. With Algorithm 4.2, the Romberg table is shown in Table 4.8.   □

TABLE 4.8   0

1.57079633	2.09439511				
1.89611890	2.00455976	1.99857073			
1.97423160	2.00026917	1.99998313	2.00000555		
1.99357034	2.00001659	1.99999975	2.00000001	1.99999999	
1.99839336	2.00000103	2.00000000	2.00000000	2.00000000	2.00000000

Exercise Set 4.6

1.   Use Romberg integration to calculate $R_{3,3}$ for the following definite integrals. Compare your results to those obtained in Exercises 1 and 2 of Section 4.4.

a) $\displaystyle\int_1^3 \frac{dx}{x}$

b) $\displaystyle\int_0^2 x^3\,dx$

c) $\int_0^3 x\sqrt{1 + x^2} \, dx$

d) $\int_0^1 \sin \pi x \, dx$

e) $\int_0^{2\pi} x \sin x \, dx$

f) $\int_0^1 x^2 e^x \, dx$

2. Use the Romberg integration procedure to find an approximation to $\int_1^3 e^x \sin x \, dx$ that is accurate within 10^{-6}. Compare your answers to the exact result and to the result obtained in Exercise 3 of Section 4.4.

3. Use the Romberg integration procedure to find approximations to $\int_0^{1.5} (1 + x)^{-1} \, dx$, completing the table for $n = 6$. Compare this result with the values obtained in Exercise 6 of Section 4.4 and Exercise 10 of Section 4.3.

4. Approximate $\int_1^{10} \ln x \, dx$ using the Romberg integration procedure with $n = 10$.

5. Let $N = 10$. Apply the Romberg integration procedure to the integral $\int_a^b f(x) \, dx$ until $|R_{k,k} - R_{k-1,k-1}| \leq 10^{-5}$ or until $k > N$ for the following functions and values of a and b.

   a) $f(x) = x^{1/3}$, $a = 0, b = 1$;

   b) $f(x) = \begin{cases} x^3 + 1, & 0 \leq x \leq .1, \\ 1.001 + .03(x - .1) + .3(x - .1)^2 + 2(x - .1)^3, & .1 < x \leq .2, \\ 1.009 + .15(x - .2) + .9(x - .2)^2 + 2(x - .2)^3, & .2 < x \leq .3, \end{cases}$

   with $a = 0, b = .3$.

6. Show that the approximation obtained from $R_{k,2}$ is the same as that given by the Simpson's composite rule described in Theorem 4.4 with $h = h_k$.

7. Show that, for any k,

$$\sum_{i=1}^{2^{k-1}-1} f\left(a + \frac{i}{2} h_{k-1}\right) = \sum_{i=1}^{2^{k-2}} f\left(a + \left(i - \frac{1}{2}\right)h_{k-1}\right) + \sum_{i=1}^{2^{k-2}-1} f(a + ih_{k-1}).$$

8. Use the result of Exercise 7 to verify Eq. (4.58); that is, show that, for all k,

$$R_{k,1} = \frac{1}{2}\left[R_{k-1,1} + h_{k-1}\sum_{i=1}^{2^{k-2}} f\left(a + \left(i - \frac{1}{2}\right)h_{k-1}\right)\right].$$

9. Compute an approximation to $\int_0^{48} \sqrt{1 + (\cos x)^2} \, dx$ that is accurate to within 10^{-4} by using Romberg integration.

4.7 Gaussian Quadrature*

The closed Newton–Cotes formulas presented in Section 4.3 were derived by integrating the Lagrange interpolating polynomials. Since the error term in the Lagrange interpolating polynomial of degree n involves the $(n + 1)$st derivative of the function being approximated, it has been remarked previously that the formula is exact when approximating any polynomial of degree less than or equal to n. Consequently, the closed Newton–Cotes formulas have degree of precision at least n. In fact, the degree of precision of the odd formulas is exactly n, while the even formulas have degree of precision $(n + 1)$. An analogous situation occurs for the open Newton–Cotes formulas.

---

* This section requires some material concerning orthogonal functions, which is discussed in Section 7.2.

All the Newton–Cotes formulas require that the values of the function whose integral is to be approximated be known at evenly spaced points, which might be the expected situation if tabulated data for the function was being used. If the function is given explicitly, however, the points for evaluating the function could be chosen in another manner, which leads to increased accuracy of approximation. **Gaussian quadrature** is concerned with choosing the points for evaluation in an optimal manner. It presents a procedure for choosing values $x_1, x_2, \ldots, x_n$ in the interval $[a, b]$ and constants $c_1, c_2, \ldots, c_n$, that are expected to minimize the error obtained in performing the approximation

$$(4.63) \qquad \int_a^b f(x)\, dx \approx \sum_{i=1}^n c_i f(x_i)$$

for an arbitrary function f. In order to measure this accuracy, it is generally assumed that the best choice of these values will be the choice that maximizes the degree of precision for the formula.

Since the values of $c_1, c_2, \ldots, c_n$ are completely arbitrary and those of $x_1, x_2, \ldots, x_n$ are restricted only in the sense that the function whose integral is being approximated must be defined at these points, there are at most $2n$ parameters involved, n given by the constants $c_1, c_2, \ldots, c_n$ and n given by $x_1, x_2, \ldots, x_n$.

If the coefficients of a polynomial are also considered as parameters, the class of polynomials of degree at most $(2n - 1)$ contains $2n$ parameters and is the largest class of polynomials for which it is reasonable to expect Eq. (4.63) to be exact. In fact, for the proper choice of the values and constants, exactness on this set can be obtained. This implies that Eq. (4.63) can be designed to have degree of precision $(2n - 1)$.

Before beginning the study of Gaussian quadrature, it is necessary to discuss some of the material of orthogonal sets of functions that is presented in Section 7.2. The set of functions $\{\phi_0, \phi_1, \ldots, \phi_n\}$ is said to be **orthogonal** on $[a, b]$ with respect to the weight function $w(x) \geq 0$, provided

$$\int_a^b \phi_k(x)\phi_j(x)w(x)\, dx$$

is zero when $j \neq k$ and positive when $j = k$.

It is easily shown (see Exercise 14 of Section 7.2) that if $\{\phi_0, \phi_1, \ldots, \phi_n\}$ is an orthogonal set of polynomials defined on $[a, b]$ and ϕ_i is of degree i for each $i = 0, 1, \ldots, n$, then for any polynomial Q of degree at most n, there exist unique constants $\alpha_0, \alpha_1, \ldots, \alpha_n$ with $Q(x) = \sum_{i=0}^n \alpha_i \phi_i(x)$. This result will be used in the proof of the next theorem, a result that is of primary importance in determining the optimal choice of the values $x_1, x_2, \ldots, x_n$.

Theorem 4.7   If $\{\phi_0, \phi_1, \ldots, \phi_n\}$ is a set of orthogonal polynomials defined on $[a, b]$ with respect to the continuous weight function w and ϕ_k is a polynomial of degree k for each $k = 0, 1, \ldots, n$, then ϕ_k has k distinct roots, and these roots lie in the interval (a, b).

Proof   Since ϕ_0 is a polynomial of degree zero, a constant $C \neq 0$ exists with $\phi_0(x) = C$. This implies that for $k \geq 1$

$$0 = \int_a^b \phi_k(x)\phi_0(x)w(x)\, dx = C \int_a^b \phi_k(x)w(x)\, dx.$$

Since w is a weight function, $w(x) \geq 0$, but $w(x) \not\equiv 0$. Thus, ϕ_k must change sign at least once in (a, b). Suppose ϕ_k changes sign precisely j times in (a, b), at the points $\{r_i\}_{i=1}^{j}$, where $a < r_1 < r_2 < \cdots < r_j < b$, and that $j < k$. Without loss of generality, it may be assumed that $\phi_k(x) > 0$ on (a, r_1) (see Exercise 9). Consequently, $\phi_k(x) < 0$ on (r_1, r_2), $\phi_k(x) > 0$ on (r_2, r_3), and in general ϕ_k is of opposite sign on each of the adjacent intervals $(a, r_1), (r_1, r_2), \ldots, (r_j, b)$.

Define the jth-degree polynomial, P, by

$$P(x) = \prod_{i=1}^{j} (x - r_i).$$

Note that the sign of P agrees with that of ϕ_k on each of the subintervals $(a, r_1), (r_1, r_2), \ldots, (r_j, b)$ and, consequently, that $P(x)\phi_k(x) > 0$ on each of these intervals. Since $w(x) \geq 0$ on (a, b) but $w(x) \not\equiv 0$, this implies that

(4.64)
$$\int_a^b P(x)\phi_k(x)w(x)\,dx > 0.$$

However, P is a polynomial of degree $j < k$, so

$$P(x) = \sum_{i=0}^{j} \alpha_i \phi_i(x)$$

for some collection of constants $\alpha_0, \alpha_1, \ldots, \alpha_j$, which implies

$$\int_a^b P(x)\phi_k(x)w(x)\,dx = \sum_{i=0}^{j} \alpha_i \int_a^b \phi_i(x)\phi_k(x)w(x)\,dx = 0$$

in contradiction with inequality (4.64).

The only assumption made in this procedure was that ϕ_k changes sign precisely j times in (a, b), where $j < k$; so this statement must be erroneous. This implies that ϕ_k changes sign at least k times in (a, b). The Intermediate Value Theorem (Theorem 1.12) implies that a root exists at each sign change; so ϕ_k must have k distinct roots in (a, b). □

To apply the result of Theorem 4.7, suppose that f is a function for which an approximation to $\int_{-1}^{1} f(x)\,dx$ is needed. The set of Legendre polynomials, $\{P_0, P_1, \ldots, P_n\}$, are defined in Example 4 of Section 7.2 to be orthogonal on $[-1, 1]$ with respect to $w(x) \equiv 1$. Choose $x_1, x_2, \ldots, x_n$ to be the n distinct roots of P_n, which Theorem 4.7 assures us lie in $(-1, 1)$.

Suppose P is an arbitrary polynomial of degree k, where $k \leq 2n - 1$. Dividing P by P_n gives

$$P(x) = Q(x)P_n(x) + R(x),$$

where both Q and R are polynomials of degree less than n. Suppose a quadrature formula with degree of precision at least $(n - 1)$ is given in the form

$$\int_{-1}^{1} f(x)\,dx \approx \sum_{i=1}^{n} c_i f(x_i)$$

where
$$c_i = \int_{-1}^{1} \prod_{\substack{j=1 \\ j \neq i}}^{n} \frac{(x - x_j)}{(x_i - x_j)}\,dx \qquad \text{for each } i = 1, 2, \ldots, n,$$

and the values $x_1, \ldots, x_n$ are as mentioned above. Since Q and R are of degree less than n,

$$\int_{-1}^{1} P(x)\, dx = \int_{-1}^{1} Q(x)P_n(x)\, dx + \int_{-1}^{1} R(x)\, dx$$

$$= 0 + \int_{-1}^{1} R(x)\, dx = \sum_{i=1}^{n} c_i R(x_i).$$

But since $x_1, x_2, \ldots, x_n$ are all roots of P_n, $P(x_i) = Q(x_i)P_n(x_i) + R(x_i) = R(x_i)$; so

$$\int_{-1}^{1} P(x)\, dx = \sum_{i=1}^{n} c_i P(x_i),$$

and the formula is exact for all polynomials of degree at most $2n - 1$.

The constants $c_1, c_2, \ldots, c_n$ could be obtained by solving the n linear equations in n unknowns:

$$\beta_0 = \int_{-1}^{1} P_0^2(x)\, dx = \sum_{i=1}^{n} c_i P_0^2(x_i),$$

(4.65)
$$\beta_1 = \int_{-1}^{1} P_1^2(x)\, dx = \sum_{i=1}^{n} c_i P_1^2(x_i),$$

$$\vdots \qquad\qquad \vdots \qquad\qquad \vdots$$

$$\beta_{n-1} = \int_{-1}^{1} P_{n-1}^2(x)\, dx = \sum_{i=1}^{n} c_i P_{n-1}^2(x_i),$$

but in actuality both these constants and the values of the roots of P_n are extensively tabulated; see, for example, Stroud and Secrest [86].

Since the simple linear transformation $t = [1/(b - a)](2x - a - b)$ will translate any interval $[a, b]$ into $[-1, 1]$ provided $b > a$, the Legendre polynomials can be used to approximate

(4.66)
$$\int_{a}^{b} f(x)\, dx = \int_{-1}^{1} f\left(\frac{(b - a)t + b + a}{2}\right) \frac{(b - a)}{2}\, dt$$

for any function which can be evaluated at the required points.

EXAMPLE 1   Consider the problem of finding approximations to $\int_{1}^{1.5} e^{-x^2}\, dx$. The Newton–Cotes formulas given in Section 4.3 can be used with the resulting approximations being:

Newton–Cotes Closed Formulas

$n = 1, h = \frac{1}{2}$:

$$\int_{1}^{1.5} e^{-x^2}\, dx \approx \tfrac{1}{4}[e^{-1} + e^{-(3/2)^2}] = .1183197;$$

$n = 2, h = \frac{1}{4}$:

$$\int_1^{1.5} e^{-x^2}\, dx \approx \tfrac{1}{12}[e^{-1} + 4e^{-(5/4)^2} + e^{-(3/2)^2}] = .1093104;$$

$n = 3, h = \frac{1}{6}$:

$$\int_1^{1.5} e^{-x^2}\, dx \approx \tfrac{1}{16}[e^{-1} + 3e^{-(7/6)^2} + 3e^{-(4/3)^2} + e^{-(3/2)^2}] = .1093404;$$

$n = 4, h = \frac{1}{8}$:

$$\int_1^{1.5} e^{-x^2}\, dx \approx \tfrac{1}{180}[7e^{-1} + 32e^{-(9/8)^2} + 32e^{-(5/4)^2} + 32e^{-(11/8)^2} + 7e^{-(3/2)^2}]$$

$$= .1093643.$$

Newton–Cotes Open Formulas

$n = 0, h = \frac{1}{4}$:

$$\int_1^{1.5} e^{-x^2}\, dx \approx \tfrac{1}{2}e^{-(5/4)^2} = .1048057;$$

$n = 1, h = \frac{1}{6}$:

$$\int_1^{1.5} e^{-x^2}\, dx \approx \tfrac{1}{4}[e^{-(7/6)^2} + e^{-(4/3)^2}] = .1063473;$$

$n = 2, h = \frac{1}{8}$:

$$\int_1^{1.5} e^{-x^2}\, dx \approx \tfrac{1}{6}[2e^{-(9/8)^2} - e^{-(5/4)^2} + 2e^{-(11/8)^2}] = .1094116.$$

$n = 3, h = \frac{1}{10}$:

$$\int_1^{1.5} e^{-x^2}\, dx \approx \tfrac{1}{48}[11e^{-(11/10)^2} + e^{-(6/5)^2} + e^{-(13/10)^2} + 11e^{-(7/5)^2}] = .1093971.$$

The Gaussian quadrature procedure applied to this problem requires that the integral be first translated into a problem whose interval of integration is $[-1, 1]$. Using Eq. (4.66),

$$\int_1^{1.5} e^{-x^2}\, dx = \frac{1}{4} \int_{-1}^{1} e^{(-(t+5)^2/16)}\, dt.$$

Table 4.9 lists the roots of the Legendre polynomials for various values of n, together with the coefficients that are associated with the values of the function to be evaluated at these roots.

TABLE 4.9

n	Roots	Coefficients
2	.5773502692	1.0000000000
	−.5773502692	1.0000000000
3	.7745966692	.5555555556
	.0000000000	.8888888889
	−.7745966692	.5555555556
4	.8611363116	.3478548451
	.3399810436	.6521451549
	−.3399810436	.6521451549
	−.8611363116	.3478548451
5	.9061798459	.2369268850
	.5384693101	.4786286705
	.0000000000	.5688888889
	−.5384693101	.4786286705
	−.9061798459	.2369268850

Using these values, the Gaussian quadrature approximations for this problem are:

$n = 2$:

$$\int_1^{1.5} e^{-x^2}\, dx \approx \tfrac{1}{4}[e^{-(5+.5773502692)^2/16} + e^{-(5-.5773502692)^2/16}] = .1094003,$$

and

$n = 3$:

$$\int_1^{1.5} e^{-x^2}\, dx \approx \tfrac{1}{4}[(.5555555556)e^{-(5+.7745966692)^2/16} + (.8888888889)e^{-(5)^2/16}$$

$$+ (.5555555556)e^{-(5-.7745966692)^2/16}]$$

$$= .1093642.$$

For comparison the values obtained by using the Romberg procedure with $n = 4$ are listed in Table 4.10.

TABLE 4.10

.1183197			
.1115627	.1093104		
.1099114	.1093610	.1093643	
.1095009	.1093641	.1093643.	.1093643

The correct answer, to 7 decimal places, is .1093643.   □

Other sets of orthogonal polynomials can be used in the Gaussian quadrature technique; examples of some of these are given in the exercises.

Exercise Set 4.7

1. Use Gaussian quadrature with $n = 2$ to approximate the following definite integrals. Compare the approximations to the results of the corresponding exercises in Sections 4.4 and 4.6.

   a) $\displaystyle\int_1^3 \frac{dx}{x}$ b) $\displaystyle\int_0^2 x^3 \, dx$

   c) $\displaystyle\int_0^3 x\sqrt{1 + x^2} \, dx$ d) $\displaystyle\int_0^1 \sin \pi x \, dx$

   e) $\displaystyle\int_0^{2\pi} x \sin x \, dx$ f) $\displaystyle\int_0^1 x^2 e^x \, dx$

2. Use Gaussian quadrature with $n = 3$ to approximate the definite integrals in Exercise 1. Compare these results with previously obtained approximations.

3. Use the Gaussian quadrature procedure with $n = 2, 3,$ and 4 to find approximations to $\int_1^3 e^x \sin x \, dx$. Compare your answers with the results obtained in Exercise 2 in Section 4.6 and in Exercise 3 of Section 4.4, as well as to the exact value.

4. Use the Gaussian quadrature procedure with $n = 2, 3,$ and 4 to find approximations to $\int_0^{1.5} (1 + x)^{-1} \, dx$. Compare your answers with the results obtained in Exercise 3 of Section 4.6, Exercise 6 of Section 4.4, and Exercise 10 of Section 4.3, as well as to the exact value.

5. Verify the entries for the values of $n = 2$ and 3 in Table 4.9 by finding the roots of the respective Legendre polynomials. Using the equations given in (4.65), find the coefficients associated with the values.

6. In Section 7.2, Exercise 12, a set $\{L_0, L_1, L_2, L_3\}$ of Laguerre polynomials is given that is orthogonal on the interval $(0, \infty)$ with respect to the weight function e^{-x}. These polynomials can be used to give approximations to $\int_0^\infty e^{-x} f(x) \, dx$, provided this improper integral exists. The derivation parallels that given for the Legendre polynomials presented after the proof of Theorem 4.7. Show that this set of polynomials gives a formula with degree of precision three to approximate

$$\int_0^\infty e^{-x} f(x) \, dx.$$

7. Find the roots of the Laguerre polynomials L_1, L_2 and L_3 discussed in Exercise 6, and use the corresponding coefficients obtained from the equations in (4.65) to find approximations to:

$$\int_0^\infty e^{-x} \sin x \, dx \qquad \text{when } n = 2 \quad \text{and} \quad n = 3.$$

8. Approximate $\int_0^\infty \sqrt{x} e^{-x} \, dx$ using the Laguerre polynomials as in Exercise 7, and compare the approximation to that obtained in Exercise 16(c) of Section 4.4.

9. In the proof of Theorem 4.7, the statement was made that there was no loss of generality in assuming that $\phi_k(x)$ was positive on (a, r_1). Show that this statement is correct.

4.8   Multiple Integrals

The techniques discussed in the previous sections of the chapter can be modified in a straightforward manner for use in the approximation of multiple integrals. The first type of multiple integral problem we will consider involves the approximation of a double integral

$$\iint_R f(x, y)\, dA$$

where R is a rectangular region in the plane; that is,

$$R = \{(x, y)\,|\, a \le x \le b,\, c \le y \le d\}$$

for some constants a, b, c, and d.

To illustrate the approximation technique, we will employ the composite Simpson's rule although any of the other composite Newton–Cotes formulas could be used in its place.

Suppose that integers n and m are chosen to determine the step sizes $h = (b - a)/2n$ and $k = (d - c)/2m$. Writing the double integral as an iterated integral

$$\iint_R f(x, y)\, dA = \int_a^b \left(\int_c^d f(x, y)\, dy \right) dx,$$

we first use the composite Simpson's rule to evaluate

$$\int_c^d f(x, y)\, dy,$$

treating x as a constant. Letting $y_0 = c$ and $y_j = c + jk$ for each $j = 1, 2, \ldots, 2m$, gives

$$\int_c^d f(x, y)\, dy = \frac{k}{3}\left[f(x, y_0) + 2\sum_{j=1}^{m-1} f(x, y_{2j}) + 4\sum_{j=1}^{m} f(x, y_{2j-1}) + f(x, y_{2m}) \right]$$

$$- \frac{(d - c)k^4}{180}\frac{\partial^4 f(x, \mu)}{\partial y^4}$$

for some μ in $[c, d]$. Thus,

(4.67)
$$\int_a^b \int_c^d f(x, y)\, dy\, dx = \frac{k}{3}\int_a^b f(x, y_0)\, dx + \frac{2k}{3}\sum_{j=1}^{m-1}\int_a^b f(x, y_{2j})\, dx$$

$$+ \frac{4k}{3}\sum_{j=1}^{m}\int_a^b f(x, y_{2j-1})\, dx + \frac{k}{3}\int_a^b f(x, y_{2m})\, dx$$

$$- \frac{(d - c)}{180}k^4 \int_a^b \frac{\partial^4 f(x, \mu)}{\partial y^4}\, dx$$

Simpson's composite rule is now employed on each integral in Eq. (4.67). Letting $x_0 = a$ and $x_i = a + ih$ for each $i = 1, 2, \ldots, 2n$ produces for each $j = 0, 1, \ldots, 2m$:

$$\int_a^b f(x, y_j)\, dx = \frac{h}{3}\left[f(x_0, y_j) + 2\sum_{i=1}^{n-1} f(x_{2i}, y_j) + 4\sum_{i=1}^{n} f(x_{2i-1}, y_j) + f(x_{2n}, y_j) \right]$$

$$- \frac{(b - a)}{180}h^4 \frac{\partial^4 f}{\partial x^4}(\xi_j, y_j)$$

for some ξ_j in (a, b). The resulting approximation has the form

$$\int_a^b \int_c^d f(x, y) \, dy \, dx \approx \frac{hk}{9} \left[f(x_0, y_0) + 2 \sum_{i=1}^{n-1} f(x_{2i}, y_0) + 4 \sum_{i=1}^{n} f(x_{2i-1}, y_0) \right.$$

$$+ f(x_{2n}, y_0) + 2 \sum_{j=1}^{m-1} f(x_0, y_{2j}) + 4 \sum_{j=1}^{m-1} \sum_{i=1}^{n-1} f(x_{2i}, y_{2j})$$

$$+ 8 \sum_{j=1}^{m-1} \sum_{i=1}^{n} f(x_{2i-1}, y_{2j}) + 2 \sum_{j=1}^{m-1} f(x_{2n}, y_{2j})$$

$$+ 4 \sum_{j=1}^{m} f(x_0, y_{2j-1}) + 8 \sum_{j=1}^{m} \sum_{i=1}^{n-1} f(x_{2i}, y_{2j-1})$$

$$+ 16 \sum_{j=1}^{m} \sum_{i=1}^{n} f(x_{2i-1}, y_{2j-1}) + 4 \sum_{j=1}^{m} f(x_{2n}, y_{2j-1})$$

$$+ f(x_0, y_{2m}) + 2 \sum_{i=1}^{n-1} f(x_{2i}, y_{2m}) + 4 \sum_{i=1}^{n} f(x_{2i-1}, y_{2m})$$

$$\left. + f(x_{2n}, y_{2m}) \right].$$

The error term E is given by

$$E = \frac{-k(b-a)h^4}{540} \left[\frac{\partial^4 f(\xi_0, y_0)}{\partial x^4} + 2 \sum_{j=1}^{m-1} \frac{\partial^4 f(\xi_{2j}, y_{2j})}{\partial x^4} + 4 \sum_{j=1}^{m} \frac{\partial^4 f(\xi_{2j-1}, y_{2j-1})}{\partial x^4} \right.$$

$$\left. + \frac{\partial^4 f(\xi_{2m}, y_{2m})}{\partial x^4} \right] - \frac{(d-c)k^4}{180} \int_a^b \frac{\partial^4 f(x, \mu)}{\partial y^4} \, dx.$$

If $\partial^4 f/\partial x^4$ and $\partial^4 f/\partial y^4$ are continuous on R, the error term can be reexpressed as

$$E = \frac{-k(b-a)h^4}{540} \left[6m \frac{\partial^4 f}{\partial x^4} (\bar{\eta}, \bar{\mu}) \right] - \frac{(d-c)(b-a)}{180} k^4 \frac{\partial^4 f}{\partial y^4} (\hat{\eta}, \hat{\mu})$$

$$= \frac{-(d-c)(b-a)}{180} \left[h^4 \frac{\partial^4 f}{\partial x^4} (\bar{\eta}, \bar{\mu}) + k^4 \frac{\partial^4 f}{\partial y^4} (\hat{\eta}, \hat{\mu}) \right]$$

for some $(\bar{\eta}, \bar{\mu})$ and $(\hat{\eta}, \hat{\mu})$ in R. The proof of this statement results from applying the Weighted Mean Value Theorem to $\partial^4 f/\partial y^4$ and applying a generalization of the Intermediate Value Theorem to $\partial^4 f/\partial x^4$.

EXAMPLE 1 The composite Simpson's rule applied to approximate

$$\int_{1.4}^{2.0} \int_{1.0}^{1.5} \ln(x + 2y) \, dy \, dx$$

with $n = 2$ and $m = 1$ uses the step sizes $h = .15$ and $k = .25$. The region of integration R is shown in Fig. 4.7 together with the nodes for (x_i, y_j) $i = 0, 1, 2, 3, 4$; $j = 0, 1, 2$; and $w_{i,j}$, the coefficients of $f(x_i, y_i) = \ln(x_i + 2y_i)$ in the sum. The approximation is

$$\int_{1.4}^{2.0} \int_{1.0}^{1.5} \ln(x + 2y) \, dy \, dx \approx \frac{(.15)(.25)}{9} \sum_{i=0}^{4} \sum_{j=0}^{2} w_{i,j} \ln(x_i + 2y_j) = .4295524387.$$

Since

$$\frac{\partial^4 f}{\partial x^4}(x, y) = \frac{-6}{(x + 2y)^4} \quad \text{and} \quad \frac{\partial^4 f}{\partial y^4}(x, y) = \frac{-96}{(x^2 + 2y)^4},$$

the error is bounded by

$$|E| \leq \frac{(.5)(.6)}{180}\left[(.15)^4 \max_{(x, y) \text{ in } R} \frac{6}{(x + 2y)^4} + (.25)^4 \max_{(x, y) \text{ in } R} \frac{96}{(x + 2y)^4}\right] \leq 4.72 \times 10^{-6}.$$

The actual value of the integral to ten decimal places is

$$\int_{1.4}^{2.0} \int_{1.0}^{1.5} \ln(x + 2y) \, dy \, dx = .4295545265;$$

so the approximation is accurate to within 2.1×10^{-6}.   □

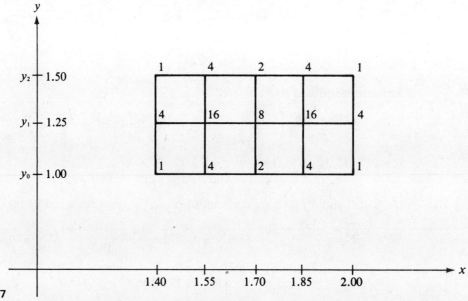

FIGURE 4.7

The same techniques can be applied for the approximation of triple integrals, as well as higher integrals for functions of more than three variables. The number of functional evaluations required for the approximation is the product of the number of functional evaluations required when the method is applied to each variable. In order to reduce the number of functional evaluations, more efficient methods such as Gaussian quadrature, Romberg integration, or adaptive quadrature can be incorporated in place of the Newton–Cotes formulas. The following example illustrates the use of Gaussian quadrature for the integral considered in Example 1.

EXAMPLE 2   Consider the double integral $\int_{1.4}^{2.0} \int_{1.0}^{1.5} \ln(x + 2y) \, dy \, dx$. Before employing a Gaussian quadrature technique to approximate this integral, we must translate the region of integration

$$R = \{(x, y) \mid 1.4 \leq x \leq 2.0, 1.0 \leq y \leq 1.5\}$$

into $\qquad \hat{R} = \{(u, v) | -1 \le u \le 1, -1 \le v \le 1\}.$

A linear transformation that accomplishes this is

$$u = \frac{1}{2.0 - 1.4}(2x - 1.4 - 2.0), \qquad v = \frac{1}{1.5 - 1.0}(2y - 1.0 - 1.5).$$

Employing this change of variables gives an integral on which Gaussian quadrature can be applied

$$\int_{1.4}^{2.0} \int_{1.0}^{1.5} \ln(x + 2y)\, dy\, dx = .075 \int_{-1}^{1} \int_{-1}^{1} \ln(.3u + .5v + 4.2)\, dv\, du.$$

The Gaussian quadrature formula for $n = 3$ in both u and v requires that we use the nodes,

$$u_1 = v_1 = 0, \qquad u_0 = v_0 = -.774596692, \qquad \text{and} \qquad u_2 = v_2 = .7745966692.$$

The associated weights are $w_1 = .8888888889$ and $w_0 = w_2 = .5555555556.$ (See Table 4.9.) Thus,

$$\int_{1.4}^{2.0} \int_{1.0}^{1.5} \ln(x + 2y)\, dy\, dx \simeq \sum_{i=0}^{2} \sum_{j=0}^{2} w_i w_j \ln(3u_i + 5v_j + 4.2) = .4295545313.$$

Although this result requires only 6 functional evaluations compared to 15 for the composite Simpson's rule considered in Example 1, this result is accurate to within 4.8×10^{-9}. $\qquad\qquad\qquad\qquad\qquad\qquad\qquad\qquad\qquad\qquad\qquad\square$

The use of approximation methods for double integrals is not limited to those integrals with rectangular regions of integration. The techniques previously discussed can easily be modified to approximate integrals of the form

(4.68) $$\int_{a}^{b} \int_{c(x)}^{d(x)} f(x, y)\, dy\, dx$$

or

(4.69) $$\int_{c}^{d} \int_{a(y)}^{b(y)} f(x, y)\, dx\, dy.$$

In fact, integrals on regions not of this type can also be approximated by performing appropriate partitions of the region. (See Exercise 4.)

To describe the technique involved with approximating an integral in the form

$$\int_{a}^{b} \int_{c(x)}^{d(x)} f(x, y)\, dy\, dx,$$

we will use Simpson's rule to integrate with respect to both variables. The step size for the variable x is $h = (b - a)/2$, but the step size for y varies with x (see Fig. 4.8).

$$k(x) = \frac{d(x) - c(x)}{2}.$$

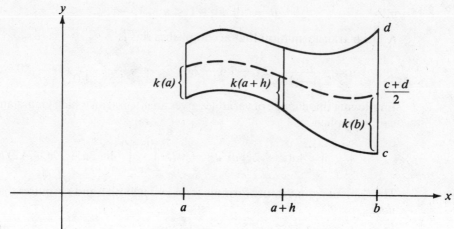

FIGURE 4.8

Consequently,

$$\int_a^b \int_{c(x)}^{d(x)} f(x, y) \, dy \, dx \approx \int_a^b \frac{k(x)}{3} \left[f(x, c(x)) + 4f(x, c(x) + k(x)) + f(x, d(x)) \right] dx$$

$$\approx \frac{h}{3} \left\{ \frac{k(a)}{3} \left[f(a, c(a)) + 4f(a, c(a) + k(a)) + f(a, d(a)) \right] \right.$$

$$+ \frac{4k(a + h)}{3} \left[f(a + h, c(a + h)) + 4f(a + h, c(a + h) \right.$$

$$+ k(a + h)) + f(a + h, d(a + h)) \right]$$

$$+ \left. \frac{k(b)}{3} \left[f(b, c(b)) + 4f(b, c(b) + k(b)) + f(b, d(b)) \right] \right\}.$$

The following algorithm applies the composite Simpson's rule to an integral in the form (4.68). Integrals in the form (4.69) can, of course, be handled similarly.

Multiple Integral Algorithm 4.4

To approximate the integral $I = \int_a^b \int_{c(x)}^{d(x)} f(x, y) \, dy \, dx$:

INPUT endpoints a, b; positive integers m, n.

OUTPUT approximation J.

Step 1 Set $h = (b - a)/(2n)$.

Step 2 Set $J_1 = 0$; *(End terms.)*
$\qquad J_2 = 0$; *(Even terms.)*
$\qquad J_3 = 0$. *(Odd terms.)*

Step 3 For $i = 0, 1, \ldots, 2n$
$\qquad$ set $x = a + ih$; *(Composite Simpson's method for fixed x.)*
$\qquad HX = (d(x) - c(x))/(2m)$;
$\qquad K_1 = f(x, c(x)) + f(x, d(x))$;
$\qquad K_2 = 0$; *(Even terms.)*
$\qquad K_3 = 0$; *(Odd terms.)*

for $j = 1, \ldots, 2m - 1$
    set $y = c(x) + jHX$;
        $Z = f(x, y)$;
    if j is even then set $K_2 = K_2 + Z$
            else set $K_3 = K_3 + Z$;
    set $L = (K_1 + 2K_2 + 4K_3)HX/3$;

$$\left(L \simeq \int_{c(x_i)}^{d(x_i)} f(x_i, y) \, dy \text{ by Composite Simpson's method.} \right)$$

if $i = 0$ or $i = 2m$
    then set $J_1 = J_1 + L$
    else if i is even then set $J_2 = J_2 + L$
            else set $J_3 = J_3 + L$.

Step 4 Set $J = (J_1 + 2J_2 + 4J_3)h/3$.

Step 5 OUTPUT (J);
    STOP.

EXAMPLE 3 Algorithm 4.4 applied to the function $f(x, y) = e^{(y/x)}$ on the region $R = \{(x, y) | .1 \leq x \leq .5, \ x^3 \leq y \leq x^2\}$ (see Fig. 4.9) with $n = m = 5$ produces the approximation

$$\int_{.1}^{.5} \int_{x^3}^{x^2} e^{(y/x)} \, dy \, dx \approx .0333054.$$

The exact value of this integral to seven decimal places is .0333056. $\square$

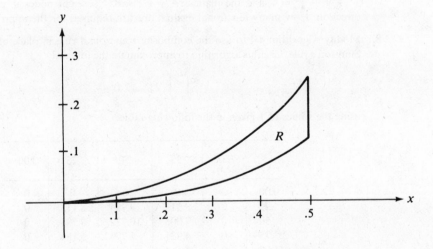

FIGURE 4.9

Exercise Set 4.8

1. Use Algorithm 4.4 with $n = m = 3$ to approximate $\int_0^1 \int_0^1 e^{-x^2 y^2} \, dy \, dx$.

2. Use Algorithm 4.4 with $n = 3$ and $m = 5$ to approximate $\int_0^1 \int_0^\pi y \sin \sqrt{x} \, dy \, dx$.

3. Use Algorithm 4.4 with $n = m = 7$ to approximate

$$\iint\limits_{R} e^{-(x+y)} \, dA$$

for the region R in the plane bounded by the curves $y = x^2$ and $y = \sqrt{x}$.

4. Use Algorithm 4.4 to approximate

$$\iint\limits_{R} \sqrt{xy + y^2} \, dA,$$

where R is the region in the plane bounded by the lines $x + y = 6$, $3y - x = 2$, and $3x - y = 2$. First partition R into two regions R_1 and R_2 on which Algorithm 4.4 can be applied. Use $n = m = 3$ on both R_1 and R_2.

5. Use Simpson's rule to approximate the triple integral $\int_0^1 \int_1^2 \int_0^{.5} e^{xyz} \, dx \, dy \, dz$.

6. Modify Algorithm 4.4 to approximate triple integrals. Use this modification and six nodes in each coordinate direction to approximate

$$\iiint\limits_{S} xy \sin(yz) \, dz \, dy \, dx,$$

where S is the solid bounded by the coordinate planes and the planes $x = \pi$, $y = \pi/2$, $z = \pi/3$. Compare this approximation to the exact result.

7. Use the Algorithm generated in Exercise 6 to approximate

$$\iiint\limits_{S} \sqrt{xyz} \, dV,$$

when S is the region in the first octant bounded by the cylinder $x^2 + y^2 = 4$, the sphere $x^2 + y^2 + z^2 = 4$, and the plane $x + y + z = 8$. Use eight nodes in each coordinate direction. How many functional evaluations are required for the approximation?

8. Modify Algorithm 4.4 to use the composite trapezoidal rule in place of the composite Simpson's rule. Use this algorithm to approximate the integral

$$\int_0^1 \int_0^1 f(x, y) \, dy \, dx,$$

using the values of f given in the following table.

x \ y	0.00	.25	.50	.75	1.00
0.00	0	0	0	0	0
.25	0	8.113	8.994	8.113	0
.50	0	7.005	10.722	8.704	0
.75	0	4.921	6.779	5.184	0
1.00	0	0	0	0	0

9. Modify Algorithm 4.4 to use the composite trapezoidal rule in one variable and the composite Simpson's rule in the other variable. Use the algorithm to approximate

$$\int_0^{1.5} \int_0^{1.5} f(x, y) \, dy \, dx,$$

using the values of f given in the following table:

y x	0.00	.25	.50	.75	1.00	1.25	1.50
0.00	0	0	0	0	0	0	0
.50	0	.2886	.7582	1.2480	1.4325	1.2754	.9235
1.00	0	.2160	.7163	1.1791	1.3534	1.2049	.8724
1.50	0	.0928	.3078	.5067	.5816	.5178	.3749

10. Modify Algorithm 4.4 to use Gaussian Quadrature in place of the composite Simpson's Rule. Use this algorithm with $n = m = 4$ to approximate

$$\iint_R e^{-(x+y)} \, dA$$

for the region R in the plane bounded by the curves $y = x^2$ and $y = \sqrt{x}$. Compare this approximation to the result obtained in Exercise 3, as well as to the exact result.

11. A plane lamina is defined to be a thin sheet of continuously distributed mass. If ρ is a function describing the density of lamina having the shape of a region R in the xy-plane, then the center of mass of the lamina $(\bar{x}, \bar{y})$ is defined by

$$\bar{x} = \frac{\iint_R x\rho(x, y) \, dA}{\iint_R \rho(x, y) \, dA}, \qquad \bar{y} = \frac{\iint_R y\rho(x, y) \, dA}{\iint_R \rho(x, y) \, dA}.$$

Use Algorithm 4.4 with $n = m = 7$ to find the center of mass of the lamina described by $R = \{(x, y) | 0 \le x \le 1, \ 0 \le y \le \sqrt{1 - x^2}\}$ with density function $\rho(x, y) = e^{-(x^2 + y^2)}$. Compare the approximation to the exact result.

12. The area of a surface described by $z = f(x, y)$ for (x, y) in R is given by

$$\iint_R \sqrt{[f_x(x, y)]^2 + [f_y(x, y)]^2} \, dA.$$

Use Algorithm 4.4 with $n = m = 4$ to find an approximation to the area of the surface on the hemisphere $x^2 + y^2 + z^2 = 9, z \ge 0$ that lies above the region in the plane described by $R = \{(x, y) | 0 \le x \le 1, 0 \le y \le 1\}$.

5

Initial-Value Problems for Ordinary Differential Equations

The motion of a swinging pendulum under certain simplifying assumptions can be described by the second-order differential equation

$$\frac{d^2\theta}{dt^2} + \frac{g}{L}\sin\theta = 0,$$

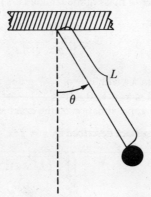

where L is the length of the pendulum, g is the gravitational constant of the earth, and θ is the angle the pendulum makes with the vertical or equilibrium position. If, in addition, we specify the position of the pendulum when the motion begins, $\theta(t_0) = \theta_0$, and its velocity at that point, $\theta'(t_0) = \theta'_0$, we have what is called an initial-value problem.

For small values of θ, the approximation $\theta \approx \sin\theta$ can be used to simplify this problem to the linear equation

$$\frac{d^2\theta}{dt^2} + \frac{g}{L}\theta = 0,$$

which can be solved by a standard differential-equation technique. For larger values of θ, however, approximations of the type discussed in this chapter must be used.

Although any textbook on the subject of ordinary differential equations details a number of methods for finding solutions to first-order initial-value problems, in actual practice few of the problems originating from the study of physical phenomena can be solved exactly.

The first part of this chapter is concerned with approximating the solution $y(t)$ to a problem of the form

$$(5.1) \qquad \frac{dy}{dt} = f(t, y) \qquad \text{for } a \leq t \leq b,$$

subject to an initial condition

$$(5.2) \qquad y(a) = \alpha.$$

Later in the chapter, we deal with the extension of these methods to a system of first-order differential equations in the form

$$(5.3) \qquad \frac{dy_1}{dt} = f_1(t, y_1, y_2, \ldots, y_n),$$

$$\frac{dy_2}{dt} = f_2(t, y_1, y_2, \ldots, y_n),$$

$$\vdots \qquad \qquad \vdots$$

$$\frac{dy_n}{dt} = f_n(t, y_1, y_2, \ldots, y_n),$$

for $a \leq t \leq b$, subject to the initial conditions

$$(5.4) \qquad y_1(a) = \alpha_1, \quad y_2(a) = \alpha_2, \quad \ldots, \quad y_n(a) = \alpha_n$$

and the relationship of a system of this type to the general nth-order initial-value problem of the form

$$(5.5) \qquad y^{(n)} = f(t, y, y', y'', \ldots, y^{(n-1)})$$

for $a \leq t \leq b$, subject to the initial conditions

$$(5.6) \qquad y(a) = \alpha_0, \quad y'(a) = \alpha_1, \quad \ldots, \quad y^{(n-1)}(a) = \alpha_{n-1}.$$

5.1 Elementary Theory of Initial-Value Problems

Before attempting to solve an initial-value problem, we would like to know whether a solution exists, and if so, whether it is unique. Moreover, since problems obtained by observing physical phenomena generally only approximate the actual situation, it is of interest to know whether small changes in the statement of the problem will introduce correspondingly small changes in the solution. This is also important because of the possibility of rounding errors when numerical methods are used.

To discuss these problems we need some definitions and results from the theory of ordinary differential equations. The first definition is simply the extension of the definition given in Exercise 9 of Section 1.1 to a function involving two variables.

Definition 5.1   A function $f(t, y)$ is said to satisfy a **Lipschitz condition** in the variable y on a set $D \subset R^2$ provided a constant $L > 0$ exists with the property that

$$|f(t, y_1) - f(t, y_2)| \leq L|y_1 - y_2|$$

whenever $(t, y_1), (t, y_2) \in D$.

The constant L is called a **Lipschitz constant** for f.

EXAMPLE 1   If $D = \{(t, y) | 1 \leq t \leq 2, -3 \leq y \leq 4\}$ and $f(t, y) = t|y|$, then for each $(t, y_1), (t, y_2) \in D$

$$|f(t, y_1) - f(t, y_2)| = \big| t|y_1| - t|y_2| \big| = |t| \, \big| |y_1| - |y_2| \big| \leq 2|y_1 - y_2|.$$

Thus f satisfies a Lipschitz condition on D in the variable y with Lipschitz constant 2. In fact, the smallest value possible for the Lipschitz constant for this problem is $L = 2$. $\square$

Definition 5.2   A set $D \subset R^2$ is said to be **convex** if whenever (t_1, y_1) and (t_2, y_2) belong to D, the point $((1 - \lambda)t_1 + \lambda t_2, (1 - \lambda)y_1 + \lambda y_2)$ also belongs to D for each $\lambda, 0 \leq \lambda \leq 1$.

In geometric terms, Definition 5.2 states that a set is convex provided that whenever two points belong to the set the entire straight-line segment between the points must also belong to the set. The sets we consider in this chapter will generally be of the form $D = \{(t, y) | a \leq t \leq b, -\infty < y < \infty\}$ for some constants a and b. It is easy to verify (see Exercise 5) that such sets are convex.

Theorem 5.3   Suppose that $f(t, y)$ is defined on a convex set $D \subset R^2$. If a constant $L > 0$ exists with

(5.7)
$$\left| \frac{\partial f}{\partial y}(t, y) \right| \leq L \qquad \text{for all } (t, y) \in D,$$

then f satisfies a Lipschitz condition on D in the variable y with Lipschitz constant L.

The proof of Theorem 5.3 is discussed in Exercise 4, and is similar to the proof of the corresponding result for functions of one variable discussed in Exercise 9 of Section 1.1.

As the next theorem will show, it is often of significant interest to determine whether the function involved in an initial-value problem satisfies a Lipschitz condition in its second variable, and condition (5.7) is generally much easier to apply than the definition. It should be remarked, however, that Theorem 5.3 gives only *sufficient* conditions for a Lipschitz condition to hold; a reexamination of Example 1 will demonstrate that these conditions are definitely *not necessary*.

The following theorem is a version of the fundamental existence and uniqueness theorem for first-order ordinary differential equations. Although the theorem can be proved with the hypothesis reduced somewhat, this form of the theorem is sufficient for our purposes. (The proof of the theorem, in approximately this form, can be found in Birkhoff and Rota [13], pages 103, 112–115.)

Theorem 5.4 Suppose that $D = \{(t, y) | a \leq t \leq b, -\infty < y < \infty\}$, and that $f(t, y)$ is continuous on D. If f satisfies a Lipschitz condition on D in the variable y, then the initial-value problem

$$y' = f(t, y), \qquad a \leq t \leq b, \quad y(a) = \alpha,$$

has a unique solution $y(t)$ for $a \leq t \leq b$.

EXAMPLE 2 Consider the initial-value problem

$$y' = 1 + t \sin(ty), \qquad 0 \leq t \leq 2, \quad y(0) = 0.$$

Holding t constant and applying the Mean Value Theorem (Theorem 1.8, p. 5) to the function

$$f(t, y) = 1 + t \sin(ty),$$

we find that whenever $y_1 < y_2$, a number ξ, $y_1 < \xi < y_2$, exists with

$$t^2 \cos(\xi t) = \frac{\partial}{\partial y} f(t, \xi) = \frac{f(t, y_2) - f(t, y_1)}{y_2 - y_1}.$$

Thus, for all $y_1 < y_2$,

$$|f(t, y_2) - f(t, y_1)| = |y_2 - y_1||t^2 \cos(\xi t)| \leq 4|y_2 - y_1|,$$

and f satisfies a Lipschitz condition in the variable y with Lipschitz constant four.

Since, additionally, $f(t, y)$ is continuous when $0 \leq t \leq 2$ and $-\infty < y < \infty$, Theorem 5.4 implies that a unique solution exists to this initial-value problem. It might be interesting for the reader who has completed a course in differential equations to try to find the exact solution to this problem. $\square$

Now that we have, to some extent, taken care of the question of when initial-value problems have unique solutions, we can move to the other question posed earlier in the section. Is there a way to determine whether a particular problem has the property that small changes or perturbations in the statement of the problem introduce correspondingly small changes in the solution? As usual, we need to first give a workable definition to express this concept.

Definition 5.5 The initial-value problem

$$(5.8) \qquad \frac{dy}{dt} = f(t, y), \qquad a \le t \le b, \quad y(a) = \alpha,$$

is said to be a **well-posed problem** if:

i) a unique solution, $y(t)$, to the problem exists;

ii) a number $\varepsilon > 0$ exists, with the property that a unique solution, $z(t)$, to the problem

$$(5.9) \qquad \frac{dz}{dt} = f(t, z) + \delta(t), \qquad a \le t \le b, \quad z(a) = \alpha + \varepsilon_0,$$

exists whenever $|\varepsilon_0| < \varepsilon$ and $|\delta(t)| < \varepsilon$ for all $a \le t \le b$;

iii) a constant $k > 0$ exists with the property that

$$|z(t) - y(t)| < k\varepsilon \qquad \text{for all } a \le t \le b.$$

The problem specified by Eq. (5.9) is often called a **perturbed problem** associated with the original problem (5.8). The following theorem specifies conditions which ensure that an initial-value problem is well-posed. The proof of this theorem can be found in the previously mentioned reference, Birkhoff and Rota [13], pages 103–107.

Theorem 5.6 Suppose $D = \{(t, y) | a \le t \le b \text{ and } c \le y \le d\}$. The initial-value problem

$$(5.10) \qquad \frac{dy}{dt} = f(t, y), \qquad a \le t \le b, \quad y(a) = \alpha,$$

is well-posed provided f is continuous and satisfies a Lipschitz condition in the variable y on the set D.

EXAMPLE 3 Let $D = \{(t, y) | 0 \le t \le 1, \ -\infty < y < \infty\}$ and consider the initial-value problem

$$(5.11) \qquad \frac{dy}{dt} = -y + t + 1, \qquad 0 \le t \le 1, \quad y(0) = 1.$$

Since

$$\left| \frac{\partial(-y + t + 1)}{\partial y} \right| = 1,$$

Theorem 5.3 implies that $f(t, y) = -y + t + 1$ satisfies a Lipschitz condition on D with Lipschitz constant 1. Since, additionally, f is continuous on D, Theorem 5.6 implies that Eq. (5.11) is a well-posed problem.

Consider the perturbed problem

$$(5.12) \qquad \frac{dz}{dt} = -z + t + 1 + \delta, \qquad 0 \le t \le 1, \quad z(0) = 1 + \varepsilon_0,$$

where δ and ε_0 are constants. The solutions to (5.11) and (5.12) can be shown to be

$$y(t) = e^{-t} + t \quad \text{and} \quad z(t) = (1 + \varepsilon_0 - \delta)e^{-t} + t + \delta,$$

respectively. It is easy to verify that if $|\delta| < \varepsilon$ and $|\varepsilon_0| < \varepsilon$, then

$$|y(t) - z(t)| = |(\delta - \varepsilon_0)e^{-t} - \delta| \le |\varepsilon_0| + |\delta||1 - e^{-t}| \le 2\varepsilon,$$

for all t, which corroborates the result obtained by the use of Theorem 5.6. $\square$

Exercise Set 5.1

1. For each choice of $f(t, y)$ given in (a)–(d):

   i) Does f satisfy a Lipschitz condition on

   $$D = \{(t, y)|0 \le t \le 1, -\infty < y < \infty\}?$$

   ii) Determine whether the initial-value problem

   $$y' = f(t, y), \quad 0 \le t \le 1, \quad y(0) = 1,$$

   is well-posed.

   a) $f(t, y) = t^2 y + 1$ b) $f(t, y) = ty$

   c) $f(t, y) = \dfrac{1}{y^2 + 1}$ d) $f(t, y) = \sqrt{y + 1}$

2. Use Theorem 5.4 to show that

   $$y' = y \cos t, \quad 0 \le t \le 1, \quad y(0) = 1$$

   has a unique solution. Find the solution.

3. Use Theorem 5.4 to show that

   $$y' - \frac{2}{t}y = t^2 e^t, \quad 1 \le t \le 2, \quad y(1) = 0$$

   has a unique solution. Find the solution.

4. Prove Theorem 5.3 by applying the Mean Value Theorem (Theorem 1.8, p. 5) to $f(t, y)$, holding t fixed.

5. Show that, for any constants a and b, the set $D = \{(t, y)|a \le t \le b, -\infty < y < \infty\}$ is convex.

5.2   Euler's Method

Although Euler's method is seldom used in practice, the simplicity of its derivation can be used to illustrate the techniques involved in the construction of some of the more advanced techniques, without the cumbersome algebra, which accompanies these constructions.

The object of the method is to obtain an approximation to the well-posed initial-value problem

(5.13) $\dfrac{dy}{dt} = f(t, y), \quad a \le t \le b, \quad y(a) = \alpha.$

In actuality, a continuous approximation to the solution $y(t)$ will not be obtained; instead approximations to y will be generated at various points, called **mesh points**, in the interval $[a, b]$. Once the approximate solution is obtained at these points, the approximate solution at other points in the interval can be obtained by using one of the interpolation procedures discussed in Chapter 3.

In this section we make the stipulation that the mesh points be equally distributed throughout the interval $[a, b]$. This condition can be ensured by choosing a positive integer N and selecting the points $\{t_0, t_1, t_2, \ldots, t_N\}$ where

$$t_i = a + ih \qquad \text{for each } i = 0, 1, 2, \ldots, N.$$

The common distance between the points, $h = (b - a)/N$, is called the **step size**.

To derive Euler's method, we will utilize Taylor's Theorem (Theorem 1.13, p. 7). Another method of derivation is presented in Exercise 8.

Suppose that $y(t)$, the unique solution to (5.13), has two continuous derivatives on $[a, b]$, so that for each $i = 0, 1, 2, \ldots, N - 1$, $y(t_{i+1})$ can be written

$$(5.14) \qquad y(t_{i+1}) = y(t_i + h) = y(t_i) + hy'(t_i) + \frac{h^2}{2} y''(\xi_i)$$

for some point $\xi_i, t_i < \xi_i < t_{i+1}$.

Using the notation $h = t_{i+1} - t_i$ implies that a number $\theta_i, 0 < \theta_i < 1$, exists with

$$(5.15) \qquad y(t_{i+1}) = y(t_i) + hy'(t_i) + \frac{h^2}{2} y''(t_i + \theta_i h),$$

and, since $y(t)$ satisfies the differential equation (5.13),

$$y(t_{i+1}) = y(t_i) + hf(t_i, y(t_i)) + \frac{h^2}{2} y''(t_i + \theta_i h).$$

Rewriting this equation gives

$$(5.16) \qquad \frac{y(t_{i+1}) - y(t_i)}{h} = f(t_i, y(t_i)) + \frac{h}{2} y''(t_i + \theta_i h)$$

for each $i = 0, 1, \ldots, N - 1$.

When h is sufficiently small the continuity of y'' suggests that the term $(h/2)y''(t_i + \theta_i h)$ is also small, and that

$$(5.17) \qquad \frac{y(t_{i+1}) - y(t_i)}{h} \approx f(t_i, y(t_i)).$$

Euler's method uses this assumption by defining

$$(5.18) \qquad \begin{aligned} w_0 &= \alpha, \\ w_{i+1} &= w_i + hf(t_i, w_i) \end{aligned}$$

and assuming that $w_i \approx y(t_i)$ for each $i = 1, 2, \ldots, N$.

Equation (5.18) is called the **difference equation** associated with Euler's method. As we will see later in this chapter, the theory and solution of difference equations parallels, in many ways, the theory and solution of differential equations. The algorithmic form of Euler's method is presented as follows.

Euler's Algorithm 5.1

To approximate the solution of the initial-value problem

$$y' = f(t, y), \qquad a \le t \le b, \quad y(a) = \alpha,$$

at $(N + 1)$ equally spaced numbers in the interval $[a, b]$:

INPUT endpoints a, b; integer N; initial condition α.

OUTPUT approximation w to y at the $(N + 1)$ values of t.

Step 1 Set $h = (b - a)/N$;
$\qquad t = a$;
$\qquad w = \alpha$;
$\qquad$ OUTPUT (t, w).

Step 2 For $i = 1, 2, \ldots, N$ do Steps 3, 4.

$\qquad$ Step 3 Set $w = w + hf(t, w)$;   (*Compute* w_i.)
$\qquad\qquad\qquad t = a + ih$.   (*Compute* t_i.)

$\qquad$ Step 4 OUTPUT (t, w).

Step 5 STOP.

To visualize the geometric interpretation of Euler's method we first introduce the notation $y_i = y(t_i)$ for each $i = 0, 1, 2, \ldots, N$, and note that, when w_i is a close approximation to y_i, the assumption that the problem is well-posed implies that

$$f(t_i, w_i) \approx y'(t_i) = f(t_i, y(t_i)).$$

One step in the method consequently appears as in Fig. 5.1, and a series of steps appear in Fig. 5.2.

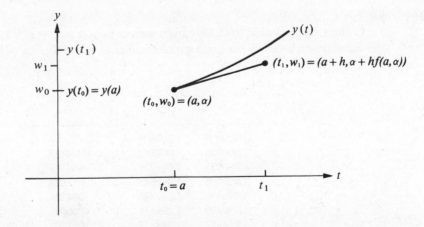

FIGURE 5.1

EXAMPLE 1 To find approximations to the initial-value problem.

(5.19) $\qquad\qquad y' = -y + t + 1, \qquad 0 \le t \le 1, \quad y(0) = 1.$

suppose $N = 10$ so that $h = .1$ and $t_i = .1i$.

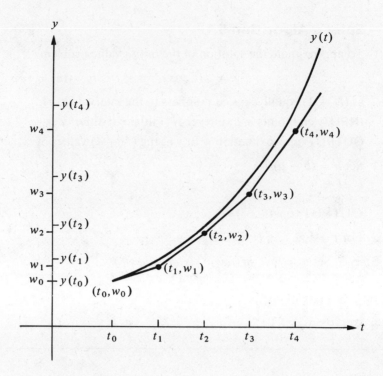

FIGURE 5.2

Using the fact that $f(t, y) = -y + t + 1$, we have

$$w_0 = 1$$

and

$$w_i = w_{i-1} + h(-w_{i-1} + t_{i-1} + 1)$$
$$= w_{i-1} + .1(-w_{i-1} + .1(i-1) + 1)$$
$$= .9w_{i-1} + .01(i-1) + .1$$

for $i = 1, 2, \ldots, 10$.

The actual solution to (5.19) can be seen to be $y(t) = t + e^{-t}$. Table 5.1 shows the comparison between the approximate values at t_i and the actual values. □

TABLE 5.1

t_i	w_i	y_i	Error $= |w_i - y_i|$
0	1.000000	1.000000	0
.1	1.000000	1.004837	.004837
.2	1.010000	1.018731	.008731
.3	1.029000	1.040818	.011818
.4	1.056100	1.070320	.014220
.5	1.090490	1.106531	.016041
.6	1.131441	1.148812	.017371
.7	1.178297	1.196585	.018288
.8	1.230467	1.249329	.018862
.9	1.287420	1.306570	.019150
1.0	1.348678	1.367879	.019201

Note that the error grows slightly as the value of t_i increases. This controlled error growth is a consequence of the stability of Euler's method, which implies that the errors due to rounding are expected to grow in no worse than a linear manner.

In order to obtain an error bound for a general case of Euler's method, it is convenient to present two computational lemmas, which will be needed in the development.

Lemma 5.7   For all $x \geq -1$ and any positive integer m,

(5.20) $$0 \leq (1 + x)^m \leq e^{mx}.$$

Proof   Applying Taylor's Theorem (Theorem 1.13, p. 7) with $f(x) = e^x$, $x_0 = 0$, and $n = 1$, gives

$$e^x = 1 + x + \tfrac{1}{2}x^2 e^{\xi}$$

where ξ is between x and zero. Thus,

$$0 \leq 1 + x \leq 1 + x + \tfrac{1}{2}x^2 e^{\xi} = e^x$$

and, since $1 + x \geq 0$,

$$0 \leq (1 + x)^m \leq (e^x)^m = e^{mx}. \qquad \square$$

Lemma 5.8   If m and n are positive real numbers and $\{a_i\}_{i=0}^{k}$ is a sequence satisfying $a_0 \geq -n/m$ and

(5.21) $$a_{i+1} \leq (1 + m)a_i + n \qquad \text{for each } i = 0, 1, 2, \ldots, k,$$

then

(5.22) $$a_{i+1} \leq e^{(i+1)m}\left(\frac{n}{m} + a_0\right) - \frac{n}{m}.$$

Proof   For a fixed integer i, inequality (5.21) implies that

$$
\begin{aligned}
a_{i+1} &\leq (1 + m)a_i + n \\
&\leq (1 + m)[(1 + m)a_{i-1} + n] + n \\
&\leq (1 + m)\{(1 + m)[(1 + m)a_{i-2} + n] + n\} + n \\
&\;\;\vdots \\
&\leq (1 + m)^{i+1}a_0 + [1 + (1 + m) + (1 + m)^2 + \cdots + (1 + m)^i]n.
\end{aligned}
$$

But

$$1 + (1 + m) + (1 + m)^2 + \cdots + (1 + m)^i = \sum_{j=0}^{i}(1 + m)^j$$

is a geometric series with ratio $(1 + m)$ and, as such, sums to

$$\frac{1 - (1 + m)^{i+1}}{1 - (1 + m)}.$$

Thus, $\quad a_{i+1} \leq (1 + m)^{i+1}a_0 + \dfrac{(1 + m)^{i+1} - 1}{m}n = (1 + m)^{i+1}\left(\dfrac{n}{m} + a_0\right) - \dfrac{n}{m},$

and, by Lemma 5.7,

$$a_{i+1} \leq e^{(i+1)m}\left(\frac{n}{m} + a_0\right) - \frac{n}{m}. \qquad \square$$

Theorem 5.9   Let $y(t)$ denote the unique solution to the well-posed initial-value problem

$$y' = f(t, y), \qquad a \le t \le b, \quad y(a) = \alpha,$$

and $w_0, w_1, \ldots, w_N$ be the approximations generated by Euler's method for some positive integer N.

If f satisfies a Lipschitz condition with constant L on

$$D = \{(t, y) | a \le t \le b, -\infty < y < \infty\},$$

and a constant M exists with the property that

$$|y''(t)| \le M \qquad \text{for all } t \in [a, b],$$

then

(5.23)
$$|y(t_i) - w_i| \le \frac{hM}{2L} [e^{L(t_i - a)} - 1]$$

for each $i = 0, 1, 2, \ldots, N$.

Proof   When $i = 0$ the result is clearly true.

From Eq. (5.16), we have for $i = 0, 1, \ldots, N - 1$,

$$y(t_{i+1}) = y(t_i) + hf(t_i, y(t_i)) + \frac{h^2}{2} y''(t_i + \theta_i h);$$

and from the equations in (5.18)

$$w_{i+1} = w_i + hf(t_i, w_i).$$

Consequently, using the notation $y_i = y(t_i)$ and $y_{i+1} = y(t_{i+1})$:

$$y_{i+1} - w_{i+1} = y_i - w_i + h[f(t_i, y_i) - f(t_i, w_i)] + \frac{h^2}{2} y''(t_i + \theta_i h)$$

and $$|y_{i+1} - w_{i+1}| \le |y_i - w_i| + h|f(t_i, y_i) - f(t_i, w_i)| + \frac{h^2}{2} |y''(t_i + \theta_i h)|.$$

The assumptions that f satisfies a Lipschitz condition in the second variable with constant L and that $|y''(t)| \le M$ imply that

$$|y_{i+1} - w_{i+1}| \le |y_i - w_i|(1 + hL) + \frac{h^2 M}{2}.$$

Referring to Lemma 5.8 and letting $a_j = |y_j - w_j|$ for each $j = 0, 1, \ldots, N$, while $m = hL$ and $n = h^2 M/2$, we see that

$$|y_{i+1} - w_{i+1}| \le e^{(i+1)hL} \left(|y_0 - w_0| + \frac{h^2 M}{2hL} \right) - \frac{h^2 M}{2hL}$$

Since $|y_0 - w_0| = 0$ and $(i + 1)h = t_{i+1} - t_0 = t_{i+1} - a$, we have the desired result:

$$|y_{i+1} - w_{i+1}| \le \frac{hM}{2L} (e^{(t_{i+1} - a)L} - 1). \qquad \square$$

The weakness of Theorem 5.9 lies in the requirement that a bound be known for the second derivative of the solution. Although this condition often prohibits us from obtaining a realistic error bound, it should be noted that if $\partial f / \partial t$ and $\partial f / \partial y$ both exist, then

$$y''(t) = \frac{dy'}{dt}(t) = \frac{df}{dt}(t, y(t))$$

$$= \frac{\partial f}{\partial t}(t, y(t)) + \frac{\partial f}{\partial y}(t, y(t)) f(t, y(t)),$$

and a bound for $y''(t)$ might be possible to obtain for certain problems without actually knowing $y(t)$.

EXAMPLE 2   Returning to the initial-value problem

$$y' = -y + t + 1, \qquad 0 \le t \le 1, \quad y(0) = 1,$$

considered in Example 1, we see that with $f(t, y) = -y + t + 1$, $\partial f / \partial y = -1$; so Theorem 5.3 implies that f satisfies a Lipschitz condition with $L = 1$.

Also, since in this case we know that the exact solution is $y(t) = t + e^{-t}$, we have $y''(t) = e^{-t}$,

and
$$|y''(t)| \le e^{-0} = 1 \qquad \text{for all } t \in [0, 1].$$

Using inequality (5.23) with $h = .1$ and $M = L = 1$ gives the error bound

$$|y_i - w_i| \le .05(e^{t_i} - 1).$$

The following table lists the actual error found in Example 1, together with this error bound.

t_i	.1	.2	.3	.4	.5	.6	.7	.8	.9	1.0
Actual error	.00484	.00873	.01182	.01422	.01604	.01737	.01829	.01886	.01915	.01920
Error bound	.00526	.01107	.01749	.02459	.03244	.04111	.05069	.06128	.07298	.08591

Note that, even though the true bound for the second derivative of the solution was used, the error bound is considerably larger than the actual error. $\square$

The principal importance of the error-bound formula given in Theorem 5.9 is the fact that the bound depends linearly upon the step size h. Consequently, diminishing the step size should give correspondingly greater accuracy to the approximations.

Neglected in the result of Theorem 5.9 is the effect that round-off error plays in the choice of step size. It is clear that as h becomes smaller more calculations are necessary, and when finite-digit arithmetic is used, which is certainly the usual case, more round-off error would be expected. In actuality then, the difference-equation form

$$w_0 = \alpha,$$

$$w_i = w_{i-1} + hf(t_{i-1}, w_{i-1}) \qquad \text{for each } i = 1, 2, \ldots, N,$$

is not used to calculate the approximation to the solution y_i at a mesh point t_i, but instead an equation of the form

(5.24)
$$u_0 = \hat{\alpha},$$
$$u_i = u_{i-1} + hf(t_{i-1}, u_{i-1}) + \delta_i \qquad \text{for each } i = 1, 2, \ldots, N,$$

is used, where $\hat{\alpha}$ is the approximation to the initial value α and δ_i denotes the round-off error associated with the computation of $u_{i-1} + hf(t_{i-1}, u_{i-1})$. By using methods similar to those in the proof of Theorem 5.9 (see Exercise 10), we can obtain the following result, which gives an error bound for the finite-digit approximations to y_i obtained using Euler's method.

Theorem 5.10 Let $y(t)$ denote the unique solution to the well-posed initial-value problem

(5.25)
$$y' = f(t, y), \qquad a \le t \le b, \quad y(a) = \alpha$$

and $u_0, u_1, \ldots, u_N$ be the approximations obtained using (5.24). If $|\delta_i| < \delta$ for each $i = 0, 1, \ldots, N$, where δ_0 denotes $\hat{\alpha} - \alpha$, and the hypotheses of Theorem 5.9 hold for (5.25), then

(5.26)
$$|y(t_i) - u_i| \le \frac{1}{L}\left(\frac{hM}{2} + \frac{\delta}{h}\right)[e^{L(t_i - a)} - 1] + |\delta_0|e^{L(t_i - a)}$$

for each $i = 0, 1, \ldots, N$.

The error bound (5.26) is no longer linear in h and, in fact, since

$$\lim_{h \to 0}\left(\frac{hM}{2} + \frac{\delta}{h}\right) = +\infty,$$

the error would be expected to become large for sufficiently small values of h. Ordinary techniques of calculus can be used to determine a lower bound for the step size h. Letting $E(h) = (hM/2) + (\delta/h)$ implies that $E'(h) = (M/2) - (\delta/h^2)$.

If $h > \sqrt{2\delta/M}$, then $E'(h) < 0$ and $E(h)$ is decreasing.
If $h < \sqrt{2\delta/M}$, then $E'(h) > 0$ and $E(h)$ is increasing.

The minimal value of $E(h)$ occurs when

$$h = \sqrt{2\delta/M};$$

decreasing h beyond this value would tend to increase the total error in the approximation.

Normally, however, the value of δ is sufficiently small that this lower bound for h will not affect the operation of Euler's method.

Exercise Set 5.2

1. Use Euler's method to approximate the solution for each of the following initial-value problems:

a) $y' = \left(\frac{y}{t}\right)^2 + \left(\frac{y}{t}\right)$, $1 \le t \le 1.2$, $y(1) = 1$ with $h = .1$;

b) $y' = \sin t + e^{-t}$, $0 \le t \le 1$, $y(0) = 0$ with $h = .5$;

c) $y' = \dfrac{1}{t}(y^2 + y)$, $1 \le t \le 3$, $y(1) = -2$ with $h = .5$.

2.   Approximate the solution to

$$y' = t^2, \qquad 0 \le t \le 2, \quad y(0) = 0,$$

using Euler's method with $h = .5$. Compare the results to a graph of the solution $y = \frac{1}{3}t^3$.

3.   a)   Use Euler's method with $h = .1$ to approximate the solution to

$$y' = \frac{2}{t}y + t^2 e^t, \qquad 1 \le t \le 2, \quad y(1) = 0.$$

     Compute the value of h necessary for $|y(t_i) - w_i| \le .1$, using Eq. (5.23).

   b)   For the initial-value problem

$$y' = \frac{1}{t^2} - \frac{y}{t} - y^2, \qquad 1 \le t \le 2, \quad y(1) = -1,$$

     compute an approximation to the solution $y(t) = -1/t$, using Euler's method with $h = .05$. Compare the approximation to the actual solution.

   c)   Approximate the solution $y(t) = e^{-t} + t$ to the initial-value problem

$$y' = -y + t + 1, \qquad 0 \le t \le 5, \quad y(0) = 1,$$

     using Euler's method with $h = .2$ and $h = .1$.

4.   Consider the initial-value problem

$$y' = -10y, \qquad 0 \le t \le 2 \quad y(0) = 1,$$

which has solution $y(t) = e^{-10t}$. What happens when Euler's method is applied to this problem with $h = .1$? Does this behavior violate Theorem 5.9?

5.   Obtain an approximation to the solution of

$$y' = 1 + t \sin(ty), \qquad 0 \le t \le 2, \quad y(0) = 0,$$

using Euler's method with $h = .1$.

6.   In a circuit with impressed voltage $\mathscr{E}$, and resistance R, inductance L, capacitance C in parallel, the current i satisfies the differential equation

$$\frac{di}{dt} = C\frac{d^2\mathscr{E}}{dt^2} + \frac{1}{R}\frac{d\mathscr{E}}{dt} + \frac{1}{L}\mathscr{E}.$$

Suppose $C = .3$ farads, $R = 1.4$ ohms, $L = 1.7$ henries, and the voltage is given by

$$\mathscr{E}(t) = e^{-.06\pi t}\sin(2t - \pi).$$

If $i(0) = 0$, find the current i for the values $t = .1j, j = 0, 1, \ldots, 100$ using Euler's method.

7.   In a book entitled "Looking at History Through Mathematics," Rashevsky [69] (pages 103–110) considers a model for a problem involving the production of nonconformists in society. Suppose that a society has a population of $x(t)$ individuals at time t, in years, and that all nonconformists who mate with other nonconformists have offspring who are also nonconformists, while a fixed proportion r of all other offspring are also nonconformist. If the birth and death rates for all individuals are assumed to be the constants b and d,

respectively, and if conformists and nonconformists mate at random, the problem can be expressed by the differential equations

$$\frac{dx(t)}{dt} = (b - d)x(t) \quad \text{and} \quad \frac{dx_n(t)}{dt} = (b - d)x_n(t) + rb(x(t) - x_n(t)),$$

where $x_n(t)$ denotes the number of nonconformists in the population at time t.

a) If the variable $p(t) = x_n(t)/x(t)$ is introduced to represent the proportion of non-conformists in the society at time t, show that these equations can be combined and simplified to the single differential equation

$$\frac{dp(t)}{dt} = rb(1 - p(t)).$$

b) Assuming that $p(0) = .01$, $b = .02$, $d = .015$, and $r = .1$, use Algorithm 5.1 to approximate the solution $p(t)$ from $t = 0$ to $t = 50$ when the step size is $h = 1$ year.
c) Solve the differential equation for $p(t)$ exactly, and compare your result in part (b) when $t = 50$ with the exact value at that time.

8. For the initial-value problem $y' = f(t, y)$, $a \leq t \leq b$, with $y(a) = \alpha$.

a) Derive Euler's method by integrating the differential equation from t_i to t_{i+1} and using an appropriate quadrature formula to approximate the integral.
b) Obtain a difference method called the **trapezoidal method** by using the trapezoidal rule as the quadrature formula.

9. Let $E(h) = \dfrac{hM}{2} + \dfrac{\delta}{h}$.

a) For the initial-value problem

$$y' = -y + 1, \quad 0 \leq t \leq 1, \quad y(0) = 0,$$

compute the value of h to minimize $E(h)$. Assume $\delta = 5 \times 10^{-(n+1)}$ if you will be using n-digit arithmetic in part (c).
b) For the optimal h computed in part (a), use Eq. (5.26) to compute the minimal error obtainable.
c) Compare the actual error obtained using $h = .1$ and $h = .01$ to the minimal error in part (b). Can you explain the results?

10. Prove Theorem 5.10.

5.3 Higher-Order Taylor Methods

Since the object of numerical techniques is generally to determine sufficiently accurate approximations with minimal effort, it is necessary to have a means of comparing the efficiency of various approximation methods with respect to their use in computation. For difference-equation techniques for solving ordinary differential equations, such as Euler's method, the first measurement device we will consider is called the **local truncation error** of the method. The local truncation error at a specified step measures the amount by which the exact solution to the differential equation fails to satisfy the difference equation being used for the approximation. For Euler's method the local truncation error at the ith step for the problem

$$y' = f(t, y), \qquad a \le t \le b, \quad y(a) = \alpha,$$

is
$$\tau_i = \frac{y_i - y_{i-1}}{h} - f(t_{i-1}, y_{i-1}) \qquad \text{for each } i = 1, 2, \ldots, N,$$

where, as usual, $y_i = y(t_i)$ denotes the exact value of the solution at t_i. This error is called a local error because it measures the accuracy of the method at a specific step, assuming that the method was exact at the previous step. As such, it depends on the differential equation, the step size, and the particular step in the approximation.

By considering Eq. (5.16), in the previous section, we see that

$$\tau_i = \frac{h}{2} y''(t_i + \theta_i h) \qquad \text{for some } \theta_i, 0 < \theta_i < 1$$

and, when $y''(t)$ is known to be bounded by a constant M on $[a, b]$, this implies

$$|\tau_i| \le \frac{h}{2} M.$$

Recalling Definition 1.18, p. 17, we infer that the truncation error in Euler's method is $O(h)$. This result implies that one way to select difference-equation methods for solving ordinary differential equations is in such a manner that their local truncation errors are $O(h^p)$ for as large a value of p as possible, while keeping the number and complexity of calculations of the methods within a reasonable bound.

Since Euler's method was derived by using Taylor's Theorem with $n = 2$ to approximate the solution of the differential equation, our first attempt to find methods of improving the convergence properties of difference methods will be to extend this technique of derivation to larger values of n.

Assuming that the solution $y(t)$ to the initial-value problem

$$y' = f(t, y), \qquad a \le t \le b, \quad y(a) = \alpha,$$

has $(n + 1)$ continuous derivatives and expanding that solution, $y(t)$, in terms of its nth-degree Taylor polynomial about t_i, we obtain:

(5.27)
$$y(t_{i+1}) = y(t_i) + hy'(t_i) + \frac{h^2}{2} y''(t_i) + \cdots$$

$$+ \frac{h^n}{n!} y^{(n)}(t_i) + \frac{h^{n+1}}{(n+1)!} y^{(n+1)}(t_i + \theta_i h)$$

for some $\theta_i, 0 < \theta_i < 1$.

Successive differentiation of the solution, $y(t)$, gives

$$y'(t) = f(t, y(t));$$
$$y''(t) = f'(t, y(t));$$

and, in general,

$$y^{(k)}(t) = f^{(k-1)}(t, y(t)).$$

Substituting these results into Eq. (5.27) gives:

$$(5.28) \quad y(t_{i+1}) = y(t_i) + hf(t_i, y(t_i)) + \frac{h^2}{2} f'(t_i, y(t_i)) + \cdots$$

$$+ \frac{h^n}{n!} f^{(n-1)}(t_i, y(t_i)) + \frac{h^{n+1}}{(n+1)!} f^{(n)}(t_i + \theta_i h, y(t_i + \theta_i h))$$

for some $\theta_i, 0 < \theta_i < 1$.

The difference method corresponding to Eq. (5.28) is obtained by neglecting the remainder term involving θ_i. This method is called the **Taylor method of order n:**

$$(5.29) \quad \begin{aligned} w_0 &= \alpha, \\ w_{i+1} &= w_i + hT^{(n)}(t_i, w_i) \qquad \text{for each } i = 0, 1, \ldots, N-1, \end{aligned}$$

where $\quad T^{(n)}(t_i, w_i) = f(t_i, w_i) + \frac{h}{2} f'(t_i, w_i) + \cdots + \frac{h^{n-1}}{n!} f^{(n-1)}(t_i, w_i).$

Note that, with this terminology, Euler's method is Taylor's method of order one.

EXAMPLE 1   To apply Taylor's method of orders two and four to the initial-value problem

$$y' = -y + t + 1, \qquad 0 \le t \le 1, \quad y(0) = 1,$$

which was studied in Example 1 of Section 5.2, we must find the first three derivatives of $f(t, y(t)) = -y + t + 1$,

$$f'(t, y(t)) = \frac{d}{dt}(-y + t + 1) = -y' + 1 = y - t - 1 + 1 = y - t,$$

$$f''(t, y(t)) = \frac{d}{dt}(y - t) = y' - 1 = -y + t + 1 - 1 = -y + t,$$

and $f'''(t, y(t)) = \frac{d}{dt}(-y + t) = -y' + 1 = y - t - 1 + 1 = y - t;$

so

$$T^{(2)}(t_i, w_i) = f(t_i, w_i) + \frac{h}{2} f'(t_i, w_i)$$

$$= -w_i + t_i + 1 + \frac{h}{2}(w_i - t_i)$$

$$= \left(1 - \frac{h}{2}\right)(t_i - w_i) + 1$$

and $T^{(4)}(t_i, w_i) = f(t_i, w_i) + \frac{h}{2} f'(t_i, w_i) + \frac{h^2}{6} f''(t_i, w_i) + \frac{h^3}{24} f'''(t_i, w_i)$

$$= -w_i + t_i + 1 + \frac{h}{2}(w_i - t_i) + \frac{h^2}{6}(-w_i + t_i) + \frac{h^3}{24}(w_i - t_i)$$

$$= \left(1 - \frac{h}{2} + \frac{h^2}{6} - \frac{h^3}{24}\right)(t_i - w_i) + 1.$$

The Taylor methods of orders two and four are consequently

$$w_0 = 1,$$

(5.30) $$w_{i+1} = w_i + h\left[\left(1 - \frac{h}{2}\right)(t_i - w_i) + 1\right] \quad \text{for } i = 0, 1, \ldots, N-1,$$

and

$$w_0 = 1,$$

(5.31) $$w_{i+1} = w_i + h\left[\left(1 - \frac{h}{2} + \frac{h^2}{6} - \frac{h^3}{24}\right)(t_i - w_i) + 1\right]$$

for $i = 0, 1, \ldots, N-1$, respectively.

Using $h = .1$ implies $N = 10$ and $t_i = .1i$ for each $i = 1, 2, \ldots, 10$; so (5.30) becomes

$$w_0 = 1,$$

(5.32) $$w_{i+1} = w_i + .1\left[\left(1 - \frac{.1}{2}\right)(.1i - w_i) + 1\right]$$

$$= .905w_i + .0095i + .1,$$

and (5.31) becomes

$$w_0 = 1,$$

(5.33) $$w_{i+1} = w_i + .1\left[\left(1 - \frac{.1}{2} + \frac{.01}{6} - \frac{.001}{24}\right)(.1i - w_i) + 1\right]$$

$$= .9048375w_i + .00951625i + .1$$

for each $i = 0, 1, \ldots, 9$.

Table 5.2 lists the actual values of the solution $y(t) = t + e^{-t}$ together with the approximations obtained from Euler's method in Example 1 of Section 5.2, the results from the Taylor methods of orders two and four and the actual errors involved with these methods. $\square$

As might be expected from our study of Euler's method, Taylor's method of order n has local truncation error $O(h^n)$, provided that the solution of the differential equation is sufficiently well-behaved. This is easily seen by noticing that Eq. (5.28) can be rewritten

$$y_{i+1} - y_i - hf(t_i, y_i) - \frac{h^2}{2} f'(t_i, y_i) - \cdots - \frac{h^n}{n!} f^{(n-1)}(t_i, y_i)$$

$$= \frac{h^{n+1}}{(n+1)!} f^{(n)}(t_i + \theta_i h, y(t_i + \theta_i h))$$

so the local truncation error at the $(i + 1)$st step is

$$\tau_{i+1} = \frac{y_{i+1} - y_i}{h} - T^{(n)}(t_i, y_i) = \frac{h^n}{(n+1)!} f^{(n)}(t_i + \theta_i h, y(t_i + \theta_i h))$$

for each $i = 0, 1, \ldots, N-1$. If $y \in C^{n+1}[a, b]$, this implies that $y^{(n+1)}(t) = f^{(n)}(t, y(t))$ is bounded on $[a, b]$, and that $\tau_i = O(h^n)$ for each $i = 1, 2, \ldots, N$.

TABLE 5.2

t	Exact value	Euler's method	Error in Euler's method	Taylor's method of order two	Error in Taylor's method of order two	Taylor's method of order four	Error in Taylor's method of order four
0	1.0000000000	1.000000	0	1.000000	0	1.0000000000	0
.1	1.0048374180	1.000000	4.837×10^{-3}	1.005000	1.626×10^{-4}	1.0048375000	8.200×10^{-8}
.2	1.0187307531	1.010000	8.731×10^{-3}	1.019025	2.942×10^{-4}	1.0187309014	1.483×10^{-7}
.3	1.0408182207	1.029000	1.182×10^{-2}	1.041218	3.998×10^{-4}	1.0408184220	2.013×10^{-7}
.4	1.0703200460	1.056100	1.422×10^{-2}	1.070802	4.820×10^{-4}	1.0703202889	2.429×10^{-7}
.5	1.1065306597	1.090490	1.604×10^{-2}	1.107076	5.453×10^{-4}	1.1065309344	2.747×10^{-7}
.6	1.1488116361	1.131441	1.737×10^{-2}	1.149404	5.924×10^{-4}	1.1488119344	2.983×10^{-7}
.7	1.1965853038	1.178297	1.829×10^{-2}	1.197211	6.257×10^{-4}	1.1965856187	3.149×10^{-7}
.8	1.2493289641	1.230467	1.887×10^{-2}	1.249976	6.470×10^{-4}	1.2493292897	3.256×10^{-7}
.9	1.3065696597	1.287420	1.915×10^{-2}	1.307228	6.583×10^{-4}	1.3065699912	3.315×10^{-7}
1.0	1.3678794412	1.348678	1.920×10^{-2}	1.368541	6.616×10^{-4}	1.3678797744	3.332×10^{-7}

Exercise Set 5.3

1. Use Taylor's method of order two to approximate the solution to each of the following initial-value problems:

   a) $y' = \sin t + e^{-t}$, $0 \le t \le 1$, $y(0) = 0$ with $h = .5$;

   b) $y' = \dfrac{1}{t}(y^2 + y)$, $1 \le t \le 3$, $y(1) = -2$ with $h = .5$;

   c) $y' = t^2$, $0 \le t \le 2$, $y(0) = 0$ with $h = .5$.

2. Use Taylor's method of order four to approximate the solution to:

   a) $y' = \sin t + e^{-t}$, $0 \le t \le 1$, $y(0) = 0$ with $h = .5$;
   b) $y' = t^2$, $0 \le t \le 2$, $y(0) = 0$ with $h = .5$.

3. Use the Taylor methods of orders two and four with $h = .1$, to approximate the solutions to:

   a) $y' = t + y$, $0 \le t \le 2$, $y(0) = -1$;
   b) $y' = 1 - y$, $0 \le t \le 2$, $y(0) = 0$.

4. For the initial-value problem

$$y' = \frac{2}{t}y + t^2 e^t, \qquad 1 \le t \le 2, \quad y(1) = 0,$$

   obtain an approximation to the solution using Taylor's method of order four with $h = .05$. Compare your answers to the correct values of y.

5. Obtain an approximation to $y(5)$ using Taylor's method of order four with $h = .2$ for the initial-value problem

$$y' = -y + t + 1, \qquad 0 \le t \le 5, \quad y(0) = 2.$$

6. Use the Taylor method of order two with $h = .1$ to approximate the solution to

$$y' = 1 + t \sin(ty), \qquad 0 \le t \le 2, \quad y(0) = 0.$$

   Compare your answers to those obtained in Exercise 5, Section 5.2.

7. A projectile of mass $m = .11$ kg shot vertically upward with initial velocity $v(0) = 8$ meter/sec is slowed due to the force of gravity $F_b = -mg$ and due to air resistance $F_r = -kv^2$, where $g = 9.8$ meter/sec^2 and $k = .002$ kg/meter. The differential equation for the velocity v is given by:

$$mv' = -mg - kv^2.$$

   a) Find the velocity after $.1, .2, \ldots, 1.0$ seconds, using the Taylor methods of orders two and four.
   b) To the nearest tenth of a second, determine when the projectile reaches its maximum height and begins falling. Use the Taylor method of order four.

8. Can the difference method

$$w_{i+1} = w_{i-1} + 2hf(t_i, w_i) \qquad \text{for each } i = 1, 2, \dots, N - 1,$$

with $w_0 = \alpha$ and $w_1 = y(t_1)$ be used with the differential equation $y' = f(t, y)$, $a < t < b$, $y(a) = \alpha$? Find the local truncation error.

5.4   Runge–Kutta Methods

The Taylor methods outlined in the previous section have the desirable property of high-order local truncation error but the disadvantage of requiring computation and evaluation of the derivatives of $f(t, y)$. This can be a very complicated and time-consuming procedure for many problems and, consequently, the Taylor methods are seldom used in practice. The **Runge–Kutta methods** utilize the high-order local truncation error of the Taylor methods while eliminating the computation and evaluation of the derivatives of $f(t, y)$. Before presenting the ideas behind their derivation, we need to state Taylor's Theorem in two variables. The proof of this result can be found in any standard book on advanced calculus (see, for example, Fulks [40], page 260).

Theorem 5.11   Suppose that $f(t, y)$ and all of its partial derivatives of order less than or equal to $n + 1$ are continuous on $D = \{(t, y) | a \le t \le b, c \le y \le d\}$. Let $(t_0, y_0) \in D$. For every $(t, y) \in D$, there exists ξ between t and t_0 and η between y and y_0 with

$$f(t, y) = P_n(t, y) + R_n(t, y),$$

where

$$P_n(t, y) = f(t_0, y_0) + \left[(t - t_0) \frac{\partial f}{\partial t}(t_0, y_0) + (y - y_0) \frac{\partial f}{\partial y}(t_0, y_0) \right]$$

$$+ \left[\frac{(t - t_0)^2}{2} \frac{\partial^2 f}{\partial t^2}(t_0, y_0) + (t - t_0)(y - y_0) \frac{\partial^2 f}{\partial t \, \partial y}(t_0, y_0) \right.$$

$$+ \left. \frac{(y - y_0)^2}{2} \frac{\partial^2 f}{\partial y^2}(t_0, y_0) \right]$$

$$+ \cdots + \left[\frac{1}{n!} \sum_{j=0}^{n} \binom{n}{j} (t - t_0)^{n-j} (y - y_0)^j \frac{\partial^n f}{\partial t^{n-j} \, \partial y^j}(t_0, y_0) \right]$$

and

$$R_n(t, y) = \frac{1}{(n+1)!} \sum_{j=0}^{n+1} \binom{n+1}{j} (t - t_0)^{n+1-j} (y - y_0)^j \frac{\partial^{n+1} f(\xi, \eta)}{\partial t^{n+1-j} \, \partial y^j}.$$

P_n is called the **Taylor polynomial of degree n in two variables** for the function f about (t_0, y_0), and $R_n(t, y)$ is called the **truncation error** associated with $P_n(t, y)$.

EXAMPLE 1   The Taylor polynomial of degree three for $f(t, y) = \sin(ty)$ about $(0, \pi)$ is found from

$$P_3(t, y) = f(t_0, y_0) + (t - t_0)\frac{\partial f}{\partial t}(t_0, y_0) + (y - y_0)\frac{\partial f}{\partial y}(t_0, y_0)$$

$$+ \left[\frac{(t - t_0)^2}{2}\frac{\partial^2 f}{\partial t^2}(t_0, y_0) + (t - t_0)(y - y_0)\frac{\partial^2 f}{\partial t\,\partial y}(t_0, y_0)\right.$$

$$+ \frac{(y - y_0)^2}{2}\frac{\partial^2 f}{\partial y^2}(t_0, y_0)\right]$$

$$+ \left[\frac{(t - t_0)^3}{6}\frac{\partial^3 f}{\partial t^3}(t_0, y_0) + \frac{(t - t_0)^2}{2}(y - y_0)\frac{\partial^3 f}{\partial t^2\,\partial y}(t_0, y_0)\right.$$

$$+ \frac{(t - t_0)(y - y_0)^2}{2}\frac{\partial^2 f}{\partial t\,\partial y^2}(t_0, y_0) + \frac{(y - y_0)^3}{6}\frac{\partial^3 f}{\partial y^3}(t_0, y_0)\right].$$

Evaluating each of these partial derivatives at $(t_0, y_0) = (0, \pi)$ reduces $P_3(t, y)$ to

$$P_3(t, y) = \pi t + t(y - \pi) - \frac{\pi^3}{6}t^3.$$

This polynomial will give a close approximation to $\sin(ty)$ provided t is close to zero and y is close to π. For example,

$$P_3(.01, \pi + .01) = .03151076,$$

while
$$\sin(.01(\pi + .01)) = .03151071. \qquad \square$$

The first step in deriving a Runge–Kutta method is to determine values for $a_1, \alpha_1,$ and β_1 with the property that $a_1 f(t + \alpha_1, y + \beta_1)$ approximates

$$T^{(2)}(t, y) = f(t, y) + \frac{h}{2}f'(t, y)$$

with error no greater than $O(h^2)$, the local truncation error for the Taylor method of order two.

Since

$$f'(t, y) = \frac{df}{dt}(t, y) = \frac{\partial f}{\partial t}(t, y) + \frac{\partial f}{\partial y}(t, y)\cdot y'(t) \qquad \text{and} \qquad y'(t) = f(t, y),$$

this implies

(5.34)
$$T^{(2)}(t, y) = f(t, y) + \frac{h}{2}\frac{\partial f}{\partial t}(t, y) + \frac{h}{2}\frac{\partial f}{\partial y}(t, y)\cdot f(t, y).$$

Expanding $f(t + \alpha_1, y + \beta_1)$ in its Taylor polynomial of degree one about (t, y) implies that:

(5.35)
$$a_1 f(t + \alpha_1, y + \beta_1) = a_1 f(t, y) + a_1\alpha_1\frac{\partial f}{\partial t}(t, y)$$

$$+ a_1\beta_1\frac{\partial f}{\partial y}(t, y) + a_1\cdot R_1(t + \alpha_1, y + \beta_1),$$

where

$$(5.36) \qquad R_1(t + \alpha_1, y + \beta_1) = \frac{\alpha_1^2}{2} \frac{\partial^2 f}{\partial t^2}(\xi, \eta) + \alpha_1 \beta_1 \frac{\partial^2 f}{\partial t\, \partial y}(\xi, \eta) + \frac{\beta_1^2}{2} \frac{\partial^2 f}{\partial y^2}(\xi, \eta)$$

for some ξ between t and $t + \alpha_1$, and η between y and $y + \beta_1$.

Matching the coefficients of f and its derivatives in equations (5.34) and (5.35) gives the three equations

$$f(t, y): \quad a_1 = 1; \qquad \frac{\partial f}{\partial t}(t, y): \quad a_1 \alpha_1 = \frac{h}{2};$$

and

$$\frac{\partial f}{\partial t}(t, y): \quad a_1 \beta_1 = \frac{h}{2} f(t, y).$$

The parameters a_1, α_1, and β_1 are uniquely determined to be

$$a_1 = 1, \qquad \alpha_1 = \frac{h}{2}, \qquad \text{and} \qquad \beta_1 = \frac{h}{2} f(t, y);$$

so

$$T^{(2)}(t, y) = f\left(t + \frac{h}{2}, y + \frac{h}{2} f(t, y)\right) - R_1\left(t + \frac{h}{2}, y + \frac{h}{2} f(t, y)\right)$$

and from Eq. (5.36),

$$R_1\left(t + \frac{h}{2}, y + \frac{h}{2} f(t, y)\right) = \frac{h^2}{8} \frac{\partial^2 f}{\partial t^2}(\xi, \eta) + \frac{h^2}{4} f(t, y) \frac{\partial^2 f}{\partial t\, \partial y}(\xi, \eta)$$

$$+ \frac{h^2}{8} (f(t, y))^2 \frac{\partial^2 f}{\partial y^2}(\xi, \eta).$$

If all the second-order partial derivatives of f are bounded,

$$R_1\left(t + \frac{h}{2}, y + \frac{h}{2} f(t, y)\right)$$

will be $O(h^2)$, the order of the local truncation error of Taylor's method of order two.

The difference method resulting from replacing $T^{(2)}(t, y)$ in Taylor's method of order two by $f(t + (h/2), y + (h/2)f(t, y))$ is a specific Runge–Kutta method known as the **Midpoint method**.

Midpoint Method

$$w_0 = \alpha,$$

$$(5.37) \qquad w_{i+1} = w_i + hf\left(t_i + \frac{h}{2}, w_i + \frac{h}{2} f(t_i, w_i)\right)$$

$$\text{for each } i = 0, 1, \ldots, N - 1.$$

Since only three parameters were present in $a_1 f(t + \alpha_1, y + \beta_1)$, and all were needed in the match of $T^{(2)}$, we would expect to need a more complicated form to satisfy the conditions required for any of the higher-order Taylor methods. In fact, the most appropriate four parameter form for approximating

$$T^{(3)}(t, y) = f(t, y) + \frac{h}{2} f'(t, y) + \frac{h^2}{6} f''(t, y),$$

is

(5.38) $a_1 f(t, y) + a_2 f(t + \alpha_2, y + \delta_2 f(t, y));$

and even with this, there is insufficient flexibility to match the term

$$\frac{h^2}{6} \left[\frac{\partial f}{\partial y} (t, y) \right]^2 f(t, y)$$

resulting from the expansion of $(h^2/6) f''(t, y)$. Consequently, the best that can be obtained from using (5.38) are methods with $O(h^2)$ error bound. The fact that (5.38) has four parameters, however, gives a flexibility in their choice so that a number of $O(h^2)$ methods can be derived. The two most important are called the **modified Euler method**, which corresponds to choosing $a_1 = a_2 = \frac{1}{2}$ and $\alpha_2 = \delta_2 = h$ and has difference-equation form:

Modified Euler Method

$$w_0 = \alpha,$$

(5.39)
$$w_{i+1} = w_i + \frac{h}{2} [f(t_i, w_i) + f(t_{i+1}, w_i + hf(t_i, w_i))]$$

for each $i = 0, 1, 2, \ldots, N - 1,$

and the **Heun's method**, which corresponds to $a_1 = \frac{1}{4}$, $a_2 = \frac{3}{4}$, and $\alpha_2 = \delta_2 = \frac{2}{3} h$ and has difference-equation form:

Heun Method

$$w_0 = \alpha,$$

(5.40)
$$w_{i+1} = w_i + \frac{h}{4} [f(t_i, w_i) + 3f(t_i + \tfrac{2}{3} h, w_i + \tfrac{2}{3} hf(t_i, w_i))]$$

for each $i = 0, 1, 2, \ldots, N - 1.$

Both of these methods are classified as Runge–Kutta methods of order two, the order of their local truncation error.

EXAMPLE 2   Applying any of the Runge–Kutta methods of order two to our usual example,

$$y' = -y + t + 1, \qquad 0 \le t \le 1, \quad y(0) = 1,$$

would give the same difference equation as that of Taylor's method of order two,

$$w_0 = 1,$$
$$w_{i+1} = .905 w_i + .0095 i + .1,$$

because of the nature of the differential equation.

In order to compare the various results from these methods we will use, instead, the initial-value problem

$$y' = -y + t^2 + 1, \qquad 0 \le t \le 1, \quad y(0) = 1,$$

which has $y(t) = -2e^{-t} + t^2 - 2t + 3$ as its exact solution. Table 5.3 lists the results of these calculations. $\square$

TABLE 5.3

t_i	Exact values	Midpoint method	Error for Midpoint method	Modified Euler method	Error for modified Euler method	Heun's method	Error for Heun's method
0	1.0000000	1.0000000	0	1.0000000	0	1.0000000	0
.1	1.0003252	1.0002500	7.52×10^{-5}	1.0005000	1.75×10^{-4}	1.0003333	8.10×10^{-6}
.2	1.0025385	1.0024263	1.12×10^{-4}	1.0029025	3.64×10^{-4}	1.0025850	4.65×10^{-5}
.3	1.0083636	1.0082458	1.18×10^{-4}	1.0089268	5.63×10^{-4}	1.0084728	1.09×10^{-4}
.4	1.0193599	1.0192624	9.75×10^{-5}	1.0201288	7.69×10^{-4}	1.0195512	1.91×10^{-4}
.5	1.0369387	1.0368825	5.62×10^{-5}	1.0379166	9.78×10^{-4}	1.0372272	2.88×10^{-4}
.6	1.0623767	1.0623787	2.00×10^{-6}	1.0635645	1.19×10^{-3}	1.0627739	3.97×10^{-4}
.7	1.0968294	1.0969027	7.33×10^{-5}	1.0982259	1.40×10^{-3}	1.0973437	5.14×10^{-4}
.8	1.1413421	1.1414969	1.55×10^{-4}	1.1429444	1.60×10^{-3}	1.1419794	6.37×10^{-4}
.9	1.1968607	1.1971047	2.44×10^{-4}	1.1986647	1.80×10^{-3}	1.1976247	7.64×10^{-4}
1.0	1.2642411	1.2645798	3.39×10^{-4}	1.2662416	2.00×10^{-3}	1.2651337	8.93×10^{-4}

Although $T^{(3)}(t, y)$ can be approximated with error $O(h^3)$ by an expression of the form

$$f(t + \alpha_1, y + \delta_1 f(t + \alpha_2, y + \delta_2 f(t, y))),$$

involving four parameters, the algebra involved in the determination of α_1, δ_1, α_2, and δ_2 is quite involved and will not be presented. In fact, the Runge–Kutta method of order three resulting from this expression is not generally used in practice. The most common Runge–Kutta method in use is of order four and, in difference form, is given by:

(5.41)

$$w_0 = \alpha,$$

$$k_1 = hf(t_i, w_i),$$

$$k_2 = hf\left(t_i + \frac{h}{2}, w_i + \frac{1}{2}k_1\right),$$

$$k_3 = hf\left(t_i + \frac{h}{2}, w_i + \frac{1}{2}k_2\right),$$

$$k_4 = hf(t_{i+1}, w_i + k_3),$$

$$w_{i+1} = w_i + \tfrac{1}{6}(k_1 + 2k_2 + 2k_3 + k_4),$$

for each $i = 0, 1, \ldots, N - 1$. This method has local truncation error $O(h^4)$, provided the solution $y(t)$ has five continuous derivatives. The reason for introducing the terminology k_1, k_2, k_3, k_4 in the method is to eliminate the need for successive nesting in the second variable of the function $f(t, y)$ (see Exercise 12).

Runge–Kutta (Order Four) Algorithm 5.2

To approximate the solution of the initial-value problem

$$y' = f(t, y), \qquad a \le t \le b, \quad y(a) = \alpha,$$

at $(N + 1)$ equally spaced numbers in the interval $[a, b]$:

INPUT   endpoints a, b; integer N; initial condition α.

OUTPUT   approximation w to y at the $(N + 1)$ values of t.

Step 1   Set $h = (b - a)/N$;
  $t = a$;
  $w = \alpha$;
  OUTPUT (t, w).

Step 2   For $i = 1, 2, \ldots, N$ do Steps 3–5.

  Step 3   Set $K_1 = hf(t, w)$;
    $K_2 = hf(t + h/2, w + K_1/2)$;
    $K_3 = hf(t + h/2, w + K_2/2)$;
    $K_4 = hf(t + h, w + K_3)$.

  Step 4   Set $w = w + (K_1 + 2K_2 + 2K_3 + K_4)/6$;   (*Compute w_i.*)
    $t = a + ih$.   (*Compute t_i.*)

  Step 5   OUTPUT (t, w).

Step 6   STOP.

EXAMPLE 3   Using the Runge–Kutta method of order four to obtain approximations to the solution of the initial-value problem

$$y' = -y + t + 1, \qquad 0 \le t \le 1, \quad y(0) = 1$$

with $h = .1, N = 10$, and $t_i = .1i$ gives the results and errors listed in Table 5.4.   □

TABLE 5.4

t_i	Exact values	Runge–Kutta values of order four	Error
0	1.0000000000	1.0000000000	0
.1	1.0048374180	1.0048375000	8.200×10^{-8}
.2	1.0187307531	1.0187309014	1.483×10^{-7}
.3	1.0408182207	1.0408184220	2.013×10^{-7}
.4	1.0703200460	1.0703202889	2.429×10^{-7}
.5	1.1065306597	1.1065309344	2.747×10^{-7}
.6	1.1488116360	1.1488119344	2.984×10^{-7}
.7	1.1965853038	1.1965856187	3.149×10^{-7}
.8	1.2493289641	1.2493292897	3.256×10^{-7}
.9	1.3065696597	1.3065699912	3.315×10^{-7}
1.0	1.3678794412	1.3678797744	3.332×10^{-7}

The main computational effort in applying the Runge–Kutta methods is the evaluation of f. In the second-order methods the truncation error is $O(h^2)$, but the cost is two functional evaluations per step. The Runge–Kutta method of order four requires four evaluations per step, but the truncation error is $O(h^4)$. Butcher [23] has established the following relationship between the number of evaluations per step and the order of the local truncation error:

Evaluations per step	2	3	4	5	6	7	$n \ge 8$
Best possible local truncation error	$O(h^2)$	$O(h^3)$	$O(h^4)$	$O(h^4)$	$O(h^5)$	$O(h^6)$	$O(h^{n-2})$

Consequently, the lower-order methods, with smaller step size, are used in preference to the possible higher-order methods using a larger step size.

One measure to compare the lower-order Runge–Kutta methods is described as follows: Since the Runge–Kutta method of order four requires four evaluations per step, it should give more accurate answers than Euler's method with one-quarter the mesh size (where by mesh size we mean the difference between consecutive mesh points) if it is to be superior. Similarly, if the Runge–Kutta method of order four is to be superior to the second-order Runge–Kutta methods, then it should give more accuracy with step size h than a second-order method with step size $\frac{1}{2}h$, because the fourth-order method requires twice as many evaluations per step. An illustration of the superiority of the Runge–Kutta fourth-order method by this measure is shown in the following example.

EXAMPLE 4 For the problem

$$y' = -y + 1, \qquad 0 \le t \le 1, \quad y(0) = 0,$$

Euler's method with $h = .025$, the Modified Euler's method (5.39) with $h = .05$, and the Runge–Kutta fourth-order method (5.41) with $h = .1$ will be compared at the mesh points .1, .2, .3, .4, and .5. The results are given in Table 5.5. For this example the fourth-order method is clearly superior. □

TABLE 5.5

t	Euler's method $h = .025$	Modified Euler's method $h = .05$	Fourth-order Runge–Kutta method $h = .1$	Actual value
.1	.096312	.095123	.09516250	.095162582
.2	.183348	.181198	.18126910	.181269247
.3	.262001	.259085	.25918158	.259181779
.4	.333079	.329563	.32967971	.329679954
.5	.397312	.393337	.39346906	.393469340

Exercise Set 5.4

1. Use the modified Euler's method to approximate the solution to each of the following initial-value problems:

   a) $y' = \left(\dfrac{y}{t}\right)^2 + \left(\dfrac{y}{t}\right), \quad 1 \le t \le 1.2, \quad y(1) = 1$ with $h = .1$;

   b) $y' = \sin t + e^{-t}, \quad 0 \le t \le 1, \quad y(0) = 0$ with $h = .5$;

   c) $y' = \dfrac{1}{t}(y^2 + y), \quad 1 \le t \le 3, \quad y(1) = -2$ with $h = .5$;

   d) $y' = t^2, \quad 0 \le t \le 2, \quad y(0) = 0$ with $h = .5$.

2. Use the Runge–Kutta fourth-order method to approximate the solution to each of the following initial-value problems:

   a) $y' = \left(\dfrac{y}{t}\right)^2 + \left(\dfrac{y}{t}\right), \quad 1 \le t \le 1.2, \quad y(1) = 1$ with $h = .1$;

   b) $y' = \sin t + e^{-t}, \quad 0 \le t \le 1, \quad y(0) = 0$ with $h = .5$;

   c) $y' = \dfrac{1}{t}(y^2 + y), \quad 1 \le t \le 3, \quad y(1) = -2$ with $h = .5$;

   d) $y' = t^2, \quad 0 \le t \le 2, \quad y(0) = 0$ with $h = .5$.

3. Use the Midpoint method, the Modified Euler method, and Heun's method with $h = .1$ to approximate the solution to:

   $$y' = 1 + t \sin(ty), \qquad 0 \le t \le 2, \quad y(0) = 0.$$

   Compare your answers to those obtained in Exercise 6, Section 5.3 and Exercise 5, Section 5.2.

4. Use the Runge–Kutta method of order four with $h = .05$ to approximate the solution to:

$$y' = \frac{2}{t}y + t^2 e^t, \quad 1 \le t \le 2, \quad y(1) = 0.$$

Compare your answers to those obtained by the Taylor method of order four in Exercise 4, Section 5.3. Which of the two fourth-order methods requires the longer computation time and which requires the longer implementation time?

5. Approximate $y(5)$ using the Runge–Kutta fourth-order method with $h = .2$ for the initial-value problem

$$y' = -y + t + 1, \quad 0 \le t \le 5, \quad y(0) = 2;$$

and compare to Exercise 5, Section 5.3.

6. Use the Runge–Kutta method of order four for the initial-value problem

$$y' = 50t^2 - 50y + 2t, \quad 0 \le t \le 1, \quad y(0) = \tfrac{1}{3},$$

with (a) $h = .1$, (b) $h = .025$, and (c) $h = .01$. Find the exact solution. Are the results consistent with the local truncation error?

7. Use Euler's method with $h = .025$, a Runge–Kutta second-order method with $h = .05$, and the Runge–Kutta fourth-order method with $h = .1$ for the problems:

a) $y' = \dfrac{1}{t^2} - \dfrac{y}{t} - y^2, \quad 1 \le t \le 2, \quad y(1) = -1;$

b) $y' = t + y, \quad 0 \le t \le 2, \quad y(0) = -1;$

c) $y' = -2y + 2t^2 + 2t, \quad 0 \le t \le 1, \quad y(0) = 1.$

Which method appears to be superior for each problem?

8. Show that the Midpoint method, the Modified Euler method, and Heun's method give the same approximations to the initial-value problem

$$y' = -y + t + 1, \quad 0 \le t \le 1, \quad y(0) = 1,$$

for any choice of h. Why?

9. A liquid of low viscosity, such as water, flows from an inverted conical tank with circular orifice at the rate

$$\frac{dx}{dt} = -.6\pi r^2 \sqrt{2g}\,\frac{\sqrt{x}}{A(x)},$$

where r is the radius of the orifice, x is the height of the liquid level from the vertex of the cone, and $A(x)$ is the area of the cross section of the tank x units above the orifice. Suppose $r = .1$ feet, $g = 32$ feet/sec^2, and the tank has an initial water level of 8 feet and initial volume of $512\,\pi/3$ cubic feet.

a) Find $A(x)$.
b) Compute the water level after 10 seconds, using the Runge–Kutta fourth-order method with $h = .1$.

10. The irreversible chemical reaction in which two molecules of solid potassium dichromate ($K_2Cr_2O_7$), two molecules of water (H_2O), and three atoms of solid sulfur (S) combine to yield three molecules of the gas sulfur dioxide (SO_2), four molecules of solid potassium hydroxide (KOH), and two molecules of solid chromic oxide (Cr_2O_3) can be represented symbolically by the stoichiometric equation:

$$2K_2Cr_2O_7 + 2H_2O + 3S \longrightarrow 4KOH + 2Cr_2O_3 + 3SO_2.$$

If n_1 molecules of $K_2Cr_2O_7$, n_2 molecules of H_2O and n_3 molecules of S are originally available, the following differential equation describes the amount $x(t)$ of KOH after time t:

$$\frac{dx}{dt} = k\left(n_1 - \frac{x}{2}\right)^2 \left(n_2 - \frac{x}{2}\right)^2 \left(n_3 - \frac{3x}{4}\right)^3$$

where k is the velocity constant of the reaction. If $k = 6.22 \times 10^{-19}$, $n_1 = n_2 = 1000$, and $n_3 = 1500$, how many units of potassium hydroxide will have been formed after two seconds? Use the Runge–Kutta fourth-order method with $h = .1$.

11. Show that the difference method

$$w_0 = \alpha,$$

$$w_{i+1} = w_i + a_1 f(t_i, w_i) + a_2 f(t_i + \alpha_2, w_i + \delta_2 f(t_i, w_i))$$

for each $i = 0, 1, \ldots, N - 1$, cannot have local truncation error $O(h^3)$ for any choice of constants $a_1, a_2, \alpha_2,$ and δ_2.

12. The Runge–Kutta method of order four can be written in the form

$$w_0 = \alpha,$$

$$w_{i+1} = w_i + \frac{h}{6} f(t_i, w_i) + \frac{h}{3} f(t_i + \alpha_1 h, w_i + \delta_1 f(t_i, w_i))$$

$$+ \frac{h}{3} f(t_i + \alpha_2 h, w_i + \delta_2 hf(t_i + \gamma_2 h, w_i + \gamma_3 hf(t_i, w_i)))$$

$$+ \frac{h}{6} f(t_i + \alpha_3 h, w_i + \delta_3 hf(t_i + \gamma_4 h, w_i + \gamma_5 hf(t_i + \gamma_6 h, w_i + \gamma_7 hf(t_i, w_i)))).$$

Find the values of the constants

$$\alpha_1, \quad \alpha_2, \quad \alpha_3, \quad \delta_1, \quad \delta_2, \quad \delta_3, \quad \gamma_2, \quad \gamma_3, \quad \gamma_4, \quad \gamma_5, \quad \gamma_6, \quad \gamma_7.$$

5.5 Error Control and the Runge–Kutta–Fehlberg Method

An ideal difference-equation method

$$(5.42) \qquad w_{i+1} = w_i + h_i \phi(t_i, h_i, w_i), \qquad i = 0, 1, \ldots, N - 1,$$

for approximating the solution, $y(t)$, to the initial-value problem

$$(5.43) \qquad y' = f(t, y), \qquad a \leq t \leq b, \quad y(a) = \alpha,$$

would have the property that, whenever a tolerance $\varepsilon > 0$ was given, the minimal number of mesh points would be used to ensure that the global error, $|y(t_i) - w_i|$, did not exceed ε for any $i = 0, 1, 2, \ldots, N$. Having a minimal number of mesh points and also controlling the global error of a difference method is, not unsurprisingly, inconsistent with the points being equally spaced in the interval. In this section we will examine a technique that can be used to control the error of a difference method in an efficient manner by the appropriate choice of mesh points.

Since we cannot generally determine the global error of a difference method, we will work instead with the local truncation error of the method. In Section 5.9 we will present a result, Theorem 5.19, that demonstrates, under suitable hypotheses on the initial-value problem and difference method, a connection between the local truncation error of the method and the global error. This result essentially states that a bound on the local truncation error will produce a corresponding bound on the global error:

To illustrate the technique of estimating the local truncation error, suppose that a local-error estimate is desired for Euler's method,

$$w_0 = \alpha,$$

(5.44)

$$w_{i+1} = w_i + hf(t_i, w_i),$$

a method with local truncation error of order $O(h)$ given by

$$\tau_{i+1} = \frac{y(t_{i+1}) - y(t_i)}{h} - f(t_i, y(t_i)).$$

The modified Euler method,

$$\tilde{w}_0 = \alpha$$

(5.45)

$$\tilde{w}_{i+1} = \tilde{w}_i + \frac{h}{2} [f(t_i, \tilde{w}_i) + f(t_{i+1}, \tilde{w}_i + hf(t_i, \tilde{w}_i))],$$

has local truncation error $\tilde{\tau}_{i+1}$ of order $O(h^2)$.

Suppose that $w_i \approx y(t_i) \approx \tilde{w}_i$, then

$$
\begin{aligned}
y(t_{i+1}) - w_{i+1} &= y(t_{i+1}) - w_i - hf(t_i, w_i) \\
&\approx y(t_{i+1}) - y(t_i) - hf(t_i, y(t_i)) \\
&= h\tau_{i+1}.
\end{aligned}
$$

So

$$\tau_{i+1} \approx \frac{1}{h} [y(t_{i+1}) - w_{i+1}]$$

$$= \frac{1}{h} [y(t_{i+1}) - \tilde{w}_{i+1}] + \frac{1}{h} (\tilde{w}_{i+1} - w_{i+1})$$

$$\approx \tilde{\tau}_{i+1} + \frac{1}{h} (\tilde{w}_{i+1} - w_{i+1}).$$

But τ_{i+1} is $O(h)$ while $\tilde{\tau}_{i+1} = O(h^2)$, and so the most significant portion of τ_{i+1} must be attributed to $(\tilde{w}_{i+1} - w_{i+1})/h$. Consequently the equation

(5.46)

$$\tau_{i+1} \approx \frac{1}{h} (\tilde{w}_{i+1} - w_{i+1})$$

can be used to approximate the local truncation error of Euler's method.

We will now determine how the estimation of the local truncation error of a difference method can be used to advantage in approximating the optimal step size to control the global error. Suppose two difference methods are available for the approximation of the solution to the initial-value problem (5.43) and that one of these methods,

$$(5.47) \qquad \begin{aligned} w_0 &= \alpha, \\ w_{i+1} &= w_i + h_i \phi(t_i, h_i, w_i), \end{aligned}$$

has local truncation error τ_{i+1} of order $O(h^n)$, while the other method

$$(5.48) \qquad \begin{aligned} \tilde{w}_0 &= \alpha, \\ \tilde{w}_{i+1} &= \tilde{w}_i + \tilde{h}_i \tilde{\phi}(\tilde{t}_i, \tilde{h}_i, \tilde{w}_i), \end{aligned}$$

has local truncation error $\tilde{\tau}_{i+1}$ of order $O(\tilde{h}^{n+1})$.

By using the same analysis as in Euler's method, we have

$$(5.49) \qquad \tau_{i+1} \approx \frac{1}{h}(\tilde{w}_{i+1} - w_{i+1}).$$

However, τ_{i+1} is of order $O(h^n)$; so a constant k exists with

$$(5.50) \qquad \tau_{i+1} \approx kh^n.$$

To estimate k we use the two relationships for τ_{i+1}, (5.49) and (5.50):

$$(5.51) \qquad kh^n \approx \frac{1}{h}(\tilde{w}_{i+1} - w_{i+1}).$$

The design of the procedure is to use this estimate to choose an appropriate step size. To accommodate this, let us consider the truncation error with h replaced by qh where q is positive but bounded above and away from zero. By (5.50) and (5.51),

$$\tau_{i+1}(qh) \approx k(qh)^n = q^n(kh^n) \approx \frac{q^n}{h}(\tilde{w}_{i+1} - w_{i+1}).$$

To bound $\tau_{i+1}(qh)$ by ε, we choose q so that

$$\frac{q^n}{h}|\tilde{w}_{i+1} - w_{i+1}| \approx |\tau_{i+1}(qh)| \le \varepsilon,$$

that is, so that

$$(5.52) \qquad q \le \left(\frac{\varepsilon h}{|\tilde{w}_{i+1} - w_{i+1}|} \right)^{1/n}.$$

One popular technique that utilizes inequality (5.52) for error control is called the **Runge–Kutta–Fehlberg method** presented by Fehlberg [37], and consists of using a Runge–Kutta method with local truncation error of order five,

$$\tilde{w}_{i+1} = w_i + \frac{16}{135}k_1 + \frac{6656}{12825}k_3 + \frac{28561}{56430}k_4 - \frac{9}{50}k_5 + \frac{2}{55}k_6,$$

to estimate the local error in a Runge–Kutta method of order four,

$$w_{i+1} = w_i + \frac{25}{216}k_1 + \frac{1408}{2565}k_3 + \frac{2197}{4104}k_4 - \frac{1}{5}k_5,$$

where $k_1 = hf(t_i, w_i)$,

$$k_2 = hf\left(t_i + \frac{h}{4}, w_i + \tfrac{1}{4}k_1\right),$$

$$k_3 = hf\left(t_i + \frac{3h}{8}, w_i + \tfrac{3}{32}k_1 + \tfrac{9}{32}k_2\right),$$

$$k_4 = hf\left(t_i + \frac{12h}{13}, w_i + \tfrac{1932}{2197}k_1 - \tfrac{7200}{2197}k_2 + \tfrac{7296}{2197}k_3\right),$$

$$k_5 = hf\left(t_i + h, w_i + \tfrac{439}{216}k_1 - 8k_2 + \tfrac{3680}{513}k_3 - \tfrac{845}{4104}k_4\right),$$

$$k_6 = hf\left(t_i + \frac{h}{2}, w_i - \tfrac{8}{27}k_1 + 2k_2 - \tfrac{3544}{2565}k_3 + \tfrac{1859}{4104}k_4 - \tfrac{11}{40}k_5\right).$$

A clear advantage to this method is that only six evaluations of f are required per step whereas arbitrary Runge–Kutta methods of order four and five used together would require ten evaluations of f per step.

In the error-control theory, an initial value of h at the ith step was used to find the first values of w_{i+1} and $\tilde{w}_{i+1}$, which led to the determination of q for that step, and then the calculations were repeated. This procedure requires twice the number of functional evaluations per step as without the error control. In practice, the value of q to be used is chosen somewhat differently in order to make the increased cost worthwhile. The value of q determined at the ith step is used for two purposes:

1)   To reject the initial choice of h at the ith step if necessary and repeat the calculations using qh, and

2)   To predict an appropriate initial choice of h for the $(i + 1)$st step.

Because of the penalty (in terms of functional evaluations) that must be paid if many of the steps are repeated, q tends to be chosen rather conservatively; in fact, for the Runge–Kutta–Fehlberg method with $n = 4$, the usual choice is

$$q = \left(\frac{\varepsilon h}{2|\tilde{w}_{i+1} - w_{i+1}|}\right)^{1/4} = .84\left(\frac{\varepsilon h}{|\tilde{w}_{i+1} - w_{i+1}|}\right)^{1/4}.$$

The following algorithm uses the Runge–Kutta–Fehlberg method with error control. In Step 9, we eliminate large changes in step size. We do this to avoid spending too much time with very small step sizes in regions with irregularities in the derivatives of y and to avoid large step sizes, which may result in skipping sensitive regions nearby. In some instances the step size increase procedure is omitted from the algorithm and the step size decrease procedure is modified so that it is incorporated only when needed to bring the error under control.

Runge–Kutta–Fehlberg Algorithm 5.3

To approximate the solution of the initial-value problem

$$y' = f(t, y), \qquad a \le t \le b, \quad y(a) = \alpha,$$

with local truncation error within a given tolerance.

INPUT   endpoints a, b; initial condition α; tolerance TOL; maximum step size $hmax$; minimum step size $hmin$.

OUTPUT t, w, h where w approximates $y(t)$ and step size h was used or a message that minimum step size exceeded.

Step 1   Set $t = a$;
   $w = \alpha$;
   $h = hmax$
OUTPUT (t, w).

Step 2   While $(t \leq b)$ do Steps 3–11.

   Step 3   Set $K_1 = hf(t, w)$;
      $K_2 = hf(t + \frac{1}{4}h, w + \frac{1}{4}K_1)$;
      $K_3 = hf(t + \frac{3}{8}h, w + \frac{3}{32}K_1 + \frac{9}{32}K_2)$;
      $K_4 = hf(t + \frac{12}{13}h, w + \frac{1932}{2197}K_1 - \frac{7200}{2197}K_2 + \frac{7296}{2197}K_3)$;
      $K_5 = hf(t + h, w + \frac{439}{216}K_1 - 8K_2 + \frac{3680}{513}K_3 - \frac{845}{4104}K_4)$;
      $K_6 = hf(t + \frac{1}{2}h, w - \frac{8}{27}K_1 + 2K_2 - \frac{3544}{2565}K_3 + \frac{1859}{4104}K_4 - \frac{11}{40}K_5)$.

   Step 4   Set $R = |\frac{1}{360}K_1 - \frac{128}{4275}K_3 - \frac{2197}{75240}K_4 + \frac{1}{50}K_5 + \frac{2}{55}K_6|/h$.
      $(R = |\tilde{w}_{i+1} - w_{i+1}|/h.)$

   Step 5   Set $\delta = .84(TOL/R)^{1/4}$.

   Step 6   If $R \leq TOL$ then do Steps 7 and 8.

      Step 7   Set $t = t + h$;   (*Approximation accepted.*)
         $w = w + \frac{25}{216}K_1 + \frac{1408}{2565}K_3 + \frac{2197}{4104}K_4 - \frac{1}{5}K_5$.

      Step 8   OUTPUT (t, w, h).

   Step 9   If $\delta \leq .1$ then set $h = .1h$
         else if $\delta \geq 4$ then set $h = 4h$
            else set $h = \delta h$.
      (*Calculate new h.*)

   Step 10   If $h > hmax$ then set $h = hmax$.

   Step 11   If $h < hmin$ then
         OUTPUT ('minimum h exceeded');
         (*Procedure completed unsuccessfully.*)
         STOP.

Step 12   (*The procedure is complete.*)
      STOP.

EXAMPLE 1   Algorithm 5.3 will be used to approximate the solution to the initial-value problem

$$y' = -y + t + 1, \qquad 0 \leq t \leq 1, \quad y(0) = 1.$$

The input included tolerance $TOL = 5 \times 10^{-5}$, a maximum step size of $hmax = .1$ and a minimum step size of $hmin = .02$. Initially, h was set to $(TOL)^{1/4}$, and if a maximum step size had not been imposed, h would have been increased beyond .1. Six digits of precision were used for all computations, and the results are shown in Table 5.6.   □

TABLE 5.6	i	t_i	h_i	w_i	r_i	$y(t_i)$	$|y(t_i) - w_i|$
	1	.08408963	.08408963	1.003437	9.674×10^{-8}	1.003437	0
	2	.1840896	.1	1.015948	2.398×10^{-7}	1.015950	2×10^{-6}
	3	.2840896	.1	1.036785	1.420×10^{-7}	1.036787	2×10^{-6}
	4	.3840897	.1	1.065155	1.863×10^{-7}	1.065159	4×10^{-6}
	5	.4840897	.1	1.100341	1.257×10^{-7}	1.100347	6×10^{-6}
	6	.5840897	.1	1.141696	1.490×10^{-7}	1.141702	6×10^{-6}
	7	.6840897	.1	1.188632	2.002×10^{-7}	1.188638	6×10^{-6}
	8	.7840898	.1	1.240616	1.839×10^{-7}	1.240624	8×10^{-6}
	9	.8840898	.1	1.297170	1.350×10^{-7}	1.297179	9×10^{-6}
	10	.9840898	.1	1.357858	8.615×10^{-8}	1.357868	1×10^{-5}

Exercise Set 5.5

1. Use the Runge–Kutta–Fehlberg Algorithm 5.3 with $TOL = 10^{-3}$ to approximate the solution to the following initial-value problems:

   a) $y' = \left(\dfrac{y}{t}\right)^2 + \left(\dfrac{y}{t}\right)$, $1 \le t \le 1.2$, $y(1) = 1$ with $hmax = .1$;

   b) $y' = \sin t + e^{-t}$, $0 \le t \le 1$, $y(0) = 0$ with $hmax = .5$;

   c) $y' = \dfrac{1}{t}(y^2 + y)$, $1 \le t \le 3$, $y(1) = -2$ with $hmax = .5$;

   d) $y' = t^2$, $0 \le t \le 2$, $y(0) = 0$ with $hmax = .5$.

2. Use the Runge–Kutta–Fehlberg Algorithm 5.3 to approximate the solutions to the following initial-value problems.

   a) $y' = 1 - y$, $0 \le t \le 1$, $y(0) = 0$; use $TOL = 10^{-6}$.
   b) $y' = -y + t + 1$, $0 \le t \le 5$, $y(0) = 2$; use $TOL = 10^{-4}$.

   c) $y' = \dfrac{2}{t}y + t^2 e^t$, $1 \le t \le 2$, $y(1) = 0$; use $TOL = 10^{-4}$.

   d) $y' = 1 + y^2$, $0 \le t \le \pi/4$, $y(0) = 0$; use $TOL = 10^{-4}$.

3. Use the Runge–Kutta–Fehlberg Algorithm 5.3 to approximate the solutions to the following initial-value problems.

   a) $y' = 2|t - 2|y$, $0 \le t \le 3$, $y(0) = e^{-4}$; use $TOL = 10^{-4}$.
   b) $y' = 1 + t \sin(ty)$, $0 \le t \le 2$, $y(0) = 0$; use $TOL = 10^{-4}$.
   c) $y' = 50t^2 - 50y + 2t$, $0 \le t \le 1$, $y(0) = \frac{1}{3}$; use $TOL = 10^{-3}$.
   d) $y' = -2y + 2t^2 + 2t$, $0 \le t \le 1$, $y(0) = 1$; use $TOL = 10^{-6}$.

4. In the theory of the spread of contagious disease (see Bailey [6] or [7]), a relatively elementary differential equation can be used to predict the number of infective individuals in the population at any time, provided appropriate simplification assumptions are made. In

particular, let us assume that all individuals in a fixed population have an equally likely change of being infected and once infected remain in that state. If we let $x(t)$ denote the number of susceptible individuals at time t and $y(t)$ denote the number of infectives, it is reasonable to assume that the rate at which the number of infectives changes is proportional to the product of $x(t)$ and $y(t)$ since the rate depends upon both the number of infectives and the number of susceptibles present at that time. If the population is large enough to assume that $x(t)$ and $y(t)$ are continuous variables, the problem can be expressed as

$$\frac{dy}{dt}(t) = kx(t)y(t)$$

where k is a constant and $x(t) + y(t) = m$, the total population. This equation can be rewritten involving only $y(t)$ as

$$\frac{dy}{dt}(t) = ky(t)(m - y(t)).$$

a) Assuming that $m = 100{,}000$, $y(0) = 1000$, $k = 2 \times 10^{-6}$, and that time is measured in days, find an approximation to the number of infective individuals at the end of 30 days by using Algorithm 5.3 and $TOL = 1$.

b) The differential equation in part (a) is called a **Bernoulli equation** and can be transformed into a linear differential equation in $z(t)$ by letting $z(t) = (y(t))^{-1}$. Use this technique to find the exact solution to the equation, under the same assumptions as in part (a), and compare the true value of $y(t)$ to the approximation given there. What is $\lim_{t \to \infty} y(t)$? Does this agree with your intuition?

5. In the previous exercise, all infected individuals remained in the population to spread the disease. A more realistic proposal is to introduce a third variable $z(t)$ to represent the number of individuals who are removed from the affected population at a given time t, either by isolation, recovery and consequent immunity, or death. This quite naturally complicates the problem, but it can be shown (see Bailey [7]) that an approximate solution can be given in the form

$$x(t) = x(0)e^{-(k_1/k_2)z(t)} \qquad \text{and} \qquad y(t) = m - x(t) - z(t),$$

where k_1 is the infective rate, k_2 is the removal rate, and $z(t)$ is determined from the differential equation

$$\frac{dz}{dt}(t) = k_2(m - z(t) - x(0)e^{-(k_1/k_2)z(t)}).$$

The authors are not aware of any technique for solving this problem directly, and so a numerical procedure must be applied. Use Algorithm 5.3 to find an approximation to $z(30)$, $y(30)$, and $x(30)$ assuming that $m = 100{,}000$, $x(0) = 99{,}000$, $k_1 = 2 \times 10^{-6}$, and $k_2 = 10^{-4}$, and taking $TOL = 1$.

5.6   Multistep Methods

The methods discussed previously in the chapter are called **one-step methods** since the approximation for the mesh point t_{i+1} involves information from only one of the previous mesh points, t_i. Although these methods generally use functional evaluation information at points between t_i and t_{i+1}, they do not retain that information for direct use in future approximations. All the information utilized by these methods is consequently obtained within the interval over which the solution is being approximated.

Since the approximate solution is available at each of the mesh points $t_0, t_1, \ldots, t_i$ before the approximation at t_{i+1} is obtained, and because the error $|w_j - y(t_j)|$ tends to increase with j, it seems reasonable to develop methods that use this more accurate previous data when approximating the solution at t_{i+1}.

Methods utilizing the approximation at more than one previous mesh point to determine the approximation at the next point are called **multistep** methods. The precise definition of these methods follows, together with the definition of the two types of multistep methods.

Definition 5.12   A **multistep method** for solving the initial-value problem

(5.53) $$y' = f(t, y), \qquad a \le t \le b, \quad y(a) = \alpha,$$

is one whose difference equation for finding the approximation w_{i+1} at the mesh point t_{i+1} can be represented by the following equation, where m is an integer greater than 1:

(5.54) $$\begin{aligned} w_{i+1} = {}& a_{m-1}w_i + a_{m-2}w_{i-1} + \cdots + a_0 w_{i+1-m} \\ &+ h[b_m f(t_{i+1}, w_{i+1}) + b_{m-1} f(t_i, w_i) \\ &+ \cdots + b_0 f(t_{i+1-m}, w_{i+1-m})] \end{aligned}$$

for $i = m - 1, m, \ldots, N - 1$, where the starting values

$$w_0 = \alpha_0, w_1 = \alpha_1, w_2 = \alpha_2, \ldots, w_{m-1} = \alpha_{m-1}$$

are specified and $h = (b - a)/N$.

When $b_m = 0$, the method is called an **explicit** or **open method** and Eq. (5.54) gives w_{i+1} explicitly in terms of previously determined values. When $b_m \ne 0$, the method is called an **implicit** or **closed method** since w_{i+1} occurs on both sides of Eq. (5.54) and is determined only in an implicit manner.

EXAMPLE 1   The equations

(5.55)
$$w_0 = \alpha_0, \quad w_1 = \alpha_1, \quad w_2 = \alpha_2, \quad w_3 = \alpha_3,$$

$$w_{i+1} = w_i + \frac{h}{24}\,[55f(t_i, w_i) - 59f(t_{i-1}, w_{i-1})$$

$$+\, 37f(t_{i-2}, w_{i-2}) - 9f(t_{i-3}, w_{i-3})]$$

for each $i = 3, 4, \ldots, N - 1$, define an explicit four-step method, known as the **fourth-order Adams–Bashforth technique**. The equations

(5.56)
$$w_0 = \alpha_0, \quad w_1 = \alpha_1, \quad w_2 = \alpha_2,$$

$$w_{i+1} = w_i + \frac{h}{24}\,[9f(t_{i+1}, w_{i+1}) + 19f(t_i, w_i)$$

$$-\, 5f(t_{i-1}, w_{i-1}) + f(t_{i-2}, w_{i-2})]$$

for each $i = 2, 3, \ldots, N - 1$, define an implicit three-step method known as the **fourth-order Adams–Moulton technique**. $\square$

The starting values in either (5.55) or (5.56) must be specified, generally by assuming $\alpha_0 = \alpha$, and generating the remaining values by either a Runge–Kutta method or some other one-step technique.

To apply an implicit method such as (5.56) directly, we must solve the implicit equation for w_{i+1}. It is not clear that this can be done in general, or that a unique solution for w_{i+1} will always be obtained, as we will see later in this section.

To begin the derivation of the multistep methods, note that the solution to the initial-value problem (5.53), if integrated over the interval $[t_i, t_{i+1}]$, has the property that

$$(5.57) \qquad y(t_{i+1}) - y(t_i) = \int_{t_i}^{t_{i+1}} y'(t)\, dt = \int_{t_i}^{t_{i+1}} f(t, y(t))\, dt.$$

Consequently,

$$(5.58) \qquad y(t_{i+1}) = y(t_i) + \int_{t_i}^{t_{i+1}} f(t, y(t))\, dt.$$

·Since we cannot integrate $f(t, y(t))$ without knowing $y(t)$, the solution to the problem, we instead integrate an interpolating polynomial, P, that is determined by some of the previously obtained data points $(t_0, w_0), (t_1, w_1), \ldots, (t_i, w_i)$. When we assume, in addition, that $y(t_i) = w_i$, Eq. (5.58) becomes:

$$(5.59) \qquad y(t_{i+1}) \approx w_i + \int_{t_i}^{t_{i+1}} P(t)\, dt.$$

Although any form of the interpolating polynomial can be used for this purpose, the polynomial most useful in practice is the Newton backward interpolating polynomial discussed in Section 3.4. Since the derivation is quite cumbersome, we will derive only a simple two-step method using the Newton backward interpolating polynomial. We refer the reader to Henrici [46], pages 276–283, for a treatment of higher-order methods. The derivation involved is very similar to the numerical-integration development in Chapter 4. The major difference is that the interpolating polynomial for numerical integration uses, principally, points within the interval for its interpolation. In this case, most of the points are taken outside the interval of integration.

EXAMPLE 2 To derive a two-step explicit method, we first replace $f(t, y(t))$ in

$$(5.60) \qquad y(t_{i+1}) = y(t_i) + \int_{t_i}^{t_{i+1}} f(t, y(t))\, dt$$

by the backward difference polynomial of degree one through the points $(t_{i-1}, y(t_{i-1}))$ and $(t_i, y(t_i))$,

$$P(t) = P(t_i + sh) = f(t_i, y(t_i)) - \binom{-s}{1} \nabla f(t_i, y(t_i))$$

and its error term

$$R(t) = \frac{(t - t_{i-1})(t - t_i)}{2}\, f''(\xi_{i+1}, y(\xi_{i+1}))$$

where $t_{i-1} < \xi_{i+1} < t_{i+1}$. Equation (5.60) then becomes

$$
y(t_{i+1}) = y(t_i) + \int_{t_i}^{t_{i+1}} P(t)\, dt + \int_{t_i}^{t_{i+1}} R(t)\, dt
$$

$$
= y(t_i) + f(t_i, y(t_i)) \int_{t_i}^{t_{i+1}} dt - \nabla f(t_i, y(t_i)) \int_{t_i}^{t_{i+1}} \binom{-s}{1}\, dt + \int_{t_i}^{t_{i+1}} R(t)\, dt
$$

$$
= y(t_i) + hf(t_i, y(t_i)) + [f(t_i, y(t_i)) - f(t_{i-1}, y(t_{i-1}))]h \int_0^1 s\, ds
$$

$$
+ \int_{t_i}^{t_{i+1}} R(t)\, dt
$$

or

(5.61) $y(t_{i+1}) = y(t_i) - \dfrac{h}{2} f(t_{i-1}, y(t_{i-1})) + \tfrac{3}{2}hf(t_i, y(t_i))$

$$
+ \frac{1}{2} \int_{t_i}^{t_{i+1}} (t - t_i)(t - t_{i-1}) f''(\xi_{i+1}, y(\xi_{i+1}))\, dt.
$$

Since $(t - t_i)(t - t_{i-1})$ does not change sign on the interval $[t_i, t_{i+1}]$, the Mean Value Theorem for Integrals (Theorem 1.10, p. 5) ensures that a number μ_{i+1}, $t_i < \mu_{i+1} < t_{i+1}$ exists with

$$
\int_{t_i}^{t_{i+1}} (t - t_{i-1})(t - t_i) f''(\xi_{i+1}, y(\xi_{i+1}))\, dt
$$

$$
= f''(\mu_{i+1}, y(\mu_{i+1})) \int_{t_i}^{t_{i+1}} (t - t_{i-1})(t - t_i)\, dt
$$

$$
= \tfrac{5}{6}h^3 f''(\mu_{i+1}, y(\mu_{i+1})).
$$

The fact that $y' = f(t, y)$ allows Eq. (5.61) to be rewritten as

(5.62) $y(t_{i+1}) = y(t_i) - \dfrac{h}{2} f(t_{i-1}, y(t_{i-1})) + \tfrac{3}{2}hf(t_i, y(t_i)) + \tfrac{5}{12}y'''(\mu_{i+1})h^3.$

The corresponding difference method

(5.63)
$$
w_0 = \alpha_0, \quad w_1 = \alpha_1,
$$
$$
w_{i+1} = w_i - \frac{h}{2} f(t_{i-1}, w_{i-1}) + \tfrac{3}{2}hf(t_i, w_i), \qquad i = 1, 2, \ldots, N - 1,
$$

is called the **Adams–Bashforth two-step method**. $\square$

Example 2 shows one typical way to derive multistep methods. Another procedure based on Taylor series is considered in Exercise 8, and a derivation using a Lagrange interpolating polynomial is discussed in Exercise 9(a).

The local truncation error for multistep methods can be defined analogously to that of one-step methods. This error provides a measure of how much the solution to the differential equation fails to solve the difference equation. Note that this definition implies that the local truncation error for the Adams–Bashforth two-step method given in Example 2 is of order two, provided $y \in C^3[a, b]$.

Definition 5.13   If $y(t)$ is the solution to the initial-value problem

(5.64)
$$y' = f(t, y), \qquad a \le t \le b, \quad y(a) = \alpha,$$

and

(5.65) $w_{i+1} = a_m w_i + a_{m-1} w_{i-1} + \cdots + a_0 w_{i-m}$
$$+ h[b_{m+1} f(t_{i+1}, w_{i+1}) + b_m f(t_i, w_i) + \cdots + b_0 f(t_{i-m}, w_{i-m})]$$

is the $(i + 1)$st step in a multistep method, the **local truncation error** at this step, τ_{i+1}, is

(5.66)
$$\tau_{i+1} = \frac{y(t_{i+1}) - a_m y(t_i) - \cdots - a_0 y(t_{i-m})}{h}$$
$$- [b_{m+1} f(t_{i+1}, y(t_{i+1})) + \cdots + b_0 f(t_{i-m}, y(t_{i-m}))]$$

for each $i = m, m + 1, \ldots, N - 1$.

Some of the explicit multistep methods together with their required starting values and local truncation errors are given as follows. The derivation of these techniques is similar to the procedure in Example 2.

Adams–Bashforth Three-Step Method

$$w_0 = \alpha, \quad w_1 = \alpha_1, \quad w_2 = \alpha_2,$$
(5.67)
$$w_{i+1} = w_i + \frac{h}{12}[23f(t_i, w_i) - 16f(t_{i-1}, w_{i-1}) + 5f(t_{i-2}, w_{i-2})]$$

where $i = 2, 3, \ldots, N - 1$; local truncation error $\tau_{i+1} = \frac{3}{8}y^4(\mu_i)h^3$.

Adams–Bashforth Four-Step Method

$$w_0 = \alpha, \quad w_1 = \alpha_1, \quad w_2 = \alpha_2, \quad w_3 = \alpha_3,$$
(5.68)
$$w_{i+1} = w_i + \frac{h}{24}[55f(t_i, w_i) - 59f(t_{i-1}, w_{i-1}) + 37f(t_{i-2}, w_{i-2})$$
$$- 9f(t_{i-3}, w_{i-3})],$$

where $i = 3, 4, \ldots, N - 1$; local truncation error $\tau_{i+1} = \frac{251}{720}y^{(5)}(\mu_i)h^4$.

Adams–Bashforth Five-Step Method

$$w_0 = \alpha, \quad w_1 = \alpha_1, \quad w_2 = \alpha_2, \quad w_3 = \alpha_3, \quad w_4 = \alpha_4,$$
(5.69)
$$w_{i+1} = w_i + \frac{h}{720}[1901f(t_i, w_i) - 2774f(t_{i-1}, w_{i-1})$$
$$+ 2616f(t_{i-2}, w_{i-2}) - 1274f(t_{i-3}, w_{i-3}) + 251f(t_{i-4}, w_{i-4})]$$

where $i = 4, 5, \ldots, N - 1$; local truncation error $\tau_{i+1} = \frac{95}{288}y^{(6)}(\mu_i)h^5$.

Implicit methods are derived by using the additional point $(t_{i+1}, f(t_{i+1}, y(t_{i+1})))$ as an interpolation node in the approximation of the integral

$$\int_{t_i}^{t_{i+1}} f(t, y(t))\, dt.$$

Some of the more common implicit methods are listed as follows.

Adams–Moulton Two-Step Method

(5.70)
$$w_0 = \alpha_0, \quad w_1 = \alpha_1,$$
$$w_{i+1} = w_i + \frac{h}{12}[5f(t_{i+1}, w_{i+1}) + 8f(t_i, w_i) - f(t_{i-1}, w_{i-1})]$$

where $i = 1, 2, \ldots, N - 1$; local truncation error $\tau_{i+1} = -\frac{1}{24}y^{(4)}(\mu_i)h^3$.

Adams–Moulton Three-Step Method

(5.71)
$$w_0 = \alpha_0, \quad w_1 = \alpha_1, \quad w_2 = \alpha_2,$$
$$w_{i+1} = w_i + \frac{h}{24}[9f(t_{i+1}, w_{i+1}) + 19f(t_i, w_i) - 5f(t_{i-1}, w_{i-1})$$
$$+ f(t_{i-2}, w_{i-2})]$$

where $i = 2, 3, \ldots, N - 1$; local truncation error $\tau_{i+1} = -\frac{19}{720}y^{(5)}(\mu_i)h^4$.

Adams–Moulton Four-Step Method

(5.72)
$$w_0 = \alpha_0, \quad w_1 = \alpha_1, \quad w_2 = \alpha_2, \quad w_3 = \alpha_3,$$
$$w_{i+1} = w_i + \frac{h}{720}[251f(t_{i+1}, w_{i+1}) + 646f(t_i, w_i)$$
$$- 264f(t_{i-1}, w_{i-1}) + 106f(t_{i-2}, w_{i-2}) - 19f(t_{i-3}, w_{i-3})]$$

where $i = 3, 4, \ldots, N - 1$; local truncation error $\tau_{i+1} = -\frac{3}{160}y^{(6)}(\mu_i)h^5$.

It is interesting to compare an m-step Adams–Bashforth explicit method to an $(m - 1)$-step Adams–Moulton implicit method. Both require m evaluations of f per step, and both have the terms $y^{(m+1)}(\mu_i)h^m$ in their local truncation errors. In general, the coefficients of the terms involving f and in the local truncation error are smaller for the Adams–Moulton methods. This leads to greater stability for the implicit methods and smaller rounding errors. This is illustrated in the next example.

EXAMPLE 3   The initial-value problem

(5.73)
$$y' = -y + t + 1, \quad 0 \leq t \leq 1, \quad y(0) = 1,$$

will be considered, the approximations being given by the Adams–Bashforth four-step method, (5.68), and the Adams–Moulton three-step method, (5.71), both using $h = .1$ and the values from the exact solution $y(t) = e^{-t} + t$ as their starting values.

The Adams–Bashforth method has the difference equation

$$w_{i+1} = w_i + \frac{h}{24} \left[55f(t_i, w_i) - 59f(t_{i-1}, w_{i-1}) + 37f(t_{i-2}, w_{i-2}) \right.$$

$$\left. - 9f(t_{i-3}, w_{i-3}) \right], \quad \text{for } i = 3, 4, \ldots, 9,$$

which, when simplified using (5.73), $h = .1$, and $t_i = .1i$, becomes

$$w_{i+1} = \tfrac{1}{24} [18.5w_i + 5.9w_{i-1} - 3.7w_{i-2} + .9w_{i-3} + .24i + 2.52]$$

for $i = 3, 4, \ldots, 9$.

The Adams–Moulton method has the difference equation

$$w_{i+1} = w_i + \frac{h}{24} \left[9f(t_{i+1}, w_{i+1}) + 19f(t_i, w_i) - 5f(t_{i-1}, w_{i-1}) \right.$$

$$\left. + f(t_{i-2}, w_{i-2}) \right], \quad \text{for } i = 2, 3, \ldots, 9,$$

which, when simplified, reduces to

$$w_{i+1} = \tfrac{1}{24} [-.9w_{i+1} + 22.1w_i + .5w_{i-1} - .1w_{i-2} + .24i + 2.52],$$

for $i = 2, 3, \ldots, 9$. In order to use this method, however, we can solve explicitly for w_{i+1}, which gives

$$w_{i+1} = \frac{1}{24.9} [22.1w_i + .5w_{i-1} - .1w_{i-2} + .24i + 2.52] \quad \text{for } i = 2, 3, \ldots, 9.$$

Using the exact values from $y(t) = e^{-t} + t$ for α_0, α_1, and α_2 in the Adams–Bashforth case and for α_0 and α_1 in the Adams–Moulton case gives the results in Table 5.7.   ☐

TABLE 5.7

t_i	Adams–Bashforth w_i	Error	Adams–Moulton w_i	Error
.3	starting value	—	1.0408180061	2.146×10^{-7}
.4	1.0703229200	2.874×10^{-6}	1.0703196614	3.846×10^{-7}
.5	1.1065354755	4.816×10^{-6}	1.1065301384	5.213×10^{-7}
.6	1.1488184077	6.772×10^{-6}	1.1488110076	6.285×10^{-7}
.7	1.1965933934	8.090×10^{-6}	1.1965845932	7.106×10^{-7}
.8	1.2493381564	9.192×10^{-6}	1.2493281927	7.714×10^{-7}
.9	1.3065796139	9.954×10^{-6}	1.3065688456	8.141×10^{-7}
1.0	1.3678899580	1.052×10^{-5}	1.3678785994	8.418×10^{-7}

In Example 3 the Adams–Moulton method gave considerably better results than the Adams–Bashforth method of the same order. Although this is generally the case, the implicit methods have the inherent weakness of having to first convert the method algebraically to an explicit representation for w_{i+1}. That this procedure can become difficult, if not impossible, can be seen by considering the rather elementary initial-value problem

$$y' = e^y, \quad 0 \le t \le .25, \quad y(0) = 1.$$

Since $f(t, y) = e^y$, the three-step Adams–Moulton method has

$$w_{i+1} = w_i + \frac{h}{24}[9e^{w_{i+1}} + 19e^{w_i} - 5e^{w_{i-1}} + e^{w_{i-2}}]$$

as its difference equation and this cannot be solved exactly for w_{i+1}.

In actual practice, implicit multistep methods are not used as described above; rather, they are used to improve upon approximations obtained by explicit methods. The combination of an explicit and implicit technique is called a **predictor-corrector method**.

If a fourth-order method for solving an initial-value problem is needed, the first step is to calculate the starting values w_0, w_1, w_2, and w_3 for the four-step Adams–Bashforth method by using the Runge–Kutta method of order four, given in Algorithm 5.2. The next step is to calculate an approximation, $w_4^{(0)}$, to $y(t_4)$ using the Adams–Bashforth method

$$w_4^{(0)} = w_3 + \frac{h}{24}[55f(t_3, w_3) - 59f(t_2, w_2) + 37f(t_1, w_1) - 9f(t_0, w_0)].$$

This approximation is then improved by use of the three-step Adams–Moulton method to

$$w_4^{(1)} = w_3 + \frac{h}{24}[9f(t_4, w_4^{(0)}) + 19f(t_3, w_3) - 5f(t_2, w_2) + f(t_1, w_1)].$$

The value $w_4^{(1)}$ is then used as the approximation to $y(t_4)$ and the technique of using the Adams–Bashforth method as a predictor and the Adams–Moulton method as a corrector repeated to find $w_5^{(0)}$ and $w_5^{(1)}$, the initial and final approximations to $y(t_5)$, etc.

In theory, improved approximations to $y(t_{i+1})$ can be obtained by iterating the Adams–Moulton formula

$$w_{i+1}^{(k+1)} = w_i + \frac{h}{24}[9f(t_{i+1}, w_{i+1}^{(k)}) + 19f(t_i, w_i) - 5f(t_{i-1}, w_{i-1}) + f(t_{i-2}, w_{i-2})].$$

In practice, since $\{w_{i+1}^{(k+1)}\}$ will converge to the actual approximation given by the implicit formula rather than to the solution $y(t_{i+1})$, it is more efficient to use a reduction in the step size if improved accuracy is needed.

The algorithm given below is based on the fourth-order Adams–Bashforth method as predictor and one iteration of the Adams–Moulton method as corrector, with the starting values obtained via the fourth-order Runge–Kutta method.

Adams Fourth-Order Predictor-Corrector Algorithm 5.4

To approximate the solution of the initial-value problem

$$y' = f(t, y), \qquad a \le t \le b, \quad y(a) = \alpha,$$

at $(N + 1)$ equally spaced numbers in the interval $[a, b]$.

INPUT endpoints a, b; integer N; initial condition α.

OUTPUT approximation w to y at the $(N + 1)$ values of t.

Step 1 Set $h = (b - a)/N$;
$$t_0 = a;$$
$$w_0 = \alpha;$$
OUTPUT (t_0, w_0).

Step 2 For $i = 1, 2, 3$ do Steps 3–5.   (*Compute starting values using Runge–Kutta method.*)

Step 3 Set $K_1 = hf(t_{i-1}, w_{i-1})$;
$$K_2 = hf(t_{i-1} + h/2, w_{i-1} + K_1/2);$$
$$K_3 = hf(t_{i-1} + h/2, w_{i-1} + K_2/2);$$
$$K_4 = hf(t_{i-1} + h, w_{i-1} + K_3).$$

Step 4 Set $w_i = w_{i-1} + (K_1 + 2K_2 + 2K_3 + K_4)/6$;
$$t_i = a + ih.$$

Step 5 OUTPUT (t_i, w_i).

Step 6 For $i = 4, \ldots, N$ do Steps 7–10.

Step 7 Set $t = a + ih$;
$$w = w_3 + h[55f(t_3, w_3) - 59f(t_2, w_2) + 37f(t_1, w_1)$$
$$- 9f(t_0, w_0)]/24 \quad (Predict\ w_i.)$$
$$w = w_3 + h[9f(t, w) + 19f(t_3, w_3) - 5f(t_2, w_2)$$
$$+ f(t_1, w_1)]/24. \quad (Correct\ w_i.)$$

Step 8 OUTPUT (t, w).

Step 9 For $j = 0, 1, 2$
set $t_j = t_{j+1}$;   (*Prepare for next iteration.*)
$$w_j = w_{j+1}.$$

Step 10 Set $t_3 = t$;
$$w_3 = w.$$

Step 11 STOP.

EXAMPLE 4   Table 5.8 lists the results obtained by using Algorithm 5.4 for the initial-value problem

$$y' = -y + t + 1, \qquad 0 \le t \le 1, \quad y(0) = 1,$$

TABLE 5.8

t_i	w_i	$\lvert y(t_i) - w_i \rvert$
.4	1.0703199182	1.278×10^{-7}
.5	1.1065302684	3.923×10^{-7}
.6	1.1488110326	6.035×10^{-7}
.7	1.1965845314	7.724×10^{-7}
.8	1.2493280604	9.043×10^{-7}
.9	1.3065686568	1.003×10^{-6}
1.0	1.3678783660	1.075×10^{-6}

with $N = 10$. Although the results here are not as accurate as those in Example 3, which used only the corrector (i.e., the Adams–Moulton method), it should be recalled that in order for the method to be applied in that example it was first necessary to change to an explicit representation for w_{i+1}. Moreover, the exact starting values were used in that case. □

As in the case of the one-step methods when two approximations are available for the same value the possibility exists for approximating the error in one of the methods. This error approximation can then be used to adjust the step size to control the local error and, because of Theorem 5.20, the global error as well.

Since the Adams–Bashforth four-step method comes from the relation

$$y(t_{i+1}) = y(t_i) + \frac{h}{24}\left[55f(t_i, y(t_i)) - 59f(t_{i-1}, y(t_{i-1}))\right.$$

$$\left. + 37f(t_{i-2}, y(t_{i-2})) - 9f(t_{i-3}, y(t_{i-3}))\right] + \tfrac{251}{720}y^{(5)}(\mu_i)h^5$$

for some $t_{i-3} < \mu_i < t_{i+1}$, the assumption that the approximations $w_0, w_1, \ldots, w_i$ are all exact gives

(5.74)
$$\frac{y(t_{i+1}) - w_{i+1}^{(0)}}{h} = \tfrac{251}{720}y^{(5)}(\mu_i)h^4.$$

A similar analysis of the Adams–Moulton three-step method leads to

(5.75)
$$\frac{y(t_{i+1}) - w_{i+1}}{h} \approx -\tfrac{19}{720}y^{(5)}(\tilde{\mu}_i)h^4, \qquad \text{for some } t_{i-2} < \tilde{\mu}_i < t_{i+1}.$$

If we subtract Eq. (5.75) from Eq. (5.74) and assume that, for small values of h, $y^{(5)}$ can be evaluated in both equations at the same point, μ, we can infer that:

$$\frac{w_{i+1} - w_{i+1}^{(0)}}{h} \approx \frac{h^4}{720}\left[251y^{(5)}(\mu) + 19y^{(5)}(\mu)\right] = \tfrac{3}{8}h^4 y^{(5)}(\mu).$$

Using this result to eliminate the term involving $h^4 y^{(5)}$ from (5.75) gives the approximation to the error

(5.76)
$$\frac{|y(t_{i+1}) - w_{i+1}|}{h} \approx \frac{19|w_{i+1} - w_{i+1}^{(0)}|}{270h}.$$

Since a number of assumptions were used to obtain the approximation in Eq. (5.76), it is generally the practice to use the conservative estimate σ for the error where

$$\sigma = \frac{|w_{i+1} - w_{i+1}^{(0)}|}{10h} \approx \frac{|y(t_{i+1}) - w_{i+1}|}{h}.$$

For a given tolerance $\varepsilon > 0$ and an initial step size h, if $\varepsilon/10 < \sigma < \varepsilon$, the value of w_{i+1} is accepted and the procedure continues by calculating $w_{i+2}^{(0)}$ using the same step size h. When σ is not in this range, however, the step size will be changed by multiplying by a constant $q > 0$. To determine an appropriate value for q, recall from Eq. (5.75) that

$$\sigma \approx \frac{|y(t_{i+1}) - w_{i+1}|}{h} \approx \tfrac{19}{720}y^{(5)}(\mu)h^4,$$

assuming the step size h. If the step size is changed to qh and $\hat{w}_{i+1}$ denotes the new approximation at $t_i + qh$, then

$$\frac{|y(t_i + qh) - \hat{w}_{i+1}|}{qh} \approx \tfrac{19}{720}|y^5(\hat{\mu})h^4q^4| \approx \sigma q^4.$$

In order to control the error $|y(t_i + qh) - \hat{w}_{i+1}|/(qh)$, q is chosen to ensure that σq^4 does not exceed ε. The procedure, outlined in the next algorithm, uses the relation

$$\sigma q^4 = \frac{\varepsilon}{2}$$

for this purpose, which implies that q is chosen as

$$q = \sqrt[4]{\frac{\varepsilon}{2\sigma}}.$$

This choice of q ensures that the relative error at each step is between $\varepsilon/10$ and ε. In actual practice, the value of q is generally given an upper bound, in this case 4, so that the step size is not made too large. It should be noted that since the multistep methods require equal step sizes for the starting values, any change in step size necessitates recalculating new starting values at that point. This will be done by calling a Runge–Kutta subalgorithm (see Algorithm 5.2).

Adams Variable Step-Size Predictor-Corrector Algorithm 5.5

To approximate the solution of the initial-value problem

$$y' = f(t, y), \qquad a \le t \le b, \quad y(a) = \alpha.$$

with local truncation error within a given tolerance.

INPUT   endpoints a, b; initial condition α; tolerance TOL; maximum step size $hmax$; minimum step size $hmin$.

OUTPUT i, t_i, w_i, h where at the ith step w_i approximates $y(t_i)$, and step size h was used or a message that the minimum step size was exceeded.

Step 1   Set up a subalgorithm for the Runge–Kutta fourth-order method to be called $RK4(h, v_0, x_0, v_1, x_1, v_2, x_2, v_3, x_3)$, which accepts as input a step size h and starting values $v_0 \approx y(x_0)$ and returns $\{(x_j, v_j)\,|\,j = 1, 2, 3\}$ defined by the following:

for $j = 1, 2, 3$
  set $K_1 = hf(x_{j-1}, v_{j-1})$;
    $K_2 = hf(x_{j-1} + h/2, v_{j-1} + K_1/2)$;
    $K_3 = hf(x_{j-1} + h/2, v_{j-1} + K_2/2)$;
    $K_4 = hf(x_{j-1} + h, v_{j-1} + K_3)$;
    $v_j = v_{j-1} + (K_1 + 2K_2 + 2K_3 + K_4)/6$;
    $x_j = x_0 + jh$.

Step 2   Set $t_0 = a$;
    $w_0 = \alpha$;
    $h = hmax$;
  OUTPUT (t_0, w_0).

Step 3 Call $RK4(h, w_0, t_0, w_1, t_1, w_2, t_2, w_3, t_3)$;
   Set $NFLAG = 1$;   (*Indicates computation from RK4.*)
      $MFLAG = 1$;   (*MFLAG = 0 indicates acceptable computation.*)
      $i = 4$;
      $t = t_3 + h$.

Step 4 While ($t_{i-1} \leq b$ or $MFLAG = 1$) do Steps 5–27.   (*Insures that algorithm stops with an acceptable value.*)

Step 5 Set $WP = w_{i-1} + \dfrac{h}{24}\left[55f(t_{i-1}, w_{i-1}) - 59f(t_{i-2}, w_{i-2})\right.$

$$\left. + 37f(t_{i-3}, w_{i-3}) - 9f(t_{i-4}, w_{i-4})\right];$$ (*Predict w_i.*)

$$WC = w_{i-1} + \dfrac{h}{24}\left[9f(t, WP) + 19f(t_{i-1}, w_{i-1})\right.$$

$$\left. - 5f(t_{i-2}, w_{i-2}) + f(t_{i-3}, w_{i-3})\right];$$ (*Correct w_i.*)

$$\sigma = |WC - WP|/(10h).$$

Step 6 If $\sigma \leq TOL$ then do Steps 7–10.

   Step 7 Set $w_i = WC$;   (*Result accepted.*)
      $t_i = t$;
      $MFLAG = 0$.

   Step 8 If $NFLAG = 1$ then for $j = i - 3, i - 2, i - 1, i$
               OUTPUT (j, t_j, w_j, h)
               (*Previous results also accepted.*)
            else OUTPUT (i, t_i, w_i, h).
               (*Previous results already accepted.*)

   Step 9 Set $i = i + 1$;
         $NFLAG = 0$.

   Step 10 If $\sigma \leq .1\ TOL$ then do Steps 11–16.
                  (*Increase h.*)

      Step 11 Set $HOLD = h$.

      Step 12 Set $q = (TOL/(2\sigma))^{1/4}$.

      Step 13 If $q > 4$ then set $h = 4h$
                     else set $h = qh$.

      Step 14 If $h > hmax$ then set $h = hmax$.

      Step 15 If $t_{i-1} + 3h \geq b$ then set $h = HOLD$.
                  (*Avoid terminating with change in step size.*)

      Step 16 If $h \neq HOLD$ then do Steps 17–19.

         Step 17 Set $NFLAG = 1$.

         Step 18 Call $RK4(h, w_{i-1}, t_{i-1}, w_i, t_i, w_{i+1}, t_{i+1},$
               $w_{i+2}, t_{i+2})$.

Step 19 Set $i = i + 3$.

else do Steps 20 through 26.
(*Result rejected.*)

Step 20 Set $MFLAG = 1$.

Step 21 Set $q = (TOL/(2\sigma))^{1/4}$.

Step 22 If $q < .1$ then set $h = .1h$
else set $h = qh$.

Step 23 If $h < hmin$ then
OUTPUT ('*hmin* exceeded');
(*Procedure fails.*)
STOP.

Step 24 If $NFLAG = 1$ then set $i = i - 3$.
(*Previous results also rejected.*)

Step 25 Call $RK4(h, w_{i-1}, t_{i-1}, w_i, t_i, w_{i+1}, t_{i+1}, w_{i+2}, t_{i+2})$.

Step 26 Set $i = i + 3$;
$NFLAG = 1$.

Step 27 Set $t = t_{i-1} + h$.

Step 28 STOP.

EXAMPLE 5 Table 5.9 lists the results obtained using Algorithm 5.5 to find approximations to the solution of the initial-value problem

$$y' = -y + t + 1, \qquad 0 \le t \le 1, \quad y(0) = 1.$$

Included in the input was tolerance $TOL = 5 \times 10^{-6}$, maximum step size $hmax = .2$ and minimum step size $hmin = .02$. The computations were performed using seven digits of precision. $\square$

TABLE 5.9

i	t_i	h_i	w_i	$y(t_i)$	$|y(t_i) - w_i|$
0	0	—	1.000000	1.000000	0
1		.1999999	result rejected		
1	.0982566	.0982566	1.004672	1.004672	0
2	.1965131	.0982566	1.018102	1.018103	1×10^{-6}
3	.2947697	.0982566	1.039471	1.039472	1×10^{-6}
4	.3930263	.0982566	1.068035	1.068037	2×10^{-6}
5	.4912829	.0982566	1.103120	1.103123	3×10^{-6}
6	.5895395	.0982566	1.144118	1.144121	3×10^{-6}
7	.6877961	.0982566	1.190474	1.190478	4×10^{-6}
8	.7860527	.0982566	1.241687	1.241691	4×10^{-6}
9	.8843092	.0982566	1.297304	1.297307	3×10^{-6}
10	.9825658	.0982566	1.356910	1.356914	4×10^{-6}

Other multistep methods can be derived by using integration of interpolating polynomials over intervals of the form $[t_j, t_{i+1}]$, for $j \leq i - 1$, to obtain an approximation to $y(t_{i+1})$. One particular method that results when a Newton backward polynomial is integrated over $[t_{i-3}, t_{i+1}]$ is an explicit method called **Milne's method**:

$$(5.77) \quad w_{i+1} = w_{i-3} + \frac{4h}{3} [2f(t_i, w_i) - f(t_{i-1}, w_{i-1}) + 2f(t_{i-2}, w_{i-2})],$$

which has local truncation error $(14/45)h^4 y^{(5)}(\xi_i)$ where $t_{i-3} < \xi_i < t_{i+1}$.

This method is occasionally used as a predictor for an implicit method called **Simpson's method**,

$$(5.78) \quad w_{i+1} = w_{i-1} + \frac{h}{3} [f(t_{i+1}, w_{i+1}) + 4f(t_i, w_i) + f(t_{i-1}, w_{i-1})].$$

which has local truncation error $-(h^4/90)y^{(5)}(\xi_i)$ for $t_{i-1} < \xi_i < t_{i+1}$ and is obtained by integrating a Newton backward polynomial over $[t_{i-1}, t_{i+1}]$.

Although the local truncation error involved with a predictor–corrector method of the Milne–Simpson type is generally smaller than that of the Adams–Bashforth–Moulton method, the technique has limited use because of problems in stability, which do not occur with the Adams procedure. More elaboration on this difficulty is contained in Section 5.9.

Exercise Set 5.6

1. Use Algorithm 5.4 to approximate the solutions to the following initial value problems:

   a) $y' = \left(\frac{y}{t}\right)^2 + \left(\frac{y}{t}\right)$, $1 \leq t \leq 1.2$, $y(1) = 1$ with $h = .05$;

   b) $y' = \sin t + e^{-t}$, $0 \leq t \leq 1$, $y(0) = 0$ with $h = .25$;

   c) $y' = \frac{1}{t}(y^2 + y)$, $1 \leq t \leq 3$, $y(1) = -2$ with $h = .5$;

   d) $y' = t^2$, $0 \leq t \leq 2$, $y(0) = 0$ with $h = .5$.

2. Use Algorithm 5.5 with $TOL = 10^{-3}$ to approximate the solutions to the following initial-value problems:

   a) $y' = \left(\frac{y}{t}\right)^2 + \left(\frac{y}{t}\right)$, $1 \leq t \leq 1.2$, $y(1) = 1$ with $hmax = .05$.

   b) $y' = \sin t + e^{-t}$, $0 \leq t \leq 1$, $y(0) = 0$ with $hmax = .25$;

   c) $y' = \frac{1}{t}(y^2 + y)$, $1 \leq t \leq 3$, $y(1) = -2$ with $hmax = .5$;

   d) $y' = t^2$, $0 \leq t \leq 2$, $y(0) = 0$ with $hmax = .5$.

3. Use the Adams–Bashforth method (5.55) and the Adams–Moulton method (5.56) to approximate the solutions to the following initial-value problems.

   a) $y' = 1 - y$, $0 \leq t \leq 1$, $y(0) = 0$;   use $h = .1$.
   b) $y' = -y + t + 1$, $0 \leq t \leq 5$, $y(0) = 2$;
      use $h = .2$ and compare results to Exercise 5, Section 5.4.

c) $y' = \dfrac{2}{t} y + t^2 e^t$, $1 \le t \le 2$, $y(1) = 0$;

 use $h = .05$, and compare results to Exercise 4. Section 5.4.

d) $y' = 50t^2 - 50y + 2t$, $0 \le t \le 1$, $y(0) = \frac{1}{3}$;

 use $h = .1, .025, .01$, and compare results to Exercise 6, Section 5.4.

4. Use an implementation of Algorithm 5.5 for the following problems, and compare your results to those obtained in Exercises 2 and 3 of Section 5.5.

 a) $y' = 1 - y$, $0 \le t \le 1$, $y(0) = 0$;   use $TOL = 10^{-6}$.

 b) $y' = -y + t + 1$, $0 \le t \le 5$, $y(0) = 2$;   use $TOL = 10^{-4}$.

 c) $y' = \dfrac{2}{t} y + t^2 e^t$, $1 \le t \le 2$, $y(1) = 0$;   use $TOL = 10^{-4}$.

 d) $y' = 1 + y^2$, $0 \le t \le \pi/4$, $y(0) = 0$;   use $TOL = 10^{-4}$.

 e) $y' = 2|t - 2|y$, $0 \le t \le 3$, $y(0) = e^{-4}$;   use $TOL = 10^{-4}$.

 f) $y' = 1 + t \sin(ty)$, $0 \le t \le 2$, $y(0) = 0$;   use $TOL = 10^{-4}$.

 g) $y' = 50t^2 - 50y + 2t$, $0 \le t \le 1$, $y(0) = \frac{1}{3}$;   use $TOL = 10^{-3}$.

 h) $y' = -2y + 2t^2 + 2t$, $0 \le t \le 1$, $y(0) = 1$;   use $TOL = 10^{-6}$.

5. The three-step Adams–Moulton method for the differential equation

$$y' = e^y, \qquad 0 \le t \le .20, \quad y(0) = 1,$$

gives the difference equation:

$$w_{i+1} = w_i + \frac{h}{24} [9e^{w_{i+1}} + 19e^{w_i} - 5e^{w_{i-1}} - e^{w_{i-2}}].$$

 a) Show that the difference equation has a unique solution for certain values of $h > 0$.

 b) With $h = .01$, obtain w_i by functional iteration for $i = 3, \ldots, 20$ using exact starting values w_0, w_1, w_2. At each step use w_i to initially approximate w_{i+1}. The solution is

$$y(t) = \ln \left| \frac{e}{1 - et} \right|.$$

 c) Will Newton's method speed the convergence over functional iteration?

6. Use the Milne–Simpson predictor-corrector method to approximate the solution to

$$y' = -5y, \qquad 0 < t < 2. \quad y(0) = 1,$$

with $h = .1$. Repeat the procedure with $h = .05$. Are the answers consistent with the local truncation error?

7. An electric circuit consists of a capacitor of constant capacitance $C = 1.1$ farads in series with a resistor of constant resistance $R_0 = 2.1$ ohms. A voltage $\mathscr{E}(t) = 110 \sin t$ is applied at time $t = 0$. When the resistor heats up the resistance becomes a function of the current i; that is,

$$R(t) = R_0 + ki \qquad \text{where } k = .9,$$

and the differential equation for i becomes

$$\left(1 + \frac{2k}{R_0} i\right) \frac{di}{dt} + \frac{1}{R_0 C} i = \frac{1}{R_0} \frac{d\mathscr{E}}{dt}.$$

Find the current i after 2 seconds, using Algorithm 5.5 with $TOL = 10^{-3}$, if $i(0) = 0$.

8. Derive formula (5.67) by the following method. Set

$$y(t_{i+1}) = y(t_i) + ahf(t_i, y(t_i)) + bhf(t_{i-1}, y(t_{i-1})) + chf(t_{i-2}, y(t_{i-2})).$$

Expand $y(t_{i+1})$, $f(t_{i-2}, y(t_{i-2}))$, and $f(t_{i-1}, y(t_{i-1}))$ in Taylor series about $(t_i, y(t_i))$, and equate the coefficients of h, h^2 and h^3 in order to obtain a, b, and c.

9. a) Derive (5.63) by using the Lagrange form of the interpolating polynomial.
   b) Derive (5.67) by using Newton's backward difference form of the interpolating polynomial.

10. Derive formula (5.70) and its local truncation error by using an appropriate form of an interpolating polynomial.

11. Derive the local truncation errors of Milne's and Simpson's methods.

12. Write an algorithm for the Adams–Bashforth and Adams–Moulton predictor corrector method, incorporating a change of step size by halving or doubling h depending on the error estimate.

13. Derive Simpson's method by applying Simpson's rule to the integral

$$y(t_{i+1}) - y(t_{i-1}) = \int_{t_{i-1}}^{t_{i+1}} f(t, y(t)) \, dt.$$

5.7 Extrapolation Methods

The technique of extrapolation was introduced in Sections 4.2 and 4.5 to derive accurate approximations from low-order formulas applied to numerical differentiation and integration. Extrapolation can also be used to develop efficient procedures for approximating the solution to initial-value problems associated with ordinary differential equations. To review and describe the extrapolation technique, we will first use the procedure with Euler's method. Extrapolation will then be incorporated into a difference scheme to obtain an extrapolation procedure whose idea is credited to Gragg [43].

For the initial-value problem

(5.79) $$y' = f(t, y), \qquad a \le t \le b, \quad y(a) = \alpha,$$

Euler's method with the step size $h > 0$ is given by

$$w_0 = \alpha,$$

and

(5.80) $$w_{i+1} = w_i + hf(t_i, w_i), \qquad \text{for each } i = 0, 1, \ldots, N - 1,$$

where $N = (b - a)/h$ and $t_i = a + ih$. A careful analysis of the error $y(t_i) - w_i$ (see Gear [41], page 15) leads to the existence of a function $\delta(t)$ with the property that

(5.81) $$y(t_i) = w_i + h\delta(t_i) + O(h^2)$$

for each $i = 1, 2, \ldots, N$, the important point being that δ is independent of h.

Since the extrapolation process involves approximations using different step sizes, the notation $w(t, h)$ will be used to represent the approximation to $y(t)$ using the step size h. To illustrate the procedure, select two step sizes h_0 and $h_1 < h_0$.

Since extrapolation can only be applied to approximate y at a particular value of t, we will consider approximating $y(b)$. Suppose

$$q_0 = \frac{b-a}{h_0} \quad \text{and} \quad q_1 = \frac{b-a}{h_1}$$

are both integers and Eq. (5.80) is used first with $h = h_0$ and $N = q_0$ to obtain $w(b, h_0) = w_{q_0}$ and then with $h = h_1$ and $N = q_1$ to obtain $w(b, h_1) = w_{q_1}$. This produces

(5.82) $$y(b) = w(b, h_0) + h_0 \delta(b) + O(h_0^2)$$

and

(5.83) $$y(b) = w(b, h_1) + h_1 \delta(b) + O(h_1^2).$$

Multiplying Eq. (5.83) by h_0 and Eq. (5.82) by h_1, and subtracting the resulting equations eliminates the term involving $\delta(b)$. Solving for $y(b)$ gives the $O(h_0^2)$ approximation for $y(b)$:

$$y(b) = \frac{h_0 w(b, h_1) - h_1 w(b, h_0)}{h_0 - h_1} + O(h_0^2).$$

EXAMPLE 1   We will apply Euler's method with extrapolation to the initial-value problem

$$y' = -y + t + 1, \quad 0 \le t \le 1, \quad y(0) = 1,$$

using $h_0 = .1$ and $h_1 = .05$. The difference formula for $h = h_0$ is

$$w_0 = 1$$
$$w_{i+1} = .9w_i + .1t_i + .1, \quad \text{for each } i = 0, 1, \ldots, 9,$$

and for $h = h_1$ is

$$w_0 = 1,$$
$$w_{i+1} = .95w_i + .05t_i + .05, \quad \text{for each } i = 0, 1, \ldots, 19.$$

Because of the choice of h_0 and h_1, extrapolation can be used at the common nodes $t_i = .1i$ for $i = 0, 1, \ldots, 10$. The results and the values of the actual solution $y(t) = t + e^{-t}$ are given in Table 5.10. $\square$

TABLE 5.10	$t = t_i$	$w(t, .1)$	$w(t, .05)$	$2w(t, .05) - w(t, .1)$	$y(t)$	Error
	0	1.000000	1.000000	1.000000	1.000000	0
	.1	1.000000	1.002500	1.005000	1.004837	1.63×10^{-4}
	.2	1.010000	1.014506	1.019012	1.018731	2.81×10^{-4}
	.3	1.029000	1.035092	1.041184	1.040818	3.66×10^{-4}
	.4	1.056100	1.063420	1.070740	1.070320	4.20×10^{-4}
	.5	1.090490	1.098737	1.106984	1.106531	4.53×10^{-4}
	.6	1.131441	1.140360	1.149279	1.148812	4.67×10^{-4}
	.7	1.178297	1.187675	1.197053	1.196585	4.68×10^{-4}
	.8	1.230467	1.240126	1.249785	1.249329	4.56×10^{-4}
	.9	1.287420	1.297214	1.307008	1.306570	4.38×10^{-4}
	1.0	1.348678	1.358485	1.368292	1.367879	4.13×10^{-4}

To apply extrapolation to a difference method, the method must have a particular type of error expansion. Suppose that a difference method of the form

$$(5.84) \quad \begin{aligned} w_0 &= \alpha \\ w_{i+1} &= w_i + h\phi(t_i, w_i, h), \quad \text{for each } i = 0, 1, \ldots, N-1, \end{aligned}$$

has the property that, if $y \in C^{2p+2}[a, b]$, then

$$(5.85) \qquad y(t) = w(t, h) + \sum_{k=1}^{p} \delta_k(t)h^{2k} + O(h^{2p+2})$$

for some $p > 0$, where the functions δ_k are independent of h for each $k = 1, 2, \ldots, p$. Select a basic step size h and integers $q_0 < q_1 < q_2$, and set $h_j = h/q_j$ for each $j = 0, 1, 2$. Compute $w(b, h_0)$, $w(b, h_1)$ and $w(b, h_2)$ from (5.84), and use (5.85) to obtain

$$(5.86) \quad y(b) = w(b, h_0) + \delta_1(b)h_0^2 + \delta_2(b)h_0^4 + \cdots + \delta_p(b)h_0^{2p} + O(h^{2p+2}),$$

$$(5.87) \quad y(b) = w(b, h_1) + \delta_1(b)h_1^2 + \delta_2(b)h_1^4 + \cdots + \delta_p(b)h_1^{2p} + O(h^{2p+2}),$$

and

$$(5.88) \quad y(b) = w(b, h_2) + \delta_1(b)h_2^2 + \delta_2(b)h_2^4 + \cdots + \delta_p(b)h_2^{2p} + O(h^{2p+2}).$$

Multiplying (5.87) by h_0^2, (5.86) by h_1^2, and subtracting the resulting equations eliminates the term involving $\delta_1(b)$. Solving for $y(b)$ gives

$$(5.89) \qquad y(b) = \frac{h_0^2 w(b, h_1) - h_1^2 w(b, h_0)}{h_0^2 - h_1^2} - \delta_2(b)h_0^2 h_1^2$$

$$- \cdots - \delta_p(b)h_0^2 h_1^2 \frac{h_0^{2p-2} - h_1^{2p-2}}{h_0^2 - h_1^2} + O(h^{2p+2}).$$

Using equations (5.88) and (5.87) in a similar manner results in

$$(5.90) \qquad y(b) = \frac{h_1^2 w(b, h_2) - h_2^2 w(b, h_1)}{h_1^2 - h_2^2} - \delta_2(b)h_1^2 h_2^2$$

$$- \cdots - \delta_p(b)h_1^2 h_2^2 \frac{h_1^{2p-2} - h_2^{2p-2}}{h_1^2 - h_2^2} + O(h^{2p+2}).$$

Since the algebraic expressions become even more complicated in subsequent calculations, we will simplify the notation by letting

$$y_{i,1} = w(t, h_{i-1}) \qquad \text{for each } i = 1, 2, 3$$

and

$$y_{i,2} = \frac{h_{i-2}^2 y_{i,1} - h_{i-1}^2 y_{i-1,1}}{h_{i-2}^2 - h_{i-1}^2} \qquad \text{for each } i = 2, 3.$$

Combining equations (5.89) and (5.90) to eliminate $\delta_2(b)$ yields

$$y(b) = \frac{h_0^2 y_{3,2} - h_2^2 y_{2,2}}{h_0^2 - h_2^2} + \delta_4(b)h_0^2 h_1^2 h_2^2 + \cdots + O(h^{2p+2}).$$

If $y_{3,3}$ is defined by

$$y_{3,3} = \frac{h_0^2 y_{3,2} - h_2^2 y_{2,2}}{h_0^2 - h_2^2},$$

the resulting table

$$y_{1,1} = w(t, h_0)$$

$$y_{2,1} = w(t, h_1) \qquad y_{2,2} = \frac{h_0^2 y_{2,1} - h_1^2 y_{1,1}}{h_0^2 - h_1^2}$$

$$y_{3,1} = w(t, h_2) \qquad y_{3,2} = \frac{h_1^2 y_{3,1} - h_2^2 y_{2,1}}{h_1^2 - h_2^2} \qquad y_{3,3} = \frac{h_0^2 y_{3,2} - h_2^2 y_{2,2}}{h_0^2 - h_2^2}.$$

---

is similar to an iterated interpolation table presented in Section 3.3. (In particular, see Exercise 5 of Section 3.3). The reason for this similarity is that the construction actually consists of forming the linear interpolating polynomials $P_{0,1}(h^2)$ through $(h_0^2, w(t, h_0))$ and $(h_1^2, w(t, h_1))$ and $P_{1,2}(h^2)$ through $(h_1^2, w(t, h_1))$ and $(h_2^2, w(t, h_2))$. Evaluating at $h = 0$ gives $P_{0,1}(0) = y_{2,2}$ and $P_{1,2}(0) = y_{3,2}$. Performing linear iterated interpolation on $y_{2,2}$ and $y_{3,2}$ gives $P_{0,1,2}(0) = y_{3,3}$. This process can be continued by selecting h_3 and computing $y_{4,1}, y_{4,2}, y_{4,3}$ and $y_{4,4}$.

The use of the extrapolation process relies on finding a difference method with error expansion of the form (5.85). A modification of the midpoint method

$$w_{i+1} = w_{i-1} + 2hf(t_i, w_i)$$

with an end-correction can be shown to have the required error expansion. A verification of this can be found in Gragg [43].

The following algorithm uses the extrapolation technique with a sequence of integers $q_0 = 2, q_1 = 3, q_2 = 4, q_3 = 6, q_4 = 8, q_5 = 12, q_6 = 16, q_7 = 18$. A basic step size h is selected, and the method progresses by using $h_j = (1/q_j)h$ for each $j = 0, \ldots, 7$. The error is controlled by requiring that the approximations $y_{1,1}, y_{2,2}, \ldots$ be computed until $|y_{i,i} - y_{i-1,i-1}|$ is less than a given tolerance. If the tolerance is not achieved by $i = 8$, then h is reduced, and the process is reapplied. If $y_{i,i}$ is found to be acceptable, then w_1 is set to $y_{i,i}$, and computations begin for the approximation of w_2.

Extrapolation Algorithm 5.6

To approximate the solution of the initial value problem

$$y' = f(t, y), \qquad a \le t \le b, \quad y(a) = \alpha,$$

with local error within a given tolerance:

INPUT endpoints a, b; initial condition α; tolerance TOL; level limit $p \le 8$; maximum step size $hmax$; minimum step size $hmin$.

OUTPUT T, W, h where W approximates $y(t)$ and step size h was used or a message that minimum step size exceeded.

Step 1 Initialize the array $NK = (2, 3, 4, 6, 8, 12, 16, 18)$.

Step 2 Set $TO = a$;
$WO = \alpha$;
$h = hmax$.

Step 3 For $i = 1, 2, \ldots, 7$
        for $j = 1, \ldots, i$
            set $Q_{i,j} = (NK_{i+1}/NK_j)^2$. $(Q_{i,j} = h_j^2/h_{i+1}^2.)$

Step 4 While $(TO < b)$ do Steps 5–21.

    Step 5 Set $k = 1$;
            $FLAG = 0$.   (*When desired accuracy achieved FLAG is set to* 1.)

    Step 6 While $(k \leq p$ and $FLAG = 0)$ do Steps 7–14.

        Step 7 Set $HK = h/NK_k$;
             $T = TO$;
             $W2 = WO$;
             $W3 = W2 + HK \cdot f(T, W2)$;   (*Euler first step.*)
             $T = TO + HK$.

        Step 8 For $j = 1, \ldots, NK_k - 1$
             set $W1 = W2$;
                $W2 = W3$;
                $W3 = W1 + 2 \cdot HK \cdot f(T, W2)$;   (*Midpoint method.*)
                $T = TO + (j + 1) \cdot HK$.

        Step 9 Set $y_k = [W3 + W2 + HK \cdot f(T, W3)]/2$.
             (*Smoothing to compute* $y_{k,1}$.)

        Step 10 If $k \geq 2$ then do Steps 11–13.

    (*Note*: $y_{k-1} \equiv y_{k-1,1}$, $y_{k-2} \equiv y_{k-2,2}, \ldots, y_1 \equiv y_{k-1,k-1}$ *since only previous row of table is saved.*)

        Step 11 Set $j = k$;
             $v = y_1$.   (*Save* $y_{k-1,k-1}$.)

        Step 12 While $(j \geq 2)$ do

$$\text{set } y_{j-1} = y_j + \frac{y_j - y_{j-1}}{Q_{k-1,j-1} - 1};$$

             (*Extrapolation to compute* $y_{j-1} \equiv y_{k,k-j+2}$).

$$\left(Note: \quad y_{j-1} = \frac{h_{j-1}^2 y_j - h_k^2 y_{j-1}}{h_{j-1}^2 - h_k^2}. \right)$$

             $j = j - 1$.

        Step 13 If $|y_1 - v| \leq TOL$ then set $FLAG = 1$.
             (y_1 *accepted as new w.*)

        Step 14 Set $k = k + 1$.

    Step 15 Set $k = k - 1$.

    Step 16 If $FLAG = 0$ then do Steps 17 and 18.

        Step 17 Set $h = h/2$.   (*New value for w rejected, decrease h.*)

Step 18 If $h < hmin$ then
          OUTPUT ('Minimum h exceeded, method fails'.);
          STOP.

               else do Steps 19–21.

Step 19 Set $WO = y_1$;   (*New value for w accepted.*)
          $TO = TO + h$.

Step 20 OUTPUT (TO, WO, h);

Step 21 If ($k \le 3$ and $h < hmax/2$) then set $h = 2h$.
          (*Increase h if possible.*)

Step 22 STOP.

EXAMPLE 2 Consider the initial-value problem

(5.91) $$y' = -y + t + 1, \qquad 0 \le t \le 1, \quad y(0) = 1,$$

which has solution $y(t) = t + e^{-t}$. Extrapolation Algorithm 5.6 will be applied to (5.91) with $h = .1$ and $TOL = 10^{-5}$. In the computation of w_1, the following table is obtained.

$y_{1,1} = 1.004874$		
$y_{2,1} = 1.004854$	$y_{2,2} = 1.004837$	
$y_{3,1} = 1.004845$	$y_{3,2} = 1.004834$	$y_{3,3} = 1.004832$

The computation stopped with $w_1 = y_{3,3}$ because $|y_{3,3} - y_{2,2}| \le 10^{-5}$. All computations were performed using six digits of precision. The complete set of approximations is given in Table 5.11. In each case $y_{3,3}$ was accepted because $|y_{3,3} - y_{2,2}| \le 10^{-5}$. However, each w_i is accurate to within 2×10^{-5} of the actual value. $\square$

TABLE 5.11

i	t_i	w_i
0	0.0	1.000000
1	.1	1.004832
2	.2	1.018721
3	.3	1.040806
4	.4	1.070306
5	.5	1.106516
6	.6	1.148796
7	.7	1.196569
8	.8	1.249311
9	.9	1.306552
10	1.0	1.367859

The proof that the method presented in Algorithm 5.6 converges involves results from summability theory, and can be found in the original paper of Gragg [43]. A number of other extrapolation procedures are available, some of which utilize variable step-size techniques, and research in this area is quite active. For additional procedures based on the extrapolation process, see Bulirsch and Stoer [19], [20], [21], or the text by Stetter [84]. The methods used by Bulirsch and Stoer involve interpolation with rational functions instead of the polynomial interpolation used in the Gragg procedure.

Exercise Set 5.7

1. Use Algorithm 5.6 with $TOL = 10^{-3}$ to approximate the solutions to the following initial value problems:

   a) $y' = \left(\dfrac{y}{t}\right)^2 + \left(\dfrac{y}{t}\right)$, $1 \le t \le 1.2$, $y(1) = 1$ with $hmax = .1$;

   b) $y' = \sin t + e^{-t}$, $0 \le t \le 1$, $y(0) = 0$ with $hmax = .5$;

   c) $y' = \dfrac{1}{t}(y^2 + y)$, $1 \le t \le 3$, $y(1) = -2$ with $hmax = .5$;

   d) $y' = t^2$, $0 \le t \le 2$, $y(0) = 0$ with $hmax = .5$.

2. Use the Extrapolation Algorithm 5.6 for the following initial-value problems, and compare the results with the corresponding exercise in Section 5.5.

   a) $y' = 1 - y$, $0 \le t \le 1$, $y(0) = 0$;   use $h = .1$ and $TOL = 10^{-6}$.
   b) $y' = -y + t + 1$, $0 \le t \le 5$, $y(0) = 2$;   use $h = .2$ and $TOL = 10^{-4}$.
   c) $y' = \dfrac{2}{t}y + t^2 e^t$, $1 \le t \le 2$, $y(1) = 0$;   use $h = .1$ and $TOL = 10^{-4}$.
   d) $y' = 1 + y^2$, $0 \le t \le \pi/4$, $y(0) = 0$;   use $h = \pi/20$ and $TOL = 10^{-4}$.
   e) $y' = 2|t - 2|y$, $0 \le t \le 3$, $y(0) = e^{-4}$;   use $h = .2$ and $TOL = 10^{-4}$.
   f) $y' = 1 + t \sin(ty)$, $0 \le t \le 2$, $y(0) = 0$;   use $h = .1$ and $TOL = 10^{-4}$.
   g) $y' = 50t^2 - 50y + 2t$, $0 \le t \le 1$, $y(0) = \frac{1}{3}$;   use $h = .1$ and $TOL = 10^{-3}$.
   h) $y' = -2y + 2t^2 + 2t$, $0 \le t \le 1$, $y(0) = 1$;   use $h = .1$ and $TOL = 10^{-6}$.

3. Let $P(t)$ be the number of individuals in a population at time t, measured in years. If the average birth rate b is constant and the average death rate d is proportional to the size of the population (due to overcrowding), then the growth rate of the population is given by the **logistic equation**

$$\frac{dP(t)}{dt} = bP(t) - k[P(t)]^2$$

where $d = kP(t)$. Suppose $P(0) = 50{,}976$, $b = 2.9 \times 10^{-2}$, and $k = 1.4 \times 10^{-7}$. Find the population after five years, using Algorithm 5.6.

5.8 Higher-Order Equations and Systems of Differential Equations

This section contains an introduction to the numerical solution of higher-order differential equations subject to initial conditions. The techniques we will discuss are limited to those that transform a higher-order equation into a system of first-order differential equations. Before discussing the transformation procedure, some remarks are needed concerning systems which involve first-order differential equations.

An **mth-order system** of first-order initial-value problems can be expressed in the form

$$\frac{du_1}{dt} = f_1(t, u_1, u_2, \ldots, u_m),$$

$$\frac{du_2}{dt} = f_2(t, u_1, u_2, \ldots, u_m),$$

(5.92)

$$\vdots$$

$$\frac{du_m}{dt} = f_m(t, u_1, u_2, \ldots, u_m).$$

for $a \le t \le b$, with the initial conditions

$$u_1(a) = \alpha_1,$$

$$u_2(a) = \alpha_2,$$

(5.93)

$$\vdots$$

$$u_m(a) = \alpha_m.$$

The object is to find m functions $u_1, u_2, \ldots, u_m$ that satisfy the system of differential equations as well as the initial conditions.

In order to discuss existence and uniqueness of solutions to systems of equations, it is first necessary to extend the definition of Lipschitz condition to functions of several variables.

Definition 5.14   The function $f(t, y_1, \ldots, y_m)$, defined on the set

$$D = \{(t, u_1, \ldots, u_m) | a \le t \le b, -\infty < u_i < \infty, \text{ for each } i = 1, 2, \ldots, m\}$$

is said to satisfy a **Lipschitz condition** on D in the variables $u_1, u_2, \ldots, u_m$ if a constant $L > 0$ exists with the property that

(5.94) $$|f(t, u_1, \ldots, u_m) - f(t, z_1, \ldots, z_m)| \le L \sum_{j=1}^{m} |u_j - z_j|$$

for all $(t, u_1, u_2, \ldots, u_m)$ and $(t, z_1, \ldots, z_m)$ in D.

By using the Mean Value Theorem, it can be shown that if f and its first partial derivatives are continuous in D and if

$$\left| \frac{\partial f(t, u_1, \ldots, u_m)}{\partial u_i} \right| \le L$$

for each $i = 1, 2, \ldots, m$ and all $(t, u_1, \ldots, u_m)$ in D, then f satisfies a Lipschitz condition on D with Lipschitz constant L (see Birkhoff–Rota [13], page 102). A basic existence and uniqueness theorem is given below. Its proof can be found in Birkhoff–Rota [13], pages 112–113.

Theorem 5.15   Suppose

$$D = \{(t, u_1, u_2, \ldots, u_m) | a \leq t \leq b, -\infty < u_i < \infty, \text{ for each } i = 1, 2, \ldots, m\}$$

and let $f_i(t, u_1, \ldots, u_m)$, for each $i = 1, 2, \ldots, m$, be continuous on D and satisfy a Lipschitz condition there. The system of first-order differential equations (5.92), subject to the initial conditions (5.93), has a unique solution $u_1(t), \ldots, u_m(t)$ for $a \leq t \leq b$.

To convert a general mth-order differential equation of the form

$$y^{(m)}(t) = f(t, y, y', \ldots, y^{(m-1)}), \qquad a \leq t \leq b,$$

with initial conditions $y(a) = \alpha_1, y'(a) = \alpha_2, \ldots, y^{(m-1)}(a) = \alpha_m$ into a system of equations in the form (5.92) and (5.93), let $u_1(t) = y(t), u_2(t) = y'(t), \ldots, u_m(t) = y^{(m-1)}(t)$. Using this notation, we obtain the first-order system

$$\frac{du_1}{dt} = \frac{dy}{dt} = u_2,$$

$$\frac{du_2}{dt} = \frac{dy'}{dt} = u_3,$$

$$\vdots$$

$$\frac{du_m}{dt} = \frac{dy^{(m-1)}}{dt} = y^{(m)} = f(t, y, y', \ldots, y^{(m-1)})$$

$$= f(t, u_1, u_2, \ldots, u_m),$$

with initial conditions

$$u_1(a) = y(a) = \alpha_1,$$

$$u_2(a) = y'(a) = \alpha_2,$$

$$\vdots$$

$$u_m(a) = y^{(m-1)}(a) = \alpha_m.$$

EXAMPLE 1   Consider the second-order differential equation

(5.95) $$y'' - 2y' + 2y = e^{2t} \sin t, \qquad 0 \leq t \leq 1,$$

with initial conditions $y(0) = -.4, y'(0) = -.6$. With $u_1(t) = y(t)$ and $u_2(t) = y'(t)$, (5.95) is transformed into the system

(5.96) $$\begin{aligned} u_1'(t) &= u_2(t), \\ u_2'(t) &= e^{2t} \sin t - 2u_1(t) + 2u_2(t), \end{aligned}$$

with initial conditions

(5.97) $$\begin{aligned} u_1(0) &= -.4, \\ u_2(0) &= -.6. \end{aligned}$$

Clearly, $f_1(t, u_1, u_2) = u_2$ satisfies a Lipschitz condition on

$$D = \{(t, u_1, u_2) | 0 \leq t \leq 1, -\infty < u_1, u_2 < \infty\}$$

with $L = 1$, and is continuous on D. With $f_2(t, u_1, u_2) = e^{2t} \sin t - 2u_1 + 2u_2$, we have

$$|f_2(t, u_1, u_2) - f_2(t, z_1, z_2)| = |e^{2t} \sin t - 2u_1 + 2u_2 - e^{2t} \sin t + 2z_1 - 2z_2|$$
$$\leq 2|u_1 - z_1| + 2|u_2 - z_2|,$$

so f_2 also satisfies a Lipschitz condition on D, but with $L = 2$. Moreover, f_2 is continuous on D, so the system has a unique solution for $0 \leq t \leq 1$. It is easily verified that

$$u_1(t) = .2e^{2t}(\sin t - 2 \cos t),$$

$$u_2(t) = .2e^{2t}(4 \sin t - 3 \cos t),$$

is this unique solution; so $y(t) = .2e^{2t}(\sin t - 2 \cos t)$ is the unique solution to the second-order equation satisfying $y(0) = -.4$ and $y'(0) = -.6$.   □

Methods to solve systems of first-order differential equations are simply generalizations of the methods for a single first-order equation presented earlier in this chapter. For example, the classical Runge–Kutta method of order four given by

$$w_0 = \alpha,$$

$$k_1 = hf(t_i, w_i),$$

$$k_2 = hf\left(t_i + \frac{h}{2}, w_i + \tfrac{1}{2}k_1\right),$$

$$k_3 = hf\left(t_i + \frac{h}{2}, w_i + \tfrac{1}{2}k_2\right),$$

$$k_4 = hf(t_{i+1}, w_i + k_3),$$

and $w_{i+1} = w_i + \tfrac{1}{6}[k_1 + 2k_2 + 2k_3 + k_4]$ for each $i = 0, 1, \ldots, N - 1,$

used to solve the first-order initial-value problem

$$y' = f(t, y), \quad a \leq t \leq b, \quad y(a) = \alpha,$$

can be generalized as follows. Let an integer $N > 0$ be chosen and set $h = (b - a)/N$. Partition the interval $[a, b]$ into N subintervals with the mesh points

$$t_j = a + jh \quad \text{for each } j = 0, 1, \ldots, N.$$

Use the notation w_{ij} to denote an approximation to $u_i(t_j)$ for each $j = 0, 1, \ldots, N$, and $i = 1, 2, \ldots, m$; that is, w_{ij} will approximate the ith solution $u_i(t)$ of (5.92) at the jth mesh point t_j. For the initial conditions, set

(5.98)
$$w_{1,0} = \alpha_1,$$

$$w_{2,0} = \alpha_2,$$

$$\vdots$$

$$w_{m,0} = \alpha_m.$$

If we assume that the values $w_{1,j}, w_{2,j}, \ldots, w_{m,j}$ have been computed, we obtain $w_{1,j+1}, w_{2,j+1}, \ldots, w_{m,j+1}$ by first calculating

(5.99) $k_{1,i} = hf_i(t_j, w_{1,j}, w_{2,j}, \ldots, w_{m,j})$ for each $i = 1, 2, \ldots, m$;

(5.100) $k_{2,i} = hf_i\left(t_j + \dfrac{h}{2}, w_{1,j} + \frac{1}{2}k_{1,1}, w_{2,j} + \frac{1}{2}k_{1,2}, \ldots, w_{m,j} + \frac{1}{2}k_{1,m}\right)$

$$\text{for each } i = 1, 2, \ldots, m;$$

(5.101) $k_{3,i} = hf_i\left(t_j + \dfrac{h}{2}, w_{1,j} + \frac{1}{2}k_{2,1}, w_{2,j} + \frac{1}{2}k_{2,2}, \ldots, w_{m,j} + \frac{1}{2}k_{2,m}\right)$

$$\text{for each } i = 1, 2, \ldots, m;$$

(5.102) $k_{4,i} = hf_i(t_j + h, w_{1,j} + k_{3,1}, w_{2,j} + k_{3,2}, \ldots, w_{m,j} + k_{3,m})$

$$\text{for each } i = 1, 2, \ldots, m;$$

and then

(5.103) $$w_{i,j+1} = w_{i,j} + \tfrac{1}{6}[k_{1,i} + 2k_{2,i} + 2k_{3,i} + k_{4,i}]$$

$$\text{for each } i = 1, 2, \ldots, m.$$

It should be noted that $k_{1,1}, k_{1,2}, \ldots, k_{1,m}$ must all be computed before $k_{2,1}$ can be determined. In general, each $k_{l,1}, k_{l,2}, \ldots, k_{l,m}$ must be computed before any of the expressions $k_{l+1,i}$.

EXAMPLE 2   Consider the first-order system from Example 1, given by

(5.104)
$$u_1' = u_2,$$
$$u_2' = e^{2t} \sin t - 2u_1 + 2u_2,$$

for $0 \leq t \leq 1$, with initial conditions

(5.105)
$$u_1(0) = -.4,$$
$$u_2(0) = -.6.$$

The classical Runge–Kutta fourth-order method will be used to approximate the solution to this problem using $h = .1$. The initial conditions give $w_{1,0} = -.4$ and $w_{2,0} = -.6$. Using equations (5.99) through (5.102) with $j = 1$,

$k_{1,1} = hf_1(t_0, w_{1,0}, w_{2,0}) = hw_{2,0} = -.06,$

$k_{1,2} = hf_2(t_0, w_{1,0}, w_{2,0})$

$\quad = h[e^{2t_0} \sin t_0 - 2w_{1,0} + 2w_{2,0}] = -.04,$

$k_{2,1} = hf_1\left(t_0 + \dfrac{h}{2}, w_{1,0} + \frac{1}{2}k_{1,1}, w_{2,0} + \frac{1}{2}k_{1,2}\right)$

$\quad = h[w_{2,0} + \frac{1}{2}k_{1,2}] = -.062,$

$k_{2,2} = hf_2\left(t_0 + \dfrac{h}{2}, w_{1,0} + \frac{1}{2}k_{1,1}, w_{2,0} + \frac{1}{2}k_{1,2}\right)$

$\quad = h[e^{2(t_0 + .05)} \sin(t_0 + .05) - 2(w_{1,0} + \frac{1}{2}k_{1,1}) + 2(w_{2,0} + \frac{1}{2}k_{1,2})]$

$\quad = -.03247644757,$

$$k_{3,1} = h[w_{2,0} + \tfrac{1}{2}k_{2,2}] = -.06162382238,$$

$$k_{3,2} = h[e^{2(t_0+.05)}\sin(t_0 + .05) - 2(w_{1,0} + \tfrac{1}{2}k_{2,1}) + 2(w_{2,0} + \tfrac{1}{2}k_{2,2})]$$
$$= -.03152409237,$$

$$k_{4,1} = h[w_{2,0} + k_{3,2}] = -.06315240924,$$

$$k_{4,2} = h[e^{2(t_0+.1)}\sin(t_0 + .1) - 2(w_{1,0} + k_{3,1}) + 2(w_{2,0} + k_{3,2})]$$
$$= -.02178637298;$$

so $$w_{1,1} = w_{1,0} + \tfrac{1}{6}[k_{1,1} + 2k_{2,1} + 2k_{3,1} + k_{4,1}] = -.4617333423,$$

and $$w_{2,1} = w_{2,0} + \tfrac{1}{6}[k_{1,2} + 2k_{2,2} + 2k_{3,2} + k_{4,2}] = -.6316312421.$$

The value $w_{1,1}$ approximates $u_1(.1) = y(.1) = .2e^{2(.1)}[\sin .1 - 2\cos .1]$ and $w_{2,1}$ approximates $u_2(.1) = y'(.1) = .2e^{2(.1)}[4\sin .1 - 3\cos .1]$.

The set of values $w_{1,j}$ and $w_{2,j}$ for $j = 0, 1, \ldots, 10$, are presented in Table 5.12 and the actual value of $u_1(t) = .2e^{2t}[\sin t - 2\cos t]$ is also given. Values of $u_2(t)$ are not given since $u_2 = u_1'$. □

TABLE 5.12

t_j	$w_{1,j}$	$w_{2,j}$	$y(t_j) = u_1(t_j)$	$\lvert y(t_j) - w_{1,j}\rvert$
0	$-.40000000$	$-.60000000$	$-.40000000$	0
.1	$-.46173334$	$-.63163124$	$-.46173297$	3.7×10^{-7}
.2	$-.52555988$	$-.64014895$	$-.52555905$	8.3×10^{-7}
.3	$-.58860144$	$-.61366381$	$-.58860005$	1.39×10^{-6}
.4	$-.64661231$	$-.53658203$	$-.64661028$	2.03×10^{-6}
.5	$-.69356666$	$-.38873810$	$-.69356395$	2.71×10^{-6}
.6	$-.72115190$	$-.14438087$	$-.72114849$	3.41×10^{-6}
.7	$-.71815295$	$.22899702$	$-.71814890$	4.05×10^{-6}
.8	$-.66971133$	$.77199180$	$-.66970677$	4.56×10^{-6}
.9	$-.55644290$	$.15347815$	$-.55643814$	4.76×10^{-6}
1.0	$-.35339886$	$.25787663$	$-.35339436$	4.50×10^{-6}

The other one-step methods can similarly be extended to systems. If methods such as the Runge–Kutta–Fehlberg method are extended with error control, then each component of the numerical solution $(w_{1j}, w_{2j}, \ldots, w_{mj})$ must be examined for accuracy. If any of the components fail to be sufficiently accurate, the entire numerical solution $(w_{1j}, w_{2j}, \ldots, w_{mj})$ must be recomputed.

The multistep methods and predictor–corrector techniques can also be extended easily to systems. Again, if error control is used, each component must be accurate.

The extension of the extrapolation technique to systems can also be done, but the notation becomes quite involved. An extension of this type is considered in Exercise 4.

Convergence theorems and error estimates for systems are very similar to those which will be considered in Section 5.9 for the single equations, except that the bounds are given in terms of vector norms, a topic considered in Chapter 8. [A good reference for these theorems is Gear [41], pages 45–72.]

Exercise Set 5.8

1. Change the Adams Fourth-Order Predictor–Corrector Algorithm 5.4 to obtain approximate solutions to systems of first-order differential equations, and solve the following initial-value problems:

    a) $u'_1(t) = 3u_1(t) + 2u_2(t)$, $0 \le t \le 1$, $u_1(0) = 0$;
        $u'_2(t) = 4u_1(t) + u_2(t)$, $0 \le t \le 1$, $u_2(0) = 1$.
        Use $h = .1$ and compare to the actual solution

        $$u_1(t) = \tfrac{1}{3}(e^{5t} - e^{-t}) \quad \text{and} \quad u_2(t) = \tfrac{1}{3}(e^{5t} + 2e^{-t}).$$

    b) $y''(t) + 2ty'(t) + t^2 y(t) = e^t$, $0 \le t \le 2$, $y(0) = 1$, $y'(0) = -1$.
        Use $h = .1$ and compare to the approximation generated by the first five terms of the power series of the actual solution.

    c) $y''(t) - 2y'(t) + 2y(t) = e^{2t} \sin t$, $0 \le t \le 1$, $y(0) = -.4$, $y'(0) = -6$.
        Use $h = .1$ and compare to Example 2.

    d) $t^2 y'' - 2ty' + 2y = t^3 \ln t$, $1 \le t \le 2$, $y(1) = 1$, $y'(1) = 0$.
        Use $h = .05$ and compare to the actual solution

        $$y(t) = \tfrac{7}{4}t + \frac{t^3}{2} \ln t - \tfrac{3}{4}t^3.$$

    e) $y''' + 2y'' - y' - 2y = e^t$, $0 \le t \le 3$, $y(0) = 1$, $y'(0) = 2$, $y''(0) = 0$.
        Use $h = .2$ and compare to the actual solution.

    f) $u'_1 + 4u_1 + 2u_2 = \cos t + 4 \sin t$, $0 \le t \le 5$, $u_1(0) = 0$;
        $u'_2 - 3u_1 - u_2 = -3 \sin t$, $0 \le t \le 5$, $u_2(0) = -1$.
        Use $h = .1$ and compare to the actual solution

        $$u_1(t) = 2e^{-t} - 2e^{-2t} + \sin t, \qquad u_2(t) = -3e^{-t} + 2e^{-2t}.$$

    g) $u'_1 = u_2$, $0 \le t \le 1$, $u_1(0) = 3$;
        $u'_2 = -u_1 + 2e^{-t} + 1$, $0 \le t \le 1$, $u_2(0) = 0$.
        Use $h = .05$ and compare to the actual solution

        $$u_1(t) = \cos t + \sin t + e^{-t} + 1, \quad u_2(t) = -\sin t + \cos t - e^{-t}.$$

    h) $y'' = 2y^3$, $1 < t < 1.9$, $y(1) = -1$, $y'(1) = -1$.
        Use $h = .05$ and compare to the actual solution $y(t) = (t - 2)^{-1}$.

2. Repeat Exercise 1, using the Runge–Kutta fourth-order method. (Omit part (c).)

3. Change the Runge–Kutta–Fehlberg Algorithm 5.3 to approximate the solution to systems, and repeat Exercise 1.

4. Change the Extrapolation Algorithm 5.6 to approximate the solution to systems, and repeat Exercise 1.

5. The study of mathematical models for predicting the population dynamics of competing species has its origin in independent works published in the early part of this century by A. J. Lotka and V. Volterra. Consider the problem of predicting the population of two species, one of which is a predator, whose population at time t is $x_2(t)$, feeding on the other (which is called a prey) whose population is $x_1(t)$. We will assume that the prey always has an adequate food supply and that its birth rate at any time is proportional to the number of prey alive at that time; that is, birth rate (prey) $= k_1 x_1(t)$. The death rate of the prey depends upon both the number of prey and predators alive at that time. For simplicity we will assume the death rate (prey) $= k_2 x_1(t)x_2(t)$. The birth rate of the predator, on the other hand, depends upon its food supply, $x_1(t)$, as well as on the number of predators available for reproduction purposes. For this reason we assume that the birth rate(predator) $= k_3 x_1(t)x_2(t)$. The death rate of the predator will be taken as simply proportional to the number of predators alive at the time; that is, death rate (predator) $= k_4 x_2(t)$.

Since $x_1'(t)$ and $x_2'(t)$ represent the change in the prey and predator populations, respectively, with respect to time, the problem is expressed by the system of nonlinear differential equations

$$x_1'(t) = k_1 x_1(t) - k_2 x_1(t) x_2(t) \quad \text{and} \quad x_2'(t) = k_3 x_1(t) x_2(t) - k_4 x_2(t).$$

Use the fourth-order Runge-Kutta method with $h = .5$ to solve this system, assuming that the initial population of the prey is 1000 and of the predators is 200, and that the constants are $k_1 = 3$, $k_2 = .002$, $k_3 = .0006$, and $k_4 = .5$. Sketch a graph of the solutions to this problem, plotting both populations with time, and describe the physical phenomena represented. Is there a stable solution to this population model; if so, for what values of x_1 and x_2 is the solution stable?

6. In Exercise 5, we considered the problem of predicting the population in a predator–prey model. Another problem of this type is concerned with two species competing for the same food supply. If the number of the species alive at time t are denoted by $x_1(t)$ and $x_2(t)$, it is often assumed that, while the birth rate of each of the species is simply proportional to the number of the species alive at that time, the death rate of each species depends upon the population of both species. We will assume that the population of a particular pair of species is described by the equations

$$\frac{dx_1(t)}{dt} = x_1(t)[4 - .0003 x_1(t) - .0004 x_2(t)]$$

and

$$\frac{dx_2(t)}{dt} = x_2(t)[2 - .0002 x_1(t) - .0001 x_2(t)].$$

If it is known that the initial population of each species is 10,000, find the solution to this system. Is there a stable solution to this population model; if so, for what values of x_1 and x_2 is the solution stable?

5.9 Stability

A number of different methods have been presented in this chapter for approximating the solution to an initial-value problem. Although numerous other techniques are available for this purpose, we have chosen the methods described here because generally they satisfied three criteria: first, their development is clear enough so that the first-year student in numerical analysis can understand how and why they work; second, most of the more advanced and complex techniques have their bases in one or more of the procedures described here; and third, one or more of the methods will give satisfactory results for most of the problems that are encountered by undergraduate students in science and engineering. The topic we will discuss in this section is the reason these methods might be expected to give satisfactory results when some similar methods would not.

Before we begin this discussion, we need to present two definitions concerned with the convergence of one-step difference-equation methods to the solution of the differential equation as the step size decreases.

Definition 5.16 A difference-equation method with local truncation error τ_i at the ith step is said to be **consistent** with the differential equation it approximates if

$$\lim_{h \to 0} \max_{1 \le i \le N} |\tau_i| = 0.$$

It should be noted that this definition is essentially a "local" definition since, for each of the values τ_i, we are comparing the exact value $f(t_i, y_i)$ to the difference-equation approximation of y'. A more realistic means of analyzing the effects of making h small is to determine the "global" effect of the method, which is the maximum error of the method over the entire range of the approximation, assuming only that the method gives the exact result at the initial value. The definition describing a method that has convergence in this sense is given below.

Definition 5.17   A difference-equation method is said to be **convergent** with respect to the differential equation it approximates if

$$\lim_{h \to 0} \max_{1 \le i \le N} |y_i - w_i| = 0,$$

where $y_i = y(t_i)$ denotes the exact value of the solution of the differential equation and w_i the approximation obtained from the difference method at the ith step.

Examining inequality (5.23) in the error-bound formula for Euler's method, it can be seen that under the hypotheses of Theorem 5.9.

$$\max_{1 \le i \le N} |y_i - w_i| \le \frac{Mh}{2L} |e^{L(b-a)} - 1|;$$

so Euler's method is convergent with respect to a differential equation satisfying the conditions of this theorem and the rate of convergence is $O(h)$.

For one-step methods, this definition of the consistency of a method is satisfied precisely when the difference equation for the method approaches the differential equation when the step size goes to zero; that is, the local truncation error approaches zero as the step size approaches zero. The definition of convergence has a similar connotation since a method is convergent precisely when the solution to the difference equation approaches the solution to the differential equation as the step size goes to zero.

The other error-bound type of problem that exists when using difference methods to approximate solutions to differential equations results from not using exact results. In practice, neither the initial conditions nor the arithmetic that is subsequently performed is represented exactly because of the roundoff error associated with finite-digit arithmetic. In Section 5.2 we saw that this consideration can lead to difficulties even for Euler's method, which is both consistent and convergent. In order to at least partially analyze this situation, we will try to determine which methods are stable, in the sense that small changes or perturbations in the initial conditions produce correspondingly small changes in the subsequent approximations; that is, a **stable** method is one that depends *continuously* upon the initial data.

Since the concept of stability of a one-step difference equation is somewhat analogous to the condition of a differential equation being well-posed, it is not surprising that the Lipschitz condition appears here as it did in the corresponding theorem for differential equations, Theorem 5.6 (p. 184). The proof of this result is not difficult and is considered in Exercise 1.

Theorem 5.18   If the initial-value problem

(5.106) $y' = f(t, y),$ $a \leq t \leq b,$ $y(a) = \alpha$

is approximated by a one-step difference method in the form

$$w_0 = \alpha,$$

(5.107)

$$w_{i+1} = w_i + h\phi(t_i, w_i, h)$$

and $\phi(t, w, h)$ satisfies a Lipschitz condition in the variable w on the set

$$D = \{(t, w, h) | a \leq t \leq b, -\infty < w < \infty, 0 \leq h \leq h_0\},$$

then the method is stable.

Similar theorems are also available that give sufficient conditions for a one-step method to be consistent and convergent. The proof of the following theorem can be found within the material presented by Gear [41] on pages 57–58.

Theorem 5.19   If the initial-value problem

$$y' = f(t, y),   a \leq t \leq b,   y(a) = \alpha,$$

is approximated by a one-step difference method in the form

$$w_0 = \alpha,$$

$$w_{i+1} = w_i + h\phi(t_i, w_i, h)$$

and $\phi(t, w, h)$ is continuous and satisfies a Lipschitz condition in the variable w on the set

$$D = \{(t, w, h) | a \leq t \leq b, -\infty < w < \infty, 0 \leq h \leq h_0\},$$

then

i)   the difference method is convergent if and only if it is consistent; i.e., if and only if

$$\phi(t, y, 0) = f(t, y)   \text{for all } a \leq t \leq b;$$

ii)   if the local truncation error, τ_i, satisfies

$$|\tau_i| \leq \tau(h)   \text{for all } i = 1, 2, \ldots, N$$

whenever $0 \leq h \leq h_0$, then

$$|y(t_i) - w_i| \leq \frac{\tau(h)}{L} e^{L(t_i - a)}$$

for each $i = 1, 2, \ldots, N$ where L denotes the Lipschitz constant.

Note that part (ii) of this theorem justifies the remark made in Section 5.5 about controlling the global error of a method by controlling its local error and implies that when the local truncation error has the rate of convergence $O(h^n)$, the global error will have the same rate of convergence.

EXAMPLE   1   Consider the modified Euler method given by

$$w_0 = \alpha,$$

$$w_{i+1} = w_i + \frac{h}{2}\left[f(t_i, w_i) + f(t_{i+1}, w_i + hf(t_i, w_i))\right] \qquad \text{for } i = 0, 1, \ldots, N - 1.$$

For this method

$$\phi(t, w, h) = \tfrac{1}{2}f(t, w) + \tfrac{1}{2}f(t + h, w + hf(t, w)).$$

Letting $h = 0$,

$$\phi(t, w, 0) = \tfrac{1}{2}f(t, w) + \tfrac{1}{2}f(t + 0, w + 0 \cdot f(t, w)) = f(t, w),$$

so the consistency condition expressed in Theorem 5.19, part (i), holds.

If f satisfies a Lipschitz condition on $\{(t, w) | a \le t \le b, -\infty < w < \infty\}$ with constant L, then, since

$$\phi(t, w, h) - \phi(t, \overline{w}, h) = \tfrac{1}{2}f(t, w) + \tfrac{1}{2}f(t + h, w + hf(t, w))$$
$$- \tfrac{1}{2}f(t, \overline{w}) - \tfrac{1}{2}f(t + h, \overline{w} + hf(t, \overline{w})),$$

the Lipschitz condition on f leads to

$$|\phi(t, w, h) - \phi(t, \overline{w}, h)| \le \tfrac{1}{2}L|w - \overline{w}| + \tfrac{1}{2}L|w + hf(t, w) - \overline{w} - hf(t, \overline{w})|$$
$$\le L|w - \overline{w}| + \tfrac{1}{2}L|hf(t, w) - hf(t, \overline{w})|$$
$$\le L|w - \overline{w}| + \tfrac{1}{2}hL^2|w - \overline{w}|$$
$$= (L + \tfrac{1}{2}hL^2)|w - \overline{w}|.$$

Therefore ϕ satisfies a Lipschitz condition in w on the set

$$\{(t, w, h) | a \le t \le b, -\infty < w < \infty, 0 \le h \le h_0\}$$

for any $h_0 > 0$ with constant

$$L' = (L + \tfrac{1}{2}h_0 L^2).$$

Finally, if f is continuous on $\{(t, w) | a \le t \le b, -\infty < w < \infty\}$, then ϕ is continuous on

$$\{(t, w, h) | a \le t \le b, -\infty < w < \infty, 0 \le h \le h_0\};$$

so Theorems 5.18 and 5.19 imply that the modified Euler method is convergent and stable. Moreover, we have seen that for this method the local truncation error is $O(h^2)$; so the convergence of the modified Euler method also has rate $O(h^2)$.   □

The problems involved with consistency, convergence, and stability of multi-step methods are compounded because of the number of approximations involved at each step. In the one-step methods, the approximation w_{i+1} depends directly only on the previous approximation w_i, while the multistep methods use at least two of the previous approximations, and the methods which are commonly used involve more. Before discussing the nature of the problems with the multistep methods, it is convenient to present those methods in a slightly different context.

The general multistep method for approximating the solution to the initial-value problem

$$(5.108) \qquad y' = f(t, y), \qquad a \le t \le b, \quad y(a) = \alpha,$$

can be written in the form

$$w_0 = \alpha, \quad w_1 = \alpha_1, \quad \ldots, \quad w_{m-1} = \alpha_{m-1},$$

$$(5.109) \qquad \begin{aligned} w_{i+1} + a_{m-1}w_i + a_{m-2}w_{i-1} + \cdots + a_0 w_{i+1-m} \\ = hF(t_i, h, w_{i+1}, w_i, \ldots, w_{i+1-m}) \end{aligned}$$

for each $i = m - 1, m, \ldots, N - 1$, where $a_0, a_1, \ldots, a_{m-1}$ are constants and, as usual, $h = (b - a)/N$ and $t_i = a + ih$.

If F does not involve w_{i+1}, the method is explicit; otherwise the method is implicit. The fourth-order Adams–Bashforth method can be expressed in the form (5.109) by choosing $m = 4$, $a_0 = 0$, $a_1 = 0$, $a_2 = 0$, $a_3 = -1$, and

$$\begin{aligned} F(t_i, h, w_{i+1}, w_i, \ldots, w_{i-3}) = \tfrac{1}{24}[55f(t_i, w_i) - 59f(t_{i-1}, w_{i-1}) + 37f(t_{i-2}, w_{i-2}) \\ - 9f(t_{i-3}, w_{i-3})] \end{aligned}$$

since the method is

$$\begin{aligned} w_{i+1} - w_i = \frac{h}{24}\big[55f(t_i, w_i) - 59f(t_{i-1}, w_{i-1}) \\ + 37f(t_{i-2}, w_{i-2}) - 9f(t_{i-3}, w_{i-3})\big]. \end{aligned}$$

Likewise, the fourth-order Adams–Moulton method can be expressed in the form (5.109) by choosing $m = 3$, $a_0 = 0$, $a_1 = 0$, $a_2 = -1$, and

$$\begin{aligned} F(t_i, h, w_{i+1}, \ldots, w_{i-2}) = \tfrac{1}{24}[9f(t_{i+1}, w_{i+1}) + 19f(t_i, w_i) - 5f(t_{i-1}, w_{i-1}) \\ + f(t_{i-2}, w_{i-2})]. \end{aligned}$$

Throughout the analysis two assumptions will be made concerning the function F: first, if $f \equiv 0$ (that is, the differential equation is homogeneous), then $F \equiv 0$ also; and second, that F satisfies a type of Lipschitz condition with respect to the sequence $\{w_j\}_{j=i+1-m}^{i+1}$, in the sense that, for a fixed function f, a constant C exists with

$$|F(t_i, h, w_{i+1}, \ldots, w_{i+1-m}) - F(t_i, h, v_{i+1}, \ldots, v_{i+1-m})| \le C \sum_{j=0}^{m} |w_{i+1-j} - v_{i+1-j}|$$

for each $i = m - 1, m, \ldots, N$, and sequences $\{w_j\}_{j=0}^{N}$ and $\{v_j\}_{j=0}^{N}$.

The Adams–Bashforth and Adams–Moulton methods satisfy both of these conditions, provided f satisfies a Lipschitz condition. (See Exercise 2.)

The local truncation error for a multistep method expressed in the form (5.109) is

$$\begin{aligned} \tau_{i+1} = \frac{y(t_{i+1}) + a_{m-1}y(t_i) + \cdots + a_0 y(t_{i+1-m})}{h} \\ - F(t_i, h, y(t_{i+1}), y(t_i), \ldots, y(t_{i+1-m})) \end{aligned}$$

for each $i = m - 1, m, \ldots, N - 1$; and, as in the one-step methods, measures how much the solution y to the differential equation fails to satisfy the difference equation.

For the four-step Adams–Bashforth method, we have seen that

$$\tau_{i+1} = \tfrac{251}{720} y^{(5)}(\mu_i) h^4 \qquad \text{for } t_{i-3} < \mu_i < t_{i+1},$$

while the local truncation error for the Adams–Moulton method we have been considering is

$$\tau_{i+1} = \frac{-19}{720} y^{(5)}(\mu_i) h^4, \qquad t_{i-2} < \mu_i < t_{i+1},$$

provided, of course that $y \in C^5[a, b]$.

The concept of convergence for multistep methods is the same as that for one-step methods; i.e., a multistep method is **convergent** if the solution to the difference equation approaches the solution to the differential equation as the step size approaches zero. This means that $\lim_{h \to 0} \max_{0 \le i \le N} |w_i - y(t_i)| = 0$.

For consistency, however, a slightly different situation occurs. Again, we will want a method to be **consistent** provided that the difference equation approaches the differential equation as the step size approaches zero; that is, the local truncation error must approach zero at each step as the step size approaches zero. The additional condition occurs because of the number of starting values required for multistep methods. Since it would be expected that at most the first starting value, w_0, is exact, it will be necessary to require that the errors in all the starting values approach zero as the step size approaches zero; that is, both

$$\lim_{h \to 0} |\tau_i| = 0, \qquad \text{for all } i = m, m + 1, \ldots, N,$$

and

$$\lim_{h \to 0} |\alpha_i - y(t_i)| = 0, \qquad \text{for all } i = 0, 1, \ldots, m - 1,$$

must be true in order for a multistep method in the form (5.109) to be consistent.

The following theorem for multistep method is similar to Theorem 5.19(ii) and gives a relationship between the local truncation error and global error of a multistep method. It provides the theoretical justification for attempting to control global error by controlling local truncation error. The proof of a slightly more general form of this theorem can be found in Isaacson and Keller [52].

Theorem 5.20 Suppose the initial-value problem

$$y' = f(t, y), \qquad a \le t \le b, \quad y(a) = \alpha,$$

is approximated by an Adams predictor-corrector method with an m step Adams–Bashforth predictor equation

$$w_{i+1} = w_i + h[b_{m-1} f(t_i, w_i) + \cdots + b_0 f(t_{i+1-m}, w_{i+1-m})]$$

with local truncation error τ_{i+1} and an $m - 1$ step Adams–Moulton corrector equation

$$w_{i+1} = w_i + h[b'_{m-1} f(t_{i+1}, w_{i+1}) + b'_{m-2} f(t_i, w_i) \\ + \cdots + b'_0 f(t_{i+2-m}, w_{i+2-m})]$$

with local truncation error τ'_{i+1}. In addition, suppose that $f(t, y)$ and $f_y(t, y)$ are continuous on $D = \{(t, y) | a \le t \le b \text{ and } -\infty < y < \infty\}$ and that f_y is also bounded. Then the local truncation error σ_{i+1} of the predictor-corrector method is

$$\sigma_{i+1} = \tau'_{i+1} + h\tau_{i+1}b'_{m-1}\frac{\partial f}{\partial y}(t_{i+1}, \theta_{i+1})$$

where θ_{i+1} is a number between zero and $h\tau_{i+1}$. Moreover, there exist constants k_1 and k_2 such that

$$|y(t_i) - w_i| \le \left[\max_{0 \le j \le m-1} |w_j - y(t_j)| + \sigma k_1\right]e^{k_2(t_i - a)}$$

where $\sigma = \max_{m \le j \le N}|\sigma_j|$.

Before discussing some of the connections between consistency, convergence, and stability for multistep methods, we must consider in more detail the difference equation for a multistep method.

Associated with the difference equation given in (5.109)

$$w_0 = \alpha, \quad w_1 = \alpha_1, \quad \dots, \quad w_{m-1} = \alpha_{m-1},$$

and

$$w_{i+1} + a_{m-1}w_i + a_{m-2}w_{i-1} + \cdots + a_0 w_{i+1-m}$$
$$= hF(t_i, h, w_{i+1}, w_i, \dots, w_{i+1-m}),$$

is a polynomial, called the **characteristic polynomial** of the method, given by

$$(5.110) \qquad p(\lambda) = \lambda^m + a_{m-1}\lambda^{m-1} + a_{m-2}\lambda^{m-2} + \cdots + a_1\lambda + a_0.$$

It is not difficult to see (Exercise 3) that p has a zero at β precisely when $w_n = \beta^n$ for each n is a solution to the difference equation for the case when $f \equiv 0$. For this reason, it is the zeros of p that are of interest to us.

Definition 5.21   Let $\lambda_1, \lambda_2, \dots, \lambda_m$ denote the (not necessarily distinct) roots of the characteristic polynomial equation

$$p(\lambda) = \lambda^m + a_{m-1}\lambda^{m-1} + \cdots + a_1\lambda + a_0 = 0$$

associated with the multistep difference method

$$w_0 = \alpha, \quad w_1 = \alpha_1, \quad \dots, \quad w_{m-1} = \alpha_{m-1},$$

and

$$w_{i+1} + a_{m-1}w_i + a_{m-2}w_{i-1} + \cdots + a_0 w_{i+1-m}$$
$$= hF(t_i, h, w_{i+1}, w_i, \dots, w_{i+1-m}).$$

If $|\lambda_i| \le 1$ for each $i = 1, 2, \dots, m$, and all roots with absolute value 1 are simple roots, then the difference method is said to satisfy the **root condition**.

EXAMPLE 2   We have seen that the fourth-order Adams–Bashforth method can be expressed as

$$w_{i+1} - w_i = hF(t_i, h, w_{i+1}, w_i, \dots, w_{i-3}),$$

where $F(t_i, h, w_{i+1}, w_i, \dots, w_{i-3}) = \frac{1}{24}[55f(t_i, w_i) - 59f(t_{i-1}, w_{i-1})$
$$+ 37f(t_{i-2}, w_{i-2}) - 9f(t_{i-3}, w_{i-3})];$$

so $m = 4$, $a_0 = 0$, $a_1 = 0$, $a_2 = 0$, and $a_3 = -1$.

The characteristic polynomial for the Adams–Bashforth method is consequently,

$$p(\lambda) = \lambda^4 - \lambda^3 = \lambda^3(\lambda - 1),$$

which has roots $\lambda_1 = 0$, $\lambda_2 = 0$, $\lambda_3 = 0$, and $\lambda_4 = 1$, and satisfies the root condition.

The Adams–Moulton method has a similar characteristic polynomial, $P(\lambda) = \lambda^3 - \lambda^2$, and also satisfies the root condition.   □

The importance of the root condition is its connection with consistency, convergence, and stability of a method. Stability in a multistep method means that, for sufficiently small step sizes, small perturbations in any or all of the starting values produce only small perturbations in the subsequent values. The next theorem details these connections. For the proof of this result and the theory on which it is based the reader is urged to see Isaacson and Keller [52].

Theorem 5.22   A multistep method of the form

$$w_0 = \alpha, \quad w_1 = \alpha_1, \quad \ldots, \quad w_{m-1} = \alpha_{m-1}$$

and
$$w_{i+1} + a_{m-1}w_i + a_{m-2}w_{i-1} + \cdots + a_0 w_{i+1-m}$$
$$= hF(t_i, h, w_{i+1}, w_i, \ldots, w_{i+1-m})$$

is stable if and only if it satisfies the root condition. Moreover, if the difference method is consistent with the differential equation, then the method is stable if and only if it is convergent.

EXAMPLE 3   The implicit multistep method given by

$$w_{i+1} - w_{i-1} = \frac{h}{3}\left[f(t_{i+1}, w_{i+1}) + 4f(t_i, w_i) + f(t_{i-1}, w_{i-1})\right]$$

was introduced in Section 5.6 as the fourth-order Simpson's method. Since the characteristic polynomial for this method, $p(\lambda) = \lambda^2 - 1$, has zeros $\lambda_1 = 1$ and $\lambda_2 = -1$, the method satisfies the root condition and is a stable method. This does not, however, imply that this method will always give accurate approximations, as can be seen by considering the initial-value problem

(5.111) $y' = -6y + 6, \qquad 0 \le t \le 1.5, \quad y(0) = 2.$

With $h = .1$ and $t_i = .1i$, the method can be explicitly expressed as

(5.112) $w_{i+1} = \dfrac{-.8w_i + .8w_{i-1} + 1.2}{1.2}$ for each $i = 1, 2, \ldots, 14.$

For comparison purposes the Adams–Moulton three-step method

$$v_{i+1} = v_i + \frac{h}{24}\left[9f(t_{i+1}, v_{i+1}) + 19f(t_i, v_i) - 5f(t_{i-1}, v_{i-1}) + f(t_{i-2}, v_{i-2})\right]$$

will also be used. For problem (5.111) the Adams–Moulton difference equation is

$$v_{i+1} = \frac{12.6v_i + 3v_{i-1} - .6v_{i-2} + 14.4}{29.4} \qquad \text{for each } i = 2, \ldots, 14.$$

Using exact values for the starting values w_0, w_1, v_0, v_1, v_2, the results w_i and v_i along with the exact solution $y_i = e^{-6t_i} + 1$, and absolute errors $|y_i - w_i|$ and $|y_i - v_i|$ are shown in Table 5.13. It is easy to see that the errors for the fourth-order Simpson's method tend to grow in magnitude, but that the errors for the fourth-order Adams–Moulton method diminish. □

TABLE 5.13

t_i	w_i	$|w_i - y_i|$	v_i	$|v_i - y_i|$	y_i
0	—	—	—	—	2.000000
.1	—	—	—	—	1.548810
.2	1.300792	4.015×10^{-4}	—	—	1.301194
.3	1.165344	4.578×10^{-5}	1.164674	6.237×10^{-4}	1.165298
.4	1.090298	4.189×10^{-4}	1.090107	6.094×10^{-4}	1.090717
.5	1.050029	2.432×10^{-4}	1.049273	5.131×10^{-4}	1.049786
.6	1.026845	4.768×10^{-4}	1.026950	3.719×10^{-4}	1.027322
.7	1.015455	4.606×10^{-4}	1.014738	2.565×10^{-4}	1.014994
.8	1.007593	6.361×10^{-4}	1.008060	1.688×10^{-4}	1.008229
.9	1.005240	7.248×10^{-4}	1.004407	1.078×10^{-4}	1.004515
1.0	1.001567	9.108×10^{-4}	1.002409	6.867×10^{-5}	1.002478
1.1	1.002447	1.087×10^{-3}	1.001317	4.196×10^{-5}	1.001359
1.2	.999413	1.332×10^{-3}	1.000720	2.575×10^{-5}	1.000745
1.3	1.002021	1.613×10^{-3}	1.000393	1.526×10^{-5}	1.000409
1.4	.9982603	1.964×10^{-3}	1.000214	9.537×10^{-6}	1.000224
1.5	1.002507	2.384×10^{-3}	1.000116	6.676×10^{-6}	1.000123

One of the reasons for the growth of error in Simpson's method is that, while Simpson's method is stable, two roots of its characteristic equation have magnitude one. The root $\lambda = -1$ caused the oscillation in the values w_i for $i \geq 11$. The Adams–Moulton method has only one characteristic root with magnitude one and has exhibited more stable behavior for increasing i. (See Exercises 10 and 11.) For this reason, methods that satisfy the root condition and have only one root with magnitude one are called **strongly stable**, while those with more than one root with magnitude one are called **weakly stable**.

The reason for choosing the Adams–Bashforth–Moulton as our standard fourth-order predictor–corrector technique in Section 5.6 over the Milne–Simpson method of the same order is that both the Adams–Bashforth and Adams–Moulton methods are strongly stable and, hence, more likely to give accurate approximations to a wider class of problems than is the predictor–corrector based upon the Milne and Simpson techniques, both of which are weakly stable.

Exercise Set 5.9

1. To prove Theorem 5.18, show that the hypotheses imply that there exists a constant $K > 0$ such that

$$|u_i - v_i| \leq K|u_0 - v_0| \qquad \text{for each } 1 \leq i \leq N,$$

whenever $\{u_i\}_{i=1}^{N}$ and $\{v_i\}_{i=1}^{N}$ satisfy (5.107).

2. Show that a constant C exists so that:

$$F(t_i, h, w_{i+1}, \ldots, w_{i+1-m}) = 0, \qquad \text{if } f \equiv 0;$$

and $\quad |F(t_i, h, w_{i+1}, \ldots, w_{i+1-m}) - F(t_i, h, v_{i+1}, \ldots, v_{i+1-m})| \le C \sum_{j=0}^{m} |w_{i+1-j} - v_{i+1-j}|$

for the Adams–Bashforth and Adams–Moulton methods.

3. Show that $w_n = \beta^n$ satisfies:

$$w_{i+1} + a_{m-1} w_i + \cdots + a_0 w_{i+1-m} = 0, \qquad i = m-1, \ldots, N$$

if and only if $p(\beta) = 0$, where

$$p(\lambda) = \lambda^m + a_{m-1} \lambda^{m-1} + \cdots + a_0.$$

4. Show that the Runge–Kutta fourth-order method is consistent.

5. Consider the differential equation

$$y' = f(t, y), \qquad a \le t \le b, \quad y(a) = \alpha.$$

   a) Show that

   $$y'(t_i) = \frac{-3y(t_i) + 4y(t_{i+1}) - y(t_{i+2})}{2h} + \frac{h^2}{3} y'''(\xi_i),$$

   for some ξ_i, where $t_i < \xi_i < t_{i+2}$.

   b) Part (a) suggests the difference method

   $$w_{j+2} = 4w_{j+1} - 3w_j - 2hf(t_j, w_j) \qquad j = 0, 1, \ldots, N-2.$$

   Use this method to solve

   $$y' = 1 - y, \qquad 0 \le t \le 1, \quad y(0) = 0,$$

   with $h = .1$. Use the starting values $w_0 = 0$ and $w_1 = y(t_1) = 1 - e^{-.1}$.

   c) Repeat part (b) with $h = .01$ and $w_1 = 1 - e^{-.01}$.

   d) Analyze this method for consistency, stability, and convergence.

6. Given the multistep method

   $$w_{i+1} = -\tfrac{3}{2} w_i + 3w_{i-1} - \tfrac{1}{2} w_{i-2} + 3hf(t_i, w_i) \qquad \text{for each } i = 2, \ldots, N-1,$$

   with starting values w_0, w_1, w_2:

   a) Find the local truncation error.

   b) Comment on consistency, stability, and convergence.

7. Obtain an approximate solution to the differential equation

   $$y' = -y, \qquad 0 \le t \le 10, \quad y(0) = 1$$

   using Milne's method with $h = .1$ and then $h = .01$, with starting values $w_0 = 1$ and $w_1 = e^{-h}$ in both cases. How does decreasing h from $h = .1$ to $h = .01$ affect the number of correct digits in the approximate solutions at $t = 1$ and $t = 10$?

8. Investigate stability for the difference method

   $$w_{i+1} + 4w_i - 5w_{i-1} = 2h[f(t_i, w_i) + 2hf(t_{i-1}, w_{i-1})],$$

   for $i = 1, 2, \ldots, N-1$, with starting values w_0, w_1.

9. Consider the problem $y' = 0$, $0 \le t \le 10$, $y(0) = 0$, which has solution $y \equiv 0$. If the difference method of Exercise 5 is applied to the problem, then

$$w_{i+1} = 4w_i - 3w_{i-1}, \qquad \text{for } i = 1, 2, \ldots, N - 1,$$

$$w_0 = 0,$$

$$w_1 = \alpha_1.$$

Suppose $w_1 = \alpha_1 = \varepsilon$, where ε is a small rounding error. Compute w_i for $i = 2, 3, \ldots, 6$ exactly to find how the error ε is propagated.

10. The fourth-order Adams–Moulton method applied to the differential equation

$$y' = \lambda y, \qquad t \ge 0, \quad y(0) = 1$$

is given by

$$w_{i+1} = w_i + \frac{9h\lambda}{24} w_{i+1} + \frac{19h\lambda}{24} w_i - \frac{5h\lambda}{24} w_{i-1} + \frac{h\lambda}{24} w_{i-2}.$$

a) Find the characteristic polynomial for this difference equation.
b) If β_1, β_2, and β_3 are the zeros of the characteristic polynomial, then the general solution to the difference equation is

$$w_i = c_1 \beta_1^i + c_2 \beta_2^i + c_3 \beta_3^i.$$

If $|\beta_j| > 1$ for some j, then β_j^i grows exponentially as i increases, so that the Adams–Moulton method becomes unstable. Find the zeros of the characteristic polynomial for various values of $h\lambda < 0$ to experimentally determine a value of h beyond which there is a zero β_j with $|\beta_j| > 1$.

11. Repeat Exercise 10 using Simpson's method.

5.10 Stiff Differential Equations

Significant difficulties can occur when standard numerical techniques are applied to approximate the solution of a differential equation when the exact solution contains terms of the form $e^{\lambda t}$, where λ is a complex number with negative real part. This term decays to zero with increasing t. However, most approximating methods have very poor round-off characteristics when applied to this type of problem since the round-off error tends to conceal the decay. The problem is particularly acute when the exact solution consists of a steady-state term that does not grow significantly with t, together with a transient term that decays rapidly to zero. In such a problem, the numerical method should approximate the steady state portion of the solution, but unless extreme care is taken, the round-off error associated with the decaying transient portion can dominate the calculations and produce meaningless results.

Problems involving rapidly decaying transient solutions occur naturally in a wide variety of applications, including the study of spring and damping systems, the analysis of control systems, and problems in chemical kinetics. These are all examples of a class of problems called **stiff systems** of differential equations.

Although stiffness is usually associated with systems of differential equations, the round-off error characteristics of a particular numerical method applied to a stiff system can be predicted by examining the round-off error produced when the method is applied to a simple *test equation*,

$$y' = \lambda y, \qquad y(0) = y_0$$

when λ is a negative real number. (A more complete discussion of the round-off error associated with stiff systems requires examining the test equation when λ is a complex number with negative imaginary part; see Gear [41], page 222.)

Suppose we first consider Euler's method applied to the test equation. Letting $h = (b - a)/N$ and $t_j = jh, j = 1, 2, \ldots, N$, equation (5.18) implies that

$$w_0 = y_0,$$

$$\begin{aligned} w_{j+1} &= w_j + h(\lambda w_j) \\ &= (1 + h\lambda)w_j \\ &= (1 + h\lambda)^{j+1}w_0, \qquad \text{for } j = 0, 1, \ldots, N - 1. \end{aligned}$$

Since the exact solution is $y(t) = y_0 e^{\lambda t}$, the absolute error is

$$|y(t_j) - w_j| = |e^{\lambda h j} - (1 + h\lambda)^j||y_0|,$$

and the accuracy is determined by how well the term $1 + h\lambda$ approximates $e^{h\lambda} = 1 + h\lambda + (h\lambda)^2/2! + (h\lambda)^3/3! + \cdots$.

If a rounding error δ_0 is introduced in the initial condition for Euler's method,

$$w_0 = y_0 + \delta_0,$$

then at the jth step the rounding error is

$$\delta_j = (1 + h\lambda)^j\delta_0.$$

Unless $|1 + h\lambda| \leq 1$, the error grows as the jth power of $(1 + h\lambda)$. In the case $\lambda > 0$, the solution $y(t)$ grows exponentially, and, since $1 + h\lambda < e^{h\lambda}$, the rounding error may not be serious. In the stiff situation, $\lambda < 0$ so $y(t)$ is decaying exponentially to zero with t and any growth in rounding error soon dominates the approximation. Clearly, then, Euler's method requires that $|1 + h\lambda| \leq 1$ if the solution to a stiff system is being approximated.

The situation is similar for other one-step methods. In general, a function Q exists with the property that the difference method, when applied to the test equation, gives

$$(5.113) \qquad\qquad w_{i+1} = Q(h\lambda)w_i.$$

All of the explicit second-order Runge–Kutta methods have $Q(h\lambda) = 1 + h\lambda + \frac{1}{2}(h\lambda)^2$, and the classical fourth-order Runge–Kutta method has $Q(h\lambda) = 1 + h\lambda + \frac{1}{2}(h\lambda)^2 + \frac{1}{6}(h\lambda)^3 + \frac{1}{24}(h\lambda)^4$. (See Exercise 6.) The accuracy of the method depends upon how well $Q(h\lambda)$ approximates $e^{h\lambda}$, and the rounding error grows intolerably unless $|Q(h\lambda)| \leq 1$.

When a multistep method is applied to the test equation, the result is

$$w_{j+1} = a_{m-1}w_j + \cdots + a_0 w_{j+1-m} + h\lambda(b_m w_{j+1} + b_{m-1}w_j + \cdots + b_0 w_{j+1-m})$$

for $j = m - 1, \ldots, N - 1$ or

$$(1 - h\lambda b_m)w_{j+1} - (a_{m-1} + h\lambda b_{m-1})w_j - \cdots - (a_0 + h\lambda b_0)w_{j+1-m} = 0.$$

(See Eq. (5.65).)

Associated with this homogeneous difference equation is a characteristic polynomial

$$Q(z, h\lambda) = (1 - h\lambda b_m)z^m - (a_{m-1} + h\lambda b_{m-1})z^{m-1} - \cdots - (a_0 + h\lambda b_0).$$

Suppose $w_0, \ldots, w_{m-1}$ are given, and, for fixed $h\lambda$, let $\beta_1, \ldots, \beta_m$ be the zeros of the polynomial $Q(z, h\lambda)$. If $\beta_1, \ldots, \beta_m$ are distinct, then constants $c_1, \ldots, c_m$ exist with

$$(5.114) \qquad w_j = \sum_{k=1}^{m} c_k(\beta_k)^j \qquad \text{for } j = 0, \ldots, N.$$

(If $Q(z, h\lambda)$ has multiple roots, w_j is similarly defined (See Henrici [46], pages 119–145).) If w_j is to approximate $y(t_j) = e^{\lambda h j} = (e^{h\lambda})^j$, then all zeros β_k must satisfy $|\beta_k| < 1$; since if one zero β_k exceeds one in modulus, then certain choices of y_0 result in $c_k \neq 0$, and the term $c_k(\beta_k)^j$ will not decay to zero.

EXAMPLE 1   The test differential equation

$$y' = -30y, \qquad 0 < t < 1.5 \quad y(0) = \tfrac{1}{3}$$

has exact solution $y = \tfrac{1}{3}e^{-30t}$. Using $h = .1$ for Euler's Algorithm 5.1, Runge–Kutta Fourth Order Algorithm 5.2, and the Adams Predictor-Corrector Algorithm 5.4 gives the results at $t = 1.5$ in Table 5.14.   □

TABLE 5.14

Exact solution	9.54173×10^{-21}
Euler's method	-1.09225×10^4
Runge–Kutta method	3.95730×10^1
Predictor–Corrector method	8.03840×10^5

The inaccuracies in Example 1 are due to the fact that $|Q(h\lambda)| > 1$ for Euler's method and the Runge–Kutta method and that $Q(z, h\lambda)$ has roots with modulus exceeding one for the predictor-corrector method. In order to apply these methods to this problem, the step size must be reduced. To describe the amount of step-size reduction that is required, we need the following definition.

Definition 5.23   The **region R of absolute stability** for a one-step method is defined by $R = \{h\lambda \in \mathbb{C} \,|\, |Q(h\lambda)| < 1\}$ and for a multistep method by $R = \{h\lambda \in \mathbb{C} \,|\, |\beta_k| < 1$ for all roots β_k of $Q(z, h\lambda)\}$.

From equations (5.113) and (5.114), it is clear that a method can be applied effectively to a stiff equation only if $h\lambda$ is in the region of absolute stability of the method, which for a given problem places a restriction on the size of h. Even though the exponential term in the exact solution decays quickly to zero, λh must remain within the region of absolute stability throughout the interval of t values or rounding error will dominate the approximation. This means that while normally h could be increased because of truncation error considerations, the absolute stability criterion forces h to remain small.

Since the region of absolute stability of a method is generally the critical factor in producing accurate approximations for stiff systems, numerical methods have been sought with as large a region of absolute stability as possible. A numerical method is said to be A-**stable** if its region R of absolute stability contains the left half-plane $\{h\lambda \in \mathbb{C} \mid \text{Re}(h\lambda) < 0\}$. The implicit **Trapezoidal method**, given by

$$w_0 = y_0$$

$$(5.115) \qquad w_{j+1} = w_j + \frac{h}{2}[f(t_{j+1}, w_{j+1}) + f(t_j, w_j)] \qquad 0 \le j \le N-1$$

is an A-stable method (See Exercise 7.).

EXAMPLE 2 The trapezoidal method and the classical Runge–Kutta fourth-order method will be applied to the test equation

$$y' = -30y, \qquad 0 < t < 1.5, \quad y(0) = \tfrac{1}{3}$$

with $h = .3, .1, .05, .01$. The results in Table 5.15 compare the values obtained with the exact solution at $t = 1.5$. The Runge–Kutta method is more accurate than the trapezoidal method when $h\lambda$ is inside the region of absolute stability since it is a fourth-order method and the trapezoidal method is only second order.   ☐

TABLE 5.15

		Values at $t = 1.5$	
h	Trapezoidal	Runge–Kutta	Exact
.3	-3.47861×10^{-2}	7.10213×10^{10}	9.54173×10^{-21}
.1	-1.09227×10^{-11}	3.95748×10^{1}	
.05	1.47890×10^{-26}	4.25227×10^{-18}	
.01	6.77706×10^{-21}	9.57905×10^{-21}	

All the techniques commonly used for stiff systems are implicit multistep methods. Generally, w_{i+1} is obtained by solving a nonlinear equation or nonlinear system iteratively, often by Newton's method. For example, to obtain w_{j+1} from

$$w_{j+1} = w_j + \frac{h}{2}[f(t_j, w_j) + f(t_{j+1}, w_{j+1})]$$

in the trapezoidal method applied to $y' = f(t, y)$, select $w_{j+1}^{(0)}$ and generate $w_{j+1}^{(k)}$ by

$$w_{j+1}^{(k)} = w_{j+1}^{(k-1)} - \frac{w_{j+1}^{(k-1)} - w_j - \dfrac{h}{2}(f(t_j, w_j) + f(t_{j+1}, w_{j+1}^{(k-1)}))}{1 - \dfrac{h}{2}f_y(t_{j+1}, w_{j+1}^{(k-1)})}$$

until $|w_{j+1}^{(k)} - w_{j+1}^{(k-1)}|$ is sufficiently small.

We have presented here only a small amount of what the reader should know if he expects to encounter stiff systems frequently. We recommend in this case, Gear [41], especially the program DIFSUB on pages 158–166, Lambert [58], or Shampine and Gear [80] be consulted. Some packaged programs for stiff systems are available, for example, the previously mentioned program of Gear and SDBASIC due to Enright [33]. Enright, Hull and Lindberg [34] compare some of the common packages for stiff systems.

In closing this chapter, it seems appropriate also to give some recommendations on methods to use for solving nonstiff initial-value problems. Most of these recommendations are based on the results presented in papers by Hull, Enright, Fellen, and Sedgwick [51], published in 1972, by Enright, Hull, and Lindberg [34] published in 1975, and by Hull and Enright [50], published in 1976. Additional discussion of these topics can also be found in references [41] and [58]. Any updates of these results are likely to appear in the *SIAM Journal of Numerical Analysis*.

When the function being evaluated is relatively simple, that is, does not require many time-consuming manipulations, an extrapolation procedure such as Algorithm 6.6 is most efficient, while the Adams-based predictor–corrector method is favored when the evaluation of the function is complicated. The use of the Runge–Kutta procedures should be restricted to finding starting values for the Adams method, or to problems where the function is easy to evaluate and the accuracy needed is small, about 10^{-4}.

Exercise Set 5.10

1. Discuss consistency, stability, and convergence for the Trapezoidal method

$$w_{i+1} = w_i + \frac{h}{2}[f(t_{i+1}, w_{i+1}) + f(t_i, w_i)] \qquad \text{for } i = 0, 1, \ldots, N-1$$

with $w_0 = \alpha$ applied to the differential equation

$$y' = f(t, y), \qquad a \le t \le b, \quad y(a) = \alpha.$$

2. Repeat Exercise 9 of Section 5.9 using the Trapezoidal method.

3. Solve the following "stiff" initial-value problems using (i) Euler's method, (ii) Runge–Kutta fourth-order method, (iii) Adams–Bashforth and Adams–Moulton predictor-corrector method, (iv) Trapezoidal method, and (v) extrapolation.

   a) $y' = -20(y - t^2) + 2t, \quad 0 \le t \le 1, \quad y(0) = \frac{1}{3}$.
   Use $h = .075$, and $h = .01$ for $0 \le t \le .2$, and $h = .075$ for $.2 \le t \le 1$. The actual solution is given by $y(t) = t^2 + \frac{1}{3}e^{-20t}$.

   b) $y' = -20y + 20 \sin t + \cos t, \quad 0 \le t \le 1, \quad y(0) = 1$.
   Use $h = .01$ for $0 \le t \le .2$ and $h = .075$ for $.2 \le t \le 1$. The actual solution is $y(t) = e^{-20t} + \sin t$.

   c) $y' = (50/y) - 50y, \quad 0 \le t \le 1, \quad y(0) = \sqrt{2}$.
   Use $h = .01$ for $0 \le t \le .2$ and $h = .1$ for $.2 \le t \le 1$. The actual solution is $y(t) = \sqrt{1 + e^{-100t}}$.

   d) $u_1'(t) = 32u_1 + 66u_2 + \frac{2}{3}t + \frac{2}{3}$,
   $u_2'(t) = -66u_1 - 133u_2 - \frac{1}{3}t - \frac{1}{3}, \quad 0 \le t \le 1$,
   $u_1(0) = \frac{1}{3}, \quad u_2(0) = \frac{1}{3}$.

Use $h = .005$ for $0 \leq t \leq .2$ and $h = .1$ for $.2 \leq t \leq 1$. The actual solution is:

$$u_1(t) = \tfrac{2}{3}t + \tfrac{2}{3}e^{-t} - \tfrac{1}{3}e^{-100t} \quad \text{and} \quad u_2(t) = -\tfrac{1}{3}t - \tfrac{1}{3}e^{-t} + \tfrac{2}{3}e^{-100t}.$$

4. In Exercise 7 of Section 5.2 the differential equation

$$\frac{dp(t)}{dt} = rb(1 - p(t))$$

was obtained as a model for studying the proportion $p(t)$ of nonconformists in a society whose birth rate was b and where r represented the rate at which offspring would become nonconformists when at least one of their parents was a conformist. That exercise required that an approximation for $p(t)$ be found by using Euler's method, Algorithm 5.1, for integral values of t when given $p(0) = .01$, $b = .02$, and $r = .1$, and then the approximation for $p(50)$ be compared with the actual value. Explain the reason for the large error in this approximation. Use the Trapezoidal method to obtain another approximation for $p(50)$, again assuming that $h = 1$ year.

5. The backward Euler method is defined by

$$w_{i+1} = w_i + hf(t_{i+1}, w_{i+1}) \quad \text{for } i = 0, \ldots, N - 1.$$

a) Show that $Q(h\lambda) = 1/(1 - h\lambda)$ for the backward Euler method.
b) Apply the backward Euler method to the differential equations given in Exercise 3. Use Newton's method to solve for w_{i+1}.

6. Show that the fourth-order Runge–Kutta method,

$$w_{i+1} = w_i + \tfrac{1}{6}(k_1 + 2k_2 + 2k_3 + k_4),$$

$$k_1 = hf(t_i, w_i),$$

$$k_2 = hf(t_i + h/2, w_i + k_1/2),$$

$$k_3 = hf(t_i + h/2, w_i + k_2/2).$$

$$k_4 = hf(t_i + h, w_i + k_3),$$

when applied to the differential equation $y' = \lambda y$ can be written in the form

$$w_{i+1} = (1 + h\lambda + \tfrac{1}{2}(h\lambda)^2 + \tfrac{1}{6}(h\lambda)^3 + \tfrac{1}{24}(h\lambda)^4)w_i.$$

7. a) Show that the Trapezoidal method (5.115) is A-stable.
    b) Show that the backward Euler method described in Exercise 5 is A-stable.

Direct Methods for Solving Linear Systems

Kirchhoff's laws of electrical circuits state that the net flow of current through each junction of a circuit is zero, and that the net voltage drop around each closed loop of the circuit is zero. Suppose that a potential of V volts is applied between the points A and G in the circuit in Fig. 6.1 and v_b, v_c, v_d, v_e, and v_f are the potentials at the points B, C, D, E, and F, respectively. Using G as a reference point, Kirchhoff's laws imply that these potentials satisfy the following system of linear equations:

$$-31v_b + 10v_c \qquad\qquad + 6v_f = 15 \text{ V},$$
$$2v_b - 8v_c + 3v_d + 3v_e \qquad = 0,$$
$$v_c - 3v_d + 2v_e \qquad = 0,$$
$$2v_c + 4v_d - 9v_e + v_f = 0,$$
$$12v_b \qquad + 15v_d \qquad - 47v_f = 0.$$

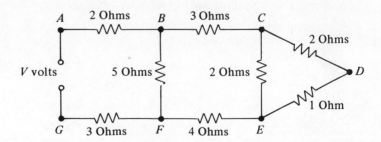

FIGURE 6.1

The solution of systems of this type will be considered in this chapter.

Linear systems of equations are associated with many problems in engineering and science, as well as with applications of mathematics to the social sciences and the quantitative study of business and economic problems.

In this chapter, direct techniques are considered to solve the linear system

(6.1)

$$
\begin{aligned}
E_1: &\quad a_{11}x_1 + a_{12}x_2 + \cdots + a_{1n}x_n = b_1, \\
E_2: &\quad a_{21}x_1 + a_{22}x_2 + \cdots + a_{2n}x_n = b_2, \\
&\ \ \vdots \qquad \vdots \qquad \vdots \qquad\qquad \vdots \qquad \vdots \\
E_n: &\quad a_{n1}x_1 + a_{n2}x_2 + \cdots + a_{nn}x_n = b_n
\end{aligned}
$$

for $x_1, \ldots, x_n$, given the $a_{i,j}$ for each $i, j = 1, 2, \ldots, n$, and b_i, for each $i = 1, 2, \ldots, n$. Direct techniques are methods that give an answer in a fixed number of steps, subject only to rounding errors. In the presentation it is also necessary to introduce some elementary notions from the subject of linear algebra.

Methods of approximating the solution to linear systems by iterative methods will be discussed in Chapter 8.

6.1   Linear Systems of Equations

To solve a linear system such as (6.1), three operations are permitted on the equations:

1)   Equation E_i can be multiplied by any nonzero constant λ and the resulting equation used in place of E_i. This operation will be denoted $(\lambda E_i) \to (E_i)$.
2)   Equation E_j can be multiplied by any constant λ, added to equation E_i, and the resulting equation used in place of E_i. This operation will be denoted $(E_i + \lambda E_j) \to (E_i)$.
3)   Equations E_i and E_j can be transposed in order. This operation will be denoted $(E_i) \leftrightarrow (E_j)$.

By a sequence of the operations given above, a linear system can be transformed to a more easily solved linear system with the same set of solutions. The sequence of operations will be illustrated in the next example.

EXAMPLE 1   The four equations

(6.2)

$$
\begin{aligned}
E_1: &\quad x_1 + x_2 \qquad\ \ + 3x_4 = 4, \\
E_2: &\quad 2x_1 + x_2 - x_3 + x_4 = 1, \\
E_3: &\quad 3x_1 - x_2 - x_3 + 2x_4 = -3, \\
E_4: &\quad -x_1 + 2x_2 + 3x_3 - x_4 = 4,
\end{aligned}
$$

will be solved for the unknowns x_1, x_2, x_3, and x_4. The first step is to use equation E_1 to eliminate the unknown x_1 from E_2, E_3, and E_4 by performing $(E_2 - 2E_1) \to (E_2)$, $(E_3 - 3E_1) \to (E_3)$, and $(E_4 + E_1) \to (E_4)$. The resulting system is:

(6.3)

$$
\begin{aligned}
E_1: &\quad x_1 + x_2 \qquad\ \ + 3x_4 = 4, \\
E_2: &\quad\ -x_2 - x_3 - 5x_4 = -7, \\
E_3: &\quad\ -4x_2 - x_3 - 7x_4 = -15, \\
E_4: &\quad\ 3x_2 + 3x_3 + 2x_4 = 8
\end{aligned}
$$

where the new equations are, for simplicity, again labeled E_1, E_2, E_3, and E_4.

In this system, E_2 is used to eliminate x_2 from E_3 and E_4 by the operations $(E_3 - 4E_2) \rightarrow (E_3)$ and $(E_4 + 3E_2) \rightarrow (E_4)$, resulting in the system

$$
\begin{aligned}
E_1: \quad & x_1 + x_2 && + 3x_4 = 4, \\
E_2: \quad & -x_2 - x_3 - 5x_4 = -7, \\
E_3: \quad & 3x_3 + 13x_4 = 13, \\
E_4: \quad & -13x_4 = -13.
\end{aligned}
$$

(6.4)

The system of equations (6.4) is now in **triangular** or **reduced form** and can easily be solved for the unknowns by a **backward-substitution** process. Noting that E_4 implies $x_4 = 1$, E_3 can be solved for x_3:

$$
x_3 = \tfrac{1}{3}(13 - 13x_4) = \tfrac{1}{3}(13 - 13) = 0.
$$

Continuing, E_2 gives:

$$
x_2 = -(-7 + 5x_4 + x_3) = -(-7 + 5 + 0) = 2;
$$

and E_1 gives

$$
x_1 = 4 - 3x_4 - x_2 = 4 - 3 - 2 = -1.
$$

The solution to (6.4) is therefore $x_1 = -1$, $x_2 = 2$, $x_3 = 0$, and $x_4 = 1$. It can easily be verified that these values also solve the equations in (6.2). $\square$

Now that it is clear how the procedure should work, it is important that it be justified.

Theorem 6.1   Any operation of the form $(\lambda E_i) \rightarrow (E_i)$ for $\lambda \neq 0$, $(E_i + \lambda E_j) \rightarrow (E_i)$, or $(E_i) \leftrightarrow (E_j)$ will not change the solution of the linear system

$$
\begin{aligned}
E_1: \quad & a_{11}x_1 + a_{12}x_2 + \cdots + a_{1n}x_n = b_1, \\
E_2: \quad & a_{21}x_1 + a_{22}x_2 + \cdots + a_{2n}x_n = b_2, \\
& \ \ \vdots \qquad\ \ \vdots \qquad\quad \vdots \qquad\ \ \vdots \\
E_n: \quad & a_{n1}x_1 + a_{n2}x_2 + \cdots + a_{nn}x_n = b_n.
\end{aligned}
$$

(6.5)

Proof   We will give the proof that $(\lambda E_i) \rightarrow (E_i)$, $\lambda \neq 0$, does not change the solution set. The other two verifications are in Exercise 5.

Suppose $x_1', x_2', \ldots, x_n'$ is a solution to the original system (6.5). To show that $x_1', \ldots, x_n'$ solves the new system

$$
\begin{aligned}
& a_{11}x_1 + a_{12}x_2 + \cdots + a_{1n}x_n = b_1, \\
& \ \ \vdots \qquad\ \ \vdots \qquad\quad \vdots \qquad\ \ \vdots \\
& \lambda a_{i1}x_1 + \lambda a_{i2}x_2 + \cdots + \lambda a_{in}x_n = \lambda b_i, \\
& \ \ \vdots \qquad\ \ \vdots \qquad\quad \vdots \qquad\ \ \vdots \\
& a_{n1}x_1 + a_{n2}x_2 + \cdots + a_{nn}x_n = b_n,
\end{aligned}
$$

(6.6)

it clearly suffices to show that E_i holds. Since

$$
a_{i1}x_1' + a_{i2}x_2' + \cdots + a_{in}x_n' = b_i,
$$

multiplying this equation by λ gives

$$
\lambda a_{i1}x_1' + \lambda a_{i2}x_2' + \cdots + \lambda a_{in}x_n' = \lambda b_i;
$$

so E_i is also satisfied by $x_1', \ldots, x_n'$.

On the other hand, suppose $x'_1, \ldots, x'_n$ is a solution for the system (6.6). To show that it is also a solution to system (6.5) requires only a verification that equation E_i of system (6.5) holds. Since $x'_1, \ldots, x'_n$ solves system (6.6),

$$\lambda a_{i1} x'_1 + \lambda a_{i2} x'_2 + \cdots + \lambda a_{in} x'_n = \lambda b_i.$$

But $\lambda \neq 0$, so division by λ is defined and yields

$$a_{i1} x'_1 + a_{i2} x'_2 + \cdots + a_{in} x'_n = b_i. \qquad \square$$

When performing the calculations of Example 1, we did not need to write out the full equations at each step or to carry the variables x_1, x_2, x_3, and x_4 through the calculations since they always remained in the same column. The only variation from system to system occurred in the coefficients of the unknowns and in the values on the right side of the equations. For this reason, a linear system is often replaced by a matrix, which contains all the information about the system that is necessary to determine its solution, but in a compact form.

Definition 6.2   An *n* **by** *m* **matrix** is a rectangular array of elements with n rows and m columns in which not only is the value of an element important, but also its position in the array.

The notation for an $n \times m$ (n by m) matrix will be a capital letter such as A for the matrix and lowercase letters with double subscripts, such as a_{ij}, to refer to the entry at the intersection of the ith row and jth column; that is,

$$A = (a_{ij}) = \begin{bmatrix} a_{11} & a_{12} & \cdots & a_{1m} \\ a_{21} & a_{22} & \cdots & a_{2m} \\ \vdots & \vdots & & \vdots \\ a_{n1} & a_{n2} & \cdots & a_{nm} \end{bmatrix}.$$

EXAMPLE 2   The matrix

$$A = \begin{bmatrix} 2 & -1 & 7 \\ 3 & 1 & 0 \end{bmatrix}$$

is a 2×3 matrix with $a_{11} = 2$, $a_{12} = -1$, $a_{13} = 7$, $a_{21} = 3$, $a_{22} = 1$, and $a_{23} = 0$. $\square$

The $1 \times n$ matrix

$$A = [a_{11} \quad a_{12} \quad \cdots \quad a_{1n}]$$

is called an *n*-**dimensional row vector**, and an $n \times 1$ matrix

$$A = \begin{bmatrix} a_{11} \\ a_{21} \\ \vdots \\ a_{n1} \end{bmatrix}$$

is called an *n*-**dimensional column vector**. Usually the unnecessary subscript is omitted for vectors and a boldface lowercase letter used for notation. Thus,

$$\mathbf{x} = \begin{bmatrix} x_1 \\ x_2 \\ \vdots \\ x_n \end{bmatrix}$$

denotes a column vector and

$$\mathbf{y} = [y_1 \quad y_2 \quad \cdots \quad y_n]$$

a row vector.

An $(n + 1)$ by n matrix can be used to represent the linear system

$$
\begin{aligned}
a_{11}x_1 + a_{12}x_2 + \cdots + a_{1n}x_n &= b_1, \\
a_{21}x_1 + a_{22}x_2 + \cdots + a_{2n}x_n &= b_2, \\
\vdots \qquad \vdots \qquad\qquad \vdots \qquad &\ \vdots \\
a_{n1}x_1 + a_{n2}x_2 + \cdots + a_{nn}x_n &= b_n
\end{aligned}
$$

by first constructing

$$A = \begin{bmatrix} a_{11} & a_{12} & \cdots & a_{1n} \\ a_{21} & a_{22} & \cdots & a_{2n} \\ \vdots & \vdots & & \vdots \\ a_{n1} & a_{n2} & \cdots & a_{nn} \end{bmatrix} \quad \text{and} \quad \mathbf{b} = \begin{bmatrix} b_1 \\ b_2 \\ \vdots \\ b_n \end{bmatrix}$$

and then combining these matrices to form the **augmented matrix**:

$$[A, \mathbf{b}] = \begin{bmatrix} a_{11} & a_{12} & \cdots & a_{1n} & \vdots & b_1 \\ a_{21} & a_{22} & \cdots & a_{2n} & \vdots & b_2 \\ \vdots & \vdots & & \vdots & \vdots & \vdots \\ a_{n1} & a_{n2} & \cdots & a_{nn} & \vdots & b_n \end{bmatrix},$$

where the broken line is used to separate the coefficients of the unknowns from the values on the righthand side of the equations.

EXAMPLE 3   Repeating the operations involved in Example 1 with the matrix notation results in considering first the augmented matrix associated with system (6.2):

(6.7)
$$\begin{bmatrix} 1 & 1 & 0 & 3 & \vdots & 4 \\ 2 & 1 & -1 & 1 & \vdots & 1 \\ 3 & -1 & -1 & 2 & \vdots & -3 \\ -1 & 2 & 3 & -1 & \vdots & 4 \end{bmatrix}.$$

Performing the operations associated with $(E_2 - 2E_1) \rightarrow (E_2)$, $(E_3 - 3E_1) \rightarrow (E_3)$, and $(E_4 + E_1) \rightarrow (E_4)$ in system (6.2) is accomplished by manipulating the respective rows of the augmented matrix (6.7), which becomes the matrix corresponding to the system (6.3):

(6.8)
$$\begin{bmatrix} 1 & 1 & 0 & 3 & \vdots & 4 \\ 0 & -1 & -1 & -5 & \vdots & -7 \\ 0 & -4 & -1 & -7 & \vdots & -15 \\ 0 & 3 & 3 & 2 & \vdots & 8 \end{bmatrix}.$$

Performing the final manipulations results in the augmented matrix corresponding to system (6.4):

(6.9)
$$\left[\begin{array}{cccc:c} 1 & 1 & 0 & 3 & 4 \\ 0 & -1 & -1 & -5 & -7 \\ 0 & 0 & 3 & 13 & 13 \\ 0 & 0 & 0 & -13 & -13 \end{array}\right].$$

□

This matrix can now be transformed into its corresponding linear system (6.4) and solutions for x_1, x_2, x_3 and x_4 obtained. The procedure involved in this process is called **Gaussian elimination with backward substitution**. In subsequent sections, we will consider conditions on the linear system under which the method can be used.

Exercise Set 6.1

1. Solve the following linear systems using the elimination method.

   a) $x_1 + 3x_2 - x_3 + x_4 = 5,$
      $2x_2 + x_3 + x_4 = 7,$
      $x_3 + 2x_4 = 6,$
      $3x_4 = 9.$

   b) $x_1 - 2x_2 = 3,$
      $2x_1 + x_2 = 4.$

   c) $x_1 + x_2 - x_3 = 3,$
      $2x_1 - x_2 + 3x_3 = 0,$
      $-x_1 - 2x_2 + x_3 = -5.$

   d) $3x - y + 2z = -3,$
      $x + y + z = -4,$
      $2x + y - z = -3.$

   e) $x_1 + \frac{1}{2}x_2 + \frac{1}{3}x_3 = \frac{11}{6},$
      $\frac{1}{2}x_1 + \frac{1}{3}x_2 + \frac{1}{4}x_3 = \frac{13}{12},$
      $\frac{1}{3}x_1 + \frac{1}{4}x_2 + \frac{1}{5}x_3 = \frac{47}{60}.$

2. For each of the following systems, obtain a solution graphically, if possible, and then try to solve the system by the methods in this section.

   a) $x_1 + 2x_2 = 3,$
      $x_1 - x_2 = 0.$

   b) $x_1 + 2x_2 = 3,$
      $-2x_1 - 4x_2 = 6.$

   c) $x_1 + 2x_2 = 3,$
      $2x_1 + 4x_2 = 6.$

   d) $0 \cdot x_1 + x_2 = 3,$
      $2x_1 - x_2 = 7.$

3. Use the procedures of this section to try to solve the following linear systems. Explain why the procedure fails in each case for which no solutions can be obtained.

   a) $x_1 + x_2 + x_4 = 2,$
      $2x_1 + x_2 - x_3 + x_4 = 1,$
      $-x_1 + 2x_2 + 3x_3 - x_4 = 4,$
      $3x_1 - x_2 - x_3 + 2x_4 = -3.$

   b) $x_1 + x_2 - x_3 = 3,$
      $-x_1 + x_2 + x_3 = 2,$
      $x_1 + 3x_2 - x_3 = 8,$

   c) $x_1 + x_2 - x_3 = 3,$
      $-x_1 + x_2 + x_3 = 2,$
      $x_1 + 3x_2 - x_3 = 6.$

4. What can be said, from a geometrical standpoint, about the following sets of equations?

a) $2x + y = -1,$
$\quad\ x + y = \ \ 2,$
$\quad\ x - 3y = \ \ 5.$

b) $2x + y + z = \ \ \ 1,$
$\quad 2x + 4y - z = -1.$

5. Complete the proof of Theorem 6.1.

6. Suppose that in a biological system there are n species of animals and m sources of food. Let x_j represent the population of the jth species for each $j = 1, \ldots, n$; b_i represent the available daily supply of the ith food; and a_{ij} represent the amount of the ith food consumed on the average by a member of the jth species. The linear system

$$a_{11}x_1 + a_{12}x_2 + \cdots + a_{1n}x_n = b_1,$$
$$a_{21}x_1 + a_{22}x_2 + \cdots + a_{2n}x_n = b_2,$$
$$\vdots \qquad \vdots \qquad \quad \vdots \qquad \vdots$$
$$a_{m1}x_1 + a_{m2}x_2 + \cdots + a_{mn}x_n = b_m$$

represents an equilibrium where there is a daily supply of food to precisely meet the average daily consumption of each species.

a) Let $A = (a_{ij}) = \begin{bmatrix} 1 & 2 & 0 & 3 \\ 1 & 0 & 2 & 2 \\ 0 & 0 & 1 & 1 \end{bmatrix},$

$\mathbf{x} = (x_j) = [1000, 500, 350, 400]$, and $\mathbf{b} = (b_i) = [3500, 2700, 900]$. Is there sufficient food to satisfy the average daily consumption?

b) What is the maximum number of animals of each species that could be added to the system with the supply of food still meeting the consumption?

c) If species 1 became extinct, how much of an increase of each of the remaining species could be supported?

d) If species 2 also became extinct, how much of an increase of each of the remaining species could be supported?

6.2   Gaussian Elimination and Backward Substitution

The general Gaussian elimination procedure applied to the linear system

$$E_1: \quad a_{11}x_1 + a_{12}x_2 + \cdots + a_{1n}x_n = b_1,$$
$$E_2: \quad a_{21}x_1 + a_{22}x_2 + \cdots + a_{2n}x_n = b_2,$$

(6.10)
$$\vdots \qquad \vdots \qquad \vdots \qquad \qquad \vdots \qquad \vdots$$
$$E_n: \quad a_{n1}x_1 + a_{n2}x_2 + \cdots + a_{nn}x_n = b_n$$

is handled in a manner similar to the procedure followed in Example 3. We form the augmented matrix $\tilde{A}$:

(6.11)
$$\tilde{A} = [A, \mathbf{b}] = \begin{bmatrix} a_{11} & a_{12} & \cdots & a_{1n} & \vdots & a_{1,n+1} \\ a_{21} & a_{22} & \cdots & a_{2n} & \vdots & a_{2,n+1} \\ \vdots & \vdots & & \vdots & & \vdots \\ a_{n1} & a_{n2} & \cdots & a_{nn} & \vdots & a_{n,n+1} \end{bmatrix}$$

where A denotes the matrix formed by the coefficients and the entries in the $(n + 1)$st column are the values of $\mathbf{b}$, that is, $a_{i,n+1} = b_i$ for each $i = 1, 2, \ldots, n$.

Provided $a_{11} \neq 0$, the operations corresponding to $(E_j - (a_{j1}/a_{11})E_1) \rightarrow (E_j)$ are performed for each $j = 2, 3, \ldots, n$ to eliminate the coefficient of x_1 in each of these rows. Although the entries in rows $2, 3, \ldots, n$ are expected to change, for ease of notation, we will again denote the entry in the ith row and the jth column by a_{ij}. With this in mind, we follow a sequential procedure for $i = 2, 3, \ldots, n - 1$ and perform the operation $(E_j - (a_{ji}/a_{ii})E_i) \rightarrow (E_j)$ for each $j = i + 1, i + 2, \ldots, n$, provided $a_{ii} \neq 0$. This will eliminate (that is, change the coefficient to zero) x_i in each row below the ith for all values of $i = 1, 2, \ldots, n - 1$. The resulting matrix will have the form:

$$\tilde{\tilde{A}} = \begin{bmatrix} a_{11} & a_{12} & \cdots & a_{1n} & \vdots & a_{1,n+1} \\ 0 & a_{22} & \cdots & a_{2n} & \vdots & a_{2,n+1} \\ \vdots & & \ddots & \vdots & \vdots & \vdots \\ 0 & \cdots\cdots & 0 & a_{nn} & \vdots & a_{n,n+1} \end{bmatrix}$$

where, as was mentioned above, the values of a_{ij} are not expected to agree with those in the original matrix $\tilde{A}$. This matrix represents a linear system with the same solution set as system (6.10). Since the equivalent linear system is triangular:

$$a_{11}x_1 + a_{12}x_2 + \cdots + a_{1n}x_n = a_{1,n+1},$$
$$a_{22}x_2 + \cdots + a_{2n}x_n = a_{2,n+1},$$
$$\vdots \qquad \vdots$$
$$a_{nn}x_n = a_{n,n+1},$$

the backward substitution can be performed. Solving the nth equation for x_n gives:

$$x_n = \frac{a_{n,n+1}}{a_{nn}}.$$

Solving the $(n - 1)$st equation for x_{n-1} and using x_n yields

$$x_{n-1} = \frac{[a_{n-1,n+1} + a_{n-1,n}x_n]}{a_{n-1,n-1}},$$

and continuing this process, we obtain:

$$x_i = \frac{[a_{i,n+1} - a_{in}x_n - a_{i,n-1}x_{n-1} - \cdots - a_{i,i+1}x_{i+1}]}{a_{ii}}$$
$$= \frac{[a_{i,n+1} - \sum_{j=i+1}^{n} a_{ij}x_j]}{a_{ii}} \qquad \text{for each } i = n - 1, n - 2, \ldots, 2, 1.$$

The Gaussian elimination procedure can be presented more precisely, although more intricately, by forming a sequence of augmented matrices $\tilde{A}^{(1)}$, $\tilde{A}^{(2)}, \ldots, \tilde{A}^{(n)}$ where $\tilde{A}^{(1)}$ is the matrix $\tilde{A}$ given in (6.11) and $\tilde{A}^{(k)}$ for each $k = 2, 3, \ldots, n$ has entries $a_{ij}^{(k)}$ where:

$$a_{ij}^{(k)} = \begin{cases} a_{ij}^{(k-1)} & \text{when } i = 1, 2, \ldots, k - 1 \text{ and } j = 1, 2, \ldots, n + 1, \\ 0 & \text{when } i = k, k + 1, \ldots, n \text{ and } j = 1, 2, \ldots, k - 1, \\ a_{ij}^{(k-1)} - \dfrac{a_{i,k-1}^{(k-1)}}{a_{k-1,k-1}^{(k-1)}} a_{k-1,j}^{(k-1)} & \text{when } i = k, k + 1, \ldots, n \text{ and } j = k, k + 1, \ldots, n + 1. \end{cases}$$

Thus,

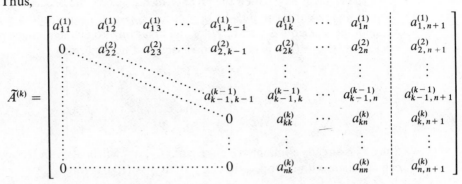

$$\tilde{A}^{(k)} = \begin{bmatrix} a_{11}^{(1)} & a_{12}^{(1)} & a_{13}^{(1)} & \cdots & a_{1,k-1}^{(1)} & a_{1k}^{(1)} & \cdots & a_{1n}^{(1)} & \vdots & a_{1,n+1}^{(1)} \\ 0 & a_{22}^{(2)} & a_{23}^{(2)} & \cdots & a_{2,k-1}^{(2)} & a_{2k}^{(2)} & \cdots & a_{2n}^{(2)} & \vdots & a_{2,n+1}^{(2)} \\ & & & & a_{k-1,k-1}^{(k-1)} & a_{k-1,k}^{(k-1)} & \cdots & a_{k-1,n}^{(k-1)} & \vdots & a_{k-1,n+1}^{(k-1)} \\ & & & & 0 & a_{kk}^{(k)} & \cdots & a_{kn}^{(k)} & \vdots & a_{k,n+1}^{(k)} \\ 0 & \cdots & & & 0 & a_{nk}^{(k)} & \cdots & a_{nn}^{(k)} & \vdots & a_{n,n+1}^{(k)} \end{bmatrix}$$

represents the equivalent linear system for which the variable x_{k-1} has just been eliminated from equations $E_k, E_{k+1}, \ldots, E_n$.

The procedure will not work if any of the elements $a_{11}^{(1)}, a_{22}^{(2)}, a_{33}^{(3)}, \ldots, a_{n-1,n-1}^{(n-1)}, a_{nn}^{(n)}$ are zero for, in this case, the step

$$\left(E_i - \frac{a_{i,k}^{(k)}}{a_{kk}^{(k)}} E_k \right) \to E_i$$

either cannot be performed (this occurs if one of $a_{11}^{(1)}, \ldots, a_{n-1,n-1}^{(n-1)}$ is zero), or the backward substitution cannot be accomplished (in the case $a_{nn}^{(n)} = 0$). This does not necessarily mean that the linear system is not solvable, but that the technique of solution must be altered. An illustration is given in the following example.

EXAMPLE 1   Consider the linear system

$$E_1: \quad x_1 - x_2 + 2x_3 - x_4 = -8,$$
$$E_2: \quad 2x_1 - 2x_2 + 3x_3 - 3x_4 = -20,$$
$$E_3: \quad x_1 + x_2 + x_3 \qquad\quad = -2,$$
$$E_4: \quad x_1 - x_2 + 4x_3 + 3x_4 = \quad 4.$$

The augmented matrix is:

$$\tilde{A} = \tilde{A}^{(1)} = \begin{bmatrix} 1 & -1 & 2 & -1 & \vdots & -8 \\ 2 & -2 & 3 & -3 & \vdots & -20 \\ 1 & 1 & 1 & 0 & \vdots & -2 \\ 1 & -1 & 4 & 3 & \vdots & 4 \end{bmatrix},$$

and, performing the operations

$$(E_2 - 2E_1) \to (E_2), \quad (E_3 - E_1) \to (E_3), \quad \text{and} \quad (E_4 - E_1) \to (E_4),$$

we write:

$$\tilde{A}^{(2)} = \begin{bmatrix} 1 & -1 & 2 & -1 & \vdots & -8 \\ 0 & 0 & -1 & -1 & \vdots & -4 \\ 0 & 2 & -1 & 1 & \vdots & 6 \\ 0 & 0 & 2 & 4 & \vdots & 12 \end{bmatrix}.$$

Since the element $a_{22}^{(2)}$, called the **pivot element**, is zero, the procedure cannot continue in its present form. But the operation $(E_i) \leftrightarrow (E_j)$ is permitted, so a search is made of the elements $a_{32}^{(2)}$ and $a_{42}^{(2)}$ for the first nonzero element. Since $a_{32}^{(2)} \neq 0$, the operation $(E_2) \leftrightarrow (E_3)$ is performed to obtain a new matrix

$$\tilde{A}^{(2)'} = \begin{bmatrix} 1 & -1 & 2 & -1 & \vdots & -8 \\ 0 & 2 & -1 & 1 & \vdots & 6 \\ 0 & 0 & -1 & -1 & \vdots & -4 \\ 0 & 0 & 2 & 4 & \vdots & 12 \end{bmatrix}.$$

Since x_2 is already eliminated from E_3 and E_4, $\tilde{A}^{(3)}$ will be $\tilde{A}^{(2)'}$ and the computation can continue with the operation $(E_4 + 2E_3) \to (E_4)$, giving

$$\tilde{A}^{(4)} = \begin{bmatrix} 1 & -1 & 2 & -1 & \vdots & -8 \\ 0 & 2 & -1 & 1 & \vdots & 6 \\ 0 & 0 & -1 & -1 & \vdots & -4 \\ 0 & 0 & 0 & 2 & \vdots & 4 \end{bmatrix}.$$

Finally, the backward substitution can be applied:

$$x_4 = \tfrac{4}{2} = 2,$$

$$x_3 = \frac{[-4 - (-1)x_4]}{(-1)} = 2,$$

$$x_2 = \frac{[6 - x_4 - (-1)x_3]}{2} = 3,$$

$$x_1 = \frac{[-8 - (-1)x_4 - 2x_3 - (-1)x_2]}{(1)} = -7. \qquad \square$$

Example 1 illustrates what is done if $a_{kk}^{(k)} = 0$ for some $k = 1, 2, \ldots, n - 1$. The kth column of $\tilde{A}^{(k-1)}$ from the kth row to the nth row is searched for the first nonzero entry. If $a_{pk}^{(k)} \neq 0$ for some $p, k + 1 \leq p \leq n$, then the operation $(E_k) \leftrightarrow (E_p)$ is performed to obtain $\tilde{A}^{(k-1)'}$. The procedure can then be continued to form $\tilde{A}^{(k)}$, and so on. If $a_{pk}^{(k)} = 0$ for $p = k, k + 1, \ldots, n$, it can be shown (see Theorem 6.13, p. 285) that the linear system does not have a unique solution and the procedure stops. Finally, if $a_{nn}^{(n)} = 0$, the linear system does not have a unique solution and again the procedure stops. To summarize the entire Gaussian elimination with backward substitution, we present the following algorithm.

Gaussian Elimination with Backward Substitution Algorithm 6.1

To solve the $n \times n$ linear system

$$E_1: \quad a_{11}x_1 + a_{12}x_2 + \cdots + a_{1n}x_n = a_{1,n+1}$$

$$E_2: \quad a_{21}x_1 + a_{22}x_2 + \cdots + a_{2n}x_n = a_{2,n+1}$$

$$\vdots \qquad \vdots \qquad \vdots \qquad \qquad \vdots \qquad \qquad \vdots$$

$$E_n: \quad a_{n1}x_1 + a_{n2}x_2 + \cdots + a_{nn}x_n = a_{n,n+1}:$$

INPUT number of unknowns and equations n; augmented matrix $A = (a_{ij})$ where $1 \le i \le n$ and $1 \le j \le n + 1$.

OUTPUT solution $x_1, x_2, \ldots, x_n$ or message that linear system has no unique solution.

Step 1 For $i = 1, \ldots, n - 1$ do Steps 2–4.   (*Elimination Process.*)

Step 2 Let p be the smallest integer with $i \le p \le n$ and $a_{p,i} \ne 0$.
    If no integer p can be found
        then OUTPUT ('no unique solution exists');
            STOP.

Step 3 If $p \ne i$ then perform $(E_p) \leftrightarrow (E_i)$.

Step 4 For $j = i + 1, \ldots, n$ do Steps 5 and 6.

    Step 5 Set $m_{ji} = a_{ji}/a_{ii}$.

    Step 6 Perform $(E_j - m_{ji}E_i) \to (E_j)$.

Step 7 If $a_{nn} = 0$ then OUTPUT ('no unique solution exists');
            STOP.

Step 8 Set $x_n = a_{n,n+1}/a_{nn}$.   (*Start backward substitution.*)

Step 9 For $i = n - 1, \ldots, 1$ set $x_i = \left[a_{i,n+1} - \sum_{j=i+1}^{n} a_{ij}x_j \right] \Big/ a_{ii}$.

Step 10 OUTPUT $(x_1, \ldots, x_n)$;   (*Procedure completed successfully.*)
            STOP.

EXAMPLE 2   The purpose of this example is to show what can happen if Algorithm 6.1 fails. The computations will be done simultaneously on two linear systems:

$$
\begin{aligned}
x_1 + x_2 + x_3 + x_4 &= 7, \\
x_1 + x_2 \qquad\; + 2x_4 &= 8, \\
2x_1 + 2x_2 + 3x_3 \qquad &= 10, \\
-x_1 - x_2 - 2x_3 + 2x_4 &= 0,
\end{aligned}
\qquad \text{and} \qquad
\begin{aligned}
x_1 + x_2 + x_3 + x_4 &= 7, \\
x_1 + x_2 \qquad\; + 2x_4 &= 5, \\
2x_1 + 2x_2 + 3x_3 \qquad &= 10, \\
-x_1 - x_2 - 2x_3 + 2x_4 &= 0.
\end{aligned}
$$

These systems produce matrices

$$
\tilde{A} = \begin{bmatrix}
1 & 1 & 1 & 1 & \vdots & 7 \\
1 & 1 & 0 & 2 & \vdots & 8 \\
2 & 2 & 3 & 0 & \vdots & 10 \\
-1 & -1 & -2 & 2 & \vdots & 0
\end{bmatrix}
\qquad \text{and} \qquad
\tilde{A} = \begin{bmatrix}
1 & 1 & 1 & 1 & \vdots & 7 \\
1 & 1 & 0 & 2 & \vdots & 5 \\
2 & 2 & 3 & 0 & \vdots & 10 \\
-1 & -1 & -2 & 2 & \vdots & 0
\end{bmatrix}.
$$

Since $a_{11} \neq 0$, the steps in eliminating x_1 from E_2, E_3, and E_4 give

$$m_{ji} = m_{j1} = a_{j1}/1 = a_{j1}.$$
$$j = 2, \quad m_{12} = 1,$$
$$j = 3, \quad m_{13} = 2,$$
$$j = 4, \quad m_{14} = -1,$$
$$(E_3 - E_1) \rightarrow (E_2),$$
$$(E_3 - 2E_1) \rightarrow (E_3),$$
$$(E_4 + E_1) \rightarrow (E_4).$$

The matrices become:

$$\tilde{A} = \begin{bmatrix} 1 & 1 & 1 & 1 & \vdots & 7 \\ 0 & 0 & -1 & 1 & \vdots & 1 \\ 0 & 0 & 1 & -2 & \vdots & -4 \\ 0 & 0 & -1 & 3 & \vdots & 7 \end{bmatrix} \quad \text{and} \quad \tilde{A} = \begin{bmatrix} 1 & 1 & 1 & 1 & \vdots & 7 \\ 0 & 0 & -1 & 1 & \vdots & -2 \\ 0 & 0 & 1 & -2 & \vdots & -4 \\ 0 & 0 & -1 & 3 & \vdots & 7 \end{bmatrix}.$$

At this point $a_{22} = a_{23} = a_{24} = 0$ and the algorithm requires that the procedure be halted, and no solution to either system is obtained.

To examine more closely the reason for difficulty, perform $(E_4 + E_3) \rightarrow (E_4)$ to obtain:

$$\begin{bmatrix} 1 & 1 & 1 & 1 & \vdots & 7 \\ 0 & 0 & -1 & 1 & \vdots & 1 \\ 0 & 0 & 1 & -2 & \vdots & -4 \\ 0 & 0 & 0 & 1 & \vdots & 3 \end{bmatrix} \quad \text{and} \quad \begin{bmatrix} 1 & 1 & 1 & 1 & \vdots & 7 \\ 0 & 0 & -1 & 1 & \vdots & -2 \\ 0 & 0 & 1 & -2 & \vdots & -4 \\ 0 & 0 & 0 & 1 & \vdots & 3 \end{bmatrix}.$$

Writing the equations for each system gives:

$$x_1 + x_2 + x_3 + x_4 = 7, \quad \text{and} \quad x_1 + x_2 + x_3 + x_4 = 7,$$
$$-x_3 + x_4 = 1, \qquad\qquad -x_3 + x_4 = -2,$$
$$x_3 - 2x_4 = -4, \qquad\qquad x_3 - 2x_4 = -4,$$
$$x_4 = 3, \qquad\qquad x_4 = 3.$$

Performing backward substitution in each system yields:

$$x_4 = 3, \qquad\qquad x_4 = 3,$$
$$\text{and}$$
$$x_3 = -4 + 2x_4 = 2, \qquad x_3 = -4 + 2x_4 = 2.$$

If the backward substitution is continued to the second equation in each case, the difference in the two systems becomes apparent because

$$-x_3 + x_4 = 1 \quad \text{implies} \quad 1 = 1,$$

while

$$-x_3 + x_4 = -2 \quad \text{implies} \quad 1 = -2.$$

The first linear system has an infinite number of solutions $x_4 = 3$, $x_3 = 2$, x_2 arbitrary, and $x_1 = 2 - x_2$, while the other leads to a contradiction, and no solution exists. In both cases, however, there is no unique solution as we conclude from Algorithm 6.1. $\qquad\square$

Although Algorithm 6.1 can be viewed as the construction of the augmented matrices $\tilde{A}^{(1)}, \ldots, \tilde{A}^{(n)}$, the computations can be performed in a computer using only one n by $(n + 1)$ array for storage by simply replacing at each step the previous value of a_{ij} by the new one. It is also advantageous to store the multipliers m_{ji} in the locations of a_{ji} since a_{ji} has the value zero for each $i = 1, 2, \ldots, n - 1$, and $j = i + 1, i + 2, \ldots, n$. Thus, A can be overwritten by the multipliers below the main diagonal and by the nonzero entries of $\tilde{A}^{(n)}$ on and above the main diagonal. These values can be used to solve other linear systems involving the original matrix A. (See Section 6.6.)

When comparing techniques for solving linear systems, we need to consider other concepts in addition to the amount of storage required. One of these is the effect of roundoff error and another is the amount of time required to complete the calculations. Both of these topics depend upon the number of arithmetic operations that need to be performed in solving a routine problem. In general, the amount of time required to perform a multiplication or division on a computer is about the same, and is considerably greater than that required to perform an addition or subtraction. The actual differences in execution time, however, depend upon the particular computing system being used. In order to demonstrate the procedure involved with counting operations for a given method, we will count the operations required to solve a typical linear system of n equations in n unknowns using Algorithm 6.1. We will keep the count of the additions/subtractions separate from the count of the multiplications/divisions because of the time differential.

No arithmetic operations are performed until Steps 5 and 6 in the algorithm. Step 5 requires that $(n - i)$ divisions be performed. The replacement of the equation E_j by $(E_j - m_{ji} E_i)$ in Step 6 requires that m_{ji} be multiplied by each term in E_i resulting in a total of $(n - i)(n - i + 1)$ multiplications. After this is completed, each term of the resulting equation is subtracted from the corresponding term in E_j. This requires $(n - i)(n - i + 1)$ subtractions. For each $i = 1, 2, \ldots, n - 1$, the operations required in Steps 5 and 6 are

Multiplications/divisions: $(n - i) + (n - i)(n - i + 1)$;

Additions/subtractions: $(n - i)(n - i + 1)$.

The total number of operations required by these steps is consequently

Multiplications/divisions:

$$\sum_{i=1}^{n-1} (n - i)(n - i + 2) = (n^2 + 2n) \sum_{i=1}^{n-1} 1 - 2(n + 1) \sum_{i=1}^{n-1} i + \sum_{i=1}^{n-1} i^2$$

$$= (n^2 + 2n)(n - 1) - 2(n + 1)\frac{(n)(n - 1)}{2}$$

$$+ \frac{(n - 1)(n)(2n - 1)}{6}$$

$$= \frac{2n^3 + 3n^2 - 5n}{6}.$$

Additions/subtractions:

$$\sum_{i=1}^{n-1}(n-i)(n-i+1) = (n^2+n)\sum_{i=1}^{n-1}1 - (2n+1)\sum_{i=1}^{n-1}i + \sum_{i=1}^{n-1}i^2$$

$$= (n^2+n)(n-1) - (2n+1)\frac{(n)(n-1)}{2}$$

$$+ \frac{(n-1)(n)(2n-1)}{6}$$

$$= \frac{n^3-n}{3}.$$

The only other steps in Algorithm 6.1 that require arithmetic operations are Steps 8 and 9. Step 8 requires one division and Step 9 requires $(n-i)$ multiplications and $(n-i-1)$ additions for each summation term, and then one subtraction and one division. The total number of operations in Steps 8 and 9 is

Multiplications/divisions:

$$1 + \sum_{i=1}^{n-1}((n-i)+1) = \frac{n^2+n}{2}.$$

Additions/subtractions:

$$\sum_{i=1}^{n-1}((n-i-1)+1) = \frac{n^2-n}{2}.$$

The total arithmetic operations in Algorithm 6.1 is therefore

Multiplications/divisions:

$$\frac{2n^3+3n^2-5n}{6} + \frac{n^2+n}{2} = \frac{n^3+3n^2-n}{3}.$$

Additions/subtractions:

$$\frac{n^3-n}{3} + \frac{n^2-n}{2} = \frac{2n^3+3n^2-5n}{6}.$$

Since the total number of multiplications and divisions is approximately $\frac{1}{3}n^3$, and similarly for additions and subtractions, the amount of computation and the time required will increase with n in proportion to n^3. The following table indicates the growth of computations with increasing n.

n	Multiplications/divisions	Additions/subtractions
3	17	11
10	430	375
50	44,150	42,875
100	343,300	338,250

Exercise Set 6.2

1. Use Algorithm 6.1 to solve the following linear systems, if possible, and determine whether row interchanges are necessary.

a) $x_1 - x_2 + 3x_3 = 2,$
$3x_1 - 3x_2 + x_3 = -1,$
$x_1 + x_2 = 3.$

b) $\frac{1}{4}x_1 + \frac{1}{5}x_2 + \frac{1}{6}x_3 = 9,$
$\frac{1}{3}x_1 + \frac{1}{4}x_2 + \frac{1}{5}x_3 = 8,$
$\frac{1}{2}x_1 + x_2 + 2x_3 = 8.$

c) $2x_1 = 3,$
$x_1 + 1.5x_2 = 4.5,$
$- 3x_2 + .5x_3 = -6.6,$
$2x_1 - 2x_2 + x_3 + x_4 = .8.$

d) $2x_1 - x_2 + x_3 = -1,$
$3x_1 + 3x_2 + 9x_3 = 0,$
$3x_1 + 3x_2 + 5x_3 = 4.$

e) $x_1 + \frac{1}{2}x_2 + \frac{1}{3}x_3 + \frac{1}{4}x_4 = \frac{1}{6},$
$\frac{1}{2}x_1 + \frac{1}{3}x_2 + \frac{1}{4}x_3 + \frac{1}{5}x_4 = \frac{1}{7},$
$\frac{1}{3}x_1 + \frac{1}{4}x_2 + \frac{1}{5}x_3 + \frac{1}{6}x_4 = \frac{1}{8},$
$\frac{1}{4}x_1 + \frac{1}{5}x_2 + \frac{1}{6}x_3 + \frac{1}{7}x_4 = \frac{1}{9}.$

f) $x_2 + 4x_3 = 0,$
$x_1 - x_2 - x_3 = .375,$
$x_1 - x_2 + 2x_3 = 0.$

g) $2x_1 - 1.5x_2 + 3x_3 = 1,$
$-x_1 + 2x_3 = 3,$
$4x_1 - 4.5x_2 + 5x_3 = -1.$

2. Use Algorithm 6.1 to solve the following linear systems, if possible, and determine whether row interchanges are necessary.

a) $4.01x_1 + 1.23x_2 + 1.43x_3 - .73x_4 = 5.94,$
$1.23x_1 + 7.41x_2 + 2.41x_3 + 3.02x_4 = 14.07,$
$1.43x_1 + 2.41x_2 + 5.79x_3 - 1.11x_4 = 8.52,$
$-.73x_1 + 3.02x_2 - 1.11x_3 + 6.41x_4 = 7.59.$

b) $3.333x_1 + 15920x_2 - 10.333x_3 = 15913,$
$2.222x_1 + 16.71x_2 + 9.612x_3 = 28.544,$
$1.5611x_1 + 5.1791x_2 + 1.6852x_3 = 8.4254.$

3. **Gauss–Jordan Method** This method, used to solve the linear equation $Ax = b$, can be described as follows. Use the ith equation to eliminate not only x_i from the equations $E_{i+1}, E_{i+2}, \ldots, E_n$, as was done in the Gauss elimination method, but also from $E_1,$ $E_2, \ldots, E_{i-1}$. Upon reducing $[A \vdots b]$ to:

$$\begin{bmatrix} a_{11}^{(1)} & 0 & \cdots & 0 & \vdots & a_{1,n+1}^{(1)} \\ 0 & a_{22}^{(2)} & \ddots & \vdots & \vdots & a_{2,n+1}^{(2)} \\ \vdots & \ddots & \ddots & 0 & \vdots & \vdots \\ 0 & \cdots & 0 & a_{nn}^{(n)} & \vdots & a_{n,n+1}^{(n)} \end{bmatrix}$$

the solution is obtained by setting

$$x_i = \frac{a_{i,n+1}^{(i)}}{a_{ii}^{(i)}}$$

for each $i = 1, 2, \ldots, n$. This procedure circumvents the backward substitution in the Gaussian elimination. Construct an algorithm for the Gauss–Jordan procedure patterned after that of Algorithm 6.1.

4. Use the Gauss–Jordan method to solve the linear systems in Exercise 1.

5.   Use the Gauss–Jordan method to solve the linear systems in Exercise 2.

6.   a)   Show that the Gauss–Jordan method requires

$$\frac{n^3}{2} + n^2 - \frac{n}{2} \text{ multiplications/divisions} \quad \text{and} \quad \frac{n^3}{2} - \frac{n}{2} \text{ additions/subtractions.}$$

   b)   Make a table comparing the required operations for the Gauss–Jordan and Gaussian elimination methods for $n = 3, 10, 50, 100$. Which method requires less computation?

6.3   Linear Algebra and Matrix Inversion

The concept of a matrix was introduced in the previous two sections as a convenient method of expressing and manipulating a linear system. The material in this section is the algebra associated with matrices and the way it can be used to solve problems involving linear systems.

Definition 6.3   Two matrices A and B are said to be **equal** if both are of the same size, say $m \times n$, and if $a_{ij} = b_{ij}$ for each $i = 1, 2, \ldots, m$, and $j = 1, 2, \ldots, n$.

EXAMPLE 1   Let

$$A = \begin{bmatrix} 2 & -1 & 7 \\ 3 & 1 & 0 \end{bmatrix}, \quad B = \begin{bmatrix} 2 & -1 & 7 \\ 3 & 1 & 0 \end{bmatrix}, \quad C = \begin{bmatrix} 2 & 3 \\ -1 & 1 \\ 7 & 0 \end{bmatrix}.$$

Then $A = B$ but $A \neq C$.   □

Two important operations performed on matrices are the sum of two matrices and the multiplication of a matrix by a real number.

Definition 6.4   If A and B are both $m \times n$ matrices, then the **sum** of A and B, denoted $A + B$, is the $m \times n$ matrix whose entries are $a_{ij} + b_{ij}$, for each $i = 1, 2, \ldots, m$, and $j = 1, 2, \ldots, n$.

Definition 6.5   If A is an $m \times n$ matrix and λ a real number, then the **product** of λ and A, denoted λA, is the $m \times n$ matrix whose entries are λa_{ij} for each $i = 1, 2, \ldots, m$ and $j = 1, 2, \ldots, n$.

EXAMPLE 2   Let

$$A = \begin{bmatrix} 2 & -1 & 7 \\ 3 & 1 & 0 \end{bmatrix}, \quad B = \begin{bmatrix} 4 & 2 & -8 \\ 0 & 1 & 6 \end{bmatrix}, \quad C = \begin{bmatrix} 1 & 2 & 3 \\ 4 & 5 & 6 \end{bmatrix}$$

and $\lambda = -2, \mu = 3$. Then

$$A + B = \begin{bmatrix} 2 + 4 & -1 + 2 & 7 - 8 \\ 3 + 0 & 1 + 1 & 0 + 6 \end{bmatrix} = \begin{bmatrix} 6 & 1 & -1 \\ 3 & 2 & 6 \end{bmatrix};$$

$$B + A = \begin{bmatrix} 6 & 1 & -1 \\ 3 & 2 & 6 \end{bmatrix};$$

$$(A + B) + C = \begin{bmatrix} (2 + 4) + 1 & (-1 + 2) + 2 & (7 - 8) + 3 \\ (3 + 0) + 4 & (1 + 1) + 5 & (0 + 6) + 6 \end{bmatrix} = \begin{bmatrix} 7 & 3 & 2 \\ 7 & 7 & 12 \end{bmatrix};$$

$$A + (B + C) = \begin{bmatrix} 7 & 3 & 2 \\ 7 & 7 & 12 \end{bmatrix};$$

$$\lambda(A + B) = \begin{bmatrix} -2(6) & -2(1) & -2(-1) \\ -2(3) & -2(2) & -2(6) \end{bmatrix} = \begin{bmatrix} -12 & -2 & 2 \\ -6 & -4 & -12 \end{bmatrix};$$

$$\lambda A + \lambda B = \begin{bmatrix} -4 & 2 & -14 \\ -6 & -2 & 0 \end{bmatrix} + \begin{bmatrix} -8 & -4 & 16 \\ 0 & -2 & -12 \end{bmatrix} = \begin{bmatrix} -12 & -2 & 2 \\ -6 & -4 & -12 \end{bmatrix};$$

$$(\lambda + \mu)A = (-2 + 3)A = \begin{bmatrix} 2 & -1 & 7 \\ 3 & 1 & 0 \end{bmatrix};$$

$$\lambda A + \mu A = -2A + 3A = \begin{bmatrix} -4 & 2 & -14 \\ -6 & -2 & 0 \end{bmatrix} + \begin{bmatrix} 6 & -3 & 21 \\ 9 & 3 & 0 \end{bmatrix}$$

$$= \begin{bmatrix} 2 & -1 & 7 \\ 3 & 1 & 0 \end{bmatrix};$$

$$\lambda(\mu A) = (-2)(3A) = \begin{bmatrix} -12 & 6 & -42 \\ -18 & -6 & 0 \end{bmatrix};$$

$$(\lambda\mu)A = [(-2)(3)]A = \begin{bmatrix} -12 & 6 & -42 \\ -18 & -6 & 0 \end{bmatrix};$$

and

$$1A = \begin{bmatrix} 2 & -1 & 7 \\ 3 & 1 & 0 \end{bmatrix} = A. \qquad \square$$

Letting the matrix all of whose entries are zero be denoted simply by O, and $-A$ be the matrix whose entries are $-a_{ij}$, we can list the following general properties of matrix addition and scalar multiplication. These properties are sufficient to classify the set of all $m \times n$ matrices with real entries as a **vector space** over the field of real numbers. (See Noble and Daniel, [63], pages 107–109.)

Theorem 6.6 Let $A, B,$ and C be $m \times n$ matrices and λ and μ real numbers. The following properties of addition and scalar multiplication hold:

a) $A + B = B + A,$
b) $(A + B) + C = A + (B + C),$
c) $A + O = O + A = A,$
d) $A + (-A) = -A + A = O,$
e) $\lambda(A + B) = \lambda A + \lambda B,$
f) $(\lambda + \mu)A = \lambda A + \mu A,$
g) $\lambda(\mu A) = (\lambda\mu)A,$
h) $1A = A.$

Properties (a), (b), (e), (f), (g), and (h) were demonstrated in Example 2 for the special matrices given there. All of the properties can be easily verified. (See Exercise 4.)

Definition 6.7   Let A be an $m \times n$ matrix and B an $n \times p$ matrix. The **product** of A and B, denoted AB, is an $m \times p$ matrix C whose entries c_{ij} are given by

(6.12) $$c_{ij} = \sum_{k=1}^{n} a_{ik}b_{kj} = a_{i1}b_{1j} + a_{i2}b_{2j} + \cdots + a_{in}b_{nj}$$

for each $i = 1, 2, \ldots, m$, and $j = 1, 2, \ldots, p$.

The computation of c_{ij} can be viewed as the multiplication of the entries of the ith row of A with corresponding entries in the jth column of B, followed by a summation. That is,

$$[a_{i1} \quad a_{i2} \quad \cdots \quad a_{in}] \begin{bmatrix} b_{1j} \\ b_{2j} \\ \vdots \\ b_{nj} \end{bmatrix} = c_{ij},$$

where $$c_{ij} = a_{i1}b_{1j} + a_{i2}b_{2j} + \cdots + a_{in}b_{nj} = \sum_{k=1}^{n} a_{ik}b_{kj}$$

This explains why the number of columns of A must equal the number of rows of B in order for the product AB to be defined.

The following example should serve to further clarify the matrix multiplication process.

EXAMPLE 3   Let

$$A = \begin{bmatrix} 2 & 1 & -1 \\ 3 & 1 & 2 \\ 0 & -2 & -3 \end{bmatrix}, \qquad B = \begin{bmatrix} 3 & 2 \\ -1 & 1 \\ 6 & 4 \end{bmatrix},$$

$$C = \begin{bmatrix} 2 & 1 & 0 \\ -1 & 3 & 2 \end{bmatrix}, \qquad \text{and} \qquad D = \begin{bmatrix} 1 & -1 & 1 \\ 2 & -1 & 2 \\ 3 & 0 & 3 \end{bmatrix}$$

Then,

$$AD = \begin{bmatrix} 2 & 1 & -1 \\ 3 & 1 & 2 \\ 0 & -2 & -3 \end{bmatrix} \begin{bmatrix} 1 & -1 & 1 \\ 2 & -1 & 2 \\ 3 & 0 & 3 \end{bmatrix} = \begin{bmatrix} 1 & -3 & 1 \\ 11 & -4 & 11 \\ -13 & 2 & -13 \end{bmatrix}$$

and $$DA = \begin{bmatrix} 1 & -1 & 1 \\ 2 & -1 & 2 \\ 3 & 0 & 3 \end{bmatrix} \begin{bmatrix} 2 & 1 & -1 \\ 3 & 1 & 2 \\ 0 & -2 & -3 \end{bmatrix} = \begin{bmatrix} -1 & -2 & -6 \\ 1 & -5 & -10 \\ 6 & -3 & -12 \end{bmatrix},$$

which shows that $AD \neq DA$.

Further,

$$BC = \begin{bmatrix} 3 & 2 \\ -1 & 1 \\ 6 & 4 \end{bmatrix} \begin{bmatrix} 2 & 1 & 0 \\ -1 & 3 & 2 \end{bmatrix} = \begin{bmatrix} 4 & 9 & 4 \\ -3 & 2 & 2 \\ 8 & 18 & 8 \end{bmatrix}$$

and

$$CB = \begin{bmatrix} 2 & 1 & 0 \\ -1 & 3 & 2 \end{bmatrix} \begin{bmatrix} 3 & 2 \\ -1 & 1 \\ 6 & 4 \end{bmatrix} = \begin{bmatrix} 5 & 5 \\ 6 & 9 \end{bmatrix}.$$

Here BC and CB are not even the same size.

Finally,

$$AB = \begin{bmatrix} 2 & 1 & -1 \\ 3 & 1 & 2 \\ 0 & -2 & -3 \end{bmatrix} \begin{bmatrix} 3 & 2 \\ -1 & 1 \\ 6 & 4 \end{bmatrix} = \begin{bmatrix} -1 & 1 \\ 20 & 15 \\ -16 & -14 \end{bmatrix}$$

while

$$BA = \begin{bmatrix} 3 & 2 \\ -1 & 1 \\ 6 & 4 \end{bmatrix} \begin{bmatrix} 2 & 1 & -1 \\ 3 & 1 & 2 \\ 0 & -2 & -3 \end{bmatrix} \qquad \text{cannot be computed.} \qquad \Box$$

Definition 6.8   The **identity matrix of order** n, $I_n = (\delta_{ij})$, is the matrix with entries

$$\delta_{ij} = \begin{cases} 1 & \text{if } i = j, \\ 0 & \text{if } i \neq j. \end{cases}$$

When the size of I_n is clear, this matrix is sometimes simply written as I.

EXAMPLE 4   The identity matrix of order three is

$$I_3 = \begin{bmatrix} 1 & 0 & 0 \\ 0 & 1 & 0 \\ 0 & 0 & 1 \end{bmatrix}.$$

If A is any other 3×3 matrix, then

$$AI_3 = \begin{bmatrix} a_{11} & a_{12} & a_{13} \\ a_{21} & a_{22} & a_{23} \\ a_{31} & a_{32} & a_{33} \end{bmatrix} \begin{bmatrix} 1 & 0 & 0 \\ 0 & 1 & 0 \\ 0 & 0 & 1 \end{bmatrix} = \begin{bmatrix} a_{11} & a_{12} & a_{13} \\ a_{21} & a_{22} & a_{23} \\ a_{31} & a_{32} & a_{33} \end{bmatrix} = A. \qquad \Box$$

The identity matrix can be seen to commute with A; that is, the order of multiplication does not matter, $I_3 A = A = AI_3$.

In Example 3 it was seen that the property $AB = BA$ does not generally hold for matrix multiplication. Some of the properties involving matrix multiplication that do hold are presented in the next theorem.

Theorem 6.9   Let A be an $n \times m$ matrix, B an $m \times k$ matrix, C a $k \times p$ matrix, D an $m \times k$ matrix, and λ a real number. The following properties hold:

a) $A(BC) = (AB)C$;

b) $I_m B = B$ and $BI_k = B$;

c) $A(B + D) = AB + AD$;

d) $\lambda(AB) = (\lambda A)B = A(\lambda B)$.

Proof   The verification of (a) is presented to show the method involved. (The other verifications are left to Exercise 6.) To show $A(BC) = (AB)C$, compute the i, j-entry of each side of the equation. BC is an $m \times p$ matrix with i, j-entry

$$(BC)_{ij} = \sum_{l=1}^{k} b_{il} c_{lj}.$$

Thus, $A(BC)$ is an $n \times p$ matrix with entries:

$$[A(BC)]_{ij} = \sum_{s=1}^{m} a_{is}(BC)_{sj} = \sum_{s=1}^{m} a_{is}\left(\sum_{l=1}^{k} b_{sl} c_{lj}\right) = \sum_{s=1}^{m} \sum_{l=1}^{k} a_{is} b_{sl} c_{lj}.$$

Similarly, AB is an $n \times k$ matrix with entries:

$$(AB)_{ij} = \sum_{s=1}^{m} a_{is} b_{sj},$$

so $(AB)C$ is an $n \times p$ matrix with entries:

$$[(AB)C]_{ij} = \sum_{l=1}^{k} (AB)_{il} c_{lj} = \sum_{l=1}^{k}\left(\sum_{s=1}^{m} a_{is} b_{sl}\right) c_{lj} = \sum_{l=1}^{k} \sum_{s=1}^{m} a_{is} b_{sl} c_{lj} = \sum_{s=1}^{m} \sum_{l=1}^{k} a_{is} b_{sl} c_{lj}.$$

This implies $[A(BC)]_{ij} = [(AB)C]_{ij}$ for each $i = 1, 2, \ldots, n$, and $j = 1, 2, \ldots, p$, and $A(BC) = (AB)C$.   □

With the definition of matrix multiplication, the relationship of linear systems to linear algebra can be discussed. The linear system

$$a_{11}x_1 + a_{12}x_2 + \cdots + a_{1n}x_n = b_1,$$
$$a_{21}x_1 + a_{22}x_2 + \cdots + a_{2n}x_n = b_2,$$
$$\vdots \qquad \vdots \qquad \qquad \vdots \qquad \vdots$$
$$a_{n1}x_1 + a_{n2}x_2 + \cdots + a_{nn}x_n = b_n,$$

can be viewed as the matrix equation

$$A\mathbf{x} = \mathbf{b},$$

where

$$A = \begin{bmatrix} a_{11} & a_{12} & \cdots & a_{1n} \\ a_{21} & a_{22} & \cdots & a_{2n} \\ \vdots & \vdots & & \vdots \\ a_{n1} & a_{n2} & \cdots & a_{nn} \end{bmatrix}, \qquad \mathbf{x} = \begin{bmatrix} x_1 \\ x_2 \\ \vdots \\ x_n \end{bmatrix}, \qquad \text{and} \qquad \mathbf{b} = \begin{bmatrix} b_1 \\ b_2 \\ \vdots \\ b_n \end{bmatrix}.$$

Also related to linear systems is the concept of the **inverse of a matrix**.

Definition 6.10   An $n \times n$ matrix A is said to be **nonsingular** if an $n \times n$ matrix A^{-1} exists with $AA^{-1} = A^{-1}A = I$. The matrix A^{-1} is called the **inverse** of A. A matrix without an inverse is called **singular**.

EXAMPLE 5   Let

$$A = \begin{bmatrix} 1 & 2 & -1 \\ 2 & 1 & 0 \\ -1 & 1 & 2 \end{bmatrix} \qquad \text{and} \qquad B = \tfrac{1}{9}\begin{bmatrix} -2 & 5 & -1 \\ 4 & -1 & 2 \\ -3 & 3 & 3 \end{bmatrix}.$$

Since

$$AB = \begin{bmatrix} 1 & 2 & -1 \\ 2 & 1 & 0 \\ -1 & 1 & 2 \end{bmatrix} \cdot \tfrac{1}{9} \begin{bmatrix} -2 & 5 & -1 \\ 4 & -1 & 2 \\ -3 & 3 & 3 \end{bmatrix} = \begin{bmatrix} 1 & 0 & 0 \\ 0 & 1 & 0 \\ 0 & 0 & 1 \end{bmatrix}$$

and

$$BA = \tfrac{1}{9} \begin{bmatrix} -2 & 5 & -1 \\ 4 & -1 & 2 \\ -3 & 3 & 3 \end{bmatrix} \begin{bmatrix} 1 & 2 & -1 \\ 2 & 1 & 0 \\ -1 & 1 & 2 \end{bmatrix} = \begin{bmatrix} 1 & 0 & 0 \\ 0 & 1 & 0 \\ 0 & 0 & 1 \end{bmatrix},$$

A and B are nonsingular; $B = A^{-1}$ and $A = B^{-1}$.

The usefulness of the inverse can be shown by considering the linear system

$$x_1 + 2x_2 - x_3 = 2,$$
$$2x_1 + x_2 = 3,$$
$$-x_1 + x_2 + 2x_3 = 4.$$

First, convert the system to the matrix equation

$$\begin{bmatrix} 1 & 2 & -1 \\ 2 & 1 & 0 \\ -1 & 1 & 2 \end{bmatrix} \begin{bmatrix} x_1 \\ x_2 \\ x_3 \end{bmatrix} = \begin{bmatrix} 2 \\ 3 \\ 4 \end{bmatrix},$$

and then multiply both sides by the inverse given above:

$$\tfrac{1}{9} \begin{bmatrix} -2 & 5 & -1 \\ 4 & -1 & 2 \\ -3 & 3 & 3 \end{bmatrix} \left(\begin{bmatrix} 1 & 2 & -1 \\ 2 & 1 & 0 \\ -1 & 1 & 2 \end{bmatrix} \begin{bmatrix} x_1 \\ x_2 \\ x_3 \end{bmatrix} \right) = \tfrac{1}{9} \begin{bmatrix} -2 & 5 & -1 \\ 4 & -1 & 2 \\ -3 & 3 & 3 \end{bmatrix} \begin{bmatrix} 2 \\ 3 \\ 4 \end{bmatrix};$$

so

$$\left(\begin{bmatrix} -\tfrac{2}{9} & \tfrac{5}{9} & -\tfrac{1}{9} \\ \tfrac{4}{9} & -\tfrac{1}{9} & \tfrac{2}{9} \\ -\tfrac{3}{9} & \tfrac{3}{9} & \tfrac{3}{9} \end{bmatrix} \begin{bmatrix} 1 & 2 & -1 \\ 2 & 1 & 0 \\ -1 & 1 & 2 \end{bmatrix} \right) \begin{bmatrix} x_1 \\ x_2 \\ x_3 \end{bmatrix} = \tfrac{1}{9} \begin{bmatrix} 7 \\ 13 \\ 15 \end{bmatrix}$$

and

$$I_3 \begin{bmatrix} x_1 \\ x_2 \\ x_3 \end{bmatrix} = \begin{bmatrix} x_1 \\ x_2 \\ x_3 \end{bmatrix} = \begin{bmatrix} \tfrac{7}{9} \\ \tfrac{13}{9} \\ \tfrac{15}{9} \end{bmatrix},$$

which gives the solution $x_1 = \tfrac{7}{9}$, $x_2 = \tfrac{13}{9}$, and $x_3 = \tfrac{15}{9}$.   □

In general, the technique in Example 5 for solving $A\mathbf{x} = \mathbf{b}$ is not recommended because of the number of operations involved. (See Exercise 7(f), 7(g), and 7(h).)

To find a method of computing A^{-1}, assuming its existence, let us look again at matrix multiplication. Let B_j be the jth column of the $n \times n$ matrix B,

$$B_j = \begin{bmatrix} b_{1j} \\ b_{2j} \\ \vdots \\ b_{nj} \end{bmatrix}.$$

Form the product

$$AB_j = \begin{bmatrix} a_{11} & a_{12} & \cdots & a_{1n} \\ a_{21} & a_{22} & \cdots & a_{2n} \\ \vdots & \vdots & & \vdots \\ a_{n1} & a_{n2} & \cdots & a_{nn} \end{bmatrix} \begin{bmatrix} b_{1j} \\ b_{2j} \\ \vdots \\ b_{nj} \end{bmatrix} = \begin{bmatrix} \sum_{k=1}^{n} a_{1k}b_{kj} \\ \sum_{k=1}^{n} a_{2k}b_{kj} \\ \vdots \\ \sum_{k=1}^{n} a_{nk}b_{kj} \end{bmatrix}.$$

If $AB = C$, then the jth column of C is given by:

$$C_j = \begin{bmatrix} c_{1j} \\ c_{2j} \\ \vdots \\ c_{nj} \end{bmatrix} = \begin{bmatrix} \sum_{k=1}^{n} a_{1k}b_{kj} \\ \sum_{k=1}^{n} a_{2k}b_{kj} \\ \vdots \\ \sum_{k=1}^{n} a_{nk}b_{kj} \end{bmatrix}.$$

Hence, the jth column of the product AB is the product of A and the jth column of B.

Suppose that A^{-1} exists and that $A^{-1} = B = (b_{ij})$; then $AB = I$ and

$$AB_j = \begin{bmatrix} 0 \\ \vdots \\ 0 \\ 1 \\ 0 \\ \vdots \\ 0 \end{bmatrix} \qquad \text{where the value 1 appears in the } j\text{th row.}$$

To actually find B, we must solve n linear systems in which the jth column of the inverse is the solution of the linear system with righthand side the jth column of I. The next example demonstrates the method involved.

EXAMPLE 6   Let

$$A = \begin{bmatrix} 1 & 2 & -1 \\ 2 & 1 & 0 \\ -1 & 1 & 2 \end{bmatrix}.$$

To compute A^{-1}, we must solve the three linear systems

$$\begin{aligned} x_1 + 2x_2 - x_3 &= 1, & x_1 + 2x_2 - x_3 &= 0, \\ 2x_1 + x_2 \phantom{{}+ 2x_3} &= 0, & 2x_1 + x_2 \phantom{{}+ 2x_3} &= 1, \\ -x_1 + x_2 + 2x_3 &= 0; & -x_1 + x_2 + 2x_3 &= 0; \end{aligned}$$

$$\begin{aligned} x_1 + 2x_2 - x_3 &= 0, \\ 2x_1 + x_2 \phantom{{}+ 2x_3} &= 0, \\ -x_1 + x_2 + 2x_3 &= 1. \end{aligned}$$

Using Gaussian elimination, the computations are conveniently performed upon the larger augmented matrix, formed by combining the matrices

$$\begin{bmatrix} 1 & 2 & -1 & \vdots & 1 & 0 & 0 \\ 2 & 1 & 0 & \vdots & 0 & 1 & 0 \\ -1 & 1 & 2 & \vdots & 0 & 0 & 1 \end{bmatrix}.$$

To elaborate, since the actual coefficient matrix does not change, we must perform the same sequence of row operations for each linear system. First, performing $(E_2 - 2E_1) \to (E_2)$ and $(E_3 + E_1) \to (E_3)$, we have:

$$\left[\begin{array}{ccc:ccc} 1 & 2 & -1 & 1 & 0 & 0 \\ 0 & -3 & 2 & -2 & 1 & 0 \\ 0 & 3 & 1 & 1 & 0 & 1 \end{array}\right].$$

Next, performing $(E_3 + E_2) \to (E_3)$, we get:

$$\left[\begin{array}{ccc:ccc} 1 & 2 & -1 & 1 & 0 & 0 \\ 0 & -3 & 2 & -2 & 1 & 0 \\ 0 & 0 & 3 & -1 & 1 & 1 \end{array}\right].$$

Backward substitution could be performed on each of the three augmented matrices,

$$\left[\begin{array}{ccc:c} 1 & 2 & -1 & 1 \\ 0 & -3 & 2 & -2 \\ 0 & 0 & 3 & -1 \end{array}\right], \quad \left[\begin{array}{ccc:c} 1 & 2 & -1 & 0 \\ 0 & -3 & 2 & 1 \\ 0 & 0 & 3 & 1 \end{array}\right], \quad \left[\begin{array}{ccc:c} 1 & 2 & -1 & 0 \\ 0 & -3 & 2 & 0 \\ 0 & 0 & 3 & 1 \end{array}\right],$$

to find all the entries of A^{-1}, but it is often more convenient to use further row reduction. In particular, the operation $(\frac{1}{3}E_3) \to (E_3)$ yields:

$$\left[\begin{array}{ccc:ccc} 1 & 2 & -1 & 1 & 0 & 0 \\ 0 & -3 & 2 & -2 & 1 & 0 \\ 0 & 0 & 1 & -\frac{1}{3} & \frac{1}{3} & \frac{1}{3} \end{array}\right]$$

and $(E_2 - 2E_3) \to (E_2)$ and $(E_1 + E_3) \to (E_1)$ produce:

$$\left[\begin{array}{ccc:ccc} 1 & 2 & 0 & \frac{2}{3} & \frac{1}{3} & \frac{1}{3} \\ 0 & -3 & 0 & -\frac{4}{3} & \frac{1}{3} & -\frac{2}{3} \\ 0 & 0 & 1 & -\frac{1}{3} & \frac{1}{3} & \frac{1}{3} \end{array}\right].$$

Performing $(-\frac{1}{3}E_2) \to (E_2)$, we obtain:

$$\left[\begin{array}{ccc:ccc} 1 & 2 & 0 & \frac{2}{3} & \frac{1}{3} & \frac{1}{3} \\ 0 & 1 & 0 & \frac{4}{9} & -\frac{1}{9} & \frac{2}{9} \\ 0 & 0 & 1 & -\frac{1}{3} & \frac{1}{3} & \frac{1}{3} \end{array}\right];$$

and finally, $(E_1 - 2E_2) \to (E_1)$ gives us:

$$\left[\begin{array}{ccc:ccc} 1 & 0 & 0 & -\frac{2}{9} & \frac{5}{9} & -\frac{1}{9} \\ 0 & 1 & 0 & \frac{4}{9} & -\frac{1}{9} & \frac{2}{9} \\ 0 & 0 & 1 & -\frac{3}{9} & \frac{3}{9} & \frac{3}{9} \end{array}\right].$$

The final augmented matrix represents the solutions to the three linear systems

$$x_1 = -\tfrac{2}{9}, \quad x_1 = \tfrac{5}{9}, \quad x_1 = -\tfrac{1}{9},$$
$$x_2 = \tfrac{4}{9}, \quad x_2 = -\tfrac{1}{9}, \quad x_2 = \tfrac{2}{9},$$
$$x_3 = -\tfrac{3}{9}, \quad x_3 = \tfrac{3}{9}, \quad x_3 = \tfrac{3}{9},$$

so

$$A^{-1} = \left[\begin{array}{ccc} -\frac{2}{9} & \frac{5}{9} & -\frac{1}{9} \\ \frac{4}{9} & -\frac{1}{9} & \frac{2}{9} \\ -\frac{3}{9} & \frac{3}{9} & \frac{3}{9} \end{array}\right] = \frac{1}{9}\left[\begin{array}{ccc} -2 & 5 & -1 \\ 4 & -1 & 2 \\ -3 & 3 & 3 \end{array}\right]. \qquad \square$$

In the last example, we illustrated the computation of A^{-1}. As we saw in that example, it is convenient to set up the larger augmented matrix

$$[A \;\vdots\; I].$$

Upon performing the elimination in accordance with Algorithm 6.1, we obtain an augmented matrix of the form

$$[U \;\vdots\; Y],$$

where U is an $n \times n$ matrix with $u_{ij} = 0$ whenever $i > j$ and Y represents the $n \times n$ matrix obtained by performing the same operations on the identity I that were performed to take A into U. At this point there is a choice between n applications of the backward-substitution algorithm or further reduction to

$$[I \;\vdots\; A^{-1}].$$

If straightforward backward substitution is used, $\frac{4}{3}n^3 - \frac{1}{3}n$ multiplications and divisions and $\frac{4}{3}n^3 - \frac{3}{2}n^2 + \frac{1}{6}n$ additions and subtractions are required (see Exercise 7(c)). It is also shown in that exercise that the further reduction, which is a special case of the Gauss–Jordan method (see Exercise 3, Section 6.2), requires $\frac{3}{2}n^3 - \frac{1}{2}n$ multiplications and divisions and $\frac{3}{2}n^3 - 2n^2 + \frac{1}{2}n$ additions and subtractions. The following table indicates the relative amount of computations for varying sizes of n.

	Gauss Elimination		Gauss–Jordan	
n	Mult./div.	Add./sub.	Mult./div.	Add./sub.
	$(4n^3 - n)/3$	$(8n^3 - 9n^2 + n)/6$	$(3n^3 - n)/2$	$(3n^3 - 4n^2 + n)/2$
3	35	23	39	24
10	1330	1185	1495	1305
50	166,650	162,925	187,475	182,525
100	1,333,300	1,318,350	1,499,950	1,480,050

In the actual implementation of either method, special care can be given to note the operations that need *not* be performed, as, for example, a multiplication when one of the multipliers is known to be unity, or a subtraction when the subtrahend is known to be zero. For both methods, the number of multiplications and divisions required can then be reduced to n^3 and the number of additions and subtractions reduced to $n^3 - 2n^2 + n$ (see Exercises 7(d) and 7(e)).

	Both methods, eliminating unnecessary computations	
n	Mult./div.	Add./sub.
3	27	12
10	1000	810
50	125,000	120,050
100	1,000,000	980,100

A fundamental concept of linear algebra that is very useful in determining the existence and uniqueness of solutions to linear systems is the **determinant** of an $n \times n$ matrix. The only approach to defining and computing the determinant given here will be the recursive definition. The determinant of a matrix A will be denoted by det A, but it is also common to use the notation $|A|$. A **submatrix** of a matrix A is a matrix "extracted" from A by deleting certain rows and/or columns of A.

Definition 6.11

a)   If $A = [a]$ is a 1×1 matrix, then det $A = a$.

b)   The **minor**, M_{ij}, is the determinant of the $(n - 1) \times (n - 1)$ submatrix of an $n \times n$ matrix A obtained by deleting the ith row and jth column.

c)   The **cofactor**, A_{ij}, associated with M_{ij}, is defined to be $A_{ij} = (-1)^{i+j} M_{ij}$.

d)   The **determinant** of an $n \times n$ matrix A, where $n > 1$, is given by either

(6.13)
$$\det A = \sum_{j=1}^{n} a_{ij} A_{ij} \qquad \text{for any } i = 1, 2, \ldots, n,$$

or

(6.14)
$$\det A = \sum_{i=1}^{n} a_{ij} A_{ij} \qquad \text{for any } j = 1, 2, \ldots, n.$$

It appears that there are $2n$ different definitions of det A depending upon which row or column is expanded. However, it can be shown that all definitions give the same numerical result. (See Noble and Daniels, [63], page 200.) For the case of 2×2 matrices, the following example shows the equivalence of these definitions.

EXAMPLE 7

a)   Let $A = \begin{bmatrix} a_{11} & a_{12} \\ a_{21} & a_{22} \end{bmatrix}$. Using Eq. (6.13) with $i = 1$,

$$\begin{aligned} \det A &= a_{11} A_{11} + a_{12} A_{12} \\ &= a_{11}[(-1)^{1+1} \det(a_{22})] + a_{12}[(-1)^{1+2} \det(a_{21})] \\ &= a_{22} a_{11} - a_{21} a_{12}. \end{aligned}$$

Using Eq. (6.13) with $i = 2$,

$$\det A = a_{21} A_{21} + a_{22} A_{22} = -a_{21} a_{12} + a_{22} a_{11}.$$

Using Eq. (6.14) with $j = 1$,

$$\det A = a_{11} A_{11} + a_{21} A_{21} = a_{11} a_{22} - a_{21} a_{12}.$$

Using Eq. (6.14) with $j = 2$,

$$\det A = a_{12} A_{12} + a_{22} A_{22} = -a_{12} a_{21} + a_{22} a_{11}.$$

In all cases,

$$\det A = \det \begin{bmatrix} a_{11} & a_{12} \\ a_{21} & a_{22} \end{bmatrix} = a_{11} a_{22} - a_{12} a_{21}.$$

b)   Let

$$A = \begin{bmatrix} 2 & 0 & 0 \\ 3 & -1 & 1 \\ 4 & 6 & -2 \end{bmatrix}.$$

Using Eq. (6.13) with $i = 3$,

$$\det A = a_{31}A_{31} + a_{32}A_{32} + a_{33}A_{33}$$

$$= 4A_{31} + 6A_{32} - 2A_{33}$$

$$= 4(-1)^{3+1} \det \begin{bmatrix} 0 & 0 \\ -1 & 1 \end{bmatrix} + 6(-1)^{3+2} \det \begin{bmatrix} 2 & 0 \\ 3 & 1 \end{bmatrix}$$

$$- 2(-1)^{3+3} \det \begin{bmatrix} 2 & 0 \\ 3 & -1 \end{bmatrix}$$

$$= -8.$$

Clearly, it is most convenient to compute $\det A$ across the row with the most zeros or down the column with the most zeros.

c)   Let

$$A = \begin{bmatrix} 2 & -1 & 3 & 0 \\ 4 & -2 & 7 & 0 \\ -3 & -4 & 1 & 5 \\ 6 & -6 & 8 & 0 \end{bmatrix}.$$

To compute $\det A$, it is easiest to use the fourth column, Eq. (6.14) with $j = 4$;

$$\det A = a_{14}A_{14} + a_{24}A_{24} + a_{34}A_{34} + a_{44}A_{44}$$

$$= 5A_{34}$$

$$= -5 \det \begin{bmatrix} 2 & -1 & 3 \\ 4 & -2 & 7 \\ 6 & -6 & 8 \end{bmatrix}$$

$$= -5 \left\{ 2 \det \begin{bmatrix} -2 & 7 \\ -6 & 8 \end{bmatrix} - (-1)\det \begin{bmatrix} 4 & 7 \\ 6 & 8 \end{bmatrix} + 3 \det \begin{bmatrix} 4 & -2 \\ 6 & -6 \end{bmatrix} \right\}$$

$$= -30. \qquad \square$$

The following properties of determinants are useful in relating linear systems and Gaussian elimination to determinants.

Theorem 6.12   Suppose A is an $n \times n$ matrix:

a)   If any row or column of A has only zero entries, then $\det A = 0$.
b)   If $\tilde{A}$ is obtained from A by the operation $(E_i) \leftrightarrow (E_j)$, with $i \neq j$, then $\det \tilde{A} = -\det A$.
c)   If A has two rows the same, then $\det A = 0$.
d)   If $\tilde{A}$ is obtained from A by the operation $(\lambda E_i) \rightarrow (E_i)$, then $\det \tilde{A} = \lambda \det A$.
e)   If $\tilde{A}$ is obtained from A by the operation $(E_i + \lambda E_j) \rightarrow (E_i)$, with $i \neq j$, then $\det \tilde{A} = \det A$.
f)   If B is also an $n \times n$ matrix, then $\det AB = \det A \det B$.

Proof

a)   Suppose the ith row of A is zero. Expanding along the ith row, we write:

$$\det A = \sum_{j=1}^{n} a_{ij} A_{ij} = \sum_{j=1}^{n} 0 \cdot A_{ij} = 0.$$

b)   The proof can be found in any standard linear algebra text. (See, for example, Noble and Daniels [63], page 201.) The case when n is three is considered in Exercise 5.

c)   Suppose the ith and jth rows are identical. Form $\tilde{A}$ by interchanging the ith and jth rows. By (b), $\det \tilde{A} = -\det A$. Since the operation $(E_i) \leftrightarrow (E_j)$ did not change A, in this case $\det \tilde{A} = \det A$, which implies $\det A = -\det A$ and $\det A = 0$.

d)   Expanding $\det \tilde{A}$ along the ith row gives

$$\det \tilde{A} = \sum_{k=1}^{n} \lambda a_{ik} A_{ik} = \lambda \sum_{k=1}^{n} a_{ik} A_{ik} = \lambda \det A.$$

e)   Expanding $\det \tilde{A}$ along the ith row gives

$$\det A = \sum_{k=1}^{n} (a_{ik} + \lambda a_{jk}) A_{ik} = \sum_{k=1}^{n} a_{ik} A_{ik} + \lambda \sum_{k=1}^{n} a_{jk} A_{ik} = \det A + \lambda \det B,$$

where the matrix B is exactly the same as A except that the ith row of B is the jth row of A. Thus, $\det B = 0$ by (c) and $\det \tilde{A} = \det A$.

f)   The proof of this result can be found in Noble and Daniels [63], page 204. $\square$

We now present the key that relates nonsingularity, Gaussian elimination, linear systems, and determinants. The proof of this theorem is not difficult but is laborious, and will not be presented here. (See Noble and Daniels [63].)

Theorem 6.13   The following statements are equivalent for any $n \times n$ matrix A:

a)   The equation $A\mathbf{x} = \mathbf{0}$ has the unique solution $\mathbf{x} = \mathbf{0}$.

b)   The linear system $A\mathbf{x} = \mathbf{b}$ has a unique solution for any n-dimensional column vector $\mathbf{b}$.

c)   The matrix A is nonsingular; that is, A^{-1} exists.

d) $\det A \neq 0$.

e)   Algorithm 6.1 (Gaussian elimination with row interchanges) can be performed on the linear system $A\mathbf{x} = \mathbf{b}$ for any n-dimensional column vector $\mathbf{b}$.

Exercise Set 6.3

1.   Determine which of the following matrices are nonsingular and compute their inverses;

a)
$$A = \begin{bmatrix} 4 & 2 & 6 \\ 3 & 0 & 7 \\ -2 & -1 & -3 \end{bmatrix}$$

b)
$$B = \begin{bmatrix} 2 & 3 & 1 & 2 \\ -2 & 4 & -1 & 5 \\ 3 & 7 & \frac{3}{2} & 1 \\ 6 & 9 & 3 & 7 \end{bmatrix}$$

c)
$$C = \begin{bmatrix} 4 & 0 & 0 & 0 \\ 6 & 7 & 0 & 0 \\ 9 & 11 & 1 & 0 \\ 5 & 4 & 1 & 1 \end{bmatrix}$$

2. Compute det A, det B, det AB, and det BA for

$$A = \begin{bmatrix} 4 & 6 & 1 & -1 \\ 2 & 1 & 0 & \frac{1}{2} \\ 3 & 0 & 0 & 1 \\ 1 & -1 & 1 & 1 \end{bmatrix} \quad \text{and} \quad B = \begin{bmatrix} 1 & 2 & 3 & 4 \\ 0 & 2 & -1 & 1 \\ 0 & 0 & 3 & 2 \\ 0 & 0 & 0 & -1 \end{bmatrix}.$$

3. Show that if A and B are $n \times n$ matrices and $AB = I$, then $B = A^{-1}$.

4. Prove Theorem 6.6.

5. Let A be a 3×3 matrix. Show that if $\tilde{A}$ is the matrix obtained from A using any of the operations

$$(E_1) \leftrightarrow (E_2), \qquad (E_1) \leftrightarrow (E_3), \qquad \text{or} \qquad (E_2) \leftrightarrow (E_3),$$

then det $\tilde{A} = -\det A$.

6. Verify parts (b), (c) and (d) of Theorem 6.9.

7. Suppose m linear systems

$$A\mathbf{x}^{(p)} = \mathbf{b}^{(p)}, \qquad p = 1, 2, \ldots, m,$$

are to be solved, each with the coefficient matrix A.

a) Show that Gaussian elimination applied to the augmented matrix

$$[A: \quad \mathbf{b}^{(1)}\mathbf{b}^{(2)} \cdots \mathbf{b}^{(m)}]$$

requires

$$\frac{n^3}{3} + mn^2 - \frac{n}{3} \quad \text{multiplications/divisions}$$

and

$$\frac{n^3}{3} + mn^2 - \frac{n^2}{2} - mn + \frac{n}{6} \quad \text{additions/subtractions.}$$

b) Show that the Gauss–Jordan method (See Exercise 3, Section 6.2) applied to the augmented matrix

$$[A: \quad \mathbf{b}^{(1)}\mathbf{b}^{(2)} \cdots \mathbf{b}^{(m)}]$$

requires

$$\tfrac{1}{2}n^3 + mn^2 - \tfrac{1}{2}n \quad \text{multiplications/divisions}$$

and

$$\tfrac{1}{2}n^3 + (m - 1)n^2 + (\tfrac{1}{2} - m)n \quad \text{additions/subtractions.}$$

c) For the special case $\mathbf{b}^{(p)} = \begin{bmatrix} 0 \\ \vdots \\ 0 \\ 1 \\ \vdots \\ 0 \end{bmatrix} \leftarrow p\text{th row}$

for each $p = 1, \ldots, m$, with $m = n$, the solution $\mathbf{x}^{(p)}$ is the pth column of A^{-1}. Show that Gaussian elimination requires

$$\tfrac{4}{3}n^3 - \frac{n}{3} \quad \text{multiplications/divisions}$$

and $\qquad\qquad \frac{4}{3}n^3 - \frac{3}{2}n^2 + \frac{1}{6}n$ additions/subtractions

for this application, and that the Gauss–Jordan method requires

$$\frac{3}{2}n^3 - \frac{1}{2}n \quad \text{multiplications/divisions}$$

and $\qquad\qquad \frac{3}{2}n^3 - 2n^2 + \frac{1}{2}n$ additions/subtractions.

    d) Construct an algorithm using Gaussian elimination to find A^{-1}, but do not perform multiplications when one of the multipliers is known to be unity, and do not perform additions/subtractions when one of the elements involved is known to be zero. Show that the required computations are reduced to n^3 multiplications/divisions and $n^3 - 2n^2 + n$ additions/subtractions.

    e) Repeat part (d), using the Gauss–Jordan method.

    f) Show that solving the linear system $A\mathbf{x} = \mathbf{b}$, when A^{-1} is known, still requires n^2 multiplications/divisions and $n^2 - n$ additions/subtractions.

    g) Show that solving m linear systems $A\mathbf{x}^{(p)} = \mathbf{b}^{(p)}$ for $p = 1, 2, \ldots, m$, by the method $\mathbf{x}^{(p)} = A^{-1}\mathbf{b}^{(p)}$ requires mn^2 multiplications and $m(n^2 - n)$ additions, if A^{-1} is known.

    h) Let A be an $n \times n$ matrix. Compare the number of operations required to solve m linear systems involving A by Gaussian elimination and by first inverting A and then multiplying $A\mathbf{x} = \mathbf{b}$ by A^{-1}, for $n = 3, 10, 50, 100$. Is it ever advantageous to compute A^{-1} for the purpose of solving linear systems?

8. Use the algorithm developed in Exercise 7 to find the inverses of the nonsingular matrices in Exercise 1.

9. It is often useful to partition matrices into a collection of submatrices by drawing vertical and horizontal lines. For example, the matrices

$$A = \begin{bmatrix} 1 & 2 & -1 \\ 3 & -4 & -3 \\ 6 & 5 & 0 \end{bmatrix} \quad \text{and} \quad B = \begin{bmatrix} 2 & -1 & 7 & 0 \\ 3 & 0 & 4 & 5 \\ -2 & 1 & -3 & 1 \end{bmatrix}$$

can be partitioned into

$$\left[\begin{array}{cc|c} 1 & 2 & -1 \\ 3 & -4 & -3 \\ \hline 6 & 5 & 0 \end{array}\right] = \left[\begin{array}{c|c} A_{11} & A_{12} \\ \hline A_{21} & A_{22} \end{array}\right]$$

and

$$\left[\begin{array}{ccc|c} 2 & -1 & 7 & 0 \\ 3 & 0 & 4 & 5 \\ \hline -2 & 1 & -3 & 1 \end{array}\right] = \left[\begin{array}{c|c} B_{11} & B_{12} \\ \hline B_{21} & B_{22} \end{array}\right]$$

    a) Show that the product of A and B in this case is

$$AB = \left[\begin{array}{c|c} A_{11}B_{11} + A_{12}B_{21} & A_{11}B_{12} + A_{12}B_{22} \\ \hline A_{21}B_{11} + A_{22}B_{21} & A_{21}B_{12} + A_{22}B_{22} \end{array}\right].$$

    b) If B were instead partitioned into

$$B = \left[\begin{array}{ccc|c} 2 & -1 & 7 & 0 \\ \hline 3 & 0 & 4 & 5 \\ -2 & 1 & -3 & 1 \end{array}\right] = \left[\begin{array}{c|c} B_{11} & B_{12} \\ \hline B_{21} & B_{22} \end{array}\right]$$

would the result in (a) hold?

    c) Make a conjecture concerning the conditions necessary for the result in (a) to hold in the general case.

10. The study of food chains is an important topic in the determination of the spread and accumulation of environmental pollutants in living matter. Suppose that a food chain has three links. The first link consists of vegetation of types $v_1, v_2, \ldots, v_n$, which provide all the food requirements for herbivores of species $h_1, h_2, \ldots, h_m$ in the second link. The third link consists of carnivorous animals $c_1, c_2, \ldots, c_k$, which depend entirely upon the herbivores in the second link for their food supply.

The coordinate a_{ij} of the matrix

$$A = \begin{bmatrix} a_{11} & a_{12} & \cdots & a_{1m} \\ a_{21} & a_{22} & \cdots & a_{2m} \\ \vdots & \vdots & & \vdots \\ a_{n1} & a_{n2} & \cdots & a_{nm} \end{bmatrix}$$

represents the total number of plants of type v_i eaten by the herbivores in the species h_j, while b_{ij} in

$$B = \begin{bmatrix} b_{11} & b_{12} & \cdots & b_{1k} \\ b_{21} & b_{22} & \cdots & b_{2k} \\ \vdots & \vdots & & \vdots \\ b_{m1} & b_{m2} & \cdots & b_{mk} \end{bmatrix}$$

describes the number of herbivores in species h_i which are devoured by the animals of type c_j.

a) Show that the number of plants of type v_i which eventually end up in the animals of species c_j would be given by the entry in the ith row and jth column of the matrix AB.
b) What physical significance is associated with the matrices A^{-1}, B^{-1}, and $(AB)^{-1} = B^{-1}A^{-1}$?

11. In a paper entitled "Population Waves," Bernadelli [11] (see also Searle [78]) hypothesizes a type of simplified beetle, which has a natural life span of three years. The female of this species has a survival rate of $\frac{1}{2}$ in the first year of life, a survival rate of $\frac{1}{3}$ from the second to third years, and gives birth to an average of six new females before expiring at the end of the third year. A matrix can be used to show the contribution an individual female beetle makes, in a probabilistic sense, to the female population of the species, by letting a_{ij} in the matrix $A = (a_{ij})$, denote the contribution that a single female beetle of age j will make to the next years' female population of age i; that is,

$$A = \begin{bmatrix} 0 & 0 & 6 \\ \frac{1}{2} & 0 & 0 \\ 0 & \frac{1}{3} & 0 \end{bmatrix}.$$

a) The contribution that a female beetle will make to the population two years hence could be determined from the entries of A^2, of three years hence from A^3, and so on. Construct A^2 and A^3, and try to make a general statement about the contribution of a female beetle to the population in n years' time for any positive integral value of n.
b) Use your conclusions from (a) to describe what will occur in future years to a population of these beetles that initially consists of 6000 female beetles in each of the three age groups.
c) Construct A^{-1} and describe its significance regarding the population of this species.

12. The solution by **Cramer's rule** to the linear system

$$a_{11}x_1 + a_{12}x_2 + a_{13}x_3 = b_1,$$
$$a_{21}x_1 + a_{22}x_2 + a_{23}x_3 = b_2,$$
$$a_{31}x_1 + a_{32}x_2 + a_{33}x_3 = b_3,$$

has

$$x_1 = \frac{1}{D} \det \begin{bmatrix} b_1 & a_{12} & a_{13} \\ b_2 & a_{22} & a_{23} \\ b_3 & a_{32} & a_{33} \end{bmatrix} \equiv \frac{D_1}{D},$$

$$x_2 = \frac{1}{D} \det \begin{bmatrix} a_{11} & b_1 & a_{13} \\ a_{21} & b_2 & a_{23} \\ a_{31} & b_3 & a_{33} \end{bmatrix} \equiv \frac{D_2}{D},$$

and

$$x_3 = \frac{1}{D} \det \begin{bmatrix} a_{11} & a_{12} & b_1 \\ a_{21} & a_{22} & b_2 \\ a_{31} & a_{32} & b_3 \end{bmatrix} \equiv \frac{D_3}{D},$$

where

$$D = \det \begin{bmatrix} a_{11} & a_{12} & a_{13} \\ a_{21} & a_{22} & a_{23} \\ a_{31} & a_{32} & a_{33} \end{bmatrix}.$$

a) Find the solution to the linear system

$$2x_1 + 3x_2 - x_3 = 4,$$
$$x_1 - 2x_2 + x_3 = 6,$$
$$x_1 - 12x_2 + 5x_3 = 10,$$

by Cramer's Rule.

b) Show that the linear system

$$2x_1 + 3x_2 - x_3 = 4,$$
$$x_1 - 2x_2 + x_3 = 6,$$
$$- x_1 - 12x_2 + 5x_3 = 9.$$

does not have a solution. Compute D_1, D_2, and D_3.

c) Show that the linear system

$$2x_1 + 3x_2 - x_3 = 4,$$
$$x_1 - 2x_2 + x_3 = 6,$$
$$-x_1 - 12x_2 + 5x_3 = 10,$$

has an infinite number of solutions. Compute D_1, D_2, and D_3.

d) Prove that if a 3×3 linear system with $D = 0$ is to have solutions, then $D_1 = D_2 = D_3 = 0$.

6.4 Pivoting Strategies

During the derivation of Algorithm 6.1, it was found that obtaining a zero for a pivot element $a_{k,k}^{(k)}$ necessitated a row interchange of the form $(E_k) \leftrightarrow (E_p)$ where $k + 1 \leq p \leq n$ was the smallest integer with $a_{p,k}^{(k)} \neq 0$. In practice it is often desirable to perform row interchanges involving the pivot elements even when they are not zero. The reason for this is that when the calculations are performed using finite-digit arithmetic, as would be the case for calculator or computer-generated solutions, a pivot element that is small compared to the entries below it in the same column can lead to substantial roundoff error. An illustration of this difficulty is given in the following example.

EXAMPLE 1 The linear system

$$E_1: \quad .003000x_1 + 59.14x_2 = 59.17,$$

$$E_2: \quad 5.291x_1 - 6.130x_2 = 46.78,$$

has the exact solution $x_1 = 10.00$ and $x_2 = 1.000$.

To illustrate the difficulties of roundoff error, Gaussian elimination will be performed on this system, using four-digit arithmetic with rounding.

The first pivot element is $a_{11}^{(1)} = .003000$, and its associated multiplier is

$$m_{21} = \frac{5.291}{.003000} = 1763.6\bar{6},$$

which rounds to 1764. Performing the operation $(E_2 - m_{21}E_1) \rightarrow (E_2)$ and the appropriate rounding, we get the new system

$$.003000x_1 + 59.14x_2 = 59.17$$

$$-104300x_2 = -104400.$$

Backward substitution implies

$$x_2 = 1.001 \quad \text{and} \quad x_1 = \frac{59.17 - (59.14)(1.001)}{.003000} = -10.00.$$

The large error in the numerical solution for x_1 resulted from the small error of .001 in solving for x_2. This error was magnified by a factor of 20,000 in the solution of x_1 because of the order in which the calculations were performed. □

The example certainly illustrates that difficulties can arise in some cases when the pivot element $a_{k,k}^{(k)}$ is small relative to the entries $a_{i,j}^{(k)}$ for $k \le i \le n$ and $k \le j \le n$. Pivoting strategies in general are accomplished by selecting a new element for the pivot $a_{p,q}^{(k)}$ and then interchanging the kth and pth rows, followed by the interchange of the kth and qth columns, if necessary. Perhaps the simplest such strategy is to select the element in the same column that is below the diagonal and has the largest absolute value; that is, determine p such that

$$|a_{p,k}^{(k)}| = \max_{k \le i \le n} |a_{i,k}^{(k)}|,$$

and perform $(E_k) \leftrightarrow (E_p)$. In this case no column interchange is considered.

EXAMPLE 2 Reconsidering the system

$$E_1: \quad .003000x_1 + 59.14x_2 = 59.17,$$

$$E_2: \quad 5.291x_1 - 6.130x_2 = 46.78,$$

and using the column pivoting procedure described above results in first finding

$$\max\{|a_{11}^{(1)}|, |a_{21}^{(1)}|\} = \max\{|.003000|, |5.291|\} = |5.291| = |a_{21}^{(1)}|.$$

The operation $(E_2) \leftrightarrow (E_1)$ is performed to give the system

$$E_1: \quad 5.291x_1 - 6.130x_2 = 46.78,$$
$$E_2: \quad .003000x_1 + 59.14x_2 = 59.17.$$

The multiplier for this system is

$$m_{21} = \frac{a_{21}^{(1)}}{a_{11}^{(1)}} = .0005670,$$

and the operation $(E_2 - m_{21}E_1) \to (E_2)$ reduces the system to

$$5.291x_1 - 6.130x_2 = 46.78,$$
$$59.14x_2 = 59.14.$$

The four-digit answers resulting from the backward substitution are the correct values $x_1 = 10.00$ and $x_2 = 1.000$. $\square$

This technique is known as **maximal column pivoting** or **partial pivoting**, and is described in detail below. The actual row interchanging is simulated in the algorithm by interchanging the content of $NROW$ in Step 5.

Gaussian Elimination with
Maximal Column Pivoting Algorithm 6.2

To solve the $n \times n$ linear system

$$E_1: \quad a_{11}x_1 + a_{12}x_2 + \cdots + a_{1n}x_n = a_{1,n+1}$$
$$E_2: \quad a_{21}x_1 + a_{22}x_2 + \cdots + a_{2n}x_n = a_{2,n+1}$$
$$\vdots \qquad \vdots \qquad \vdots \qquad\qquad \vdots \qquad\quad \vdots$$
$$E_n: \quad a_{n1}x_1 + a_{n2}x_2 + \cdots + a_{nn}x_n = a_{n,n+1}:$$

INPUT number of unknowns and equations n; augmented matrix $A = (a_{ij})$ where $1 \le i \le n$ and $1 \le j \le n + 1$.

OUTPUT solution $x_1, \ldots, x_n$ or message that linear system has no unique solution.

Step 1 For $i = 1, \ldots, n$ set $NROW(i) = i$.   (*Initialize row pointer.*)

Step 2 For $i = 1, \ldots, n - 1$ do Steps 3–6.   (*Elimination process.*)

Step 3 Let p be the smallest integer with $i \le p \le n$ and
$$|a(NROW(p), i)| = \max_{i \le j \le n} |a(NROW(j), i)|.$$

(*Notation*: $a(NROW(i), j) \equiv a_{NROW_i, j}$.)

Step 4 If $a(NROW(p), i) = 0$ then OUTPUT (no unique solution exists');
STOP.

Step 5 If $NROW(i) \neq NROW(p)$ then set $NCOPY = NROW(i)$;
$$NROW(i) = NROW(p);$$
$$NROW(p) = NCOPY.$$
(*Simulated row interchange.*)

Step 6 For $j = i + 1, \ldots, n$ do Steps 7 and 8.

   Step 7 Set $m(NROW(j), i) = a(NROW(j), i)/a(NROW(i), i)$.

   Step 8 Perform $(E_{NROW(j)} - m(NROW(j), i)E_{NROW(i)}) \rightarrow (E_{NROW(j)})$.

Step 9 If $a(NROW(n), n) = 0$ then OUTPUT ('no unique solution exists');
STOP.

Step 10 Set $x_n = a(NROW(n), n + 1)/a(NROW(n), n)$. (*Start backward substitution.*)

Step 11 For $i = n - 1, \ldots, 1$

$$\text{set } x_i = \frac{a(NROW(i), n + 1) - \sum_{j=i+1}^{n} a(NROW(i), j)x_j}{a(NROW(i), i)}.$$

Step 12 OUTPUT $(x_1, \ldots, x_n)$;
(*Procedure completed successfully.*)
STOP.

The procedures detailed in this algorithm are sufficient to guarantee that each multiplier m_{ji} has magnitude not exceeding one. While the maximal column pivoting strategy is sufficient for most linear systems, situations do arise for which this strategy is inadequate.

EXAMPLE 3 The linear system

$$E_1: \quad 30.00x_1 + 591,400x_2 = 591,700,$$

$$E_2: \quad 5.291x_1 - \quad 6.130x_2 = 46.78,$$

is the same system as that presented in the previous example except that all entries in the first equation are multiplied by 10^4. The procedure described in Algorithm 6.2 with 4-digit arithmetic would lead to the same results as were obtained in Example 1, since the maximal value in the first column is 30.00 and the multiplier

$$m_{21} = \frac{5.291}{30.00} = .1764$$

leads to the system

$$30.00x_1 + 591,400x_2 = 591,700,$$

$$-104,300x_2 = -104,400,$$

which has solutions $x_2 = 1.001$ and $x_1 = -10.00$. □

A pivoting procedure that will handle this situation and many of the other systems for which problems occur when Algorithm 6.2 is applied, is known as **scaled column pivoting**. Instead of simply choosing the largest element below the diagonal to use for the pivot, the scaled column pivoting procedure first scales the values relative to the largest entry in the respective row, and subsequently chooses as the pivot element the one with the largest scaled value. It can be seen that this procedure applied to Example 3 results in scaled values of

$$\frac{30.00}{591,400} = .00005073 \quad \text{and} \quad \frac{5.291}{6.130} = .8631.$$

Using the pivoting strategy, E_1 and E_2 are interchanged and the results are the same as those in Example 2; that is, $x_1 = 10.00$ and $x_2 = 1.000$, the correct values.

Gaussian Elimination with Scaled Column Pivoting Algorithm 6.3

The only steps in this algorithm that differ from those of Algorithm 6.2 are:

Step 1 For $i = 1, \ldots, n$ set $s_i = \max_{1 \le j \le n} |a_{ij}|$;

if $s_i = 0$ then OUTPUT ('no unique solution exists');
STOP.

set $NROW(i) = i$.

Step 2 For $i = 1, \ldots, n - 1$ do Steps 3–6.   (*Elimination process.*)

Step 3 Let p be the smallest integer with $i \le p \le n$ and

$$\frac{|a(NROW(p), i)|}{s(NROW(p))} = \max_{i \le j \le n} \frac{|a(NROW(j), i)|}{s(NROW(j))}.$$

The technique in Algorithm 6.3 can be viewed as a type of **equilibration**, a term used to refer to the process of changing to one the largest entry in absolute value in each row or column. The row equilibration in Algorithm 6.3 can also be achieved by multiplying the matrix equation $A\mathbf{x} = \mathbf{b}$ by the diagonal matrix D^{-1}, whose ith diagonal entry is $(s_i)^{-1}$. The coefficient matrix in the resulting equation $D^{-1}A\mathbf{x} = D^{-1}\mathbf{b}$ is row equilibrated.

The most complete pivoting strategy, called **maximal** (or **total**) **pivoting**, involves selecting the pivot element $a_{pq}^{(k)}$ to be $\max_{k \le i, j \le n} |a_{i,j}^{(k)}|$, which is accomplished by both row and column interchanges. It is useful to incorporate a column as well as row equilibration technique into maximal pivoting. To do this, define

$$r_i = \max_{1 \le j \le n} \left| \frac{a_{ji}}{s_j} \right|$$

for each $i = 1, 2, \ldots, n$, and let C be the diagonal matrix whose ith diagonal entry is r_i. Let $\mathbf{y} = C\mathbf{x}$ and solve the matrix equation $D^{-1}AC^{-1}\mathbf{y} = D^{-1}\mathbf{b}$ for $\mathbf{y}$ using maximal pivoting. The solution $\mathbf{x}$ is then found from $\mathbf{x} = C^{-1}\mathbf{y}$. Because of the number of comparisons and operations necessary, this strategy is recommended for only the most stubborn systems where the extensive amount of execution time needed for this method can be justified. For problems involving maximal pivoting, see Exercises 5–8.

Exercise Set 6.4

1. Apply Algorithms 6.1, 6.2, and 6.3 to the following linear systems using four-digit rounding arithmetic. Which algorithm yields the most accurate results?

   a) $58.09x_1 + 1.003x_2 = 68.12,$
       $321.8x_1 + 5.550x_2 = 377.3;$

   b) $1.003x_1 + 58.09x_2 = 68.12,$
       $5.550x_1 + 321.8x_2 = 377.3;$

   c) $1.003x_1 + 58.09x_2 = 68.12,$
       $321.8x_1 + 5.550x_2 = 377.3.$

2. Use Algorithms 6.1, 6.2, and 6.3 to solve the following linear systems using three-digit rounding arithmetic. Which algorithm yields the most accurate results?

   a) $.03x_1 + 58.9x_2 = 59.2,$
       $5.31x_1 - 6.10x_2 = 47.0;$

   b) $58.9x_1 + .03x_2 = 59.2,$
       $-6.10x_1 + 5.31x_2 = 47.0;$

   c) $x_1 + \frac{1}{2}x_2 + \frac{1}{3}x_3 = \frac{11}{16},$
       $5x_1 + \frac{10}{3}x_2 + \frac{5}{2}x_3 = \frac{65}{6},$
       $\frac{100}{3}x_1 + 25x_2 + 20x_3 = \frac{235}{3};$

   d) $.832x_1 + .448x_2 + .193x_3 = 1.00,$
       $.784x_1 + .421x_2 - .207x_3 = 0.00,$
       $.784x_1 - .421x_2 + .279x_3 = 0.00.$

3. Repeat Exercise 2 using three-digit chopping arithmetic.

4. Apply Algorithms 6.1, 6.2, and 6.3 to the following linear system using five-digit rounding arithmetic. Which algorithm yields the most accurate result?

$$3.3330x_1 + 15920x_2 - 10.333x_3 = 15913$$

$$2.2220x_1 + 16.710x_2 + 9.6120x_3 = 28.544$$

$$1.5611x_1 + 5.1791x_2 + 1.6852x_3 = 8.4254$$

5. Construct an algorithm for the maximal pivoting procedure discussed after Algorithm 6.3. Do not include equilibration.

6. Repeat Exercise 5 incorporating equilibration.

7. Use the maximal pivoting algorithm developed in Exercise 5 to obtain solutions to Exercises 1, 2, 3, and 4. Compare the answers obtained with the results of those exercises.

8. Use the equilibrated maximal pivoting algorithm developed in Exercise 6 to obtain solutions to Exercises 1, 2, 3, and 4. Compare the answers obtained with the results of those exercises.

9. Algorithm 6.3 can be changed to incorporate the matrix multiplications $D^{-1}A$ with maximal column pivoting instead of choosing the pivot based upon $\max_{i \le j \le n}|a_{ji}/s_j|$. Which approach requires the most multiplications/divisions?

6.5 Special Types of Matrices

In this section some additional material on matrices is presented, which will be useful in the succeeding chapters. The first type of matrix we will consider is produced naturally when Gaussian elimination is performed on a linear system.

Definition 6.14   An **upper-triangular** $n \times n$ matrix U has for each j entries

$$u_{ij} = 0 \qquad \text{for each } i = j + 1, j + 2, \ldots, n;$$

and a **lower-triangular** matrix L has for each j entries

$$l_{ij} = 0 \qquad \text{for each } i = 1, 2, \ldots, j - 1.$$

A **diagonal** matrix D is a matrix that is both upper and lower triangular; that is, the only nonzero entries are on the diagonal.

EXAMPLE 1   Reconsider Examples 1 and 3 of Section 6.1, in which the linear system

$$
\begin{aligned}
x_1 + \; x_2 \qquad\quad + 3x_4 &= \;\; 4, \\
2x_1 + \; x_2 - \; x_3 + \; x_4 &= \;\; 1, \\
3x_1 - \; x_2 - \; x_3 + 2x_4 &= -3, \\
-x_1 + 2x_2 + 3x_3 - \; x_4 &= \;\; 4.
\end{aligned}
$$

was reduced to the equivalent system

$$
\left[
\begin{array}{cccc:c}
1 & 1 & 0 & 3 & 4 \\
0 & -1 & -1 & -5 & -7 \\
0 & 0 & 3 & 13 & 13 \\
0 & 0 & 0 & -13 & -13
\end{array}
\right].
$$

Let U be the 4×4 upper-triangular matrix

$$
U = \begin{bmatrix}
1 & 1 & 0 & 3 \\
0 & -1 & -1 & -5 \\
0 & 0 & 3 & 13 \\
0 & 0 & 0 & -13
\end{bmatrix},
$$

which is the result of Gaussian elimination performed on A. For $i = 1, 2, 3$, define m_{ji} for each $j = i + 1, i + 2, \ldots, 4$, to be the number used in the elimination step $(E_j - m_{ji}E_i) \rightarrow E_j$; that is, $m_{21} = 2, m_{31} = 3, m_{41} = -1, m_{32} = 4, m_{42} = -3$. and $m_{43} = 0$. If L is defined to be the 4×4 lower-triangular matrix with entries l_{ij} given by

$$
l_{ij} = \begin{cases}
0, & \text{when } i = 1, 2, \ldots, j - 1, \\
1, & \text{when } i = j, \\
m_{ij}, & \text{when } i = j + 1, j + 2, \ldots, n,
\end{cases}
$$

then
$$
L = \begin{bmatrix}
1 & 0 & 0 & 0 \\
2 & 1 & 0 & 0 \\
3 & 4 & 1 & 0 \\
-1 & -3 & 0 & 1
\end{bmatrix};
$$

and it is easy to verify that

$$
LU = \begin{bmatrix} 1 & 0 & 0 & 0 \\ 2 & 1 & 0 & 0 \\ 3 & 4 & 1 & 0 \\ -1 & -3 & 0 & 1 \end{bmatrix} \begin{bmatrix} 1 & 1 & 0 & 3 \\ 0 & -1 & -1 & -5 \\ 0 & 0 & 3 & 13 \\ 0 & 0 & 0 & -13 \end{bmatrix}
$$

$$
= \begin{bmatrix} 1 & 1 & 0 & 3 \\ 2 & 1 & -1 & 1 \\ 3 & -1 & -1 & 2 \\ -1 & 2 & 3 & -1 \end{bmatrix} = A.
$$

□

The results obtained in Example 1 are true in general and are given in the following theorem.

Theorem 6.15 If the Gaussian elimination procedure (Algorithm 6.1) can be performed on the system $A\mathbf{x} = \mathbf{b}$ without row interchanges, then the matrix A can be factored into the product of a lower-triangular matrix with an upper-triangular matrix

$$
A = LU,
$$

where $U = (u_{ij})$ and $L = (l_{ij})$ are defined for each j by:

$$
u_{ij} = \begin{cases} a_{ij}^{(i)}, & \text{when } i = 1, 2, \ldots, j, \\ 0, & \text{when } i = j + 1, j + 2, \ldots, n, \end{cases}
$$

and

$$
l_{ij} = \begin{cases} 0, & \text{when } i = 1, 2, \ldots, j - 1, \\ 1, & \text{when } i = j, \\ m_{ij}, & \text{when } i = j + 1, j + 2, \ldots, n, \end{cases}
$$

where $a_{ij}^{(i)}$ is the i, j-element of $A^{(n)}$, the final matrix obtained in the Gaussian elimination method, and m_{ij} is the multiplier.

Proof See Exercise 9.   □

If row interchanges must be performed in order for the procedure to work, then A can be factored into LU, where U is the same as in Theorem 6.15, but in general, L is not lower-triangular. This topic is further elaborated upon in the exercises (in particular, see Exercises 10 and 11).

To evaluate the determinant of an arbitrary matrix can require considerable manipulation. A matrix in triangular form, however, has an easily calculated determinant.

Theorem 6.16 If $A = (a_{ij})$ is an $n \times n$ matrix that is either upper triangular, lower triangular, or diagonal, then $\det A = \prod_{i=1}^{n} a_{ii}$.

The proof of this result consists of expanding the matrix and each submatrix about either the first row or first column and will be left to the reader.

The problem of computing the determinant of a matrix can be simplified by first reducing the matrix to triangular form and then using Theorem 6.16 to find the determinant of the triangular matrix.

EXAMPLE 2   Let

$$A = \begin{bmatrix} 1 & 1 & 0 & 3 \\ 2 & 1 & -1 & 1 \\ -1 & 2 & 3 & -1 \\ 3 & -1 & -1 & 2 \end{bmatrix}.$$

By the three permitted types of operation, the matrix A will be reduced to an upper-triangular matrix. First perform $(E_2 - 2E_1) \to (E_2)$, $(E_3 + E_1) \to (E_3)$, and $(E_4 - 3E_1) \to (E_4)$, to obtain:

$$\tilde{A}_1 = \begin{bmatrix} 1 & 1 & 0 & 3 \\ 0 & -1 & -1 & -5 \\ 0 & 3 & 3 & 2 \\ 0 & -4 & -1 & -7 \end{bmatrix}.$$

By Theorem 6.12(e), $\det A = \det \tilde{A}_1$. Forming $\tilde{A}_2$ from $\tilde{A}_1$ by the operations $(E_3 + 3E_2) \to (E_3)$ and $(E_4 - 4E_2) \to (E_4)$,

$$\tilde{A}_2 = \begin{bmatrix} 1 & 1 & 0 & 3 \\ 0 & -1 & -1 & -5 \\ 0 & 0 & 0 & -13 \\ 0 & 0 & 3 & 13 \end{bmatrix},$$

and, again, $\det \tilde{A}_2 = \det \tilde{A}_1 = \det A$. Let $\tilde{A}_3$ be formed from $\tilde{A}_2$ by $(E_3) \leftrightarrow (E_4)$, thus:

$$\tilde{A}_3 = \begin{bmatrix} 1 & 1 & 0 & 3 \\ 0 & -1 & -1 & -5 \\ 0 & 0 & 3 & 13 \\ 0 & 0 & 0 & -13 \end{bmatrix}.$$

By Theorem 6.16, $\det \tilde{A}_3 = (1)(-1)(3)(-13) = 39$, and since $\tilde{A}_3$ was formed from $\tilde{A}_2$ by a row interchange,

$$\det A = \det \tilde{A}_2 = -\det \tilde{A}_3 = -39. \qquad \square$$

Definition 6.17   The **transpose** of an $m \times n$ matrix A, denoted A^t, is an $n \times m$ matrix whose entries are $(A^t)_{ij} = (A)_{ji}$. A matrix whose transpose is itself is said to be **symmetric**.

EXAMPLE 3   The matrices

$$A = \begin{bmatrix} 7 & 2 & 0 \\ 3 & 5 & -1 \\ 0 & 5 & -6 \end{bmatrix}, \qquad B = \begin{bmatrix} 2 & 4 & 7 \\ 3 & -5 & -1 \end{bmatrix}, \qquad C = \begin{bmatrix} 6 & 4 & 3 \\ 4 & -2 & 0 \\ 3 & 0 & 1 \end{bmatrix}$$

have transposes

$$
A^t = \begin{bmatrix} 7 & 3 & 0 \\ 2 & 5 & 5 \\ 0 & -1 & -6 \end{bmatrix}, \quad
B^t = \begin{bmatrix} 2 & 3 \\ 4 & -5 \\ 7 & -1 \end{bmatrix}, \quad
C^t = \begin{bmatrix} 6 & 4 & 3 \\ 4 & -2 & 0 \\ 3 & 0 & 1 \end{bmatrix}.
$$

Since C^t and C have the same entries, C is symmetric. $\square$

Theorem 6.18 The following operations involving the transpose of a matrix
hold whenever the operation is possible:

1. $(A^t)^t = A$,
2. $(A + B)^t = A^t + B^t$,
3. $(AB)^t = B^t A^t$,
4. If A^{-1} exists, $(A^{-1})^t = (A^t)^{-1}$,
5. $\det A^t = \det A$.

The proof of these statements is elementary and is left to the exercises.

Definition 6.19 An $n \times n$ matrix is called a **band matrix** if integers p and q,
$1 < p, q < n$, exist, with the property that $a_{ij} = 0$ whenever $i + p \leq j$ or
$j + q \leq i$. The **band width** for a matrix of this type is defined to be $w = p + q - 1$.

The matrix A defined in Example 3 is a band matrix with $p = q = 2$ and band
width 3.

The definition of band matrix forces those matrices to concentrate all their
nonzero entries about the diagonal. Two special cases of band matrices that occur
often in practice have $p = q = 2$ and $p = q = 4$. The matrix of band width 3,
occurring when $p = q = 2$, has already been encountered in connection with
the study of cubic spline approximations in Section 3.6. These matrices are often
called **tridiagonal** since they have the form

$$
A = \begin{bmatrix}
a_{11} & a_{12} & 0 & \cdots\cdots\cdots\cdots & 0 \\
a_{21} & a_{22} & a_{23} & & \vdots \\
0 & a_{32} & a_{33} & a_{34} & 0 \\
\vdots & & & & a_{n-1,n} \\
0 & \cdots\cdots\cdots & 0 & a_{n,n-1} & a_{nn}
\end{bmatrix}.
$$

Matrices of this type will be discussed again in Chapter 10 in connection with the
study of piecewise linear approximations to boundary-value problems. The case of
$p = q = 4$ will also be used for the solution of boundary-value problems, when the
approximating functions assume the form of cubic splines.

Another useful class of matrices is described in the following definition.

Definition 6.20   The $n \times n$ matrix A is said to be **strictly diagonally dominant** in case

(6.15)
$$|a_{ii}| > \sum_{\substack{j=1, \\ j \neq i}}^{n} |a_{ij}|$$

holds for each $i = 1, 2, \ldots, n$.

The matrix A described in Example 3 can be seen to be strictly diagonally dominant since $|7| > |2| + |0|$, $|5| > |3| + |-1|$, and $|-6| > |0| + |5|$. It is interesting to note that A^t is not strictly diagonally dominant, however.

Theorem 6.21   If A is a strictly diagonally dominant $n \times n$ matrix, then A is nonsingular. Moreover, Gaussian elimination can be performed on any linear system of the form $A\mathbf{x} = \mathbf{b}$ to obtain its unique solution without row or column interchanges, and the computations are stable with respect to the growth of rounding errors.

The proof that Gaussian elimination can be performed without interchanges is left to the reader (as Exercise 13). The demonstration of stability for this procedure can be found in Wendroff [89].

To show that A is nonsingular, consider the linear system described by $A\mathbf{x} = \mathbf{0}$, and suppose that a nonzero solution $\mathbf{x} = (x_i)$ to this system exists. In this case, for some k, $0 < |x_k| = \max_{1 \leq j \leq n} |x_j|$. Since $\sum_{j=1}^{n} a_{ij} x_j = 0$ for each $i = 1, 2, \ldots, n$, when $i = k$

$$a_{kk} x_k = - \sum_{\substack{j=1, \\ j \neq k}}^{n} a_{kj} x_j.$$

This implies

$$|a_{kk}||x_k| \leq \sum_{\substack{j=1, \\ j \neq k}}^{n} |a_{kj}||x_j| \quad \text{or} \quad |a_{kk}| \leq \sum_{\substack{j=1, \\ j \neq k}}^{n} |a_{kj}| \frac{|x_j|}{|x_k|} \leq \sum_{\substack{j=1, \\ j \neq k}}^{n} |a_{kj}|,$$

in contradiction to the strict diagonal dominance of A. Consequently, the only solution to $A\mathbf{x} = \mathbf{0}$ is $\mathbf{x} = \mathbf{0}$, a condition shown in Theorem 6.13 to be equivalent to the nonsingularity of A.

Note that Theorem 6.21 can be used to give elementary proofs of Theorem 3.10 and of Theorem 3.11 of Section 3.6 when applied to the matrices given in those theorems.

The final special class of matrices to be discussed in this section is called **positive definite**.

Definition 6.22   A symmetric $n \times n$ matrix A is called **positive definite** if $\mathbf{x}^t A \mathbf{x} > 0$ for every n-dimensional column vector $\mathbf{x} \neq \mathbf{0}$.

To be precise, the definition should specify that the 1×1 matrix generated by the operation $\mathbf{x}^t A \mathbf{x}$ has a positive value for its only entry, since the operation is performed as follows:

$$\mathbf{x}^t A \mathbf{x} = [x_1, x_2, \ldots, x_n] \begin{bmatrix} a_{11} & a_{12} & \cdots & a_{1n} \\ a_{21} & a_{22} & \cdots & a_{2n} \\ \vdots & \vdots & & \vdots \\ a_{n1} & a_{n2} & \cdots & a_{nn} \end{bmatrix} \begin{bmatrix} x_1 \\ x_2 \\ \vdots \\ x_n \end{bmatrix}$$

$$= [x_1, x_2, \ldots, x_n] \begin{bmatrix} \sum_{j=1}^n a_{1j} x_j \\ \sum_{j=1}^n a_{2j} x_j \\ \vdots \\ \sum_{j=1}^n a_{nj} x_j \end{bmatrix} = \left[\sum_{i=1}^n \sum_{j=1}^n a_{ij} x_i x_j \right].$$

EXAMPLE 4   The matrix

$$A = \begin{bmatrix} 2 & -1 & 0 \\ -1 & 2 & -1 \\ 0 & -1 & 2 \end{bmatrix}$$

is positive definite, for suppose $\mathbf{x}$ is any three-dimensional column vector, then

$$\mathbf{x}^t A \mathbf{x} = [x_1, x_2, x_3] \begin{bmatrix} 2 & -1 & 0 \\ -1 & 2 & -1 \\ 0 & -1 & 2 \end{bmatrix} \begin{bmatrix} x_1 \\ x_2 \\ x_3 \end{bmatrix}$$

$$= [x_1, x_2, x_3] \begin{bmatrix} 2x_1 - x_2 \\ -x_1 + 2x_2 - x_3 \\ -x_2 + 2x_3 \end{bmatrix}$$

$$= [2x_1^2 - 2x_1 x_2 + 2x_2^2 - 2x_2 x_3 + 2x_3^2]$$

$$= [x_1^2 + (x_1 - x_2)^2 + (x_2 - x_3)^2 + x_3^2]$$

and

$$x_1^2 + (x_1 - x_2)^2 + (x_2 - x_3)^2 + x_3^2 > 0,$$

unless $x_1 = x_2 = x_3 = 0$.   □

It should be clear from the example that using the definition to determine whether a matrix is positive definite can be extremely tedious. Fortunately, there are more easily verified criteria, which will be presented in Chapter 8, for identifying members of this important class. Conditions that can be used to eliminate certain matrices from consideration are discussed in Exercise 7.

The next theorem parallels Theorem 6.21, which dealt with strictly diagonally dominant matrices.

Theorem 6.23   If A is a positive definite $n \times n$ matrix, then A is nonsingular. Moreover, Gaussian elimination can be performed on any linear system of the form $A\mathbf{x} = \mathbf{b}$ to obtain its unique solution without row or column interchanges, and the computations are stable with respect to the growth of rounding errors.

Proof   If $\mathbf{x} \neq \mathbf{0}$ is a vector that satisfies $A\mathbf{x} = \mathbf{0}$, then $\mathbf{x}^t A\mathbf{x} = 0$. This contradicts the assumption that A is positive definite. Consequently, $A\mathbf{x} = \mathbf{0}$ has only the zero solution and Theorem 6.13 implies that A is nonsingular.

To show the validity of the initial statement concerning Gaussian elimination, let $\hat{A}$ denote a leading principal submatrix of A for an arbitrary k, $k = 1, 2, \ldots, n$ (the definition of leading principal submatrix is given in Exercise 12). If $\hat{A}$ is not positive definite, then a k-dimensional column vector $\hat{\mathbf{x}} \neq \mathbf{0}$ exists with $\hat{\mathbf{x}}^t \hat{A} \hat{\mathbf{x}} \leq \mathbf{0}$. Construct an n-dimensional column vector $\mathbf{x} \neq \mathbf{0}$ from $\hat{\mathbf{x}}$ by placing zeros in the last $(n - k)$ coordinates. Since

$$\mathbf{x}^t A\mathbf{x} = \hat{\mathbf{x}}^t \hat{A} \hat{\mathbf{x}} \leq 0,$$

A is not positive definite and we have a contradiction. Consequently, every principal submatrix of A is positive definite and hence nonsingular, and the result follows from Exercise 12.

The proof of stability of the Gaussian elimination procedure can be found in Wendroff [89], beginning on page 120. □

Exercise Set 6.5

1. Determine which of the following matrices are singular and which are positive definite.

   a)
   $$A = \begin{bmatrix} 4 & 2 & 6 \\ 3 & 0 & 7 \\ -2 & -1 & -3 \end{bmatrix}$$

   b)
   $$B = \begin{bmatrix} 2 & 3 & 1 & 2 \\ -2 & 4 & -1 & 5 \\ 3 & 7 & 1.5 & 1 \\ 6 & 9 & 3 & 7 \end{bmatrix}$$

   c)
   $$C = \begin{bmatrix} 4 & 0 & 0 & 0 \\ 6 & 7 & 0 & 0 \\ 9 & 11 & 1 & 0 \\ 5 & 4 & 1 & 1 \end{bmatrix}$$

   d)
   $$D = \begin{bmatrix} 2 & -1 & 0 \\ -1 & 4 & 2 \\ 0 & 2 & 2 \end{bmatrix}$$

2. Verify the statements in Theorem 6.18 for the matrices

   $$A = \begin{bmatrix} 4 & 6 & 1 & -1 \\ 2 & 1 & 0 & .5 \\ 3 & 0 & 0 & 1 \\ 1 & -1 & 1 & 1 \end{bmatrix}, \quad B = \begin{bmatrix} 1 & 2 & 3 & 4 \\ 0 & 2 & -1 & 1 \\ 0 & 0 & 3 & 2 \\ 0 & 0 & 0 & -1 \end{bmatrix}.$$

3. Prove the following statements or provide counterexamples to show they are not true:

   a) The product of two symmetric matrices is symmetric.
   b) The inverse of a nonsingular symmetric matrix is a nonsingular symmetric matrix.
   c) If A and B are $n \times n$ matrices, then $(AB)^t = A^t B^t$.

4. Prove the statements presented in Theorem 6.18.

5. Construct a matrix A which is nonsymmetric but for which $\mathbf{x}^t A\mathbf{x} > 0$ for all $\mathbf{x} \neq \mathbf{0}$.

6. An $n \times n$ matrix A is called **diagonally dominant** if

   $$|a_{ii}| \geq \sum_{\substack{j=1, \\ j \neq i}}^{n} |a_{ij}| \qquad \text{for each } i = 1, 2, \ldots, n.$$

   a) Find a singular diagonally dominant matrix with no zero entries.
   b) Does a singular diagonally dominant matrix exist that has no zero entries and for which strict inequality holds in inequality (6.15) for all but one row?

7. Suppose A is an $n \times n$ positive definite matrix. Show that:

   a) $a_{ii} > 0$ for each $i = 1, 2, \ldots, n$.
   b) $\max_{1 \leq i \leq n} |a_{ii}| \geq \max_{1 \leq i, j \leq n} |a_{ij}|$.
   c) $(a_{ij})^2 < a_{ii}a_{jj}$ for each $i, j = 1, 2, \ldots, n$, with $i \neq j$.

8. Show that the matrix

$$A = \begin{bmatrix} 4 & -1 & 1 \\ -1 & 4.25 & 2.75 \\ 1 & 2.75 & 3.5 \end{bmatrix}$$

   is positive definite.

9. Let $m_{ji} = a_{ji}^{(i)}/a_{ii}^{(i)}$ for each $j = i + 1, i + 2, \ldots, n$, and define

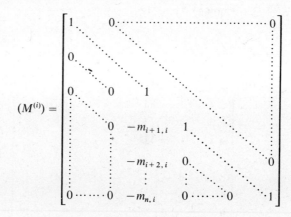

   for each $i = 1, 2, \ldots, n - 1$.

   a) Show that $\tilde{A}^{(2)} = M^{(1)}\tilde{A}^{(1)}$; that is, the operations $(E_j - m_{j1}E_1) \to (E_j)$ for each $j = 2, 3, \ldots, n$ are equivalent to the matrix multiplications $M^{(1)}A$ and $M^{(1)}\mathbf{b}$.
   b) Assuming no row interchanges are necessary, show that $\tilde{A}^{(k)} = M^{(k-1)}M^{(k-2)} \cdots M^{(1)}\tilde{A}^{(1)}$ for each $k = 2, \ldots, n$.
   c) Show that $M^{(i)}$ is nonsingular for each i, and that:

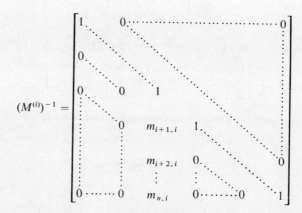

   d) Let $M = M^{(n-1)}M^{(n-2)} \cdots M^{(1)}$. Show that M is lower triangular and nonsingular and that $U = MA$ where U is the upper-triangular matrix formed in Gauss elimination.

e) Show that

$$L = M^{-1} = \begin{bmatrix} 1 & 0 & \cdots\cdots\cdots\cdots & 0 \\ m_{21} & 1 & & \vdots \\ \vdots & & \ddots & 0 \\ m_{n1} & \cdots\cdots & m'_{n,n-1} & 1 \end{bmatrix}$$

and that A can be factored into the product of a lower-triangular matrix and an upper-triangular matrix. *

10. A **permutation matrix** P is a matrix that has precisely one entry whose value is one in each column and each row, and all of whose other entries are zero. For example,

$$P = \begin{bmatrix} 1 & 0 & 0 \\ 0 & 0 & 1 \\ 0 & 1 & 0 \end{bmatrix}$$

is a 3×3 permutation matrix.

a) Show that the operation $(E_i) \leftrightarrow (E_j)$ performed on an $n \times n$ matrix A is equivalent to the multiplication PA, where P is the $n \times n$ permutation matrix formed by applying $(E_i) \leftrightarrow (E_j)$ on I_n.

b) What effect does the multiplication AP have when P is the permutation matrix obtained in (a)?

c) Show that P^{-1} can be obtained from P for any permutation matrix by interchanging rows and columns; that is, $(P^{-1})_{ij} = (P)_{ji}$.

d) If the row interchanges required in Algorithm 6.1 to reduce A are also applied to I, resulting in the matrix P, show that $LU = PA$. This implies $A = (P^{-1}L)U$, but $P^{-1}L$ is not necessarily lower triangular.

11. Let

$$A = \begin{bmatrix} 1 & 1 & 0 & 3 \\ 2 & 1 & -1 & 1 \\ -1 & 2 & 3 & -1 \\ 3 & -1 & -1 & 2 \end{bmatrix}.$$

Use the discussion in Exercise 10 to obtain a factorization of A in the form $A = (P^{-1}L)U$.

12. Given an $n \times n$ matrix A, a **leading principal submatrix** of A is defined to be a submatrix of the form

$$\begin{bmatrix} a_{11} & a_{12} & \cdots & a_{1k} \\ a_{21} & a_{22} & \cdots & a_{2k} \\ \vdots & \vdots & & \vdots \\ a_{k1} & a_{k2} & \cdots & a_{kk} \end{bmatrix}$$

where $1 \le k \le n$. Show that Gaussian elimination can be performed on A without row interchanges if and only if all leading principal submatrices of A are nonsingular. [*Hint*: Fix $1 \le k \le n$ and recall, from Exercise 9 that $A^{(k)} = M^{(k-1)} \cdots M^{(1)}A$. Partition each of these matrices vertically between the kth and $(k + 1)$st columns and horizontally between the kth and $(k + 1)$st rows, to obtain:

$$\begin{bmatrix} A_{11}^{(k)} & A_{12}^{(k)} \\ A_{21}^{(k)} & A_{22}^{(k)} \end{bmatrix} = \begin{bmatrix} M_{11}^{(k-1)} & 0 \\ M_{21}^{(k-1)} & M_{22}^{(k-1)} \end{bmatrix} \begin{bmatrix} M_{11}^{(k-2)} & 0 \\ M_{21}^{(k-2)} & M_{22}^{(k-2)} \end{bmatrix} \cdots \begin{bmatrix} M_{11}^{(1)} & 0 \\ M_{21}^{(1)} & M_{22}^{(1)} \end{bmatrix} \begin{bmatrix} A_{11} & A_{12} \\ A_{21} & A_{22} \end{bmatrix}.$$

Show that A_{11} nonsingular implies $A_{11}^{(k)}$ nonsingular. If Gaussian elimination can be performed without row interchanges, then $A = LU$. Use this, with the consideration of the partitioned product, to obtain the result.]

13. Use Exercise 12 and mathematical induction to prove that the Gaussian elimination procedure can be performed to find the unique solution to $A\mathbf{x} = \mathbf{b}$ without row or column interchanges whenever A is a strictly diagonally dominant matrix.

24. In a paper by Dorn and Burdick [32], it is reported that the average wing length which resulted from mating three mutant varieties of fruit flies (*Drosophila melanogaster*) can be expressed in the symmetric matrix form

$$A = \begin{bmatrix} 1.59 & 1.69 & 2.13 \\ 1.69 & 1.31 & 1.72 \\ 2.13 & 1.72 & 1.85 \end{bmatrix},$$

where a_{ij} denotes the average wing length of an offspring resulting from the mating of a male of type i with a female of type j.

a) What physical significance is associated with the symmetry of this matrix?

b) Is this matrix positive definite? If so, prove it; if not, find a nonzero vector $\mathbf{x}$ for which $\mathbf{x}^t A \mathbf{x} \leq 0$.

6.6   Direct Factorization of Matrices

The discussion centering around Theorem 6.15 on page 296 concerned factoring a matrix A in terms of a lower triangular matrix L and an upper triangular matrix U. It was shown that this factorization existed whenever the linear system $A\mathbf{x} = \mathbf{b}$ could be solved uniquely by Gaussian elimination without row or column interchanges. The system $LU\mathbf{x} = A\mathbf{x} = \mathbf{b}$ can then be transformed into the system $U\mathbf{x} = L^{-1}\mathbf{b}$ and, since U is upper triangular, backward substitution can be applied. Although the specific forms of L and U can be obtained from the Gaussian elimination process, it is desirable to find a more direct method for their determination so that if many systems are to be solved using A, only a forward and backward substitution need be performed (see Steps 11–14 of Algorithm 6.5). To illustrate a procedure for calculating the entries of these matrices, let us consider an example.

EXAMPLE 1   Consider the strictly diagonally dominant 4×4 matrix

$$A = \begin{bmatrix} 6 & 2 & 1 & -1 \\ 2 & 4 & 1 & 0 \\ 1 & 1 & 4 & -1 \\ -1 & 0 & -1 & 3 \end{bmatrix}.$$

Theorems 6.21 (p. 299) and 6.15 (p. 296) guarantee that A can be factored in the form $A = LU$, where:

$$L = \begin{bmatrix} l_{11} & 0 & 0 & 0 \\ l_{21} & l_{22} & 0 & 0 \\ l_{31} & l_{32} & l_{33} & 0 \\ l_{41} & l_{42} & l_{43} & l_{44} \end{bmatrix} \quad \text{and} \quad U = \begin{bmatrix} u_{11} & u_{12} & u_{13} & u_{14} \\ 0 & u_{22} & u_{23} & u_{24} \\ 0 & 0 & u_{33} & u_{34} \\ 0 & 0 & 0 & u_{44} \end{bmatrix}.$$

The 16 known entries in A can be used to partially determine the ten unknown entries in L and the like number in U. If a procedure leading to a unique solution is desired, however, four additional conditions on the entries of L and U are needed. The method to be used in this example arbitrarily requires that $l_{11} = l_{22} = l_{33} = l_{44} = 1$; this is known as **Doolittle's method**. Later in this section, methods requiring that all the diagonal elements of U be one—**Crout's Method**—and requiring that $l_{ii} = u_{ii}$ for each value of i—**Choleski's Method**—will be considered. The portion of the multiplication of L by U,

$$L \cdot U = \begin{bmatrix} 1 & 0 & 0 & 0 \\ l_{21} & 1 & 0 & 0 \\ l_{31} & l_{32} & 1 & 0 \\ l_{41} & l_{42} & l_{43} & 1 \end{bmatrix} \begin{bmatrix} u_{11} & u_{12} & u_{13} & u_{14} \\ 0 & u_{22} & u_{23} & u_{24} \\ 0 & 0 & u_{33} & u_{34} \\ 0 & 0 & 0 & u_{44} \end{bmatrix},$$

that determines the first row of A, results in the four equations

$$u_{11} = 6, \qquad u_{12} = 2, \qquad u_{13} = 1, \qquad u_{14} = -1.$$

The portion of the multiplication that determines the remaining entries in the first column of A provides equations

$$\begin{aligned} l_{21}u_{11} &= 2, & l_{21} &= \tfrac{1}{3}, \\ l_{31}u_{11} &= 1, & \text{and} \quad l_{31} &= \tfrac{1}{6}, \\ l_{41}u_{11} &= -1, & l_{41} &= -\tfrac{1}{6}. \end{aligned}$$

At this stage the matrices L and U assume the form

$$L = \begin{bmatrix} 1 & 0 & 0 & 0 \\ \tfrac{1}{3} & 1 & 0 & 0 \\ \tfrac{1}{6} & l_{32} & 1 & 0 \\ -\tfrac{1}{6} & l_{42} & l_{43} & 1 \end{bmatrix} \quad \text{and} \quad U = \begin{bmatrix} 6 & 2 & 1 & -1 \\ 0 & u_{22} & u_{23} & u_{24} \\ 0 & 0 & u_{33} & u_{34} \\ 0 & 0 & 0 & u_{44} \end{bmatrix}.$$

The portion of the multiplication that determines the remaining entries in the second row of A leads to the equations

$$\begin{aligned} \tfrac{2}{3} + u_{22} &= 4, & u_{22} &= \tfrac{10}{3}, \\ \tfrac{1}{3} + u_{23} &= 1, & \text{so} \quad u_{23} &= \tfrac{2}{3}, \\ -\tfrac{1}{3} + u_{24} &= 0, & u_{24} &= \tfrac{1}{3}; \end{aligned}$$

and that which determines the remaining entries in the second column of A give:

$$\begin{aligned} \tfrac{2}{6} + \tfrac{10}{3}l_{32} &= 1, & l_{32} &= \tfrac{1}{5}, \\ & & \text{so} & \\ -\tfrac{2}{6} + \tfrac{10}{3}l_{42} &= 0, & l_{42} &= \tfrac{1}{10}. \end{aligned}$$

This process is continued, alternating columns and rows between L and U, to finally obtain:

$$L = \begin{bmatrix} 1 & 0 & 0 & 0 \\ \tfrac{1}{3} & 1 & 0 & 0 \\ \tfrac{1}{6} & \tfrac{1}{5} & 1 & 0 \\ -\tfrac{1}{6} & \tfrac{1}{10} & -\tfrac{9}{37} & 1 \end{bmatrix} \quad \text{and} \quad U = \begin{bmatrix} 6 & 2 & 1 & -1 \\ 0 & \tfrac{10}{3} & \tfrac{2}{3} & \tfrac{1}{3} \\ 0 & 0 & \tfrac{37}{10} & -\tfrac{9}{10} \\ 0 & 0 & 0 & \tfrac{191}{74} \end{bmatrix}. \qquad \square$$

A general procedure for factoring matrices into a product of triangular matrices is contained in the following algorithm. Although new matrices L and U are constructed, the actual values generated can replace the corresponding entries of A that are no longer needed. Thus, the new matrix has entries $a_{ij} = l_{ij}$ for each $i = 2, 3, \ldots, n$ and $j = 1, 2, \ldots, i - 1$; and $a_{ij} = u_{ij}$ for each $i = 1, 2, \ldots, n$ and $j = i, i + 1, \ldots, n$.

Direct Factorization Algorithm 6.4

To factor the $n \times n$ matrix $A = (a_{ij})$ into the product of the lower triangular matrix $L = (l_{ij})$ and $u = (u_{ij})$, that is, $A = LU$ where the main diagonal of either L or U is given:

INPUT dimension n; the entries a_{ij}, $1 \le i, j \le n$, of A; the diagonal $l_{11}, \ldots, l_{nn}$ of L or the diagonal $u_{11}, \ldots, u_{nn}$ of U.

OUTPUT the entries l_{ij}, $1 \le j \le i$, $1 \le i \le n$ of L and the entries u_{ij}, $i \le j \le n$, $1 \le i \le n$ of U.

Step 1   Select l_{11} and u_{11} satisfying $l_{11}u_{11} = a_{11}$.
If $l_{11}u_{11} = 0$ then OUTPUT ('Factorization impossible');
STOP.

Step 2   For $j = 2, \ldots, n$ set $u_{1j} = a_{1j}/l_{11}$;   *(First row of U.)*
$l_{j1} = a_{j1}/u_{11}$.   *(First column of L.)*

Step 3   For $i = 2, \ldots, n - 1$ do Steps 4 and 5.

Step 4   Select l_{ii} and u_{ii} satisfying $l_{ii}u_{ii} = a_{ii} - \sum_{k=1}^{i-1} l_{ik}u_{ki}$.

If $l_{ii}u_{ii} = 0$ then OUTPUT ('Factorization impossible');
STOP.

Step 5   For $j = i + 1, \ldots, n$

set $u_{ij} = \dfrac{1}{l_{ii}}\left[a_{ij} - \sum_{k=1}^{i-1} l_{ik}u_{kj} \right]$;   *(ith row of U.)*

$l_{ji} = \dfrac{1}{u_{ii}}\left[a_{ji} - \sum_{k=1}^{i-1} l_{jk}u_{ki} \right]$.   *(ith column of L.)*

Step 6   Select l_{nn} and u_{nn} satisfying $l_{nn}u_{nn} = a_{nn} - \sum_{k=1}^{n-1} l_{nk}u_{kn}$.

(Note: If $l_{nn}u_{nn} = 0$, then $A = LU$ but A is singular.)

Step 7   OUTPUT (l_{ij} for $j = 1, \ldots, i$ and $i = 1, \ldots, n$);
OUTPUT (u_{ij} for $j = i, \ldots, n$ and $i = 1, \ldots, n$);
STOP.

A difficulty that can arise when using this algorithm to obtain the factorization of the coefficient matrix of a linear system of equations is caused by the fact that no pivoting is used to reduce the effect of round-off error. It has been seen in previous calculations that round-off error can be quite significant when finite-digit arithmetic is used and any efficient algorithm must take this effect into consideration.

Although column interchange is difficult to incorporate into the factorization algorithm, the algorithm can be easily altered to include a row-interchange technique equivalent to the maximal column pivoting procedure described in Algorithm 6.2. This interchange is sufficient in most cases.

The following algorithm incorporates the factorization procedure of Algorithm 6.4 together with maximal column pivoting and forward and backward substitution to obtain a solution to a linear system of equations. The actual process involves writing the linear system $A\mathbf{x} = \mathbf{b}$ as $LU\mathbf{x} = \mathbf{b}$. The forward substitution solves the system $L\mathbf{z} = \mathbf{b}$ and the backward substitution solves the system $U\mathbf{x} = \mathbf{z} = L^{-1}\mathbf{b}$. It should be noted that the nonzero entries of L and U can be stored in the corresponding entries of A except for the diagonal of L or U, which must be input.

Direct Factorization with
Maximal Column Pivoting Algorithm 6.5

To solve the $n \times n$ linear system $A\mathbf{x} = \mathbf{b}$ in the form

$$E_1: \quad a_{11}x_1 + a_{12}x_2 + \cdots + a_{1n}x_n = a_{1,n+1}$$
$$E_2: \quad a_{21}x_1 + a_{22}x_2 + \cdots + a_{2n}x_n = a_{2,n+1}$$
$$\vdots \qquad \vdots \qquad \vdots \qquad \qquad \vdots \qquad \vdots$$
$$E_n: \quad a_{n1}x_1 + a_{n2}x_2 + \cdots + a_{nn}x_n = a_{n,n+1}$$

by factoring A into LU, solving $L\mathbf{z} = \mathbf{b}$ and $U\mathbf{x} = \mathbf{z}$ where the main diagonal of either L or U is given:

INPUT the dimension n; the entries a_{ij}, $1 \le j \le n + 1$, $1 \le i \le n$ of the augmented form of A; the diagonal $l_{11}, \ldots, l_{nn}$ of L or the diagonal $u_{11}, \ldots, u_{nn}$ of U.

OUTPUT the solution $x_1, \ldots, x_n$ or a message that linear system has no unique solution.

Step 1   Let p be the smallest integer such that $1 \le p \le n$ and $|a_{p1}| = \max\limits_{1 \le j \le n} |a_{j1}|$;
  (*Find first pivot element.*)
  If $|a_{p1}| = 0$, then OUTPUT ('No unique solution exists');
       STOP.

Step 2   If $p \ne 1$ then interchange rows p and 1 in A.

Step 3   Select l_{11} and u_{11} satisfying $l_{11}u_{11} = a_{11}$.

Step 4   For $j = 2, \ldots, n$ set $u_{1j} = a_{1j}/l_{11}$;   (*First row of U.*)
            $l_{j1} = a_{j1}/u_{11}$.   (*First column of L.*)

Step 5   For $i = 2, \ldots, n - 1$ do Steps 6–9.

Step 6   Let p be the smallest integer such that $i \le p \le n$ and

$$\left| a_{pi} - \sum_{k=1}^{i-1} l_{pk}u_{ki} \right| = \max_{i \le j \le n} \left| a_{ji} - \sum_{k=1}^{i-1} l_{jk}u_{ki} \right|.$$

(*Find ith pivot element.*)
If the maximum is zero then OUTPUT ('No unique solution exists').

Step 7 If $p \neq i$ then interchange rows p and i in both matrices A and L.

Step 8 Select l_{ii} and u_{ii} satisfying $l_{ii}u_{ii} = a_{ii} - \sum_{k=1}^{i-1} l_{ik}u_{ki}$.

Step 9 For $j = i + 1, \ldots, n$

$$\text{set } u_{ij} = \frac{1}{l_{ii}}\left[a_{ij} - \sum_{k=1}^{i-1} l_{ik}u_{kj}\right]; \quad (\textit{ith row of } U.)$$

$$l_{ji} = \frac{1}{u_{ii}}\left[a_{ji} - \sum_{k=1}^{i-1} l_{jk}u_{ki}\right]. \quad (\textit{ith column of } L.)$$

Step 10 Set $HOLD = a_{nn} - \sum_{k=1}^{n-1} l_{nk}u_{kn}$;

If $HOLD = 0$ then OUTPUT ('No unique solution exists.');
STOP.

Select l_{nn} and u_{nn} satisfying $l_{nn}u_{nn} = a_{nn} - \sum_{k=1}^{n-1} l_{nk}u_{kn}$.

(*Steps* 11 *and* 12 *solve lower triangular system* $L\mathbf{z} = \mathbf{b}$.)

Step 11 Set $z_1 = a_{1,n+1}/l_{11}$.

Step 12 For $i = 2, \ldots, n$ set $z_i = \frac{1}{l_{ii}}\left[a_{i,n+1} - \sum_{j=1}^{i-1} l_{ij}z_j\right]$.

(*Steps* 13 *and* 14 *solve upper triangular system* $U\mathbf{x} = \mathbf{z}$.)

Step 13 Set $x_n = z_n/u_{nn}$.

Step 14 For $i = n - 1, \ldots, 1$ set $x_i = \frac{1}{u_{ii}}\left[z_i - \sum_{j=i+1}^{n} u_{ij}x_j\right]$.

Step 15 OUTPUT $(x_1, \ldots, x_n)$;
STOP.

When the matrix is known to be positive definite, a significant improvement in the matrix factorization technique can be made with regard to the number of arithmetic operations required.

Theorem 6.24 If A is a positive definite $n \times n$ matrix, then A has a factorization of the form $A = LL^t$, where L is a lower-triangular matrix. The factorization can be achieved by applying Algorithm 6.4 with $l_{ii} = u_{ii}$ for each $i = 1, 2, \ldots, n$.

Proof Since A is positive definite, Theorems 6.23 and 6.15 imply that A can be factored in the form $A = LU$, where L is lower triangular and U is upper triangular. Positive definiteness also implies (see Exercise 7 of Section 6.5) that $a_{ii} > 0$ for each $i = 1, 2, \ldots, n$. With the notation of Algorithm 6.4, the procedure is started by choosing $l_{11} = u_{11} = \sqrt{a_{11}}$.
Since A is symmetric,

$$l_{j1} = \frac{a_{j1}}{u_{11}} = \frac{a_{1j}}{l_{11}} = u_{1j}$$

for each $j = 2, 3, \ldots, n$ so the entries in the first row of U agree with the corresponding entries in the first column of L.

The proof proceeds by mathematical induction. Assume that $k < n$ and that the entries in the first k rows of U agree with the corresponding entries in the first k columns of L. The proof of the theorem will be complete if it can be established that the entries in the $(k + 1)$st row of U agree with the entries in the $(k + 1)$st column of L.

Since L is lower triangular and U is upper triangular, it is clear that

$$l_{j, k+1} = 0 = u_{k+1, j} \qquad \text{whenever } j = 1, 2, \ldots, k.$$

Choose $l_{k+1, k+1} = u_{k+1, k+1} = (a_{k+1, k+1} - \sum_{j=1}^{k} l_{k+1, j}^2)^{1/2}$, which can be shown to be a real number (see Exercise 19).

Since A is symmetric and $l_{ji} = u_{ij}$ for each $i = 1, 2, \ldots, k$, and $j = 1, 2, \ldots, n$,

$$l_{j, k+1} = \frac{1}{u_{k+1, k+1}} \left[a_{j, k+1} - \sum_{i=1}^{k} l_{ji} u_{i, k+1} \right] = \frac{1}{l_{k+1, k+1}} \left[a_{k+1, j} - \sum_{i=1}^{k} u_{ij} l_{k+1, i} \right] = u_{k+1, j}$$

for each $j = k + 2, k + 3, \ldots, n$. $\square$

For a positive definite matrix, this theorem can be used to simplify the Factorization Algorithm 6.4. If a linear system is to be solved involving a positive definite matrix, Steps 1–6 of the following algorithm can be substituted for Steps 1–10 of Algorithm 6.5 to take advantage of the simplification that results provided u_{ij} is replaced by l_{ji} in Steps 13 and 14. The factorization procedure is described in the following algorithm.

Choleski's Algorithm 6.6

To factor the positive definite $n \times n$ matrix A into LL^t where L is lower triangular.

INPUT the dimension n; entries a_{ij}, $1 \le i, j \le n$ of A.

OUTPUT the entries l_{ij}, $1 \le j \le i$, $1 \le i \le n$ of L.   (*The entries of $U = L^t$ are $u_{ij} = l_{ji}$, $i \le j \le n$, $1 \le i \le n$.*)

Step 1   Set $l_{11} = \sqrt{a_{11}}$.

Step 2   For $j = 2, \ldots, n$ set $l_{j1} = a_{j1}/l_{11}$.

Step 3   For $i = 2, \ldots, n - 1$ do Steps 4 and 5.

Step 4   Set $l_{ii} = \left[a_{ii} - \sum_{k=1}^{i-1} l_{ik}^2 \right]^{1/2}$.

Step 5   For $j = i + 1, \ldots, n$

set $l_{ji} = \frac{1}{l_{ii}} \left[a_{ji} - \sum_{k=1}^{i-1} l_{jk} l_{ik} \right]$.

Step 6   Set $l_{nn} = \left[a_{nn} - \sum_{k=1}^{n-1} l_{nk}^2 \right]^{1/2}$.

Step 7   OUTPUT (l_{ij} *for* $j = 1, \ldots, i$ *and* $i = 1, \ldots, n.$);
STOP.

It is not difficult to verify (see Exercise 17) that the solution of a typical linear system involving a positive definite matrix by using Choleski's Algorithm requires

n square roots,

$$\frac{n^3 + 9n^2 + 2n}{6}$$ multiplications/divisions,

and $$\frac{n^3 + 6n^2 - 7n}{6}$$ additions/subtractions.

This is about half the arithmetic operations that are required in the Gauss Elimination Algorithm 6.1, a method that is typical with regard to the number of arithmetic operations required. The computational advantage of the Choleski method depends on the number of operations that are required for determining the values of the n square roots which, since it is a linear factor of n, will decrease in significance as n increases.

EXAMPLE 2   To illustrate the steps involved in applying Choleski's method, consider the matrix

$$A = \begin{bmatrix} 4 & -1 & 1 \\ -1 & 4.25 & 2.75 \\ 1 & 2.75 & 3.5 \end{bmatrix},$$

which can be shown to be positive definite (see Exercise 8 of Section 6.5).

Tracing the steps of Algorithm 6.6,

Step 1 $l_{11} = \sqrt{a_{11}} = \sqrt{4} = 2.$

Step 2 $l_{21} = \dfrac{a_{21}}{l_{11}} = \dfrac{-1}{2} = -.5, \quad l_{31} = \dfrac{a_{31}}{l_{11}} = \dfrac{1}{2} = .5.$

Step 3 $i = 2.$

Step 4 $l_{22} = (a_{22} - l_{21}^2)^{1/2} = (4.25 - (-.5)^2)^{1/2} = 2.$

Step 5 $l_{32} = \dfrac{1}{l_{22}}(a_{32} - l_{31}l_{21}) = \dfrac{1}{2}(2.75 - .5(-.5)) = 1.5.$

Step 6 $l_{33} = (a_{33} - l_{31}^2 - l_{32}^2)^{1/2} = (3.5 - (.5)^2 - (1.5)^2)^{1/2} = 1.$

Since L is lower triangular and $U = L^t$,

$$l_{12} = l_{13} = l_{23} = 0,$$

$$u_{11} = l_{11} = 2, \quad u_{12} = l_{21} = -.5, \quad u_{13} = l_{31} = .5,$$

$$u_{21} = l_{12} = 0, \quad u_{22} = l_{22} = 2, \quad u_{23} = l_{32} = 1.5,$$

$$u_{31} = l_{31} = 0, \quad u_{32} = l_{23} = 0, \quad u_{33} = l_{33} = 1.$$

So

$$L = \begin{bmatrix} 2 & 0 & 0 \\ -.5 & 2 & 0 \\ .5 & 1.5 & 1 \end{bmatrix}, \quad U = \begin{bmatrix} 2 & -.5 & .5 \\ 0 & 2 & 1.5 \\ 0 & 0 & 1 \end{bmatrix}. \qquad \square$$

The factorization algorithms can be simplified considerably in the case of band matrices because of the large number of zeros appearing in these matrices in regular patterns. It is particularly interesting to observe the form which the Crout or the Doolittle method assumes in this case. To illustrate the situation, suppose that a tridiagonal matrix

$$A = \begin{bmatrix} a_{11} & a_{12} & 0 & \cdots & \cdots & 0 \\ a_{21} & a_{22} & a_{23} & & & 0 \\ 0 & & & & & a_{n-1,n} \\ 0 & \cdots & 0 & & a_{n,n-1} & a_{nn} \end{bmatrix}$$

can be factored into the triangular matrices

$$L = \begin{bmatrix} l_{11} & 0 & \cdots & 0 \\ l_{21} & l_{22} & & \vdots \\ \vdots & & & 0 \\ l_{n1} & l_{n2} & \cdots & l_{nn} \end{bmatrix} \quad \text{and} \quad U = \begin{bmatrix} u_{11} & u_{12} & \cdots & u_{1n} \\ 0 & u_{22} & \cdots & u_{2n} \\ \vdots & & & \vdots \\ 0 & \cdots & 0 & u_{nn} \end{bmatrix}.$$

Since A has only $(3n - 2)$ nonzero entries, there are only $(3n - 2)$ conditions to be applied to determine the entries of L and U, provided, of course, that the zero entries of A are also obtained. Suppose that the matrices can actually be found in the form

$$L = \begin{bmatrix} l_{11} & 0 & \cdots & \cdots & 0 \\ l_{21} & l_{22} & & & \vdots \\ & & & & \\ 0 & & & & 0 \\ \vdots & & & & \\ 0 & \cdots & 0 & l_{n,n-1} & l_{nn} \end{bmatrix} \quad \text{and} \quad U = \begin{bmatrix} 1 & u_{12} & 0 & \cdots & 0 \\ 0 & 1 & & & 0 \\ \vdots & & & & \\ & & & & u_{n-1,n} \\ 0 & \cdots & \cdots & 0 & 1 \end{bmatrix}.$$

In this form, there are $(2n - 1)$ undetermined entries of L and $(n - 1)$ undetermined entries of U, which totals the number of conditions mentioned above, and the zero entries of A are obtained automatically.

The multiplication involved with $A = LU$ gives, in addition to the zero entries, the equations

(6.16) $a_{11} = l_{11};$

(6.17) $a_{i,i-1} = l_{i,i-1}$ for each $i = 2, 3, \ldots, n;$

(6.18) $a_{ii} = l_{i,i-1}u_{i-1,i} + l_{ii}$ for each $i = 2, 3, \ldots, n;$

and

(6.19) $a_{i,i+1} = l_{ii}u_{i,i+1}$ for each $i = 1, 2, \ldots, n - 1.$

A solution to this system of equations can be found by first obtaining all the nonzero off-diagonal terms in L, using Eq. (6.17), and then using equations (6.19) and (6.18) to alternately obtain the remainder of the entries in U and L, which can be stored in the corresponding entries of A.

A complete algorithm for solving an $n \times n$ system of linear equations whose coefficient matrix is tridiagonal follows.

Crout Reduction for Tridiagonal Linear Systems Algorithm 6.7

To solve the $n \times n$ linear system

$$E_1: \quad a_{11}x_1 + a_{12}x_2 \qquad\qquad\qquad\qquad\qquad = a_{1,n+1}$$
$$E_2: \quad a_{21}x_1 + a_{22}x_2 + a_{23}x_3 \qquad\qquad\qquad = a_{2,n+1}$$
$$\vdots \qquad\qquad\qquad \vdots \qquad\qquad\qquad\qquad \vdots$$
$$E_{n-1}: \qquad\qquad a_{n-1,n-2}x_{n-2} + a_{n-1,n-1}x_{n-1} + a_{n-1,n}x_n = a_{n-1,n+1}$$
$$E_n: \qquad\qquad\qquad\qquad a_{n,n-1}x_{n-1} + a_{nn}x_n \qquad = a_{n,n+1},$$

which is assumed to have a unique solution:

INPUT the dimension n; the entries of A.

OUTPUT the solution $x_1, \ldots, x_n$.

Step 1 Set $l_{11} = a_{11}$;
$u_{12} = a_{12}/l_{11}$.

Step 2 For $i = 2, \ldots, n - 1$ set $l_{i,i-1} = a_{i,i-1}$;   (*ith row of L.*)
$l_{ii} = a_{ii} - l_{i,i-1}u_{i-1,i}$;
$u_{i,i+1} = a_{i,i+1}/l_{ii}$.   ((*i + 1)st column of U.*)

Step 3 Set $l_{n,n-1} = a_{n,n-1}$;   (*nth row of L.*)
$l_{nn} = a_{nn} - l_{n,n-1}u_{n-1,n}$.

(*Steps 4, 5 solve* $Lz = b$.)

Step 4 Set $z_1 = a_{1,n+1}/l_{11}$.

Step 5 For $i = 2, \ldots, n$ set $z_i = \dfrac{1}{l_{ii}}[a_{i,n+1} - l_{i,i-1}z_{i-1}]$.

(*Steps 6, 7 solve* $Ux = z$.)

Step 6 Set $x_n = z_n$.

Step 7 For $i = n - 1, \ldots, 1$ set $x_i = z_i - u_{i,i+1}x_{i+1}$.

Step 8 OUTPUT $(x_1, \ldots, x_n)$;
STOP.

This algorithm requires only $(5n - 4)$ multiplication/divisions and $(3n - 3)$ addition/subtractions, and consequently has considerable computational advantage over the methods that do not consider the tridiagonality of the matrix, especially for large values of n.

EXAMPLE 3   To illustrate the procedure involved in Algorithm 6.7, consider the tridiagonal system of equations

$$2x_1 - x_2 \qquad\qquad = 1,$$
$$-x_1 + 2x_2 - x_3 \qquad = 0,$$
$$- x_2 + 2x_3 - x_4 = 0,$$
$$- x_3 + 2x_4 = 1,$$

whose augmented matrix is

$$\begin{bmatrix} 2 & -1 & 0 & 0 & \vdots & 1 \\ -1 & 2 & -1 & 0 & \vdots & 0 \\ 0 & -1 & 2 & -1 & \vdots & 0 \\ 0 & 0 & -1 & 2 & \vdots & 1 \end{bmatrix}.$$

Tracing the steps of Algorithm 6.7:

Step 1 $l_{11} = 2,\quad u_{12} = \dfrac{a_{12}}{l_{11}} = -\dfrac{1}{2}.$

Step 2 $i = 2.$

$$l_{21} = a_{21} = -1,\quad l_{22} = a_{22} - l_{21}u_{12} = 2 - (-1)(-\tfrac{1}{2}) = \tfrac{3}{2},$$

$$u_{23} = \frac{a_{23}}{l_{22}} = -\frac{1}{(\tfrac{3}{2})} = -\frac{2}{3}$$

$i = 3.$

$$l_{32} = a_{32} = -1,\quad l_{33} = a_{33} - l_{32}u_{23} = 2 - (-1)\left(-\frac{2}{3}\right) = \frac{4}{3},$$

$$u_{34} = \frac{a_{34}}{l_{33}} = -\frac{3}{4}.$$

Step 3 $l_{43} = a_{43} = -1,\quad l_{44} = a_{44} - l_{43}u_{34} = 2 - (-1)(-\tfrac{3}{4}) = \tfrac{5}{4}.$

Step 4 $z_1 = \dfrac{a_{15}}{l_{11}} = \dfrac{1}{2}.$

Step 5 $z_2 = \dfrac{1}{l_{22}}[a_{25} - l_{21}z_1] = \dfrac{0 - (-1)(\tfrac{1}{2})}{(\tfrac{3}{2})} = \dfrac{1}{3},$

$$z_3 = \frac{1}{l_{33}}[a_{35} - l_{32}z_2] = \frac{0 - (-1)(\tfrac{1}{3})}{(\tfrac{4}{3})} = \frac{1}{4},$$

$$z_4 = \frac{1}{l_{44}}[a_{45} - l_{43}z_3] = \frac{1 - (-1)(\tfrac{1}{4})}{(\tfrac{5}{4})} = 1.$$

Step 6 $x_4 = 1.$

Step 7 $x_3 = z_3 - u_{34}x_4 = \tfrac{1}{4} - (-\tfrac{3}{4}) = 1,$

$$x_2 = z_2 - u_{23}x_3 = \tfrac{1}{3} - (-\tfrac{2}{3}) = 1,$$

$$x_1 = z_1 - u_{12}x_2 = \tfrac{1}{2} - (-\tfrac{1}{2}) = 1.$$

The algorithm factored the matrix involved into

$$
\begin{bmatrix}
2 & -1 & 0 & 0 \\
-1 & 2 & -1 & 0 \\
0 & -1 & 2 & -1 \\
0 & 0 & -1 & 2
\end{bmatrix}
=
\begin{bmatrix}
2 & 0 & 0 & 0 \\
-1 & \frac{3}{2} & 0 & 0 \\
0 & -1 & \frac{4}{3} & 0 \\
0 & 0 & -1 & \frac{5}{4}
\end{bmatrix}
\begin{bmatrix}
1 & -\frac{1}{2} & 0 & 0 \\
0 & 1 & -\frac{2}{3} & 0 \\
0 & 0 & 1 & -\frac{3}{4} \\
0 & 0 & 0 & 1
\end{bmatrix},
$$

and gave the correct solution, $x_1 = x_2 = x_3 = x_4 = 1$.   □

Algorithm 6.7 can be applied whenever $l_{ii} \neq 0$ for each $i = 1, 2, \ldots, n$. Two conditions either of which will ensure that this is true, are that the coefficient matrix of the system is positive definite or that it is strictly diagonally dominant. An additional condition that ensures that this algorithm can be applied is given in the next theorem, whose proof is discussed in Exercise 12.

Theorem 6.25   Suppose that $A = (a_{ij})$ is tridiagonal with $a_{i, i-1} a_{i, i+1} \neq 0$ for each $i = 2, 3, \ldots, n - 1$. If $|a_{11}| > |a_{12}|$ and $|a_{ii}| \geq |a_{i, i-1}| + |a_{i, i+1}|$ for each $i = 2, 3, \ldots, n - 1$, and $|a_{nn}| > |a_{n, n-1}|$, then A is nonsingular and the values of l_{ii} described in Algorithm 6.7 are nonzero for each $i = 1, 2, \ldots, n$.

The Choice of a Method for Solving a Linear System

When the linear system is small enough to be efficiently accommodated in the main memory of a computer, it is generally most efficient to use a direct technique that minimizes the effect of rounding error. Specifically, Gaussian Elimination with Scaled Column Pivoting Algorithm 6.3 is appropriate.

Large linear systems with primarily zero entries occurring in regular patterns can generally be efficiently solved by using an iterative procedure such as those discussed in Chapter 8. Systems of this type arise naturally when finite difference techniques are used to solve boundary value problems, a common application in the numerical solution of partial-differential equations, for example.

It can be very difficult to solve a large linear system that has primarily nonzero entries, or one where the zero entries are not in a predictable pattern. The matrix associated with such a system can be placed in secondary storage in partitioned form (see Exercise 9 of Section 6.3) and portions read into main memory only as needed for calculation. Methods that require secondary storage can be either iterative or direct, but they generally require techniques from the fields of data structures and graph theory. A study of the problems involved in theory and implementation is well beyond the scope of this text. The reader is referred to Bunch and Rose [22], and Rose and Willoughby [70] for a discussion of the current techniques.

Exercise Set 6.6

1. Factor the following matrices into the LU decomposition using Algorithm 6.4 with $l_{ii} = 1$ for all i:

a)
$$
\begin{bmatrix}
2 & -1 & 1 \\
3 & 3 & 9 \\
3 & 3 & 5
\end{bmatrix}
$$

b)
$$
\begin{bmatrix}
2 & 0 & 0 & 0 \\
1 & 1.5 & 0 & 0 \\
0 & -3 & .5 & 0 \\
2 & -2 & 1 & 1
\end{bmatrix}
$$

c) $\begin{bmatrix} 2 & -1.5 & 3 \\ -1 & 0 & 2 \\ 4 & -4.5 & 5 \end{bmatrix}$ d) $\begin{bmatrix} 1.012 & -2.132 & 3.104 \\ -2.132 & 4.096 & -7.013 \\ 3.104 & -7.013 & .014 \end{bmatrix}$

2. Solve the following linear systems using Algorithm 6.5 with $l_{ii} = 1$ for each i:

   a) $2x_1 - x_2 + x_3 = -1$
   $3x_1 + 3x_2 + 9x_3 = 0$
   $3x_1 + 3x_2 + 5x_3 = 4$

   b) $2x_1 - 1.5x_2 + 3x_3 = 1$
   $-x_1 \qquad + 2x_3 = 3$
   $4x_1 - 4.5x_2 + 5x_3 = -1$

   c) $2x_1 \qquad\qquad = 3$
   $x_1 + 1.5x_2 \qquad = 4.5$
   $\quad - 3x_2 + .5x_3 \qquad = -6.6$
   $2x_1 - 2x_2 + x_3 + x_4 = .8$

   d) $1.012x_1 - 2.132x_2 + 3.104x_3 = 1.984$
   $-2.132x_1 + 4.096x_2 - 7.013x_3 = -5.049$
   $3.104x_1 - 7.013x_2 + .014x_3 = -3.895$

3. Use Algorithm 6.6 to find a factorization of the form $A = LL^t$ for the following matrices:

   a) $A = \begin{bmatrix} 2 & -1 & 0 \\ -1 & 2 & -1 \\ 0 & -1 & 2 \end{bmatrix}$

   b) $A = \begin{bmatrix} 4 & 1 & 1 & 1 \\ 1 & 3 & -1 & 1 \\ 1 & -1 & 2 & 0 \\ 1 & 1 & 0 & 2 \end{bmatrix}$

   c) $A = \begin{bmatrix} 4 & 1 & -1 & 0 \\ 1 & 3 & -1 & 0 \\ -1 & -1 & 5 & 2 \\ 0 & 0 & 2 & 4 \end{bmatrix}$

   d) $A = \begin{bmatrix} 6 & 2 & 1 & -1 \\ 2 & 4 & 1 & 0 \\ 1 & 1 & 4 & -1 \\ -1 & 0 & -1 & 3 \end{bmatrix}$

4. Modify Algorithm 6.4 so that it can be used to solve linear systems. Repeat Exercise 2 using the modified algorithm, and compare your answers to those obtained in Exercise 2.

5. Solve the following linear systems using Algorithm 6.7:

   a) $x_1 - x_2 \qquad = 0$
   $-2x_1 + 4x_2 - 2x_3 = -1$
   $\quad - x_2 + 2x_3 = 1.5$

   b) $3x_1 + x_2 \qquad = -1$
   $2x_1 + 4x_2 + x_3 = 7$
   $\quad 2x_2 + 5x_3 = 9$

   c) $2x_1 - x_2 \qquad = 3$
   $-x_1 + 2x_2 - x_3 = -3$
   $\quad - x_2 + 2x_3 = 1$

   d) $.5x_1 + .25x_2 \qquad = .35$
   $.35x_1 + .8x_2 + .4x_3 \qquad = .77$
   $.25x_2 + x_3 + .5x_4 = -.5$
   $x_3 - 2x_4 = -2.25$

6. Modify Algorithm 6.6 so that it can be used to solve linear systems. Use the modified algorithm to solve the following linear systems:

   a) $2x_1 - x_2 \qquad = 3$
   $-x_1 + 2x_2 - x_3 = -3$
   $\quad - x_2 + 2x_3 = 1$

   b) $4x_1 + x_2 + x_3 + x_4 = .65$
   $x_1 + 3x_2 - x_3 + x_4 = .05$
   $x_1 - x_2 + 2x_3 \qquad = 0$
   $x_1 + x_2 \qquad + 2x_4 = .5$

   c) $4x_1 + x_2 - x_3 \qquad = 7$
   $x_1 + 3x_2 - x_3 \qquad = 8$
   $-x_1 - x_2 + 5x_3 + 2x_4 = -4$
   $2x_3 + 4x_4 = 6$

7. Tridiagonal matrices are usually labelled by using the notation

$$A = \begin{bmatrix} a_1 & c_1 & 0 & \cdots & \cdots & 0 \\ b_2 & a_2 & c_2 & & & \vdots \\ 0 & b_3 & & & & 0 \\ \vdots & & & & & c_{n-1} \\ 0 & \cdots & \cdots & 0 & b_n & a_n \end{bmatrix}$$

in order to emphasize that it is not necessary to consider the entire matrix. Rewrite Algorithm 6.7, using this notation, and changing the notation of the l_{ij} and u_{ij} in a similar manner.

8. Derive an algorithm for the factorization of tridiagonal matrices directly from Algorithm 6.4, using $u_{ii} = 1$ for each $i = 1, 2, \ldots, n$, and compare this procedure with Algorithm 6.7.

9. Derive an algorithm for the factorization of tridiagonal matrices using Algorithm 6.4 with $l_{ii} = 1$ for each $i = 1, 2, \ldots, n$.

10. Let A be the 10×10 tridiagonal matrix given by $a_{ii} = 2$, $a_{i,i+1} = a_{i,i-1} = -1$ for each $i = 2, \ldots, 9$ with $a_{11} = a_{10,10} = 2$, $a_{12} = a_{10,9} = -1$. Let $\mathbf{b}$ be the ten-dimensional column vector given by $b_1 = b_{10} = 1$ and $b_i = 0$ for each $i = 2, 3, \ldots, 9$. Solve $A\mathbf{x} = \mathbf{b}$, using Algorithm 6.7.

11. Repeat Exercise 10 using a method based on the algorithm obtained in Exercise 9.

12. Prove Theorem 6.25. [*Hint:* Show that $|u_{i,i+1}| < 1$ for each $i = 1, 2, \ldots, n-1$, and that $|l_{ii}| > 0$ for each $i = 1, 2, \ldots, n$, and deduce that det $A = \det L \cdot \det U \neq 0$.]

13. A block (or partitioned) tridiagonal matrix is a matrix of the form

$$A = \begin{bmatrix} A_1 & C_1 & 0 & \cdots & \cdots & 0 \\ B_2 & A_2 & C_2 & & & \vdots \\ 0 & B_3 & & & & 0 \\ \vdots & & & & & C_{N-1} \\ 0 & \cdots & \cdots & 0 & B_N & A_N \end{bmatrix}$$

where each A_i is an $n_i \times n_i$ matrix, each B_i is an $n_i \times n_{i-1}$ matrix and each C_i is an $n_i \times n_{i+1}$ matrix for some collection of positive integers $n_1, n_2, \ldots, n_N$.

a) Factor A into LU, where

$$L = \begin{bmatrix} L_1 & 0 & \cdots & \cdots & 0 \\ B_2 & L_2 & & & \vdots \\ 0 & & & & \vdots \\ \vdots & & & & 0 \\ 0 & \cdots & \cdots & 0 & B_n & L_n \end{bmatrix}, \quad U = \begin{bmatrix} I_1 & \Gamma_1 & 0 & \cdots & \cdots & 0 \\ 0 & I_2 & \Gamma_2 & & & \vdots \\ \vdots & & & & & 0 \\ & & & & & \Gamma_{n-1} \\ 0 & \cdots & \cdots & 0 & I_n \end{bmatrix},$$

and each L_i is an $n_i \times n_i$ matrix, each Γ_i is an $n_i \times n_{i+1}$ matrix, and I_i denotes the $n_i \times n_i$ identity matrix.

b) Derive a block tridiagonal matrix algorithm, similar to Algorithm 6.7, to solve a linear system of the form $A\mathbf{x} = \mathbf{y}$ where $\mathbf{y}$ is row partitioned the same as A.

14. Consider the block tridiagonal matrix defined in Exercise 13. Show that if the leading principal submatrices of A of the form

$$A^{(k)} = \begin{bmatrix} A_1 & C_1 & 0 & \cdots & \cdots & 0 \\ B_2 & A_2 & C_2 & & & \vdots \\ 0 & & & & & 0 \\ \vdots & & & & & C_{k-1} \\ 0 & \cdots & \cdots & 0 & B_k & A_k \end{bmatrix}$$

are all nonsingular for each $k = 1, 2, \ldots, n$, then the matrices $L_1, L_2, \ldots, L_n$ are also nonsingular, and the algorithm obtained in Exercise 13(b) can be performed.

15. Let A be the block tridiagonal matrix defined by

$$A = \begin{bmatrix} A_1 & C_1 & 0 \\ B_2 & A_2 & C_2 \\ 0 & B_3 & A_3 \end{bmatrix}, \quad \text{where} \quad A_i = \begin{bmatrix} 4 & -1 & 0 \\ -1 & 4 & -1 \\ 0 & -1 & 4 \end{bmatrix}$$

for each $i = 1, 2, 3$, and

$$B_{i+1} = C_i = \begin{bmatrix} -1 & 0 & 0 \\ 0 & -1 & 0 \\ 0 & 0 & -1 \end{bmatrix}$$

for $i = 1, 2$. Given the nine-dimensional column vector $\mathbf{b}$ with $b_1 = b_9 = 2$ and $b_i = 1$ for each $i = 2, 3, \ldots, 8$, solve $A\mathbf{x} = \mathbf{b}$ by the algorithm obtained in Exercise 13.

16. Show that Algorithm 6.5 requires as many multiplications/divisions and additions/subtractions as Gaussian elimination.

17. Construct the operation count for an $n \times n$ linear system, using Algorithm 6.6.

18. Construct the operation count for an $n \times n$ linear system, using Algorithm 6.7.

19. Show that for any positive definite matrix A, Algorithm 6.5 can be performed with $l_{ii} > 0$ and $u_{ii} > 0$ for each $i = 1, 2, \ldots, n$.

Approximation Theory

Hooke's law states that when a force is applied to a spring constructed of uniform material, the length of the spring will be a linear function of the force applied. We can write the linear function as $F(l) = k \cdot l + E$ where $F(l)$ represents the force required to stretch the spring l units, the constant E represents the length of the spring with no force applied, and the constant k is called the spring constant.

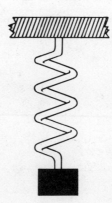

Suppose we want to determine the spring constant for a spring that has an initial length of 5.3 inches. We consecutively apply forces of 2, 4, and 6 pounds to the spring and find that its length increases to 7.0, 9.4, and 12.3 inches, respectively. A quick examination shows that the points (0, 5.3), (2, 7.0), (4, 9.4), and (6, 12.3) do not quite lie in a straight line. Although we could simply use one random pair of these data points to approximate the spring constant, it would seem more reasonable to find the line that *best* approximates all the data points, in order to determine the constant. This type of approximation is known as least-squares approximation and will be considered in this chapter.

The study of approximation theory involves two general types of problems. One problem arises when a function is given explicitly but it is desirable to find a "simpler" type of function, such as a polynomial, that can be used to determine approximate values of the given function. The other problem in approximation theory is concerned with fitting functions to given data and finding the "best" function in a certain class that can be used to represent the data.

Both problems have been touched upon in Chapter 3. The Taylor polynomial of degree n about the point x_0 was discussed as an excellent approximation to an $(n + 1)$-times differentiable function f in a small neighborhood of a point x_0. The Lagrange interpolating polynomials, or more generally osculatory polynomials, were discussed both as approximating polynomials and as polynomials to fit certain data. Cubic splines were also discussed in that chapter. In this chapter, limitations to these techniques will be pointed out and other avenues of approach will be discussed.

7.1 Discrete Least-Squares Approximation

Consider the problem of estimating the values of a function at nontabulated points, given the experimental data in Table 7.1.

TABLE 7.1

i	x_i	y_i
1	2	2
2	4	11
3	6	28
4	8	40

The methods described in Chapter 3 require that either a Lagrange polynomial of degree 3 or a cubic spline be constructed that assumes the value y_i at x_i for each $i = 1, 2, 3, 4$. Sketching the graph of the values given in Table 7.1, however, indicates that it would be reasonable (see Fig. 7.1) to assume that the actual

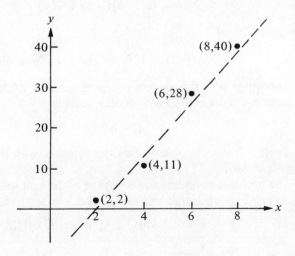

FIGURE 7.1

relationship is a linear one and that no line fits the data exactly because of error in the data-collection procedure.

If this is indeed the case, it would be unreasonable to require that the approximating function agree exactly with the given data; in fact, such an approximating function would introduce oscillations that were not present originally. A better approach for a problem of this type would be to find the "best" (in some sense) line that could be used as an approximating function even though it might not agree precisely with the data at any point.

The least squares approach to this problem involves determining the best approximating line when the error involved is the sum of the squares of the differences between the values on the approximating line and the given values. Letting $ax_i + b$ denote the ith value on the approximating line and y_i the ith given value, requires that constants a and b be found that minimize the quantity

$$\sum_{i=1}^{n} [y_i - (ax_i + b)]^2.$$

This particular problem reduces to finding constants a and b that minimize

$$\sum_{i=1}^{4} [y_i - (ax_i + b)]^2 = [2 - (2a + b)]^2 + [11 - (4a + b)]^2$$
$$+ [28 - (6a + b)]^2 + [40 - (8a + b)]^2.$$

If $\sum_{i=1}^{4} [y_i - (ax_i + b)]^2$ is considered to be a function of two variables a and b, an elementary result from multivariate calculus implies that, for a minimum to occur at (a, b), it is necessary for

$$0 = \frac{\partial}{\partial a} \sum_{i=1}^{4} [y_i - (ax_i + b)]^2 \quad \text{and} \quad 0 = \frac{\partial}{\partial b} \sum_{i=1}^{4} [y_i - (ax_i + b)]^2$$

Consequently,

$$0 = 2(2 - 2a - b)(-2) + 2(11 - 4a - b)(-4)$$
$$+ 2(28 - 6a - b)(-6) + 2(40 - 8a - b)(-8),$$

and

$$0 = 2(2 - 2a - b)(-1) + 2(11 - 4a - b)(-1)$$
$$+ 2(28 - 6a - b)(-1) + 2(40 - 8a - b)(-1),$$

which simplifies to

$$30a + 5b = 134 \quad \text{and} \quad 20a + 4b = 81.$$

The solution to this system of equations is $a = 6.55$ and $b = -12.5$, so the best linear equation in the least squares sense is

$$y = 6.55x - 12.5.$$

Table 7.2 lists the observed values together with the values obtained using this approximation.

The problem of finding the equation of the best linear approximation in the absolute sense requires that values of a and b be found to minimize

$$\text{Max} \{|y_i - (ax_i + b)|\}.$$
$$i = 1, 2, 3, 4$$

TABLE 7.2

i	x_i	y_i	$6.55x - 12.5$
1	2	2	.6
2	4	11	13.7
3	6	28	26.8
4	8	40	39.9

This is commonly called a **minimax** problem and cannot be handled by elementary techniques.

Another approach to determining the best linear approximation involves finding values of a and b to minimize

$$\sum_{i=1}^{4} |y_i - (ax_i + b)| = |2 - (2a + b)| + |11 - (4a + b)|$$

$$+ |28 - (6a + b)| + |40 - (8a + b)|.$$

This quantity is called the **absolute deviation**. In order to minimize the absolute deviation, it is necessary that:

$$0 = \frac{\partial}{\partial a} \sum_{i=1}^{4} |y_i - (ax_i + b)| \quad \text{and} \quad 0 = \frac{\partial}{\partial b} \sum_{i=1}^{4} |y_i - (ax_i + b)|.$$

The difficulty with this procedure is that the absolute-value function is not differentiable at zero, and solutions to this pair of equations cannot necessarily be obtained.

The preceding remarks indicate that the least squares method is the most convenient procedure for determining best linear approximations; fortunately, there are important theoretical considerations that also favor this method. The minimax approach will generally assign too much weight to a bit of data that is badly in error, while the method using absolute deviation simply averages the error at the various points and does not give sufficient weight to a point that is considerably out of line with the approximation. The least-squares approach puts substantially more weight on a point that is out of line with the rest of the data but will not allow that point to completely dominate the approximation.

An additional reason for considering the least-squares approach involves the study of the statistical distribution of error. If the data is known or assumed to have its mean distributed in a linear manner, the values obtained from a linear least-squares procedure are unbiased estimates for the equation that describes the mean. Moreover, the values obtained can be used to calculate an unbiased estimator for the variance associated with the distribution. (An easily readable presentation of the theory involved can be found in Larson [59], pages 359–370.)

The general problem of fitting the best least-squares line to a collection of data involves minimizing $\sum_{i=1}^{n} [y_i - (ax_i + b)]^2$ with respect to the parameters a and b. In order for a minimum to occur, it is necessary that

$$0 = \frac{\partial}{\partial a} \sum_{i=1}^{n} [y_i - (ax_i + b)]^2 = 2 \sum_{i=1}^{n} (y_i - ax_i - b)(-x_i)$$

and

$$0 = \frac{\partial}{\partial b} \sum_{i=1}^{n} (y_i - ax_i - b)^2 = 2 \sum_{i=1}^{n} (y_i - ax_i - b)(-1).$$

These equations simplify to what is known as the **normal equations**:

$$a \sum_{i=1}^{n} x_i^2 + b \sum_{i=1}^{n} x_i = \sum_{i=1}^{n} x_i y_i \quad \text{and} \quad a \sum_{i=1}^{n} x_i + b \cdot n = \sum_{i=1}^{n} y_i.$$

The solution to this system of equations is

(7.1)
$$a = \frac{n(\sum_{i=1}^{n} x_i y_i) - (\sum_{i=1}^{n} x_i)(\sum_{i=1}^{n} y_i)}{n(\sum_{i=1}^{n} x_i^2) - (\sum_{i=1}^{n} x_i)^2}$$

and

(7.2)
$$b = \frac{(\sum_{i=1}^{n} x_i^2)(\sum_{i=1}^{n} y_i) - (\sum_{i=1}^{n} x_i y_i)(\sum_{i=1}^{n} x_i)}{n(\sum_{i=1}^{n} x_i^2) - (\sum_{i=1}^{n} x_i)^2}.$$

EXAMPLE 1   Consider the data presented in Table 7.3.

TABLE 7.3

x_i	y_i
1	1.3
2	3.5
3	4.2
4	5.0
5	7.0
6	8.8
7	10.1
8	12.5
9	13.0
10	15.6
11	16.1

To find the least-squares line approximating this data, extend the table and sum the columns, as shown in Table 7.4.

TABLE 7.4

x_i	y_i	x_i^2	$x_i y_i$
1	1.3	1	1.3
2	3.5	4	7.0
3	4.2	9	12.6
4	5.0	16	20.0
5	7.0	25	35.0
6	8.8	36	52.8
7	10.1	49	70.7
8	12.5	64	100.0
9	13.0	81	117.0
10	15.6	100	156.0
11	16.1	121	177.1
66	97.1	506	749.5

TABLE 7.5

x_i	y_i	$1.517x_i - .276$
1	1.3	1.24
2	3.5	2.76
3	4.2	4.28
4	5.0	5.79
5	7.0	7.31
6	8.8	8.83
7	10.1	10.34
8	12.5	11.86
9	13.0	13.38
10	15.6	14.89
11	16.1	16.41

Equations (7.1) and (7.2) imply that

$$a = \frac{11(749.5) - 66(97.1)}{11(506) - (66)^2} = 1.517$$

and

$$b = \frac{506(97.1) - 66(749.5)}{11(506) - (66)^2} = -.276.$$

The graph of this line together with the data points is shown in Fig. 7.2.

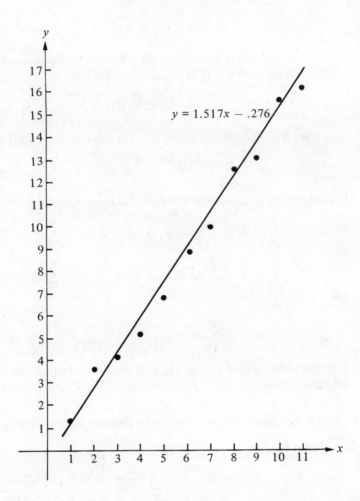

FIGURE 7.2

The approximate values given by the least-squares technique and the data points are given in Table 7.5.   □

The general problem of approximating a set of data, $\{(x_i, y_i) | i = 0, 1, \ldots, M\}$, with a polynomial $P_n(x) = \sum_{k=0}^{n} a_k x^k$ of degree $n < M$ using the least-squares procedure is handled in a similar manner and requires choosing the constants $a_0, a_1, \ldots, a_n$ to minimize

$$
\begin{aligned}
E &= \sum_{i=0}^{M} (y_i - P(x_i))^2 \\
&= \sum_{i=0}^{M} y_i^2 - 2 \sum_{i=0}^{M} P(x_i) y_i + \sum_{i=0}^{M} (P(x_i))^2 \\
&= \sum_{i=0}^{M} y_i^2 - 2 \sum_{i=0}^{M} \left(\sum_{j=0}^{n} a_j x_i^j \right) y_i + \sum_{i=0}^{M} \left(\sum_{j=0}^{n} a_j x_i^j \right)^2 \\
&= \sum_{i=0}^{M} y_i^2 - 2 \sum_{j=0}^{n} a_j \left(\sum_{i=0}^{M} y_i x_i^j \right) + \sum_{j=0}^{n} \sum_{k=0}^{n} a_j a_k \left(\sum_{i=0}^{M} x_i^{j+k} \right).
\end{aligned}
$$

As in the linear case, in order for E to be minimized, it is necessary that $\partial E / \partial a_j = 0$ for each $j = 0, 1, \ldots, n$. Thus, for each j,

$$
0 = \frac{\partial E}{\partial a_j} = -2 \sum_{i=0}^{M} y_i x_i^j + 2 \sum_{k=0}^{n} a_k \sum_{i=0}^{M} x_i^{j+k}.
$$

This gives $n + 1$ equations in the $n + 1$ unknowns, a_j, called the **normal equations**,

$$
\sum_{k=0}^{n} a_k \sum_{i=0}^{M} x_i^{j+k} = \sum_{i=0}^{M} y_i x_i^j, \qquad j = 0, 1, \ldots, n.
$$

It is helpful to write out the equations as follows:

$$
a_0 \sum_{i=0}^{M} x_i^0 + a_1 \sum_{i=0}^{M} x_i^1 + a_2 \sum_{i=0}^{M} x_i^2 + \cdots + a_n \sum_{i=0}^{M} x_i^n = \sum_{i=0}^{M} y_i x_i^0,
$$

$$
a_0 \sum_{i=0}^{M} x_i^1 + a_1 \sum_{i=0}^{M} x_i^2 + a_2 \sum_{i=0}^{M} x_i^3 + \cdots + a_n \sum_{i=0}^{M} x_i^{n+1} = \sum_{i=0}^{M} y_i x_i^1
$$

$$
\vdots
$$

$$
a_0 \sum_{i=0}^{M} x_i^n + a_1 \sum_{i=0}^{M} x_i^{n+1} + a_2 \sum_{i=0}^{M} x_i^{n+2} + \cdots + a_n \sum_{i=0}^{M} x_i^{2n} = \sum_{i=0}^{M} y_i x_i^n.
$$

It can be shown, (see Exercise 11), that the normal equations always have a unique solution provided that the x_i, for $i = 0, 1, \ldots, M$, are distinct.

EXAMPLE 2 Fit the data in Table 7.6 with the discrete least squares polynomial of degree two.

For this problem, $n = 2$, $M = 4$, and the three normal equations are:

$$
\begin{aligned}
5a_0 + \quad 2.5a_1 + \quad 1.875a_2 &= 8.7680 \\
2.5a_0 + \quad 1.875a_1 + 1.5625a_2 &= 5.4514 \\
1.875a_0 + 1.5625a_1 + 1.3828a_2 &= 4.4015
\end{aligned}
$$

TABLE 7.6

i	0	1	2	3	4
x_i	0	.25	.5	.75	1.00
y_i	1.0000	1.2840	1.6487	2.1170	2.7183

The solution to this system is

$$a_0 = 1.0052, \qquad a_1 = .8641, \qquad a_2 = .8437.$$

Thus, the least-squares polynomial of degree two fitting the above data is $P_2(x) = 1.0052 + .8641x + .8437x^2$, whose graph is shown in Fig. 7.3. At the given values of x_i, we have the approximations shown in Table 7.7.

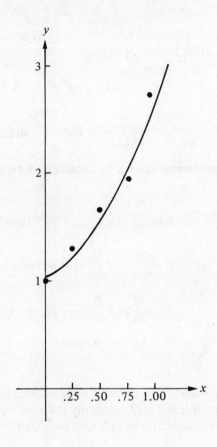

FIGURE 7.3

TABLE 7.7

i	0	1	2	3	4
x_i	0	.25	.50	.75	1.00
y_i	1.0000	1.2840	1.6487	2.1170	2.7183
$P(x_i)$	1.0052	1.2740	1.6482	2.1279	2.7130
$y_i - P(x_i)$	$-.0052$	.0100	.0005	$-.0109$	.0053

The error

$$\sum_{i=0}^{4} |y_i - P(x_i)|^2 = 2.76 \times 10^{-4}$$

is the least that can be obtained by using a quadratic polynomial. $\square$

Although least squares using polynomials is the most extensively used procedure, occasionally it is appropriate to assume that the data is exponentially related. This requires the approximating function to be of the form

$$(7.3) \hspace{3cm} y = be^{ax}$$

or

$$(7.4) \hspace{3cm} y = bx^a$$

for some constants a and b. The difficulty with applying the least-squares procedure in a situation of this type comes from attempting to minimize

$$E = \sum_{i=0}^{M} (y_i - be^{ax_i})^2 \hspace{1cm} \text{in the case of Eq. (7.3)}$$

or

$$E = \sum_{i=0}^{M} (y_i - bx_i^a)^2 \hspace{1cm} \text{in the case of Eq. (7.4).}$$

The normal equations associated with these procedures are obtained from either

$$0 = \frac{\partial E}{\partial b} = 2 \sum_{i=0}^{M} (y_i - be^{ax_i})(-e^{ax_i})$$

and

$$0 = \frac{\partial E}{\partial a} = 2 \sum_{i=0}^{M} (y_i - be^{ax_i})(-bx_i e^{ax_i}), \hspace{1cm} \text{in the case of Eq. (7.3),}$$

or

$$0 = \frac{\partial E}{\partial b} = 2 \sum_{i=0}^{M} (y_i - bx_i^a)(-x_i^a)$$

and

$$0 = \frac{\partial E}{\partial a} = 2 \sum_{i=0}^{M} (y_i - bx_i^a)(-b(\ln x_i)x_i^a), \hspace{1cm} \text{in the case of Eq. (7.4).}$$

No exact solution to either of these systems can generally be found.

The method that is usually followed when the data is suspected to be exponentially related is to consider the logarithm of the approximating equation:

$$(7.5) \hspace{2cm} \ln y = \ln b + ax \hspace{1cm} \text{in the case of Eq. (7.3),}$$

and

$$(7.6) \hspace{2cm} \ln y = \ln b + a \ln x \hspace{1cm} \text{in the case of Eq. (7.4).}$$

In either case, a linear problem now appears and solutions for $\ln b$ and a can be obtained from appropriately modifying Eqs. (7.1) and (7.2). It should be remembered, however, that the approximation obtained in this manner is not the least-squares approximation for the original problem and that this approximation can in some cases differ significantly from the least-squares approximation to the original problem. The application in Exercise 10 describes such a problem. This application will be reconsidered as an exercise in Section 9.2, where the exact solution to the exponential least-squares problem is approximated by using methods suitable for nonlinear systems of equations.

EXAMPLE 3   Consider the collection of data in Table 7.8.

TABLE 7.8

i	x_i	y_i
1	1.00	5.10
2	1.25	5.79
3	1.50	6.53
4	1.75	7.45
5	2.00	8.46

If x_i is graphed with $\ln y_i$, the data appears to have a linear relation, or, equivalently, if (x_i, y_i) is graphed on semilog paper, a linear shape will appear. Thus, it is reasonable to assume an approximation of the form

$$y = be^{ax} \qquad \text{or} \qquad \ln y = \ln b + ax.$$

Extending the table and summing the columns gives the data in Table 7.9.

TABLE 7.9

i	x_i	y_i	$\ln y_i$	x_i^2	$x_i \ln y_i$
1	1.00	5.10	1.629	1.0000	1.629
2	1.25	5.79	1.756	1.5625	2.195
3	1.50	6.53	1.876	2.2500	2.814
4	1.75	7.45	2.008	3.0625	3.514
5	2.00	8.46	2.135	4.0000	4.270
	7.50		9.404	11.875	14.422

Using Eqs. (7.1) and (7.2),

$$a = \frac{(5)(14.422) - (7.5)(9.404)}{(5)(11.875) - (7.5)^2} = .5056$$

and

$$\ln b = \frac{(11.875)(9.404) - (14.422)(7.5)}{(5)(11.875) - (7.5)^2} = 1.122$$

TABLE 7.10

i	x_i	y_i	$3.071 e^{.5056 x_i}$
1	1.00	5.10	5.09
2	1.25	5.79	5.78
3	1.50	6.53	6.56
4	1.75	7.45	7.44
5	2.00	8.46	8.44

Since $b = e^{1.122} = 3.071$, the approximation assumes the form

$$y = 3.071e^{.5056x},$$

which, at the data points, gives the values in Table 7.10.   □

Exercise Set 7.1

1. Compute the least-squares polynomial of degree one for the data of Example 2.

2. The data for Example 2 is actually given by the function $f(x) = e^x$. Considering the graph of $f(x) = e^x$ on [0, 1], do you believe that choosing a higher-degree least-squares polynomial would improve upon the results of Example 2? Support your claim by considering least-squares polynomials of degree three and four.

3. Compute the Lagrange interpolating polynomial of degree two for $f(x) = e^x$ on the nodes $x_0 = 0, x_1 = .5,$ and $x_2 = 1.0$. Compare this second-degree polynomial and that of Example 2 to determine which better approximates $f(x) = e^x$ on [0, 1].

4. Find the least-squares polynomials of degree 1, 2, 3, and 4 for the data in the table below.

i	0	1	2	3	4	5
x_i	0	.15	.31	.5	.6	.75
y_i	1.0	1.004	1.031	1.117	1.223	1.422

Which degree gives the best least-squares approximation; that is, which gives the smaller error? Can you justify your answer?

5. Listed below are homework grades and final examination grades for 30 numerical analysis students. Find the equation of the least-squares line for this data, and use this line to determine the homework grade required to predict minimal A(90%) and D(60%) grades on the final.

Homework	Final	Homework	Final	Homework	Final
302	45	343	83	234	51
325	72	290	74	337	53
285	54	326	76	351	100
339	54	233	57	339	67
334	79	254	45	343	83
322	65	323	83	314	42
331	99	337	99	344	79
279	63	337	70	185	59
316	65	304	62	340	75
347	99	319	66	316	45

6. The following table lists the college grade-point averages of 20 mathematics and computer science majors, together with the scores that these students received on the ACT (American College Testing Program) test while in high school. Plot this data, and find the equation of the least-squares line for this data.

ACT score	Grade-point average	ACT score	Grade-point average
28	3.84	29	3.75
25	3.21	28	3.65
28	3.23	27	3.87
27	3.63	29	3.75
28	3.75	21	1.66
33	3.20	28	3.12
28	3.41	28	2.96
29	3.38	26	2.92
23	3.53	30	3.10
27	2.03	24	2.81

7. In order to determine a relationship between the number of fish and the number of species of fish in samples taken for a portion of the Great Barrier Reef, P. Sale and R. Dybdahl [72] fit a linear least-squares polynomial to the following collection of data, which was collected in samples over a two-year period: Let x be the number of fish in the sample, and y be the number of species in the sample.

x	y	x	y	x	y
13	11	29	12	60	14
15	10	30	14	62	21
16	11	31	16	64	21
21	12	36	17	70	24
22	12	40	13	72	17
23	13	42	14	100	23
25	13	55	22	130	34

Determine the linear least-squares polynomial for this data.

8. The following set of data, presented in March, 1970, to the Senate Antitrust Subcommittee, shows the comparative crash-survivability characteristics of cars in various classes. Find the best least-squares line that approximates this data. (The table shows the percent of accident-involved vehicles in which the most severe injury was fatal or serious.)

Type	Average weight	Percent occurrence
1. Domestic "luxury" regular	4800 lb	3.1
2. Domestic "intermediate" regular	3700 lb	4.0
3. Domestic "economy" regular	3400 lb	5.2
4. Domestic compact	2800 lb	6.4
5. Foreign compact	1900 lb	9.6

9. In order to determine a functional relationship between the attenuation coefficient and the thickness of a sample of taconite, V. P. Singh [82] fits a collection of data by using a linear least-squares polynomial. The following collection of data is taken from a graph in that paper. Find the best linear least-squares polynomial fitting this data.

Thickness (cm)	Attenuation coefficient db/cm
.040	26.5
.041	28.1
.055	25.2
.056	26.0
.062	24.0
.071	25.0
.071	26.4
.078	27.2
.082	25.6
.090	25.0
.092	26.8
.100	24.8
.105	27.0
.120	25.0
.123	27.3
.130	26.9
.140	26.2

10. In a paper dealing with the efficiency of energy utilization of the larvae of the Modest Sphinx moth (*Pachysphinx modesta*), L. Schroeder [75] used the following data to determine a relation between W, the live weight of the larvae in grams, and R, the oxygen consumption of the larvae in ml/hr. For biological reasons, it is assumed that a relationship in the form of Eq. (7.4) exists between W and R.

a) Find the logarithmic linear least-squares polynomial by using

$$\ln R = \ln b + a \ln W.$$

b) Compute the error associated with the approximation in (a):

$$E = \sum_{i=1}^{37} (R_i - bW_i^a)^2.$$

c) Modify the logarithmic least squares equation in (a) by adding the quadratic term $c(\ln W_i)^2$ and determine the logarithmic quadratic least-squares polynomial.

d) Determine the formula for and compute the error associated with the approximation in (c).

W	R	W	R	W	R	W	R	W	R
0.017	0.154	0.025	0.23	0.020	0.181	0.020	0.180	0.025	0.234
0.087	0.296	0.111	0.357	0.085	0.260	0.119	0.299	0.233	0.537
0.174	0.363	0.211	0.366	0.171	0.334	0.210	0.428	0.783	1.47
1.11	0.531	0.999	0.771	1.29	0.87	1.32	1.15	1.35	2.48
1.74	2.23	3.02	2.01	3.04	3.59	3.34	2.83	1.69	1.44
4.09	3.58	4.28	3.28	4.29	3.40	5.48	4.15	2.75	1.84
5.45	3.52	4.58	2.96	5.30	3.88			4.83	4.66
5.96	2.40	4.68	5.10					5.53	6.94

11. Show that the normal equations discussed on page 324 resulting from discrete least-squares approximation yield a symmetric and nonsingular matrix and hence have a unique solution. [*Hint*:   Let $A = (a_{ij})$ where

$$a_{ij} = \sum_{k=1}^{M} x_k^{i+j-2},$$

and $x_1, x_2, \ldots, x_n$ are distinct points with $n < M$. Suppose A is singular and that $\mathbf{c} \neq \mathbf{0}$ is such that $\mathbf{c}^t A \mathbf{c} = 0$. Show that the nth-degree polynomial whose coefficients are the coordinates of $\mathbf{c}$ has more than n roots, and use this to establish a contradiction.]

(7.2) Orthogonal Polynomials and Least-Squares Approximation

The previous section considered the problem of least-squares approximation to fit a collection of data. The other approximation problem mentioned in the introduction concerns the approximation of functions.

Suppose $f \in C[a, b]$ and that a polynomial of degree at most n, P_n, is required that will minimize

(7.7) $$\int_a^b (f(x) - P_n(x))^2 \, dx.$$

To determine a least-squares approximating polynomial, that is, a polynomial to minimize expression (7.7), let

$$P_n(x) = a_n x^n + a_{n-1} x^{n-1} + \cdots + a_1 x + a_0 = \sum_{k=0}^{n} a_k x^k,$$

and define

$$E(a_0, a_1, \ldots, a_n) = \int_a^b \left(f(x) - \sum_{k=0}^{n} a_k x^k \right)^2 dx.$$

The problem is to find real coefficients $a_0, \ldots, a_n$ that will minimize E. From the calculus of functions of several variables, a necessary condition for the numbers $a_0, \ldots, a_n$ to minimize E is that

$$\frac{\partial E}{\partial a_j} = 0 \qquad \text{for each } j = 0, 1, \ldots, n.$$

Since $$E = \int_a^b [f(x)]^2 \, dx - 2 \sum_{k=0}^{n} a_k \int_a^b x^k f(x) \, dx + \int_a^b \left(\sum_{k=0}^{n} a_k x^k \right)^2 dx,$$

$$\frac{\partial E}{\partial a_j} = -2 \int_a^b x^j f(x) \, dx + 2 \sum_{k=0}^{n} a_k \int_a^b x^{j+k} \, dx.$$

Hence, in order to find P_n, the $(n + 1)$ linear equations

(7.8) $$\sum_{k=0}^{n} a_k \int_a^b x^{j+k} \, dx = \int_a^b x^j f(x) \, dx, \qquad j = 0, 1, \ldots, n,$$

must be solved for the $(n + 1)$ unknowns $a_j, j = 0, 1, \ldots, n$. These equations are called the **normal equations**. It can be shown that the normal equations always have a unique solution provided $f \in C[a, b]$ and $a \neq b$. (See Exercise 9.)

EXAMPLE 1   Find the least-squares approximating polynomial of degree two for the function $f(x) = \sin \pi x$ on the interval $[0, 1]$. The normal equations for $P_2(x) = a_2 x^2 + a_1 x + a_0$ are given by:

$$a_0 \int_0^1 1 \, dx + a_1 \int_0^1 x \, dx + a_2 \int_0^1 x^2 \, dx = \int_0^1 \sin \pi x \, dx,$$

$$a_0 \int_0^1 x \, dx + a_1 \int_0^1 x^2 \, dx + a_2 \int_0^1 x^3 \, dx = \int_0^1 x \sin \pi x \, dx,$$

$$a_0 \int_0^1 x^2 \, dx + a_1 \int_0^1 x^3 \, dx + a_2 \int_0^1 x^4 \, dx = \int_0^1 x^2 \sin \pi x \, dx.$$

Performing the integration yields

$$a_0 + \tfrac{1}{2} a_1 + \tfrac{1}{3} a_2 = \frac{2}{\pi},$$

$$\tfrac{1}{2} a_0 + \tfrac{1}{3} a_1 + \tfrac{1}{4} a_2 = \frac{1}{\pi},$$

$$\tfrac{1}{3} a_0 + \tfrac{1}{4} a_1 + \tfrac{1}{5} a_2 = \frac{\pi^2 - 4}{\pi^3}.$$

The three equations in three unknowns can be solved to obtain

$$a_0 = \frac{12\pi^2 - 120}{\pi^3} \approx -.050465,$$

$$a_1 = \frac{720 - 60\pi^2}{\pi^3} \approx 4.12251,$$

$$a_2 = \frac{60\pi^2 - 720}{\pi^3} \approx -4.12251.$$

Consequently, the least-squares polynomial approximation of degree two for $f(x) = \sin \pi x$ on $[0, 1]$ is $P_2(x) = -4.12251x^2 + 4.12251x - .050465$; see Fig. 7.4.   □

Example 1 serves to illustrate the degree of difficulty in obtaining a least-squares polynomial approximation. One must solve an $(n + 1) \times (n + 1)$ linear system for the coefficients $a_0, \ldots, a_n$ of P_n. Also, since the coefficients in the linear system are of the form

$$\int_a^b x^{j+k} \, dx = \frac{b^{j+k+1} - a^{j+k+1}}{j + k + 1},$$

the linear system does not have a convenient numerical solution because of the presence of rounding errors. Another disadvantage to the technique used above is similar to the situation that occurred when the Lagrange polynomials were first introduced in Section 3.2; that is, the calculations that were performed in obtaining the best nth-degree polynomial, P_n, do not lessen the amount of work required to obtain P_{n+1}, the next higher degree polynomial.

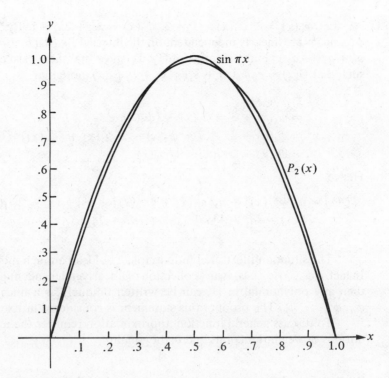

FIGURE 7.4

Another technique to obtain least-squares approximations will now be considered. This technique turns out to be efficient in computation and also useful, in the sense that, once P_n is known, it is easy to determine P_{n+1}. To facilitate further discussion, some new concepts will be introduced.

For notation, **let Π_n be the set of all polynomials of degree at most n.** Throughout $\phi_0, \ldots, \phi_n$ will denote polynomials and ϕ_j will be of degree j. It is not actually necessary to restrict this study to polynomials; all that is needed is for the functions $\phi_0, \ldots, \phi_n$ to be continuous and to satisfy the following property.

Definition 7.1   The functions $\phi_0, \ldots, \phi_n$ are said to be **linearly independent** on $[a, b]$ where $b > a$ if, whenever

$$c_0\phi_0(x) + c_1\phi_1(x) + \cdots + c_n\phi_n(x) = 0 \qquad \text{for all } x \in [a, b],$$

then $c_0 = c_1 = \cdots = c_n = 0$. Otherwise the functions are said to be **linearly dependent**.

Theorem 7.2   If $\phi_j(x)$ is a polynomial of degree j, for each $j = 0, 1, \ldots, n$, then $\phi_0, \ldots, \phi_n$ are linearly independent on any interval $[a, b]$ where $a < b$.

Proof   Suppose $c_0, \ldots, c_n$ are real numbers for which

$$c_0\phi_0(x) + c_1\phi_1(x) + \cdots + c_n\phi_n(x) = 0 \qquad \text{for all } x \in [a, b].$$

Since $P(x) = \sum_{k=0}^{n} c_k\phi_k(x)$ vanishes on $[a, b]$, P has an infinite number of roots. P is a polynomial of degree at most n, so Corollary 2.15, page 56, implies $P \equiv 0$ and that the coefficient of each power of x vanishes. Since each ϕ_j is of degree exactly j, this implies that $c_j = 0$ for each $j = 0, 1, \ldots, n$.   □

EXAMPLE 2 Let $\phi_0(x) = 2$, $\phi_1(x) = x - 3$, $\phi_2(x) = x^2 + 2x + 7$. By Theorem 7.2, ϕ_0, ϕ_1, and ϕ_2 are linearly independent on the interval $[a, b]$ if $b > a$. Suppose $Q(x) = a_0 + a_1 x + a_2 x^2$ is any member of Π_2. To show that there exist constants c_0, c_1, c_2 such that $Q(x) = c_0 \phi_0(x) + c_1 \phi_1(x) + c_2 \phi_2(x)$, note that

$$1 = \tfrac{1}{2}\phi_0(x),$$
$$x = \phi_1(x) + 3 = \phi_1(x) + \tfrac{3}{2}\phi_0(x),$$
$$x^2 = \phi_2(x) - 2x - 7 = \phi_2(x) - 2[\phi_1(x) + \tfrac{3}{2}\phi_0(x)] - 7[\tfrac{1}{2}\phi_0(x)]$$
$$= \phi_2(x) - 2\phi_1(x) - \tfrac{13}{2}\phi_0(x).$$

Hence,

$$Q(x) = a_0[\tfrac{1}{2}\phi_0(x)] + a_1[\phi_1(x) + \tfrac{3}{2}\phi_0(x)] + a_2[\phi_2(x) - 2\phi_1(x) - \tfrac{13}{2}\phi_0(x)]$$
$$= [\tfrac{1}{2}a_0 + \tfrac{3}{2}a_1 - \tfrac{13}{2}a_2]\phi_0(x) + [a_1 - 2a_2]\phi_1(x) + a_2\phi_2(x). \qquad \square$$

The situation illustrated in Example 2 holds in a much more general setting. In fact, if $\phi_0, \phi_1, \ldots, \phi_n$ is any collection of linearly independent polynomials in Π_n, then any polynomial in Π_n can be written uniquely as a linear combination of $\phi_0, \phi_1, \ldots, \phi_n$. The proof of this statement is considered in Exercise 12.

To discuss general function approximation requires the introduction of the notions of weight functions and orthogonality.

Definition 7.3 An integrable function w is called a **weight function** on $[a, b]$ if $w(x) \geq 0$ for $x \in [a, b]$, but $w(x) \not\equiv 0$ on any subinterval of $[a, b]$.

The purpose of a weight function is to assign varying degrees of importance to approximation errors on certain portions of the interval.

Suppose that $\phi_0, \phi_1, \ldots, \phi_n$ are linearly independent functions on $[a, b]$, that w is a weight function for $[a, b]$, and that for $f \in C[a, b]$ a linear combination

$$P(x) = \sum_{k=0}^{n} a_k \phi_k(x)$$

is sought to minimize

$$(7.9) \qquad E(a_0, \ldots, a_n) = \int_a^b w(x)\left[f(x) - \sum_{k=0}^{n} a_k \phi_k(x)\right]^2 dx$$

This problem reduces to the situation considered at the beginning of this section in the special case when $w(x) \equiv 1$ and $\phi_k(x) = x^k$ for each $k = 0, 1, \ldots, n$.

The normal equations associated with Eq. (7.9) are derived from the fact that for each $j = 0, 1, \ldots, n$,

$$0 = \frac{\partial E}{\partial a_j} = 2 \int_a^b w(x)\left[f(x) - \sum_{k=0}^{n} a_k \phi_k(x)\right]\phi_j(x)\, dx.$$

The system of normal equations can be written:

$$\int_a^b w(x)f(x)\phi_j(x)\, dx = \sum_{k=0}^{n} a_k \int_a^b w(x)\phi_k(x)\phi_j(x)\, dx \qquad \text{for } j = 0, 1, \ldots, n.$$

Suppose that the functions $\phi_0, \phi_1, \ldots, \phi_n$ can be chosen so that

$$(7.10) \qquad \int_a^b w(x)\phi_k(x)\phi_j(x)\,dx = \begin{cases} 0 & \text{when } j \neq k \\ \alpha_k > 0, & \text{when } j = k \end{cases}$$

Then, for each $j = 0, 1, \ldots, n$.

$$\int_a^b w(x)f(x)\phi_j(x)\,dx = a_j \int_a^b w(x)[\phi_j(x)]^2\,dx = a_j\alpha_j$$

and

$$a_j = \frac{1}{\alpha_j} \int_a^b w(x)f(x)\phi_j(x)\,dx.$$

Hence the least-squares approximation problem is greatly simplified when the functions $\phi_0, \phi_1, \ldots, \phi_n$ are chosen to satisfy Eq. (7.10). The remainder of this section will be devoted to studying collections of this type.

Definition 7.4 $\{\phi_0, \phi_1, \ldots, \phi_n\}$ is said to be an **orthogonal set of functions** for the interval $[a, b]$ with respect to the weight function w if

$$\int_a^b w(x)\phi_j(x)\phi_k(x)\,dx = \begin{cases} 0, & \text{whenever } j \neq k, \\ \alpha_k > 0, & \text{whenever } j = k. \end{cases}$$

If, in addition, $\alpha_k = 1$ for each $k = 0, 1, \ldots, n$, the set is said to be **orthonormal**.

This definition, together with the remarks preceding it, produces the following theorem.

Theorem 7.5   If $\{\phi_0, \ldots, \phi_n\}$ is an orthogonal set of functions on an interval $[a, b]$ with respect to the weight function w, then the least-squares approximation to f on $[a, b]$ with respect to w is

$$P(x) = \sum_{k=0}^n a_k \phi_k(x)$$

where

$$a_k = \frac{\int_a^b w(x)\phi_k(x)f(x)\,dx}{\int_a^b w(x)[\phi_k(x)]^2\,dx} = \frac{1}{\alpha_k} \int_a^b w(x)\phi_k(x)f(x)\,dx.$$

EXAMPLE 3   For each positive integer n, the set of functions $\mathscr{T}_n = \{\phi_0, \phi_1, \ldots, \phi_{2n-1}\}$ where

$$\phi_0(x) \equiv \frac{1}{\sqrt{2\pi}},$$

$$\phi_k(x) = \frac{1}{\sqrt{\pi}} \cos(kx) \qquad \text{for each } k = 1, 2, \ldots, n,$$

and $\qquad \phi_{n+k}(x) = \frac{1}{\sqrt{\pi}} \sin(kx) \qquad \text{for each } k = 1, 2, \ldots, n-1,$

form an orthonormal set of functions on $[-\pi, \pi]$ with respect to $w(x) \equiv 1$. (See Exercise 3.) (Some sources include an additional function in the set, $\phi_{2n}(x) = (1/\sqrt{\pi})\sin(nx)$. We will not follow this convention.) Given $f \in C[-\pi, \pi]$, the least-squares approximation (called a **trigonometric polynomial**) by functions in $\mathcal{T}_n$ is defined by

$$S_n(x) = \sum_{k=0}^{2n-1} a_k \phi_k(x),$$

where $\qquad a_k = \int_{-\pi}^{\pi} f(x)\phi_k(x)\, dx \qquad$ for each $k = 0, 1, \ldots, 2n - 1.$

The limit of S_n when $n \to \infty$ is called the **Fourier Series** of f. Fourier series are an extremely useful tool for describing the solution of various ordinary and partial-differential equations that occur in physical situations.

To determine the trigonometric polynomial from $\mathcal{T}_n$ that approximates

$$f(x) = |x| \qquad \text{for } -\pi < x < \pi$$

requires finding

$$a_0 = \int_{-\pi}^{\pi} |x|\, \frac{1}{\sqrt{2\pi}}\, dx$$

$$= -\frac{1}{\sqrt{2\pi}} \int_{-\pi}^{0} x\, dx + \frac{1}{\sqrt{2\pi}} \int_{0}^{\pi} x\, dx$$

$$= \frac{2}{\sqrt{2\pi}} \int_{0}^{\pi} x\, dx = \frac{\sqrt{2}\pi^2}{2\sqrt{\pi}},$$

$$a_k = \frac{1}{\sqrt{\pi}} \int_{-\pi}^{\pi} |x| \cos kx\, dx = \frac{2}{\sqrt{\pi}} \int_{0}^{\pi} x \cos kx\, dx$$

$$= \frac{2}{\sqrt{\pi}k^2} [(-1)^k - 1] \qquad \text{for each } k = 1, 2, \ldots, n,$$

and

$$a_{n+k} = \frac{1}{\sqrt{\pi}} \int_{-\pi}^{\pi} |x| \sin kx\, dx$$

$$= 0 \qquad \text{for each } k = 1, 2, \ldots, n - 1.$$

The trigonometric polynomial from $\mathcal{T}_n$ approximating f is therefore

$$S_n(x) = \frac{\pi}{2} - \frac{2}{\pi} \sum_{k=1}^{n} \frac{(-1)^k - 1}{k^2} \cos kx.$$

The Fourier series for f is

$$S(x) = \frac{\pi}{2} - \frac{2}{\pi} \sum_{k=1}^{\infty} \frac{(-1)^k - 1}{k^2} \cos kx. \qquad \square$$

For the remainder of this section, only orthogonal sets of polynomials will be considered. The next theorem, which is based on what is called the **Gram–Schmidt process**, describes how to construct orthogonal polynomials on $[a, b]$ with respect to a weight function w.

Theorem 7.6 The set of polynomials $\{\phi_0, \phi_1, \ldots, \phi_n\}$ defined in the following way is orthogonal on $[a, b]$ with respect to the weight function w.

$$\phi_0(x) \equiv 1,$$

$$\phi_1(x) = x - B_1 \qquad \text{for each } a \leq x \leq b,$$

where

$$B_1 = \frac{\int_a^b xw(x)[\phi_0(x)]^2 \, dx}{\int_a^b w(x)[\phi_0(x)]^2 \, dx};$$

when $k \geq 2$,

$$\phi_k(x) = (x - B_k)\phi_{k-1}(x) - C_k\phi_{k-2}(x) \qquad \text{for each } a \leq x \leq b,$$

where

$$B_k = \frac{\int_a^b xw(x)[\phi_{k-1}(x)]^2 \, dx}{\int_a^b w(x)[\phi_{k-1}(x)]^2 \, dx}$$

and

$$C_k = \frac{\int_a^b xw(x)\phi_{k-1}(x)\phi_{k-2}(x) \, dx}{\int_a^b w(x)[\phi_{k-2}(x)]^2 \, dx}.$$

Proof   Each ϕ_k is of the form $1 \cdot x^k + $ lower-order terms, so all denominators for B_k and C_k are nonzero.

We will show by induction on k that

$$\int_a^b w(x)\phi_k(x)\phi_i(x) \, dx = 0 \qquad \text{for each } i < k.$$

For $k = 1$,

$$\int_a^b w(x)\phi_1(x)\phi_0(x) \, dx = \int_a^b w(x)(x - B_1) \, dx$$

$$= \int_a^b xw(x) \, dx - B_1 \int_a^b w(x) \, dx$$

$$= \int_a^b xw(x)[\phi_0(x)]^2 \, dx$$

$$- \left[\frac{\int_a^b xw(x)[\phi_0(x)]^2 \, dx}{\int_a^b w(x)[\phi_0(x)]^2 \, dx} \right] \int_a^b w(x)[\phi_0(x)]^2 \, dx$$

$$= 0.$$

Assume that the result is true for $k = n - 1$. Then,

$$\int_a^b w(x)\phi_n(x)\phi_{n-1}(x) \, dx = \int_a^b w(x)[(x - B_n)\phi_{n-1}(x) - C_n\phi_{n-2}(x)] \cdot \phi_{n-1}(x) \, dx$$

$$= \int_a^b w(x)(x - B_n)[\phi_{n-1}(x)]^2 \, dx$$

$$= \int_a^b xw(x)[\phi_{n-1}(x)]^2 \, dx - B_n \int_a^b w(x)[\phi_{n-1}(x)]^2 \, dx$$

$$= 0.$$

In a similar manner (see Exercise 14), it can be shown that

(7.11)
$$\int_a^b w(x)\phi_n(x)\phi_{n-2}(x) \, dx = 0.$$

For $i < n - 2$

$$\int_a^b w(x)\phi_n(x)\phi_i(x) \, dx = \int_a^b w(x)[(x - B_n)\phi_{n-1}(x) - C_n\phi_{n-2}(x)]\phi_i(x) \, dx$$

$$= \int_a^b w(x)x\phi_{n-1}(x)\phi_i(x) \, dx$$

$$= \int_a^b w(x)\phi_{n-1}(x)[\phi_{i+1}(x) + B_{i+1}\phi_i(x) + C_{i+1}\phi_{i-1}(x)] \, dx$$

$$= 0. \qquad \qquad \square$$

Corollary 7.7   For any $n > 0$, the polynomials $\phi_0, \ldots, \phi_n$ given in Theorem 7.6 are linearly independent on $[a, b]$ and

$$\int_a^b w(x)\phi_n(x)Q_k(x) \, dx = 0,$$

for any polynomial Q_k of degree $k < n$.

Proof   To show that $\phi_0, \ldots, \phi_n$ are linearly independent suppose

$$0 = c_0\phi_0(x) + \cdots + c_n\phi_n(x) \qquad \text{for every } x \in [a, b].$$

For any $k = 0, 1, \ldots, n$, multiply by $w(x)\phi_k(x)$ to obtain

$$0 = \sum_{j=0}^n c_j w(x)\phi_j(x)\phi_k(x).$$

Then

$$0 = \sum_{j=0}^n c_j \int_a^b w(x)\phi_j(x)\phi_k(x) \, dx = c_k \int_a^b w(x)[\phi_k(x)]^2 \, dx,$$

and it follows that $c_k = 0$. Since this is true for each $k = 0, 1, \ldots, n$, $\{\phi_0, \ldots, \phi_n\}$ is a linearly independent set.

Let $Q_k(x)$ be a polynomial of degree k. In Exercise 12, it is shown that there exist numbers $c_0, \ldots, c_k$ such that

$$Q_k(x) = \sum_{j=0}^k c_j\phi_j(x).$$

Thus,

$$\int_a^b w(x)Q_k(x)\phi_n(x) \, dx = \sum_{j=0}^k c_j(x) \int_a^b w(x)\phi_j(x)\phi_n(x) \, dx = 0$$

since ϕ_n is orthogonal to ϕ_j for each $j = 0, 1, \ldots, k$. $\qquad \square$

EXAMPLE 4   One of the most common sets of orthogonal polynomials is the set of **Legendre polynomials** $\{P_n\}$, which are orthogonal on $[-1, 1]$ with respect to the weight function $w(x) \equiv 1$. The classical definition of the Legendre polynomials requires that $P_n(1) = 1$ for each n, and a recursive relation can be used to generate the polynomials when $n \geq 2$. This normalization will not be needed in our discussion, and the least-squares approximating polynomials generated in either case will be the same. Using the procedure of Theorem 7.6,

$$P_0(x) \equiv 1$$

$$B_1 = \frac{\int_{-1}^{1} x \, dx}{\int_{-1}^{1} dx} = 0$$

so $$P_1(x) = (x - B_1)P_0(x) = x.$$

Also, $$B_2 = \frac{\int_{-1}^{1} x^3 \, dx}{\int_{-1}^{1} x^2 \, dx} = 0 \quad \text{and} \quad C_2 = \frac{\int_{-1}^{1} x^2 \, dx}{\int_{-1}^{1} 1 \, dx} = \frac{1}{3},$$

so $$P_2(x) = (x - B_2)P_1(x) - C_2 P_0(x) = (x - 0)x - \tfrac{1}{3} \cdot 1 = x^2 - \tfrac{1}{3}.$$

Finally,

$$B_3 = \frac{\int_{-1}^{1} x \cdot (x^2 - \tfrac{1}{3})^2 \, dx}{\int_{-1}^{1} (x^2 - \tfrac{1}{3})^2 \, dx} = 0$$

and $$C_3 = \frac{\int_{-1}^{1} x \cdot x \cdot (x^2 - \tfrac{1}{3}) \, dx}{\int_{-1}^{1} x^2 \, dx} = \frac{\tfrac{8}{45}}{\tfrac{2}{3}} = \frac{4}{15},$$

so $$P_3(x) = (x - B_3)P_2(x) - C_3 P_1(x)$$
$$= x \cdot (x^2 - \tfrac{1}{3}) - \tfrac{4}{15}x = x^3 - \tfrac{3}{5}x. \qquad \square$$

EXAMPLE 5   Another set of orthogonal polynomials, a set that will be used later in this chapter, is called the set of **Chebyshev polynomials**, $\{T_n\}$. They can be derived by means of Theorem 7.6 for the interval $(-1, 1)$, using the weight function $w(x) = (1 - x^2)^{-1/2}$. We will derive the Chebyshev polynomials differently here, and then show that they satisfy the required orthogonality properties.

For $x \in [-1, 1]$, define

$$T_n(x) = \cos[n \arccos x] \qquad \text{for each } n \geq 0.$$

Introducing the substitution $\theta = \arccos x$ changes this equation to

$$T_n(\theta) = \cos(n\theta) \qquad \text{where } \theta \in [0, \pi].$$

A recurrence relation can be derived by noting that

$$T_{n+1}(\theta) = \cos((n + 1)\theta) = \cos(n\theta)\cos \theta - \sin(n\theta)\sin \theta$$

and $$T_{n-1}(\theta) = \cos((n - 1)\theta) = \cos(n\theta)\cos \theta + \sin(n\theta)\sin \theta,$$

so $$T_{n+1}(\theta) = 2 \cos(n\theta)\cos \theta - T_{n-1}(\theta).$$

Returning to the variable x, we obtain:

(7.12) $$T_{n+1}(x) = 2xT_n(x) - T_{n-1}(x) \qquad \text{for each } 1 \leq n.$$

Since

$$T_0(x) = \cos(0 \cdot \arccos x) = 1 \qquad \text{and} \qquad T_1(x) = \cos(1 \arccos x) = x,$$

the Chebyshev polynomials are now easily obtained in a sequential manner by using Eq. (7.12):

$$T_2(x) = 2xT_1(x) - T_0(x) = 2x^2 - 1,$$

$$T_3(x) = 2xT_2(x) - T_1(x) = 4x^3 - 3x,$$

$$T_4(x) = 2xT_3(x) - T_2(x) = 8x^4 - 8x^2 + 1, \quad \text{etc.}$$

To show the orthogonality of the Chebyshev polynomials, consider

$$\int_{-1}^{1} \frac{T_n(x)T_m(x)}{\sqrt{1-x^2}} \, dx = \int_{-1}^{1} \frac{\cos(n \arccos x)\cos(m \arccos x)}{\sqrt{1-x^2}} \, dx.$$

Reintroducing the substitution $\theta = \arccos x$, we obtain the integral

$$\int_{-1}^{1} \frac{T_n(x)T_m(x)}{\sqrt{1-x^2}} \, dx = \int_{\pi}^{0} \frac{\cos(n\theta)\cos(m\theta)}{\sin \theta} (-\sin \theta \, d\theta)$$

$$= -\int_{\pi}^{0} \cos(n\theta)\cos(m\theta) \, d\theta$$

$$= \int_{0}^{\pi} \cos(n\theta)\cos(m\theta) \, d\theta.$$

Suppose $n \neq m$, since $\cos n\theta \cos m\theta = \frac{1}{2}[\cos(n + m)\theta + \cos(n - m)\theta]$,

$$\int_{-1}^{1} \frac{T_n(x)T_m(x) \, dx}{\sqrt{1-x^2}} = \frac{1}{2} \int_{0}^{\pi} \cos((n + m)\theta) \, d\theta + \frac{1}{2} \int_{0}^{\pi} \cos((n - m)\theta) \, d\theta$$

$$= \left[\frac{1}{2(n + m)} \sin((n + m)\theta) + \frac{1}{2(n - m)} \sin((n - m)\theta) \right] \Bigg|_{0}^{\pi}$$

$$= 0.$$

It can also be shown that

(7.13) $$\int_{-1}^{1} \frac{T_n^2(x)}{\sqrt{1-x^2}} \, dx = \frac{\pi}{2} \qquad \text{for each } n \geq 1$$

by a similar technique (see Exercise 6.)   □

   The main motivation for considering *weighted* least squares is that the weight function puts different emphasis upon distinct parts of an interval. For example, the weight function

$$w(x) = \frac{1}{\sqrt{1-x^2}}$$

places less emphasis near the center of the interval $(-1, 1)$ and more emphasis when $|x|$ is near one. However, the integrals to be evaluated in the weighted least-squares techniques of approximation can be troublesome to evaluate, and numerical integration techniques are often needed to determine the coefficients.

The process discussed in Theorem 7.7 provides, for any interval $[a, b]$ and weight function $w(x)$, a method of obtaining a recursive formula of the type

$$(7.14) \qquad \phi_k(x) = (x - B_k)\phi_{k-1}(x) - C_k\phi_{k-2}(x) \qquad \text{whenever } k \geq 2.$$

If this formula is known and the values of the coefficients

$$a_k = \frac{\int_a^b w(x)\phi_k(x)f(x)\,dx}{\int_a^b w(x)[\phi_k(x)]^2\,dx}$$

have been obtained, then the least-squares approximation

$$P_n(x) = \sum_{k=0}^{n} a_k \phi_k(x)$$

for the function f can be easily evaluated at any point by using the following algorithm.

Least-Squares Evaluation Algorithm 7.1

To evaluate the least squares approximation

$$P_n(x) = \sum_{k=0}^{n} a_k \phi_k(x)$$

to the function f at the number $\bar{x}$ where

$$a_k = \frac{\int_a^b w(x)\phi_k(x)f(x)\,dx}{\int_a^b w(x)[\phi_k(x)]^2\,dx} \qquad \text{for each } k = 0, 1, \ldots, n$$

and $\{\phi_k\}$ is a set of monic orthogonal polynomials with respect to the weight function $w(x)$:

INPUT endpoints a, b; n; coefficients $a_0, \ldots, a_n$; $\bar{x}$. *(Note: $a_0, \ldots, a_n$ can be computed using a numerical integration method from Chapter 4.)*

OUTPUT $b_0 = P_n(\bar{x})$.

Step 1 Set $b_{n+2} = 0$;
  $b_{n+1} = 0$;
  $B_{n+1} = 0$;
  $C_{n+2} = 0$;
  $C_{n+1} = 0$.

Step 2 For $k = n, n - 1, \ldots, 2$
    set

$$B_k = \frac{\int_a^b xw(x)[\phi_{k-1}(x)]^2\,dx}{\int_a^b w(x)[\phi_{k-1}(x)]^2\,dx};$$

$$C_k = \frac{\int_a^b xw(x)\phi_{k-1}(x)\phi_{k-2}(x)\,dx}{\int_a^b w(x)[\phi_{k-2}(x)]^2\,dx};$$

$$b_k = (\bar{x} - B_{k+1})b_{k+1} - C_{k+2}b_{k+2} + a_k.$$

Step 3   Set $B_1 = \dfrac{\int_a^b xw(x)[\phi_0(x)]^2 \, dx}{\int_a^b w(x)[\phi_0(x)]^2 \, dx}$

$\qquad\qquad b_1 = (\bar{x} - B_2)b_2 - C_3 b_3 + a_1.$

Step 4   Set $b_0 = (\bar{x} - B_1)b_1 - C_2 b_2 + a_0.$

Step 5   OUTPUT (b_0);
$\qquad\qquad$ STOP.

To verify that the procedure described by Algorithm 7.1 is correct, note that

$$a_k = b_k - (x - B_{k+1})b_{k+1} + C_{k+2}b_{k+2},$$

so
$$P_n(x) = \sum_{k=0}^{n} a_k \phi_k(x)$$

$$= \sum_{k=0}^{n} [b_k - (x - B_{k+1})b_{k+1} + C_{k+2}b_{k+2}]\phi_k(x)$$

$$= b_0 \phi_0(x) + b_1[\phi_1(x) - (x - B_1)\phi_0(x)]$$

$$+ \sum_{k=2}^{n} b_k[\phi_k(x) - (x - B_k)\phi_{k-1}(x) + C_k \phi_{k-2}(x)]$$

$$= b_0 \phi_0(x)$$

$$= b_0.$$

EXAMPLE 6   Reconsider the Chebyshev polynomials defined in Example 5 on $[-1, 1]$ by

$$T_0(x) = 1, \qquad T_1(x) = x$$

and
$$T_{n+1}(x) = 2xT_n(x) - T_{n-1}(x) \qquad \text{for each } n \geq 1.$$

Note that $T_n(x)$ has leading coefficient 2^{n-1} for each $n \geq 1$. This follows inductively since

$$T_2(x) = 2x^2 - 1 = 2^{2-1}x^2 - 1.$$

$$T_3(x) = 2xT_2(x) - T_1(x) = 2^{3-1}x^3 + \text{lower-order terms},$$

$$T_4(x) = 2xT_3(x) - T_2(x) = 2^{4-1}x^4 + \text{lower-order terms}, \quad \text{etc.}$$

It is often desirable to use polynomials with leading coefficient 1, which requires defining

$$\tilde{T}_n(x) = 2^{1-n}T_n(x) \qquad \text{for each } n \geq 1.$$

The recursion formula for $\tilde{T}_n$ will be

$$\tilde{T}_0(x) \equiv 1, \qquad \tilde{T}_1(x) = x, \qquad \tilde{T}_2(x) = x^2 - \tfrac{1}{2},$$

and

(7.15) $\qquad \tilde{T}_{n+1}(x) = x\tilde{T}_n(x) - \tfrac{1}{4}\tilde{T}_{n-1}(x) \qquad \text{for each } n \geq 2.$

To use Algorithm 7.1 to evaluate

$$P_4(x) = \sum_{k=0}^{4} \frac{(-1)^k}{k^2 + 1} \tilde{T}_k(x)$$

at $x = \frac{1}{2}$, first note that, since $a_k = (-1)^k/(k^2 + 1)$ for each $k = 0, 1, 2, 3, 4$, the recursive equation given in Eq. (7.15) implies that

$$B_1 = B_2 = B_3 = B_4 = 0, \qquad C_2 = \tfrac{1}{2}, \qquad C_3 = C_4 = \tfrac{1}{4}.$$

By Algorithm 7.1, $n = 4$, so $B_5 = 0, C_6 = C_5 = 0$ and $b_6 = b_5 = 0$. Hence,

$$b_4 = (x - B_5)b_5 - C_6 b_6 + a_4 = .05882353,$$
$$b_3 = (x - B_4)b_4 - C_5 b_5 + a_3 = -.07058824,$$
$$b_2 = (x - B_3)b_3 - C_4 b_4 + a_2 = .14999999,$$
$$b_1 = (x - B_2)b_2 - C_3 b_3 + a_1 = -.40735294,$$

and $\qquad P_4(.5) = b_0 = (x - B_1)b_1 - C_2 b_2 + a_0 = .7213235.$ $\qquad\square$

Exercise Set 7.2

1. Find the least-squares polynomial approximation of degree one to $f(x)$ on the indicated interval if

   a) $f(x) = x^2 - 2x + 3$, $[0, 1]$ b) $f(x) = x^3 - 1$, $[0, 2]$
   c) $f(x) = 1/x$, $[1, 3]$ d) $f(x) = e^{-x}$, $[0, 1]$
   e) $f(x) = \cos \pi x$, $[0, 1]$ f) $f(x) = \ln x$, $[1, 2]$

2. Find the least-squares polynomial approximation of degree two to the functions and intervals in Exercise 1.

3. Show that the functions $\phi_0(x) = 1/\sqrt{2\pi}$, $\phi_1(x) = (1/\sqrt{\pi})\cos x, \ldots, \phi_n(x) = (1/\sqrt{\pi})\cos nx$, $\phi_{n+1}(x) = (1/\sqrt{\pi})\sin x, \ldots, \phi_{2n-1}(x) = (1/\sqrt{\pi})\sin(n - 1)x$ are orthogonal on $[-\pi, \pi]$. [*Hint*: Use the trigonometric identities for $\cos(mx \pm nx)$ and $\sin(mx \pm nx)$ to simplify the integrals involved.]

4. Find the trigonometric least-squares approximating polynomial for $f(x) = x$ on $[-\pi, \pi]$, using Example 3 with $n = 2$.

5. Find the general least-squares trigonometric polynomial, $S_n(x)$, for

$$f(x) = \begin{cases} -1 & \text{if } -\pi < x < 0, \\ 1 & \text{if } 0 < x < \pi, \end{cases}$$

6. Show that

$$\int_{-1}^{1} \frac{T_n^2(x)}{\sqrt{1 - x^2}} \, dx = \frac{\pi}{2} \qquad \text{for each Chebyshev polynomial } T_n(x).$$

7. Calculate the value of the Legendre polynomial $P_4(x)$ at $x = .75$,

   a) by direct substitution in $P_4(x)$,   b) by using Algorithm 7.1.

8. Obtain the least-squares approximating polynomial of degree three for each of the following functions, using Theorem 7.6 and Algorithm 7.1. [Note: The integration required in this exercise is quite time consuming and might best be done numerically.]

   a) $f(x) = \cos x, 0 \le x \le 1$; $w(x) = 1$
   b) $f(x) = \ln x, 1 \le x \le 2$; $w(x) = 1$
   c) $f(x) = x^4, 0 \le x \le 1$; $w(x) = e^x$
   d) $f(x) = e^{-.1x}, 0 \le x \le 1$; $w(x) = 1$.

9. Show that the normal equations (7.8) have a unique solution. [*Hint*: Show that the only solution for the function $f(x) \equiv 0$ is $a_j = 0, j = 0, 1, \ldots, n$. Multiply (7.8) by a_j and sum over all j. Interchange the integral sign and the summation sign to obtain $\int_a^b [P(x)]^2 \, dx = 0$. Thus, $P(x) = 0$ or $a_j = 0$ for $j = 0, \ldots, n$. Hence, the coefficient matrix is nonsingular, and there is a unique solution to Eq. (7.8)].

10. Use the technique presented in Theorem 7.6 to calculate L_1, L_2 and L_3 where $\{L_0, L_1, L_2, L_3\}$ is an orthogonal set of polynomials defined on $(0, \infty)$ with respect to the weight function $w(x) = e^{-x}$ and $L_0(x) \equiv 1$. The polynomials obtained from this procedure are called the **Laguerre polynomials**.

11. Calculate P_4 and P_5, the fourth- and fifth-degree Legendre polynomials introduced in Example 4.

12. Suppose $\{\phi_0, \phi_1, \ldots, \phi_n\}$ is any linearly independent set of Π_n. Show that, for any element $Q \in \Pi_n$, there exist unique constants $c_0, c_1, \ldots, c_n$ such that

$$Q(x) = \sum_{k=0}^{n} c_k \phi_k(x).$$

13. Show that, if $\{\phi_0, \ldots, \phi_n\}$ is an orthogonal set of functions on $[a, b]$ with respect to the weight function w, then $\{\phi_0, \ldots, \phi_n\}$ is a linearly independent set.

14. Show that the induction step $\int_a^b w(x)\phi_n(x)\phi_{n-2}(x) \, dx = 0$ holds in Theorem 7.6.

7.3 Chebyshev Polynomials and Economization of Power Series

In this section we will continue the study of the Chebyshev polynomials initiated in Examples 5 and 6 of Section 7.2. This study will lead to the following two results: first, an optimal placing of interpolating points to minimize the error in Lagrange interpolation; second, a means of reducing the degree of an approximating polynomial with minimal loss of accuracy.

Recall that the definition of the Chebyshev polynomial is

(7.16) $T_n(x) = \cos[n \arccos x]$ for $x \in [-1, 1]$ and $n = 0, 1, 2, \ldots$.

The substitution $\theta = \arccos x$ implies that

(7.17) $T_n(x) = \cos(n\theta)$ for $\theta \in [0, \pi]$ and $n = 0, 1, 2, \ldots$.

With the aid of Eq. (7.17) and various trigonometric identities, it was shown in Example 5 of Section 7.2 that:

(7.18) $T_{n+1}(x) = 2xT_n(x) - T_{n-1}(x)$ for each $n = 1, 2, \ldots$ and $x \in [-1, 1]$.

Using the fact that $T_0(x) = 1$ and $T_1(x) = x$, we can infer from the recurrence relation in Eq. (7.18) that, for each $n \geq 1$, T_n is a polynomial of degree n with leading coefficient 2^{n-1}.

The first result in this section concerns the roots of the polynomial T_n.

Theorem 7.8   The Chebyshev polynomial T_n of degree $n \geq 1$ has n simple zeros in $[-1, 1]$ at

(7.19)
$$\bar{x}_k = \cos\left(\frac{2k-1}{2n}\pi\right) \qquad \text{for each } k = 1, 2, \ldots, n.$$

Moreover, T_n assumes its absolute extrema at the points

(7.20)
$$\bar{x}'_k = \cos\left(\frac{k\pi}{n}\right) \qquad \text{for each } k = 0, \ldots, n,$$

with

(7.21)
$$T_n(\bar{x}'_k) = (-1)^k \qquad \text{for each } k = 0, 1, \ldots, n.$$

Proof   If

$$\bar{x}_k = \cos\left(\frac{2k-1}{2n}\pi\right) \qquad \text{for } k = 1, 2, \ldots, n,$$

then

$$T_n(\bar{x}_k) = \cos(n \arccos \bar{x}_k) = \cos\left(n \arccos\left(\cos\left(\frac{2k-1}{2n}\pi\right)\right)\right) = \cos\left(\frac{2k-1}{2}\pi\right) = 0;$$

so $\bar{x}_k$ is a zero of T_n for each $k = 1, 2, \ldots, n$. Since T_n is a polynomial of degree n, all zeros of T_n must be of this form.

To show the second part of this theorem, note first that

$$T'_n(x) = \frac{d}{dx}\left[\cos(n \arccos x)\right] = \frac{n \sin(n \arccos x)}{\sqrt{1 - x^2}},$$

and that, when $1 \leq k \leq n - 1$,

$$T'_n(\bar{x}'_k) = \frac{n \sin\left(n \arccos\left(\cos\left(\frac{k\pi}{n}\right)\right)\right)}{\sqrt{1 - \cos^2\left(\frac{k\pi}{n}\right)}} = \frac{n \sin(k\pi)}{\sin\left(\frac{k\pi}{n}\right)} = 0.$$

Moreover, since T_n is a polynomial of degree n, T'_n is a polynomial of degree $(n-1)$ and all zeros of T'_n occur at these points. The only other possibilities for extrema of the function T_n occur at the endpoints of the interval $[-1, 1]$; that is, at $\bar{x}'_0 = 1$ and $\bar{x}'_n = -1$.

Since

$$T_n(\bar{x}'_k) = \cos\left(n \arccos\left(\cos\left(\frac{k\pi}{n}\right)\right)\right) = \cos(k\pi) = (-1)^k,$$

a maximum occurs at each even value of k and a minimum at each odd value. $\square$

In Example 6 of Section 7.2, the monic Chebyshev polynomials (polynomials with leading coefficient 1), $\tilde{T}_n$, were defined by

(7.22)
$$\tilde{T}_0(x) = 1, \qquad \tilde{T}_n(x) = 2^{1-n} T_n(x) \qquad \text{for each } n \geq 1.$$

It was also demonstrated in that example that these polynomials satisfied the recurrence relation

(7.23) $\tilde{T}_2(x) = x\tilde{T}_1(x) - \frac{1}{2}\tilde{T}_0(x);$ $\tilde{T}_{n+1}(x) = x\tilde{T}_n(x) - \frac{1}{4}\tilde{T}_{n-1}(x),$

for each $n \geq 2$.

Because of the linear relationship between $\tilde{T}_n$ and T_n, Theorem 7.8 implies that the zeros of $\tilde{T}_n$ also occur at

$$\bar{x}_k = \cos\left(\frac{2k-1}{2n}\pi\right) \quad \text{for each } k = 1, 2, \ldots, n,$$

and the extreme values of the $\tilde{T}_n$ occur at

$$\bar{x}'_k = \cos\left(\frac{k\pi}{n}\right) \quad \text{for each } k = 0, 1, 2, \ldots, n.$$

At these extreme values, for $n \geq 1$,

(7.24) $$\tilde{T}_n(\bar{x}'_k) = \frac{(-1)^k}{2^{n-1}} \quad \text{for each } k = 0, 1, 2, \ldots, n.$$

If $\tilde{\Pi}_n$ denotes **the set of all monic polynomials of degree** n, the relation expressed in Eq. (7.24) leads to an important minimization property that distinguishes the polynomials $\tilde{T}_n$ from the other members of $\tilde{\Pi}_n$.

Theorem 7.9   The polynomials of the form $\tilde{T}_n$, when $n \geq 1$, have the property that

(7.25) $$\frac{1}{2^{n-1}} = \max_{x \in [-1, 1]} |\tilde{T}_n(x)| \leq \max_{x \in [-1, 1]} |P_n(x)| \quad \text{for all } P_n \in \tilde{\Pi}_n.$$

Equality in (7.25) can occur only if $P_n = \tilde{T}_n$.

Proof   Suppose $P_n \in \tilde{\Pi}_n$ and

$$\max_{x \in [-1, 1]} |P_n(x)| \leq \frac{1}{2^{n-1}} = \max_{x \in [-1, 1]} |\tilde{T}_n(x)|.$$

Let $Q = \tilde{T}_n - P_n$. Since $\tilde{T}_n$ and P_n are both monic polynomials of degree n, Q is a polynomial of degree at most $(n - 1)$. Moreover, at the extreme points of $\tilde{T}_n$,

$$Q(\bar{x}'_k) = \tilde{T}_n(\bar{x}'_k) - P_n(\bar{x}'_k) = \frac{(-1)^k}{2^{n-1}} - P_n(\bar{x}'_k).$$

The fact that

$$|P_n(\bar{x}'_k)| \leq \frac{1}{2^{n-1}}$$

implies, that, for $k = 1, 2, \ldots, n$,

$$Q(\bar{x}'_k) \leq 0, \quad \text{when } k \text{ is odd,}$$

and

$$Q(\bar{x}'_k) \geq 0, \quad \text{when } k \text{ is even.}$$

Since Q is continuous, the Intermediate Value Theorem can be used to show that the $(n-1)$st-degree polynomial Q must have at least n zeros in the interval $[-1, 1]$, which is clearly impossible unless $Q \equiv 0$. This implies $P_n = \tilde{T}_n$, which establishes the result. $\square$

Theorem 7.9 can be used to answer the question of where to place interpolating nodes to minimize the error in Lagrange interpolation. Theorem 3.3 (p. 83), applied to the interval $[-1, 1]$, states that, if $x_0, \ldots, x_n$ are distinct numbers in the interval $[-1, 1]$ and $f \in C^{n+1}[-1, 1]$, then, for each $x \in [-1, 1]$, $\xi(x)$ in $(-1, 1)$ exists with

$$f(x) = P(x) + \frac{f^{(n+1)}(\xi(x))}{(n+1)!} (x - x_0)(x - x_1) \cdots (x - x_n)$$

where P is the Lagrange interpolating polynomial. To minimize the error by shrewd placement of the nodes $x_0, \ldots, x_n$ would, in general, be equivalent to finding the $x_0, \ldots, x_n$, which would minimize the quantity

$$|(x - x_0)(x - x_1) \cdots (x - x_n)|$$

throughout the interval $[-1, 1]$. This can be accomplished by choosing the $x_0, x_1, \ldots, x_n$ in a manner that minimizes the maximum value of $|(x - x_0)(x - x_1) \cdots (x - x_n)|$. Since $(x - x_0)(x - x_1) \cdots (x - x_n)$ is a monic polynomial of degree $(n + 1)$, Theorem 7.9 implies that this minimum is obtained if and only if

$$(x - x_0)(x - x_1) \cdots (x - x_n) = \tilde{T}_{n+1}(x).$$

When x_k is chosen to be the $(k + 1)$st zero of $\tilde{T}_{n+1}$ for each $k = 0, 1, \ldots, n$, that is, when x_k is chosen to be

(7.26) $$\bar{x}_{k+1} = \cos \frac{2k + 1}{2(n+1)} \pi,$$

the maximum value of $|(x - x_0)(x - x_1) \cdots (x - x_n)|$ will be minimized. Since $\max_{x \in [-1, 1]} |\tilde{T}_{n+1}(x)| = 2^{-n}$, this implies that

$$\begin{aligned} \frac{1}{2^n} &= \max_{x \in [-1, 1]} |(x - \bar{x}_1)(x - \bar{x}_2) \cdots (x - \bar{x}_{n+1})| \\ &\leq \max_{x \in [-1, 1]} |(x - x_0)(x - x_1) \cdots (x - x_n)|, \end{aligned}$$

for any choice of $x_0, x_1, \ldots, x_n$ in the interval $[-1, 1]$.

This technique for choosing points that will minimize the interpolating error can be easily extended to a general closed interval $[a, b]$ by using the change of variable $\tilde{x} = \frac{1}{2}[(b - a)x + a + b]$ to transform the numbers $\bar{x}_k$ in the interval $[-1, 1]$ into the corresponding numbers $\tilde{x}_k$ in the interval $[a, b]$.

EXAMPLE 1   Let $f(x) = xe^x$ on $[0, 1.5]$. Two interpolation polynomials of degree at most three will be constructed. First, the equally spaced nodes $x_0 = 0$, $x_1 = .5$, $x_2 = 1$, $x_3 = 1.5$ are used to give

$$L_0(x) = -1.3333x^3 + 4.0000x^2 - 3.6667x + 1,$$

$$L_1(x) = 4.0000x^3 - 10.000x^2 + 6.0000x,$$

$$L_2(x) = -4.0000x^3 + 8.0000x^2 - 3.0000x,$$

$$L_3(x) = 1.3333x^3 - 2.0000x^2 + .66667x.$$

TABLE 7.11

x	$f(x) = xe^x$
0	.0000
.5	.824361
1.0	2.71828
1.5	6.72253

With the values listed in Table 7.11, the interpolating polynomial is given by

$$P_3(x) = 1.3875x^3 + .057570x^2 + 1.2730x.$$

For the second interpolating polynomial, shift the zeros $\bar{x}_k = \cos((2k + 1)/8)\pi$. $k = 0, 1, 2, 3$ of $\tilde{T}_4$ from $[-1, 1]$ to $[0, 1.5]$, using the transformation

$$\tilde{x}_k = .75 + .75\bar{x}_k$$

to obtain

$$\tilde{x}_0 = 1.44291, \quad \tilde{x}_1 = 1.03701, \quad \tilde{x}_2 = .46299, \quad \text{and} \quad \tilde{x}_3 = .05709.$$

The Lagrange coefficient polynomials are then computed as:

$$\tilde{L}_0(x) = 1.8142x^3 - 2.8249x^2 + 1.0264x - .049728,$$

$$\tilde{L}_1(x) = -4.3799x^3 + 8.5977x^2 - 3.4026x + .16705,$$

$$\tilde{L}_2(x) = 4.3799x^3 - 11.112x^2 + 7.1738x - .37415,$$

$$\tilde{L}_3(x) = -1.8142x^3 + 5.3390x^2 - 4.7976x + 1.2568.$$

TABLE 7.12

x	$f(x) = xe^x$
$\tilde{x}_0 = 1.44291$	6.10783
$\tilde{x}_1 = 1.03701$	2.92517
$\tilde{x}_2 = .46299$	.73560
$\tilde{x}_3 = .05709$	.060444

The functional values required for these polynomials are given in Table 7.12, and the interpolation polynomial of degree at most three is given by

$$\tilde{P}_3(x) = 1.3811x^3 + .044445x^2 + 1.3030x - .014357.$$

TABLE 7.13

x	$f(x) = xe^x$	$P_3(x)$	$|xe^x - P_3(x)|$	$\tilde{P}_3(x)$	$|xe^x - \tilde{P}_3(x)|$
.15	.1743	.1969	.0226	.1868	.0125
.25	.3210	.3435	.0225	.3358	.0148
.35	.4967	.5121	.0154	.5064	.0097
.65	1.245	1.233	.0120	1.231	.0140
.75	1.588	1.572	.0160	1.571	.0170
.85	1.989	1.976	.0130	1.973	.0160
1.15	3.632	3.650	.0180	3.643	.0110
1.25	4.363	4.391	.0280	4.381	.0180
1.35	5.208	5.237	.0290	5.224	.0160

For comparison, Table 7.13 lists various values of x, together with the values of $f(x)$, $P_3(x)$, and $\tilde{P}_3(x)$. It can be seen from this table that, although the error using P_3 is less than using $\tilde{P}_3$ near the middle of the table, the maximum error involved with using $\tilde{P}_3$ is considerably less.   □

The following corollary can be easily obtained from Theorem 7.9. The proof is considered in Exercise 3.

Corollary 7.10   If P is the interpolating polynomial of degree at most n with nodes at the roots of $T_{n+1}(x)$, then

$$\max_{x \in [-1, 1]} |f(x) - P(x)| \le \frac{1}{2^n (n + 1)!} \max_{x \in [-1, 1]} |f^{n+1}(x)|$$

for each $f \in C^{n+1}[-1, 1]$.

Chebyshev polynomials can also be used to reduce the degree of an approximating polynomial with a minimal loss of accuracy. This is a particularly useful technique to employ when the approximating polynomial being used is a Taylor polynomial. Although Taylor polynomials are very accurate near the point about which they are expanded, their accuracy deteriorates rapidly when they are employed in situations farther from this point. For this reason, a high-degree Taylor polynomial may be needed to achieve a prescribed error tolerance. Because the Chebyshev polynomials have minimum maximum-absolute value that is spread uniformly on an interval, they can be used to reduce the degree of the Taylor polynomial without exceeding the error tolerance. The following example illustrates the technique involved.

EXAMPLE 2   The function $f(x) = e^x$ can be approximated on the interval $[-1, 1]$ by the fourth-degree Taylor polynomial expanded about zero,

$$P_4(x) = 1 + x + \frac{x^2}{2} + \frac{x^3}{6} + \frac{x^4}{24},$$

with an error

$$|R_4(x)| = \frac{|f^{(5)}(\xi(x))||x^5|}{120} \le \frac{e}{120} \approx .023 \qquad \text{for } -1 \le x \le 1.$$

Suppose that an actual error of .05 is tolerable, and consider the situation that occurs if we replaced the term in the Taylor polynomial involving x^4 by the equivalent Chebyshev polynomials of degree less than or equal to four.

Before continuing with this example, we list the explicit representation of x^k in terms of $T_0, T_1, \ldots, T_k$ for some of the smaller positive integers k. The verification of Table 7.14 is considered in Exercise 4.

TABLE 7.14

k	T_k	x^k
0	1	T_0
1	x	T_1
2	$2x^2 - 1$	$\frac{1}{2}T_0 + \frac{1}{2}T_2$
3	$4x^3 - 3x$	$\frac{3}{4}T_1 + \frac{1}{4}T_3$
4	$8x^4 - 8x^2 + 1$	$\frac{3}{8}T_0 + \frac{1}{2}T_2 + \frac{1}{8}T_4$
5	$16x^5 - 20x^3 + 5x$	$\frac{5}{8}T_1 + \frac{5}{16}T_3 + \frac{1}{16}T_5$
6	$32x^6 - 48x^4 + 18x^2 - 1$	$\frac{5}{16}T_0 + \frac{15}{32}T_2 + \frac{3}{16}T_4 + \frac{1}{32}T_6$
7	$64x^7 - 112x^5 + 56x^3 - 7x$	$\frac{35}{64}T_1 + \frac{21}{64}T_3 + \frac{7}{64}T_5 + \frac{1}{64}T_7$

Thus,

$$
\begin{aligned}
P_4(x) &= 1 + x + \tfrac{1}{2}x^2 + \tfrac{1}{6}x^3 + \tfrac{1}{24}[\tfrac{3}{8}T_0(x) + \tfrac{1}{2}T_2(x) + \tfrac{1}{8}T_4(x)] \\
&= 1 + x + \tfrac{1}{2}x^2 + \tfrac{1}{6}x^3 + \tfrac{1}{64}T_0(x) + \tfrac{1}{48}T_2(x) + \tfrac{1}{192}T_4(x) \\
&= 1 + x + \tfrac{1}{2}x^2 + \tfrac{1}{6}x^3 + \tfrac{1}{64} + \tfrac{1}{48}(2x^2 - 1) + \tfrac{1}{192}T_4(x) \\
&= (1 + \tfrac{1}{64} - \tfrac{1}{48}) + x + (\tfrac{1}{2} + \tfrac{1}{24})x^2 + \tfrac{1}{6}x^3 + \tfrac{1}{192}T_4(x) \\
&= \tfrac{191}{192} + x + \tfrac{13}{24}x^2 + \tfrac{1}{6}x^3 + \tfrac{1}{192}T_4(x).
\end{aligned}
$$

But $\qquad \max\limits_{x \in [-1,\,1]} |T_4(x)| = 1, \qquad |\tfrac{1}{192}T_4(x)| \le \tfrac{1}{192} = .0053,$

and $\qquad |R_4(x)| + |\tfrac{1}{192}T_4(x)| \le .023 + .0053 = .0283$

is still less than the tolerance .05. Consequently, the fourth-degree term $(1/192)T_4(x)$ can be omitted from the approximating polynomial and the desired accuracy will still be obtained. The third-degree polynomial,

$$P_3(x) = \tfrac{191}{192} + x + \tfrac{13}{24}x^2 + \tfrac{1}{6}x^3$$

will give the desired accuracy on the interval $[-1, 1]$.

To attempt to eliminate the third-degree term requires replacing x^3 with $\frac{3}{4}T_1 + \frac{1}{4}T_3$ and results in

$$
\begin{aligned}
P_3(x) &= \tfrac{191}{192} + x + \tfrac{13}{24}x^2 + \tfrac{1}{6}[\tfrac{3}{4}T_1(x) + \tfrac{1}{4}T_3(x)] \\
&= \tfrac{191}{192} + \tfrac{9}{8}x + \tfrac{13}{24}x^2 + \tfrac{1}{24}T_3(x).
\end{aligned}
$$

However, $\max_{x \in [-1,\,1]} |\tfrac{1}{24}T_3(x)| = .0417$, which, when combined with the possible error of .0283 obtained previously, gives an error bound of .07, and exceeds the tolerance of .05. $P_3(x)$ is therefore the lowest-degree polynomial suitable for this approximation. Table 7.15 lists the function and the approximating polynomials at some representative points where $P_2(x)$ represents the polynomial

$$P_2(x) = \tfrac{191}{192} + \tfrac{9}{8}x + \tfrac{13}{24}x^2.$$

TABLE 7.15

x	e^x	$P_4(x)$	$P_3(x)$	$P_2(x)$
$-.75$	.47237	.47412	.47917	.45573
$-.25$	.77880	.77881	.77604	.74740
0	1.00000	1.00000	.99479	.99479
.25	1.28403	1.28402	1.28125	1.30990
.75	2.11700	2.11475	2.11979	2.14323

Even though the error bound for the polynomial

$$P_2(x) = \tfrac{191}{192} + \tfrac{9}{8}x + \tfrac{13}{24}x^2$$

was greater than the tolerance of .05, it should be noted that the tabulated entries are well within that bound. □

Exercise Set 7.3

1. Use the zeros of $\tilde{T}_3$ to construct an interpolating polynomial of degree two for the functions and intervals given.

   a) $f(x) = e^x$, $[-1, 1]$ b) $f(x) = \sin x$, $[0, \pi]$

   c) $f(x) = \ln x$, $[1, 2]$ d) $f(x) = 1 + x$, $[0, 1]$

2. Use the zeros of $\tilde{T}_4$ to construct an interpolating polynomial of degree at most three for the functions and intervals given in Exercise 1.

3. Prove Corollary 7.10.

4. Derive the entries in Table 7.14.

5. Find the sixth-degree Taylor polynomial for xe^x, and use Chebyshev economization to obtain a lesser-degree polynomial approximation while keeping the error less than .01 on $[-1, 1]$.

6. Find the third-degree interpolating polynomial P_3 to

   $$f(x) = \sqrt{1 + x}, \; -1 \le x \le 1,$$

   using the roots of T_3 as interpolating nodes. Compute $P_3(.1)$, and compare to Example 2 of Section 3.1.

7. Find the third-degree interpolating polynomial P_3 to $f(x) = \sqrt{1 + x}$, $0 \le x \le 1$, using the roots of T_3 and the transformation of the interval $[0, 1]$ onto $[-1, 1]$. Compute $P_3(.1)$, and compare to Exercise 6 and Example 2 of Section 3.1.

8. Show that for any positive integers i and j

   $$T_i(x)T_j(x) = \tfrac{1}{2}[T_{i+j}(x) - T_{|i-j|}(x)].$$

7.4 Rational Function Approximation

The class of algebraic polynomials has some distinct advantages for use in approximation. There are a sufficient number of polynomials to approximate any continuous function on a closed interval to within an arbitrary tolerance, polynomials are easily evaluated at arbitrary values, and the derivatives and integrals of polynomials exist and are easily determined. The disadvantage to using polynomials for approximation is their tendency for oscillation. This tendency often causes error bounds in polynomial approximation to significantly exceed the average approximation error since error bounds are determined by the maximum approximation error. In order to find techniques that decrease approximation error bounds, we will consider methods that more evenly spread the approximation error over the approximation interval. These techniques require the introduction of a new class of approximating functions, the class of **rational functions**.

A rational function r of degree N is a function of the form

$$r(x) = \frac{p(x)}{q(x)}$$

where p and q are polynomials whose degrees sum to N.

Since every polynomial is also a rational function (simply let $q(x) \equiv 1$), approximation by rational functions will give results with no greater error bounds than approximation by polynomials. In fact, rational functions whose numerator and denominator have the same or nearly the same degree generally produce approximation results superior to polynomial methods for the same amount of computation effort. (This statement is based on the assumption that the amount of computation effort required for division is approximately the same as for multiplication. This assumption is valid in many, but not all, computing systems.) Rational functions have the added advantage of permitting efficient approximation of functions that have infinite discontinuities near, but outside, the interval of approximation. Polynomial approximation is generally unacceptable in this situation.

Suppose that r is a rational function of degree $N = n + m$ of the form

$$r(x) = \frac{p(x)}{q(x)} = \frac{p_0 + p_1 x + \cdots + p_n x^n}{q_0 + q_1 x + \cdots + q_m x^m}$$

that will be used to approximate a function f on a closed interval I containing zero. For r to be defined at zero requires that $q_0 \neq 0$. In fact, we can assume that $q_0 = 1$, for if this is not the case we can simply replace $p(x)$ by $p(x)/q_0$ and $q(x)$ by $q(x)/q_0$. Consequently, there are $N + 1$ parameters $q_1, q_2, \ldots, q_m, p_0, p_1, \ldots, p_n$ available for the approximation of f by r.

The **Padé approximation technique** chooses the $N + 1$ parameters so that $f^{(k)}(0) = r^{(k)}(0)$ for each $k = 0, 1, \ldots, N$. Pade approximation is the extension of Taylor polynomial approximation to rational functions. In fact, when $n = N$ and $m = 0$, the Padé approximation is the Taylor polynomial of degree N expanded about zero, that is, the Maclaurin polynomial of degree N.

Suppose that f has the Maclaurin expansion $f(x) = \sum_{i=0}^{\infty} a_i x^i$. Since

$$f(x) - r(x) = f(x) - \frac{p(x)}{q(x)} = \frac{f(x)q(x) - p(x)}{q(x)},$$

(7.27) $$f(x) - r(x) = \frac{\sum_{i=0}^{\infty} a_i x^i \sum_{i=0}^{m} q_i x^i - \sum_{i=0}^{n} p_i x^i}{q(x)}$$

The function $f - r$ and its first N derivatives will be zero if the right-hand side of Eq. (7.27) can be written as $x^{N+1} Q(x)$ for some continuous function Q (see Exercise 10 of Section 2.4). But this can occur only if the coefficients of x^k in the numerator of the right-hand side of Eq. (7.27) are zero for each $k = 0, 1, \ldots, N$. Consequently, $(f - r)^{(k)}(0) = 0$; that is $f^{(k)}(0) = r^{(k)}(0)$ for $k = 0, 1, \ldots, N$ provided that

(7.28) $$(a_0 + a_1 x + \cdots)(1 + q_1 x + \cdots + q_m x^m) - (p_0 + p_1 x + \cdots + p_n x^n)$$

has no terms of degree less than or equal to N. To simplify notation, let us define $p_{n+1} = p_{n+2} = \cdots = p_N = 0$ and $q_{m+1} = q_{m+2} = \cdots = q_N = 0$. We can then express the coefficient of x^k in expression (7.28) as

$$\sum_{i=0}^{k} a_i q_{k-i} - p_k;$$

so the rational function for Padé approximation results from the solution of the $N + 1$ linear equations

$$\sum_{i=0}^{k} a_i q_{k-i} - p_k = 0, \qquad k = 0, 1, \ldots, N$$

in the $N + 1$ unknowns $q_1, q_2, \ldots, q_m, p_0, p_1, \ldots, p_n$.

EXAMPLE 1   The Maclaurin series expansion for e^{-x} is

$$\sum_{i=0}^{\infty} \frac{(-1)^i}{i!} x^i.$$

To find the Padé approximation to e^{-x} of degree 5 with $n = 3$ and $m = 2$ requires choosing p_0, p_1, p_2, p_3, q_1, and q_2 so that the coefficients of x^k for $k = 0, 1, \ldots, 5$ are zero in the expression

$$\left(1 - x + \frac{x^2}{2} - \frac{x^3}{6} + \cdots\right)(1 + q_1 x + q_2 x^2) - (p_0 + p_1 x + p_2 x^2 + p_3 x^3).$$

Expanding and collecting terms produces

$$x^5: \quad -\frac{1}{120} + \frac{1}{24}q_1 - \frac{1}{6}q_2 = 0; \qquad x^2: \quad \frac{1}{2} - q_1 + q_2 = p_2;$$

$$x^4: \quad \frac{1}{24} - \frac{1}{6}q_1 + \frac{1}{2}q_2 = 0; \qquad x^1: \quad -1 + q_1 = p_1;$$

$$x^3: \quad -\frac{1}{6} + \frac{1}{2}q_1 - q_2 = p_3; \qquad x^0: \quad 1 = p_0.$$

The solution to this system is:

$$p_0 = 1, \quad p_1 = -\tfrac{3}{5}, \quad p_2 = \tfrac{3}{20}, \quad p_3 = -\tfrac{1}{60}, \quad q_1 = \tfrac{2}{5}, \quad \text{and} \quad q_2 = \tfrac{1}{20};$$

so the Padé approximation is

$$r(x) = \frac{1 - \frac{3}{5}x + \frac{3}{20}x^2 - \frac{1}{60}x^3}{1 + \frac{2}{5}x + \frac{1}{20}x^2}.$$

Table 7.16 lists values of $r(x)$ and $P_5(x)$, the fifth-degree Taylor polynomial about $x = 0$. The Padé approximation is clearly superior in this example.   □

TABLE 7.16	x	e^{-x}	$P_5(x)$	$|e^{-x} - P_5(x)|$	$r(x)$	$|e^{-x} - r(x)|$
	.2	.81873075	.81873067	8.64×10^{-8}	.81873075	7.55×10^{-9}
	.4	.67032005	.67031467	5.38×10^{-6}	.67031963	4.11×10^{-7}
	.6	.54881164	.54875200	5.96×10^{-5}	.54880763	4.00×10^{-6}
	.8	.44932896	.44900267	3.26×10^{-4}	.44930966	1.93×10^{-5}
	1.0	.36787944	.36666667	1.21×10^{-3}	.36781609	6.33×10^{-5}

It is also interesting to compare the number of arithmetic operations required for calculations of $P_5(x)$ and $r(x)$. Using nested multiplication, $P_5(x)$ can be expressed as

$$P_5(x) = 1 - x(1 - x(\tfrac{1}{2} - x(\tfrac{1}{6} - x(\tfrac{1}{24} - \tfrac{1}{120}x)))).$$

Assuming that the coefficients of 1, x, x^2, x^3, x^4, and x^5 are represented as decimals, a single calculation of $P_5(x)$ requires five multiplications and five additions/subtractions. Using nested multiplication, $r(x)$ can be expressed as

$$r(x) = \frac{1 - x(\tfrac{3}{5} - x(\tfrac{3}{20} - \tfrac{1}{60}x))}{1 + x(\tfrac{2}{5} + \tfrac{1}{20}x)};$$

so a single calculation of $r(x)$ requires five multiplications, five additions/subtractions, and one division. Hence, computational effort appears to favor the polynomial approximation. However, by re-expressing $r(x)$ by continued division, we can write

$$r(x) = \frac{1 - \tfrac{3}{5}x + \tfrac{3}{20}x^2 - \tfrac{1}{60}x^3}{1 + \tfrac{2}{5}x + \tfrac{1}{20}x^2}$$

$$= \frac{-\tfrac{1}{3}x^3 + 3x^2 - 12x + 20}{x^2 + 8x + 20}$$

$$= -\tfrac{1}{3}x + \tfrac{17}{3} + \frac{(-\tfrac{152}{3}x - \tfrac{280}{3})}{(x^2 + 8x + 20)}$$

$$= -\tfrac{1}{3}x + \tfrac{17}{3} + \frac{-\tfrac{152}{3}}{\left(\dfrac{x^2 + 8x + 20}{x + \tfrac{35}{19}}\right)}$$

or

(7.29)
$$r(x) = -\tfrac{1}{3}x + \tfrac{17}{3} + \frac{-\tfrac{152}{3}}{\left(x + \tfrac{117}{19} + \dfrac{\tfrac{3125}{361}}{(x + \tfrac{35}{19})}\right)}.$$

Written in this form, a single calculation of $r(x)$ requires one multiplication, five addition/subtractions, and two divisions. Consequently, if the amount of computation required for division is approximately the same as for multiplication, the computational effort required for an evaluation of $P_5(x)$ should significantly exceed that required for an evaluation of $r(x)$.

Expressing a rational function approximation in a form such as Eq. (7.29) is called **continued fraction** approximation. This is a classical approximation technique that is of current interest because of the computational efficiency of this representation. It is, however, a specialized technique, and one we will not discuss further. A rather extensive treatment of this subject, and of rational approximation in general, can be found in Ralston and Rabinowitz [67], pages 285–322.

Although the rational function approximation in Example 1 gave results superior to the polynomial approximation of the same degree, the approximation has a wide variation in accuracy; the approximation at .2 is accurate to within 8×10^{-9} while, at 1.0, the approximation and the function agree only to within 7×10^{-5}. This accuracy variation is not unexpected because the Padé approximation is based on a Taylor polynomial representation of e^{-x}, and this representation has a wide variation of accuracy in $[.2, 1.0]$.

To obtain more uniformly accurate rational function approximations, we will use a class of polynomials that exhibit more uniform behavior on the interval $[-1, 1]$, the set of Chebyshev polynomials. The general Chebyshev rational function approximation method proceeds in the same manner as Padé approximation except that each x^k term in the Padé approximation is replaced by the kth degree Chebyshev polynomial T_k. Consequently, we approximate a function f by an Nth degree rational function r written in the form

$$r(x) = \frac{\sum_{k=0}^{n} p_k T_k(x)}{\sum_{k=0}^{m} q_k T_k(x)} \qquad \text{where } N = n + m \text{ and } q_0 = 1.$$

Writing $f(x)$ in a series involving Chebyshev polynomials gives

$$f(x) - r(x) = \sum_{k=0}^{\infty} a_k T_k(x) - \frac{\sum_{k=0}^{n} p_k T_k(x)}{\sum_{k=0}^{m} q_k T_k(x)}$$

or

$$(7.30) \qquad f(x) - r(x) = \frac{\left[\sum_{k=0}^{\infty} a_k T_k(x)\right]\left[\sum_{k=0}^{m} q_k T_k(x)\right] - \sum_{k=0}^{n} p_k T_k(x)}{\sum_{k=0}^{m} q_k T_k(x)}.$$

The coefficients $q_1, q_2, \ldots, q_m$ and $p_0, p_1, \ldots, p_n$ are chosen so that the numerator on the right-hand side of Eq. (7.30) has zero coefficients for $T_k(x)$ when $k = 0$, $1, \ldots, N$.

Two problems arise with the Chebyshev procedure that make it more difficult to implement than the Padé method. One problem occurs because the product of the polynomial $q(x)$ and the series for $f(x)$ involve products of Chebyshev polynomials. This problem is easily resolved by making use of the relationship

$$(7.31) \qquad T_i(x)T_j(x) = \tfrac{1}{2}[T_{i+j}(x) + T_{|i-j|}(x)].$$

(See Exercise 8 of Section 7.3.) The other problem is more difficult to resolve and involves the computation of the Chebyshev series for $f(x)$. In theory, this is not a difficult problem for if

$$f(x) = \sum_{k=0}^{\infty} a_k T_k(x),$$

then the orthogonality of the Chebyshev polynomials (see Example 5 of Section 7.2 for a verification) implies that

$$a_0 = \frac{1}{\pi} \int_{-1}^{1} \frac{f(x)}{\sqrt{1 - x^2}} \, dx \qquad \text{and} \qquad a_k = \frac{2}{\pi} \int_{-1}^{1} \frac{f(x)T_k(x)}{\sqrt{1 - x^2}} \, dx \quad \text{when } k \geq 1.$$

Practically, however, these integrals can seldom be evaluated in closed form and a numerical integration technique is required for each evaluation.

EXAMPLE 2   The first five terms of the Chebyshev expansion for e^{-x} can be shown to be

$$\tilde{P}_5(x) = 1.266066 T_0(x) - 1.130318 T_1(x) + .271495 T_2(x) - .044337 T_3(x)$$
$$+ .005474 T_4(x) - .000543 T_5(x).$$

To determine the Chebyshev rational approximation of degree 5 with $n = 3$ and $m = 2$ requires choosing $p_0, p_1, p_2, p_3, q_1,$ and q_2 so that for $k = 0, 1, 2, 3, 4, 5$ the coefficients of $T_k(x)$ are zero in the expansion

$$\tilde{P}_5(x)[T_0(x) + q_1 T_1(x) + q_2 T_2(x)] - [p_0 T_0(x) + p_1 T_1(x) + p_2 T_2(x)].$$

Using the relation (7.31) and collecting terms gives the equations

$$T_0: \quad 1.266066 - .565159 q_1 + .1357485 q_2 = p_0$$

$$T_1: \quad -1.130318 + 1.401814 q_1 - .583275 q_2 = p_1$$

$$T_2: \quad .271495 - .587328 q_1 + 1.268803 q_2 = p_2$$

$$T_3: \quad -.044337 + .138485 q_1 - .565431 q_2 = p_3$$

$$T_4: \quad .005474 - .022440 q_1 + .135748 q_2 = 0$$

$$T_5: \quad -.000543 + .002737 q_1 - .022169 q_2 = 0.$$

The solution to this system produces the rational function

$$r_T(x) = \frac{1.055265 T_0(x) - .613016 T_1(x) + .077478 T_2(x) - .004506 T_3(x)}{T_0(x) + .378331 T_1(x) + .022216 T_2(x)}.$$

Using Table 7.14 to convert to an expression involving powers of x gives

$$r_T(x) = \frac{.977787 - .599499x + .154956x^2 - .018022x^3}{.977784 + .378331x + .044432x^2}.$$

Table 7.17 lists values of $r_T(x)$ and, for comparison purposes, the values of $r(x)$ obtained in Example 1. Notice that the approximation given by $r(x)$ is superior to that of $r_T(x)$ for $x = .2$ and $.4$, but that the maximum error for $r(x)$ is 6.33×10^{-5} as compared to 9.13×10^{-6} for $r_T(x)$.   □

TABLE 7.17

x	e^{-x}	$r(x)$	$|e^{-x} - r(x)|$	$r_T(x)$	$|e^{-x} - r_T(x)|$
.2	.81873075	.81873075	7.55×10^{-9}	.81872510	5.66×10^{-6}
.4	.67032005	.67031963	4.11×10^{-7}	.67031310	6.95×10^{-6}
.6	.54881164	.54880763	4.00×10^{-6}	.54881292	1.28×10^{-6}
.8	.44932896	.44930966	1.93×10^{-5}	.44933809	9.13×10^{-6}
1.0	.36787944	.36781609	6.33×10^{-5}	.36787155	7.89×10^{-6}

The Chebyshev method does not produce the best rational function approximation in the sense of the approximation whose maximum approximation error is minimal. The method can, however, be used as a starting point for an iterative method known as the second Remes' algorithm that converges to the best approximation. A discussion of the techniques involved with this procedure and an improvement on this algorithm can be found in the previously mentioned book by Ralston and Rabinowitz [67] pages 292–305.

Exercise Set 7.4

1. Determine the Padé approximation of degree 5 with $n = 2$ and $m = 3$ for $f(x) = e^x$. Compare the results at $x_i = .2i$, $i = 1, 2, 3, 4, 5$ from this approximation with those from the fifth-degree Maclaurin polynomial.

2. Repeat Exercise 1 using instead the Padé approximation of degree 5 with $n = 3$ and $m = 2$. Compare the results at each x_i with those computed in Exercise 1. Compare the rational function obtained with the rational function determined in Example 1.

3. Determine the Padé approximation of degree 6 with $n = m = 3$ for $f(x) = \sin x$. Compare the results at $x_i = .1i$, $i = 0, 1, \ldots, 5$ with the exact results and with the results of the sixth-degree Maclaurin polynomial.

4. Determine the Padé approximations of degree 6 with a) $n = 2, m = 4$ and b) $n = 4, m = 2$ for $f(x) = \sin x$. Compare the results at each x_i to those obtained in Exercise 3.

5. Table 7.16 lists results of the Padé approximation of degree 5 with $n = 2$ and $m = 3$, the Maclaurin polynomial of degree 5, and the exact values of $f(x) = e^{-x}$ when $x_i = .2i$, $i = 1, 2, 3, 4, 5$. Compare these results with those produced from the other Padé approximations of degree 5:

   a) $n = 0$, $m = 5$,
   c) $n = 3$, $m = 2$,

   b) $n = 1$, $m = 4$,
   d) $n = 4$, $m = 1$.

6. Express the following rational functions in continued-fraction form.

   a) $\dfrac{x^2 + 3x + 2}{x^2 - x + 1}$

   b) $\dfrac{4x^4 + 3x^2 - 7x + 5}{x^5 - 2x^4 + 3x^3 + 6}$

   c) $\dfrac{3x^5 - 4x^2 + 8x - 1}{x^4 + 5x^2 + x - 2}$

7. Use three-digit rounding arithmetic to compute the values of the rational fractions in Exercise 6 for $x = 1.23$

   a) precisely as listed in the exercise,
   b) using nested multiplication in the numerator and denominator,
   c) in continued-fraction form.

   Compare to the exact result.

8. Determine the first four terms of the Chebyshev expansion for

   a) $\sin x$,

   b) e^x.

9. Find the Chebyshev rational approximation of degree 4 with $n = m = 2$ for $f(x) = \sin x$. Compare the results at $x_i = .1$, $i = 0, 1, \ldots, 5$ from this approximation with those obtained in Exercise 3 using a sixth-degree Padé approximation.

10. Find the Chebyshev rational approximation of degree 4 with $n = m = 2$ for $f(x) = e^x$. Compare the results at $x_i = .2i$, $i = 1, 2, 3, 4, 5$ from this approximation with those obtained in Exercises 1 and 2 where fifth-degree Padé approximations were used.

7.5   Trigonometric Polynomial Approximation

In Section 7.2 we discussed briefly some facts concerning trigonometric polynomials and Fourier series. In particular, for each positive integer n, we defined the set

$$\mathcal{T}_n = \{\phi_0, \phi_1, \ldots, \phi_{2n-1}\}$$

where $\qquad \phi_0(x) = \dfrac{1}{\sqrt{2\pi}},$

$$\phi_k(x) = \frac{1}{\sqrt{\pi}}\cos(kx) \qquad \text{for each } k = 1, 2, \ldots, n,$$

and $\qquad \phi_{n+k}(x) = \dfrac{1}{\sqrt{\pi}}\sin(kx) \qquad \text{for each } k = 1, 2, \ldots, n - 1.$

(Some sources include an additional function in the set, $\phi_{2n}(x) = (1/\sqrt{\pi})\sin(nx)$. We will not follow this convention.)

The set $\mathcal{T}_n$ is orthogonal on $[-\pi, \pi]$ with respect to the weight function $w(x) \equiv 1$. We also found that the least-squares approximation to a function $f \in C[-\pi, \pi]$ by functions in $\mathcal{T}_n$ is

$$S_n(x) = \sum_{k=0}^{2n-1} a_k \phi_k(x),$$

where $\qquad a_k = \displaystyle\int_{-\pi}^{\pi} f(x)\phi_k(x)\, dx, \qquad \text{for each } k = 0, 1, \ldots, 2n - 1.$

There is a discrete analog to this situation that is useful for the least-squares approximation and interpolation of large amounts of data when the data is given at equally spaced points.

Suppose that a collection of $2m$ paired data points $\{(x_j, y_j)\}_{j=0}^{2m-1}$ is given with the first elements in the pairs equally spaced within an interval. For convenience we will assume that the interval is $[-\pi, \pi]$ and that

$$x_j = -\pi + \left(\frac{j}{m}\right)\pi \qquad \text{for each } j = 0, 1, \ldots, 2m - 1.$$

If this were not the case, a simple linear transformation could be used to translate the data into this form. (Example 1 contains a demonstration of the technique).

For a fixed $n < m$, we consider the set of functions $\hat{\mathcal{T}}_n = \{\hat{\phi}_0, \hat{\phi}_1, \ldots, \hat{\phi}_{2n-1}\}$ where

(7.32) $\qquad \hat{\phi}_0(x) = \dfrac{1}{\sqrt{2m}},$

(7.33) $\qquad \hat{\phi}_k(x) = \dfrac{1}{\sqrt{m}}\cos kx \qquad \text{for each } k = 1, 2, \ldots, n,$

and

(7.34) $\qquad \hat{\phi}_{n+k}(x) = \dfrac{1}{\sqrt{m}}\sin kx \qquad \text{for each } k = 1, 2, \ldots, n - 1.$

The object is to determine the trigonometric polynomial S_n composed of functions in $\hat{\mathcal{T}}_n$ with the property that

$$E(S_n) = \sum_{j=0}^{2m-1} \{y_j - S_n(x_j)\}^2$$

$$= \sum_{j=0}^{2m-1} \left\{ y_j - \left[\frac{\alpha_0}{\sqrt{2m}} + \frac{\alpha_n}{\sqrt{m}}\cos nx_j + \frac{1}{\sqrt{m}}\sum_{k=1}^{n-1}(\alpha_k \cos kx_j + \alpha_{n+k}\sin kx_j) \right] \right\}^2$$

is a minimum. As usual in least squares approximations, the constants $\alpha_0, \alpha_1, \ldots,$ α_{2n-1} are determined by the condition that this minimum can occur only if $\partial E(S_n)/\partial \alpha_k = 0$ for each $k = 0, 1, \ldots, 2n - 1$.

The determination of the constants is simplified by the fact that the set $\hat{\mathscr{T}}_n$ is orthonormal with respect to summation over the equally-spaced points $\{x_j\}_{j=0}^{2m-1}$ in $[-\pi, \pi]$. By this we mean that for each $k \neq l$,

$$\sum_{j=0}^{2m-1} \hat{\phi}_k(x_j)\hat{\phi}_l(x_j) = 0,$$

and for each k,

$$\sum_{j=0}^{2m-1} [\hat{\phi}_k(x_j)]^2 = 1.$$

The demonstration of this orthonormality is not particularly difficult, but it is quite a tedious process that relies upon the trigonometric identities

$$\sin a \cos b = \tfrac{1}{2}[\sin(a + b) + \sin(a - b)],$$

$$\sin a \sin b = \tfrac{1}{2}[\cos(a - b) - \cos(a + b)],$$

and

$$\cos a \cos b = \tfrac{1}{2}[\cos(a + b) + \cos(a - b)]$$

together with the relations

$$\sum_{j=0}^{2m-1} \cos \frac{kj\pi}{m} = \sum_{j=0}^{2m-1} \sin \frac{kj\pi}{m} = 0,$$

relations that are true for each pair of positive integers k and m. (See Exercises 7a, 7b for details and hints for the verification of these results.) The use of these relations produces the least-squares approximation S_n. It is most easily expressed in the form

$$(7.35) \qquad S_n(x) = \frac{a_0}{2} + a_n \cos nx + \sum_{k=1}^{n-1} (a_k \cos kx + a_{n+k} \sin kx),$$

where

$$(7.36) \qquad a_k = \frac{1}{m} \sum_{j=0}^{2m-1} y_j \cos kx_j \qquad \text{for each } k = 0, 1, \ldots, n$$

and

$$(7.37) \qquad a_{n+k} = \frac{1}{m} \sum_{j=0}^{2m-1} y_j \sin kx_j \qquad \text{for each } k = 1, 2, \ldots, n - 1.$$

EXAMPLE 1   To determine the least squares approximation from $\hat{\mathscr{T}}_3$ for the data $\{(x_j, y_j)\}_{j=0}^{11}$, where $x_j = j/6$ and $y_j = e^{-x_j}$, requires that the data points $\{x_j\}_{j=0}^{11}$ first be translated from $[0, 2]$ to $[-\pi, \pi]$. It is easily verified that the required transformation is

$$z_j = \pi(x_j - 1),$$

and that the translated data is of the form

$$\{(z_j, e^{-1 - z_j/\pi})\}_{j=0}^{11}.$$

The least squares trigonometric polynomial is consequently

$$S_3(z) = \left[\frac{a_0}{2} + a_3 \cos 3z + \sum_{k=1}^{2} (a_k \cos kz + a_{3+k} \sin kz) \right]$$

where
$$a_k = \tfrac{1}{6} \sum_{j=0}^{11} e^{-1-z_j/\pi} \cos kz_j \qquad \text{for } k = 0, 1, 2, 3$$

and
$$a_{3+k} = \tfrac{1}{6} \sum_{j=0}^{11} e^{-1-z_j/\pi} \sin kz_j \qquad \text{for } k = 1, 2.$$

Evaluating these sums produces the approximation.

$$S_3(z) = .93872 - .30726 \cos z + .19106 \cos 2z - .16791 \cos 3z$$
$$- .48720 \sin z + .24285 \sin 2z. \qquad \Box$$

The least-squares approximation on the $2m$ data points $\{(x_j, y_j)\}_{j=0}^{2m-1}$ by functions in $\hat{\mathscr{T}}_m$ is simply the interpolatory trigonometric polynomial for the data points. It is this form of the trigonometric polynomial that is most useful in application. The only modification that is imposed is the change of S_m to

$$(7.38) \qquad S_m(x) = \frac{a_0 + a_m \cos mx}{2} + \sum_{j=1}^{m-1} [a_j \cos jx + a_{m+j} \sin jx]$$

The interpolation of large amounts of equally spaced data by trigonometric polynomials can produce very accurate results. It is the appropriate approximation technique used in areas such as those involving digital filters, antenna field patterns, quantum mechanics, optics, and many areas involving simulation problems. Until quite recently however, the method had limited application due to the number of arithmetic calculations required for the determination of the constants in the approximation. The interpolation of $2m$ data points requires approximately $(2m)^2$ multiplications and $(2m)^2$ additions by the direct calculation technique. The approximation of thousands of data points is not unusual in areas requiring trigonometric interpolation; so the direct methods for evaluating the constants require multiplication and addition operations numbering in the millions. The round-off error associated with this number of calculations generally dominates the approximation and makes the results meaningless.

In 1965, a paper written by J. W. Cooley and J. W. Tukey was published in the journal *Mathematics of Computation* [28], which described a method of calculating the constants in the interpolating trigonometric polynomial that requires only $2m \ln(2m)$ multiplications and $2m \ln(2m)$ additions, provided m is chosen in an appropriate manner. For a problem with a thousand data points, this method reduces the number of calculations to less than 7000 compared to a million for the direct technique. The method had actually been discovered a number of years before the Cooley–Tukey paper appeared but had gone unnoticed by most researchers until that time. (Brigham [16], pages 8–9, contains a short but interesting historical summary of the method.)

The method described by Cooley and Tukey has come to be known either as the **Cooley–Tukey Algorithm** or the **Fast Fourier Transform (FFT) Algorithm** and has led to a revolution in the use of interpolatory trigonometric polynomials in many scientific areas. The method consists of organizing the problem so that the number of data points being used can be easily factored, particularly into powers of two. For example, the problem is arranged so that 128, 256, 512, 1024, etc. equally spaced data points are used for determining the interpolatory trigonometric polynomial. In such a situation, the calculations for the constants can be arranged so that a portion of the calculation of one constant can be used to more efficiently calculate a number of the other constants. The details of this procedure can be seen by closely following the calculation steps listed in the following algorithm. Verification of the validity of the procedure can be found in Hamming [44], who presents the method from a mathematical approach, or in Bracewell [14] where the presentation is based on methods more likely to be familiar to engineers. Aho, Hopcroft, and Ullman [4], pages 252–269, is a good reference for a discussion of the computational aspects of the method.

The fast Fourier transform algorithm computes the coefficients c_k in the formula

$$F(x) = \sum_{k=0}^{N-1} c_k e^{ikx}$$

where $\quad c_k = \dfrac{2}{N} \sum_{j=0}^{N-1} y_j e^{2\pi ijk/N} \quad$ for each $k = 0, 1, \ldots, N - 1$ and $i = \sqrt{-1}$.

The discrete least-squares approximation of degree $m = N/2$ is

$$S_m(x) = \frac{a_0 + a_m \cos mx}{2} + \sum_{k=1}^{m-1} (a_k \cos kx + b_k \sin kx),$$

where $\quad a_k = \dfrac{2}{N} \sum_{j=0}^{N-1} y_j \cos k\left(-\pi + \dfrac{2j\pi}{N}\right) \quad$ for $k = 0, 1, \ldots, m$

and $\quad b_k = \dfrac{2}{N} \sum_{j=0}^{N-1} y_j \sin k\left(-\pi + \dfrac{2j\pi}{N}\right) \quad$ for $k = 1, \ldots, m - 1$.

By Euler's formula,

$$e^{ikx} = \cos kx + i \sin kx.$$

So $\quad a_k + ib_k = \dfrac{2}{N} \sum_{j=0}^{N-1} y_j \left[\cos k\left(-\pi + \dfrac{2j\pi}{N}\right) + i \sin k\left(-\pi + \dfrac{2j\pi}{N}\right)\right]$

$$= \frac{2}{N} \sum_{j=0}^{N-1} y_j e^{ik(-\pi + 2j\pi/N)}$$

$$= \frac{2e^{-ik\pi}}{N} \sum_{j=0}^{N-1} y_j e^{2\pi ijk/N}.$$

Thus, if the complex numbers $c_0, \ldots, c_{N-1}$ have been computed, then

$$a_k = \text{Re}(c_k e^{-k\pi i}) \qquad \text{for } k = 0, 1, \ldots, m$$

and $\qquad b_k = \text{Im}(c_k e^{-k\pi i}) \qquad \text{for } k = 1, \ldots, m - 1,$

where $\qquad c_k \equiv \text{Re}(c_k) + i \cdot \text{Im}(c_k).$

---**Fast Fourier Transform Algorithm 7.2**---------

To compute the discrete approximation

$$F(x) = \sum_{k=0}^{N-1} c_k e^{ik\pi x} = \sum_{k=0}^{N-1} c_k(\cos k\pi x + i \sin k\pi x) \qquad \text{where } i = \sqrt{-1},$$

for the data $\{(x_j, y_j)\}_{j=0}^{N-1}$ where $N = 2^p$ and $x_j = -\pi + 2j\pi/N$ for $j = 0, 1, \ldots, N - 1$.

INPUT N; $y_0, y_1, \ldots, y_{N-1}$.

OUTPUT complex numbers $c_0, \ldots, c_{N-1}$.

Step 1 Set $M = N/2$;
$\qquad p = \log_2 N$;
$\qquad q = p - 1$;
$\qquad \zeta = e^{2\pi i/N}$.

Step 2 For $j = 0, 1, \ldots, N - 1$ set $c_j = y_j$.

Step 3 For $j = 1, 2, \ldots, M$ $\qquad$ set $\xi_j = \zeta^j$;
$\qquad\qquad\qquad\qquad\qquad \xi_{j+M} = -\xi_j$.

Step 4 Set $K = 0$;
$\qquad \xi_0 = 1$.

Step 5 For $L = 1, 2, \ldots, p$ do Steps 6–12.

Step 6 While $K < N - 1$ do Steps 7–11.

Step 7 For $j = 1, 2, \ldots, M$ do Steps 8–10.

Step 8 Let $K = k_{p-1} \cdot 2^{p-1} + k_{p-2} \cdot 2^{p-2} + \cdots + k_1 2 + k_0$;
$\qquad$ set $K_1 = K/2^q = k_{p-1} \cdot 2^{p-q-1} + \cdots + k_{q+1} \cdot 2 + k_q$;
$\qquad K_2 = k_q \cdot 2^{p-1} + k_{q+1} \cdot 2^{p-2} + \cdots + k_{p-1} \cdot 2^q$.

Step 9 Set $\eta = c_{K+M} \xi_{K_2}$;
$\qquad c_{K+M} = c_K - \eta$;
$\qquad c_K = c_K + \eta$.

Step 10 Set $K = K + 1$.

Step 11 Set $K = K + M$.

Step 12 Set $K = 0$;
$\qquad M = M/2$;
$\qquad q = q - 1$.

Step 13 While $K < N - 1$ do Steps 14–16.

Step 14 Let $K = k_{p-1} \cdot 2^{p-1} + k_{p-2} \cdot 2^{p-2} + \cdots + k_1 \cdot 2 + k_0$
$\qquad$ set $j = k_0 \cdot 2^{p-1} + k_1 \cdot 2^{p-2} + \cdots + k_{p-2} \cdot 2 + k_{p-1}$.

Step 15 If $j > K$ then interchange c_j and c_k.

Step 16 Set $K = K + 1$.

Step 17 For $j = 0, 1, \ldots, N - 1$ set $c_j = 2c_j/N$.

Step 18 OUTPUT $(c_0, \ldots, c_{N-1})$;
        STOP.

EXAMPLE 2 To determine the trigonometric interpolating polynomial of degree 8 for the data $\{(x_j, y_j)\}_{j=0}^{15}$ where $x_j = j/8$ and $y_j = e^{-x_j}$ requires a transformation of the interval $[0, 2]$ to $[-\pi, \pi]$. The translation is given by

$$z_j = \pi(x_j - 1)$$

so that the input data to Algorithm 7.2 is

$$\{z_j, e^{-1-z_j/\pi}\}_{j=0}^{15}.$$

Table 7.18 lists $\hat{c}_k = e^{-k\pi i}c_k$ for $k = 0, 1, \ldots, 15$, generated by Algorithm 7.2.

TABLE 7.18	k	$\operatorname{Re}(\hat{c}_k)$	$\operatorname{Im}(\hat{c}_k)$
	0	0.91983150	0.00000000
	1	-0.13472420	-0.24636630
	2	$0.76563530 \times 10^{-1}$	0.12707350
	3	$-0.64875540 \times 10^{-1}$	$-0.79866820 \times 10^{-1}$
	4	$0.60761670 \times 10^{-1}$	$0.53622240 \times 10^{-1}$
	5	$-0.58912200 \times 10^{-1}$	$-0.35906370 \times 10^{-1}$
	6	$0.57990620 \times 10^{-1}$	$0.22283140 \times 10^{-1}$
	7	$-0.57547600 \times 10^{-1}$	$-0.10706750 \times 10^{-1}$
	8	$0.57414520 \times 10^{-1}$	$0.28837380 \times 10^{-6}$
	9	$-0.57547800 \times 10^{-1}$	$0.10705270 \times 10^{-1}$
	10	$0.57991160 \times 10^{-1}$	$-0.22281700 \times 10^{-1}$
	11	$-0.58912600 \times 10^{-1}$	$0.35905730 \times 10^{-1}$
	12	$0.60762590 \times 10^{-1}$	$-0.53621170 \times 10^{-1}$
	13	$-0.64877270 \times 10^{-1}$	$0.79865270 \times 10^{-1}$
	14	$0.76565200 \times 10^{-1}$	-0.12707240
	15	-0.13472890	0.24636360

The interpolating polynomial is

$$S_8(x) = .4599157 - .1347242 \cos x + .0765635 \cos 2x + \cdots$$
$$+ .02870726 \cos 8x - .2463663 \sin x + \cdots - .01070678 \sin 7x. \quad \square$$

Exercise Set 7.5

1. Compute the trigonometric least-squares polynomial of degree 3 using $m = 4$ for $f(x) = e^x \cos 2x$ on the interval $[-\pi, \pi]$.

2. Repeat Exercise 1 using $m = 8$. Which approximation is better?

3. Compute the trigonometric interpolating polynomial of degree 4 for $f(x) = x(\pi - x)$ on the interval $[-\pi, \pi]$ using

   a) equations (7.36) and (7.37),   b) Algorithm 7.2.

   Compare the accuracy by evaluating each polynomial at the nodes.

4. a) Compute the trigonometric interpolating polynomial S_4 of degree 4 for $f(x) = x^2 \sin x$ on the interval $[0, 1]$.
   b) Compute $\int_0^1 S_4(x)\, dx$.
   c) Compare the integral in (b) to $\int_0^1 x^2 \sin x\, dx$.

5. a) Compute the least-squares trigonometric polynomial S_4 of degree 4 using $m = 16$, for $f(x) = x^2 \sin x$ on the interval $[0, 1]$.
   b) Compute $\int_0^1 S_4(x)\, dx$.
   c) Compare the integral in (b) to $\int_0^1 x^2 \sin x\, dx$.

6. Using Algorithm 7.2, compute the trigonometric interpolating polynomial of degree 16 for $f(x) = x^2 \cos x$ on $[-\pi, \pi]$.

7. a) Show that for any positive integers m and k

$$\sum_{j=0}^{2m-1} \cos \frac{kj\pi}{m} = \sum_{j=0}^{2m-1} \sin \frac{kj\pi}{m} = 0.$$

   [*Hint*:   Let $w = e^{k\pi i/m}$. Then

$$0 = \sum_{j=0}^{2m-1} w^j = \sum_{j=0}^{2m-1} e^{k\pi i j/m} = \sum_{j=0}^{2m-1} \left(\cos \frac{kj\pi}{m} + i \sin \frac{kj\pi}{m} \right).]$$

   b) Show that for $l \neq k$,

$$\sum_{j=0}^{2m-1} \cos kx_j \sin lx_j = \sum_{j=0}^{2m-1} \cos kx_j \cos lx_j = \sum_{j=0}^{2m-1} \sin kx_j \sin lx_j = 0$$

   and

$$\sum_{j=0}^{2m-1} (\cos kx_j)^2 = \sum_{j=0}^{2m-1} (\sin kx_j)^2 = m \qquad \text{for } k > 0.$$

8. Count the number of complex multiplication and addition operations required in Algorithm 7.2, and compare this to the number of multiplications and additions necessary to compute the trigonometric interpolating polynomial using equations (7.36) and (7.37).

9. Show that $c_0, \ldots, c_{N-1}$ in Algorithm 7.2 are given by

$$\frac{N}{2} \begin{bmatrix} c_0 \\ c_1 \\ c_2 \\ \vdots \\ c_{N-1} \end{bmatrix} = \begin{bmatrix} 1 & 1 & 1 & \cdots & 1 \\ 1 & \zeta & \zeta^2 & \cdots & \zeta^{N-1} \\ 1 & \zeta^2 & \zeta^4 & \cdots & \zeta^{2N-2} \\ \vdots & \vdots & \vdots & & \vdots \\ 1 & \zeta^{N-1} & \zeta^{2N-2} & \cdots & \zeta^{(N-1)^2} \end{bmatrix} \begin{bmatrix} y_0 \\ y_1 \\ y_2 \\ \vdots \\ y_{N-1} \end{bmatrix}$$

   where $\zeta = e^{2\pi i/N}$.

10. Consider Exercise 9 in the case $N = 4$; i.e.,

$$2 \begin{bmatrix} c_0 \\ c_1 \\ c_2 \\ c_3 \end{bmatrix} = \begin{bmatrix} 1 & 1 & 1 & 1 \\ 1 & \zeta & \zeta^2 & \zeta^3 \\ 1 & \zeta^2 & \zeta^4 & \zeta^6 \\ 1 & \zeta^3 & \zeta^6 & \zeta^9 \end{bmatrix} \begin{bmatrix} y_0 \\ y_1 \\ y_2 \\ y_3 \end{bmatrix}.$$

Show that

$$2\begin{bmatrix} c_0 \\ c_2 \\ c_1 \\ c_3 \end{bmatrix} = \begin{bmatrix} 1 & 1 & 0 & 0 \\ 1 & \zeta^2 & 0 & 0 \\ 0 & 0 & 1 & \zeta \\ 0 & 0 & 1 & \zeta^3 \end{bmatrix}\begin{bmatrix} 1 & 0 & 1 & 0 \\ 0 & 1 & 0 & 1 \\ 1 & 0 & \zeta^2 & 0 \\ 0 & 1 & 0 & \zeta^2 \end{bmatrix}\begin{bmatrix} y_0 \\ y_1 \\ y_2 \\ y_3 \end{bmatrix}.$$

Compare the above equation to a trace of Algorithm 7.2 in the case $N = 4$.

Iterative Techniques in Matrix Algebra

At the beginning of Chapter 6 we discussed an electrical circuit problem involving seven resistors and an impressed voltage. This problem led to a 5×5 linear system whose solution gave the potential at each junction in the circuit. In a circuit involving many more resistances the situation would be similar but, of course, the associated linear system would be larger. Even in large circuits, however, we would expect each junction to have only a relatively small number of connections. If we consider the matrix associated with the 5×5 linear system discussed in Chapter 6,

$$
\begin{array}{c@{\,}c}
 & \begin{array}{ccccc} v_b & v_c & v_d & v_e & v_f \end{array} \\
\begin{array}{c} v_b \\ v_c \\ v_d \\ v_e \\ v_f \end{array} &
\left[\begin{array}{rrrrr}
-31 & 10 & 0 & 0 & 6 \\
2 & -8 & 3 & 3 & 0 \\
0 & 1 & -3 & 2 & 0 \\
0 & 2 & 4 & -9 & 1 \\
12 & 0 & 15 & 0 & -47
\end{array}\right]
\end{array}
$$

we see that a zero occurs in the matrix whenever two junctions are not directly connected. Matrices associated with large electrical circuits generally have a high percentage of zero entries. Matrices of this type are called **sparse**, the appropriate methods for solving large linear systems involving sparse matrices often involve techniques which are iterative rather than direct.

The methods presented in Chapter 6 utilized direct techniques to solve a system of $n \times n$ linear equations of the form $A\mathbf{x} = \mathbf{b}$. In this chapter, we will present some methods that can be used to iteratively solve a system of this type.

8.1   Norms of Vectors and Matrices

Before considering iterative methods for solving linear systems, it is necessary to determine a method for quantitatively measuring the distance between vectors in R^n, the set of all column vectors with real components, in order to determine whether the sequence of vectors which results from using an iterative technique converges to a solution of the system. In actuality, this measure is also needed when the solution is being obtained by the direct methods presented in Chapter 6, since these methods require the performance of a large number of arithmetic operations, and using finite-digit arithmetic will lead only to an approximation to an actual solution of the system.

To define a distance in R^n, we will use the idea of the **norm** of a vector.

Definition 8.1   A **vector norm** on R^n, the collection of all n-dimensional column vectors with real components, is a function, $\|\cdot\|$, from R^n into R with the following properties:

i) $\|\mathbf{x}\| \geq 0$ for all $\mathbf{x} \in R^n$,
ii) $\|\mathbf{x}\| = 0$ if and only if $\mathbf{x} = (0, 0, \ldots, 0)^t \equiv \mathbf{0}$,
iii) $\|\alpha x\| = |\alpha|\,\|\mathbf{x}\|$ for all $\alpha \in R$ and $\mathbf{x} \in R^n$,
iv) $\|\mathbf{x} + \mathbf{y}\| \leq \|\mathbf{x}\| + \|\mathbf{y}\|$ for all $\mathbf{x}, \mathbf{y} \in R^n$.

For our purposes we will need only two specific norms on R^n, although a third norm on R^n is presented in Exercise 2.

Since vectors in R^n are column vectors, it is convenient to use the transpose notation, presented in Section 6.5 when a vector is represented in terms of its components. For example the vector

$$\mathbf{x} = \begin{bmatrix} x_1 \\ x_2 \\ \vdots \\ x_n \end{bmatrix}$$

will generally be written $\mathbf{x} = (x_1, x_2, \ldots, x_n)^t$,

Definition 8.2   The l_2 and l_∞ norms for the vector $\mathbf{x} = (x_1, x_2, \ldots, x_n)^t$ are defined by

(8.1)
$$\|\mathbf{x}\|_2 = \left\{ \sum_{i=1}^{n} x_i^2 \right\}^{1/2}$$

and

(8.2)
$$\|\mathbf{x}\|_\infty = \max_{1 \leq i \leq n} |x_i|.$$

The l_2 norm is often called the **Euclidean norm** of the vector **x** since it represents the usual notion of distance from the origin in case **x** is in $R^1 \equiv R$, R^2, or R^3.

EXAMPLE 1   The vector $\mathbf{x} = (-1, 1, -2)^t$ in R^3 has norms

$$\|\mathbf{x}\|_2 = \sqrt{(-1)^2 + (1)^2 + (-2)^2} = \sqrt{6}$$

and

$$\|\mathbf{x}\|_\infty = \max\{|-1|, |1|, |-2|\} = 2.$$

The collection of all vectors in R^3 having l_2 norm less than 1 lie inside the sphere about $(0, 0, 0)^t$ with radius 1 (see Fig. 8.1), while those with l_∞ norm less than 1 (see Fig. 8.2) lie inside the cube with sides of length 2 centered at $(0, 0, 0)^t$. □

The justification for calling the concept defined by $\|\mathbf{x}\|_\infty = \max_{1 \le i \le n} |x_i|$ a norm on R^n is presented in the following theorem.

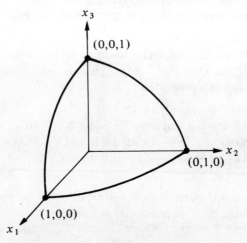

FIGURE 8.1   The vectors with l_2 norm less than one and lying in the first octant are inside this figure.

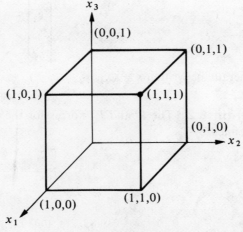

FIGURE 8.2   The vectors with l_∞ norm less than one and lying in the first octant are inside this figure.

Theorem 8.3   For each $\mathbf{x}, \mathbf{y} \in R^n$ and $\alpha \in R$,

   i) $\|\mathbf{x}\|_\infty \geq 0$,
   ii) $\|\mathbf{x}\|_\infty = 0$ if and only if $\mathbf{x} = \mathbf{0}$,
   iii) $\|\alpha\mathbf{x}\|_\infty = |\alpha|\,\|\mathbf{x}\|_\infty$
   iv) $\|\mathbf{x} + \mathbf{y}\|_\infty \leq \|\mathbf{x}\|_\infty + \|\mathbf{y}\|_\infty$,

   so $\|\cdot\|_\infty$ is a norm on R^n.

Proof   The properties follow easily from similar statements concerning the absolute value function. As an example, consider statement (iv). If $\mathbf{x} = (x_1, x_2, \ldots, x_n)^t$ and $\mathbf{y} = (y_1, y_2, \ldots, y_n)^t$, then

$$\|\mathbf{x} + \mathbf{y}\|_\infty = \max_{1 \leq i \leq n} |x_i + y_i|$$

$$\leq \max_{1 \leq i \leq n} \{|x_i| + |y_i|\}$$

$$\leq \max_{1 \leq i \leq n} |x_i| + \max_{1 \leq i \leq n} |y_i| = \|\mathbf{x}\|_\infty + \|\mathbf{y}\|_\infty.$$

The other properties are shown in a similar manner. $\square$

To show that

$$\|\mathbf{x}\|_2 = \left\{ \sum_{i=1}^{n} x_i^2 \right\}^{1/2}$$

describes a norm in R^n is more difficult. The problem occurs in proving that

$$\|\mathbf{x} + \mathbf{y}\|_2 \leq \|\mathbf{x}\|_2 + \|\mathbf{y}\|_2 \qquad \text{for each } \mathbf{x}, \mathbf{y} \in R^n.$$

Theorem 8.4   For each $\mathbf{x}, \mathbf{y} \in R^n$ and $\alpha \in R$.

   i) $\|\mathbf{x}\|_2 \geq 0$,
   ii) $\|\mathbf{x}\|_2 = 0$ if and only if $\mathbf{x} = \mathbf{0}$,
   iii) $\|\alpha\mathbf{x}\|_2 = |\alpha|\,\|\mathbf{x}\|_2$,
   iv) $\|\mathbf{x} + \mathbf{y}\|_2 \leq \|\mathbf{x}\|_2 + \|\mathbf{y}\|_2$.

Proof   For each $\mathbf{x} = (x_1, x_2, \ldots, x_n)^t$ in R^n,

$$\|\mathbf{x}\|_2 = \left\{ \sum_{i=1}^{n} x_i^2 \right\}^{1/2} \geq 0$$

and $\|\mathbf{x}\|_2 = 0$ if and only if $x_i^2 = 0$ for each $i = 1, 2, \ldots, n$.

Thus, $\|\mathbf{x}\|_2 = 0$ if and only if $\mathbf{x} = \mathbf{0}$, and properties (i) and (ii) are established.
To show that (iii) holds, simply note that:

$$\|\alpha\mathbf{x}\|_2 = \left\{ \sum_{i=1}^{n} (\alpha x_i)^2 \right\}^{1/2} = \left\{ \sum_{i=1}^{n} \alpha^2 x_i^2 \right\}^{1/2} = (\alpha^2)^{1/2} \left\{ \sum_{i=1}^{n} x_i^2 \right\}^{1/2} = |\alpha|\,\|\mathbf{x}\|_2.$$

In order to show property (iv), it is first necessary to prove the following Lemma.

Lemma 8.5 (Cauchy–Buniakowsky–Schwarz Inequality for Sums)
For each $\mathbf{x} = (x_1, x_2, \ldots, x_n)^t$ and $\mathbf{y} = (y_1, y_2, \ldots, y_n)^t$ in R^n,

$$(8.3) \qquad \sum_{i=1}^{n} |x_i y_i| \leq \left\{ \sum_{i=1}^{n} x_i^2 \right\}^{1/2} \left\{ \sum_{i=1}^{n} y_i^2 \right\}^{1/2}.$$

Proof of Lemma 8.5   If $\mathbf{y} = \mathbf{0}$ or $\mathbf{x} = \mathbf{0}$, the result is immediate, since both sides of Eq. (8.3) are zero.

Suppose $\mathbf{y} \neq \mathbf{0}$ and $\mathbf{x} \neq \mathbf{0}$. For each $\lambda \in R$,

$$0 \leq \|\mathbf{x} - \lambda \mathbf{y}\|_2^2 = \sum_{i=1}^{n} (x_i - \lambda y_i)^2 = \sum_{i=1}^{n} x_i^2 - 2\lambda \sum_{i=1}^{n} x_i y_i + \lambda^2 \sum_{i=1}^{n} y_i^2.$$

Thus,

$$2\lambda \sum_{i=1}^{n} x_i y_i \leq \sum_{i=1}^{n} x_i^2 + \lambda^2 \sum_{i=1}^{n} y_i^2 = \|\mathbf{x}\|_2^2 + \lambda^2 \|\mathbf{y}\|_2^2.$$

Since Theorem 8.4 implies that $\|\mathbf{x}\|_2 > 0$ and $\|\mathbf{y}\|_2 > 0$, choosing $\lambda = \|\mathbf{x}\|_2 / \|\mathbf{y}\|_2$ gives

$$\left(2 \frac{\|\mathbf{x}\|_2}{\|\mathbf{y}\|_2} \right) \left(\sum_{i=1}^{n} x_i y_i \right) \leq \|\mathbf{x}\|_2^2 + \frac{\|\mathbf{x}\|_2^2}{\|\mathbf{y}\|_2^2} \|\mathbf{y}\|_2^2 = 2\|\mathbf{x}\|_2^2$$

so

$$2\sum_{i=1}^{n} x_i y_i \leq 2 \frac{\|\mathbf{y}\|_2}{\|\mathbf{x}\|_2} \|\mathbf{x}\|_2^2 = 2\|\mathbf{x}\|_2\|\mathbf{y}\|_2.$$

Thus,

$$\sum_{i=1}^{n} x_i y_i \leq \|\mathbf{x}\|_2\|\mathbf{y}\|_2.$$

Replacing x_i by $-x_i$ whenever $x_i y_i < 0$ and calling the vector $\tilde{\mathbf{x}}$:

$$\sum_{i=1}^{n} |x_i y_i| \leq \|\tilde{\mathbf{x}}\|_2\|\mathbf{y}\|_2 = \|\mathbf{x}\|_2\|\mathbf{y}\|_2 = \left\{ \sum_{i=1}^{n} x_i^2 \right\}^{1/2} \left\{ \sum_{i=1}^{n} y_i^2 \right\}^{1/2} \qquad \square$$

Proof of Theorem 8.4 (*continued*)   For each $\mathbf{x}, \mathbf{y} \in R^n$,

$$\|\mathbf{x} + \mathbf{y}\|_2^2 = \sum_{i=1}^{n} (x_i + y_i)^2 = \sum_{i=1}^{n} x_i^2 + 2\sum_{i=1}^{n} x_i y_i + \sum_{i=1}^{n} y_i^2$$

(by Lemma 8.5),

$$\leq \|\mathbf{x}\|_2^2 + 2\|\mathbf{x}\|_2\|\mathbf{y}\|_2 + \|\mathbf{y}\|_2^2 = (\|\mathbf{x}\|_2 + \|\mathbf{y}\|_2)^2,$$

so

$$\|\mathbf{x} + \mathbf{y}\|_2 \leq \|\mathbf{x}\|_2 + \|\mathbf{y}\|_2. \qquad \square$$

Since the norm of a vector gives a measure for the distance between the vector and the origin, the **distance between two vectors** can be defined as the norm of the difference of the vectors.

Definition 8.6   If $\mathbf{x} = (x_1, x_2, \ldots, x_n)^t$ and $\mathbf{y} = (y_1, y_2, \ldots, y_n)^t$ are vectors in R^n, the l_2 and l_∞ distances between $\mathbf{x}$ and $\mathbf{y}$ are defined by:

$$(8.4) \qquad \|\mathbf{x} - \mathbf{y}\|_2 = \left\{ \sum_{i=1}^{n} |x_i - y_i|^2 \right\}^{1/2}$$

and

$$(8.5) \qquad \|\mathbf{x} - \mathbf{y}\|_\infty = \max_{1 \leq i \leq n} |x_i - y_i|.$$

EXAMPLE 2 The linear system

$$3.3330x_1 + 15920x_2 - 10.333x_3 = 15913,$$
$$2.2220x_1 + 16.710x_2 + 9.6120x_3 = 28.544,$$
$$1.5611x_1 + 5.1791x_2 + 1.6852x_3 = 8.4254,$$

has solution $(x_1, x_2, x_3)^t = (1.0000, 1.0000, 1.0000)^t$. If Gaussian elimination is performed in five-digit arithmetic using maximal column pivoting (Algorithm 6.2. p. 291), the solution obtained is

$$\mathbf{x} = (\tilde{x}_1, \tilde{x}_2, \tilde{x}_3)^t = (1.2001, .99991, .92538)^t.$$

Measurements of $\mathbf{x} - \tilde{\mathbf{x}}$ are given by

$$\|\mathbf{x} - \tilde{\mathbf{x}}\|_\infty = \max\{|1.0000 - 1.2001|, |1.0000 - .99991|, |1.0000 - .92538|\}$$
$$= \max\{.2001, .00009, .07462\}$$
$$= .2001,$$

and

$$\|\mathbf{x} - \tilde{\mathbf{x}}\|_2 = (|1.0000 - 1.2001|^2 + |1.0000 - .99991|^2 + |1.0000 - .92538|^2)^{1/2}$$
$$= [(.2001)^2 + (.00009)^2 + (.07462)^2]^{1/2}$$
$$= .21356. \qquad \square$$

Although the components $\tilde{\mathbf{x}}_2$ and $\tilde{\mathbf{x}}_3$ are fairly good approximations to x_2 and x_3, note that the component $\tilde{x}_1$ is a poor approximation to x_1 and that $|x_1 - \tilde{x}_1|$ dominates the norms.

The concept of distance in R^n can also be used to define a limit of a sequence of vectors in this space.

Definition 8.7 A sequence $\{\mathbf{x}^{(k)}\}_{k=1}^\infty$ of vectors in R^n is said to **converge** to $\mathbf{x}$ with respect to the norm $\|\cdot\|$, if, given any $\varepsilon > 0$, there exists an integer $N(\varepsilon)$ such that

$$\|\mathbf{x}^{(k)} - \mathbf{x}\| < \varepsilon \qquad \text{for all } k \geq N(\varepsilon).$$

It can be shown that all norms on R^n are equivalent with respect to convergence; that is, if $\|\cdot\|$ and $\|\cdot\|'$ are any two norms on R^n and $\{\mathbf{x}^{(k)}\}_{k=1}^\infty$ has the limit $\mathbf{x}$ with respect to $\|\cdot\|$, then $\{\mathbf{x}^{(k)}\}_{k=1}^\infty$ has the limit $\mathbf{x}$ with respect to $\|\cdot\|'$. The proof of this fact for the general case can be found in Ortega [64], page 8. The case for the norms $\|\ \|_2$ and $\|\ \|_\infty$ is given in Exercise 3.

EXAMPLE 3 Let $\mathbf{x}^{(k)} \in R^4$ be defined by

$$\mathbf{x}^{(k)} = (x_1^{(k)}, x_2^{(k)}, x_3^{(k)}, x_4^{(k)})^t = \left(1, 2 + \frac{1}{k}, \frac{3}{k^2}, e^{-k}\sin k\right)^t.$$

Some typical members of the sequence are

$$\mathbf{x}^{(1)} = \begin{bmatrix} 1 \\ 3 \\ 3 \\ \dfrac{\sin 1}{e} \end{bmatrix}, \qquad \mathbf{x}^{(2)} = \begin{bmatrix} 1 \\ 2.5 \\ .75 \\ \dfrac{\sin 2}{e^2} \end{bmatrix} \approx \begin{bmatrix} 1 \\ 2.5 \\ .75 \\ .1231 \end{bmatrix},$$

$$\mathbf{x}^{(10)} = \begin{bmatrix} 1 \\ 2.1 \\ .03 \\ \dfrac{\sin 10}{e^{10}} \end{bmatrix} \approx \begin{bmatrix} 1 \\ 2.1 \\ .03 \\ -2.5 \times 10^{-5} \end{bmatrix}$$

and

$$\mathbf{x}^{(100)} = \begin{bmatrix} 1 \\ 2.01 \\ 3 \times 10^{-4} \\ \dfrac{\sin 100}{e^{100}} \end{bmatrix} \approx \begin{bmatrix} 1 \\ 2.01 \\ 3 \times 10^{-4} \\ -1.88 \times 10^{-44} \end{bmatrix}$$

It appears that the vector $\mathbf{x}$ given by

$$\mathbf{x} = (1, 2, 0, 0)^t$$

is the limit of the sequence $\{\mathbf{x}^{(k)}\}_{k=1}^{\infty}$ since all component sequences converge to the components of x; that is,

$$\lim_{k \to \infty} x_1^{(k)} = \lim_{k \to \infty} 1 = 1 = x_1,$$

$$\lim_{k \to \infty} x_2^{(k)} = \lim_{k \to \infty} \left(2 + \frac{1}{k} \right) = 2 = x_2,$$

$$\lim_{k \to \infty} x_3^{(k)} = \lim_{k \to \infty} \frac{3}{k^2} = 0 = x_3,$$

and

$$\lim_{k \to \infty} x_4^{(k)} = \lim_{k \to \infty} \frac{\sin k}{e^k} = 0 = x_4.$$

To establish this fact note that

$$\mathbf{x}^{(k)} - \mathbf{x} = \left(0, \frac{1}{k}, \frac{3}{k^2}, e^{-k} \sin k \right)^t.$$

Expanding e^t in a second-degree Taylor polynomial about zero, we see that

$$e^k \geq 1 + k + \tfrac{1}{2}k^2 \geq \tfrac{1}{2}k^2.$$

It follows that for $k \geq 3$

$$0 \leq |e^{-k} \sin k| \leq \frac{2}{k^2} |\sin k| \leq \frac{2}{k^2} \leq \frac{1}{k}.$$

Hence, for $k \geq 3$

$$\|\mathbf{x}^{(k)} - \mathbf{x}\|_\infty = \max\left\{|0|, \left|\frac{1}{k}\right|, \left|\frac{3}{k^2}\right|, |e^{-k} \sin k|\right\} = \frac{1}{k}.$$

Given $\varepsilon > 0$, let N be any integer greater than $1/\varepsilon$. If $k \geq N$, then

$$\|\mathbf{x}^{(k)} - \mathbf{x}\|_\infty = \frac{1}{k} \leq \frac{1}{N} < \varepsilon,$$

and $\{\mathbf{x}^{(k)}\}_{k=1}^\infty$ converges to $\mathbf{x}$.   □

To show directly that this sequence converges to $(1, 2, 0, 0)^t$ with respect to the l_2 norm is quite complicated. It is much easier to prove the next theorem and apply it to this special case.

Theorem 8.8   For each $\mathbf{x} \in R^n$,

$$\|\mathbf{x}\|_\infty \leq \|\mathbf{x}\|_2 \leq \sqrt{n}\|\mathbf{x}\|_\infty.$$

Proof   Let x_j be the coordinate of $\mathbf{x}$ such that $\|\mathbf{x}\|_\infty = \max_{1 \leq i \leq n} |x_i| = |x_j|$. Then

$$\|\mathbf{x}\|_\infty^2 = |x_j|^2 = x_j^2 \leq \sum_{i=1}^n x_i^2 \leq \sum_{i=1}^n x_j^2 = nx_j^2 = n\|\mathbf{x}\|_\infty^2.$$

Thus, $$\|\mathbf{x}\|_\infty \leq \left\{\sum_{i=1}^n x_i^2\right\}^{1/2} = \|\mathbf{x}\|_2 \leq \sqrt{n}\|\mathbf{x}\|_\infty.$$ □

To use this result to show that the sequence $\{\mathbf{x}^{(k)}\}_{k=1}^\infty$, with

$$\mathbf{x}^{(k)} = \left(1, 2 + \frac{1}{k}, \frac{3}{k^2}, e^{-k} \sin k\right)^t,$$

converges to $(1, 2, 0, 0)^t$, suppose $\varepsilon > 0$ is given. Choose $N(\varepsilon)$ to be any integer greater than $2/\varepsilon$. Since we have already shown that $\|\mathbf{x}^{(k)} - \mathbf{x}\|_\infty < 1/k$ if $k \geq N(\varepsilon)$, Theorem 8.8 implies that

$$\|\mathbf{x}^{(k)} - \mathbf{x}\|_2 \leq \sqrt{4}\|\mathbf{x}^{(k)} - \mathbf{x}\|_\infty < \frac{2}{k} \leq \frac{2}{N(\varepsilon)} < \varepsilon.$$

It is also necessary to have a method for measuring the distance between two $n \times n$ matrices, which again requires the use of the norm concept.

Definition 8.9   A **matrix norm** on the set of all real $n \times n$ matrices is a real-valued function, $\|\cdot\|$, defined on this set, satisfying for all $n \times n$ matrices A and B and all real numbers α:

i) $\|A\| \geq 0$,
ii) $\|A\| = 0$ if and only if A is the matrix with all zero entries,
iii) $\|\alpha A\| = |\alpha|\|A\|$,
iv) $\|A + B\| \leq \|A\| + \|B\|$,
v) $\|AB\| \leq \|A\|\|B\|$.

A **distance between $n \times n$ matrices** A and B can be defined, in the usual manner, as $\|A - B\|$.

Although matrix norms can be obtained in various ways, the only norms we will consider are those that are natural consequences of the vector norms l_2 and l_∞.

The following theorem is not difficult to show, and its proof is left to Exercise 8.

Theorem 8.10 If $\|\cdot\|$ is any vector norm on R^n, then

$$\|A\| = \max_{\|\mathbf{x}\| = 1} \|A\mathbf{x}\|$$

defines a matrix norm on the set of real $n \times n$ matrices, which is called a **natural norm**.

In this text all matrix norms will be assumed to be natural matrix norms, unless specified otherwise.

The matrix norms we will consider consequently have the forms

$$\|A\|_\infty = \max_{\|\mathbf{x}\|_\infty = 1} \|A\mathbf{x}\|_\infty, \quad \text{the } l_\infty \text{ norm,}$$

and

$$\|A\|_2 = \max_{\|\mathbf{x}\|_2 = 1} \|A\mathbf{x}\|_2, \quad \text{the } l_2 \text{ norm.}$$

The l_∞ norm of a matrix has a particularly interesting representation with respect to the entries of the matrix.

Theorem 8.11 If $A = (a_{ij})$ is an $n \times n$ matrix, then

$$\|A\|_\infty = \max_{1 \le i \le n} \sum_{j=1}^{n} |a_{ij}|.$$

Proof Let $\mathbf{x}$ be an n-dimensional column vector with $1 = \|\mathbf{x}\|_\infty = \max_{1 \le i \le n} |x_i|$. Since $A\mathbf{x}$ is also an n-dimensional column vector,

$$\|A\mathbf{x}\|_\infty = \max_{1 \le i \le n} |(A\mathbf{x})_i| = \max_{1 \le i \le n} \left| \sum_{j=1}^{n} a_{ij} x_j \right|$$

$$\le \max_{1 \le i \le n} \sum_{j=1}^{n} |a_{ij}| \left(\max_{1 \le j \le n} |x_j| \right) = \max_{1 \le i \le n} \sum_{j=1}^{n} |a_{ij}| \|\mathbf{x}\|_\infty$$

$$= \max_{1 \le i \le n} \sum_{j=1}^{n} |a_{ij}|$$

so

$$\|A\mathbf{x}\|_\infty \le \max_{1 \le i \le n} \sum_{j=1}^{n} |a_{ij}|$$

for all $\mathbf{x}$ with $\|\mathbf{x}\|_\infty = 1$. Consequently,

(8.6)

$$\|A\|_\infty = \max_{\|\mathbf{x}\|_\infty = 1} \|A\mathbf{x}\|_\infty \le \max_{1 \le i \le n} \sum_{j=1}^{n} |a_{ij}|.$$

However, if p is the integer, $1 \leq p \leq n$, with

$$\sum_{j=1}^{n} |a_{pj}| = \max_{1 \leq i \leq n} \sum_{j=1}^{n} |a_{ij}|,$$

and $\mathbf{x}$ is chosen with

$$x_j = \begin{cases} 1, & \text{if } a_{p,j} \geq 0, \\ -1, & \text{if } a_{p,j} < 0, \end{cases}$$

then $\|\mathbf{x}\|_{\infty} = 1$ and $a_{p,j} x_j = |a_{p,j}|$ for all $j = 1, 2, \ldots, n$. Moreover,

$$\|A\mathbf{x}\|_{\infty} = \max_{1 \leq i \leq n} \left| \sum_{j=1}^{n} a_{ij} x_j \right| \geq \left| \sum_{j=1}^{n} a_{pj} x_j \right| = \sum_{j=1}^{n} |a_{pj}| = \max_{1 \leq i \leq n} \sum_{j=1}^{n} |a_{ij}|.$$

This implies that

$$\|A\|_{\infty} = \max_{\|\mathbf{x}\|_{\infty}=1} \|A\mathbf{x}\|_{\infty} \geq \max_{1 \leq i \leq n} \sum_{j=1}^{n} |a_{ij}|,$$

which, together with inequality (8.6), gives

$$\|A\|_{\infty} = \max_{1 \leq i \leq n} \sum_{j=1}^{n} |a_{ij}|. \qquad \square$$

EXAMPLE 4   If

$$A = \begin{bmatrix} 1 & 2 & -1 \\ 0 & 3 & -1 \\ 5 & -1 & 1 \end{bmatrix},$$

then

$$\sum_{j=1}^{3} |a_{1j}| = |1| + |2| + |-1| = 4,$$

$$\sum_{j=1}^{3} |a_{2j}| = |0| + |3| + |-1| = 4,$$

and

$$\sum_{j=1}^{3} |a_{3j}| = |5| + |-1| + |1| = 7;$$

so

$$\|A\|_{\infty} = \max\{4, 4, 7\} = 7. \qquad \square$$

To investigate the l_2-norm, it is necessary to discuss some further concepts of linear algebra.

Definition 8.12   If A is a real $n \times n$ matrix, the polynomial defined by

$$p(\lambda) = \det(A - \lambda I)$$

is called the **characteristic polynomial** of A.

It is easy to show (see Exercise 11) that p is an nth-degree polynomial with real coefficients and, consequently, has at most n distinct zeros, some of which may be complex. If λ is a zero of p, then, since $\det(A - \lambda I) = 0$, Theorem 6.13 (p. 285) implies that the linear system defined by $(A - \lambda I)\mathbf{x} = \mathbf{0}$ has a solution other than the identically zero solution. We wish to study the zeros of p and the nontrivial solutions corresponding to these systems.

Definition 8.13 If p is the characteristic polynomial of the matrix A, the zeros of p are called **eigenvalues** or characteristic values of the matrix A. If λ is an eigenvalue of A and $\mathbf{x} \neq \mathbf{0}$ has the property that $(A - \lambda I)\mathbf{x} = \mathbf{0}$, then $\mathbf{x}$ is called an **eigenvector** or characteristic vector of A corresponding to the eigenvalue λ.

EXAMPLE 5 Let

$$A = \begin{bmatrix} 1 & 0 & 1 \\ 2 & 2 & 1 \\ -1 & 0 & 0 \end{bmatrix}.$$

To compute the eigenvalues of A consider

$$p(\lambda) = \det(A - \lambda I) = \det \begin{bmatrix} 1-\lambda & 0 & 1 \\ 2 & 2-\lambda & 1 \\ -1 & 0 & -\lambda \end{bmatrix}$$

$$= (1-\lambda) \begin{vmatrix} 2-\lambda & 1 \\ 0 & -\lambda \end{vmatrix} + 1 \begin{vmatrix} 2 & 2-\lambda \\ -1 & 0 \end{vmatrix}$$

$$= (1-\lambda)(2-\lambda)(-\lambda) + (2-\lambda)$$

$$= (2-\lambda)(\lambda^2 - \lambda + 1).$$

The eigenvalues of A are the solutions of $p(\lambda) = 0$, $\lambda_1 = 2$, $\lambda_2 = \frac{1}{2} + (\sqrt{3}/2)i$, and $\lambda_3 = \frac{1}{2} - (\sqrt{3}/2)i$.

An eigenvector $\mathbf{x}$ of A associated with λ_1 is a solution of the system $(A - \lambda_1 I)\mathbf{x} = \mathbf{0}$:

$$\begin{bmatrix} -1 & 0 & 1 \\ 2 & 0 & 1 \\ -1 & 0 & -2 \end{bmatrix} \begin{bmatrix} x_1 \\ x_2 \\ x_3 \end{bmatrix} = \begin{bmatrix} 0 \\ 0 \\ 0 \end{bmatrix}.$$

Thus,
$$-x_1 + x_3 = 0,$$
$$2x_1 + x_3 = 0,$$
$$-x_1 - 2x_3 = 0,$$

which has a solution with $x_1 = x_3 = 0$ and x_2 arbitrary. In particular, $\mathbf{x} = (0, 1, 0)^t$ is an eigenvector of A corresponding to the eigenvalue $\lambda_1 = 2$.

To find an eigenvector for $\lambda_2 = \frac{1}{2} + (\sqrt{3}/2)i$ requires solving the system

$$\begin{bmatrix} 1 - \left(\frac{1}{2} + \frac{\sqrt{3}}{2}i\right) & 0 & 1 \\ 2 & 2 - \left(\frac{1}{2} + \frac{\sqrt{3}}{2}i\right) & 1 \\ -1 & 0 & -\left(\frac{1}{2} + \frac{\sqrt{3}}{2}i\right) \end{bmatrix} \begin{bmatrix} x_1 \\ x_2 \\ x_3 \end{bmatrix} = \begin{bmatrix} 0 \\ 0 \\ 0 \end{bmatrix}.$$

Using complex arithmetic, it can be shown (see Exercise 25) that one solution is

$$x_1 = -\frac{1}{2} - \frac{\sqrt{3}}{2}i, \qquad x_2 = -\frac{1}{2} + \frac{\sqrt{3}}{2}i, \qquad x_3 = 1;$$

so an eigenvector corresponding to the eigenvalue $\lambda_2 = \frac{1}{2} + (\sqrt{3}/2)i$ is

$$\mathbf{x} = \left(-\frac{1}{2} - \frac{\sqrt{3}}{2}i, -\frac{1}{2} + \frac{\sqrt{3}}{2}i, 1\right)^t.$$

In a similar manner,

$$\mathbf{x} = \left(-\frac{1}{2} + \frac{\sqrt{3}}{2}i, -\frac{1}{2} - \frac{\sqrt{3}}{2}i, 1\right)^t$$

can be shown to be an eigenvector corresponding to the eigenvalue

$$\lambda_3 = \frac{1}{2} - \frac{\sqrt{3}}{2}i. \qquad \square$$

The computation of eigenvalues and eigenvectors of matrices is an important part of numerical linear algebra and will be considered further in Sections 8.4 and 8.5.

Definition 8.14   The **spectral radius** $\rho(A)$ of a matrix A is defined by

$$\rho(A) = \max|\lambda| \qquad \text{where } \lambda \text{ is an eigenvalue of } A.$$

(Recall that, for complex $\lambda = \alpha + \beta i$, $|\lambda| = \{\alpha^2 + \beta^2\}^{1/2}$.)

For the matrix considered in Example 5

$$\rho(A) = \max\left\{2, \left|\frac{1}{2} + \frac{\sqrt{3}}{2}i\right|, \left|\frac{1}{2} - \frac{\sqrt{3}}{2}i\right|\right\} = \max\{2, 1, 1\} = 2.$$

The spectral radius is closely related to the norm of a matrix, as is shown in the following theorem.

Theorem 8.15   If A is a real $n \times n$ matrix, then

i) $[\rho(A^tA)]^{1/2} = \|A\|_2$,
ii) $\rho(A) \leq \|A\|$ for any natural norm $\|\cdot\|$.

Proof   The proof of part (i) requires more information concerning eigenvalues than we presently have available. For the details involved in the proof, see Ortega [64], page 21.

To prove part (ii), suppose λ is an eigenvalue of A with eigenvector $\mathbf{x}$ where $\|\mathbf{x}\| = 1$. (Exercise 1(b) ensures that such an eigenvector exists.) Since $(A - \lambda I)\mathbf{x} = \mathbf{0}$, $A\mathbf{x} = \lambda\mathbf{x}$, so for any natural norm

$$|\lambda| = \|\lambda\mathbf{x}\| = \|A\mathbf{x}\| \leq \|A\|.$$

Thus, $\rho(A) = \max|\lambda| \leq \|A\|.$ $\square$

An interesting and useful result, which is similar to part (ii) of Theorem 8.15, is that for any matrix A and any $\varepsilon > 0$, there exists a norm $\|\cdot\|$ with the property that $\|A\| < \rho(A) + \varepsilon$. Consequently, $\rho(A)$ is the greatest lower bound for the norms on A. The proof of this result can be found in Ortega [64], page 23.

EXAMPLE 6   If

$$A = \begin{bmatrix} 2 & 1 & 0 \\ 1 & 1 & 1 \\ 0 & 1 & 2 \end{bmatrix},$$

then, since A is symmetric, $A^t = A$ and

$$A^t A = A^2 = \begin{bmatrix} 2 & 1 & 0 \\ 1 & 1 & 1 \\ 0 & 1 & 2 \end{bmatrix} \begin{bmatrix} 2 & 1 & 0 \\ 1 & 1 & 1 \\ 0 & 1 & 2 \end{bmatrix} = \begin{bmatrix} 5 & 3 & 1 \\ 3 & 3 & 3 \\ 1 & 3 & 5 \end{bmatrix}.$$

To calculate $\rho(A^t A)$, we need the eigenvalues of $A^t A$. If

$$0 = \det(A^t A - \lambda I) = \det \begin{bmatrix} 5 - \lambda & 3 & 1 \\ 3 & 3 - \lambda & 3 \\ 1 & 3 & 5 - \lambda \end{bmatrix}$$

$$= (5 - \lambda)^2 (3 - \lambda) + 9 + 9 - (3 - \lambda) - 9(5 - \lambda) - 9(5 - \lambda)$$

$$= -\lambda^3 + 13\lambda^2 - 36\lambda$$

$$= -\lambda(\lambda - 4)(\lambda - 9),$$

then λ is 0, 4, or 9. Thus,

$$\|A\|_2 = \sqrt{\rho(A^t A)} = \sqrt{\max\{0, 4, 9\}} = 3. \qquad \square$$

In studying iterative matrix techniques, it is of particular importance to know when powers of a matrix become small; that is, when all of the entries approach zero. Matrices of this type are called **convergent**.

Definition 8.16   We call an $n \times n$ matrix A **convergent** if

$$\lim_{k \to \infty} (A^k)_{ij} = 0 \qquad \text{for each } i = 1, 2, \ldots, n \text{ and } j = 1, 2, \ldots, n.$$

EXAMPLE 7   Let

$$A = \begin{bmatrix} \frac{1}{2} & 0 \\ \frac{1}{4} & \frac{1}{2} \end{bmatrix}.$$

Computing powers of A, we obtain:

$$A^2 = \begin{bmatrix} \frac{1}{4} & 0 \\ \frac{1}{4} & \frac{1}{4} \end{bmatrix}, \qquad A^3 = \begin{bmatrix} \frac{1}{8} & 0 \\ \frac{3}{16} & \frac{1}{8} \end{bmatrix}, \qquad A^4 = \begin{bmatrix} \frac{1}{16} & 0 \\ \frac{1}{8} & \frac{1}{16} \end{bmatrix},$$

and, in general,

$$A^k = \begin{bmatrix} (\frac{1}{2})^k & 0 \\ k/2^{k+1} & (\frac{1}{2})^k \end{bmatrix}.$$

Since

$$\lim_{k \to \infty} \left(\frac{1}{2}\right)^k = 0 \quad \text{and} \quad \lim_{k \to \infty} \frac{k}{2^{k+1}} = 0,$$

A is a convergent matrix. Note that $\rho(A) = \frac{1}{2}$, since $\frac{1}{2}$ is the only eigenvalue of A. $\square$

As might be suspected, a connection exists between the spectral radius of a matrix and the convergence of the matrix.

Theorem 8.17   The following statements are equivalent:

i) A is a convergent matrix;
ii) $\lim_{n \to \infty} \|A^n\| = 0$,   for some natural norm $\|\cdot\|$;
iii) $\rho(A) < 1$.

The proof of this theorem can be found in Isaacson and Keller [52], page 14.

Exercise Set 8.1

1.   For any vector $\mathbf{x} \neq \mathbf{0}$, $\mathbf{x}/\|\mathbf{x}\|$ is a vector whose norm is one. Use this to show that

a) $\|A\mathbf{x}\| \leq \|A\| \|\mathbf{x}\|$
b)   for any eigenvalue λ an eigenvector $\mathbf{x}$ exists with $\|\mathbf{x}\| = 1$.

2.   Verify that the function $\|\cdot\|_1$ defined on R^n by

$$\|\mathbf{x}\|_1 = \sum_{i=1}^{n} |x_i|,$$

is a norm on R^n.

3.   Show that a sequence $\{\mathbf{x}^{(k)}\}_{k=1}^{\infty}$ converges to $\mathbf{x} \in R^n$ relative to the l_2 norm if and only if it converges to $\mathbf{x}$ relative to the l_∞ norm.

4.   Find $\|\mathbf{x}\|_\infty$, $\|\mathbf{x}\|_1$, and $\|\mathbf{x}\|_2$ for the following vectors:

a) $\mathbf{x} = (3, -4, 0, \frac{3}{2})^t$;
b) $\mathbf{x} = (4/(k+1), 2/k^2, k^2 e^{-k})^t$ for k a fixed positive integer.

5.   Prove that the following sequences are convergent, and find their limits.

a) $\mathbf{x}^{(k)} = (1/k, e^{1-k}, -2/k^2)^t$;
b) $\mathbf{x}^{(k)} = (ke^{-k^2}, (\cos k)/k, \sqrt{k^2 + k} - k)^t$;
c) $\mathbf{x}^{(k)} = (e^{1/k}, (k^2 + 1)/(1 - k^2), (1/k^2)(1 + 3 + 5 + \cdots + (2k - 1)))^t$.

6.   Show that $\|\cdot\|_{\text{②}}$, and $\|\cdot\|_{\text{①}}$ are matrix norms, where

$$\|A\|_{\text{②}} = \left(\sum_{i=1}^{n} \sum_{j=1}^{n} |a_{ij}|^2 \right)^{1/2} \quad \text{and} \quad \|A\|_{\text{①}} = \sum_{i=1}^{n} \sum_{j=1}^{n} |a_{ij}|$$

for any $n \times n$ matrix A.

7. Show that $\|\cdot\|_{\otimes}$, defined by $\|A\|_{\otimes} = \max_{1 \le i, j \le n} |a_{ij}|$, does not define a matrix norm.

8. Prove Theorem 8.10.

9. Define the matrix norm $\|\cdot\|_1$ by

$$\|A\|_1 = \max_{\|\mathbf{x}\|_1 = 1} \|A\mathbf{x}\|_1$$

and show that

$$\|A\|_1 = \max_{1 \le j \le n} \sum_{i=1}^{n} |a_{ij}|.$$

10. Find $\|\cdot\|_{\infty}$ and $\|\cdot\|_1$ for the following matrices:

a) $A = \begin{bmatrix} 1 & -1 \\ 2 & 1 \end{bmatrix}$
b) $B = \begin{bmatrix} -3 & 2 \\ \frac{3}{2} & \frac{7}{2} \end{bmatrix}$
c) $C = \begin{bmatrix} 10 & 15 \\ 0 & 1 \end{bmatrix}$

11. Show that the characteristic polynomial $p(\lambda) = \det(A - \lambda I)$, for the $n \times n$ matrix A, is an nth-degree polynomial. [*Hint*: Expand $\det(A - \lambda I)$ along the first row and use induction on n.]

12. Show that if A is an $n \times n$ matrix, then

$$\det A = \prod_{i=1}^{n} \lambda_i,$$

where $\lambda_1, \ldots, \lambda_n$ are the eigenvalues of A. [*Hint*: $\det A = p(0)$.]

13. Show that A is singular if and only if $\lambda = 0$ is an eigenvalue of A.

14. Compute the eigenvalues and associated eigenvectors of the following matrices:

a) $\begin{bmatrix} 2 & -1 \\ -1 & 2 \end{bmatrix}$
b) $\begin{bmatrix} 1 & 1 \\ 0 & 1 \end{bmatrix}$
c) $\begin{bmatrix} -1 & 2 & 0 \\ 0 & 3 & 4 \\ 0 & 0 & 7 \end{bmatrix}$

d) $\begin{bmatrix} 2 & 1 & 1 \\ 2 & 3 & 2 \\ 1 & 1 & 2 \end{bmatrix}$
e) $\begin{bmatrix} 2 & 1 & 0 \\ 1 & 2 & 0 \\ 0 & 0 & 3 \end{bmatrix}$
f) $\begin{bmatrix} 0 & \frac{1}{2} \\ \frac{1}{2} & 0 \end{bmatrix}$

15. Find the spectral radius for each matrix in Exercise 14.

16. Find $\|\cdot\|_2$ for each matrix in Exercise 14.

17. Show that $\|A\|_2 = \rho(A)$ if A is symmetric.

18. a) Show that the eigenvalues of the powers A^k, $k = 1, 2, 3, \ldots$ of a matrix A are the powers of the eigenvalues λ^k, $k = 1, 2, 3, \ldots$, respectively, with the same eigenvectors, where λ represents an eigenvalue of A.
   b) Show that, if A^{-1} exists, then the eigenvalues of A^{-1} are the inverses of the eigenvalues of A with the same eigenvectors.
   c) Generalize (a) and (b) to $(A^{-1})^k$ where $k = 2, 3, 4, \ldots$.
   d) Let $q(x) = q_0 + q_1 x + \cdots + q_k x^k$ be any polynomial, and define $q(A) = q_0 I + q_1 A + \cdots + q_k A^k$. Show that the eigenvalues of $q(A)$ are $q(\lambda)$ where λ represents any eigenvalue of A.

19. Find matrices A and B for which $\rho(A + B) > \rho(A) + \rho(B)$. (This shows that $\rho(A)$ cannot be a matrix norm.)

20. Show that

$$A_1 = \begin{bmatrix} 1 & 0 \\ \frac{1}{4} & \frac{1}{2} \end{bmatrix}$$

is not convergent, but

$$A_2 = \begin{bmatrix} \frac{1}{2} & 0 \\ 16 & \frac{1}{2} \end{bmatrix}$$

is convergent.

21. Which of the matrices in Exercise 14 are convergent?

22. Show that $(1/\|A^{-1}\|) \leq |\lambda| \leq \|A\|$ for any eigenvalue λ of the nonsingular matrix A where $\|\cdot\|$ is any natural norm.

23. Let S be a positive definite matrix.

a) Show that the eigenvalues of S are positive real numbers.
b) For any $\mathbf{x} \in R^n$ define $\|\mathbf{x}\| = \sqrt{\mathbf{x}^t S \mathbf{x}}$. Show that this defines a norm on R^n.

24. Let S be a real and nonsingular matrix, and let $\|\cdot\|$ be any norm on R^n. Define $\|\cdot\|'$ by $\|\mathbf{x}\|' = \|S\mathbf{x}\|$. Show that $\|\cdot\|'$ is also a norm on R^n.

25. Show that

$$\begin{bmatrix} 1 - \left(\frac{1}{2} + \frac{\sqrt{3}}{2}i\right) & 0 & 1 \\ 2 & 2 - \left(\frac{1}{2} + \frac{\sqrt{3}}{2}i\right) & 1 \\ -1 & 0 & -\left(\frac{1}{2} + \frac{\sqrt{3}}{2}i\right) \end{bmatrix} \begin{bmatrix} -\frac{1}{2} - \frac{\sqrt{3}}{2}i \\ -\frac{1}{2} + \frac{\sqrt{3}}{2}i \\ 1 \end{bmatrix} = \begin{bmatrix} 0 \\ 0 \\ 0 \end{bmatrix}.$$

26. In Exercise 14 of Section 6.5, a symmetric matrix

$$A = \begin{bmatrix} 1.59 & 1.69 & 2.13 \\ 1.69 & 1.31 & 1.72 \\ 2.13 & 1.72 & 1.85 \end{bmatrix}$$

was used to describe the average wing lengths of fruit flies which were offspring resulting from the mating of three mutants of the flies. The entry a_{ij} represents the average wing length of a fly that is the offspring of a male fly of type i and a female fly of type j.

a) Find the eigenvalues and associated eigenvectors of this matrix.
b) Use the result in Exercise 23 to answer the question posed in part (b) of Exercise 14 in Section 6.5; that is, is this matrix positive definite?

27. In Exercise 11 of Section 6.3, we assumed that the contribution a female beetle of a certain type made to the future years' female beetle population could be expressed in terms of the matrix

$$A = \begin{bmatrix} 0 & 0 & 6 \\ \frac{1}{2} & 0 & 0 \\ 0 & \frac{1}{3} & 0 \end{bmatrix}$$

where the entry in the ith row and jth column represents the probabilistic contribution of a beetle of age j onto the next year's female population of age i.

a)   Does the matrix A have any real eigenvalues? If so, determine them and any associated eigenvectors.

b)   If a sample of this species were needed for laboratory test purposes, which would have a constant proportion in each age group from year to year, what criteria could be imposed on the initial population to ensure that this would be satisfied?

8.2   Iterative Techniques for Solving Linear Systems

An iterative technique to solve the $n \times n$ linear system $A\mathbf{x} = \mathbf{b}$ starts with an initial approximation $\mathbf{x}^{(0)}$ to the solution $\mathbf{x}$, and generates a sequence of vectors $\{\mathbf{x}^{(k)}\}_{k=0}^{\infty}$ that converges to $\mathbf{x}$. Most of these iterative techniques involve a process that converts the system $A\mathbf{x} = \mathbf{b}$ into an equivalent system of the form $\mathbf{x} = T\mathbf{x} + \mathbf{c}$ for some $n \times n$ matrix T and vector $\mathbf{c}$. Having selected the initial vector $\mathbf{x}^{(0)}$, the sequence of approximate solution vectors is generated by computing

(8.7) $$\mathbf{x}^{(k)} = T\mathbf{x}^{(k-1)} + \mathbf{c}$$

for each $k = 1, 2, 3, \ldots$. This type of procedure should be reminiscent of the fixed-point iteration studied in Chapter 2.

In practice, iterative techniques are seldom used for solving linear systems of small dimension since the time required for sufficient accuracy exceeds that required for direct techniques such as the Gauss elimination method. For large systems with a high percentage of zero entries, however, these techniques are efficient in terms of computer storage and time requirements. Systems of this type arise frequently in the numerical solution of boundary-value problems and partial-differential equations.

EXAMPLE 1   The linear system $A\mathbf{x} = \mathbf{b}$ given by

$$
\begin{aligned}
E_1: & \quad 10x_1 - x_2 + 2x_3 && = 6, \\
E_2: & \quad -x_1 + 11x_2 - x_3 + 3x_4 && = 25, \\
E_3: & \quad 2x_1 - x_2 + 10x_3 - x_4 && = -11, \\
E_4: & \quad 3x_2 - x_3 + 8x_4 && = 15,
\end{aligned}
$$

has solution $\mathbf{x} = (1, 2, -1, 1)^t$. To convert $A\mathbf{x} = \mathbf{b}$ to the form $\mathbf{x} = T\mathbf{x} + \mathbf{c}$, solve equation E_i for x_i, for each $i = 1, 2, 3, 4$, to obtain

$$
\begin{aligned}
x_1 &= \phantom{-\tfrac{1}{5}x_1} \tfrac{1}{10}x_2 - \tfrac{1}{5}x_3 \phantom{+ \tfrac{1}{10}x_4} + \tfrac{3}{5}, \\
x_2 &= \tfrac{1}{11}x_1 \phantom{+ \tfrac{1}{10}x_2} + \tfrac{1}{11}x_3 - \tfrac{3}{11}x_4 + \tfrac{25}{11}, \\
x_3 &= -\tfrac{1}{5}x_1 + \tfrac{1}{10}x_2 \phantom{+ \tfrac{1}{11}x_3} + \tfrac{1}{10}x_4 - \tfrac{11}{10}, \\
x_4 &= \phantom{-\tfrac{1}{5}x_1} - \tfrac{3}{8}x_2 + \tfrac{1}{8}x_3 \phantom{+ \tfrac{1}{10}x_4} + \tfrac{15}{8}.
\end{aligned}
$$

In this example,

$$T = \begin{bmatrix} 0 & \dfrac{1}{10} & -\dfrac{1}{5} & 0 \\[2mm] \dfrac{1}{11} & 0 & \dfrac{1}{11} & \dfrac{-3}{11} \\[2mm] -\dfrac{1}{5} & \dfrac{1}{10} & 0 & \dfrac{1}{10} \\[2mm] 0 & \dfrac{-3}{8} & \dfrac{1}{8} & 0 \end{bmatrix} \quad \text{and} \quad \mathbf{c} = \begin{bmatrix} \dfrac{3}{5} \\[2mm] \dfrac{25}{11} \\[2mm] \dfrac{-11}{10} \\[2mm] \dfrac{15}{8} \end{bmatrix}$$

For an initial approximation let $\mathbf{x}^{(0)} = (0, 0, 0, 0)^t$ and generate $\mathbf{x}^{(1)}$ by:

$$x_1^{(1)} = \quad \tfrac{1}{10}x_2^{(0)} - \tfrac{1}{5}x_3^{(0)} \qquad + \tfrac{3}{5} = \quad .6000,$$

$$x_2^{(1)} = \tfrac{1}{11}x_1^{(0)} \qquad + \tfrac{1}{11}x_3^{(0)} - \tfrac{3}{11}x_4^{(0)} + \tfrac{25}{11} = \quad 2.2727,$$

$$x_3^{(1)} = -\tfrac{1}{5}x_1^{(0)} + \tfrac{1}{10}x_2^{(0)} \qquad + \tfrac{1}{10}x_4^{(0)} - \tfrac{11}{10} = -1.1000,$$

$$x_4^{(1)} = \qquad - \tfrac{3}{8}x_2^{(0)} + \tfrac{1}{8}x_3^{(0)} \qquad + \tfrac{15}{8} = \quad 1.8750.$$

Additional iterates, $\mathbf{x}^{(k)} = (x_1^{(k)}, x_2^{(k)}, x_3^{(k)}, x_4^{(k)})^t$, are generated in a similar manner and are presented in Table 8.1.

TABLE 8.1

k	0	1	2	3	4	5	6	7	8	9	10
$x_1^{(k)}$	0.0000	.6000	1.0473	.9326	1.0152	.9890	1.0032	.9981	1.0006	.9997	1.0001
$x_2^{(k)}$	0.0000	2.2727	1.7159	2.0533	1.9537	2.0114	1.9922	2.0023	1.9987	2.0004	1.9998
$x_3^{(k)}$	0.0000	-1.1000	$-.8052$	-1.0493	$-.9681$	-1.0103	$-.9945$	-1.0020	$-.9990$	-1.0004	$-.9998$
$x_4^{(k)}$	0.0000	1.8750	.8852	1.1309	.9739	1.0214	.9944	1.0036	.9989	1.0006	.9998

The decision to stop after ten iterations is based upon the fact that

$$\frac{\|\mathbf{x}^{(10)} - \mathbf{x}^{(9)}\|_\infty}{\|\mathbf{x}^{(10)}\|_\infty} = \frac{8.0 \times 10^{-4}}{1.9998} < 10^{-3}.$$

In fact, $\|\mathbf{x}^{(10)} - \mathbf{x}\|_\infty = .0002$. $\qquad\qquad\qquad\qquad\qquad\qquad\qquad\qquad$ $\square$

The method of Example 1 is called the **Jacobi iterative method**. It consists of solving the ith equation in $A\mathbf{x} = \mathbf{b}$ for x_i, to obtain

(8.8) $$x_i = \sum_{\substack{j=1 \\ j \neq i}}^{n} \left(-\frac{a_{ij}x_j}{a_{ii}} \right) + \frac{b_i}{a_{ii}} \quad \text{for } i = 1, 2, \ldots, n.$$

and generating each $x_i^{(k)}$ from $x_i^{(k-1)}$ for $k \geq 1$ by

(8.9) $$x_i^{(k)} = \frac{\sum_{\substack{j=1 \\ j \neq i}}^{n} (-a_{ij}x_j^{(k-1)}) + b_i}{a_{ii}} \quad \text{for } i = 1, 2, \ldots, n.$$

The method can be written in the form $\mathbf{x}^{(k)} = T\mathbf{x}^{(k-1)} + \mathbf{c}$ by splitting A into its diagonal and off-diagonal parts. To see this, let D be the diagonal matrix whose diagonal is the same as A, $-L$ be the strictly lower-triangular part of A, and $-U$ be the strictly upper-triangular part of A. With this notation, A is split into

$$A = \begin{bmatrix} a_{11} & a_{12} & \cdots & a_{1n} \\ a_{21} & a_{22} & \cdots & a_{2n} \\ \vdots & \vdots & & \vdots \\ a_{n1} & a_{n2} & \cdots & a_{nn} \end{bmatrix}$$

$$= \begin{bmatrix} a_{11} & 0 & \cdots & 0 \\ 0 & a_{22} & & \vdots \\ \vdots & & \ddots & 0 \\ 0 & \cdots & 0 & a_{nn} \end{bmatrix} - \begin{bmatrix} 0 & & \cdots & 0 \\ -a_{21} & & & \vdots \\ \vdots & & \ddots & \\ -a_{n1} & \cdots & -a_{n,n-1} & 0 \end{bmatrix} - \begin{bmatrix} 0 & -a_{12} & \cdots & -a_{1n} \\ & \ddots & & \vdots \\ \vdots & & & -a_{n-1,n} \\ 0 & \cdots & & 0 \end{bmatrix}$$

$$= D \qquad\qquad - L \qquad\qquad\qquad - U.$$

The equation $A\mathbf{x} = \mathbf{b}$ or $(D - L - U)\mathbf{x} = \mathbf{b}$ is then transformed into

$$D\mathbf{x} = (L + U)\mathbf{x} + \mathbf{b},$$

and finally

$$(8.10) \qquad\qquad \mathbf{x} = D^{-1}(L + U)\mathbf{x} + D^{-1}\mathbf{b}.$$

This results in the matrix form of the Jacobi iterative technique:

$$(8.11) \qquad\qquad \mathbf{x}^{(k)} = D^{-1}(L + U)\mathbf{x}^{(k-1)} + D^{-1}\mathbf{b}, \qquad k = 1, 2, \ldots$$

In practice, Eq. (8.9) is used in computation, with Eq. (8.11) being reserved for theoretical purposes.

Jacobi Iterative Algorithm 8.1

To solve $A\mathbf{x} = \mathbf{b}$ given an initial approximation $\mathbf{x}^{(0)}$.

INPUT the number of equations and unknowns n; the entries a_{ij}, $1 \leq i, j \leq n$ of the matrix A; the entries b_i, $1 \leq i \leq n$ of the inhomogeneous term $\mathbf{b}$; the entries XO_i, $1 \leq i \leq n$ of $\mathbf{x}^{(0)}$; tolerance TOL; maximum number of iterations N.

OUTPUT the approximate solution $x_1, \ldots, x_n$ or a message that the number of iterations was exceeded.

Step 1 Set $k = 1$.

Step 2 While $(k \leq N)$ do Steps 3–6.

Step 3 For $i = 1, \ldots, n$

$$\text{set } x_i = \frac{-\sum_{\substack{j=1 \\ j \neq i}}^{n} (a_{ij} XO_j) + b_i}{a_{ii}}.$$

Step 4 If $\|\mathbf{x} - \mathbf{XO}\| < TOL$ then OUTPUT $(x_1, \ldots, x_n)$;
(*Procedure completed successfully.*)
STOP.

Step 5 Set $k = k + 1$.

Step 6 For $i = 1, \ldots, n$ set $XO_i = x_i$.

Step 7 OUTPUT ('Maximum number of iterations exceeded.');
(*Procedure completed unsuccessfully.*)
STOP.

Step 3 of the algorithm requires that $a_{ii} \neq 0$ for each $i = 1, 2, \ldots, n$. If this is not the case, a re-ordering of the equations can be performed so that no $a_{ii} = 0$, unless the system is singular. It is suggested that the equations be arranged so that a_{ii} is as large as possible in order to speed convergence. (See Exercise 3(a) and 3(d).) This subject will be discussed in more detail later.

Another possible stopping criterion in Step 4 is to iterate until

$$\frac{\|\mathbf{x}^{(k)} - \mathbf{x}^{(k-1)}\|}{\|\mathbf{x}^{(k)}\|}$$

is smaller than some prescribed tolerance $\varepsilon > 0$. For this purpose, any convenient norm can be used, the most usual being the l_∞ norm.

A possible improvement in Algorithm 8.1 is suggested by an analysis of Eq. (8.9). To compute $x_i^{(k)}$ for each $i > 1$, the components of $\mathbf{x}^{(k-1)}$ are used. Since $x_1^{(k)}, \ldots, x_{i-1}^{(k)}$ have already been computed and are supposedly better approximations to the actual solutions $x_1, \ldots, x_{i-1}$ than $x_1^{(k-1)}, \ldots, x_{i-1}^{(k-1)}$, it seems reasonable to compute $x_i^{(k)}$ using these most recently calculated values; that is,

$$(8.12) \qquad x_i^{(k)} = \frac{-\sum_{j=1}^{i-1} (a_{ij} x_j^{(k)}) - \sum_{j=i+1}^{n} (a_{ij} x_j^{(k-1)}) + b_i}{a_{ii}},$$

for each $i = 1, 2, \ldots, n$, instead of Eq. (8.9). An example illustrating this procedure follows.

EXAMPLE 2 The linear system given by

$$
\begin{aligned}
10x_1 - \quad x_2 + 2x_3 \qquad\qquad &= \quad 6, \\
-x_1 + 11x_2 - \quad x_3 + 3x_4 &= \quad 25, \\
2x_1 - \quad x_2 + 10x_3 - \quad x_4 &= -11, \\
3x_2 - \quad x_3 + 8x_4 &= \quad 15,
\end{aligned}
$$

was solved in Example 1 by the Jacobi iterative method. Incorporating Eq. (8.12) into Algorithm 8.1 gives the equations to be used for each $k = 1, 2, \ldots$;

$$
\begin{aligned}
x_1^{(k)} &= \qquad\qquad \tfrac{1}{10}x_2^{(k-1)} - \tfrac{1}{5}x_3^{(k-1)} \qquad\qquad + \tfrac{3}{5}, \\
x_2^{(k)} &= \tfrac{1}{11}x_1^{(k)} \qquad\qquad + \tfrac{1}{11}x_3^{(k-1)} - \tfrac{3}{11}x_4^{(k-1)} + \tfrac{25}{11}, \\
x_3^{(k)} &= -\tfrac{1}{5}x_1^{(k)} + \tfrac{1}{10}x_2^{(k)} \qquad\qquad + \tfrac{1}{10}x_4^{(k-1)} - \tfrac{11}{10}, \\
x_4^{(k)} &= \qquad\qquad - \tfrac{3}{8}x_2^{(k)} + \tfrac{1}{8}x_3^{(k)} \qquad\qquad + \tfrac{15}{8}.
\end{aligned}
$$

Letting $\mathbf{x}^{(0)} = (0, 0, 0, 0)^t$, we generate the vector iterates in Table 8.2.

TABLE 8.2

k	0	1	2	3	4	5
$x_1^{(k)}$	0.0000	.6000	1.030	1.0065	1.0009	1.0001 .
$x_2^{(k)}$	0.0000	2.3272	2.037	2.0036	2.0003	2.0000
$x_3^{(k)}$	0.0000	−.9873	−1.014	−1.0025	−1.0003	−1.0000
$x_4^{(k)}$	0.0000	.8789	.9844	.9983	.9999	1.0000

Since

$$\frac{\|\mathbf{x}^{(5)} - \mathbf{x}^{(4)}\|_\infty}{\|\mathbf{x}^{(5)}\|_\infty} = \frac{.0008}{2.000} = 4 \times 10^{-4},$$

$\mathbf{x}^{(5)}$ is accepted as a reasonable approximation to the solution. It is interesting to note that Jacobi's method required twice as many iterations for the same accuracy. □

The technique presented in Example 2 is called the **Gauss–Seidel iterative method**. To write this method in the matrix form (8.7), multiply both sides of Eq. (8.12) by a_{ii} and collect all kth iterate terms to give

$$a_{i1}x_1^{(k)} + a_{i2}x_2^{(k)} + \cdots + a_{ii}x_i^{(k)} = -a_{i,i+1}x_{i+1}^{(k-1)} - \cdots - a_{in}x_n^{(k-1)} + b_i,$$

for each $i = 1, 2, \ldots, n$. Writing all n equations gives:

$$a_{11}x_1^{(k)} = -a_{12}x_2^{(k-1)} - a_{13}x_3^{(k-1)} - \cdots - a_{1n}x_n^{(k-1)} + b_1,$$
$$a_{21}x_1^{(k)} + a_{22}x_2^{(k)} = -a_{23}x_3^{(k-1)} - \cdots - a_{2n}x_n^{(k-1)} + b_2,$$
$$\vdots \qquad \vdots \qquad \qquad \qquad \qquad \qquad \vdots$$
$$a_{n1}x_1^{(k)} + a_{n2}x_2^{(k)} + \cdots + a_{nn}x_n^{(k)} = b_n;$$

and it follows that, in matrix form, the Gauss–Seidel method can be represented by:

$$(8.13) \qquad\qquad (D - L)\mathbf{x}^{(k)} = U\mathbf{x}^{(k-1)} + \mathbf{b}$$

or

$$(8.14) \quad \mathbf{x}^{(k)} = (D - L)^{-1}U\mathbf{x}^{(k-1)} + (D - L)^{-1}\mathbf{b} \qquad \text{for each } k = 1, 2, \ldots$$

In order for the lower-triangular matrix $D - L$ to be nonsingular, it is necessary and sufficient that $a_{ii} \neq 0$ for each $i = 1, 2, \ldots, n$.

Gauss–Seidel Iterative Algorithm 8.2

To solve $A\mathbf{x} = \mathbf{b}$ given an initial approximation $\mathbf{x}^{(0)}$.

INPUT the number of equations and unknowns n; the entries $a_{ij}, 1 \le i, j \le n$ of the matrix A; the entries $b_i, 1 \le i \le n$ of the inhomogeneous term $\mathbf{b}$; the entries $XO_i, 1 \le i \le n$ of $\mathbf{x}^{(0)}$; tolerance TOL; maximum number of iterations N.

OUTPUT the approximate solution $x_1, \ldots, x_n$ or a message that the number of iterations was exceeded.

Step 1 Set $k = 1$.

Step 2 While $(k \leq N)$ do Steps 3–6.

Step 3 For $i = 1, \ldots, n$

$$\text{set } x_i = \frac{-\sum_{j=1}^{i-1} a_{ij} x_j - \sum_{j=i+1}^{n} a_{ij} XO_j + b_i}{a_{ii}}.$$

Step 4 If $\|\mathbf{x} - \mathbf{XO}\| < TOL$ then OUTPUT $(x_1, \ldots, x_n)$;
(*Procedure completed successfully.*)
STOP.

Step 5 Set $k = k + 1$.

Step 6 For $i = 1, \ldots, n$ set $XO_i = x_i$.

Step 7 OUTPUT ('Maximum number of iterations exceeded');
(*Procedure completed unsuccessfully.*)
STOP.

The comments following Algorithm 8.1 regarding stopping criteria also apply to the Gauss–Seidel Algorithm 8.2.

The results of Examples 1 and 2 seem to imply that the Gauss–Seidel method is superior to the Jacobi method. This is generally the case but is not always true. In fact, there are linear systems for which the Jacobi method converges and the Gauss–Seidel method does not, and conversely (see Varga [87], page 74).

To study the convergence of general iteration techniques, we consider the formula

$$\mathbf{x}^{(k)} = T\mathbf{x}^{(k-1)} + \mathbf{c} \qquad \text{for each } k = 1, 2, \ldots,$$

where $\mathbf{x}^{(0)}$ is arbitrary. The study will require the following lemma.

Lemma 8.18 If the spectral radius $\rho(T)$ satisfies $\rho(T) < 1$, then $(I - T)^{-1}$ exists, and

(8.15) $$(I - T)^{-1} = I + T + T^2 + \cdots.$$

Proof For any eigenvalue λ of T, $1 - \lambda$ is an eigenvalue of $I - T$. Since $|\lambda| \leq \rho(T) < 1$, if follows that no eigenvalue of $I - T$ could be zero and, consequently, $I - T$ is nonsingular.

Let $S_m = I + T + T^2 + \cdots + T^m$. Then

$$(I - T)S_m = I - T^{m+1}$$

and, since T is convergent, Theorem 8.17 (p. 379), implies that

$$\lim_{m \to \infty} (I - T)S_m = \lim_{m \to \infty} (I - T^{m+1}) = I.$$

Thus, $\lim_{m \to \infty} S_m = (I - T)^{-1}$. $\qquad\qquad\square$

Theorem 8.19 For any $\mathbf{x}^{(0)} \in R^n$, the sequence $\{\mathbf{x}^{(k)}\}_{k=0}^{\infty}$ defined by

(8.16) $\mathbf{x}^{(k)} = T\mathbf{x}^{(k-1)} + \mathbf{c}$ for each $k \geq 1$ and $\mathbf{c} \neq 0$,

converges to the unique solution of $\mathbf{x} = T\mathbf{x} + \mathbf{c}$ if and only if $\rho(T) < 1$.

Proof From Eq. (8.16),

$$
\begin{aligned}
(8.17) \qquad \mathbf{x}^{(k)} &= T\mathbf{x}^{(k-1)} + \mathbf{c} \\
&= T(T\mathbf{x}^{(k-2)} + \mathbf{c}) + \mathbf{c} \\
&= T^2\mathbf{x}^{(k-2)} + (T + I)\mathbf{c} \\
&\;\;\vdots \\
&= T^k\mathbf{x}^{(0)} + (T^{k-1} + \cdots + T + I)\mathbf{c}
\end{aligned}
$$

Assuming $\rho(T) < 1$, we can use Theorem 8.17 and Lemma 8.18 to give

$$
\begin{aligned}
(8.18) \qquad \lim_{k \to \infty} \mathbf{x}^{(k)} &= \lim_{k \to \infty} T^k\mathbf{x}^{(0)} + \lim_{k \to \infty} \left(\sum_{j=0}^{k-1} T^j \right)\mathbf{c} \\
&= 0 \cdot \mathbf{x}^{(0)} + (I - T)^{-1}\mathbf{c} = (I - T)^{-1}\mathbf{c}.
\end{aligned}
$$

To prove the converse, let $\{\mathbf{x}^{(k)}\}$ converge to $\mathbf{x}$ for any $\mathbf{x}^{(0)}$. From Eq. (8.16) it follows that $\mathbf{x} = T\mathbf{x} + \mathbf{c}$, so for each k,

$$\mathbf{x} - \mathbf{x}^{(k)} = T(\mathbf{x} - \mathbf{x}^{(k-1)}) = \cdots = T^k(\mathbf{x} - \mathbf{x}^{(0)}).$$

Hence, for any vector $\mathbf{x}^{(0)}$,

$$\lim_{k \to \infty} T^k(\mathbf{x} - \mathbf{x}^{(0)}) = \lim_{k \to \infty} (\mathbf{x} - \mathbf{x}^{(k)}) = 0.$$

Consequently, if $\mathbf{z}$ is an arbitrary vector and $\mathbf{x}^{(0)} = \mathbf{x} - \mathbf{z}$, then

$$\lim_{k \to \infty} T^k\mathbf{z} = \lim_{k \to \infty} T^k(\mathbf{x} - (\mathbf{x} - \mathbf{z})) = 0,$$

which implies that T is convergent (see Exercise 10). By Theorem 8.17 this implies that $\rho(T) < 1$. $\qquad \square$

The proof of the following corollary is considered in Exercise 4.

Corollary 8.20 If for any matrix norm $\|T\| < 1$, then the sequence $\{\mathbf{x}^{(k)}\}_{k=0}^{\infty}$ in Eq. (8.16) converges, for any $\mathbf{x}^{(0)} \in R^n$, to a vector $\mathbf{x} \in R^n$, and the following error bounds hold:

(8.19) $\|\mathbf{x} - \mathbf{x}^{(k)}\| \leq \|T\|^k \|\mathbf{x}^{(0)} - \mathbf{x}\|;$

(8.20) $\|\mathbf{x} - \mathbf{x}^{(k)}\| \leq \dfrac{\|T\|^k}{1 - \|T\|} \|\mathbf{x}^{(1)} - \mathbf{x}^{(0)}\|.$

To apply the results above to the Jacobi or Gauss–Seidel iterative techniques, we need to write the iteration matrices for the Jacobi method, T_j, and the Gauss–Seidel method, T_g, as

$$T_j = D^{-1}(L + U), \qquad T_g = (D - L)^{-1}U.$$

Should $\rho(T_j)$ or $\rho(T_g)$ be less than one, it is clear that the sequence $\{x^{(k)}\}_{k=0}^{\infty}$ will converge to the solution $\mathbf{x}$ of $A\mathbf{x} = \mathbf{b}$. For example, the Jacobi scheme has

$$\mathbf{x}^{(k)} = D^{-1}(L + U)\mathbf{x}^{(k-1)} + D^{-1}\mathbf{b},$$

and, if $\{\mathbf{x}^{(k)}\}_{k=0}^{\infty}$ converges to $\mathbf{x}$, then

$$\mathbf{x} = D^{-1}(L + U)\mathbf{x} + D^{-1}\mathbf{b}.$$

This implies that

$$D\mathbf{x} = (L + U)\mathbf{x} + \mathbf{b} \quad \text{and} \quad (D - L - U)\mathbf{x} = \mathbf{b}.$$

Since $D - L - U = A$, $\mathbf{x}$ satisfies $A\mathbf{x} = \mathbf{b}$. We can now give easily verified sufficiency conditions for convergence of the Jacobi and Gauss–Seidel methods. (To prove convergence for the Jacobi scheme, see Exercise 5, and for the Gauss–Seidel scheme, see Ortega [64], page 120.)

Theorem 8.21 If A is strictly diagonally dominant, then for any choice of $\mathbf{x}^{(0)}$ both the Jacobi and Gauss–Seidel methods give sequences $\{\mathbf{x}^{(k)}\}_{k=0}^{\infty}$ that converge to the solution of $A\mathbf{x} = \mathbf{b}$.

The relationship of the rapidity of convergence to the spectral radius of the iteration matrix T can be seen from inequality (8.19). Since (8.19) holds for any natural matrix norm, it follows, from the statement following Theorem 8.15, that

$$(8.21) \qquad \|\mathbf{x}^{(k)} - \mathbf{x}\| \approx \rho(T)^k \|\mathbf{x}^{(0)} - \mathbf{x}\|.$$

Suppose that $\rho(T) < 1$ and that $\mathbf{x}^{(0)} = \mathbf{0}$ is to be used in an iterative technique to approximate $\mathbf{x}$ with relative error at most 10^{-t}. By the estimate (8.21), the relative error after k iterations is approximately $\rho(T)^k$ so accuracy of 10^{-t} is expected if

$$\rho(T)^k \le 10^{-t},$$

that is, if

$$k \ge \frac{t}{-\log_{10} \rho(T)}.$$

Thus, it is desirable to select the iterative technique with minimal $\rho(T) < 1$ for a particular system $A\mathbf{x} = \mathbf{b}$. In general, it is not known which of the two techniques, Jacobi or Gauss–Seidel, should be used. However, in a special case the answer is known. The proof of the theorem will not be given here, but can be found in Young [93], pages 120–127.

Theorem 8.22 (Stein–Rosenberg) If $a_{ij} \le 0$ for each $i \ne j$ and $a_{ii} > 0$ for each $i = 1, 2, \ldots, n$, then one and only one of the following statements holds:

a) $0 < \rho(T_g) < \rho(T_j) < 1$,
b) $1 < \rho(T_j) < \rho(T_g)$,
c) $\rho(T_j) = \rho(T_g) = 0$,
d) $\rho(T_j) = \rho(T_g) = 1$.

For the special case described in Theorem 8.22, we see that when one method gives convergence, then both give convergence, with the Gauss–Siedel method being faster than the Jacobi method.

Since the rate of convergence of a procedure depends on the spectral radius of the matrix associated with the method, one way to select a procedure that will lead to accelerated convergence is to choose a method whose associated matrix has minimal spectral radius. Before describing a procedure for selecting such a method we need to introduce a new means of measuring the amount by which an approximation to the solution to a linear system differs from the true solution to the system. The method makes use of the vector described in the following definition.

Definition 8.23   If $\tilde{\mathbf{x}} \in R^n$ is an approximation to the solution of the linear system defined by $A\mathbf{x} = \mathbf{b}$, the **residual vector** for $\tilde{\mathbf{x}}$ with respect to this system is defined by $\mathbf{r} = \mathbf{b} - A\tilde{\mathbf{x}}$.

In procedures such as the Jacobi or Gauss–Seidel methods, a residual vector is associated with each calculation of an approximate component to the solution vector. If we let

$$\mathbf{r}_i^{(k)} = (r_{1i}^{(k)}, r_{2i}^{(k)}, \ldots, r_{ni}^{(k)})^t$$

denote the residual vector for the Gauss–Seidel method corresponding to the approximate solution vector

$$(x_1^{(k)}, x_2^{(k)}, \ldots, x_{i-1}^{(k)}, x_i^{(k-1)}, \ldots, x_n^{(k-1)})^t,$$

the mth component of $\mathbf{r}_i^{(k)}$ is

$$r_{mi}^{(k)} = b_m - \sum_{j=1}^{i-1} a_{mj} x_j^{(k)} - \sum_{j=i}^{n} a_{mj} x_j^{(k-1)}$$

$$= b_m - \sum_{j=1}^{i-1} a_{mj} x_j^{(k)} - \sum_{j=i+1}^{n} a_{mj} x_j^{(k-1)} - a_{mi} x_i^{(k-1)}$$

for each $m = 1, 2, \ldots, n$.

In particular, the ith component of $\mathbf{r}_i^{(k)}$ is

$$r_{ii}^{(k)} = b_i - \sum_{j=1}^{i-1} a_{ij} x_j^{(k)} - \sum_{j=i+1}^{n} a_{ij} x_j^{(k-1)} - a_{ii} x_i^{(k-1)};$$

so

(8.22) $$a_{ii} x_i^{(k-1)} + r_{ii}^{(k)} = b_i - \sum_{j=1}^{i-1} a_{ij} x_j^{(k)} - \sum_{j=i+1}^{n} a_{ij} x_j^{(k-1)}.$$

Recall, however, that in the Gauss–Seidel method,

$$x_i^{(k)} = \frac{1}{a_{ii}} \left[b_i - \sum_{j=1}^{i-1} a_{ij} x_j^{(k)} - \sum_{j=i+1}^{n} a_{ij} x_j^{(k-1)} \right]$$

so Eq. (8.22) can be rewritten as

$$a_{ii} x_i^{(k-1)} + r_{ii}^{(k)} = a_{ii} x_i^{(k)}$$

or

(8.23) $$x_i^{(k)} = x_i^{(k-1)} + \frac{r_{ii}^{(k)}}{a_{ii}}.$$

In terms of the residual vectors, the Gauss–Seidel method is obtained by requiring that, at each stage, the ith coordinate of the residual vector $\mathbf{r}_{i+1}^{(k)}$ be zero. (See Exercise 9.)

Reducing one coordinate of the residual vector to zero, however, is not necessarily the most efficient way to reduce the norm of the vector $\mathbf{r}_{i+1}^{(k)}$. In fact, modifying the Gauss–Seidel procedure as given by Eq. (8.23) to:

$$(8.24) \qquad x_i^{(k)} = x_i^{(k-1)} + \omega \frac{r_{ii}^{(k)}}{a_{ii}}$$

for certain choices of positive ω will lead to significantly faster convergence.

Methods involving Eq. (8.24) are called **relaxation methods**. For choices of $0 < \omega < 1$, the procedures are called **under-relaxation methods** and can be used to obtain convergence of some systems that are not convergent by the Gauss–Seidel method. For choices $1 < \omega$ the procedures are called **over-relaxation methods**, which can be used to accelerate the convergence for systems that are convergent by the Gauss–Seidel technique. These methods are often abbreviated **SOR** for **Successive Over-Relaxation**, and are useful for solving the linear systems that occur in the numerical solution of certain partial-differential equations.

Before illustrating the advantages of the SOR method, we note that, by using Eq. (8.22), Eq. (8.24) can be reformulated for calculation purposes to read:

$$(8.25) \qquad x_i^{(k)} = (1 - \omega)x_i^{(k-1)} + \frac{\omega}{a_{ii}}\left[b_i - \sum_{j=1}^{i-1} a_{ij}x_j^{(k)} - \sum_{j=i+1}^{n} a_{ij}x_j^{(k-1)} \right].$$

EXAMPLE 3   The linear system $A\mathbf{x} = \mathbf{b}$ given by

$$4x_1 + 3x_2 \qquad = \quad 24,$$
$$3x_1 + 4x_2 - \quad x_3 = \quad 30,$$
$$- \quad x_2 + 4x_3 = -24,$$

has the solution $(3, 4, -5)^t$. Gauss–Seidel and the SOR method with $\omega = 1.25$ will be used to solve this system, using $\mathbf{x}^{(0)} = (1, 1, 1)^t$ for both methods. The equations for the Gauss–Seidel method are

$$x_1^{(k)} = -.75x_2^{(k-1)} + 6,$$
$$x_2^{(k)} = -.75x_1^{(k)} + .25x_3^{(k-1)} + 7.5,$$
$$x_1^{(k)} = .25x_2^{(k)} - 6 \qquad \text{for each } k = 1, 2, \ldots,$$

and the equations for the SOR method with $\omega = 1.25$ are

$$x_1^{(k)} = -.25x_1^{(k-1)} - .9375x_2^{(k-1)} + 7.5,$$
$$x_2^{(k)} = -.9375x_1^{(k)} - .25x_2^{(k-1)} + .3125x_2^{(k-1)} + 9.375,$$
$$x_3^{(k)} = .3125x_2^{(k)} - .25x_3^{(k-1)} - 7.5.$$

The first seven iterates for each method are listed in Tables 8.3 and 8.4.

TABLE 8.3   Gauss–Seidel

k	0	1	2	3	4	5	6	7
$x_1^{(k)}$	1	5.250000	3.1406250	3.0878906	3.0549316	3.0343323	3.0214577	3.0134110
$x_2^{(k)}$	1	3.812500	3.8828125	3.9267578	3.9542236	3.9713898	3.9821186	3.9888241
$x_3^{(k)}$	1	−5.046875	−5.0292969	−5.0183105	−5.0114441	−5.0071526	−5.0044703	−5.0027940

TABLE 8.4   SOR with $\omega = 1.25$

k	0	1	2	3	4	5	6	7
$x_1^{(k)}$	1	6.312500	2.6223145	3.1333027	2.9570512	3.0037211	2.9963276	3.0000498
$x_2^{(k)}$	1	3.5195313	3.9585266	4.0102646	4.0074838	4.0029250	4.0009262	4.0002586
$x_3^{(k)}$	1	−6.6501465	−4.6004238	−5.0966863	−4.9734897	−5.0057135	−4.9982822	−5.0003486

In order for the iterates to be accurate to seven decimal places the Gauss–Seidel method required 34 iterations, as opposed to 14 iterations for the overrelaxation method with $\omega = 1.25$.   ☐

The obvious question to ask is how the appropriate value of ω is chosen. Although no complete answer to this question is known for the general $n \times n$ linear system, the following results can be used in certain situations.

Theorem 8.24 (Kahan)   If $a_{ii} \neq 0$ for each $i = 1, 2, \ldots, n$, then $\rho(T_\omega) \geq |\omega - 1|$. This implies that $\rho(T_\omega) < 1$ only if $0 < \omega < 2$, where

$$T_\omega = (D - \omega L)^{-1}\{(1 - \omega)D + \omega U\}$$

is the iteration matrix for the SOR method.

The proof of this theorem is considered in Exercises 6 and 7. The proof of the next two results can be found in Ortega [64], p. 123–33.

Theorem 8.25 (Ostrowski–Reich)   If A is a positive definite matrix and $0 < \omega < 2$, then the SOR method converges for any choice of initial approximate solution vector $\mathbf{x}^{(0)}$.

Theorem 8.26   If A is positive definite and tridiagonal, then $\rho(T_g) = [\rho(T_j)]^2 < 1$, the optimal choice of ω for the SOR method is

$$\omega = \frac{2}{1 + \sqrt{1 - \rho(T_j)^2}},$$

and with this choice of ω, $\rho(T_\omega) = \omega - 1$.

EXAMPLE 4   In Example 3 the matrix A was given by

$$A = \begin{bmatrix} 4 & 3 & 0 \\ 3 & 4 & -1 \\ 0 & -1 & 4 \end{bmatrix}.$$

This matrix is positive definite and tridiagonal, so Theorem 8.26 applies. Since

$$T_j = D^{-1}(L + U)$$

$$= \begin{bmatrix} \frac{1}{4} & 0 & 0 \\ 0 & \frac{1}{4} & 0 \\ 0 & 0 & \frac{1}{4} \end{bmatrix} \begin{bmatrix} 0 & -3 & 0 \\ -3 & 0 & 1 \\ 0 & 1 & 0 \end{bmatrix} = \begin{bmatrix} 0 & -.75 & 0 \\ -.75 & 0 & .25 \\ 0 & .25 & 0 \end{bmatrix},$$

we have

$$T_j - \lambda I = \begin{bmatrix} -\lambda & -.75 & 0 \\ -.75 & -\lambda & .25 \\ 0 & .25 & -\lambda \end{bmatrix}$$

so
$$\det(T_j - \lambda I) = -\lambda(\lambda^2 - .625).$$

Thus,
$$\rho(T_j) = \sqrt{.625}$$

and $$\omega = \frac{2}{1 + \sqrt{1 - \rho(T_g)}} = \frac{2}{1 + \sqrt{1 - \rho(T_j)^2}} = \frac{2}{1 + \sqrt{1 - .625}} \approx 1.24.$$

This explains the comparatively rapid convergence obtained by using $\omega = 1.25$ in Example 3.   □

We close this section with an algorithm for the SOR method.

SOR Algorithm 8.3

To solve $Ax = \mathbf{b}$ given the parameter ω and an initial approximation $\mathbf{x}^{(0)}$:

INPUT the number of equations and unknowns n; the entries a_{ij}, $1 \le i, j \le n$ of the matrix A; the entries b_i, $1 \le i \le n$ of the inhomogeneous term $\mathbf{b}$; the entries XO_i, $1 \le i \le n$ of $\mathbf{x}^{(0)}$; the parameter ω; tolerance TOL; maximum number of iterations N.

OUTPUT the approximate solution $x_1, \ldots, x_n$ or a message that the number of iterations was exceeded.

Step 1 Set $k = 1$.

Step 2 While ($k \le N$) do Steps 3–6.

Step 3 For $i = 1, \ldots, n$

set $x_i = (1 - \omega)XO_i + \dfrac{\omega(-\sum_{j=1}^{i-1} a_{ij}x_j - \sum_{j=i+1}^{n} a_{ij}XO_j + b_i)}{a_{ii}}.$

Step 4 If $\|\mathbf{x} - \mathbf{XO}\| < TOL$ then OUTPUT $(x_1, \ldots, x_n)$;
(*Procedure completed successfully.*)
STOP.

Step 5   Set $k = k + 1$.

Step 6   For $i = 1, \ldots, n$ set $XO_i = x_i$.

Step 7   OUTPUT ('Maximum number of iterations exceeded');
       (*Procedure completed unsuccessfully.*)
       STOP.

Exercise Set 8.2

1. Find the first two iterations of the Jacobi method for the following linear systems:

   a) $2x_1 - x_2 + x_3 = -1$
   $3x_1 + 3x_2 + 9x_3 = 0$
   $3x_1 + 3x_2 + 5x_3 = 4$ use $\mathbf{x}^{(0)} = (1, 1, 1)^t$.

   b) $2x_1 = 3$
   $x_1 + 1.5x_2 = 4.5$
   $- 3x_2 + .5x_3 = -6.6$
   $2x_1 - 2x_2 + x_3 + x_4 = .8$ use $\mathbf{x}^{(0)} = (1, 1, 1, 1)^t$.

   c) $2x_2 + 4x_3 = 0$
   $x_1 - x_2 - x_3 = .375$
   $x_1 - x_2 + 2x_3 = 0$ use $\mathbf{x}^{(0)} = (1, 1, 1)^t$.

2. a)   Repeat Exercise 1 for the Gauss–Siedel method
   b)   Repeat Exercise 1 for the SOR method with $\omega = 1.2$.

3. Solve the following linear systems by the Jacobi method and the Gauss–Seidel method (if possible), and compare rates of convergence. As a test of convergence, iterate until $N > 25$ (divergence) or $\|\mathbf{x}^{(k+1)} - \mathbf{x}^{(k)}\|_2 \leq .01$.

   a) $2x_1 - x_2 + 10x_3 = -11,$
   $3x_2 - x_3 + 8x_4 = -11,$
   $10x_1 - x_2 + 2x_3 = 6,$
   $-x_1 + 11x_2 - x_3 + 3x_4 = 25.$

   b) $10x_1 - x_2 = 9,$
   $-x_1 + 10x_2 - 2x_3 = 7,$
   $- 4x_2 + 10x_3 = 6.$

   c) $\begin{bmatrix} 4 & -1 & 0 & -1 & 0 & 0 \\ -1 & 4 & -1 & 0 & -1 & 0 \\ 0 & -1 & 4 & 0 & 0 & -1 \\ -1 & 0 & 0 & 4 & -1 & 0 \\ 0 & -1 & 0 & -1 & 4 & -1 \\ 0 & 0 & -1 & 0 & -1 & 4 \end{bmatrix} \begin{bmatrix} x_1 \\ x_2 \\ x_3 \\ x_4 \\ x_5 \\ x_6 \end{bmatrix} = \begin{bmatrix} 0 \\ 5 \\ 0 \\ 6 \\ -2 \\ 6 \end{bmatrix}.$

   d)   Repeat (a), but rearrange the equations to improve the results.

4. Prove that

$$\|\mathbf{x}^{(k)} - \mathbf{x}\| \le \|T\|^k \|\mathbf{x}^{(0)} - \mathbf{x}\| \qquad \text{and} \qquad \|\mathbf{x}^{(k)} - \mathbf{x}\| \le \frac{\|T\|^k}{1 - \|T\|} \|\mathbf{x}^{(1)} - \mathbf{x}^{(0)}\|,$$

where T is an $n \times n$ matrix with $\|T\| < 1$ and

$$\mathbf{x}^{(k)} = T\mathbf{x}^{(k-1)} + \mathbf{c}, k = 1, 2, \ldots \qquad \text{with } \mathbf{x}^{(0)} \text{ arbitrary, } \mathbf{c} \in R^n, \text{ and } \mathbf{x} = T\mathbf{x} + \mathbf{c}.$$

Apply the bounds to Exercise 3.

5. Show that if A is strictly diagonally dominant, then $\|T_j\|_\infty < 1$.

6. Show that the iteration matrix for the SOR method is

$$T_\omega = (D - \omega L)^{-1} \{(1 - \omega)D + \omega U\} \qquad \text{and} \qquad \mathbf{c} = \omega(D - \omega L)^{-1}\mathbf{b}.$$

7. Prove Theorem 8.24. [*Hint:* If $\lambda_1, \ldots, \lambda_n$ are the eigenvalues of T_ω, then det $T_\omega = \prod_{i=1}^{n} \lambda_i$. Since det $D^{-1} = \det(D - \omega L)^{-1}$ and the determinant of a product of matrices is the product of the determinants of the factors, the result is obtained by using Exercise 6.]

8. Solve the following linear systems using the SOR method with the given values of ω and the same convergence criteria as in Exercise 3.

    a) Exercise 3(b) with $\omega = .5$ and $\omega = 1.1$.
    b) Exercise 3(c) with $\omega = 1.334$, $\omega = 1.95$, and $\omega = .95$.

9. Show that if $x_i^{(k)}$ is chosen so that

$$x_i^{(k)} = x_i^{(k-1)} + \frac{r_{ii}^{(k)}}{a_{ii}},$$

    then

$$r_{i,i+1}^{(k)} = 0.$$

10. Show that the matrix T is convergent if and only if $T^k\mathbf{x} \to 0$ for every vector $\mathbf{x}$.

11. Use the Gauss–Seidel method to solve the linear system given in the introduction to Chapter 6 involving the potential at the junctions of the electrical circuit in Figure 6.1.

8.3 Error Estimates and Iterative Refinement

It seems intuitively reasonable that if $\tilde{\mathbf{x}}$ is an approximation to the solution $\mathbf{x}$ of $A\mathbf{x} = \mathbf{b}$ and the residual vector $\mathbf{r} = A\tilde{\mathbf{x}} - \mathbf{b}$ has the property that $\|\mathbf{r}\|$ is small, then $\|\mathbf{x} - \tilde{\mathbf{x}}\|$ would be small as well. Although this is often the case, certain special systems, which occur quite often in practice, fail to have this property.

EXAMPLE 1 The linear system $A\mathbf{x} = \mathbf{b}$ given by

$$\begin{bmatrix} 1 & 2 \\ 1.0001 & 2 \end{bmatrix} \begin{bmatrix} x_1 \\ x_2 \end{bmatrix} = \begin{bmatrix} 3 \\ 3.0001 \end{bmatrix}$$

has the unique solution $\mathbf{x} = (1, 1)^t$. The approximation to this solution $\tilde{\mathbf{x}} = (3, 0)^t$ has the residual vector

$$\mathbf{r} = \mathbf{b} - A\tilde{\mathbf{x}} = \begin{bmatrix} 3 \\ 3.0001 \end{bmatrix} - \begin{bmatrix} 1 & 2 \\ 1.0001 & 2 \end{bmatrix} \begin{bmatrix} 3 \\ 0 \end{bmatrix} = \begin{bmatrix} 0 \\ .0002 \end{bmatrix},$$

so $\|\mathbf{r}\|_\infty = .0002$. Although the norm of the residual vector is small, the approximation $\tilde{\mathbf{x}} = (3, 0)^t$ is obviously quite poor; in fact, $\|\mathbf{x} - \tilde{\mathbf{x}}\|_\infty = 2$. ▢

The difficulty that arose in this example can be explained quite simply by noting that the solution to the system represents the intersection of the lines

$$l_1: \quad x_1 + 2x_2 = 3 \quad \text{and} \quad l_2: \quad 1.0001x_1 + 2x_2 = 3.0001.$$

The point $(3, 0)$ lies on l_1 and the lines are nearly parallel. This implies that $(3, 0)$ also lies close to l_2, even though it differs significantly from the intersection point $(1, 1)$. (See Fig. 8.3.)

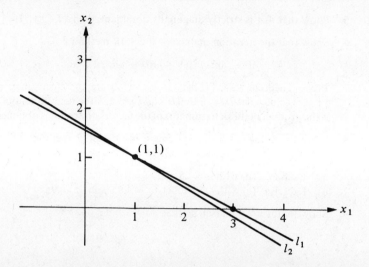

FIGURE 8.3

This example was clearly constructed to show the difficulties that might, and in fact do, arise. Had the lines not been nearly parallel, it would be expected that a small residual vector implies an accurate approximation.

In the general situation, we cannot rely on the geometry of the system to give us an indication of when problems might occur. We can, however, obtain this information by considering the norms of the matrix A and its inverse.

Theorem 8.27   If $\tilde{\mathbf{x}}$ is an approximation to the solution of $A\mathbf{x} = \mathbf{b}$ and A is a nonsingular matrix, then for any natural norm,

$$(8.26) \qquad \qquad \|\mathbf{x} - \tilde{\mathbf{x}}\| \leq \|\mathbf{r}\| \, \|A^{-1}\|$$

and

$$(8.27) \qquad \frac{\|\mathbf{x} - \tilde{\mathbf{x}}\|}{\|\mathbf{x}\|} \leq \|A\| \, \|A^{-1}\| \, \frac{\|\mathbf{r}\|}{\|\mathbf{b}\|}, \qquad \text{provided } \mathbf{x} \neq \mathbf{0} \text{ and } \mathbf{b} \neq \mathbf{0},$$

where $\mathbf{r}$ is the residual vector for $\tilde{\mathbf{x}}$ with respect to the system $A\mathbf{x} = \mathbf{b}$.

Proof   Since $\mathbf{r} = \mathbf{b} - A\tilde{\mathbf{x}} = A\mathbf{x} - A\tilde{\mathbf{x}}$ and A is nonsingular, $\mathbf{x} - \tilde{\mathbf{x}} = A^{-1}\mathbf{r}$ and Exercise 1 of Section 8.1 implies that

$$\|\mathbf{x} - \tilde{\mathbf{x}}\| = \|A^{-1}\mathbf{r}\| \leq \|A^{-1}\| \, \|\mathbf{r}\|.$$

Moreover, since $\mathbf{b} = A\mathbf{x}$, $\|\mathbf{b}\| \leq \|A\| \, \|\mathbf{x}\|$; so

$$\frac{\|\mathbf{x} - \tilde{\mathbf{x}}\|}{\|\mathbf{x}\|} \leq \frac{\|A\| \, \|A^{-1}\|}{\|\mathbf{b}\|} \, \|\mathbf{r}\|. \qquad \qquad \square$$

Inequalities (8.26) and (8.27) imply that the quantities $\|A^{-1}\|$ and $\|A\| \|A^{-1}\|$ can be used to give an indication of the connection between the residual vector and the accuracy of the approximation. In general, the relative error $\|\mathbf{x} - \tilde{\mathbf{x}}\|/\|\mathbf{x}\|$ is of most interest and, by inequality (8.27), this error is bounded by the product of $\|A\| \|A^{-1}\|$ with the relative residual for this approximation, $\|\mathbf{r}\|/\|\mathbf{b}\|$. Any convenient norm can be used for this approximation, the only requirement being that it be used consistently throughout.

Definition 8.28   The **condition number** $K(A)$ of the nonsingular matrix A relative to a norm $\|\cdot\|$ is defined to be

$$K(A) = \|A\| \|A^{-1}\|.$$

With this notation, the inequalities in Theorem 8.27 become

(8.28)
$$\|\mathbf{x} - \tilde{\mathbf{x}}\| \leq K(A) \frac{\|\mathbf{r}\|}{\|A\|}$$

and

(8.29)
$$\frac{\|\mathbf{x} - \tilde{\mathbf{x}}\|}{\|\mathbf{x}\|} \leq K(A) \frac{\|\mathbf{r}\|}{\|\mathbf{b}\|}.$$

Since for any nonsingular matrix A

$$1 = \|I\| = \|A \cdot A^{-1}\| \leq \|A\| \|A^{-1}\| = K(A),$$

the expectation is that the matrix A will be well-behaved (formally called a **well-conditioned matrix**) if $K(A)$ is close to one and not well-behaved (called **ill-conditioned**), when $K(A)$ is significantly greater than one. Behavior in this instance refers to the relative security that a small residual vector implies a correspondingly accurate approximate solution.

EXAMPLE 2   The matrix for the system considered in Example 1 was

$$A = \begin{bmatrix} 1 & 2 \\ 1.0001 & 2 \end{bmatrix},$$

which has $\|A\|_\infty = 3.0001$. This norm would not be considered to be large. However,

$$A^{-1} = \begin{bmatrix} -10000 & 10000 \\ 5000.5 & -5000 \end{bmatrix};$$

$$\|A^{-1}\|_\infty = 20,000,$$

and, for the infinity norm, $K(A) = (20,000)(3.0001) = 60,002$. The size of the condition number for this example would certainly keep us from making hasty accuracy decisions based on the residual of an approximation.   □

While in theory the condition number of a matrix depends totally upon the norms of the matrix and its inverse, in practice the calculation of the inverse is subject to round-off error and is dependent on the accuracy with which the calculations are performed. If the operations involve arithmetic with t significant digits of accuracy, the approximate condition number for the matrix A is actually the norm of the matrix times the norm of the approximation to the inverse of A, which is obtained using t-digit arithmetic. In fact, this condition number even depends upon the method used to calculate the inverse of A. If we make the assumption that the approximate solution to the linear system $A\mathbf{x} = \mathbf{b}$ is being determined using t-digit arithmetic and Gaussian elimination, it can be shown (see Forsythe and Moler [39], pages 49–51) that the residual vector $\mathbf{r}$ for the approximation $\tilde{\mathbf{x}}$ has the property

$$(8.30) \qquad \|\mathbf{r}\| \approx 10^{-t}\|A\|\,\|\tilde{\mathbf{x}}\|.$$

From this approximate equation, an estimate for the effective condition number in t-digit arithmetic can be obtained without the necessity of inverting the matrix A. In actuality, the approximation in Eq. (8.30) assumes that all the arithmetic operations in the Gaussian elimination technique are performed using t-digit arithmetic, but that the operations that are needed to determine the residual are done in double precision (that is, $2t$-digit) arithmetic. This technique does not add significantly to the computational effort, and it eliminates much of the loss of accuracy (see Section 1.2) involved with the subtraction of the nearly equal numbers that occur in the calculation of the residual.

The approximation for the t-digit condition number $K(A)$ comes from consideration of the linear system

$$A\mathbf{y} = \mathbf{r}.$$

The solution to this system can be readily approximated since the multipliers for the Gaussian elimination method have already been calculated and hopefully retained. In fact, $\tilde{\mathbf{y}}$, the approximate solution of $A\mathbf{y} = \mathbf{r}$, satisfies

$$(8.31) \qquad \tilde{\mathbf{y}} \approx A^{-1}\mathbf{r} = A^{-1}(\mathbf{b} - A\tilde{\mathbf{x}}) = A^{-1}\mathbf{b} - A^{-1}A\tilde{\mathbf{x}} = \mathbf{x} - \tilde{\mathbf{x}};$$

so $\tilde{\mathbf{y}}$ is an estimate of the error in approximating the solution to the original system. Equation (8.30) can consequently be used to deduce that

$$\|\tilde{\mathbf{y}}\| \approx \|\mathbf{x} - \tilde{\mathbf{x}}\| = \|A^{-1}\mathbf{r}\| \le \|A^{-1}\|\,\|\mathbf{r}\| \approx \|A^{-1}\|(10^{-t}\|A\|\,\|\tilde{\mathbf{x}}\|)$$
$$= 10^{-t}\|\tilde{\mathbf{x}}\|K(A).$$

This gives an approximation for the condition number involved with solving the system $A\mathbf{x} = \mathbf{b}$ using Gaussian elimination and the t-digit type of arithmetic described above:

$$(8.32) \qquad K(A) \approx \frac{\|\tilde{\mathbf{y}}\|}{\|\tilde{\mathbf{x}}\|}\,10^{t}.$$

EXAMPLE 3   The linear system given by

$$\begin{bmatrix} 3.3330 & 15920 & -10.333 \\ 2.2220 & 16.710 & 9.6120 \\ 1.5611 & 5.1791 & 1.6852 \end{bmatrix} \begin{bmatrix} x_1 \\ x_2 \\ x_3 \end{bmatrix} = \begin{bmatrix} 15913 \\ 28.544 \\ 8.4254 \end{bmatrix}$$

has the exact solution $\mathbf{x} = (1, 1, 1)^t$.

Using Gaussian elimination and 5-digit arithmetic leads to the augmented matrix

$$\begin{bmatrix} 3.3330 & 15920 & -10.333 & \vdots & 15913 \\ 0 & -10596 & 16.501 & \vdots & -10580 \\ 0 & 0 & -5.0790 & \vdots & -4.7000 \end{bmatrix}.$$

The approximate solution to this system is

$$\tilde{\mathbf{x}} = (1.2001, .99991, .92538)^t.$$

The residual vector corresponding to $\tilde{\mathbf{x}}$ is computed in double precision to be

$$\mathbf{r} = \mathbf{b} - A\tilde{\mathbf{x}}$$

$$= \begin{bmatrix} 15913 \\ 28.544 \\ 8.4254 \end{bmatrix} - \begin{bmatrix} 3.3330 & 15920 & -10.333 \\ 2.2220 & 16.710 & 9.6120 \\ 1.5611 & 5.1791 & 1.6852 \end{bmatrix} \begin{bmatrix} 1.2001 \\ .99991 \\ .92538 \end{bmatrix}$$

$$= \begin{bmatrix} -.00518 \\ .27413 \\ -.18616 \end{bmatrix};$$

so

$$\|\mathbf{r}\|_\infty = .27413.$$

Using 5-digit arithmetic for the calculations gives the approximation for the inverse of A,

$$A^{-1} = \begin{bmatrix} -1.1701 \times 10^{-4} & -1.4983 \times 10^{-1} & 8.5416 \times 10^{-1} \\ 6.2782 \times 10^{-5} & 1.2124 \times 10^{-4} & -3.0662 \times 10^{-4} \\ -8.6631 \times 10^{-5} & 1.3846 \times 10^{-1} & -1.9689 \times 10^{-1} \end{bmatrix},$$

and Theorem 8.11 can be used to show that $\|A^{-1}\|_\infty = 1.0041$ and $\|A\|_\infty = 15934$. The condition number in this case is extremely large:

$$(8.33) \qquad K(A) \approx (1.0041)(15934) = 15999.$$

The estimate for the condition number given in the preceding discussion is obtained by first solving the system

$$\begin{bmatrix} 3.3330 & 15920 & -10.333 \\ 2.2220 & 16.710 & 9.6120 \\ 1.5611 & 5.1791 & 1.6852 \end{bmatrix} \begin{bmatrix} y_1 \\ y_2 \\ y_3 \end{bmatrix} = \begin{bmatrix} -.00518 \\ .27413 \\ -.18616 \end{bmatrix},$$

which implies that $\tilde{\mathbf{y}} = (-.20008, 8.9987 \times 10^{-5}, .074607)^t$, and then using the estimate given in Eq. (8.32):

$$(8.34) \qquad K(A) \approx 10^5 \frac{\|\tilde{\mathbf{y}}\|_\infty}{\|\tilde{\mathbf{x}}\|_\infty} = \frac{10^5(.20008)}{1.2001} = 16672.$$

This estimate is quite close to the actual value and requires considerably less computational effort.

Since the actual solution $\mathbf{x} = (1, 1, 1)^t$ is known for this system, we can calculate both

$$\|\mathbf{x} - \tilde{\mathbf{x}}\|_\infty = .2001 \qquad \text{and} \qquad \frac{\|\mathbf{x} - \tilde{\mathbf{x}}\|}{\|\mathbf{x}\|} = \frac{.2001}{1} = .2001.$$

The error bounds given in Theorem 8.27 for these values are

$$\|\mathbf{x} - \tilde{\mathbf{x}}\|_\infty \le K(A) \frac{\|\mathbf{r}\|_\infty}{\|A\|_\infty} = \frac{(15999)(.27413)}{15934} = .27525$$

and

$$\frac{\|\mathbf{x} - \tilde{\mathbf{x}}\|_\infty}{\|\mathbf{x}\|_\infty} \le K(A) \frac{\|\mathbf{r}\|_\infty}{\|\mathbf{b}\|_\infty} = \frac{(15999)(.27413)}{15913} = .27561. \qquad \square$$

In Eq. (8.31) we used the estimate $\tilde{\mathbf{y}} \approx \mathbf{x} - \tilde{\mathbf{x}}$ where $\tilde{\mathbf{y}}$ is the approximate solution to the system $A\mathbf{y} = \mathbf{r}$. It would be reasonable to suspect from this result that $\tilde{\mathbf{x}} + \tilde{\mathbf{y}}$ is a more accurate approximation to the solution of the linear system $A\mathbf{x} = \mathbf{b}$ than the original approximation $\tilde{\mathbf{x}}$. The method utilizing this assumption is called **iterative refinement**, or iterative improvement, and consists of performing iterations on the system whose right-hand side is the residual vector for successive approximations until satisfactory accuracy results. The procedure is generally used only on systems when it is suspected that the matrix involved is ill-conditioned, since this technique will not significantly improve the approximation for a well-conditioned system.

Iterative Refinement Algorithm 8.4

To approximate the solution to the linear system $A\mathbf{x} = \mathbf{b}$ when A is suspected to be ill-conditioned.

INPUT the number of equations and unknowns n; the entries $a_{ij}, 1 \le l, j \le n$ of the matrix A; the entries b_i, $1 \le i \le n$ of the inhomogeneous term $\mathbf{b}$; the maximum number of iterations N; tolerance TOL.

OUTPUT the approximation $xx_1, \ldots, xx_n$ or a message that the number of iterations was exceeded.

Step 0 Solve the system $A\mathbf{x} = \mathbf{b}$ for $x_1, \ldots, x_n$ by Gaussian elimination saving the multipliers m_{ji}, $j = i + 1, i + 2, \ldots, n, i = 1, 2, \ldots, n - 1$ and noting row interchanges.

Step 1 Set $k = 1$.

Step 2 While ($k \le N$) do Steps 3–8.

Step 3 For $i = 1, 2, \ldots, n$ (*Calculate* $\mathbf{r}$.)

$$\text{set } r_i = b_i - \sum_{j=1}^{n} a_{ij} x_j.$$

(*Perform the computation in double-precision arithmetic.*)

Step 4 Solve the linear system $A\mathbf{y} = \mathbf{r}$ by using Gaussian elimination in the same order as in Step 0.

Step 5 For $i = 1, \ldots, n$ set $xx_i = x_i + y_i$.

Step 6 If $\|\mathbf{x} - \mathbf{xx}\|_\infty < TOL$ then OUTPUT (**xx**);
                             (*Procedure completed successfully.*)
                             STOP.

Step 7 Set $k = k + 1$.

Step 8   For $i = 1, \ldots, n$ set $x_i = xx_i$.

Step 9   OUTPUT ('Maximum number of iterations exceeded.');
(*Procedure completed unsuccessfully.*)
STOP.

A recommended stopping procedure in Step 6 is to iterate until $|y_i^{(k)}| \le 10^{-t}$ for each $i = 1, 2, \ldots, n$ if t-digit arithmetic is being used. It should be emphasized that the iterative refinement technique will not give satisfactory results for all systems involving ill-conditioned matrices. In particular, if $K(A) \ge 10^t$, the procedure will probably fail, and the only alternative is to use increased precision for the calculations, if possible.

EXAMPLE 4   In Example 3 we found the approximation to the problem we have been considering, using five-digit arithmetic and Gaussian elimination, to be

$$\tilde{\mathbf{x}}^{(1)} = (1.2001, .99991, .92538)^t$$

and the solution to $A\mathbf{y} = \mathbf{r}^{(1)}$ to be

$$\mathbf{y}^{(1)} = (-.20008, 8.9987 \times 10^{-5}, .074607)^t.$$

By Step 5 in the algorithm, this implies that

$$\tilde{\mathbf{x}}^{(2)} = \tilde{\mathbf{x}}^{(1)} + \mathbf{y}^{(1)} = (1.0000, 1.0000, .99999)^t,$$

and the actual error in this approximation is

$$\|\mathbf{x} - \tilde{\mathbf{x}}^{(2)}\|_\infty = 1 \times 10^{-5}.$$

Using the suggested stopping technique for the algorithm, we compute $\mathbf{r}^{(2)} = \mathbf{b} - A\tilde{\mathbf{x}}^{(2)}$, and solve the system $A\mathbf{y}^{(2)} = \mathbf{r}^{(2)}$, which gives

$$\mathbf{y}^{(2)} = (1.5002 \times 10^{-9}, 2.0951 \times 10^{-10}, 1.0000 \times 10^{-5})^t.$$

Since $\|\mathbf{y}^{(2)}\|_\infty \le 10^{-5}$, we conclude that

$$\tilde{\mathbf{x}}^{(3)} = \tilde{\mathbf{x}}^{(2)} + \tilde{\mathbf{y}}^{(2)} = (1.0000, 1.0000, 1.0000)^t$$

is sufficiently accurate. This is certainly correct.   □

Throughout this section it has been assumed that, in the linear system $A\mathbf{x} = \mathbf{b}$, A and $\mathbf{b}$ could be represented exactly. Realistically, the entries a_{ij} and b_j might be altered or perturbed by an amount δa_{ij} and δb_j, causing the linear system

$$(A + \delta A)\mathbf{x} = \mathbf{b} + \delta \mathbf{b}$$

to be solved in place of $A\mathbf{x} = \mathbf{b}$. Normally, if $\|\delta A\|$ and $\|\delta \mathbf{b}\|$ are small, on the order of 10^{-t}, the t-digit arithmetic should yield a solution $\tilde{\mathbf{x}}$ for which $\|\mathbf{x} - \tilde{\mathbf{x}}\|$ is correspondingly small. However, in the case of ill-conditioned systems, we have seen that even if A and $\mathbf{b}$ are represented exactly, rounding errors can cause $\|\mathbf{x} - \tilde{\mathbf{x}}\|$ to be large. The following theorem relates the perturbations of linear systems to the condition number of a matrix. (The proof of this result can be found in Ortega [64], page 33.)

Theorem 8.29   Suppose A is nonsingular and

$$\|\delta A\| < \frac{1}{\|A^{-1}\|}.$$

The solution $\tilde{\mathbf{x}}$ to $(A + \delta A)\tilde{\mathbf{x}} = \mathbf{b} + \delta \mathbf{b}$ approximates the solution $\mathbf{x}$ of $A\mathbf{x} = \mathbf{b}$ with error estimate

(8.35) $$\frac{\|\mathbf{x} - \tilde{\mathbf{x}}\|}{\|\mathbf{x}\|} \leq \frac{K(A)}{1 - K(A)(\|\delta A\|/\|A\|)}\left(\frac{\|\delta \mathbf{b}\|}{\|\mathbf{b}\|} + \frac{\|\delta A\|}{\|A\|}\right).$$

The estimate in inequality (8.35) states that if the matrix A is well-conditioned, that is, $K(A)$ is not too large, then small changes in A and $\mathbf{b}$ produce correspondingly small changes in the solution $\mathbf{x}$. On the other hand, if A is ill-conditioned, then small changes in A and $\mathbf{b}$ may produce large changes in $\mathbf{x}$.

This theorem is independent of the particular numerical procedure used to solve $A\mathbf{x} = \mathbf{b}$. It can be shown, by means of Wilkinson's backward error analysis (see Wilkinson [90] or [91]), that if Gaussian elimination with pivoting is used to solve $A\mathbf{x} = \mathbf{b}$ in t-digit arithmetic, the numerical solution $\tilde{\mathbf{x}}$ is the actual solution of a linear system

(8.36) $$(A + \delta A)\tilde{\mathbf{x}} = \mathbf{b}$$

where

(8.37) $$\|\delta A\|_\infty \leq f(n)10^{1-t} \max_{i,j,k}|a_{ij}^{(k)}|.$$

Wilkinson has found in practice that $f(n) \approx n$ and, at worst, $f(n) \leq 1.01(n^3 + 3n^2)$.

Exercise Set 8.3

1. Solve the following linear systems using Gaussian elimination and iterative refinement. Estimate $K(A)$.

   a) $x_1 + \frac{1}{2}x_2 + \frac{1}{3}x_3 = \frac{11}{6}$,

   $5x_1 + \frac{10}{3}x_2 + \frac{5}{2}x_3 = \frac{65}{6}$,

   $\frac{100}{3}x_1 + 25x_2 + 20x_3 = \frac{235}{3}$.   (three-digit arithmetic)

   b) $1.003x_1 + 58.09x_2 = 68.12$,

   $5.550x_1 + 321.8x_2 = 377.8$.   (four-digit arithmetic)

   c) $3.9x_1 + 1.6x_2 = 5.5$,

   $6.8x_1 + 2.9x_2 = 9.7$.   (two-digit arithmetic)

   d) $\frac{1}{4}x_1 + \frac{1}{5}x_2 + \frac{1}{6}x_3 = 9$,

   $\frac{1}{3}x_1 + \frac{1}{4}x_2 + \frac{1}{5}x_3 = 8$,

   $\frac{1}{2}x_1 + x_2 + 2x_3 = 8$.   (single-precision computer arithmetic)

   e) $4.56x_1 + 2.18x_2 = 6.74$,

   $2.79x_1 + 1.38x_2 = 4.13$.   (three-digit arithmetic)

2. Compute the condition numbers of the following matrices relative to $\|\cdot\|_\infty$.

a) $\begin{bmatrix} 1 & 2 \\ 1.0001 & 2 \end{bmatrix}$

b) $\begin{bmatrix} 3.9 & 1.6 \\ 6.8 & 2.9 \end{bmatrix}$

Compare to 1(c).

c) $\begin{bmatrix} 1.003 & 58.09 \\ 5.550 & 321.8 \end{bmatrix}$

Compare to 1(b).

3. Show that if B is singular, then

$$\frac{1}{K(A)} \le \frac{\|A - B\|}{\|A\|}.$$

[*Hint*: There exists a vector $\mathbf{x} \neq \mathbf{0}$, with $\|\mathbf{x}\| = 1$, such that $B\mathbf{x} = \mathbf{0}$. Derive the estimate using $\|A\mathbf{x}\| \ge \|\mathbf{x}\|/\|A^{-1}\|$].

4. Using Exercise 3, estimate the condition numbers for the following matrices:

a) $\begin{bmatrix} 1 & 2 \\ 1.0001 & 2 \end{bmatrix}$

b) $\begin{bmatrix} 3.9 & 1.6 \\ 6.8 & 2.9 \end{bmatrix}$

Compare to 2(a).

Compare to 1(c).

5. The linear system

$$\begin{bmatrix} 1 & 2 \\ 1.0001 & 2 \end{bmatrix} \begin{bmatrix} x_1 \\ x_2 \end{bmatrix} = \begin{bmatrix} 3 \\ 3.0001 \end{bmatrix}$$

has solution $(1, 1)^t$. Change A slightly to

$$\begin{bmatrix} 1 & 2 \\ .9999 & 2 \end{bmatrix}$$

and consider the linear system

$$\begin{bmatrix} 1 & 2 \\ .9999 & 2 \end{bmatrix} \begin{bmatrix} x_1 \\ x_2 \end{bmatrix} = \begin{bmatrix} 3 \\ 3.0001 \end{bmatrix}.$$

Compute the new solution $\tilde{\mathbf{x}}$ using five-digit arithmetic, and compare the actual error to the estimate (8.35). Is A ill-conditioned?

6. The linear system $A\mathbf{x} = \mathbf{b}$ given by

$$\begin{bmatrix} 1 & 2 \\ 1.00001 & 2 \end{bmatrix} \begin{bmatrix} x_1 \\ x_2 \end{bmatrix} = \begin{bmatrix} 3 \\ 3.00001 \end{bmatrix}$$

has solution $(1, 1)^t$. Find the solution of the perturbed system, using seven-digit arithmetic

$$\begin{bmatrix} 1 & 2 \\ 1.000011 & 2 \end{bmatrix} \begin{bmatrix} x_1 \\ x_2 \end{bmatrix} = \begin{bmatrix} 3.00001 \\ 3.00003 \end{bmatrix},$$

and compare the actual error to the estimate (8.35). Is A ill-conditioned?

7. Using a computer, analyze the linear system in Exercise 1(d) with respect to the estimate (8.37). Is $f(n)$ closer to n or $1.01(n^3 + 3n^2)$?

8. The $n \times n$ matrix $H^{(n)}$ defined by

$$H^{(n)}_{ij} = \frac{1}{i + j - 1}, \qquad 1 \le i, j \le n$$

is an ill-conditioned matrix that arises in solving the normal equations for the coefficients of the least-squares polynomial (see Example 1 of Section 7.2).

a)   Show that

$$[H^{(4)}]^{-1} = \begin{bmatrix} 16 & -120 & 240 & -140 \\ -120 & 1200 & -2700 & 1680 \\ 240 & -2700 & 6480 & -4200 \\ -140 & 1680 & -4200 & 2800 \end{bmatrix},$$

and compute $K(H^{(4)})$ relative to $\|\cdot\|_\infty$.

b)   Show that

$$[H^{(5)}]^{-1} = \begin{bmatrix} 52 & -300 & 1050 & -1400 & 630 \\ -300 & 4800 & -18900 & 26880 & -12600 \\ 1050 & -18900 & 79380 & -117600 & 56700 \\ -1400 & 26880 & -117600 & 179200 & -88200 \\ 630 & -12600 & 56700 & -88200 & 44100 \end{bmatrix},$$

and compute $K(H^{(5)})$ relative to $\|\cdot\|_1$.

c)   Solve the linear system

$$H^{(4)} \begin{bmatrix} x_1 \\ x_2 \\ x_3 \\ x_4 \end{bmatrix} = \begin{bmatrix} 1 \\ 0 \\ 0 \\ 1 \end{bmatrix}$$

using three-digit arithmetic, and compare the actual error to that estimated in (8.35).

8.4   Eigenvalues and Eigenvectors

The solution of many physical problems requires the calculation, or at least estimation, of the eigenvalues and corresponding eigenvectors of a matrix associated with a linear system of equations. We have seen (see Definitions 8.12 and 8.13 of Section 8.1) that an $n \times n$ matrix A has precisely n, not necessarily distinct, eigenvalues that are the roots of the polynomial $p(\lambda) = \det(A - \lambda I)$. Theoretically the eigenvalues of A are obtained by finding the n roots or $p(\lambda)$ and then the associated linear system is solved to determine the corresponding eigenvectors. In practice, $p(\lambda)$ is difficult to obtain, and in any case we have seen that it can be difficult to determine the roots of an nth-degree polynomial, except for small values of n; so it is necessary to construct approximation techniques for finding eigenvalues. Of particular importance is the calculation of the eigenvalue of a matrix that is largest in magnitude. The following important result gives a bound for this eigenvalue.

Theorem 8.30 (Gerschgorin Circle Theorem)   Let A be an $n \times n$ matrix and R_i denote the circle in the complex plane with center a_{ii} and radius $\sum_{j=1, j\neq i}^{n} |a_{ij}|$; that is,

$$R_i = \left\{ z \in \mathbb{C} \;\middle|\; |z - a_{ii}| \le \sum_{\substack{i=1, \\ j\neq i}}^{n} |a_{ij}| \right\},$$

where $\mathbb{C}$ is used to denote the complex plane. The eigenvalues of A are contained within $R = \bigcup_{i=1}^{n} R_i$ and the union of any k of these circles that do not intersect the remaining $(n - k)$ must contain precisely k (counting multiplicities) of the eigenvalues.

Proof The result in Exercise 1 of Section 8.1 implies that an eigenvector $\mathbf{x}$ associated with λ can be found with $\|\mathbf{x}\|_\infty = 1$.

Since $A\mathbf{x} - \lambda\mathbf{x} = \mathbf{0}$, the equivalent component representation is

$$(8.38) \qquad \sum_{j=1}^{n} a_{ij}x_j = \lambda x_i \qquad \text{for each } i = 1, 2, \ldots, n;$$

and if k is an integer with $|x_k| = \|\mathbf{x}\|_\infty = 1$, Eq. (8.38), with $i = k$, implies that

$$\sum_{j=1}^{n} a_{kj}x_j = \lambda x_k.$$

Thus

$$\sum_{\substack{j=1,\\ j \neq k}}^{n} a_{kj}x_j = \lambda x_k - a_{kk}x_k,$$

and

$$|(a_{kk} - \lambda)x_k| = \left| \sum_{\substack{j=1,\\ j \neq k}}^{n} - a_{kj}x_j \right| \leq \sum_{\substack{j=1,\\ j \neq k}}^{n} |a_{kj}||x_j|.$$

Since $|x_k| \geq |x_j|$ for all $j = 1, 2, \ldots, n$,

$$|a_{kk} - \lambda| \leq \sum_{\substack{j=1,\\ j \neq k}}^{n} |a_{kj}| \left| \frac{x_j}{x_k} \right| \leq \sum_{\substack{j=1,\\ j \neq k}}^{n} |a_{kj}|.$$

Thus $\lambda \in R_k$, which proves the first assertion in the theorem. The second part of this theorem requires a clever continuity argument. A quite readable proof is contained in Ortega [64], page 48. □

EXAMPLE 1 For the matrix

$$A = \begin{bmatrix} 4 & 1 & 1 \\ 0 & 2 & 1 \\ -2 & 0 & 9 \end{bmatrix},$$

the circles in the Gerschgorin theorem are (see Fig. 8.4)

$$R_1 = \{z \in \mathbb{C} \,||z - 4| \leq 2\},$$
$$R_2 = \{z \in \mathbb{C} \,||z - 2| \leq 1\},$$

and

$$R_3 = \{z \in \mathbb{C} \,||z - 9| \leq 2\}.$$

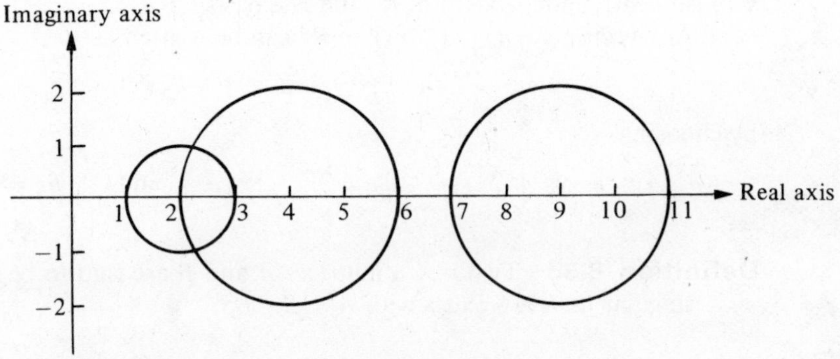

FIGURE 8.4

Since R_1 and R_2 are disjoint from R_3, there must be precisely two eigenvalues within $R_1 \cup R_2$ and one within R_3. Moreover, since $\rho(A) = \max_{1 \le i \le 3} |\lambda_i|$, $7 \le \rho(A) \le 11$. $\square$

Before considering further results concerning eigenvalues and eigenvectors, we need some definitions and results from linear algebra. All of the general results that will be needed in the remainder of this chapter are listed here for ease of reference. The proofs of these results will not be given, but they are available in most standard texts on linear algebra (see, for example, Noble and Daniel [63]).

Definition 8.31   Let $\{v^{(1)}, v^{(2)}, v^{(3)}, \ldots, v^{(k)}\}$ be a set of vectors. The set is said to be **linearly dependent** if numbers $\alpha_1, \alpha_2, \ldots, \alpha_k$ exist, not all zero, with

$$0 = \alpha_1 v^{(1)} + \alpha_2 v^{(2)} + \alpha_3 v^{(3)} + \cdots + \alpha_k v^{(k)}.$$

A set of vectors that is not linearly dependent is called **linearly independent**.

Theorem 8.32   If $\{v^{(1)}, v^{(2)}, v^{(3)}, \ldots, v^{(n)}\}$ is a set of n linearly independent vectors in R^n, then any vector $x \in R^n$ can be written uniquely as

$$x = \beta_1 v^{(1)} + \beta_2 v^{(2)} + \beta_3 v^{(3)} + \cdots + \beta_n v^{(n)}$$

for some collection of constants $\beta_1, \beta_2, \ldots, \beta_n$.

Any collection of n linearly independent vectors in R^n is called a **basis** for R^n.

EXAMPLE 2   Let $v^{(1)} = (1, 0, 0)^t$, $v^{(2)} = (-1, 1, 1)^t$, and $v^{(3)} = (0, 4, 2)^t$. If α_1, α_2, and α_3 are numbers with

$$0 = \alpha_1 v^{(1)} + \alpha_2 v^{(2)} + \alpha_3 v^{(3)},$$

then

$$(0, 0, 0)^t = \alpha_1(1, 0, 0)^t + \alpha_2(-1, 1, 1)^t + \alpha_3(0, 4, 2)^t$$
$$= (\alpha_1 - \alpha_2, \alpha_2 + 4\alpha_3, \alpha_2 + 2\alpha_3)^t,$$

so $\alpha_1 - \alpha_2 = 0,$ $\alpha_2 + 4\alpha_3 = 0,$ and $\alpha_2 + 2\alpha_3 = 0.$

Since the only solution to this system is $\alpha_1 = \alpha_2 = \alpha_3 = 0$, the set $\{v^{(1)}, v^{(2)}, v^{(3)}\}$ is linearly independent in R^3 and is a basis for R^3.

Any vector $x = (x_1, x_2, x_3)^t$ in R^3 can be written as

$$x = \beta_1 v^{(1)} + \beta_2 v^{(2)} + \beta_3 v^{(3)}$$

by choosing

$$\beta_1 = x_1 - x_2 + 2x_3, \qquad \beta_2 = 2x_3 - x_2, \qquad \text{and} \qquad \beta_3 = \tfrac{1}{2}(x_2 - x_3). \quad \square$$

Definition 8.33   Two $n \times n$ matrices A and B are said to be **similar** if a nonsingular matrix S exists with $A = S^{-1}BS$.

Theorem 8.34 Suppose A and B are similar $n \times n$ matrices and λ is an eigenvalue of A, with associated eigenvector $\mathbf{x}$. Then λ is also an eigenvalue of B and if $A = S^{-1}BS$, then $S\mathbf{x}$ is an eigenvector associated with λ and the matrix B.

Definition 8.35 A set of vectors $\{\mathbf{v}^{(1)}, \mathbf{v}^{(2)}, \ldots, \mathbf{v}^{(n)}\}$ is called **orthogonal** if $(\mathbf{v}^{(i)})^t \mathbf{v}^{(j)} = 0$ for all $i \neq j$. If, in addition, $(\mathbf{v}^{(i)})^t \mathbf{v}^{(i)} = 1$ for all $i = 1, 2, \ldots, n$, then the set is called **orthonormal**.

Theorem 8.36 If A is a matrix and $\lambda_1, \ldots, \lambda_k$ are distinct eigenvalues of A with associated eigenvectors $\mathbf{x}^{(1)}, \mathbf{x}^{(2)}, \ldots, \mathbf{x}^{(k)}$, then $\{\mathbf{x}^{(1)}, \mathbf{x}^{(2)}, \ldots, \mathbf{x}^{(k)}\}$ is linearly independent.

Theorem 8.37 An orthogonal set of vectors that does not contain the zero vector is linearly independent.

Definition 8.38 An $n \times n$ matrix P is said to be an **orthogonal matrix** if $P^{-1} = P^t$.

Theorem 8.39 If A is an $n \times n$ symmetric matrix, then there exists an orthogonal matrix P such that $D = P^{-1}AP$ where D is a diagonal matrix whose diagonal entries are the eigenvalues of A.

Corollary 8.40 If A is a symmetric $n \times n$ matrix, then there exist n eigenvectors of A that form an orthonormal set.

Theorem 8.41 The eigenvalues of a symmetric matrix are all real numbers.

This completes the material that will be needed from linear algebra. We will now consider the **power method** for finding the eigenvalues and eigenvectors of a matrix.

In order to apply the power method, we must assume that the $n \times n$ matrix A has n eigenvalues $\lambda_1, \lambda_2, \ldots, \lambda_n$ with an associated collection of eigenvectors $\{\mathbf{v}^{(1)}, \mathbf{v}^{(2)}, \mathbf{v}^{(3)}, \ldots, \mathbf{v}^{(n)}\}$, which is linearly independent.

Moreover, we assume that A has precisely one eigenvalue that is largest in magnitude and, for this reason, assume that $\lambda_1, \lambda_2, \ldots, \lambda_n$ denote the eigenvalues of A with $|\lambda_1| > |\lambda_2| \geq |\lambda_3| \geq \cdots \geq |\lambda_n| \geq 0$.

If $\mathbf{x}^{(0)}$ is any vector in R^n, the fact that $\{\mathbf{v}^{(1)}, \mathbf{v}^{(2)}, \ldots, \mathbf{v}^{(n)}\}$ is linearly independent implies that constants $\alpha_1, \alpha_2, \ldots, \alpha_n$ exist with

$$(8.39) \qquad \mathbf{x}^{(0)} = \sum_{j=1}^{n} \alpha_j \mathbf{v}^{(j)}.$$

Multiplying both sides of Eq. (8.39) by $A, A^2, \ldots, A^k$, we obtain:

$$A\mathbf{x}^{(0)} = \sum_{j=1}^{n} \alpha_j A\mathbf{v}^{(j)} = \sum_{j=1}^{n} \alpha_j \lambda_j \mathbf{v}^{(j)},$$

(8.40)
$$A^2\mathbf{x}^{(0)} = \sum_{j=1}^{n} \alpha_j \lambda_j A\mathbf{v}^{(j)} = \sum_{j=1}^{n} \alpha_j \lambda_j^2 \mathbf{v}^{(j)},$$

$$\vdots \qquad \qquad \vdots$$

$$A^k\mathbf{x}^{(0)} = \sum_{j=1}^{n} \alpha_j \lambda_j^k \mathbf{v}^{(j)}.$$

If λ_1^k is factored from each term on the right side of the equations in (8.40),

$$A^k\mathbf{x}^{(0)} = \lambda_1^k \sum_{j=1}^{n} \alpha_j \left(\frac{\lambda_j}{\lambda_1}\right)^k \mathbf{v}^{(j)},$$

the fact that $|\lambda_1| > |\lambda_j|$ for all $j = 2, 3, \ldots, n$ implies that $\lim_{k \to \infty} (\lambda_j/\lambda_1)^k = 0$, and

(8.41)
$$\lim_{k \to \infty} A^k\mathbf{x}^{(0)} = \lim_{k \to \infty} \lambda_1^k \alpha_1 \mathbf{v}^{(1)}.$$

This sequence will converge to zero if $|\lambda_1| < 1$ and diverge if $|\lambda_1| \geq 1$, provided, of course, that $\alpha_1 \neq 0$.

Advantage can be made of the relationship expressed in Eq. (8.41) by scaling the powers of $A^k\mathbf{x}^{(0)}$ in an appropriate manner. The scaling begins by choosing a unit vector $\mathbf{x}^{(0)}$ relative to $\|\cdot\|_{\infty}$ and a component x_{p_0} of $\mathbf{x}^{(0)}$ with

$$x_{p_0} = 1 = \|\mathbf{x}^{(0)}\|_{\infty}.$$

Let $\mathbf{y}^{(1)} = A\mathbf{x}^{(0)}$, and define $\mu^{(1)} = y_{p_0}^{(1)}$.

With this notation,

$$\mu^{(1)} = y_{p_0}^{(1)} = \frac{y_{p_0}^{(1)}}{x_{p_0}^{(1)}} = \frac{\alpha_1 \lambda_1 v_{p_0}^{(1)} + \sum_{j=2}^{n} \alpha_j \lambda_j v_{p_0}^{(j)}}{\alpha_1 v_{p_0}^{(1)} + \sum_{j=2}^{n} \alpha_j v_{p_0}^{(j)}} = \lambda_1 \left[\frac{\alpha_1 v_{p_0}^{(1)} + \sum_{j=2}^{n} \alpha_j (\lambda_j/\lambda_1) v_{p_0}^{(j)}}{\alpha_1 v_{p_0}^{(1)} + \sum_{j=2}^{n} \alpha_j v_{p_0}^{(j)}} \right]$$

Let p_1 be the smallest integer, $1 \leq p_1 \leq n$, such that

$$|y_{p_1}^{(1)}| = \|\mathbf{y}^{(1)}\|_{\infty}$$

and define $\mathbf{x}^{(1)}$ by

$$\mathbf{x}^{(1)} = \frac{1}{y_{p_1}^{(1)}} \mathbf{y}^{(1)}.$$

Then
$$x_{p_1}^{(1)} = 1 = \|\mathbf{x}^{(1)}\|_{\infty}.$$

In a similar manner, we can define sequences of vectors $\{\mathbf{x}^{(m)}\}_{m=0}^{\infty}$ and $\{\mathbf{y}^{(m)}\}_{m=1}^{\infty}$ and a sequence of scalars $\{\mu^{(m)}\}_{m=1}^{\infty}$ inductively by

(8.42)
$$\mathbf{y}^{(m)} = A\mathbf{x}^{(m-1)}$$

(8.43)
$$\mu^{(m)} = y_{p_{m-1}}^{(m)} = \lambda_1 \left[\frac{\alpha_1 v_{p_{m-1}}^{(1)} + \sum_{j=2}^{n} (\lambda_j/\lambda_1)^m \alpha_j v_{p_{m-1}}^{(j)}}{\alpha_1 v_{p_{m-1}}^{(1)} + \sum_{j=2}^{n} (\lambda_j/\lambda_1)^{m-1} \alpha_j v_{p_{m-1}}^{(j)}} \right],$$

and

$$(8.44) \qquad \mathbf{x}^{(m)} = \frac{\mathbf{y}^{(m)}}{y_{p_m}^{(m)}},$$

where at each step p_m is used to represent an integer for which

$$|y_{p_m}^{(m)}| = \|\mathbf{y}^{(m)}\|_\infty.$$

By examining Eq. (8.43), we see that since $|\lambda_j/\lambda_1| < 1$ for each $j = 2, 3, \ldots, n$, when $\mathbf{x}^{(0)}$ is chosen so that $\alpha_1 \neq 0$, then $\lim_{m \to \infty} \mu^{(m)} = \lambda_1$. Moreover, it can be shown (see Wilkinson [91], pages 570–572) that the sequence of vectors $\{\mathbf{x}^{(m)}\}_{m=0}^\infty$ will converge to an eigenvector of norm one associated with λ_1.

It should, however, be noted that the power method in general has the disadvantage that it is unknown at the outset whether or not the matrix has a single dominant eigenvalue, nor how $\mathbf{x}^{(0)}$ should be chosen so as to ensure that its representation in terms of the eigenvectors of the matrix will contain a nonzero contribution from the eigenvector associated with that dominant eigenvalue, should it exist.

Power Method Algorithm 8.5

To approximate the dominant eigenvalue and an associated eigenvector of the $n \times n$ matrix A given a nonzero vector $\mathbf{x}$:

INPUT dimension n; matrix A; vector $\mathbf{x}$; tolerance TOL; maximum number of iterations N.

OUTPUT approximate eigenvalue μ; approximate eigenvector $\mathbf{x}$ (with $\|\mathbf{x}\|_\infty = 1$) or a message that the maximum number of iterations was exceeded.

Step 1 Set $k = 1$;
     $\mu_0 = 0$.

Step 2 Find an integer p with $1 \leq p \leq n$ and $|x_p| = \|\mathbf{x}\|_\infty$.

Step 3 Set $\mathbf{x} = \dfrac{1}{x_p} \mathbf{x}$.

Step 4 While $(k \leq N)$ do Steps 5–11.

   Step 5 Set $\mathbf{y} = A\mathbf{x}$.

   Step 6 Set $\mu = y_p$.

   Step 7 Find an integer p with $1 \leq p \leq n$ and $|y_p| = \|\mathbf{y}\|_\infty$.

   Step 8 If $y_p = 0$ then OUTPUT ('eigenvector', $\mathbf{x}$);
                 OUTPUT ('A has eigenvalue 0, select new vector $\mathbf{x}$ and restart');
           STOP.

   Step 9 Set $\mathbf{x} = \dfrac{1}{y_p} \mathbf{y}$.

Step 10 If $|\mu - \mu_0| < TOL$ then OUTPUT $(\mu, \mathbf{x})$;
> (*Procedure completed successfully.*)
> STOP.

Step 11 Set $k = k + 1$;
$$\mu_0 = \mu.$$

Step 12 OUTPUT ('Maximum number of iterations exceeded');
STOP.

Choosing, in Step 7, the smallest integer, p_m, for which $|y_{p_m}^{(m)}| = \|\mathbf{y}^{(m)}\|_\infty$ will generally ensure that this index eventually becomes invariant. The rate at which $\{\mu^{(m)}\}_{m=1}^\infty$ converges to λ_1, is determined by the ratios $|\lambda_j/\lambda_1|^m$ for $j = 2, 3, \ldots, n$, and in particular by $|\lambda_2/\lambda_1|^m$; that is, the convergence is of order $O((\lambda_2/\lambda_1)^m)$. Thus, there is a constant k such that for large m

$$|\mu^{(m)} - \lambda_1| \approx k \left| \frac{\lambda_2}{\lambda_1} \right|^m,$$

which implies that

$$\lim_{m \to \infty} \frac{|\mu^{(m+1)} - \lambda_1|}{|\mu^{(m)} - \lambda_1|} \approx \left| \frac{\lambda_2}{\lambda_1} \right|.$$

By Definition 2.8 (p. 45) the sequence $\{\mu^{(m)}\}$ converges linearly to λ_1. The Aitken's Δ^2 procedure discussed in Theorem 2.11 (p. 61) can consequently by used to speed the convergence. Implementing the Δ^2 procedure in Algorithm 8.5 can be accomplished by modifying the algorithm as follows:

Step 1 Set $k = 1$;
$$\mu_0 = 0;$$
$$\mu_1 = 0;$$
$$\tilde{\mu}_0 = 0.$$

Step 6 Set $\mu = y_p$;

$$\tilde{\mu} = \mu_0 - \frac{(\mu_1 - \mu_0)^2}{\mu - 2\mu_1 + \mu_0}$$

Step 10 If $|\tilde{\mu} - \tilde{\mu}_0| < TOL$ and $k \geq 4$ then $OUTPUT\ (\tilde{\mu}, \mathbf{x})$;
> STOP.

Step 11 Set $k = k + 1$;
$$\mu_0 = \mu_1;$$
$$\mu_1 = \mu;$$
$$\tilde{\mu}_0 = \tilde{\mu}.$$

In actuality, it is not necessary for the matrix to have n distinct eigenvalues for the power method to converge. In fact, if the unique dominant eigenvalue, λ_1, has multiplicity r greater than one and $\mathbf{v}^{(1)}, \mathbf{v}^{(2)}, \ldots, \mathbf{v}^{(r)}$ are linearly independent eigenvectors associated with λ_1, the procedure will still converge to λ_1. The sequence of vectors $\{x^{(m)}\}_{m=0}^{\infty}$ will in this case converge to an eigenvector of λ_1 of norm one that is a linear combination of $\mathbf{v}^{(1)}$, $\mathbf{v}^{(2)}$, ..., $\mathbf{v}^{(r)}$ and depends on the choice of the initial vector $\mathbf{x}^{(0)}$.

EXAMPLE 3   The matrix

$$A = \begin{bmatrix} -4 & 14 & 0 \\ -5 & 13 & 0 \\ -1 & 0 & 2 \end{bmatrix}$$

has eigenvalues $\lambda_1 = 6$, $\lambda_2 = 3$, and $\lambda_3 = 2$. The power method described in Algorithm 8.5 will consequently converge.

Let $\mathbf{x}^{(0)} = (1, 1, 1)^t$, then

$$\mathbf{y}^{(1)} = A\mathbf{x}^{(0)} = (10, 8, 1)^t;$$

so

$$\|\mathbf{y}^{(1)}\|_{\infty} = 10, \qquad \mu^{(1)} = y_1^{(1)} = 10, \qquad \text{and} \qquad \mathbf{x}^{(1)} = \frac{\mathbf{y}^{(1)}}{\|\mathbf{y}^{(1)}\|_{\infty}} = (1, .8, .1)^t.$$

Continuing in this manner leads to the values in Table 8.5 where $\tilde{\mu}^{(m)}$ represents the sequence generated by the Aitken's Δ^2 procedure.

TABLE 8.5

m	$(\mathbf{x}^{(m)})^t$	$\mu^{(m)}$	$\tilde{\mu}^{(m)}$
0	$(1, 1, 1)$	—	
1	$(1, .8, .1)$	10	6.266667
2	$(1, .75, -.111)$	7.2	6.062473
3	$(1, .730769, -.188034)$	6.5	6.015054
4	$(1, .722200, -.220850)$	6.230769	6.004202
5	$(1, .718182, -.235915)$	6.111000	6.000855
6	$(1, .716216, -.243095)$	6.054546	6.000240
7	$(1, .715247, -.246588)$	6.027027	6.000058
8	$(1, .714765, -.248306)$	6.013453	6.000017
9	$(1, .714525, -.249157)$	6.006711	6.000003
10	$(1, .714405, -.249579)$	6.003352	6.000000
11	$(1, .714346, -.249790)$	6.001675	6.000000
12	$(1, .714316, -.249895)$	6.000837	

An approximation to the dominant eigenvalue, 6, at this stage is either $\tilde{\mu}^{(10)} = 6.000000$ or $\mu^{(12)} = 6.000837$ with approximate unit eigenvector $(1, .714316, -.249895)$. $\square$

When A is a symmetric matrix, a variation in the choice of the vectors $\mathbf{x}^{(m)}$, $\mathbf{y}^{(m)}$, and scalars $\mu^{(m)}$ can be made to significantly improve the rate of convergence of the sequence $\{\mu^{(m)}\}_{m=1}^{\infty}$ to the dominant eigenvalue λ_1. In fact, while the rate of convergence of the general power method is $O((\lambda_2/\lambda_1)^m)$, the rate of convergence of the modified procedure given below for symmetric matrices is $O((\lambda_2/\lambda_1)^{2m})$.

Power Method for Symmetric Matrices Algorithm 8.6

To approximate the dominant eigenvalue and an associated eigenvector of the $n \times n$ symmetric matrix A given a nonzero vector $\mathbf{x}$:

INPUT dimension n; matrix A; vector $\mathbf{x}$; tolerance TOL; maximum number of iterations N.

OUTPUT approximate eigenvalue μ; approximate eigenvector $\mathbf{x}$ (with $\|\mathbf{x}\|_2 = 1$) or a message that the maximum number of iterations was exceeded.

Step 1 Set $k = 1$;
$\qquad \mu_0 = 0$;
$\qquad \mathbf{x} = \mathbf{x}/\|\mathbf{x}\|_2$.

Step 2 While $(k \le N)$ do Steps 3–8.

Step 3 Set $\mathbf{y} = A\mathbf{x}$.

Step 4 Set $\mu = \mathbf{x}^t\mathbf{y}$.

Step 5 If $\|\mathbf{y}\|_2 = 0$, then OUTPUT ('eigenvector', $\mathbf{x}$);
$\qquad\qquad\qquad$ OUTPUT ('A has eigenvalue 0, select new vector $\mathbf{x}$ and
$\qquad\qquad\qquad\qquad$ restart');
$\qquad\qquad$ STOP.

Step 6 Set $\mathbf{x} = \mathbf{y}/\|\mathbf{y}\|_2$.

Step 7 If $|\mu - \mu_0| < TOL$ then OUTPUT $(\mu, \mathbf{x})$;
$\qquad\qquad\qquad$ (*Procedure completed successfully.*)
$\qquad\qquad\qquad$ STOP.

Step 8 Set $k = k + 1$;
$\qquad \mu_0 = \mu$.

Step 9 OUTPUT ('Maximum number of iterations exceeded');
$\qquad$ STOP.

EXAMPLE 4 The matrix

$$A = \begin{bmatrix} 4 & -1 & 1 \\ -1 & 3 & -2 \\ 1 & -2 & 3 \end{bmatrix}$$

is symmetric with eigenvalues $\lambda_1 = 6$, $\lambda_2 = 3$, and $\lambda_3 = 1$. Listing the results of the first ten iterations of the power methods presented in Algorithms 8.5 and 8.6 with $\mathbf{y}^{(0)} = \mathbf{x}^{(0)} = (1, 0, 0)^t$ demonstrates the significant improvement obtained by using the latter algorithm. Using Algorithm 8.5, we have the results listed in Table 8.6. The results listed in Table 8.7 are obtained using Algorithm 8.6.   □

TABLE 8.6	m	$(\mathbf{y}^{(m)})^t$	$\mu^{(m)}$	$(\mathbf{x}^{(m)})^t$
	0	—	—	$(1, 0, 0)$
	1	$(4, -1, 1)$	4	$(1, -.25, .25)$
	2	$(4.5, -2.25, 2.25)$	4.5	$(1, -.5, .5)$
	3	$(5, -3.5, 3.5)$	5	$(1, -.7, .7)$
	4	$(5.4, -4.5, 4.5)$	5.4	$(1, -.83\overline{33}, .83\overline{33})$
	5	$(5.6\overline{66}, -5.1\overline{66}, 5.1\overline{66})$	$5.6\overline{66}$	$(1, -.911765, .911765)$
	6	$(5.823529, -5.558824, 5.558824)$	5.823529	$(1, -.954545, .954545)$
	7	$(5.909091, -5.772727, 5.772727)$	5.909091	$(1, -.976923, .976923)$
	8	$(5.953846, -5.884615, 5.884615)$	5.953846	$(1, -.988372, .988372)$
	9	$(5.976744, -5.941861, 5.941861)$	5.976744	$(1, -.994163, .994163)$
	10	$(5.988327, -5.970817, 5.970817)$	5.988327	$(1, -.997076, .997076)$

TABLE 8.7	m	$(\mathbf{y}^{(m)})^t$	$\mu^{(m)}$	$(\mathbf{x}^{(m)})^t$
	0	$(1, 0, 0)$	—	$(1, 0, 0)$
	1	$(4, -1, 1)$	4	$(.942809, -.235702, .235702)$
	2	$(4.242641, -2.121320, 2.121320)$	5	$(.816497, -.408248, .408248)$
	3	$(4.082483, -2.857738, 2.857738)$	5.666667	$(.710669, -.497468, .497468)$
	4	$(3.837613, -3.198011, 3.198011)$	5.909091	$(.646997, -.539164, .539164)$
	5	$(3.666314, -3.342816, 3.342816)$	5.976744	$(.612836, -.558763, .558763)$
	6	$(3.568871, -3.406650, 3.406650)$	5.994152	$(.595247, -.568190, .568190)$
	7	$(3.517370, -3.436200, 3.436200)$	5.998536	$(.586336, -.572805, .572805)$
	8	$(3.490952, -3.450359, 3.450359)$	5.999634	$(.581852, -.575086, .575086)$
	9	$(3.477580, -3.457283, 3.457283)$	5.999908	$(.579603, -.576220, .576220)$
	10	$(3.470854, -3.460706, 3.460706)$	5.999977	$(.578477, -.576786, .576786)$

The following error bound for approximating the eigenvalues of a symmetric matrix can be used as a stopping criteria in Step 7 of Algorithm 8.6. The proof of this result can be found in Isaacson and Keller [52], pages 140–141.

Theorem 8.42   If A is an $n \times n$ symmetric matrix with eigenvalues $\lambda_1, \lambda_2, \ldots, \lambda_n$ and $\|A\mathbf{x} - \lambda\mathbf{x}\|_2 \le \varepsilon$ for some vector $\mathbf{x}$ with $\|\mathbf{x}\|_2 = 1$ and real number λ, then

$$(8.45) \qquad \min_{1 \le i \le n} |\lambda_i - \lambda| \le \varepsilon.$$

The **inverse power method** uses the power method as a basis but gives faster convergence and can often be used to find more than just the dominant eigenvalue. Assume that the matrix A has eigenvalues $\lambda_1, \ldots, \lambda_n$ with linearly independent eigenvectors $\mathbf{v}^{(1)}, \ldots, \mathbf{v}^{(n)}$. Consider the matrix $(A - qI)^{-1}$ where $q \ne \lambda_i$ for $i = 1, 2, \ldots, n$. The eigenvalues of $(A - qI)^{-1}$ are

$$\frac{1}{\lambda_1 - q}, \frac{1}{\lambda_2 - q}, \ldots, \frac{1}{\lambda_n - q}$$

with eigenvectors $\mathbf{v}^{(1)}, \mathbf{v}^{(2)}, \ldots, \mathbf{v}^{(n)}$. Applying the power method to $(A - qI)^{-1}$ gives

$$\mathbf{y}^{(m)} = (A - qI)^{-1}\mathbf{x}^{(m)}, \tag{8.46}$$

$$\mu^{(m)} = y_{p_{m-1}}^{(m)} = \frac{y_{p_{m-1}}^{(m)}}{x_{p_{m-1}}^{(m)}} = \frac{\displaystyle\sum_{j=1}^{n} \alpha_j \frac{1}{(\lambda_j - q)^m} v_{p_{m-1}}^{(j)}}{\displaystyle\sum_{j=1}^{n} \alpha_j \frac{1}{(\lambda_j - q)^{m-1}} v_{p_{m-1}}^{(j)}}, \tag{8.47}$$

and

$$\mathbf{x}^{(m)} = \frac{\mathbf{y}^{(m)}}{y_{p_m}^{(m)}} \tag{8.48}$$

where, at each step, p_m is used to represent an integer for which $|y_{p_m}^{(m)}| = \|\mathbf{y}^{(m)}\|_\infty$. The sequence $\{\mu^{(m)}\}$ in Eq. (8.47) will converge to

$$\frac{1}{\lambda_k - q} = \max_{1 \le i \le n} \frac{1}{|\lambda_i - q|}$$

where λ_k is the eigenvalue of A closest to q.

With k known, Eq. (8.47) can be written as

$$\mu^{(m)} = \frac{1}{\lambda_k - q} \left[\frac{\alpha_k v_{p_{m-1}}^{(k)} + \displaystyle\sum_{\substack{j=1 \\ j \ne k}}^{n} \alpha_j \left[\dfrac{\lambda_k - q}{\lambda_j - q}\right]^m v_{p_{m-1}}^{(j)}}{\alpha_k v_{p_{m-1}}^{(k)} + \displaystyle\sum_{\substack{j=1 \\ j \ne k}}^{n} \alpha_j \left[\dfrac{\lambda_k - q}{\lambda_j - q}\right]^{m-1} v_{p_{m-1}}^{(j)}} \right]. \tag{8.49}$$

Thus the choice of q determines the convergence, provided that $1/(\lambda_k - q)$ is a unique dominant eigenvalue of $(A - qI)^{-1}$ (although it may be a multiple eigenvalue). The closer q is to an eigenvalue λ_k of A, the faster the convergence since the convergence is of order

$$O\left(\left|\frac{\lambda_k - q}{\lambda - q}\right|^m\right)$$

where λ represents the eigenvalue of A that is second closest to q.

The determination of $\mathbf{y}^{(m)}$ is obtained from the equation

$$(A - qI)\mathbf{y}^{(m)} = \mathbf{x}^{(m)}. \tag{8.50}$$

To solve this equation, it is recommended that Gaussian elimination with pivoting be used. Although the method requires the solution of an $n \times n$ linear system at each step, the multipliers can be saved to reduce the computation. The selection of q can be based on the Gerschgorin Theorem, Theorem 8.30 (p. 404), or on any other means of localizing an eigenvalue.

The following algorithm computes q from an initial approximation to the eigenvector $\mathbf{x}^{(0)}$ (see Exercise 24) by

$$q = \frac{\mathbf{x}^{(0)t} A \mathbf{x}^{(0)}}{\mathbf{x}^{(0)t}\mathbf{x}^{(0)}}.$$

If q is close to an eigenvalue, the convergence will be quite rapid, but a pivoting technique should be used in Step 6 to avoid contamination by rounding error.

Inverse Power Method Algorithm 8.7

To approximate an eigenvalue and an associated eigenvector of the $n \times n$ matrix A given a nonzero vector $\mathbf{x}$:

INPUT dimension n; matrix A; vector $\mathbf{x}$; tolerance TOL; maximum number of iterations N.

OUTPUT approximate eigenvalue μ; approximate eigenvector $\mathbf{x}$ (with $\|\mathbf{x}\|_\infty = 1$) or a message that the maximum number of iterations was exceeded.

Step 1 Set $q = \mathbf{x}^t A \mathbf{x} / \mathbf{x}^t \mathbf{x}$.

Step 2 Set $k = 1$;
$\quad\quad \mu_0 = 0$.

Step 3 Find the integer p with $1 \le p \le n$ and $|x_p| = \|\mathbf{x}\|_\infty$.

Step 4 Set $\mathbf{x} = \dfrac{1}{x_p} \mathbf{x}$.

Step 5 While ($k \le N$) do Steps 6–11.

Step 6 Solve the linear system $(A - qI)\mathbf{y} = \mathbf{x}$.
$\quad\quad$ If the system does not have a unique solution, then
$\quad\quad\quad\quad$ OUTPUT ('q is an eigenvalue', q);
$\quad\quad\quad\quad$ STOP.

Step 7 Set $\mu = y_p$.

Step 8 Find an integer p with $1 \le p \le n$ and $|y_p| = \|\mathbf{y}\|_\infty$.

Step 9 Set $\mathbf{x} = \dfrac{1}{y_p} \mathbf{y}$.

Step 10 If $|\mu - \mu_0| < TOL$ then set $\mu = (1/\mu) + q$;
$\quad\quad\quad\quad$ OUTPUT ($\mu, \mathbf{x}$);
$\quad\quad\quad\quad$ (*Procedure completed successfully.*)
$\quad\quad\quad\quad$ STOP.

Step 11 Set $k = k + 1$;
$\quad\quad\quad$ $\mu_0 = \mu$.

Step 12 OUTPUT ('Maximum number of iterations exceeded');
$\quad\quad$ (*Procedure completed unsuccessfully.*)
$\quad\quad$ STOP.

Algorithm 8.7 with pivoting is often used to approximate an eigenvector when an approximate eigenvalue is known. In this case, q is the approximate eigenvalue.

Since the convergence of the inverse power method is linear, Aitken's Δ^2 procedure can be used to speed convergence. The following example illustrates the fast convergence of the inverse power method if q is close to an eigenvalue. To obtain the other eigenvalues for this example, see Exercise 11.

EXAMPLE 5   The matrix

$$A = \begin{bmatrix} -4 & 14 & 0 \\ -5 & 13 & 0 \\ -1 & 0 & 2 \end{bmatrix}$$

was considered in Example 3. Algorithm 8.5 gave the approximation $\mu^{(12)} = 6.000837$ using $\mathbf{x}^{(0)} = (1, 1, 1)^t$. With $\mathbf{x}^{(0)} = (1, 1, 1)^t$, we have

$$q = \frac{\mathbf{x}^{(0)t}A\mathbf{x}^{(0)}}{\mathbf{x}^{(0)t}\mathbf{x}^{(0)}} = \frac{19}{3} = 6.333333.$$

The results of applying Algorithm 8.7 are listed in Table 8.8.   ☐

TABLE 8.8

m	$\mathbf{x}^{(m)t}$	$\mu^{(m)}$
0	(1, 1, 1)	
1	(1, .720727, −.194042)	6.183183
2	(1, .715518, −.245052)	6.017244
3	(1, .714409, −.249522)	6.001719
4	(1, .714298, −.249953)	6.000175
5	(1, .714287, −.250000)	6.000021
6	(1, .714286, −.249999)	6.000005

Numerous techniques are available for obtaining approximations to the other eigenvalues of a matrix once an approximation to the dominant eigenvalue has been computed. We will restrict our presentation to **deflation techniques**. Other techniques are considered in Exercises 16 and 17.

Deflation techniques involve forming a new matrix B from the original matrix A whose eigenvalues are the same as those of A with the exception that the dominant eigenvalue of A is replaced by the eigenvalue 0 in B.

Theorem 8.43   Suppose that $\lambda_1, \lambda_2, \ldots, \lambda_n$ are eigenvalues of A with associated eigenvectors $\mathbf{v}^{(1)}, \mathbf{v}^{(2)}, \ldots, \mathbf{v}^{(n)}$, and that λ_1 has multiplicity one. If $\mathbf{x}$ is any vector with the property that $\mathbf{x}^t\mathbf{v}^{(1)} = 1$, then the matrix

(8.51) $$B = A - \lambda_1\mathbf{v}^{(1)}\mathbf{x}^t$$

has eigenvalues $0, \lambda_2, \lambda_3, \ldots, \lambda_n$ with associated eigenvectors $\mathbf{v}^{(1)}, \mathbf{w}^{(2)}, \mathbf{w}^{(3)}, \ldots, \mathbf{w}^{(n)}$ where $\mathbf{v}^{(i)}$ and $\mathbf{w}^{(i)}$ are related by the equation

(8.52) $$\mathbf{v}^{(i)} = (\lambda_i - \lambda_1)\mathbf{w}^{(i)} + \lambda_1(\mathbf{x}^t\mathbf{w}^{(i)})\mathbf{v}^{(1)}$$

for each $i = 2, 3, \ldots, n$.

The proof of this result can be found in Wilkinson [91], page 596.

A particularly useful deflation technique, called the **Wielandt's deflation** procedure, results by defining

$$
(8.53) \qquad \mathbf{x} = \frac{1}{\lambda_1 v_i^{(1)}} \begin{bmatrix} a_{i1} \\ a_{i2} \\ \vdots \\ a_{in} \end{bmatrix}
$$

where $v_i^{(1)}$ is a coordinate of $\mathbf{v}^{(1)}$ which is nonzero, and the values $a_{i1}, a_{i2}, \ldots, a_{in}$ are the entries in the ith row of A.

With this definition, we have

$$
\mathbf{x}^t \mathbf{v}^{(1)} = \frac{1}{\lambda_1 v_i^{(1)}} [a_{i1}, a_{i2}, \ldots, a_{in}](v_1^{(1)}, v_2^{(1)}, \ldots, v_n^{(1)})^t = \frac{1}{\lambda_1 v_i^{(1)}} \sum_{j=1}^n a_{ij} v_j^{(1)}
$$

where the sum is the ith coordinate of the product $A\mathbf{v}^{(1)}$. Since $A\mathbf{v}^{(1)} = \lambda_1 \mathbf{v}^{(1)}$, this implies that

$$
\mathbf{x}^t \mathbf{v}^{(1)} = \frac{1}{\lambda_1 v_i^{(1)}} (\lambda_1 v_i^{(1)}) = 1;
$$

so $\mathbf{x}$ does satisfy the hypotheses of Theorem 8.43. Moreover, (see Exercise 19) the ith row of $B = A - \lambda_1 \mathbf{v}^{(1)} \mathbf{x}^t$ consists entirely of zero entries. If $\lambda \neq 0$ is any eigenvalue with associated eigenvector $\mathbf{w}$, the relation $B\mathbf{w} = \lambda \mathbf{w}$ implies that the ith coordinate of $\mathbf{w}$ must also be zero. Consequently, the ith column of the matrix B makes no contribution to the product $B\mathbf{w} = \lambda \mathbf{w}$. Thus, the matrix B can be replaced by an $(n-1) \times (n-1)$ matrix B' that is obtained by deleting the ith row and column from B. B' will have eigenvalues $\lambda_2, \lambda_3, \ldots, \lambda_n$. If $|\lambda_2| > |\lambda_3|$, the power method can be reapplied to the matrix B' to determine this new dominant eigenvalue and an eigenvector, $\mathbf{w}^{(2)'}$, associated with λ_2, with respect to the matrix B'. To find the associated eigenvector for the original matrix A simply requires constructing $\mathbf{w}^{(2)}$ by inserting a zero coordinate between the coordinates $w_{i-1}^{(2)'}$ and $w_i^{(2)'}$ of the $(n-1)$-dimensional vector $\mathbf{w}^{(2)'}$ and then calculating $\mathbf{v}^{(2)}$ by the use of Eq. (8.52).

EXAMPLE 6   From Example 4 we know that the matrix

$$
A = \begin{bmatrix} 4 & -1 & 1 \\ -1 & 3 & -2 \\ 1 & -2 & 3 \end{bmatrix}
$$

has eigenvalues $\lambda_1 = 6$, $\lambda_2 = 3$, and $\lambda_3 = 1$. Assuming that the dominant eigenvalue $\lambda_1 = 6$ and associated unit eigenvector $\mathbf{v}^{(1)} = (1, -1, 1)^t$ have been calculated, the procedure outlined above for obtaining λ_2 proceeds as follows:

$$
\mathbf{x} = \frac{1}{6} \begin{bmatrix} 4 \\ -1 \\ 1 \end{bmatrix} = (\tfrac{2}{3}, -\tfrac{1}{6}, \tfrac{1}{6})^t,
$$

$$
\mathbf{v}^{(1)} \mathbf{x}^t = \begin{bmatrix} 1 \\ -1 \\ 1 \end{bmatrix} [\tfrac{2}{3}, -\tfrac{1}{6}, \tfrac{1}{6}] = \begin{bmatrix} \tfrac{2}{3} & -\tfrac{1}{6} & \tfrac{1}{6} \\ -\tfrac{2}{3} & \tfrac{1}{6} & -\tfrac{1}{6} \\ \tfrac{2}{3} & -\tfrac{1}{6} & \tfrac{1}{6} \end{bmatrix},
$$

and
$$B = A - \lambda_1 \mathbf{v}^{(1)} \mathbf{x}^t = \begin{bmatrix} 4 & -1 & 1 \\ -1 & 3 & -2 \\ 1 & -2 & 3 \end{bmatrix} - 6 \begin{bmatrix} \frac{2}{3} & -\frac{1}{6} & \frac{1}{6} \\ -\frac{2}{3} & \frac{1}{6} & -\frac{1}{6} \\ \frac{2}{3} & -\frac{1}{6} & \frac{1}{6} \end{bmatrix}$$

$$= \begin{bmatrix} 0 & 0 & 0 \\ 3 & 2 & -1 \\ -3 & -1 & 2 \end{bmatrix}.$$

Deleting the first row and column gives

$$B' = \begin{bmatrix} 2 & -1 \\ -1 & 2 \end{bmatrix},$$

which has eigenvalues $\lambda_2 = 3$ and $\lambda_3 = 1$. For $\lambda_2 = 3$ the eigenvector, $\mathbf{w}^{(2)'}$, can be obtained by solving the second order linear system

$$(B' - 3I)\mathbf{w}^{(2)'} = \mathbf{0},$$

resulting in

$$\mathbf{w}^{(2)'} = (1, -1)^t.$$

Thus, $\mathbf{w}^{(2)} = (0, 1, -1)^t$ and, from Eq. (8.52), we have:

$$\mathbf{v}^{(2)} = (3 - 6)(0, 1, -1)^t + 6((\tfrac{2}{3}, -\tfrac{1}{6}, \tfrac{1}{6})(0, 1, -1)^t)(1, -1, 1)^t$$
$$= (-2, -1, 1)^t. \qquad \square$$

Although this deflation process can in some cases be used to find approximations to all of the eigenvalues and eigenvectors of a matrix, the process is very susceptible to rounding error. Techniques based on similarity transformations will be presented in the next section; these methods are generally preferable when approximations to all eigenvalues and eigenvectors are needed.

We close this section with an algorithm for the calculation of the second most dominant eigenvalue and associated eigenvector for a matrix, once the dominant eigenvalue and associated eigenvector have been determined.

Wielandt's Deflation Algorithm 8.8

To approximate the second most dominant eigenvalue and an associated eigenvector of the $n \times n$ matrix A given an approximation λ to the dominant eigenvalue, an approximation $\mathbf{v}$ to a corresponding eigenvector and a vector $\mathbf{x} \in R^{n-1}$:

INPUT dimension n; matrix A; approximate eigenvalue λ; approximate eigenvector $\mathbf{v} \in R^n$; vector $\mathbf{x} \in R^{n-1}$.

OUTPUT approximate eigenvalue μ; approximate eigenvector $\mathbf{u}$ or a message that the method fails.

Step 1 Let i be the smallest integer with $1 \le i \le n$ and $|v_i| = \max\limits_{1 \le j \le n} |v_j|$.

Step 2 If $i \ne 1$ then
for $k = 1, \ldots, i - 1$
for $j = 1, \ldots, i - 1$

set $b_{kj} = a_{kj} - \dfrac{v_k}{v_i} a_{ij}.$

Step 3 If $i \neq 1$ and $i \neq n$ then
for $k = i, \ldots, n-1$
for $j = 1, \ldots, i-1$

$$\text{set } b_{kj} = a_{k+1,j} - \frac{v_{k+1}}{v_i} a_{ij};$$

$$b_{jk} = a_{j,k+1} - \frac{v_j}{v_i} a_{i,k+1}.$$

Step 4 If $i \neq n$ then
for $k = i, \ldots, n-1$
for $j = i, \ldots, n-1$

$$\text{set } b_{kj} = a_{k+1,j+1} - \frac{v_{k+1}}{v_i} a_{i,j+1}.$$

Step 5 Perform the power method on the $(n-1) \times (n-1)$ matrix $B' = (b_{kj})$ with $\mathbf{x}$ as initial approximation.
If the method fails, then OUTPUT ('Method fails');
STOP.
else let μ be the approximate eigenvalue and
$\mathbf{w}' = (w'_1, \ldots, w'_{n-1})^t$ the approximate eigenvector.

Step 6 If $i \neq 1$ then for $k = 1, \ldots, i-1$ set $w_k = w'_k$.

Step 7 Set $w_i = 0$.

Step 8 If $i \neq n$ then for $k = i+1, \ldots, n$ set $w_k = w'_{k-1}$.

Step 9 For $k = 1, \ldots, n$

$$\text{set } u_k = (\mu - \lambda)w_k + \left(\sum_{j=1}^{n} a_{ij}w_j \right) \frac{v_k}{v_i}.$$

(Compute eigenvector using Eq. 8.52).

Step 10 OUTPUT $(\mu, \mathbf{u})$;
(Procedure completed successfully.)
STOP.

Exercise Set 8.4

1. Find the first three iterations obtained by the power method applied to the following matrices:

a) $\begin{bmatrix} 1 & -1 & 0 \\ -2 & 4 & -2 \\ 0 & -1 & 1 \end{bmatrix}$ use $\mathbf{x}^{(0)} = (1, 0, 0)^t$

b) $\begin{bmatrix} 1 & -1 & 0 \\ -2 & 4 & -2 \\ 0 & -1 & 2 \end{bmatrix}$ use $\mathbf{x}^{(0)} = (1, 0, 0)^t$

c) $\begin{bmatrix} 4 & 1 & 1 & 1 \\ 1 & 3 & -1 & 1 \\ 1 & -1 & 2 & 0 \\ 1 & 1 & 0 & 2 \end{bmatrix}$ use $\mathbf{x}^{(0)} = (1, 1, 1, 1)^t$

2. Repeat Exercise 1 using the inverse power method.

3. Find the first three iterations obtained by the symmetric power method applied to the following matrices:

a) $\begin{bmatrix} 2 & -1 & 0 \\ -1 & 2 & -1 \\ 0 & -1 & 2 \end{bmatrix}$ use $\mathbf{x}^{(0)} = (1, 0, 0)^t$

b) $\begin{bmatrix} 4 & 1 & -1 & 0 \\ 1 & 3 & -1 & 0 \\ -1 & -1 & 5 & 2 \\ 0 & 0 & 2 & 4 \end{bmatrix}$ use $\mathbf{x}^{(0)} = (0, 1, 0, 0)^t$

c) $\begin{bmatrix} 4 & 1 & 1 & 1 \\ 1 & 3 & -1 & 1 \\ 1 & -1 & 2 & 0 \\ 1 & 1 & 0 & 2 \end{bmatrix}$ use $\mathbf{x}^{(0)} = (1, 0, 0, 0)^t$

4. Show that the matrix

$$A = \begin{bmatrix} 1 & 0 & 0 \\ -1 & 0 & 1 \\ -1 & -1 & 2 \end{bmatrix}$$

has fewer than three linearly independent eigenvectors.

5. a) Let B be any $n \times n$ matrix. Show that det $B = $ det B^t.
   b) Show that A and A^t have the same eigenvalues.

6. Use Theorem 8.30 to determine bounds for the eigenvalues of the following matrices.

a) $\begin{bmatrix} 1 & 0 & 0 \\ -1 & 0 & 1 \\ -1 & -1 & 2 \end{bmatrix}$ b) $\begin{bmatrix} 4 & -1 & 0 \\ -1 & 4 & -1 \\ -1 & -1 & 4 \end{bmatrix}$ c) $\begin{bmatrix} 3 & 2 & 1 \\ 2 & 3 & 0 \\ 1 & 0 & 3 \end{bmatrix}$

7. Show that any four vectors in R^3 are linearly dependent.

8. Show that if A is an $n \times n$ matrix with n distinct eigenvalues, then A has n linearly independent eigenvectors.

9. Use the power method to compute the largest eigenvalue (in absolute value) of the following matrices to within .0001, or until $N > 25$.

a) $\begin{bmatrix} 4 & 0 & 0 \\ -1 & 2 & 0 \\ -2 & -1 & -3 \end{bmatrix}$ b) $\begin{bmatrix} 1 & 1 & 3 \\ 1 & -2 & 1 \\ 3 & 1 & 3 \end{bmatrix}$ c) $\begin{bmatrix} 1 & 1 & .5 \\ 1 & 1 & .25 \\ .5 & .25 & 2 \end{bmatrix}$.

10. Repeat Exercise 9(b) and 9(c) using Algorithm 8.6. Compare the rates of convergence.

11. Apply the inverse power method to the matrix A in Example 5 using $q = 7$, 4, and 0. Explain the results.

12. Repeat Exercise 9 using Algorithm 8.7. Compare the rates of convergence. Can the other eigenvalues be obtained using Algorithm 8.7?

13. Let $\mathbf{v}_1, \ldots, \mathbf{v}_k$ be k orthogonal vectors. Show that $\{\mathbf{v}_1, \ldots, \mathbf{v}_k\}$ is a linearly independent set.

14. Let P be an orthogonal matrix. Show that the columns of P form an orthonormal set of vectors. Also, show that $\|P\|_\infty \le 1$ and $\|P^t\|_\infty \le 1$.

15. Let $\mathbf{v}_1, \ldots, \mathbf{v}_n$ be n orthonormal vectors in R^n. Let $\mathbf{x} \in R^n$. Show that $\mathbf{x} = \sum_{k=1}^{n} c_k \mathbf{v}_k$ where $c_k = \mathbf{v}_k^t \mathbf{x}$.

16. Let the eigenvalues of A be $\lambda_1 \ge \cdots \ge \lambda_{n-1} > \lambda_n$ or $\lambda_1 \le \cdots \le \lambda_{n-1} < \lambda_n$ with λ_1 dominant.

    a) Show that the eigenvalue λ_n can be approximated by applying the power method to $cI - A$ where c is any number such that $|c| > |\lambda_1|$ and c and λ_1 have the same sign.
    b) Apply this technique to the matrices in Exercise 9.

17. **Annihilation Technique** Suppose the $n \times n$ matrix A has eigenvalues $\lambda_1, \ldots, \lambda_n$ ordered by

$$|\lambda_1| > |\lambda_2| > |\lambda_3| \ge \cdots \ge |\lambda_n|$$

    with linearly independent eigenvectors $\mathbf{v}^{(1)}, \mathbf{v}^{(2)}, \ldots, \mathbf{v}^{(n)}$.

    a) Show that if the power method is applied with an initial vector $\mathbf{x}^{(0)}$ given by

$$\mathbf{x}^{(0)} = \alpha_2 \mathbf{v}^{(2)} + \alpha_3 \mathbf{v}^{(3)} + \cdots + \alpha_n \mathbf{v}^{(n)},$$

    then the sequence $\{\mu^{(m)}\}$ described in Algorithm 8.5, will converge to λ_2.
    b) Show that, for any vector $\mathbf{x} = \sum_{i=1}^{n} \alpha_i \mathbf{v}^{(i)}$, the vector $\mathbf{x}^{(0)} = (A - \lambda_1 I)\mathbf{x}$ satisfies the property given in part (a).
    c) Obtain an approximation to λ_2 for the matrices in Exercise 9.
    d) Show that this method may be continued to find λ_3 using $\mathbf{x}^{(0)} = (A - \lambda_2 I)(A - \lambda_1 I)\mathbf{x}$.

18. **Hotelling's Deflation** Assume that the largest eigenvalue λ_1 in magnitude and an associated eigenvector $\mathbf{v}^{(1)}$ have been obtained for the $n \times n$ symmetric matrix A. Show that the matrix B,

$$B = A - \frac{\lambda_1}{(\mathbf{v}^{(1)})^t \mathbf{v}^{(1)}} \mathbf{v}^{(1)} (\mathbf{v}^{(1)})^t,$$

    has the same eigenvalues $\lambda_2, \ldots, \lambda_n$ as A, except that B has eigenvalue 0 with eigenvector $\mathbf{v}^{(1)}$ instead of eigenvalue λ_1. Use this deflation method to find λ_2 for each matrix in Exercise 9. Theoretically, this method may be continued to find more eigenvalues, but rounding error soon makes the effort worthless.

19. Show that the ith row of $B = A - \lambda_1 \mathbf{v}^{(1)} \mathbf{x}^t$ is zero where λ_1 is the largest eigenvalue of A in absolute value, $\mathbf{v}^{(1)}$ is the associated eigenvector of A for λ_1, and $\mathbf{x}$ is the vector defined in Eq. (8.53).

20. Apply Wielandt's deflation to the matrices in Exercise 9.

21. Following along the line of Exercise 11 in Section 6.3 and Exercise 27 in Section 8.1, suppose that a species of beetle has a life span of four years, that a female in the first year has a survival rate of $\frac{1}{2}$, in the second year a survival rate of $\frac{1}{4}$, and in the third year a survival rate of $\frac{1}{8}$. Suppose additionally that a female gives birth, on the average, to two new females in the third year and to four new females in the fourth year. The matrix describing a single female's contribution in one year to the female population in the succeeding year is

$$A = \begin{bmatrix} 0 & 0 & 2 & 4 \\ \frac{1}{2} & 0 & 0 & 0 \\ 0 & \frac{1}{4} & 0 & 0 \\ 0 & 0 & \frac{1}{8} & 0 \end{bmatrix},$$

where again the entry in the ith row and jth column denotes the probabilistic contribution that a female of age j makes on the next year's female population of age i.

a) Use the Gerschgorin Circle Theorem 8.30 to determine a region in the complex plane containing all the eigenvalues of A.

b) Use the power method to determine the dominant eigenvalue of this matrix and its associated eigenvector.

c) Use Algorithm 8.8 to determine any remaining eigenvalues and eigenvectors of A.

d) Find the eigenvalues of A by using the characteristic polynomial of A and the Newton–Raphson method.

e) What is your long-range prediction for the population of these beetles?

22. A **persymmetric matrix** is a matrix that is symmetric about both diagonals; that is, an $N \times N$ matrix $A = (a_{ij})$ is persymmetric if $a_{ij} = a_{ji} = a_{N+1-i, N+1-j}$ for all $i = 1, 2, \ldots, N$ and $j = 1, 2, \ldots, N$. A number of problems in communication theory have solutions that involve the eigenvalues and eigenvectors of matrices that are in persymmetric form. For example, the eigenvector corresponding to the minimal eigenvalue of the 4×4 persymmetric matrix

$$A = \begin{bmatrix} 2 & -1 & 0 & 0 \\ -1 & 2 & -1 & 0 \\ 0 & -1 & 2 & -1 \\ 0 & 0 & -1 & 2 \end{bmatrix}$$

gives the unit energy-channel impulse response for a given error sequence of length 2, and subsequently the minimum weight of any possible error sequence.

a) Use the Gerschgorin Circle Theorem (Theorem 8.30) to show that if A is the matrix given above and λ is its minimal eigenvalue, then $|\lambda - 4| = \rho(A - 4I)$ where ρ denotes the spectral radius.

b) Find the minimal eigenvalue of the matrix A by finding all the eigenvalues of $A - 4I$ and computing its spectral radius. Then find the corresponding eigenvector.

c) Find the minimal eigenvalue of the matrix A by applying the power method to $A - 4I$.

d) Use the Gerschgorin Circle Theorem to show that if λ is the minimal eigenvalue of the matrix

$$B = \begin{bmatrix} 3 & -1 & -1 & 1 \\ -1 & 3 & -1 & -1 \\ -1 & -1 & 3 & -1 \\ 1 & -1 & -1 & 3 \end{bmatrix},$$

then $|\lambda - 6| = \rho(B - 6I)$.

e) Try to repeat parts (b) and (c) of this exercise using the matrix B and the result in part (d). What difficulties arise?

23. A linear dynamical system can be represented by the equations

$$\frac{d\mathbf{x}}{dt} = A(t)\mathbf{x}(t) + B(t)\mathbf{u}(t), \qquad \mathbf{y}(t) = C(t)\mathbf{x}(t) + D(t)\mathbf{u}(t)$$

where A is an $n \times n$ variable matrix, B an $n \times r$ variable matrix, C an $m \times n$ variable matrix, D an $m \times r$ variable matrix, $\mathbf{x}$ an n-dimensional vector variable, $\mathbf{y}$ an m-dimensional vector variable, and $\mathbf{u}$ an r-dimensional vector variable. In order for the system to be stable, the matrix A must have all of its eigenvalues with nonpositive real part for all t.

a) If

$$A(t) = \begin{bmatrix} -1 & 2 & 0 \\ -2.5 & -7 & 4 \\ 0 & 0 & -5 \end{bmatrix},$$

is the system stable?

b) If

$$A(t) = \begin{bmatrix} -1 & 1 & 0 & 0 \\ 0 & -2 & 1 & 0 \\ 0 & 0 & -5 & 1 \\ -1 & -1 & -2 & -3 \end{bmatrix},$$

is the system stable?

24. Show that if λ is an eigenvalue of the matrix A with eigenvector $\mathbf{x}$ then

$$\lambda = \mathbf{x}^t A \mathbf{x} / \mathbf{x}^t \mathbf{x}.$$

8.5 Householder's Method and the QL Algorithm

The deflation methods discussed in the previous section are not generally suitable for calculating all the eigenvalues of a matrix because of rounding error growth. In this section we will consider a matrix reduction technique that is designed to determine all the eigenvalues of a symmetric matrix. The method is called the **QL Algorithm**, and, because of its stability characteristics, it is the technique most often used in practice. The method can be modified for use in computing the eigenvalues of nonsymmetric matrices; these modifications are discussed briefly at the end of this section. Further remarks on this topic, as well as a complete general discussion, can be found in Wilkinson [91] and Wilkinson and Reinsch [92].

To apply the QL algorithm, the symmetric matrix must be in tridiagonal form; that is, the only nonzero entries in the matrix lie either on the diagonal or on the subdiagonals directly above or below the diagonal. Since this is not necessarily the form of a symmetric matrix, the first problem is to find a tridiagonal symmetric matrix whose eigenvalues agree with those of the original matrix. For this purpose we use a technique known as **Householder's method**.

Suppose that A is an aribtrary $n \times n$ symmetric matrix. Theorem 8.39 implies that an orthogonal matrix Q exists with the property that $T = QAQ^{-1} = QAQ^t$ where T is a diagonal matrix. If the matrix Q, and hence T, was easy to determine, the eigenvalue problem would be solved since the eigenvalues of T lie along its diagonal and agree with the eigenvalues of A. However, the matrix Q is not easily obtained although methods are available for finding the eigenvalues of A by using this diagonalization (see the Jacobi method discussion in Wilkinson [91], for example). Instead of attempting to determine the diagonal matrix T, Householder's method computes a tridiagonal matrix with the same eigenvalues as A. The QL Algorithm can be applied to the tridiagonal matrix to resolve the problem.

Householder's method begins by determining a matrix $P^{(1)}$ with $P^{(1)} = [P^{(1)}]^{-1}$ and with the additional property that the matrix $A^{(2)} = P^{(1)}AP^{(1)}$ has entries $a_{n,j}^{(2)} = a_{j,n}^{(2)} = 0$ for each $j = 1, 2, \ldots, n - 2$. The tridiagonal matrix is produced by repeating this procedure $n - 3$ times, where the kth step is described by

$$A^{(k+1)} = P^{(k)}A^{(k)}P^{(k)}$$

for a matrix $P^{(k)}$ chosen to have the properties

(8.54) $$[P^{(k)}]^{-1} = P^{(k)}$$

and

(8.55) $a_{n-k+1,j}^{(k+1)} = a_{j,n-k+1}^{(k+1)} = 0$ for each $j = 1, 2, \ldots, n - k - 1$.

The matrix $A^{(k)}$ will be tridiagonal in its last $k - 1$ rows and columns and has the form shown in (8.56) where, for convenience of notation, we have let $i = n - k + 2$.

(8.56) $$A^{(k)} = \begin{bmatrix} a_{11} & \cdots & a_{1,i-2} & a_{1,i-1} & 0 & \cdots\cdots\cdots\cdots\cdots & 0 \\ \vdots & & \vdots & \vdots & \vdots & & \\ & & & & 0 & & \\ a_{i-1,1} & \cdots & a_{i-1,i-2} & a_{i-1,i-1} & a_{i-1,i} & & \\ 0 & \cdots & 0 & a_{i,i-1} & a_{i,i} & a_{i,i+1} & 0 \\ \vdots & & & & & & a_{n-1,n} \\ 0 & \cdots\cdots\cdots\cdots\cdots\cdots\cdots & 0 & a_{n,n-1} & a_{n,n} \end{bmatrix}$$

To illustrate the construction of $P^{(k)}$, suppose that $A^{(k)}$ has been formed as shown in Eq. (8.56). Set

$$P^{(k)} = I - 2ww^t$$

where w is a vector with the property that $w^t w = 1$. Then $P^{(k)}$ has the property that (see Exercise 7)

$$[P^{(k)}]^2 = [I - 2ww^t]^2 = I.$$

and hence $[P^{(k)}]^{-1} = P^{(k)}$. This ensures that $P^{(k)}$ satisfies Eq. (8.54). To guarantee that Eq. (8.55) also holds, we choose $w_j = 0$ for each $j = i - 1, \ldots, n$ and w_j for for $j = 1, 2, \ldots, i - 2$ so that

$$[A^{(k+1)}]_{n-k+1,j} = [A^{(k+1)}]_{j,n-k+1} = 0$$ for each $j = 1, 2, \ldots, i - 2$.

The precise procedure for the choice of $w_1, w_2, \ldots, w_{i-2}$ is detailed in the following algorithm and requires solving a system of nonlinear equations in a way that roundoff error is minimized.

Computational accuracy requires (see Wilkinson and Reinsch [92]) that the larger diagonal elements be initially placed toward the upper left corner of the matrix. This is accomplished by simply interchanging appropriate rows and columns of the matrix.

Householder Algorithm 8.9

To obtain a symmetric tridiagonal matrix A_{n-1} similar to the symmetric matrix $A = A_1$ construct the following matrices $A_2, A_3, \ldots, A_{n-1}$ where $A_k = a_{i,j}^{(k)}$ for each $k = 1, 2, \ldots, n - 1$.

INPUT dimension n; matrix A.

OUTPUT A_{n-1}. (*Could over-write A.*)

Step 1 For $k = 1, 2, \ldots, n - 2$ do Steps 2–10.

Step 2 Set $i = n - k + 2$.

Step 3 Set $q = \left(\sum\limits_{j=1}^{i-2} [a_{i-1,j}^{(k)}]^2 \right)^{1/2}$.

Step 4 If $a_{i-1,i-2}^{(k)} = 0$ then set $s = q$
  else set $s = qa_{i-1,i-2}^{(k)}/|a_{i-1,i-2}^{(k)}|$.

Step 5 Set $RSQ = (q^2 + sa_{i-1,i-2}^{(k)})$.   (*Note:* $RSQ = 2r^2$)

Step 6 Set $v_{i-1} = 0$.   (*Note:* $v_i = \cdots = v_n = 0$, *but are not needed.*)
  Set $v_{i-2} = a_{i-1,i-2}^{(k)} + s$.
  If $i > 3$ then for $j = 1, \ldots, i - 3$ set $v_j = a_{i-1,j}^{(k)}$.

$$\left(Note: \mathbf{w} = \frac{1}{\sqrt{2RSQ}}\, \mathbf{v} = \frac{1}{2r}\, \mathbf{v} \right)$$

Step 7 For $j = 1, 2, \ldots, i - 1$ set $u_j = \dfrac{1}{RSQ} \sum\limits_{l=1}^{i-2} a_{jl}^{(k)} v_l$.

$$\left(Note: \mathbf{u} = \frac{1}{RSQ}\, A_k \mathbf{v} = \frac{1}{2r^2}\, A_k \mathbf{v} \right)$$

Step 8 Set $PROD = \sum\limits_{j=1}^{i-2} v_j u_j$.

$$\left(Note: PROD = \mathbf{v}^t \mathbf{u} = \frac{1}{2r^2}\, \mathbf{v}^t A_k \mathbf{v} \right)$$

Step 9 For $j = 1, 2, \ldots, i - 1$ set $z_j = u_j - \dfrac{PROD}{2RSQ}\, v_j$.

$$\left(Note: \mathbf{z} = \mathbf{u} - \frac{1}{2RSQ}\, \mathbf{v}^t \mathbf{u}\mathbf{v} = \mathbf{u} - \frac{1}{4r^2}\, \mathbf{v}^t \mathbf{u}\mathbf{v} = \mathbf{u} - \mathbf{w}\mathbf{w}^t \mathbf{u} \right.$$
$$\left. = \frac{1}{r}\, A_k \mathbf{w} - \mathbf{w}\mathbf{w}^t \frac{1}{r}\, A_k \mathbf{w} \right)$$

Step 10 For $l = 2, \ldots, i - 2$ (*Note: Compute* $A_{k+1} = A_k - \mathbf{v}\mathbf{z}^t - \mathbf{z}\mathbf{v}^t$
  $= (I - 2\mathbf{w}\mathbf{w}^t)A_k(I - 2\mathbf{w}\mathbf{w}^t).$)

  for $j = 1, \ldots, l - 1$
    set $a_{lj}^{(k+1)} = a_{lj}^{(k)} - v_l z_j - v_j z_l$;
      $a_{jl}^{(k+1)} = a_{lj}^{(k+1)}$;
    set $a_{ll}^{(k+1)} = a_{ll}^{(k)} - 2v_l z_l$;
  set $a_{11}^{(k+1)} = a_{11}^{(k)} - 2v_1 z_1$;
  for $j = 1, \ldots, i - 3$ set $a_{i-1,j}^{(k+1)} = 0$; $a_{j,i-1}^{(k+1)} = 0$;
  set $a_{i-1,i-2}^{(k+1)} = a_{i-1,i-2}^{(k)} - v_{i-2} z_{i-1}$;
    $a_{i-2,i-1}^{(k+1)} = a_{i-1,i-2}^{(k+1)}$.
  (*Note: The other elements of* A_{k+1} *are the same as* A_k.)

Step 11 OUTPUT (A_{n-1});
  (*The process is complete.* A_{n-1} *is symmetric, tridiagonal and similar to* A.)
  STOP.

EXAMPLE 1   The 4×4 matrix

$$A_1 = \begin{bmatrix} 4 & -2 & 1 & 2 \\ -2 & 3 & 0 & -2 \\ 1 & 0 & 2 & 1 \\ 2 & -2 & 1 & -1 \end{bmatrix}$$

is symmetric. To use Algorithm 8.9 to transform this matrix into a matrix that is tridiagonal, symmetric and has the same eigenvalues as A, we perform the following calculations:

Set $k = 1$.

Set $i = 5$;

$$q = \left(\sum_{j=1}^{3} [a_{4,j}^{(1)}]^2 \right)^{1/2} = 3;$$

$$s = 3a_{4,3}^{(1)}/|a_{4,3}^{(1)}| = 3;$$

$$RSQ = (q^2 + sa_{4,3}^{(1)}) = 12;$$

$$\mathbf{v} = (2, -2, 4, 0)^t;$$

$$\mathbf{u} = (1/RSQ)A\mathbf{v} = (\tfrac{4}{3}, -\tfrac{5}{6}, \tfrac{5}{6}, 1)^t;$$

$$PROD = \tfrac{23}{3};$$

$$\mathbf{z} = \mathbf{u} - (PROD/2RSQ)\mathbf{v} = (\tfrac{25}{36}, -\tfrac{7}{36}, -\tfrac{4}{9}, 1)^t;$$

$$A_2 = \begin{bmatrix} \tfrac{11}{9} & -\tfrac{2}{9} & -\tfrac{8}{9} & 0 \\ -\tfrac{2}{9} & \tfrac{20}{9} & -\tfrac{1}{9} & 0 \\ -\tfrac{8}{9} & -\tfrac{1}{9} & \tfrac{50}{9} & -3 \\ 0 & 0 & -3 & -1 \end{bmatrix}.$$

Set $k = 2$.

Set $i = 4$;

$$q = \left(\sum_{j=1}^{2} [a_{3,j}^{(2)}]^2 \right)^{1/2} = .8958064164;$$

$$s = qa_{3,2}^{(2)}/|a_{3,2}^{(2)}| = -.8958064164;$$

$$RSQ = (q^2 + sa_{3,2}^{(2)}) = .9020031820;$$

$$\mathbf{v} = (-.8888888889, -1.006917528, 0, 0)^t;$$

$$\mathbf{u} = (1/RSQ)A_2\mathbf{v} = (-.9563827706, -2.261703377, 1, 0)^t;$$

$$PROD = \mathbf{v}^t\mathbf{u} = 3.127466791;$$

$$\mathbf{z} = \mathbf{u} - (PROD/2RSQ)\mathbf{v} = (.5846153854, -.5160880660, 1, 0)^t;$$

$$A_3 = \begin{bmatrix} 2.261538463 & -.0923076911 & 0 & 0 \\ -.0923076911 & 1.182905983 & .8958064169 & 0 \\ 0 & .8958064169 & 5.555555556 & -3 \\ 0 & 0 & -3 & -1 \end{bmatrix}.$$

The process is complete. The matrix A_3 is tridiagonal and has the same eigenvalues as A, except for roundoff error. $\square$

In the remainder of this section it will be assumed that the symmetric matrix for which the eigenvalues are to be calculated is tridiagonal. When this is not the case, Householder's method will first be employed. If we let A denote a matrix of this type, we can simplify the notation somewhat by labeling the entries of A as follows:

$$A = \begin{bmatrix} a_1 & b_2 & 0 & \cdots & \cdots & 0 \\ b_2 & a_2 & b_3 & & & \vdots \\ 0 & b_3 & a_3 & & & \\ \vdots & & & & & 0 \\ \vdots & & & & & b_n \\ 0 & \cdots & \cdots & 0 & b_n & a_n \end{bmatrix}$$

The first observation that can be made is that, when $b_j = 0$ for some j, $2 \le j \le n$, the problem can be reduced to considering, instead of A, the smaller matrices

(8.57)

$$\begin{bmatrix} a_1 & b_2 & 0 & \cdots & \cdots & 0 \\ b_2 & a_2 & & & & \\ 0 & & & & & \\ \vdots & & & & & 0 \\ \vdots & & & & & b_{j-1} \\ 0 & \cdots & \cdots & 0 & b_{j-1} & a_{j-1} \end{bmatrix}$$

and

$$\begin{bmatrix} a_j & b_{j+1} & 0 & \cdots & \cdots & 0 \\ b_{j+1} & a_{j+1} & & & & \\ 0 & & & & & \\ \vdots & & & & & 0 \\ \vdots & & & & & b_n \\ 0 & \cdots & \cdots & 0 & b_n & a_n \end{bmatrix}$$

If $b_2 = 0$ or $b_n = 0$, then the 1×1 matrix $[a_1]$ or $[a_n]$ immediately produces an eigenvalue a_1 or a_n of A. If this is not the case, the QL Algorithm proceeds by forming a sequence of matrices $A^{(1)} = A$, $A^{(2)}$, $A^{(3)}$, . . . as follows:

1) $A^{(1)} = A$ is factored as a product $A^{(1)} = Q^{(1)}L^{(1)}$, where $Q^{(1)}$ is orthogonal and $L^{(1)}$ is lower triangular;

2) $A^{(2)}$ is defined as $A^{(2)} = L^{(1)}Q^{(1)}$ and factored as a product $A^{(2)} = Q^{(2)}L^{(2)}$ where $Q^{(2)}$ is orthogonal and $L^{(2)}$ is lower triangular.

In general, $A^{(i)}$ is factored as a product $A^{(i)} = Q^{(i)}L^{(i)}$ of an orthogonal matrix $Q^{(i)}$ and a lower triangular matrix $L^{(i)}$. Then $A^{(i+1)}$ is defined by the product of $L^{(i)}$ and $Q^{(i)}$ in the reverse direction $A^{(i+1)} = L^{(i)}Q^{(i)}$. Since $Q^{(i)}$ is orthogonal for each i,

(8.58) $$A^{(i+1)} = L^{(i)}Q^{(i)} = (Q^{(i)^t}A^{(i)})Q^{(i)} = Q^{(i)^t}A^{(i)}Q^{(i)}$$

and $A^{(i+1)}$ is symmetric and tridiagonal with the same eigenvalues as $A^{(i)}$. Continuing by induction, $A^{(i+1)}$ has the same eigenvalues as the original matrix A.

The success of the procedure is a result of the fact that $A^{(i+1)}$ tends to a diagonal matrix with the eigenvalues of A in increasing order of magnitude along the diagonal.

To describe the construction of the factoring matrices $Q^{(i)}$ and $L^{(i)}$, we must first introduce the notion of a **rotation matrix**.

Definition 8.44 A **rotation matrix** R is a matrix that differs from the identity matrix in at most four elements. These four elements are of the form

$$r_{ii} = r_{jj} = \cos \theta \quad \text{and} \quad r_{ij} = -r_{ji} = \sin \theta$$

for some angle θ and some $i \neq j$.

It is easy to show (see Exercise 8) that, for any rotation matrix R, the matrix AR differs from A only in the ith and jth columns and the matrix RA differs from A only in the ith and jth rows. When A is symmetric, the angle θ can be chosen so that the product RA has zero entries for $(RA)_{ij}$ and $(RA)_{ji}$. The factorization of $A^{(1)}$ into $A^{(1)} = Q^{(1)}L^{(1)}$ uses a product of $n - 1$ rotation matrices of this type to construct $L^{(1)}$.

$$L^{(1)} = P_2 P_3 \cdots P_n A^{(1)}.$$

We choose the rotation matrix P_n first so that the matrix

$$A_2^{(1)} = P_n A^{(1)}$$

has the element in the $(n - 1, n)$ position set to zero. Since the multiplication $P_n A^{(1)}$ effects both rows $n - 1$ and n of $A^{(1)}$, the matrix $A_2^{(1)}$ does not necessarily retain zero entries in positions $(n, 1), (n, 2), \ldots, (n, n - 2)$. However, since $A^{(1)}$ is tri-diagonal, it is easily seen that the $(n, 1), (n, 2), \ldots, (n, n - 3)$ entries of $A_2^{(1)}$ remain zero and only the $(n, n - 2)$ entry can become nonzero. In general, the matrix P_k is chosen so that the $(k - 1, k)$ entry in $A_{n-k+2}^{(1)} = P_k A_{n-k+1}^{(1)}$ is zero, which results in the $(k, k - 2)$ entry becoming nonzero. The matrix $A_{n-k+1}^{(1)}$ will have the form

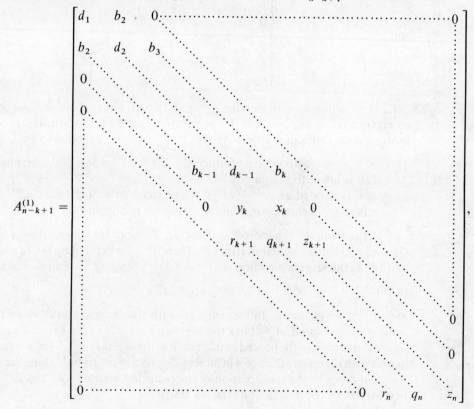

and P_k will have the form

$$P_k = \begin{bmatrix} I_{k-2} & O & O \\ \hline O & \begin{matrix} c_{k-1} & -s_{k-1} \\ s_{k-1} & c_{k-1} \end{matrix} & O \\ \hline O & O & I_{n-k} \end{bmatrix} \leftarrow \text{row } k.$$

$$\uparrow$$
$$\text{column } k$$

The constants c_{k-1} and s_{k-1} in P_k are chosen so that

$$(8.59) \qquad (c_{k-1})^2 + (s_{k-1})^2 = 1$$

(which will make P_k a rotation matrix) and so that the $(k-1, k)$ entry in $A_{n-k+2}^{(1)}$ is zero. That is,

$$(8.60) \qquad c_{k-1}b_k - s_{k-1}x_k = 0.$$

The solution to the nonlinear system (8.59) and (8.60) is

$$s_{k-1} = \frac{b_k}{\sqrt{b_k^2 + x_k^2}} \qquad \text{and} \qquad c_{k-1} = \frac{x_k}{\sqrt{b_k^2 + x_k^2}}$$

and $A_{n-k+2}^{(1)}$ has the form

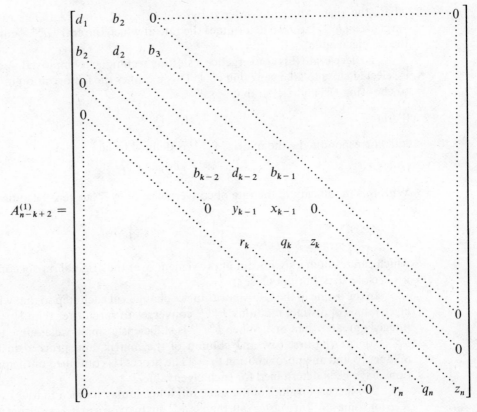

Proceeding with this construction in sequence through $P_n, P_{n-1}, \ldots, P_2$ produces the lower triangular matrix:

$$L^{(1)} \equiv A^{(1)}_{n-1} = \begin{bmatrix} x_1 & 0 & \cdots & & & & \cdots & 0 \\ q_2 & z_2 & & & & & & \vdots \\ r_3 & q_3 & z_3 & & & & & \\ 0 & & & & & & & \\ \vdots & & & & & & & 0 \\ 0 & \cdots & & \cdots & 0 & r_n & q_n & z_n \end{bmatrix}$$

The orthogonal matrix $Q^{(1)}$ is then defined as

$$Q^{(1)}_1 = P^t_n P^t_{n-1} \cdots P^t_2.$$

Consequently, the matrix $A^{(2)}$ is defined by

$$A^{(2)} = L^{(1)} Q^{(1)} = L^{(1)} P^t_n P^t_{n-1} \cdots P^t_2 = P_2 P_3 \cdots P_n A P^t_n P^t_{n-1} \cdots P^t_2$$

The matrix $A^{(2)}$ is tridiagonal, and, in general, the entries off the diagonal will be smaller in magnitude than the corresponding entries in $A^{(1)}$. The process is repeated to construct $A^{(3)}, A^{(4)}, \ldots$.

It can be shown that if the eigenvalues of A have distinct moduli, then the rate of convergence of the entry $b^{(i+1)}_{j+1}$ to zero in the matrix $A^{(i+1)}$ is

$$O\left(\left| \frac{\lambda_j}{\lambda_{j+1}} \right|^{i+1} \right)$$

where the eigenvalues are ordered by $|\lambda_1| < |\lambda_2| < \cdots < |\lambda_n|$. The rate of convergence of $b^{(i+1)}_{j+1}$ to zero determines the rate at which the entry $a^{(i+1)}_j$ converges to the jth eigenvalue λ_j.

To accelerate this convergence a shifting technique is employed. A constant s is selected close to an eigenvalue of A. This modifies the factorization in Eq. (8.58) to choosing $Q^{(i)}$ and $L^{(i)}$ so that

(8.61) $$A^{(i)} - sI = Q^{(i)} L^{(i)},$$

and correspondingly, the matrix $A^{(i+1)}$ is defined to be

(8.62) $$A^{(i+1)} = L^{(i)} Q^{(i)} + sI.$$

With this modification, the rate of convergence of $b^{(i+1)}_{j+1}$ to zero becomes

$$O\left(\left| \frac{\lambda_j - s}{\lambda_{j+1} - s} \right|^{i+1} \right),$$

which can result in a significant improvement over the original rate of convergence if s is close to λ_j but not close to λ_{j+1}.

In the listing of the QL Algorithm, we change s at each step so that when A has eigenvalues of distinct modulus, $b^{(i+1)}_2$ converges to zero faster than $b^{(i+1)}_{j+1}$ for any integer j greater than one. When $b^{(i+1)}_2$ is sufficiently small, we assume that $\lambda_1 \approx a^{(i+1)}_1$, delete the first row and column of the matrix, and proceed in the same manner to find an approximation to λ_2. The process is continued until approximations have been determined for each eigenvalue.

If A has eigenvalues of the same modulus, it may happen that $b^{(i+1)}_j$ tends to zero for some $j \neq 2$ at a faster rate than $b^{(i+1)}_2$. In this case a matrix splitting technique

bed in equations (8.61) and (8.62) can be employed to reduce the
volving a pair of matrices of reduced order.

g algorithm incorporates the shifting technique by choosing at
fting constant s_i, where s_i is the eigenvalue of the matrix

$$E^{(i)} = \begin{bmatrix} a_1^{(i)} & b_2^{(i)} \\ b_2^{(i)} & a_2^{(i)} \end{bmatrix}$$

to $a_1^{(i)}$. This shift translates the eigenvalues of A by a factor s_i. The algorithm accumulates these shifts until $b_2^{(i+1)} \approx 0$ and then adds the shifts to $a_1^{(i+1)}$ to approximate the eigenvalue λ_1.

QL Algorithm 8.10

To obtain the eigenvalues of the symmetric, tridiagonal $n \times n$ matrix

$$A \equiv \hat{A}_1 \equiv \begin{bmatrix} a_1^{(1)} & b_2^{(1)} & 0 & \cdots & \cdots & 0 \\ b_2^{(1)} & a_2^{(1)} & & & & \vdots \\ 0 & & \ddots & & & \\ \vdots & & & \ddots & & 0 \\ & & & & & b_n^{(1)} \\ 0 & \cdots & \cdots & 0 & b_n^{(1)} & a_n^{(1)} \end{bmatrix}$$

INPUT $n; a_1^{(1)}, \ldots, a_n^{(1)}; b_2^{(1)}, \ldots, b_n^{(1)}$; maximum number of iterations M.

OUTPUT eigenvalues of A or recommended splitting of A, or a message that the maximum number of iterations was exceeded.

Step 1 Set $k = 1$;
  $SHIFT = 0$.   (*Accumulated shift.*)

Step 2 While $k \le M$ do Steps 3–8.

  Step 3 (*Test for success.*)
    If $b_2^{(k)} \approx 0$ then set $\lambda = a_1^{(k)} + SHIFT$;
        OUTPUT (λ);
        set $n = n - 1$;
        $a_1^{(k)} = a_2^{(k)}$;
        for $j = 2, \ldots, n$
          set $a_j^{(k)} = a_{j+1}^{(k)}$;
            $b_j^{(k)} + b_{j+1}^{(k)}$.
    If $b_j^{(k)} \approx 0$ for $3 \le j \le n - 1$ then
          OUTPUT ('split into', $a_1^{(k)}, \ldots, a_{j-1}^{(k)}$, $b_2^{(k)}, \ldots, b_{j-1}^{(k)}$, 'and',
        $a_j^{(k)}, \ldots, a_n^{(k)}, b_{j+1}^{(k)}, \ldots, b_n^{(k)}, SHIFT$);
          STOP.
    If $b_n^{(k)} \approx 0$ then set $\lambda = a_n^{(k)} + SHIFT$;
          OUTPUT (λ);
          set $n = n - 1$.

  Step 4 (*Compute shift.*)
    Set $b = -(a_2^{(k)} + a_1^{(k)})$;
      $c = a_2^{(k)} a_1^{(k)} - [b_2^{(k)}]^2$;
      $d = (b^2 - 4c)^{1/2}$;

if $b > 0$ then set $\mu_1 = -2c/(b + d)$;
$$\mu_2 = -(b + d)/2;$$
else set $\mu_1 = (d - b)/2$;
$$\mu_2 = 2c/(d - b);$$
if $n = 2$ then set $\lambda_1 = \mu_1 + SHIFT$;
$$\lambda_2 = \mu_2 + SHIFT;$$
$$\text{OUTPUT } (\lambda_1, \lambda_2);$$
STOP.

Choose s so that $|s - a_1^{(k)}| = \min(|\mu_1 - a_1^{(k)}|, |\mu_2 - a_1^{(k)}|)$.

Step 5 *(Accumulate shift.)*
Set $SHIFT = SHIFT + s$.

Step 6 *(Perform shift.)*
For $j = 1, \ldots, n$ set $d_j = a_j^{(k)} - s$.

Step 7 *(Compute $L^{(k)}$.)*
Set $x_n = d_n$;
$$y_n = b_n;$$
for $j = n, n - 1, \ldots, 2$
set $z_j = (x_j^2 + [b_j^{(k)}]^2)^{1/2}$;
$$c_{j-1} = x_j/z_j;$$
$$s_{j-1} = b_j^{(k)}/z_j;$$
$$q_j = c_{j-1}y_j + s_{j-1}d_{j-1};$$
$$x_{j-1} = c_{j-1}d_{j-1} - s_{j-1}y_j;$$
if $j \neq 2$ then set $r_j = s_{j-1}b_{j-1}$;
$$y_{j-1} = c_{j-1}b_{j-1}.$$
($A_{n-j+2}^{(k)} = P_j A_{n-j+1}^{(k)}$ *has just been computed where*

$$
A_{n-j+2}^{(k)} =
\begin{bmatrix}
d_1 & b_2 & 0 & \cdots & & & & & & & 0 \\
b_2 & & & & & & & & & & \\
0 & & & & & & & & & & \\
0 & & & & & & & & & & \\
& & & & b_{j-2} & d_{j-2} & b_{j-1} & & & & \\
& & & & 0 & y_{j-1} & x_{j-1} & 0 & & & \\
& & & & 0 & r_j & q_j & z_j & & & \\
& & & & & & & & & 0 & \\
& & & & & & & & & & 0 \\
0 & & & & \cdots & & 0 & r_n & q_n & z_n
\end{bmatrix},
$$

$$L = A_n^{(k)} = \begin{bmatrix} x_1 & 0 & \cdots\cdots\cdots\cdots\cdots & 0 \\ q_2 & z_2 & & \\ r_3 & q_3 & z_3 & \\ 0 & & & \\ \vdots & & & 0 \\ 0 & \cdots\cdots & 0 & r_n & q_n & z_n \end{bmatrix} \qquad \text{has been computed.)}$$

Step 8 (*Compute* $A^{(k+1)}$.)

   Set $z_1 = x_1$;
   $a_n^{(k+1)} = s_{n-1}q_n + c_{n-1}z_n$;
   $b_n^{(k+1)} = s_{n-1}z_{n-1}$;
   for $j = n-1, n-2, \ldots, 2$
   set $a_j^{(k+1)} = s_{j-1}q_j + c_{j-1}c_j z_j$;
   $b_j^{(k+1)} = s_{j-1}z_{j-1}$.
   set $a_1^{(k+1)} = c_1 x_1$.
   $k = k + 1$.

Step 9 OUTPUT ('Maximum number of iterations exceeded');
   STOP.

EXAMPLE 2 Let

$$A = A_1^{(1)} = \begin{bmatrix} 4 & 2 & 0 \\ 2 & 3 & 1 \\ 0 & 1 & 2 \end{bmatrix} = \begin{bmatrix} a_1^{(1)} & b_2^{(1)} & 0 \\ b_2^{(1)} & a_2^{(1)} & b_3^{(1)} \\ 0 & b_3^{(1)} & a_3^{(1)} \end{bmatrix}.$$

To find the acceleration parameter s requires finding the eigenvalues of

$$\begin{bmatrix} 4 & 2 \\ 2 & 3 \end{bmatrix}.$$

These can be easily found to be $\mu_1 = 5.561552$ and $\mu_2 = 1.438447$. The eigenvalue closest to $a_1^{(1)} = 4$ is $s = \mu_1 = 5.561552$. Thus, $SHIFT = 5.561552$ and

$$A_1^{(1)} - sI = \begin{bmatrix} -1.561552 & 2 & 0 \\ 2 & -2.561552 & 1 \\ 0 & 1 & -3.561552 \end{bmatrix} = \begin{bmatrix} d_1 & b_2^{(2)} & 0 \\ b_2^{(2)} & d_2 & b_3^{(2)} \\ 0 & y_3 & x_3 \end{bmatrix}.$$

The computation of L in Step 7 of Algorithm 8.10 is

$$P_3 = \begin{bmatrix} 1 & 0 & 0 \\ 0 & -.9627697 & .270323 \\ 0 & -.270323 & -.9627697 \end{bmatrix},$$

$$A_2^{(1)} = P_3 A_1^{(1)} = \begin{bmatrix} -1.561552 & 2 & 0 \\ 1 & 2.195860 & 0 \\ .5406461 & -1.655216 & 3.699276 \end{bmatrix} = \begin{bmatrix} d_1 & b_2^{(1)} & 0 \\ y_2 & x_2 & 0 \\ r_3 & q_3 & z_3 \end{bmatrix},$$

$$P_2 = \begin{bmatrix} .7393092 & .6733661 & 0 \\ -.6733661 & .7393092 & 0 \\ 0 & 0 & 1 \end{bmatrix},$$

$$L = A_3^{(1)} = P_2 A_2^{(1)} = \begin{bmatrix} .1421232 & 2 & 0 \\ -2.475064 & 2.970151 & 0 \\ .5406461 & -1.655216 & 3.699276 \end{bmatrix} = \begin{bmatrix} x_1 & 0 & 0 \\ q_2 & z_2 & 0 \\ r_3 & q_3 & z_3 \end{bmatrix}.$$

Computing $A^{(2)}$ gives

$$A^{(2)} = L P_3^t P_2^t = \begin{bmatrix} .1050729 & .09570091 & 0 \\ .09570091 & -3.780732 & .8029006 \\ 0 & .8029006 & -4.008994 \end{bmatrix}.$$

Iterating again using $A^{(2)}$ gives $s = .1074284$ and

$$A^{(3)} = \begin{bmatrix} 9.885922 \times 10^{-5} & 2.583158 \times 10^{-6} & 0 \\ 2.583158 \times 10^{-6} & -3.597252 & .7013983 \\ 0 & .7013983 & -4.409784 \end{bmatrix}.$$

Accepting $b_2^{(3)} = 2.583158 \times 10^{-6}$ as sufficiently small gives

$$\lambda \approx a_1^{(3)} + SHIFT = 9.885922 \times 10^{-5} + 5.668980 = 5.669077$$

as the first eigenvalue. The other two eigenvalues are computed in Step 4 as 2.476024 and .8548984. The actual eigenvalues are 5.669079087, 2.476023608, and .8548973088. $\qquad\square$

A similar procedure can be employed to find approximations to the eigenvalues of a nonsymmetric $n \times n$ matrix A. The problem is reduced to finding the eigenvalues of a similar matrix with more easily determined eigenvalues. Then iterative matrix operations are performed so that the eigenvalues accumulate along the diagonal of a triangular matrix. The following theorem implies that this approach can be successful. The proof of this result can be found in Ralston and Rabinowitz [67].

Theorem 8.45 (Schur) Let A be an arbitrary $n \times n$ matrix. A nonsingular matrix U exists with the property that

$$T = UAU^{-1},$$

where T is an upper-triangular matrix whose diagonal entries consist of the eigenvalues of A.

The matrix U, whose existence is ensured in Theorem 8.45, can actually be chosen to satisfy a very important condition: $\|U\mathbf{x}\|_2 = \|\mathbf{x}\|_2$ for any vector $\mathbf{x}$. Matrices with this property are called **unitary matrices**. Although we will not make use of the norm-preserving property, it does significantly increase the application of the theorem.

Analogous to the symmetric case, we do not try to find the upper-triangular matrix directly. For a symmetric matrix, Theorem 8.45 guarantees a diagonal matrix, but we instead constructed a tridiagonal matrix with the same eigenvalues. For the nonsymmetric case, we find a matrix that can be considered as a combination of a tridiagonal and a triangular matrix. A matrix of this type is called **upper**

Hessenberg, it contains only zero entries below the lower subdiagonal; i.e. H is upper Hessenberg if $(H)_{ij} = 0$ for all $i \geq j + 2$. The reduction of an arbitrary nonsymmetric matrix to one that is in upper-Hessenberg form can be performed by a method similar to Householder's method or can be accomplished by a Gaussian-elimination type procedure that is modified to preserve eigenvalues.

After the upper Hessenberg matrix H has been formed with the same eigenvalues as the matrix A, a factoring procedure is employed, which, at each step, factors a matrix into a product of an orthogonal matrix, denoted by Q, and an upper triangular matrix, denoted by R. The factoring process assumes the following form. First

$$(8.63) \qquad\qquad H \equiv H_1 = Q_1 R_1;$$

then H_2 is defined by

$$(8.64) \qquad\qquad H_2 = R_1 Q_1$$

and factored into

$$(8.65) \qquad\qquad H_2 = Q_2 R_2,$$

etc. The method of factoring proceeds with the same aim as that of the QL factoring. The matrices are chosen to introduce zeros at appropriate entries of the matrix. The method is called the QR method. In practice, a shifting procedure is employed similar to that used in the QL method. However, the shifting is somewhat more complicated for nonsymmetric matrices since complex eigenvalues with the same modulus can occur. The shifting process for this procedure modifies the calculations in equations (8.63), (8.64), and (8.65) to obtain the double QR method

$$(8.66) \qquad\qquad H_1 - s_1 I = Q_1 R_1$$

$$(8.67) \qquad\qquad H_2 = R_1 Q_1 + s_1 I,$$

$$(8.68) \qquad\qquad H_2 - s_2 I = Q_2 R_2$$

and

$$(8.69) \qquad\qquad H_3 = R_2 Q_2 + s_2 I$$

where s_1 and s_2 are complex conjugates and $H_1, H_2, \ldots$ are real upper-Hessenberg matrices.

A complete description of the QR method can be found in Wilkinson [91]. Detailed algorithms and ALGOL 60 programs for this method and most other commonly employed methods are given in Wilkinson and Reinsch [92]. We refer the reader to these works if the method we have discussed does not give satisfactory results.

Both the QL and QR methods can be performed in a manner that will produce the eigenvectors of a matrix as well as its eigenvalues. The QL Algorithm 8.10 has not been designed to accomplish this, however. If the eigenvectors of a symmetric matrix are needed as well as the eigenvalues, we suggest either using the inverse power method after Algorithms 8.9 and 8.10 have been employed or using a more powerful technique such as those listed in Wilkinson and Reinsch [92], methods that are designed expressly for this purpose.

Exercise Set 8.5

1. Use Householder's method to place the following matrices in tridiagonal form.

a) $\begin{bmatrix} 12 & 10 & 4 \\ 10 & 8 & -5 \\ 4 & -5 & 3 \end{bmatrix}$ b) $\begin{bmatrix} 2 & -1 & -1 \\ -1 & 2 & -1 \\ -1 & -1 & 2 \end{bmatrix}$

2. Apply two iterations of the QL Algorithm to the following matrices.

a) $\begin{bmatrix} 2 & -1 & 0 \\ -1 & 2 & -1 \\ 0 & -1 & 2 \end{bmatrix}$ b) $\begin{bmatrix} 3 & 1 & 0 \\ 1 & 4 & 2 \\ 0 & 2 & 1 \end{bmatrix}$

c) $\begin{bmatrix} .5 & .25 & 0 & 0 \\ .25 & .8 & .4 & 0 \\ 0 & .4 & .6 & .1 \\ 0 & 0 & .1 & 1 \end{bmatrix}$

3. Use Householder's method to place the following matrices in tridiagonal form.

a) $A = \begin{bmatrix} 2 & 0 & 1 \\ 0 & 3 & -2 \\ 1 & -2 & -1 \end{bmatrix}$ b) $B = \begin{bmatrix} 6 & 1 & 0 & -1 \\ 1 & 5 & 1 & 1 \\ 0 & 1 & -4 & 2 \\ -1 & 1 & 2 & 5 \end{bmatrix}$

c) $A = \begin{bmatrix} 4 & -1 & -1 & 0 \\ -1 & 4 & 0 & -1 \\ -1 & 0 & 4 & -1 \\ 0 & -1 & -1 & 4 \end{bmatrix}$

d) $C = \begin{bmatrix} 8 & .25 & .5 & 2 & -1 \\ .25 & -4 & 0 & 1 & 2 \\ .5 & 0 & 5 & .75 & -1 \\ 2 & 1 & .75 & 5 & -.5 \\ -1 & 2 & -1 & -.5 & 6 \end{bmatrix}$

4. Use the QL Algorithm to find all the eigenvalues of the matrices in Exercise 1.

5. Use the QL Algorithm to determine all the eigenvalues of the following matrices.

a) $\begin{bmatrix} 4 & 2 & 0 & 0 & 0 \\ 2 & 4 & 2 & 0 & 0 \\ 0 & 2 & 4 & 2 & 0 \\ 0 & 0 & 2 & 4 & 2 \\ 0 & 0 & 0 & 2 & 4 \end{bmatrix}$ b) $\begin{bmatrix} 5 & -1 & 0 & 0 & 0 \\ -1 & 4.5 & .2 & 0 & 0 \\ 0 & .2 & 1 & -.4 & 0 \\ 0 & 0 & -.4 & 3 & 1 \\ 0 & 0 & 0 & 1 & 3 \end{bmatrix}$

c) $\begin{bmatrix} 4 & -3 & 0 & 0 \\ -3 & \frac{10}{3} & -\frac{5}{3} & 0 \\ 0 & -\frac{5}{3} & -\frac{99}{75} & \frac{68}{75} \\ 0 & 0 & \frac{68}{75} & \frac{149}{75} \end{bmatrix}$ d) $\begin{bmatrix} 2 & -1 & 0 \\ -1 & -1 & -2 \\ 0 & -2 & 3 \end{bmatrix}$

6. Use the inverse power method to determine the eigenvectors of the matrices in Exercise 3.

7. Let $\mathbf{w} \in R^n$ satisfy $\mathbf{w}^t\mathbf{w} = 1$. Show that $P^2 = I$ where $P = I - 2\mathbf{w}\mathbf{w}^t$.

8. Let R be a rotation matrix with $r_{ii} = r_{jj} = \cos\theta$, $r_{ij} = -r_{ji} = \sin\theta$ for $j < i$. Show that for any $n \times n$ matrix A:

$$(AR)_{p,q} = \begin{cases} a_{p,q} & \text{if } q = i, j, \\ (\cos\theta)a_{p,j} + (\sin\theta)a_{p,i} & \text{if } q = j, \\ (\cos\theta)a_{p,i} - (\sin\theta)a_{p,j} & \text{if } q = i. \end{cases}$$

$$(RA)_{p,q} = \begin{cases} a_{p,q} & \text{if } p = i, j, \\ (\cos\theta)a_{j,q} - (\sin\theta)a_{i,q} & \text{if } p = j, \\ (\sin\theta)a_{j,q} + (\cos\theta)a_{i,q} & \text{if } p = i. \end{cases}$$

9. Jacobi's method for a symmetric matrix A is described by

$$A_1 = A \qquad,$$
$$A_2 = P_1 A_1 P_1^t,$$

and in general

$$A_{i+1} = P_i A_i P_i^t$$

tends to a diagonal matrix where P_i is a rotation matrix chosen to eliminate a large off-diagonal element in A_i. If $a_{j,k}$ and $a_{k,j}$ are to be set to zero where $j < k$, then either

$$(P_i)_{jj} = (P_i)_{kk} = \sqrt{\frac{1}{2}\left(1 + \frac{b}{\sqrt{c^2 + b^2}}\right)},$$

$$(P_i)_{k,j} = \frac{c}{2(P_i)_{jj}\sqrt{c^2 + b^2}} = -(P_i)_{j,k},$$

where

$$c = 2a_{jk}\,\text{sign}(a_{jj} - a_{kk}),$$
$$b = |a_{jj} - a_{kk}| \qquad \text{if } a_{jj} \neq a_{kk},$$

or

$$(P_i)_{jj} = (P_i)_{kk} = \sqrt{2}/2,$$
$$(P_i)_{k,j} = -(P_i)_{j,k} = \sqrt{2}/2 \qquad \text{if } a_{jj} = a_{kk}.$$

Develop an algorithm to implement Jacobi's method by setting $a_{21} = 0$, then $a_{31}, a_{32}, a_{41}, a_{42}, a_{43}, \ldots, a_{n,1}, \ldots, a_{n,n-1}$ in turn go to zero. This is repeated until matrix A_k is computed with

$$\sum_{i=1}^{n} \sum_{\substack{j=1 \\ j \neq i}}^{n} |a_{ij}^{(k)}|$$

sufficiently small. The eigenvalues of A can then be approximated by the diagonal entries of A_k.

10. Apply Jacobi's method to the matrices in Exercise 1.

11. Modify the QL Algorithm to find the eigenvalues of a nonsymmetric tridiagonal matrix

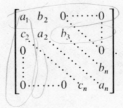

$$\begin{bmatrix} a_1 & b_2 & 0 & \cdots & 0 \\ c_2 & a_2 & b_3 & \ddots & \vdots \\ 0 & \ddots & \ddots & \ddots & 0 \\ \vdots & & \ddots & \ddots & b_n \\ 0 & \cdots & 0 & c_n & a_n \end{bmatrix}$$

12. Use the algorithm developed in Exercise 11 to find all the eigenvalues of

a) $\begin{bmatrix} 4 & 1 & 0 \\ -1 & 3 & 1 \\ 0 & 1 & 2 \end{bmatrix},$

b) $\begin{bmatrix} 6 & -2 & 0 & 0 \\ 3 & 4 & -1 & 0 \\ 0 & 0 & 5 & -1 \\ 0 & 0 & 2 & 3 \end{bmatrix},$

c) $\begin{bmatrix} 7 & 1 & 0 & 0 \\ 1 & 6 & 3 & 0 \\ 0 & -2 & 4 & 1 \\ 0 & 0 & 2 & 3 \end{bmatrix}.$

Numerical Solutions of Nonlinear Systems of Equations

The amount of pressure required to sink a large heavy object in a soft, homogeneous soil that lies above a hard base soil can be predicted by the amount of pressure required to sink smaller objects in the same soil. In particular, the amount of pressure p to sink a circular plate of radius r a distance d in the soft soil, where the hard base soil lies a distance $D > d$ below the surface, can be approximated by an equation of the form

$$p = k_1 e^{k_2 r} + k_3 r,$$

where k_1, k_2, and k_3 are constants depending on d and the consistency of the soil but not on the radius of the plate.

To determine the minimal size of plate that is required to sustain a large load, three small plates with differing radii are sunk to the same distance, and the loads required for this sinkage are measured. These calculations lead to three equations of the type given above in the three unknowns k_1, k_2, and k_3. Due to the nature of the equations, however, we cannot easily solve for the unknowns. Numerical methods are needed that can be used for solving systems of equations when the equations are not linear. See Exercise 5 of Section 9.2 for a problem of this type of application.

Fixed-point iteration methods dominated the study in Chapter 2 of numerical solutions to equations of the form $f(x) = 0$. The methods we will use for solving systems of equations of this type will basically be generalizations of Newton's method. Although other techniques are available, they are generally quite specialized with regard to the types of problem they will solve, and are probably inappropriate for inclusion in a first course in numerical analysis. A quite extensive account of such methods, however, presented on a reasonably elementary level, can be found in Acton [2], pages 361–409.

Most of the proofs of the theoretical results in this chapter will be omitted since in general they involve methods that are usually studied in a course in advanced calculus. A good general reference for this material is Ortega's book entitled *Numerical Analysis—A Second Course* [64]. A more complete reference is Ortega and Rheinboldt [65].

(9.1) Fixed Points for Functions of Several Variables

The general form of a system of nonlinear equations is:

$$
\begin{aligned}
f_1(x_1, x_2, \ldots, x_n) &= 0, \\
f_2(x_1, x_2, \ldots, x_n) &= 0, \\
&\vdots \\
f_n(x_1, x_2, \ldots, x_n) &= 0
\end{aligned}
$$

(9.1)

where each function f_i maps n-dimensional space, R^n, into the real line R. It is often desirable to represent the system alternatively by defining a function $\mathbf{F}$, mapping R^n into R^n by

$$\mathbf{F}(x_1, x_2, \ldots, x_n) = (f_1(x_1, x_2, \ldots, x_n), f_2(x_1, x_2, \ldots, x_n), \ldots, f_n(x_1, x_2, \ldots, x_n)).$$

Using vector notation to represent the variables $x_1, x_2, \ldots, x_n$, we write:

$$\mathbf{x} = (x_1, x_2, \ldots, x_n)^t, \qquad \text{or for simplicity,} \quad \mathbf{x} = (x_1, x_2, \ldots, x_n).$$

Equation (9.1) then assumes the form:

(9.2)
$$\mathbf{F}(\mathbf{x}) = \mathbf{0}.$$

EXAMPLE 1 The three-by-three nonlinear system

$$
\begin{aligned}
3x_1 - \cos(x_2 x_3) \qquad - \tfrac{1}{2} \qquad &= 0, \\
x_1^2 - 81(x_2 + .1)^2 + \sin x_3 + 1.06 \qquad &= 0, \\
e^{-x_1 x_2} \qquad + 20x_3 + \frac{10\pi - 3}{3} &= 0,
\end{aligned}
$$

can be placed in the form (9.2) by defining the three functions f_1, f_2, and f_3 from R^3 to R as

$$
\begin{aligned}
f_1(x_1, x_2, x_3) &= 3x_1 - \cos(x_2 x_3) - \tfrac{1}{2}, \\
f_2(x_1, x_2, x_3) &= x_1^2 - 81(x_2 + .1)^2 + \sin x_3 + 1.06, \\
f_3(x_1, x_2, x_3) &= e^{-x_1 x_2} + 20x_3 + \frac{10\pi - 3}{3},
\end{aligned}
$$

and $\mathbf{F}$ from $R^3 \to R^3$ by

$$\mathbf{F}(\mathbf{x}) = \mathbf{F}(x_1, x_2, x_3)$$

$$= (f_1(x_1, x_2, x_3), f_2(x_1, x_2, x_3), f_3(x_1, x_2, x_3))$$

$$= \left(3x_1 - \cos(x_2 x_3) - \tfrac{1}{2}, x_1^2 - 81(x_2 + .1)^2 \right.$$

$$\left. + \sin x_3 + 1.06, e^{-x_1 x_2} + 20x_3 + \frac{10\pi - 3}{3}\right). \qquad \square$$

Before discussing the solution of a system which is given in the form (9.1) or (9.2), it is necessary to consider some results concerning continuity and differentiability of functions from R^n into R^n. Although this study could be presented directly (see Exercise 4), we prefer instead to use an alternative method that will allow us to present the more theoretically difficult concepts of **limits** and **continuity** in terms of functions from R^n into R.

Definition 9.1   Let f be a function defined on a set $D \subset R^n$ and mapping into R. The function f is said to have the **limit** L at the point $\mathbf{x}_0$, written

$$\lim_{\mathbf{x} \to \mathbf{x}_0} f(\mathbf{x}) = L,$$

if, given any number $\varepsilon > 0$, a number $\delta > 0$ exists with the property that

$$|f(\mathbf{x}) - L| < \varepsilon$$

whenever $\mathbf{x} \in D$ and

$$0 < \|\mathbf{x} - \mathbf{x}_0\| < \delta.$$

It should be noted that the existence of a limit at a point is independent of the particular vector norm being used because of the equivalence of vector norms in R^n (see Section 8.1).

Definition 9.2   Let f be a function from a set $D \subset R^n$ into R. The function f is said to be **continuous** at $\mathbf{x}_0 \in D$ provided $\lim_{\mathbf{x} \to \mathbf{x}_0} f(\mathbf{x})$ exists and

$$\lim_{\mathbf{x} \to \mathbf{x}_0} f(\mathbf{x}) = f(\mathbf{x}_0).$$

Moreover, f is said to be **continuous** on a set D provided f is continuous at every point of D. This concept is expressed by writing $f \in C(D)$.

We can now define the limit and continuity concepts for functions from R^n into R^n by considering the coordinate functions from R^n into R.

Definition 9.3   Let $\mathbf{F}$ be a function from $D \subset R^n$ into R^n and suppose $\mathbf{F}$ has the representation

$$\mathbf{F}(\mathbf{x}) = (f_1(\mathbf{x}), f_2(\mathbf{x}), \dots, f_n(\mathbf{x})),$$

where f_i is a mapping from R^n into R for each $i = 1, 2, \ldots, n$. We define

$$\lim_{\mathbf{x} \to \mathbf{x}_0} \mathbf{F}(\mathbf{x}) = \mathbf{L} = (L_1, L_2, \ldots, L_n)$$

if and only if $\lim_{\mathbf{x} \to \mathbf{x}_0} f_i(\mathbf{x}) = L_i$ for each $i = 1, 2, \ldots, n$.

Definition 9.4   Let $\mathbf{F}$ be a function from $D \subset R^n$ into R^n with the representation $\mathbf{F}(\mathbf{x}) = (f_1(\mathbf{x}), f_2(\mathbf{x}), \ldots, f_n(\mathbf{x}))$. The function $\mathbf{F}$ is said to be **continuous** at $\mathbf{x}_0 \in D$ provided $\lim_{\mathbf{x} \to \mathbf{x}_0} \mathbf{F}(\mathbf{x})$ exists and $\lim_{\mathbf{x} \to \mathbf{x}_0} \mathbf{F}(\mathbf{x}) = \mathbf{F}(\mathbf{x}_0)$.

   $\mathbf{F}$ is said to be **continuous** on the set D if $\mathbf{F}$ is continuous at each point in D. This concept is expressed by writing $\mathbf{F} \in C(D)$.

For functions from R into R, continuity can often be shown at a point by showing that the function is differentiable at the point (see Theorem 1.6, p. 4). Although this theorem generalizes to functions of several variables, the derivative, or total derivative, of a function of several variables is quite involved and will not be presented here. Instead we will state the following theorem, which relates the continuity of a function of n variables at a point to the partial derivatives of the function at the point.

Theorem 9.5   Let f be a function from $D \subset R^n$ into R and $\mathbf{x}_0 \in D$. If constants $\delta > 0$ and $K > 0$ exist with

$$\left| \frac{\partial f(\mathbf{x})}{\partial x_j} \right| \leq K \qquad \text{for each } j = 1, 2, \ldots, n,$$

whenever $\|\mathbf{x} - \mathbf{x}_0\| < \delta$ and $\mathbf{x} \in D$, then f is continuous at $\mathbf{x}_0$.

In Chapter 2, an iterative process for solving an equation $f(x) = 0$ was developed by first transforming the equation into an equation of the form $x = g(x)$. The function g has fixed points precisely at solutions to the original equation. A similar procedure will be investigated here for functions from R^n into R^n.

Definition 9.6   A function $\mathbf{G}$ from $D \subset R^n$ into R^n is said to have a **fixed point** at $\mathbf{p} \in D$ if $\mathbf{G}(\mathbf{p}) = \mathbf{p}$.

The following theorem extends the fixed-point theorem (Theorem 2.3 p. 26) to the n-dimensional case. This theorem is a special case of the well-known **Contraction Mapping Theorem**, and its proof can be found in the previously mentioned book by Ortega [64], page 153.

Theorem 9.7   Let $D = \{(x_1, x_2, \ldots, x_n) | a_i \leq x_i \leq b_i \text{ for each } i = 1, 2, \ldots, n\}$ for some collection of constants $a_1, a_2, \ldots, a_n$, and $b_1, b_2, \ldots, b_n$. Suppose $\mathbf{G}$ is a continuous function with continuous first partial derivatives from $D \subset R^n$ into R^n with the property that $\mathbf{G}(\mathbf{x}) \in D$ whenever $\mathbf{x} \in D$. Then $\mathbf{G}$ has a fixed point in D.

If, in addition, a constant $K < 1$ exists with

$$\left|\frac{\partial g_i(\mathbf{x})}{\partial x_j}\right| \leq \frac{K}{n} \qquad \text{whenever } \mathbf{x} \in D,$$

for each $j = 1, 2, \ldots, n$, and each of the coefficient functions g_i, then the fixed point is unique.

EXAMPLE 2   Consider the nonlinear system from Example 1 given by

$$(9.3) \qquad \begin{aligned} 3x_1 - \cos(x_2 x_3) &\qquad\qquad\qquad -\tfrac{1}{2} &&= 0, \\ x_1^2 - 81(x_2 + .1)^2 + \sin x_3 + 1.06 &&&= 0, \\ e^{-x_1 x_2} &\qquad + 20x_3 + \frac{10\pi - 3}{3} &&= 0. \end{aligned}$$

If the ith equation is solved for x_i, the system can be changed into the fixed-point problem

$$x_1 = \tfrac{1}{3}\cos(x_2 x_3) + \tfrac{1}{6},$$

$$x_2 = \tfrac{1}{9}\sqrt{x_1^2 + \sin x_3 + 1.06} - .1,$$

$$x_3 = -\tfrac{1}{20}e^{-x_1 x_2} - \frac{10\pi - 3}{60}.$$

Let $\mathbf{G}: R^3 \to R^3$ be defined by $\mathbf{G}(\mathbf{x}) = (g_1(\mathbf{x}), g_2(\mathbf{x}), g_3(\mathbf{x}))$ where

$$g_1(x_1, x_2, x_3) = \tfrac{1}{3}\cos(x_2 x_3) + \tfrac{1}{6},$$

$$g_2(x_1, x_2, x_3) = \tfrac{1}{9}\sqrt{x_1^2 + \sin x_3 + 1.06} - .1,$$

$$g_3(x_1, x_2, x_3) = -\tfrac{1}{20}e^{-x_1 x_2} - \frac{10\pi - 3}{60}.$$

Theorems 9.5 and 9.7 will be used to show that $\mathbf{G}$ has a unique fixed point in

$$D = \{(x_1, x_2, x_3) \mid -1 \leq x_i \leq 1 \qquad \text{for each } i = 1, 2, 3\}.$$

For $\mathbf{x} = (x_1, x_2, x_3)$ in D,

$$|g_1(x_1, x_2, x_3)| \leq \tfrac{1}{3}|\cos(x_2 x_3)| + \tfrac{1}{6} \leq \tfrac{1}{2},$$

$$\begin{aligned} |g_2(x_1, x_2, x_3)| &= |\tfrac{1}{9}\sqrt{x_1^2 + \sin x_3 + 1.06} - .1| \\ &\leq \tfrac{1}{9}\sqrt{1 + \sin 1 + 1.06} - .1 \\ &< .090, \end{aligned}$$

and

$$\begin{aligned} |g_3(x_1, x_2, x_3)| &= \tfrac{1}{20}e^{-x_1 x_2} + \frac{10\pi - 3}{60} \\ &\leq \tfrac{1}{20}e + \frac{10\pi - 3}{60} \\ &< .61; \end{aligned}$$

so $-1 \leq g_i(x_1, x_2, x_3) \leq 1$, for each $i = 1, 2, 3$. Thus, $\mathbf{G}(\mathbf{x}) \in D$ whenever $\mathbf{x} \in D$.

Finding bounds for the partial derivatives on D gives the following:

$$\left| \frac{\partial g_1}{\partial x_1} \right| = 0,$$

$$\left| \frac{\partial g_1}{\partial x_2} \right| \le \tfrac{1}{3} |x_3| |\sin x_2 x_3| \le \tfrac{1}{3} \sin 1 < .281,$$

$$\left| \frac{\partial g_1}{\partial x_3} \right| = \tfrac{1}{3} |x_2| |\sin x_2 x_3| \le \tfrac{1}{3} \sin 1 < .281,$$

$$\left| \frac{\partial g_2}{\partial x_1} \right| = \frac{|x_1|}{9\sqrt{x_1^2 + \sin x_3 + 1.06}} < \frac{1}{9\sqrt{.218}} < .238,$$

$$\left| \frac{\partial g_2}{\partial x_2} \right| = 0,$$

$$\left| \frac{\partial g_2}{\partial x_3} \right| = \frac{|\cos x_3|}{18\sqrt{x_1^2 + \sin x_3 + 1.06}} < \frac{1}{18\sqrt{.218}} < .119,$$

$$\left| \frac{\partial g_3}{\partial x_1} \right| = \frac{|x_2|}{20} e^{-x_1 x_2} \le \tfrac{1}{20} e < .14,$$

$$\left| \frac{\partial g_3}{\partial x_2} \right| = \frac{|x_1|}{20} e^{-x_1 x_2} \le \tfrac{1}{20} e < .14,$$

and
$$\left| \frac{\partial g_3}{\partial x_3} \right| = 0.$$

Since the partial derivatives of g_1, g_2, and g_3 are bounded on D, Theorem 9.5 implies that these functions are continuous on D. Consequently, G is continuous on D. Moreover, for every $\mathbf{x} \in D$

$$\left| \frac{\partial g_i(\mathbf{x})}{\partial x_j} \right| \le .281 \qquad \text{for each } i = 1, 2, 3 \text{ and } j = 1, 2, 3,$$

and the condition in the second part of Theorem 9.7 holds with $K = .843$.

In the same manner it can also be shown that $\partial g_i / \partial x_j$ is continuous on D for each $i = 1, 2, 3$, and $j = 1, 2, 3$. (This will be considered in Exercise 3.) Consequently, $\mathbf{G}$ has a unique fixed point in D and the nonlinear system (9.3) has a unique solution in D. $\square$

After a nonlinear system of the form $\mathbf{F}(\mathbf{x}) = \mathbf{0}$ has been transformed into an equivalent fixed-point problem $\mathbf{G}(\mathbf{x}) = \mathbf{x}$, it is natural to consider a functional or fixed-point iteration process applied to $\mathbf{G}$. To do this, we select $\mathbf{x}^{(0)} = (x_1^{(0)}, x_2^{(0)}, \ldots, x_n^{(0)})$, and generate the sequence of vectors $\mathbf{x}^{(k)} = (x_1^{(k)}, x_2^{(k)}, \ldots, x_n^{(k)})$ by

(9.4) $$\mathbf{x}^{(k)} = \mathbf{G}(\mathbf{x}^{(k-1)}) \qquad \text{for each } k = 1, 2, \ldots, n,$$

or, component-wise,

(9.5)
$$x_1^{(k)} = g_1(x_1^{(k-1)}, x_2^{(k-1)}, \ldots, x_n^{(k-1)}),$$
$$x_2^{(k)} = g_2(x_1^{(k-1)}, x_2^{(k-1)}, \ldots, x_n^{(k-1)}),$$
$$\vdots \qquad \vdots$$
$$x_n^{(k)} = g_n(x_1^{(k-1)}, x_2^{(k-1)}, \ldots, x_n^{(k-1)}).$$

The following theorem is analogous to Theorem 2.4, p. 30 in the one-dimensional case and gives conditions for the iterations to converge. The proof of the theorem can be found in Ortega [64], page 153.

Theorem 9.8 Let $D = \{(x_1, x_2, \ldots, x_n)| a_i \leq x_i \leq b_i, \text{for each } i = 1, 2, \ldots, n\}$, for some collection of constants $a_1, a_2, \ldots, a_n$, and $b_1, b_2, \ldots, b_n$. Suppose **G** is a continuous function with continuous first partial derivatives from $D \subset R^n$ into R^n with the property that $\mathbf{G}(\mathbf{x}) \in D$ whenever $\mathbf{x} \in D$. If a constant $K < 1$ exists with

$$\left| \frac{\partial g_i(\mathbf{x})}{\partial x_j} \right| \leq \frac{K}{n} \qquad \text{whenever } \mathbf{x} \in D,$$

for each $j = 1, 2, \ldots, n$ and each component function g_i, then the sequence $\{\mathbf{x}^{(k)}\}_{k=0}^{\infty}$ defined in Eq. (9.4) converges to the unique fixed point $\mathbf{p} \in D$, for any $\mathbf{x}^{(0)}$ in D, and

(9.6) $$\|\mathbf{x}^{(j)} - \mathbf{p}\|_\infty \leq \frac{K^j}{1 - K} \|\mathbf{x}^{(1)} - \mathbf{x}^{(0)}\|_\infty.$$

EXAMPLE 3 The nonlinear system

$$3x_1 - \cos(x_2 x_3) \qquad\qquad -\tfrac{1}{2} \qquad\quad = 0,$$
$$x_1^2 - 81(x_2 + .1)^2 + \sin x_3 + 1.06 \qquad = 0,$$
$$e^{-x_1 x_2} \qquad\qquad + 20x_3 + \frac{10\pi - 3}{3} = 0,$$

was shown in Example 2 to have a unique solution in

$$D = \{(x_1, x_2, x_3)| -1 \leq x_i \leq 1 \qquad \text{for each } i = 1, 2, 3\},$$

by applying Theorem 9.7 to the fixed-point problem

$$x_1 = \tfrac{1}{3} \cos x_2 x_3 + \tfrac{1}{6},$$
(9.7) $$x_2 = \tfrac{1}{9}\sqrt{x_1^2 + \sin x_3 + 1.06} - .1,$$
$$x_3 = -\tfrac{1}{20}e^{-x_1 x_2} - \frac{10\pi - 3}{60}.$$

Since Theorem 9.8 holds for the equations in (9.7) with $K = .843$, the functional iteration method can be applied. The choice of $\mathbf{x}^{(0)}$ in D is arbitrary, and we will let $\mathbf{x}^{(0)} = (.1, .1, -.1)$. The sequence of vectors generated by

$$x_1^{(k)} = \tfrac{1}{3} \cos x_2^{(k-1)} x_3^{(k-1)} + \tfrac{1}{6},$$
$$x_2^{(k)} = \tfrac{1}{9}\sqrt{(x_1^{(k-1)})^2 + \sin x_3^{(k-1)} + 1.06} - .1,$$
$$x_3^{(k)} = -\tfrac{1}{20}e^{-x_1^{(k-1)} x_2^{(k-1)}} - \frac{10\pi - 3}{60},$$

will converge to the unique solution of (9.7). In this example the sequence was generated until k was found with

$$\|\mathbf{x}^{(k)} - \mathbf{x}^{(k-1)}\|_\infty < 10^{-5}.$$

The results are given in Table 9.1.

TABLE 9.1

k	$x_1^{(k)}$	$x_2^{(k)}$	$x_3^{(k)}$	$\|\mathbf{x}^{(k)} - \mathbf{x}^{(k-1)}\|_\infty$
0	.10000000	.10000000	−.10000000	—
1	.49998333	.00944115	−.52310127	.423
2	.49999593	.00002557	−.52336331	9.4×10^{-3}
3	.50000000	.00001234	−.52359814	2.3×10^{-4}
4	.50000000	.00000003	−.52359847	1.2×10^{-5}
5	.50000000	.00000002	−.52359877	3.1×10^{-7}

Using the error bound (9.6) with $K = .843$ gives

$$\|\mathbf{x}^{(5)} - \mathbf{p}\|_\infty \le \frac{(.843)^5}{1 - .843}(.423) < 1.15,$$

which does not indicate the true accuracy of $\mathbf{x}^{(5)}$. If

$$\mathbf{x}^{(3)} = (.50000000, 1.234 \times 10^{-5}, -.52359814)$$

had been chosen as an initial approximation, then $\mathbf{x}^{(5)}$ would have been the second iterate, and inequality (9.6) implies that

$$\|\mathbf{x}^{(5)} - \mathbf{p}\|_\infty \le \frac{(.843)^2}{1 - .843}(1.50 \times 10^{-5}) < 6.8 \times 10^{-5},$$

a more reasonable estimate.

Since the actual solution is

$$\mathbf{p} = \left(.5, 0, -\frac{\pi}{6}\right) \approx (.5, 0, -.5235987757),$$

the true error is

$$\|\mathbf{x}^{(5)} - \mathbf{p}\|_\infty \le 2 \times 10^{-8}. \qquad \square$$

One possible way to accelerate convergence is to use the latest estimates $x_1^{(k)}, \ldots, x_{i-1}^{(k)}$, instead of $x_1^{(k-1)}, \ldots, x_{i-1}^{(k-1)}$, to compute $x_i^{(k)}$ as in the Gauss–Seidel method for linear systems (see Section 8.2). The component equations then become:

$$x_1^{(k)} = \tfrac{1}{3}\cos(x_2^{(k-1)}x_3^{(k-1)}) + \tfrac{1}{6},$$

$$x_2^{(k)} = \tfrac{1}{9}\sqrt{(x_1^{(k)})^2 + \sin x_3^{(k-1)} + 1.06} - .1,$$

$$x_3^{(k)} = -\tfrac{1}{20}e^{-x_1^{(k)}x_2^{(k)}} - \frac{10\pi - 3}{60}.$$

With $\mathbf{x}^{(0)} = (.1, .1, -.1)$, the results of these calculations are listed in Table 9.2.

The iterate $\mathbf{x}^{(4)}$ is actually accurate to within 10^{-7} in the l_∞ norm; so the convergence was indeed accelerated for this problem by using the Seidel method. It should be remarked, however, that while the Seidel method did improve the convergence in this example it does not always give an acceleration of the convergence.

TABLE 9.2

k	$x_1^{(k)}$	$x_2^{(k)}$	$x_3^{(k)}$	$\|\mathbf{x}^{(k)} - \mathbf{x}^{(k-1)}\|_\infty$
0	.10000000	.10000000	-.10000000	—
1	.49998333	.02222979	-.52304613	.423
2	.49997747	.00002815	-.52359807	2.2×10^{-2}
3	.50000000	.00000004	-.52359877	2.8×10^{-5}
4	.50000000	.00000000	-.52359877	3.8×10^{-8}

Exercise Set 9.1

1. Show that the function $\mathbf{F}: R^3 \to R^3$ defined by

    $$\mathbf{F}(x_1, x_2, x_3) = (x_1 + 2x_3, x_1 \cos x_2, x_2^2 + x_3)$$

    is continuous at each point of R_3.

2. Give an example of a function $\mathbf{F}: R^2 \to R^2$ that is continuous at each point of R^2 except $(1, 0)$.

3. Show that the first partial derivatives in Example 2 are continuous on D.

4. Show that Definition 9.4 is equivalent to the following statement: A function $\mathbf{F}: D \subset R^n \to R^n$ is said to be continuous at $\mathbf{x}_0 \in D$ if, given any number $\varepsilon > 0$, a number $\delta > 0$ can be found with the property that

    $$\|\mathbf{F}(\mathbf{x}) - \mathbf{F}(\mathbf{x}_0)\| < \varepsilon$$

    whenever $\mathbf{x} \in D$ and $\|\mathbf{x} - \mathbf{x}_0\| < \delta$.

5. Show that the nonlinear system

    $$x_1^2 + x_2^2 - x_1 = 0,$$
    $$x_1^2 - x_2^2 - x_2 = 0,$$

    has a unique nontrivial solution. Approximate the solution graphically. Use the graphical solution as an initial approximation for an appropriate functional iteration. Determine the solution to within 10^{-3} in the l_2 norm.

6. The nonlinear system

    $$x_1^2 - 10x_1 + x_2^2 + 8 = 0,$$
    $$x_1 x_2^2 + x_1 - 10x_2 + 8 = 0,$$

    can be transformed into the fixed-point problem

    $$x_1 = g_1(x_1, x_2) = \frac{x_1^2 + x_2^2 + 8}{10},$$

    $$x_2 = g_2(x_1, x_2) = \frac{x_1 x_2^2 + x_1 + 8}{10}.$$

    a) Use Theorem 9.7 to show that $\mathbf{G} = (g_1, g_2): D \subset R^2 \to R^2$ has a unique fixed point in

        $$D = \{(x_1, x_2)| 0 \le x_1, x_2 \le 1.5\}.$$

    b) Apply functional iteration to approximate the solution.
    c) Does the Seidel method accelerate convergence?

7. Use Theorem 9.7 to show that the following functions $\mathbf{G}: D \subset R^3 \rightarrow R^3$ have unique fixed points in D:

a) $\mathbf{G}(x_1, x_2, x_3) = \left(\dfrac{\cos(x_2 x_3) + .5}{3}, \dfrac{1}{25} \sqrt{x_1^2 + .3125} - .03, -\dfrac{1}{20} e^{-x_1 x_2} - \dfrac{10\pi - 3}{60} \right),$

$D = \{(x_1, x_2, x_3) | -1 \le x_i \le 1, i = 1, 2, 3\}.$

b) $\mathbf{G}(x_1, x_2, x_3) = \left(\dfrac{7.17 + 3x_2^2 + 4x_3}{12}, \dfrac{11.54 + x_3 - x_1^2}{10}, \dfrac{7.631 - x_2^3}{7} \right),$

$D = \{(x_1, x_2, x_3) | 0 \le x_i \le 1.5, i = 1, 2, 3\}.$

c) $\mathbf{G}(x_1, x_2, x_3) = (1 - \cos(x_1 x_2 x_3), 1 - (1 - x_1)^{1/4} - .05x_3^2 + .15x_3,$

$x_1^2 + .1x_2^2 - .01x_2 + 1),$

$D = \{(x_1, x_2, x_3) | -.1 \le x_1 \le .1, -.1 \le x_2 \le .3, .5 \le x_3 \le 1.1\}.$

8. Use functional iteration and Seidel's method to approximate the fixed points in Exercise 7.

9. In Exercise 6 of Section 5.8, we considered the problem of predicting the population of two species that compete for the same food supply. In that problem, we made the assumption that the populations could be predicted by solving the system of equations:

$$\frac{dx_1(t)}{dt} = x_1(t)(4 - .0003x_1(t) - .0004x_2(t))$$

and

$$\frac{dx_2(t)}{dt} = x_2(t)(2 - .0002x_1(t) - .0001x_2(t)).$$

In this exercise we would like to consider the problem of determining equilibrium populations of the two species. The mathematical criteria that must be satisfied in order for the populations to be at equilibrium is that, simultaneously,

$$\frac{dx_1(t)}{dt} = 0 \quad \text{and} \quad \frac{dx_2(t)}{dt} = 0.$$

This clearly occurs when the first species is extinct and the second species has a population of 20,000, or when the second species is extinct and the first species has a population of 13,333. Can this equilibrium occur under any other situation?

9.2   Newton's Method

Although the problem presented in Example 2 of Section 9.1 can be transformed into a convergent fixed-point format quite easily, this is not often the situation. For this reason it is necessary to devise an algorithmic procedure that can be used to perform the transformation for a general problem.

In order to construct the algorithm that led to an appropriate fixed-point method in the one-dimensional case, we attempted to find a function ϕ with the property that the function g, given by

$$(9.8) \qquad\qquad g(x) = x - \phi(x)f(x),$$

gives quadratic convergence to the fixed point p of g. From this condition, Newton's method evolved, by choosing $\phi(x) = 1/f'(x)$, provided that $f'(p) \ne 0$ at the fixed point p.

Using a similar approach in the n-dimensional case involves a matrix

(9.9)
$$A(\mathbf{x}) = \begin{bmatrix} a_{11}(\mathbf{x}) & a_{12}(\mathbf{x}) & \cdots & a_{1n}(\mathbf{x}) \\ a_{21}(\mathbf{x}) & a_{22}(\mathbf{x}) & \cdots & a_{2n}(\mathbf{x}) \\ \vdots & \vdots & & \vdots \\ a_{n1}(\mathbf{x}) & a_{n2}(\mathbf{x}) & \cdots & a_{nn}(\mathbf{x}) \end{bmatrix}$$

where each of the entries $a_{ij}(\mathbf{x})$ is a function from R^n into R. The procedure requires that $A(\mathbf{x})$ be found so that

(9.10)
$$\mathbf{G}(\mathbf{x}) = \mathbf{x} - A(\mathbf{x})^{-1}\mathbf{F}(\mathbf{x})$$

gives quadratic convergence to the solution of $\mathbf{F}(\mathbf{x}) = \mathbf{0}$, provided, of course, that $A(\mathbf{x})$ is nonsingular at the fixed point.

The following theorem verifies that this approach can be used to motivate the choice of A.

Theorem 9.9   Suppose $\mathbf{p}$ is a solution of $\mathbf{G}(\mathbf{x}) = \mathbf{x}$ for some function $\mathbf{G} = (g_1, g_2, \ldots, g_n)$, mapping R^n into R^n. If a number $\delta > 0$ exists with the property that

i) $\partial g_i/\partial x_j$ is continuous on $N_\delta = \{\mathbf{x} \mid \|\mathbf{x} - \mathbf{p}\| < \delta\}$ for each $i = 1, 2, \ldots, n$ and $j = 1, 2, \ldots, n$,

ii) $\partial^2 g_i(\mathbf{x})/(\partial x_j \partial x_k)$ is continuous, and $|\partial^2 g_i(\mathbf{x})/(\partial x_j \partial x_k)| \leq M$ for some constant M whenever $\mathbf{x} \in N_\delta$ for each $i = 1, 2, \ldots, n, j = 1, 2, \ldots, n$, and $k = 1, 2, \ldots, n$,

iii) $\partial g_i(\mathbf{p})/\partial x_j = 0$ for each $i = 1, 2, \ldots, n$, and $j = 1, 2, \ldots, n$,

then the sequence generated by $\mathbf{x}^{(k)} = \mathbf{G}(\mathbf{x}^{(k-1)})$ converges quadratically to $\mathbf{p}$ for any choice of $\mathbf{x}^{(0)} \in N_\delta$ and

$$\|\mathbf{x}^{(k)} - \mathbf{p}\|_\infty \leq \frac{n^2 M}{2} \|\mathbf{x}^{(k-1)} - \mathbf{p}\|_\infty^2 \qquad \text{for each } k \geq 1.$$

This theorem parallels Theorem 2.9 in Section 2.4, and its proof requires being able to express $\mathbf{G}$ in terms of its Taylor series in n variables about the point $\mathbf{p}$.

In order to utilize Theorem 9.9, suppose that $A(\mathbf{x})$ is an $n \times n$ matrix of functions from R^n into R in the form of Eq. (9.9) where the specific entries will be chosen later. Assume, moreover, that $A(\mathbf{x})$ is nonsingular near a solution $\mathbf{p}$ of $\mathbf{F}(\mathbf{x}) = \mathbf{0}$, and let $b_{ij}(\mathbf{x})$ denote the entry of $A(\mathbf{x})^{-1}$ in the ith row and jth column. Since $\mathbf{G}(\mathbf{x}) = \mathbf{x} - A(\mathbf{x})^{-1}\mathbf{F}(\mathbf{x})$,

$$g_i(\mathbf{x}) = x_i - \sum_{j=1}^{n} b_{ij}(\mathbf{x}) f_j(\mathbf{x});$$

so
$$\frac{\partial g_i(\mathbf{x})}{\partial x_k} = \begin{cases} 1 - \displaystyle\sum_{j=1}^{n} \left(b_{ij}(\mathbf{x}) \frac{\partial f_j}{\partial x_k}(\mathbf{x}) + \frac{\partial b_{ij}}{\partial x_k}(\mathbf{x}) f_j(\mathbf{x}) \right), & \text{if } i = k, \\[4mm] -\displaystyle\sum_{j=1}^{n} \left(b_{ij}(\mathbf{x}) \frac{\partial f_j}{\partial x_k}(\mathbf{x}) + \frac{\partial b_{ij}(\mathbf{x})}{\partial x_k} f_j(\mathbf{x}) \right), & \text{if } i \neq k. \end{cases}$$

Theorem 9.9 implies that we need to have $\partial g_i(\mathbf{p})/\partial x_k = 0$ for each $i = 1, 2, \ldots, n$ and $k = 1, 2, \ldots, n$. This means that for $i = k$,

$$0 = 1 - \sum_{j=1}^{n} b_{ij}(\mathbf{p}) \frac{\partial f_j}{\partial x_i}(\mathbf{p}),$$

so

(9.11)
$$\sum_{j=1}^{n} b_{ij}(\mathbf{p}) \frac{\partial f_j}{\partial x_i}(\mathbf{p}) = 1,$$

and, when $k \neq i$,

$$0 = - \sum_{j=1}^{n} b_{ij}(\mathbf{p}) \frac{\partial f_j}{\partial x_k}(\mathbf{p}),$$

so

(9.12)
$$\sum_{j=1}^{n} b_{ij}(\mathbf{p}) \frac{\partial f_j}{\partial x_k}(\mathbf{p}) = 0.$$

Defining the matrix $J(\mathbf{x})$ by

(9.13)
$$J(\mathbf{x}) = \begin{bmatrix} \dfrac{\partial f_1(\mathbf{x})}{\partial x_1} & \dfrac{\partial f_1(\mathbf{x})}{\partial x_2} & \cdots & \dfrac{\partial f_1(\mathbf{x})}{\partial x_n} \\[2mm] \dfrac{\partial f_2(\mathbf{x})}{\partial x_1} & \dfrac{\partial f_2(\mathbf{x})}{\partial x_2} & \cdots & \dfrac{\partial f_2(\mathbf{x})}{\partial x_n} \\[2mm] \vdots & \vdots & & \vdots \\[2mm] \dfrac{\partial f_n(\mathbf{x})}{\partial x_1} & \dfrac{\partial f_n(\mathbf{x})}{\partial x_2} & \cdots & \dfrac{\partial f_n(\mathbf{x})}{\partial x_n} \end{bmatrix},$$

we see that conditions (9.11) and (9.12) require

$$A(\mathbf{p})^{-1} J(\mathbf{p}) = I, \qquad \text{the identity matrix,}$$

so
$$A(\mathbf{p}) = J(\mathbf{p}).$$

An appropriate choice for $A(\mathbf{x})$ is consequently $A(\mathbf{x}) = J(\mathbf{x})$ since condition (iii) in Theorem 9.9 is satisfied with this choice.

The function $\mathbf{G}$ is consequently

$$\mathbf{G}(\mathbf{x}) = \mathbf{x} - J(\mathbf{x})^{-1} \mathbf{F}(\mathbf{x}),$$

and the functional iteration procedure evolves from

(9.14)
$$\mathbf{x}^{(k)} = \mathbf{G}(\mathbf{x}^{(k-1)}) = \mathbf{x}^{(k-1)} - J(\mathbf{x}^{(k-1)})^{-1} \mathbf{F}(\mathbf{x}^{(k-1)}).$$

This method is called, quite reasonably, **Newton's method for nonlinear systems** and is generally expected to give quadratic convergence, provided that a sufficiently accurate starting value is known and $J(\mathbf{p})^{-1}$ exists.

The matrix $J(\mathbf{x})$, given in Eq. (9.13), is called the **Jacobian** matrix and has a number of applications in analysis. It might, in particular, be familiar to the reader due to its application in the multiple integration of a function of several variables over a region which requires a change of variables to be performed.

A definite weakness in the Newton's-method procedure arises from the necessity of inverting the matrix $J(\mathbf{x})$ at each step. In practice, the method is generally performed in a two-step manner. First, a vector $\mathbf{y}$ is found, which will satisfy $J(\mathbf{x}^{(k)})\mathbf{y} = -\mathbf{F}(\mathbf{x}^{(k)})$. After this has been accomplished, the new approximation, $\mathbf{x}^{(k+1)}$, can be obtained by adding $\mathbf{y}$ to $\mathbf{x}^{(k)}$. The following algorithm uses this two-step procedure.

Newton's Method for Systems Algorithm 9.1

To approximate the solution of the nonlinear system $\mathbf{F}(\mathbf{x}) = \mathbf{0}$ given an initial approximation $\mathbf{x}$:

INPUT number n of equations and unknowns; initial approximation $\mathbf{x} = (x_1, \ldots, x_n)$; tolerance TOL; maximum number of iterations N.

OUTPUT approximate solution $\mathbf{x} = (x_1, \ldots, x_n)$ or a message that the number of iterations was exceeded.

Step 1 Set $k = 1$.

Step 2 While ($k \leq N$) do Steps 3–7.

Step 3 Calculate $\mathbf{F}(\mathbf{x})$ and $J(\mathbf{x})$, where $J(\mathbf{x})_{i,j} = (\partial f_i(\mathbf{x})/\partial x_j)$ for $1 \leq i, j \leq n$.

Step 4 Solve the $n \times n$ linear system $J(\mathbf{x})\mathbf{y} = -\mathbf{F}(\mathbf{x})$.

Step 5 Set $\mathbf{x} = \mathbf{x} + \mathbf{y}$.

Step 6 If $\|\mathbf{y}\| < TOL$ then OUTPUT ($\mathbf{x}$);
                        (*Procedure completed successfully.*)
                        STOP.

Step 7 Set $k = k + 1$.

Step 8 OUTPUT ('Maximum number of iterations exceeded');
        STOP.

EXAMPLE 1 The nonlinear system

$$3x_1 - \cos(x_2 x_3) - \tfrac{1}{2} = 0,$$

$$x_1^2 - 81(x_2 + .1)^2 + \sin x_3 + 1.06 = 0,$$

$$e^{-x_1 x_2} + 20x_3 + \frac{10\pi - 3}{3} = 0,$$

was shown, in Example 2 of Section 9.1, to have a unique solution in

$$D = \{(x_1, x_2, x_3) | -1 \leq x_i \leq 1 \text{ for each } i = 1, 2, 3\}.$$

Newton's method will be used to approximate the solution when the initial approximation is $\mathbf{x}^{(0)} = (.1, .1, -.1)$.

The Jacobian matrix $J(\mathbf{x})$ for this system is given by

$$J((x_1, x_2, x_3)) = \begin{bmatrix} 3 & x_3 \sin x_2 x_3 & x_2 \sin x_2 x_3 \\ 2x_1 & -162(x_2 + .1) & \cos x_3 \\ -x_2 e^{-x_1 x_2} & -x_1 e^{-x_1 x_2} & 20 \end{bmatrix}$$

and

$$\begin{bmatrix} x_1^{(k)} \\ x_2^{(k)} \\ x_3^{(k)} \end{bmatrix} = \begin{bmatrix} x_1^{(k-1)} \\ x_2^{(k-1)} \\ x_3^{(k-1)} \end{bmatrix} + \begin{bmatrix} y_1^{(k-1)} \\ y_2^{(k-1)} \\ y_3^{(k-1)} \end{bmatrix},$$

where $\begin{bmatrix} y_1^{(k-1)} \\ y_2^{(k-1)} \\ y_3^{(k-1)} \end{bmatrix} = -(J(x_1^{(k-1)}, x_2^{(k-1)}, x_3^{(k-1)}))^{-1} \mathbf{F}(x_1^{(k-1)}, x_2^{(k-1)}, x_3^{(k-1)}).$

Thus, at the kth step, the linear system

$$\begin{bmatrix} 3 & x_3^{(k-1)} \sin x_2^{(k-1)} x_3^{(k-1)} & x_2^{(k-1)} \sin x_2^{(k-1)} x_3^{(k-1)} \\ 2x_1^{(k-1)} & -162(x_2^{(k-1)} + .1) & \cos x_3^{(k-1)} \\ -x_2^{(k-1)} e^{-x_1^{(k-1)} x_2^{(k-1)}} & -x_1^{(k-1)} e^{-x_1^{(k-1)} x_2^{(k-1)}} & 20 \end{bmatrix}$$

$$\times \begin{bmatrix} x_1^{(k)} - x_1^{(k-1)} \\ x_2^{(k)} - x_2^{(k-1)} \\ x_3^{(k)} - x_3^{(k-1)} \end{bmatrix} = - \begin{bmatrix} 3x_1^{(k-1)} - \cos x_2^{(k-1)} x_3^{(k-1)} - \frac{1}{2} \\ (x_1^{(k-1)})^2 - 81(x_2^{(k-1)} + .1)^2 + \sin x_3^{(k-1)} + 1.06 \\ e^{-x_1^{(k-1)} x_2^{(k-1)}} + 20x_3^{(k-1)} + \frac{10\pi - 3}{3} \end{bmatrix}$$

must be solved. The results obtained using this iterative procedure are shown in Table 9.3.

TABLE 9.3	k	$x_1^{(k)}$	$x_2^{(k)}$	$x_3^{(k)}$	$\|\mathbf{x}^{(k)} - \mathbf{x}^{(k-1)}\|_\infty$
	0	.10000000	.10000000	$-.10000000$	
	1	.50003702	.01946686	$-.52152047$	.422
	2	.50004593	.00158859	$-.52355711$	1.79×10^{-2}
	3	.50000034	.00001244	$-.52359845$	1.58×10^{-3}
	4	.50000000	.00000000	$-.52359877$	1.24×10^{-5}
	5	.50000000	.00000000	$-.52359877$	0

The actual solution to Example 1 is

$$\mathbf{p} = \left(.5, 0, -\frac{\pi}{6}\right) \approx (.5, 0, -.52359877).$$

Note that the convergence of Newton's method becomes very fast once an iterate is close to $\mathbf{p}$. This illustrates the quadratic convergence of the method near a solution.   □

Exercise Set 9.2

1. Solve the following nonlinear systems using Newton's method and compute the error bound given in Theorem 9.9. Iterate until $\|\mathbf{x}^{(i)} - \mathbf{x}^{(i+1)}\|_\infty < 10^{-5}$.

   a) $x_1^2 - 10x_1 + x_2^2 + 8 = 0,$
   $x_1 x_2^2 + x_1 - 10x_2 + 8 = 0.$

   Compare the convergence to that of Exercise 6, Section 9.1.

   b) $x_1^2 + x_2^2 - x_1 = 0,$
   $x_1^2 - x_2^2 - x_2 = 0.$

   Compare the convergence to that of Exercise 5, Section 9.1.

   c) $3x_1^2 - x_2^2 = 0,$
   $3x_1 x_2^2 - x_1^3 - 1 = 0.$

   d) $x_1^2 + x_2 - 37 = 0,$
   $x_1 - x_2^2 - 5 = 0,$
   $x_1 + x_2 + x_3 - 3 = 0.$

   e) $x_1 + \cos(x_1 x_2 x_3) - 1 = 0,$
   $(1 - x_1)^{1/4} + x_2 + .05x_3^2 - .15x_3 - 1 = 0,$
   $-x_1^2 - .1x_2^2 + .01x_2 + x_3 - 1 = 0.$

   Compare the convergence to that of Exercise 8(c), Section 9.1.

   f) $12x_1 - 3x_2^2 - 4x_3 = 7.17,$
   $x_1^2 + 10x_2 - x_3 = 11.54,$
   $x_2^3 + 7x_3 = 7.631.$

   Compare the convergence to that of Exercise 8(b), Section 9.1.

   g) $4x_1 - x_2 + x_3 - 2 = 0,$
   $-x_1 + 6x_2 - 3x_3 - 11 = 0,$
   $x_1 - 3x_2 + 5x_3 + 5 = 0.$

2. Can Newton's method be used to solve the following nonlinear system?

$$3x_1 - \cos(x_2 x_3) - .5 = 0,$$
$$x_1^2 - 625x_2^2 = 0,$$
$$e^{-x_1 x_2} + 20x_3 + \frac{10\pi - 3}{3} = 0.$$

3. Find the Jacobian matrix of the linear system:

$$c_{11}x_1 + c_{12}x_2 + \cdots + c_{1n}x_n - b_1 = 0,$$
$$c_{21}x_1 + c_{22}x_2 + \cdots + c_{2n}x_n - b_2 = 0,$$
$$\vdots \qquad\qquad\qquad \vdots$$
$$c_{n1}x_1 + c_{n2}x_2 + \cdots + c_{nn}x_n - b_n = 0.$$

When will this matrix have an inverse?

4. C. Chiarella, W. Charlton, and A. W. Roberts [26], in calculating the shape of a gravity-flow discharge chute which will minimize transit time of discharged granular particles, solve the following equations by Newton's method:

   i) $f_n(\theta_1, \ldots, \theta_N) = \dfrac{\sin \theta_{n+1}}{v_{n+1}} (1 - \mu w_{n+1}) - \dfrac{\sin \theta_n}{v_n} (1 - \mu w_n) = 0,$

   for each $n = 1, 2, \ldots, N - 1$;

   ii) $f_N(\theta_1, \ldots, \theta_N) = \Delta y \displaystyle\sum_{i=1}^{N} \tan \theta_i - X = 0,$

   where

   a) $\qquad v_n^2 = v_0^2 + 2gn\Delta y - 2\mu\Delta y \displaystyle\sum_{j=1}^{n} \dfrac{1}{\cos \theta_j} \qquad$ for each $n = 1, 2, \ldots, N$, and

   b) $\qquad w_n = -\Delta y v_n \displaystyle\sum_{i=1}^{N} \dfrac{1}{v_i^3 \cos \theta_i} \qquad$ for each $n = 1, 2, \ldots, N$.

   The constant v_0 is the initial velocity of the granular material, X is the x-coordinate of the end of the chute, μ is the friction force, N is the number of chute segments, and g is the gravitational constant. The variable θ_i is the angle of the ith chute segment from the vertical as shown in the following figure, and v_i is the particle velocity in the ith chute segment. Solve (i) and (ii) for $\theta = (\theta_1, \ldots, \theta_N)$ using Newton's method with $\mu = 0$, $X = 2$, $\Delta y = .2$, $N = 20$, $g = 32.2$ ft/sec^2 where the values for v_n and w_n can be obtained directly from (a) and (b). Iterate until $\|\theta^{(i+1)} - \theta^{(i)}\| < 10^{-2}$ radians.

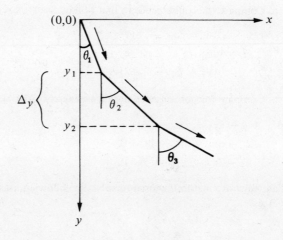

5. The amount of pressure required to sink a large heavy object in a soft homogeneous soil that lies above a hard base soil can be predicted by the amount of pressure required to sink smaller objects in the same soil. In particular, the amount of pressure p required to sink a circular plate of radius r a distance d in the soft soil, where the hard base soil lies a distance $D > d$ below the surface, can be approximated by an equation of the form

$$p = k_1 e^{k_2 r} + k_3 r,$$

where $k_1, k_2,$ and k_3 are constants, with $k_2 > 0$, depending on d and the consistency of the soil but not on the radius of the plate. (See Bekker [10], pages 89–94.)

   a) Find the values of $k_1, k_2,$ and k_3 if we assume that a plate of radius 1 inch requires a pressure of 10 lb/(in)^2 to sink 1 foot in a muddy field, a plate of radius 2 inches requires a pressure of 12 lb/(in)^2 to sink one foot, and a plate of radius 3 inches requires a pressure of 15 lb/(in)^2 to sink this distance (assuming that the mud is more than one foot deep).

b) Use your calculations from (a) to predict the minimal size of circular plate that would be required to sustain a load of 500 lb on this field with sinkage of less than 1 foot.

6. Exercise 10 of Section 7.1 dealt with determining an exponential least squares relationship of the form $R = bw^a$ to approximate a collection of data relating the weight and respiration rule of *Modest sphinx* moths. In that exercise, the problem was converted to a log-log relationship, and in part (c), a quadratic term was introduced in an attempt to improve the approximation. Instead of converting the problem, use Newton's method to determine the constants a and b that minimize $\sum_{i=1}^{n} [R_i - bw_i^a]^2$ for the data listed in Exercise 10 of Section 7.1. Compute the error associated with this approximation, and compare this to the error of the previous approximations for this problem.

7. An interesting biological experiment, see Schroeder [76], concerns the determination of the maximum water temperature, X_M, at which various species of hydra can survive without shortened life expectancy. One approach to the solution of this problem uses a weighted least squares fit of the form $f(x) = y = a/(x - b)^c$ to a collection of experimental data. The x values of the data refer to water temperatures above X_M, and the y values refer to the average life expectancy at that temperature. The constant b is the asymptote of the graph of f and as such is an approximation to X_M.

a) Show that choosing a, b, c to minimize

$$\sum_{i=1}^{n} \left[w_i y_i - \frac{a}{(x_i - b)^c} \right]^2$$

reduces to solving the nonlinear system

$$a = \left(\sum_{i=1}^{n} \frac{y_i w_i}{(x_i - b)^c} \right) \bigg/ \left(\sum_{i=1}^{n} \frac{w_i}{(x_i - b)^{2c}} \right)$$

$$0 = \sum_{i=1}^{n} \frac{y_i w_i}{(x_i - b)^c} \cdot \sum_{i=1}^{n} \frac{w_i}{(x_i - b)^{2c+1}} - \sum_{i=1}^{n} \frac{y_i w_i}{(x_i - b)^{c+1}} \cdot \sum_{i=1}^{n} \frac{w_i}{(x_i - b)^{2c}}$$

$$0 = \sum_{i=1}^{n} \frac{y_i w_i}{(x_i - b)^c} \cdot \sum_{i=1}^{n} \frac{w_i \ln(x_i - b)}{(x_i - b)^{2c}} - \sum_{i=1}^{n} \frac{w_i y_i \ln(x_i - b)}{(x_i - b)^c} \cdot \sum_{i=1}^{n} \frac{w_i}{(x_i - b)^{2c}}$$

b) Solve the nonlinear system for the species whose data is given below. Use the weights $w_i = \ln y_i$.

i	1	2	3	4
x_i	2.40	3.80	4.75	21.60
y_i	31.8	31.5	31.2	30.2

9.3 Quasi-Newton Methods

A significant weakness of Newton's method for solving systems of nonlinear equations lies in the requirement that, at each iteration, a Jacobian matrix be computed and an $n \times n$ linear system solved that involves this matrix. To illustrate the magnitude of this weakness, let us consider the amount of calculation associated with one iteration of Newton's method. The Jacobian matrix associated with a system of n nonlinear equations written in the form $\mathbf{F}(\mathbf{x}) = \mathbf{0}$ requires that the n^2

partial derivatives of the n component functions of $\mathbf{F}$ be determined and evaluated. In most situations, the exact evaluation of the partial derivatives is inconvenient and in many applications impossible. This difficulty can be overcome by using finite difference approximations to the partial derivatives. For example,

$$(9.15) \qquad \frac{\partial f_j}{\partial x_k}(\mathbf{x}^{(i)}) \approx \frac{f_j(\mathbf{x}^{(i)} + \mathbf{e}_k h) - f_j(\mathbf{x}^{(i)})}{h}$$

where h is small in absolute value and $\mathbf{e}_k$ is the vector whose only nonzero entry is a one in the kth coordinate. This approximation, however, still requires that at least n^2 scalar functional evaluations be performed to approximate the Jacobian and does not decrease the amount of calculation, in general $O(n^3)$, required for solving the linear system involving this approximate Jacobian. The total computational effort for one iteration of Newton's method is consequently at least $n^2 + n$ scalar functional evaluations (n^2 for the evaluation of the Jacobian matrix and n for the evaluation of $\mathbf{F}$) together with $O(n^3)$ arithmetic operations to solve the linear system. This amount of computational effort is prohibitive except for relatively small values of n and easily evaluated scalar functions.

In this section we will consider a generalization of the Secant method to systems of nonlinear equations, in particular a technique known as **Broyden's method** (see Broyden [18]). The method requires only n scalar functional evaluations per iteration and also reduces the number of arithmetic calculations to $O(n^2)$. It is one of a class of methods known as *least-change secant updates* that produce algorithms called **quasi-Newton**. These methods replace the Jacobian matrix in Newton's method with an approximation matrix that is updated at each iteration. The disadvantage to the methods is that the quadratic convergence of Newton's method is lost, being replaced in general by super linear convergence—

$$\lim_{i \to \infty} \frac{\|\mathbf{x}^{(i+1)} - \mathbf{p}\|}{\|\mathbf{x}^{(i)} - \mathbf{p}\|} = 0$$

where $\mathbf{p}$ denotes the solution to $\mathbf{F}(\mathbf{x}) = \mathbf{0}$ and $\mathbf{x}^{(i)}$, $\mathbf{x}^{(i-1)}$ consecutive approximations (see Exercise 17, Section 2.4). In most applications, the reduction to super linear convergence is a more than acceptable trade-off for the decrease in the amount of computation. An additional disadvantage is that, unlike Newton's method, they are not self correcting.

Procedures are also available that maintain quadratic convergence but significantly reduce the number of required functional evaluations. Methods of this type were originally proposed by Brown.* A survey and comparison of some commonly used methods of this type can be found in Moré and Cosnard.** In general, however, these methods are much more difficult to efficiently implement than the method we will now consider.

Suppose that an initial approximation $\mathbf{x}^{(0)}$ is given to the solution $\mathbf{p}$ of $\mathbf{F}(\mathbf{x}) = \mathbf{0}$. We calculate the next approximation $\mathbf{x}^{(1)}$ in the same manner as Newton's method or, if it is inconvenient to determine $J(\mathbf{x}^{(0)})$ exactly, we use the difference equations given by (9.15) to approximate the partial derivatives. At this point we

* Brown, K. M. (1969), "A quadratically convergent Newton-like method based upon Gaussian elimination." *SIAM Journal of Numerical Analysis*, **6**, No. 4, 560–569.
** Moré, J. J., and M. Y. Cosnard (1979), "Numerical solution of nonlinear equations." *ACM Transactions on Mathematical Software*, **5**, No. 1, 64–85.

depart from Newton's method and examine the Secant method for a single non-linear equation. The Secant method uses the approximation

$$\frac{f(x_1) - f(x_0)}{x_1 - x_0}$$

as a replacement for $f'(x_1)$ in Newton's method. For nonlinear systems, $\mathbf{x}^{(1)} - \mathbf{x}^{(0)}$ is a vector, and the corresponding quotient is undefined. However, the method proceeds similarly in that we replace the matrix $J(\mathbf{x}^{(1)})$ in Newton's method by a matrix A_1 with the property that

$$(9.16) \qquad A_1(\mathbf{x}^{(1)} - \mathbf{x}^{(0)}) = \mathbf{F}(\mathbf{x}^{(1)}) - \mathbf{F}(\mathbf{x}^{(0)}).$$

This equation does not define the matrix uniquely, however, because it does not describe how A_1 operates on vectors orthogonal to $\mathbf{x}^{(1)} - \mathbf{x}^{(0)}$. Since no information is available about the change in $\mathbf{F}$ in any direction other than that of $\mathbf{x}^{(1)} - \mathbf{x}^{(0)}$, we will require additionally of A_1 that

$$(9.17) \qquad A_1\mathbf{z} = J(\mathbf{x}^{(0)})\mathbf{z} \qquad \text{whenever } (\mathbf{x}^{(1)} - \mathbf{x}^{(0)})^t\mathbf{z} = 0.$$

Conditions (9.16) and (9.17) uniquely define A_1 (see Dennis and Moré [30], page 54.) as

$$A_1 = J(\mathbf{x}^{(0)}) + \frac{[\mathbf{F}(\mathbf{x}^{(1)}) - \mathbf{F}(\mathbf{x}^{(0)}) - J(\mathbf{x}^{(0)})(\mathbf{x}^{(1)} - \mathbf{x}^{(0)})](\mathbf{x}^{(1)} - \mathbf{x}^{(0)})^t}{\|\mathbf{x}^{(1)} - \mathbf{x}^{(0)}\|_2^2},$$

and it is this matrix that is used in place of $J(\mathbf{x}^{(1)})$ to determine $\mathbf{x}^{(2)}$:

$$\mathbf{x}^{(2)} = \mathbf{x}^{(1)} - A_1^{-1}\mathbf{F}(\mathbf{x}^{(1)}).$$

The method can then be repeated to determine $\mathbf{x}^{(3)}$ by using A_1 in place of $A_0 \equiv J(\mathbf{x}^{(0)})$ and with $\mathbf{x}^{(2)}$ and $\mathbf{x}^{(1)}$ in place of $\mathbf{x}^{(1)}$ and $\mathbf{x}^{(0)}$. In general, once $\mathbf{x}^{(i)}$ has been determined, $\mathbf{x}^{(i+1)}$ is computed by

$$(9.18) \qquad A_i = A_{i-1} + \frac{(\mathbf{y}_i - A_{i-1}\mathbf{s}_i)\mathbf{s}_i^t}{\|\mathbf{s}_i\|_2^2}$$

$$(9.19) \qquad \mathbf{x}^{(i+1)} = \mathbf{x}^{(i)} - A_i^{-1}\mathbf{F}(\mathbf{x}^{(i)}),$$

where the notation $\mathbf{y}_i = \mathbf{F}(\mathbf{x}^{(i)}) - \mathbf{F}(\mathbf{x}^{(i-1)})$ and $\mathbf{s}_i = \mathbf{x}^{(i)} - \mathbf{x}^{(i-1)}$ is introduced into (9.18) to simplify the equation.

If the method is performed as outlined in equations (9.18) and (9.19), the number of scalar functional evaluations is reduced from $n^2 + n$ to n (those required for evaluating $\mathbf{F}(\mathbf{x}^{(i)})$), but the method still requires $O(n^3)$ calculations to solve the associated $n \times n$ linear system (see Step 4 in Algorithm 9.1)

$$(9.20) \qquad A_i\mathbf{y}^{(i)} = -\mathbf{F}(\mathbf{x}^{(i)}).$$

Employing the method in this form would ordinarily not be justified because of the reduction to super linear convergence from the quadratic convergence of Newton's method.

A considerable improvement can be incorporated, however, by employing a matrix inversion formula of Sherman and Morrison (see, for example, Dennis and Moré [30] page 55). This result states that if A is a nonsingular matrix and $\mathbf{x}$ and $\mathbf{y}$

are vectors, then $A + \mathbf{x}\mathbf{y}^t$ is nonsingular provided that $\mathbf{y}^t A^{-1}\mathbf{x} \neq -1$. Moreover, in this case,

$$(9.21) \qquad (A + \mathbf{x}\mathbf{y}^t)^{-1} = A^{-1} - \frac{A^{-1}\mathbf{x}\mathbf{y}^t A^{-1}}{1 + \mathbf{y}^t A^{-1}\mathbf{x}}.$$

This formula permits A_i^{-1} to be computed directly from A_{i-1}^{-1}. By letting $A = A_{i-1}$, $\mathbf{x} = (\mathbf{y}_i - A_{i-1}\mathbf{s}_i)/\|\mathbf{s}_i\|_2^2$, and $\mathbf{y} = \mathbf{s}_i$, formula (9.18) together with (9.21) implies that

$$
\begin{aligned}
A_i^{-1} &= \left(A_{i-1} + \frac{(\mathbf{y}_i - A_{i-1}\mathbf{s}_i)}{\|\mathbf{s}_i\|_2^2}\mathbf{s}_i^t\right)^{-1} \\
&= A_{i-1}^{-1} - \frac{A_{i-1}^{-1}\left(\dfrac{(\mathbf{y}_i - A_{i-1}\mathbf{s}_i)}{\|\mathbf{s}_i\|_2^2}\mathbf{s}_i^t\right)A_{i-1}^{-1}}{1 + \mathbf{s}_i^t A_{i-1}^{-1}\left(\dfrac{(\mathbf{y}_i - A_{i-1}\mathbf{s}_i)}{\|\mathbf{s}_i\|_2^2}\right)} \\
&= A_{i-1}^{-1} - \frac{(A_{i-1}^{-1}\mathbf{y}_i - \mathbf{s}_i)\mathbf{s}_i^t A_{i-1}^{-1}}{\|\mathbf{s}_i\|_2^2 + \mathbf{s}_i^t A_{i-1}^{-1}\mathbf{y}_i - \|\mathbf{s}_i\|_2^2};
\end{aligned}
$$

so

$$(9.22) \qquad A_i^{-1} = A_{i-1}^{-1} + \frac{(\mathbf{s}_i - A_{i-1}^{-1}\mathbf{y}_i)\mathbf{s}_i^t A_{i-1}^{-1}}{\mathbf{s}_i^t A_{i-1}^{-1}\mathbf{y}_i}.$$

This computation involves only matrix multiplication at each step and as such requires only $O(n^2)$ arithmetic calculations. The calculation of A_i is therefore bypassed, as is the necessity of solving the linear system (9.20). Algorithm 9.2 follows directly from this construction.

Broyden Algorithm 9.2

To approximate the solution of the nonlinear system $\mathbf{F}(\mathbf{x}) = \mathbf{0}$ given an initial approximation $\mathbf{x}$:

INPUT number n of equations and unknowns; initial approximation $\mathbf{x} = (x_1, \ldots, x_n)$; tolerance TOL; maximum number of iterations N.

OUTPUT approximate solution $\mathbf{x} = (x_1, \ldots, x_n)$ or a message that the number of iterations was exceeded.

Step 1   Set $A_0 = J(\mathbf{x})$ where $J(\mathbf{x})_{i,j} = \dfrac{\partial f_i(\mathbf{x})}{\partial x_j}$ for $1 \leq i, j \leq n$;

$\mathbf{v} = \mathbf{F}(\mathbf{x})$.   (*Note:* $\mathbf{v} = \mathbf{F}(\mathbf{x}^{(0)})$.)

Step 2   Set $A = A_0^{-1}$.

Step 3   Set $k = 1$;
$\mathbf{s} = -A\mathbf{v}$;   (*Note:* $\mathbf{s} = \mathbf{s}_1$.)
$\mathbf{x} = \mathbf{x} + \mathbf{s}$.   (*Note:* $\mathbf{x} = \mathbf{x}^{(1)}$.)

Step 4   While ($k \leq N$) do Steps 5–13.

Step 5   Set $\mathbf{w} = \mathbf{v}$;   (*Save* $\mathbf{v}$.)
$\mathbf{v} = \mathbf{F}(\mathbf{x})$;   (*Note:* $\mathbf{v} = \mathbf{F}(\mathbf{x}^{(k)})$.)
$\mathbf{y} = \mathbf{v} - \mathbf{w}$.   (*Note:* $\mathbf{y} = \mathbf{y}_k$.)

Step 6 Set $\mathbf{z} = -A\mathbf{y}$. (*Note:* $\mathbf{z} = -A_{k-1}^{-1}\mathbf{y}_k$.)

Step 7 Set $p = -\mathbf{s}^t\mathbf{z}$. (*Note:* $p = \mathbf{s}_k^t A_{k-1}^{-1}\mathbf{y}_k$.)

Step 8 Set $C = pI + (\mathbf{s} + \mathbf{z})\mathbf{s}^t$.
$\quad\quad$ (*Note:* $C = \mathbf{s}_k^t A_{k-1}^{-1}\mathbf{y}_k I + (\mathbf{s}_k + A_{k-1}^{-1}\mathbf{y}_k)\mathbf{s}_k^t$.)

Step 9 Set $A = (1/p)CA$. (*Note:* $A = A_k^{-1}$.)

Step 10 Set $\mathbf{s} = -A\mathbf{v}$. (*Note:* $\mathbf{s} = -A_k^{-1}\mathbf{F}(\mathbf{x}^{(k)})$.)

Step 11 Set $\mathbf{x} = \mathbf{x} + \mathbf{s}$. (*Note:* $\mathbf{x} = \mathbf{x}^{(k+1)}$.)

Step 12 If $\|\mathbf{s}\| < TOL$ then OUTPUT (**x**);
$\quad\quad\quad\quad\quad\quad\quad\quad$ (*Procedure completed successfully.*)
$\quad\quad\quad\quad\quad\quad\quad\quad$ STOP.

Step 13 Set $k = k + 1$.

Step 14 OUTPUT ('Maximum number of iterations exceeded');
$\quad\quad\quad$ STOP.

EXAMPLE 1 The nonlinear system

$$3x_1 - \cos(x_2 x_3) - \tfrac{1}{2} = 0,$$

$$x_1^2 - 81(x_2 + .1)^2 + \sin x_3 + 1.06 = 0,$$

$$e^{-x_1 x_2} + 20x_3 + \frac{10\pi - 3}{3} = 0,$$

was solved by Newton's method in Example 1 of Section 9.2. The Jacobian matrix for this system is

$$J(x_1, x_2, x_3) = \begin{bmatrix} 3 & x_3 \sin x_2 x_3 & x_2 \sin x_2 x_3 \\ 2x_1 & -162(x_2 + .1) & \cos x_3 \\ -x_2 e^{-x_1 x_2} & -x_1 e^{-x_1 x_2} & 20 \end{bmatrix}.$$

With $\mathbf{x}^{(0)} = (.1, .1, -.1)^t$, we have

$$\mathbf{F}(\mathbf{x}^{(0)}) = \begin{bmatrix} -1.199949 \\ -2.269832 \\ 8.462026 \end{bmatrix},$$

$J(x_1^{(0)}, x_2^{(0)}, x_3^{(0)})$

$$= \begin{bmatrix} 3 & 9.999836 \times 10^{-4} & -9.999836 \times 10^{-4} \\ .2 & -323.9999 & .9950041 \\ -9.900498 \times 10^{-2} & -9.900498 \times 10^{-2} & 20 \end{bmatrix}$$

$A_0^{-1} = J(x_1^{(0)}, x_2^{(0)}, x_3^{(0)})^{-1}$

$$= \begin{bmatrix} .3333331 & 1.023852 \times 10^{-5} & 1.615703 \times 10^{-5} \\ 2.108606 \times 10^{-3} & -3.086882 \times 10^{-2} & 1.535838 \times 10^{-3} \\ 1.660522 \times 10^{-3} & -1.527579 \times 10^{-4} & 5.000774 \times 10^{-2} \end{bmatrix},$$

$$\mathbf{x}^{(1)} = \mathbf{x}^{(0)} - A_0^{-1}\mathbf{F}(\mathbf{x}^{(0)}) = \begin{bmatrix} .4998693 \\ 1.946693 \times 10^{-2} \\ -.5215209 \end{bmatrix},$$

$$\mathbf{F}(\mathbf{x}^{(1)}) = \begin{bmatrix} -3.404021 \times 10^{-4} \\ -.3443899 \\ 3.18737 \times 10^{-2} \end{bmatrix},$$

$$\mathbf{y}_1 = \mathbf{F}(\mathbf{x}^{(1)}) - \mathbf{F}(\mathbf{x}^{(0)}) = \begin{bmatrix} 1.199608 \\ 1.925442 \\ -8.430152 \end{bmatrix},$$

$$\mathbf{s}_1 = \begin{bmatrix} .3998693 \\ -8.053307 \times 10^{-2} \\ -.4215209 \end{bmatrix},$$

$$\mathbf{s}_1^t A_0^{-1}\mathbf{y}_1 = .3424604,$$

$$A_1^{-1} = A_0^{-1} + (1/.3424604)[(\mathbf{s}_1 - A_0^{-1}\mathbf{y}_1)A_0^{-1}]$$

$$= \begin{bmatrix} .3333781 & 1.11077 \times 10^{-5} & 8.944584 \times 10^{-6} \\ -2.021271 \times 10^{-3} & -3.094847 \times 10^{-2} & 2.196909 \times 10^{-3} \\ 1.022381 \times 10^{-3} & -1.650679 \times 10^{-4} & 5.010987 \times 10^{-2} \end{bmatrix},$$

and

$$\mathbf{x}^{(2)} = \mathbf{x}^{(1)} - A_1^{-1}\mathbf{F}(\mathbf{x}^{(1)}) = \begin{bmatrix} .4999863 \\ 8.737888 \times 10^{-3} \\ -.5231746 \end{bmatrix}.$$

Additional iterations are listed in Table 9.4.   □

TABLE 9.4

k	$x_1^{(k)}$	$x_2^{(k)}$	$x_3^{(k)}$	$\|\mathbf{x}^{(k)} - \mathbf{x}^{(k-1)}\|_2$
3	.5000066	8.672215×10^{-4}	$-.5236918$	7.88×10^{-3}
4	.5000005	6.087473×10^{-5}	$-.5235954$	8.12×10^{-4}
5	.5000002	-1.445223×10^{-6}	$-.5235989$	6.24×10^{-5}

Exercise Set 9.3

1. Use Algorithm 9.2 to approximate the solutions to the following systems of nonlinear equations. Iterate until $\|\mathbf{x}^{(i+1)} - \mathbf{x}^{(i)}\|_\infty < 10^{-5}$. Compare the number of iterations required for this accuracy to the number required for the corresponding approximation by Newton's method computed in Exercise 1 of Section 9.2.

   a) $x_1^2 + x_2 - 37 = 0,$

   $\qquad x_1 - x_2^2 - 5 = 0,$

   $x_1 + x_2 + x_3 - 3 = 0.$

b) $x_1 + \cos(x_1 x_2 x_3) - 1 = 0,$

$(1 - x_1)^{1/4} + x_2 + .05x_3^2 - .15x_3 - 1 = 0,$

$-x_1^2 - .1x_2^2 + .01x_2 + x_3 - 1 = 0.$

Compare the convergence to that of Exercise 8(c), Section 9.1.

c) $12x_1 - 3x_2^2 - 4x_3 = 7.17,$

$x_1^2 + 10x_2 - x_3 = 11.54,$

$x_2^3 + 7x_3 = 7.631.$

Compare the convergence to that of Exercise 8(b), Section 9.1.

d) $4x_1 - x_2 + x_3 - 2 = 0,$

$-x_1 + 6x_2 - 3x_3 - 11 = 0,$

$x_1 - 3x_2 + 5x_3 + 5 = 0.$

2. Newton's method cannot be employed to solve the following system of nonlinear equations. Does Broyden's method lead to a solution?

$$3x_1 - \cos(x_2 x_3) - .5 = 0,$$

$$x_1^2 - 625x_2^2 = 0,$$

$$e^{-x_1 x_2} + 20x_3 + \frac{10\pi - 3}{3} = 0.$$

3. Use both Algorithms 9.1 and 9.2 to attempt a solution to the system:

$$x_1 + x_2^3 - 5x_2^2 - 2x_2 = 10$$

$$x_1 + x_2^3 + x_2^2 - 14x_2 = 29$$

starting with $x_1 = 15, x_2 = -2$. Solve this problem instead by solving the first equation for x_1 and using this in the second to give a single nonlinear equation. Which method produces the best results?

4. Verify that Eq. (9.21) is correct by showing that

$$\left[A^{-1} - \frac{A^{-1}\mathbf{x}\mathbf{y}^t A^{-1}}{1 + \mathbf{y}^t A^{-1}\mathbf{x}} \right](A + \mathbf{x}\mathbf{y}^t) = I.$$

5. Exercise 4 of Section 9.2 requires that a system of 20 nonlinear equations be solved for $\boldsymbol{\theta} = (\theta_1, \theta_2, \ldots, \theta_{20})$ in order to determine the chute that will give the minimal transit time for granular particles. Use Broyden's method to approximate $\boldsymbol{\theta}$ by computing the iterations $\boldsymbol{\theta}^{(1)}, \boldsymbol{\theta}^{(2)}, \ldots,$ until $\|\boldsymbol{\theta}^{(i)} - \boldsymbol{\theta}^{(i+1)}\|_\infty < 10^{-2}$ radians. Compare the amount of computation time required for this method to the amount required to solve the corresponding problem by Newton's method.

6. Apply Broyden's method to the least squares problem described in Exercise 7 of Section 9.2.

10

Boundary-Value Problems for Ordinary Differential Equations

A common problem in civil engineering concerns the deflection of a beam of rectangular cross section, which is subject to uniform loading, while the ends of the beam are supported so that they undergo no deflection.

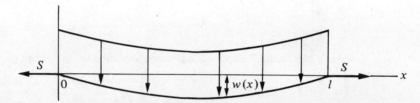

The differential equation approximating the physical situation is of the form

$$\frac{d^2w}{dx^2} - \frac{S}{EI}\,w = \frac{qx}{2EI}\,(x - l),$$

where $w = w(x)$ is the deflection a distance x from the left end of the beam, and l, q, E, S, and I represent, respectively, the length of the beam, the intensity of the uniform load, the modulus of elasticity, the stress at the endpoints, and the central moment of inertia. Associated with this differential equation are the two conditions given by the assumption that no deflection occurs at the ends of the beam, $w(0) = w(l) = 0$.

When the beam is of uniform thickness, the product EI will be constant, and the exact solution can be easily obtained. In many applications, however, the thickness is not uniform, so the moment of inertia I is a function of x, and approximation techniques are required.

Although methods for finding approximate solutions to differential equations were studied in Chapter 5, those techniques required that all conditions imposed on the differential equation occur at an initial point. For a second order equation, we need to know both $w(0)$ and $w'(0)$, which is not the case in this problem. New techniques are required for handling problems when the conditions imposed are of a boundary-value type rather than an initial-value type.

Physical problems that are position-dependent rather than time-dependent are often described in terms of differential equations with conditions imposed at more than one point. The general two-point boundary-value problems we will discuss in this chapter involve a second-order differential equation of the form

$$(10.1) \qquad y'' = f(x, y, y'), \qquad a \le x \le b,$$

together with the boundary conditions

$$(10.2) \qquad y(a) = \alpha \qquad \text{and} \qquad y(b) = \beta.$$

Most of the material concerning second-order boundary-value problems can be extended to problems with boundary conditions of the form

$$(10.3) \qquad \alpha_1 y(a) - \beta_1 y'(a) = \alpha \qquad \text{and} \qquad \alpha_2 y(b) + \beta_2 y'(b) = \beta,$$

where $|\alpha_1| + |\beta_1| \ne 0$ and $|\alpha_2| + |\beta_2| \ne 0$, but some of the techniques become quite complicated. The reader who is interested in problems of this type would be advised to consider a book specializing in boundary-value problems, such as Keller [56].

10.1   The Linear Shooting Method

The following theorem gives general conditions that ensure that the solution to a second-order boundary value problem will exist and be unique. The proof of this theorem in the general situation can be found in the book by Keller [56].

Theorem 10.1   Suppose the function f in the boundary-value problem

$$y'' = f(x, y, y'), \qquad a \le x \le b, \quad y(a) = \alpha, \quad y(b) = \beta,$$

is continuous on the set

$$D = \{(x, y, y') \mid a \le x \le b, -\infty < y < \infty, -\infty < y' < \infty\},$$

and that $\partial f / \partial y$ and $\partial f / \partial y'$ are also continuous on D. If

i) $\dfrac{\partial f}{\partial y}(x, y, y') > 0$ for all $(x, y, y') \in D$, and

ii)   a constant M exists, with

$$\left| \frac{\partial f}{\partial y'}(x, y, y') \right| \le M \qquad \text{for all } (x, y, y') \in D,$$

then the boundary-value problem has a unique solution.

EXAMPLE 1   The boundary-value problem

$$y'' + e^{-xy} + \sin y' = 0, \qquad 1 \le x \le 2, \quad y(1) = y(2) = 0,$$

has

$$f(x, y, y') = -e^{-xy} - \sin y',$$

and since

$$\frac{\partial f}{\partial y}(x, y, y') = xe^{-xy} > 0 \quad \text{and} \quad \left| \frac{\partial f}{\partial y'}(x, y, y') \right| = |-\cos y'| \le 1,$$

this problem has a unique solution.   □

When $f(x, y, y')$ can be expressed in the form

$$f(x, y, y') = p(x)y' + q(x)y + r(x),$$

the differential equation

$$y'' = f(x, y, y')$$

is called **linear**. Problems of this type occur quite often in practice, and this representation allows Theorem 10.1 to be simplified considerably.

Corollary 10.2   If the linear boundary-value problem

(10.4) $y'' = p(x)y' + q(x)y + r(x), \quad a \le x \le b, \quad y(a) = \alpha, \quad y(b) = \beta,$

satisfies:

i) $p(x)$, $q(x)$, and $r(x)$ are continuous on $[a, b]$,
ii) $q(x) > 0$ on $[a, b]$,

then the problem has a unique solution.

In order to approximate the unique solution guaranteed by the satisfaction of the hypotheses of Corollary 10.2, let us first consider the initial-value problems

(10.5) $y'' = p(x)y' + q(x)y + r(x), \quad a \le x \le b, \quad y(a) = \alpha, \quad y'(a) = 0,$

and

(10.6) $y'' = p(x)y' + q(x)y, \quad a \le x \le b, \quad y(a) = 0, \quad y'(a) = 1.$

Theorem 5.15 (p. 238) ensures that under the hypotheses in Corollary 10.2, both of these problems have a unique solution. If $y_1(x)$ denotes the solution to (10.5) and $y_2(x)$ denotes the solution to (10.6), it is not difficult to verify that

(10.7) $$y(x) = y_1(x) + \frac{\beta - y_1(b)}{y_2(b)} y_2(x)$$

is the unique solution to our boundary-value problem, provided, of course, that $y_2(b) \ne 0$. That $y_2(b) = 0$ is in conflict with the hypotheses of Corollary 10.2 is considered in Exercise 10.

The shooting method for linear equations is based on this replacement of the boundary-value problem by the two initial-value problems (10.5) and (10.6). Numerous methods are available from Chapter 5 for approximating the solutions $y_1(x)$ and $y_2(x)$, and once these approximations are available, the solution to the boundary-value problem can be approximated by using Eq. (10.7). Graphically, the method has the appearance shown in Fig. 10.1.

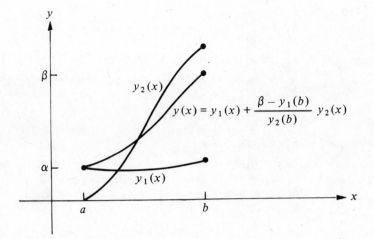

FIGURE 10.1

Algorithm 10.1 uses the fourth-order Runge–Kutta technique to find the approximations to $y_1(x)$ and $y_2(x)$, but any other technique for approximating the solutions to initial-value problems can be substituted into Step 4 (see Exercises 7 and 8). The algorithm has the additional feature of obtaining approximations for the *derivative* of the solution to the boundary-value problem as well as to the solution of the problem itself.

The use of the algorithm is not restricted to those problems for which the hypotheses of Corollary 10.2 can be verified and will, in fact, give satisfactory results for many problems that do not satisfy these hypotheses.

Linear Shooting Algorithm 10.1

To approximate the solution of the boundary-value problem

$$-y'' + p(x)y' + q(x)y + r(x) = 0, \qquad a \le x \le b, \quad y(a) = \alpha, \quad y(b) = \beta:$$

(*Note*: Equations (10.5), (10.6) *are written as first order systems and solved.*)

INPUT endpoints a, b; boundary conditions α, β; number of subintervals N.

OUTPUT approximations $w_{1,i}$ to $y(x_i)$; $w_{2,i}$ to $y'(x_i)$ for each $i = 0, 1, \ldots, N$.

Step 1 Set $h = (b - a)/N$;
  $u_{1,0} = \alpha$;
  $u_{2,0} = 0$;
  $v_{1,0} = 0$;
  $v_{2,0} = 1$.

Step 2 For $i = 0, \ldots, N - 1$ do Steps 3 and 4.
  (*Runge–Kutta method for systems is used in Steps 3 and 4.*)

Step 3 Set $x = a + ih$.

Step 4 Set $k_{1,1} = hu_{2,i}$;

   $k_{1,2} = h[p(x)u_{2,i} + q(x)u_{1,i} + r(x)]$;

   $k_{2,1} = h[u_{2,i} + \frac{1}{2}k_{1,2}]$;

   $k_{2,2} = h[p(x + h/2)(u_{2,i} + \frac{1}{2}k_{1,2})$

       $+ q(x + h/2)(u_{1,i} + \frac{1}{2}k_{1,1}) + r(x + h/2)]$;

   $k_{3,1} = h[u_{2,i} + \frac{1}{2}k_{2,2}]$;

   $k_{3,2} = h[p(x + h/2)(u_{2,i} + \frac{1}{2}k_{2,2})$

       $+ q(x + h/2)(u_{1,i} + \frac{1}{2}k_{2,1}) + r(x + h/2)]$;

   $k_{4,1} = h[u_{2,i} + k_{3,2}]$;

   $k_{4,2} = h[p(x + h)(u_{2,i} + k_{3,2}) + q(x + h)(u_{1,i} + k_{3,1}) + r(x + h)]$; .

   $u_{1,i+1} = u_{1,i} + \frac{1}{6}[k_{1,1} + 2k_{2,1} + 2k_{3,1} + k_{4,1}]$;

   $u_{2,i+1} = u_{2,i} + \frac{1}{6}[k_{1,2} + 2k_{2,2} + 2k_{3,2} + k_{4,2}]$;

   $k'_{1,1} = hv_{2,i}$;

   $k'_{1,2} = h[p(x)v_{2,i} + q(x)v_{1,i}]$;

   $k'_{2,1} = h[v_{2,i} + \frac{1}{2}k'_{1,2}]$;

   $k'_{2,2} = h[p(x + h/2)(v_{2,i} + \frac{1}{2}k'_{1,2}) + q(x + h/2)(v_{1,i} + \frac{1}{2}k'_{1,1})]$;

   $k'_{3,1} = h[v_{2,i} + \frac{1}{2}k'_{2,2}]$;

   $k'_{3,2} = h[p(x + h/2)(v_{2,i} + \frac{1}{2}k'_{2,2}) + q(x + h/2)(v_{1,i} + \frac{1}{2}k'_{2,1})]$;

   $k'_{4,1} = h[v_{2,i} + k'_{3,2}]$;

   $k'_{4,2} = h[p(x + h)(v_{2,i} + k'_{3,2}) + q(x + h)(v_{1,i} + k'_{3,1})]$;

   $v_{1,i+1} = v_{1,i} + \frac{1}{6}[k'_{1,1} + 2k'_{2,1} + 2k'_{3,1} + k'_{4,1}]$;

   $v_{2,i+1} = v_{2,i} + \frac{1}{6}[k'_{1,2} + 2k'_{2,2} + 2k'_{3,2} + k'_{4,2}]$.

Step 5 Set $w_{1,0} = \alpha$;

$$w_{2,0} = \frac{\beta - u_{1,N}}{v_{1,N}};$$

OUTPUT $(a, w_{1,0}, w_{2,0})$.

Step 6 For $i = 1, \ldots, N$

   set $W1 = u_{1,i} + w_{2,0}v_{1,i}$;

   $W2 = u_{2,i} + w_{2,0}v_{2,i}$;

   $x = a + ih$;

   OUTPUT $(x, W1, W2)$.   (*Output is* $x_i, w_{1,i}, w_{2,i}$)

Step 7 STOP.   (*Process is complete.*)

EXAMPLE 2   The boundary-value problem

(10.8) $y'' = -\dfrac{2}{x}y' + \dfrac{2}{x^2}y + \dfrac{\sin(\ln x)}{x^2}$, $1 \le x \le 2$, $y(1) = 1$, $y(2) = 2$,

has the exact solution

$$y = c_1 x + \frac{c_2}{x^2} - \tfrac{3}{10}\sin(\ln x) - \tfrac{1}{10}\cos(\ln x),$$

where $c_2 = \tfrac{1}{70}[8 - 12\sin(\ln 2) - 4\cos(\ln 2)] \approx -.03920701320$

and $c_1 = \tfrac{11}{10} - c_2 \approx 1.139207013$.

Applying Algorithm 10.1 to this problem requires approximating the solutions to the initial-value problems

$$y_1'' = -\frac{2}{x}y_1' + \frac{2}{x^2}y_1 + \frac{\sin(\ln x)}{x^2}, \qquad 1 \le x \le 2, \quad y_1(1) = 1, \quad y_1'(1) = 0,$$

and

$$y_2'' = -\frac{2}{x}y_2' + \frac{2}{x^2}y_2, \qquad 1 \le x \le 2, \quad y_2(1) = 0, \quad y_2'(1) = 1.$$

The results of the calculations using $N = 10$ and $h = .1$, are given in Table 10.1. The value listed at $u_{1,i}$ approximates $y_1(t_i)$, $v_{1,i}$ approximates $y_2(t_i)$, and w_i approximates $y(t_i)$. □

TABLE 10.1

x_i	$u_{1,i}$	$v_{1,i}$	w_i	$y(t_i)$	$|y(t_i) - w_i|$
1.0	1.0000000	.0000000	1.00000000	1.00000000	—
1.1	1.00896058	.09117986	1.09262917	1.09262930	1.43×10^{-7}
1.2	1.03245472	.16851175	1.18708471	1.18708484	1.34×10^{-7}
1.3	1.06674375	.23608704	1.28338227	1.28338236	9.78×10^{-8}
1.4	1.10928795	.29659067	1.38144589	1.38144595	6.02×10^{-8}
1.5	1.15830000	.35184379	1.48115939	1.48115942	3.06×10^{-8}
1.6	1.21248372	.40311695	1.58239245	1.58239246	1.08×10^{-8}
1.7	1.27087454	.45131840	1.68501396	1.68501396	5.43×10^{-10}
1.8	1.33273851	.49711137	1.78889854	1.78889853	5.05×10^{-9}
1.9	1.39750618	.54098928	1.89392951	1.89392951	4.41×10^{-9}
2.0	1.46472815	.58332538	2.00000000	2.00000000	—

The accuracy exhibited in Example 2 is expected because the Runge–Kutta fourth-order method gives $O(h^4)$ accuracy to the solutions of the initial-value problems. Unfortunately, there are problems hidden in this technique because of rounding errors. Should β be small in magnitude compared to $u_{1,N}$, the term $(\beta - u_{1,N})/v_{1,N}$ will be dominated by $-u_{1,N}/v_{1,N}$; and if $y_1(x)$ rapidly increases as x goes from a to b, then $u_{1,N}$ will be large. This can cause cancellation of significant digits in the computations of Steps 5 and 6. However, since $u_{1,i}$ is an approximation to $y_1(x_i)$, the behavior of y_1 can easily be monitored, and if $u_{1,i}$ increases rapidly from a to b, then the shooting technique can be employed in the other direction, that is, solving instead the initial-value problems

$$-y'' + p(x)y' + q(x)y + r(x) = 0, \qquad a \le x \le b, \quad y(b) = \beta, \quad y'(b) = 0,$$

and

$$-y'' + p(x)y' + q(x)y = 0, \qquad a \le x \le b, \quad y(b) = 0, \quad y'(b) = 1.$$

If the reverse shooting technique still gives cancellation of significant digits, and if increased precision does not yield greater accuracy, other techniques must be employed, techniques such as those presented in the last two sections of this chapter. In general, however, if $u_{1,i}$ and $v_{1,i}$ are $O(h^n)$ approximations to $y_1(x_i)$ and $y_2(x_i)$, respectively, for each $i = 0, 1, \ldots, N$, then it can be shown that $w_{1,i}$ will be an $O(h^n)$ approximation to $y_{1,i}$. In particular,

$$|w_{1,i} - y(x_i)| \le Kh^n \left| 1 + \frac{v_{1,i}}{v_{1,N}} \right|,$$

for some constant K (see Isaacson and Keller [52], page 426).

Exercise Set 10.1

1. Which of the following boundary-value problems are guaranteed by Corollary 10.2 to have a unique solution?

   a) $y'' = -\dfrac{4}{x}y' + \dfrac{2}{x^2}y - \dfrac{2\ln x}{x^2}$, $1 < x < 2$, $y(1) = -\frac{1}{2}$, $y(2) = \ln 2$;

   b) $y'' = -y - x$, $0 < x < \pi$, $y(0) = 1$, $y(\pi) = 2$;

   c) $y'' = -\sin xy$, $0 < x < \pi$, $y(0) = 0$, $y(\pi) = 1$;

   d) $y'' = \frac{1}{2}y^3$, $1 < x < 2$, $y(1) = -\frac{2}{3}$, $y(2) = -1$;

   e) $y'' = y^3 - yy'$, $1 < x < 2$, $y(1) = \frac{1}{2}$, $y(2) = \frac{1}{3}$;

   f) $y'' = -y$, $0 < x < \pi$, $y(0) = 1$, $y(\pi) = 1$;

   g) $y'' = -y$, $0 < x < \pi$, $y(0) = 1$, $y(\pi) = -1$;

   h) $y'' = -y$, $0 < x < \dfrac{\pi}{4}$, $y(0) = 1$, $y\left(\dfrac{\pi}{4}\right) = 1$;

   i) $y'' = 2y^3 - 6y - 2x^3$, $1 < x < 2$, $y(1) = 2$, $y(2) = \frac{5}{2}$.

2. Use Algorithm 10.1 to approximate the solution to the following boundary-value problem.

   a) $y'' + y = 0$, $0 < x < \pi$, $y(0) = 1$, $y(\pi) = -1$;

   use $h = \pi/3$.
   b) Repeat part (a) using $h = \pi/4$.

3. Use Algorithm 10.1 to approximate the solution to

   $$y'' + 4y = \cos x,   0 < x < \pi/4,   y(0) = 0,   y(\pi/4) = 0;$$

   use $h = \pi/12$.

4. Approximate the solution to the following boundary-value problems, using Algorithm 10.1:

   a) $y'' = -\dfrac{4}{x}y' + \dfrac{2}{x^2}y - \dfrac{2\ln x}{x^2}$, $1 < x < 2$, $y(1) = -\frac{1}{2}$, $y(2) = \ln 2$;

   use $h = .05$.

   b) $y'' = -4y' + 4y$, $0 < x < 5$, $y(0) = 1$, $y(5) = 0$;

   use $h = .2$.

   c) $y'' = -3y' + 2y + 2x + 3$, $0 < x < 1$, $y(0) = 2$, $y(1) = 1$;

   use $h = .1$.

5. Show that Corollary 10.2 does not apply to the following boundary-value problems, but that Algorithm 10.1 can still be used.

   a) $y'' = -\dfrac{4}{x}y' - \dfrac{2}{x^2}y + \dfrac{2}{x^2}\ln x$, $1 < x < 2$, $y(1) = \frac{1}{2}$, $y(2) = \ln 2$;

   use $h = .05$.

   b) $y'' = -y$, $0 < x < \pi/4$, $y(0) = 1$, $y(\pi/4) = 1$;

   use $h = \pi/40$.

6. Write the second-order initial-value problems (10.5) and (10.6) as first-order systems, and derive the equations necessary to solve the systems, using the Runge–Kutta fourth-order method for systems.

7. Construct an algorithm similar to Algorithm 10.1, using the following initial-value methods.

   a) Adams–Bashforth and Adams–Moulton predictor–corrector method;
   b) Gragg extrapolation;
   c) Runge–Kutta–Fehlberg method.

8. Using the algorithms constructed in Exercise 7, repeat Exercises 4 and 5.

9. Use either Algorithm 10.1 or the algorithms constructed in Exercise 7 to approximate the solution $y = e^{-10x}$ to the boundary-value problem

$$y'' = 100y, \qquad 0 < x < 1, \quad y(0) = 1, \quad y(1) = e^{-10}.$$

Use $h = .1$ and $.05$. Can you explain the consequences?

10. Show that if y_2 is the solution to $y'' = p(x)y' + g(x)y$ and $y_2(a) = y_2(b) = 0$, then $y_2 \equiv 0$.

11. Devise a shooting method for the general boundary-value problem:

$$y'' = p(x)y' + q(x)y + r(x), \qquad a < x < b,$$
$$\alpha_1 y(a) - \beta_1 y'(a) = \alpha, \qquad |\alpha_1| + |\beta_1| \neq 0,$$
$$\alpha_2 y(b) + \beta_2 y'(b) = \beta, \qquad |\alpha_2| + |\beta_2| \neq 0.$$

12. Let u represent the electrostatic potential between two concentric metal spheres of radii R_1 and $R_2 (R_1 < R_2)$, such that the potential of the inner sphere is kept constant at V_1 volts and the potential of the outer sphere at 0 volts. The potential in the region between the two spheres is governed by Laplace's equation (see Eq. (11.2)), which in this particular application, reduces to

$$\frac{d^2u}{dr^2} + \frac{2}{r}\frac{du}{dr} = 0, \qquad R_1 < r < R_2, \quad u(R_1) = V_1, \quad u(R_2) = 0.$$

Suppose $R_1 = 2$ in., $R_2 = 4$ in., and $V_1 = 110$ volts.

   a) Approximate $u(3)$ using the Linear Shooting Algorithm 10.1 with $N = 20$.
   b) Approximate $u(3)$ using the Linear Shooting Algorithm 10.1 with $N = 40$.
   c) Compare the results of (a) and (b) with the actual potential $u(3)$ where

$$u(r) = \frac{V_1 R_1}{r}\left(\frac{R_2 - r}{R_2 - R_1}\right).$$

10.2 The Shooting Method for Nonlinear Problems

The idea behind the shooting technique for the nonlinear second-order boundary-value problem

$$(10.9) \qquad y'' = f(x, y, y'), \qquad a \leq x \leq b, \quad y(a) = \alpha, \quad y(b) = \beta,$$

is similar to the linear case, except that the solution to a nonlinear problem cannot be simply expressed as a linear combination of the solutions to two initial-value problems. Instead, we will need to utilize the solutions to a *sequence* of initial-value problems of the form

$$(10.10) \qquad y'' = f(x, y, y'), \qquad a \leq x \leq b, \quad y(a) = \alpha, \quad y'(a) = t_k,$$

involving a parameter t_k, to approximate the solution to our boundary-value problem. We do this by choosing the parameters t_k in a manner that will ensure that

$$\lim_{k \to +\infty} y(b, t_k) = y(b) = \beta,$$

where $y(x, t_k)$ denotes the solution to the initial-value problem (10.10) and $y(x)$ denotes the solution to the boundary-value problem (10.9).

This technique is called the "shooting" method by analogy to the procedure of firing objects at a stationary target (see Fig. 10.2). We start with a parameter t_0 that determines the initial elevation at which the object is fired from the point (a, α) and along the curve described by the solution to the initial-value problem:

$$y'' = f(x, y, y'), \qquad a \le x \le b, \quad y(a) = \alpha, \quad y'(a) = t_0.$$

If $y(b, t_0)$ is not sufficiently close to β, we attempt to correct our approximation by choosing another elevation t_1 and so on, until $y(b, t_k)$ is sufficiently close to "hitting" β. (See Fig. 10.3.)

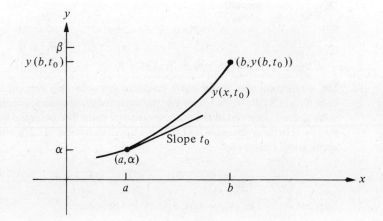

FIGURE 10.2

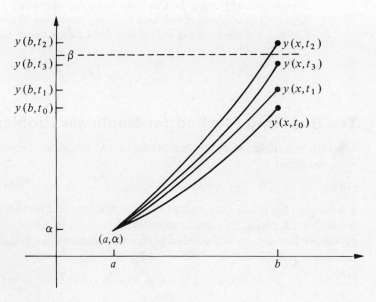

FIGURE 10.3

Of course, in practice we cannot rely on solving the sequence of initial-value problems exactly, so our approximation to the boundary-value problem will itself be based on approximations. In order to be assured that this procedure will lead to satisfactory results, we need to have conditions applied to $f(x, y, y')$ to at least ensure that the initial-value problems we are approximating, as well as the boundary-value problem, have unique solutions in the interval $[a, b]$. The following theorem is a special case of Theorem 10.1, which will guarantee that the procedure can be performed. We refer the reader again to Keller [56], page 48, for a proof of this result.

Theorem 10.3   The boundary-value problem

(10.11) $y'' = f(x, y, y')$, $a \le x \le b$, $y(a) = \alpha$, $y(b) = \beta$,

and the initial-value problem

(10.12) $y'' = f(x, y, y')$, $a \le x \le b$, $y(a) = \alpha$, $y'(a) = t$,

for an arbitrary parameter t, will have unique solutions on $[a, b]$ provided that on

$$D = \{(x, y, y') | a \le x \le b, -\infty < y < \infty, -\infty < y' < \infty\},$$

i) f, $\partial f / \partial y$, and $\partial f / \partial y'$, are continuous;
ii)   a constant M exists with

$$\left| \frac{\partial f}{\partial y'} \right| \le M; \quad \text{and}$$

iii) a constant L exists with $0 < \partial f / \partial y < L$.

To determine how the parameters t_k can be chosen, suppose that we have a boundary-value problem of the form (10.11), which satisfies the hypotheses of Theorem 10.3. If $y(x, t)$ is used to denote the solution to the initial-value problem (10.12), the question is how to choose t so that

(10.13) $y(b, t) - \beta = 0$.

Since this is a nonlinear equation of the type considered in Chapter 2, a number of methods are available. If we wish to employ the Secant method (Algorithm 2.4 of Section 2.3) to solve the problem, we need to choose initial approximations t_0 and t_1 and then generate the remaining terms of the sequence by

$$t_k = t_{k-1} - \frac{(y(b, t_{k-1}) - \beta)(t_{k-1} - t_{k-2})}{y(b, t_{k-1}) - y(b, t_{k-2})}, \quad k = 2, 3, \ldots$$

In order to use the more powerful Newton's method to generate the sequence $\{t_k\}$, only one initial value, t_0, is needed. However, the iteration has the form

(10.14) $t_k = t_{k-1} - \dfrac{(y(b, t_{k-1}) - \beta)}{(dy/dt)(b, t_{k-1})}$ where $(dy/dt)(b, t_{k-1}) \equiv \dfrac{dy(b, t_{k-1})}{dt}$

and requires the knowledge of $(dy/dt)(b, t_{k-1})$. This presents a difficulty, since an explicit representation for $y(b, t)$ is not known; we know only the values $y(b, t_0)$, $y(b, t_1), \ldots, y(b, t_{k-1})$.

Suppose we rewrite the initial-value problem (10.12), emphasizing that the solution depends on both x and t:

(10.15) $y''(x, t) = f(x, y(x, t), y'(x, t))$, $a \le x \le b$, $y(a, t) = \alpha$, $y'(a, t) = t$,

retaining the prime notation to indicate differentiation with respect to x. Since we are interested in determining $(dy/dt)(b, t)$ when $t = t_{k-1}$, we first take the partial derivative of (10.15) with respect to t.

This implies that

(10.16)

$$\frac{\partial y''}{\partial t}(x, t) = \frac{\partial}{\partial t} f(x, y(x, t), y'(x, t))$$

$$= \frac{\partial f}{\partial x}(x, y(x, t), y'(x, t)) \frac{\partial x}{\partial t} + \frac{\partial f}{\partial y}(x, y(x, t), y'(x, t)) \frac{\partial y}{\partial t}(x, t)$$

$$+ \frac{\partial f}{\partial y'}(x, y(x, t), y'(x, t)) \frac{\partial y'}{\partial t}(x, t)$$

$$= \frac{\partial f}{\partial y}(x, y(x, t), y'(x, t)) \frac{\partial y}{\partial t}(x, t) + \frac{\partial f}{\partial y'}(x, y(x, t), y'(x, t)) \frac{\partial y'}{\partial t}(x, t)$$

for $a \le x \le b$, and the initial conditions give

$$\frac{\partial y}{\partial t}(a, t) = 0 \qquad \text{and} \qquad \frac{\partial y'}{\partial t}(a, t) = 1.$$

If we simplify the notation by using $z(x, t)$ to denote $(\partial y/\partial t)(x, t)$ and assume that the order of differentiation of x and t can be reversed, we have the initial-value problem

(10.17) $z'' = \dfrac{\partial f}{\partial y}(x, y, y')z + \dfrac{\partial f}{\partial y'}(x, y, y')z'$, $a \le x \le b$, $z(a) = 0$, $z'(a) = 1$.

Newton's method therefore requires that two initial-value problems be solved for each iteration, since Eq. (10.14) becomes

(10.18) $$t_k = t_{k-1} - \frac{y(b, t_{k-1}) - \beta}{z(b, t_{k-1})}.$$

In practice, none of these initial-value problems is likely to be solved exactly; instead the solutions are approximated by one of the methods discussed in Chapter 5. Algorithm 10.2 uses the fourth-order Runge–Kutta method to approximate both solutions required by Newton's method. A similar procedure for the Secant method is considered in Exercise 4.

Nonlinear Shooting Algorithm 10.2

To approximate the solution of the nonlinear boundary-value problem

$$y'' = f(x, y, y'), \qquad a \le x \le b, \quad y(a) = \alpha, \quad y(b) = \beta:$$

(*Note*: *Equations* (10.15), (10.17) *are written as first-order systems and solved.*)

INPUT endpoints a, b; boundary conditions α, β; number of subintervals N; tolerance TOL; maximum number of iterations M.

OUTPUT approximations $w_{1,i}$ to $y(x_i)$; $w_{2,i}$ to $y'(x_i)$ for each $i = 0, 1, \ldots, N$ or a message that the maximum number of iterations was exceeded.

Step 1 Set $h = (b - a)/N$;
$k = 1$;
$TK = (\beta - \alpha)/(b - a)$.

Step 2 While $(k \leq M)$ do Steps 3–10.

Step 3 Set $w_{1,0} = \alpha$;
$w_{2,0} = TK$;
$u_1 = 0$;
$u_2 = 1$.

Step 4 For $i = 1, \ldots, N$ do Steps 5 and 6.
(Runge–Kutta method for systems is used in Steps 5 and 6.)

Step 5 Set $x = a + (i - 1)h$.

Step 6 $k_{1,1} = hw_{2,i-1}$;
$k_{1,2} = hf(x, w_{1,i-1}, w_{2,i-1})$;
$k_{2,1} = h(w_{2,i-1} + \frac{1}{2}k_{1,2})$;
$k_{2,2} = hf(x + h/2, w_{1,i-1} + \frac{1}{2}k_{1,1}, w_{2,i-1} + \frac{1}{2}k_{1,2})$;
$k_{3,1} = h(w_{2,i-1} + \frac{1}{2}k_{2,2})$;
$k_{3,2} = hf(x + h/2, w_{1,i-1} + \frac{1}{2}k_{2,1}, w_{2,i-1} + \frac{1}{2}k_{2,2})$;
$k_{4,1} = h(w_{2,i-1} + k_{3,2})$;
$k_{4,2} = hf(x + h, w_{1,i-1} + k_{3,1}, w_{2,i-1} + k_{3,2})$;
$w_{1,i} = w_{1,i-1} + (k_{1,1} + 2k_{2,1} + 2k_{3,1} + k_{4,1})/6$;
$w_{2,i} = w_{2,i-1} + (k_{1,2} + 2k_{2,2} + 2k_{3,2} + k_{4,2})/6$;
$k'_{1,1} = hu_2$;
$k'_{1,2} = h[f_y(x, w_{1,i-1}, w_{2,i-1})u_1$
$\qquad + f_{y'}(x, w_{1,i-1}, w_{2,i-1})u_2]$;
$k'_{2,1} = h[u_2 + \frac{1}{2}k'_{1,2}]$;
$k'_{2,2} = h[f_y(x + h/2, w_{1,i-1}, w_{2,i-1})(u_1 + \frac{1}{2}k'_{1,1})$
$\qquad + f_{y'}(x + h/2, w_{1,i-1}, w_{2,i-1})(u_2 + \frac{1}{2}k'_{2,1})]$;
$k'_{3,1} = h(u_2 + \frac{1}{2}k'_{2,2})$;
$k'_{3,2} = h[f_y(x + h/2, w_{1,i-1}, w_{2,i-1})(u_1 + \frac{1}{2}k'_{2,1})$
$\qquad + f_{y'}(x + h/2, w_{1,i-1}, w_{2,i-1})(u_2 + \frac{1}{2}k'_{2,2})]$;
$k'_{4,1} = h(u_2 + k'_{3,2})$;
$k'_{4,2} = h[f_y(x + h, w_{1,i-1}, w_{2,i-1})(u_1 + k'_{3,1})$
$\qquad + f_{y'}(x + h, w_{1,i-1}, w_{2,i-1})(u_2 + k'_{3,2})]$;
$u_1 = u_1 + \frac{1}{6}[k'_{1,1} + 2k'_{2,1} + 2k'_{3,1} + k'_{4,1}]$;
$u_2 = u_2 + \frac{1}{6}[k'_{1,2} + 2k'_{2,2} + 2k'_{3,2} + k'_{4,2}]$.

Step 7 If $|w_{1,N} - \beta| \leq TOL$ then do Steps 8 and 9.

Step 8 For $i = 0, 1, \ldots, N$
set $x = a + ih$;
OUTPUT $(x, w_{1,i}, w_{2,i})$.

Step 9 *(Procedure is complete.)*
STOP.

Step 10 Set $TK = TK - \left(\dfrac{w_{1,N} - \beta}{u_1} \right)$;   *(Newton's method is used to compute TK.)*

$k = k + 1$.

Step 11 OUTPUT ('Maximum number of iterations exceeded');
     STOP.

In Step 7, the best approximation to β we can expect $w_{1,N}(t_k)$ to give is $O(h^n)$ if the approximation method selected for Step 6 gives $O(h^n)$ rate of convergence.

The value t_0 selected in Step 1 is the slope of the straight line through (a, α) and (b, β). If the problem satisfies the hypotheses of Theorem 10.3, any choice of t_0 will give convergence; but in general the procedure will work for many problems for which these hypotheses are not satisfied, although a good choice of t_0 is necessary.

EXAMPLE 1 Consider the boundary-value problem

(10.19) $y'' = \tfrac{1}{8}(32 + 2x^3 - yy')$, $1 \le x \le 3$, $y(1) = 17$, $y(3) = \tfrac{43}{3}$,

which has the exact solution $y(x) = x^2 + (16/x)$.

Applying the shooting method given in Algorithm 10.2 to this problem requires approximating the initial-value problems

$$y'' = \tfrac{1}{8}(32 + 2x^3 - yy'), \quad 1 \le x \le 3, \quad y(1) = 17, \quad y'(1) = t_k,$$

and $z'' = -\tfrac{1}{8}(yz' + y'z), \quad 1 \le x \le 3, \quad z(1) = 0, \quad z'(1) = 1,$

at each step in the iteration. If the stopping technique

$$|w_{1,N}(t_k) - y(3)| \le 10^{-5}$$

is used, this problem requires four iterations and $t_4 = -14.000203$. The results obtained for this value of t are shown in Table 10.2.   □

The problem in this example was chosen because of a relative minimum which occurs for the exact solution at $x = 2$. Had the exact solution been monotonic in the interval, we would not expect to require four iterations for this degree of accuracy.

It should be mentioned that, although Newton's method used with the shooting technique requires the solution of an additional initial-value problem, it will generally be faster than the Secant method. Both methods are only locally convergent in that they require good initial approximations whenever the assumptions of Theorem 10.3 do not hold.

For a general discussion of the convergence of the shooting techniques for nonlinear problems, the reader is referred to the excellent text by Keller [56]. In this reference more general boundary conditions are discussed, and it is also noted that the shooting technique for nonlinear problems is sensitive to rounding errors, especially if the solutions $y(x)$ and $z(x)$ are rapidly increasing functions on $[a, b]$.

TABLE 10.2

x_i	w_{1i}	$y(x_i)$	$\lvert w_{1i} - y(x_i)\rvert$
1.0	17.000000	17.000000	—
1.1	15.755495	15.755455	4.06×10^{-5}
1.2	14.773389	14.773333	5.60×10^{-5}
1.3	13.997752	13.997692	5.94×10^{-5}
1.4	13.388629	13.388571	5.71×10^{-5}
1.5	12.916719	12.916667	5.23×10^{-5}
1.6	12.560046	12.560000	4.64×10^{-5}
1.7	12.301805	12.301765	4.02×10^{-5}
1.8	12.128923	12.128889	3.41×10^{-5}
1.9	12.031081	12.031053	2.84×10^{-5}
2.0	12.000023	12.000000	2.32×10^{-5}
2.1	12.029066	12.029048	1.84×10^{-5}
2.2	12.112741	12.112727	1.40×10^{-5}
2.3	12.246532	12.246522	1.01×10^{-5}
2.4	12.426673	12.426667	6.68×10^{-6}
2.5	12.650004	12.650000	3.61×10^{-6}
2.6	12.913847	12.913846	9.17×10^{-7}
2.7	13.215924	13.215926	1.43×10^{-6}
2.8	13.554282	13.554286	3.47×10^{-6}
2.9	13.927236	13.927241	5.21×10^{-6}
3.0	14.333327	14.333333	6.69×10^{-6}

Exercise Set 10.2

1. Use Algorithm 10.2 with $h = .5$ to approximate the solution to the boundary-value problem

$$y'' = -(y')^2 - y + \ln x, \qquad 1 < x < 2, \quad y(1) = 0, \quad y(2) = \ln 2.$$

Compare your answer to the actual solution $y = \ln x$.

2. Use Algorithm 10.2 with $h = .2$ to approximate the solution to the boundary-value problem

$$y'' = \frac{xy' - y}{x^2}, \qquad 1 < x < 1.6, \quad y(1) = -1, \quad y(1.6) = -.847994.$$

3. Use the nonlinear shooting algorithm with Newton's method (Algorithm 10.2) to approximate the solutions to the following boundary-value problems. The actual solution is given for comparison to your results. Do the hypotheses of Theorem 10.3 hold?

   a) $y'' = \frac{1}{2}y^3, \quad 1 < x < 2, \quad y(1) = -\frac{2}{3}, \quad y(2) = -1;$

   use $h = .05$ and compare to $y(x) = 2/(x - 4)$.

   b) $y'' = 2y^3, \quad 1 < x < 5, \quad y(1) = \frac{1}{4}, \quad y(5) = \frac{1}{8};$

   use $h = .2$ and compare to $y(x) = 1/(x + 3)$.

   c) $y'' = y^3 - yy', \quad 1 < x < 2, \quad y(1) = \frac{1}{2}, \quad y(2) = \frac{1}{3};$

   use $h = .1$ and compare to $y(x) = 1/(x + 1)$.

   d) $y'' = 2y^3 - 6y - 2x^3, \quad 1 < x < 2, \quad y(1) = 2, \quad y(2) = \frac{5}{2};$

   use $h = .05$ and compare to $y(x) = x + (1/x)$.

4. Change Algorithm 10.2 to incorporate the Secant method instead of Newton's method. Use $t_0 = (\beta - \alpha)/(b - a)$ and $t_1 = t_0 + (\beta - y(b, t_0))/(b - a)$.

5. Repeat Exercise 3 using the Secant Algorithm derived in Exercise 4, and compare the number of iterations required for the two methods.

6. Change Algorithm 10.2 to incorporate, in place of the Runge–Kutta fourth-order method,

    a) Adams–Bashforth and Adams–Moulton predictor–corrector method;
    b) Gragg extrapolation;
    c) Runge–Kutta–Fehlberg method.

7. Use the algorithms developed in Exercise 6 parts (a), (b), and (c) to approximate the solutions to the problems given in Exercise 3. Determine which method is most accurate for each problem.

10.3   Finite-Difference Methods for Boundary-Value Problems

Although the shooting methods presented in the earlier part of this chapter can be used for both linear and nonlinear boundary-value problems, they often present problems of instability. The methods we will present here have better stability characteristics, but generally require more work to obtain a specified accuracy.

Methods involving finite differences for solving boundary-value problems consist of replacing each of the derivatives in the differential equation by an appropriate difference-quotient approximation. The difference quotient is generally chosen so that a certain order of truncation error is maintained.

The linear second-order boundary-value problem,

$$(10.20) \qquad y'' = p(x)y' + q(x)y + r(x), \qquad a \le x \le b, \quad y(a) = \alpha, \quad y(b) = \beta,$$

requires that difference-quotient approximations be used for approximating both y' and y''. To accomplish this, we select an integer $N > 0$ and divide the interval $[a, b]$ into $(N + 1)$ equal subintervals, whose endpoints are the meshpoints $x_i = a + ih$, for $i = 0, 1, \ldots, N + 1$, where $h = (b - a)/(N + 1)$. Choosing the constant h in this manner will facilitate the application of a matrix algorithm from Chapter 6, which in this form will require solving a linear system involving an $N \times N$ matrix.

At the interior meshpoints, x_i, $i = 1, 2, \ldots, N$, the differential equation to be approximated is

$$(10.21) \qquad y''(x_i) = p(x_i)y'(x_i) + q(x_i)y(x_i) + r(x_i).$$

Expanding y in a third-degree Taylor polynomial about x_i evaluated at x_{i+1} and x_{i-1} we have:

$$(10.22) \qquad y(x_{i+1}) = y(x_i + h) = y(x_i) + hy'(x_i) + \frac{h^2}{2} y''(x_i)$$

$$+ \frac{h^3}{6} y'''(x_i) + \frac{h^4}{24} y^{(4)}(\xi_i^+),$$

for some ξ_i^+, $x_i < \xi_i^+ < x_{i+1}$, and

$$(10.23) \quad y(x_{i-1}) = y(x_i - h) = y(x_i) - hy'(x_i) + \frac{h^2}{2} y''(x_i) - \frac{h^3}{6} y'''(x_i)$$

$$+ \frac{h^4}{24} y^{(4)}(\xi_i^-),$$

for some ξ_i^-, $x_{i-1} < \xi_i^- < x_i$, assuming $y \in C^4[x_{i-1}, x_{i+1}]$. If these equations are added together, the terms involving $y'(x_i)$ and $y'''(x_i)$ are eliminated, and a simple algebraic manipulation gives

$$y''(x_i) = \frac{1}{h^2} [y(x_{i+1}) - 2y(x_i) + y(x_{i-1})] - \frac{h^2}{24} [y^{(4)}(\xi_i^+) + y^{(4)}(\xi_i^-)].$$

The Intermediate Value Theorem can be used to simplify this even further:

$$(10.24) \quad y''(x_i) = \frac{1}{h^2} [y(x_{i+1}) - 2y(x_i) + y(x_{i-1})] - \frac{h^2}{12} y^{(4)}(\xi_i),$$

for some point ξ_i, $x_{i-1} < \xi_i < x_{i+1}$. Equation (10.24) is called the **centered-difference formula** for $y''(x_i)$.

A centered-difference formula for $y'(x_i)$ can be obtained in a similar manner (the details are considered in Exercise 6) resulting in

$$(10.25) \quad y'(x_i) = \frac{1}{2h} [y(x_{i+1}) - y(x_{i-1})] - \frac{h^2}{6} y'''(\eta_i)$$

for some η_i where $x_{i-1} < \eta_i < x_{i+1}$.

The use of these centered-difference formulas in Eq. (10.21) results in the equation

$$\frac{y(x_{i+1}) - 2y(x_i) + y(x_{i-1})}{h^2} = p(x_i) \left[\frac{y(x_{i+1}) - y(x_{i-1})}{2h} \right] + q(x_i)y(x_i)$$

$$+ r(x_i) - \frac{h^2}{12} [2p(x_i)y'''(\eta_i) - y^{(4)}(\xi_i)].$$

A finite-difference method with truncation error of order $O(h^2)$ results by using this equation together with the boundary conditions $y(a) = \alpha$ and $y(b) = \beta$ to define

$$w_0 = \alpha, \qquad w_{N+1} = \beta,$$

and

$$(10.26) \quad \left(\frac{2w_i - w_{i+1} - w_{i-1}}{h^2} \right) + p(x_i) \left(\frac{w_{i+1} - w_{i-1}}{2h} \right) + q(x_i)w_i = -r(x_i)$$

for each $i = 1, 2, \ldots, N$.

In the form we will consider, Eq. (10.26) is rewritten as

$$- \left(1 + \frac{h}{2} p(x_i) \right) w_{i-1} + (2 + h^2 q(x_i))w_i - \left(1 - \frac{h}{2} p(x_i) \right) w_{i+1} = -h^2 r(x_i),$$

and the resulting system of equations is expressed in the tridiagonal $N \times N$-matrix form shown in (10.27).

$$
\begin{bmatrix}
2 + h^2 q(x_1) & -1 + \dfrac{h}{2} p(x_1) & 0 & \cdots & \cdots & 0 \\[2mm]
-1 - \dfrac{h}{2} p(x_2) & 2 + h^2 q(x_2) & -1 + \dfrac{h}{2} p(x_2) & & & \\[2mm]
0 & & & & & 0 \\[2mm]
& & & & & -1 + \dfrac{h}{2} p(x_{N-1}) \\[2mm]
0 & \cdots & 0 & -1 - \dfrac{h}{2} p(x_N) & & 2 + h^2 q(x_N)
\end{bmatrix}
$$

(10.27)

$$
\times
\begin{bmatrix}
w_1 \\
w_2 \\
\vdots \\
w_{N-1} \\
w_N
\end{bmatrix}
=
\begin{bmatrix}
-h^2 r(x_1) + \left(1 + \dfrac{h}{2} p(x_1)\right) w_0 \\[2mm]
-h^2 r(x_2) \\
\vdots \\
-h^2 r(x_{N-1}) \\[2mm]
-h^2 r(x_N) + \left(1 - \dfrac{h}{2} p(x_N)\right) w_{N+1}
\end{bmatrix}
$$

The following theorem gives conditions under which the tridiagonal linear system (10.26) has a unique solution. Its proof is a consequence of Theorem 6.25 (p. 314) and is considered in Exercise 7.

Theorem 10.4   Suppose that p, q, and r are continuous on $[a, b]$. If $q(x) \geq 0$ on $[a, b]$, then the tridiagonal linear system (10.27) has a unique solution provided that $h < 2/L$ where $L = \max_{a \leq x \leq b} |p(x)|$.

It should be noted that the hypotheses of Theorem 10.4 guarantee a unique solution to the boundary-value problem (10.20), but they do not guarantee that $y \in C^4[a, b]$. It is necessary to establish that $y^{(4)}$ is continuous on $[a, b]$ in order to ensure that the truncation error has order $O(h^2)$.

Linear Finite-Difference Algorithm 10.3

To approximate the solution of the boundary-value problem

$$y'' = p(x)y' + q(x)y + r(x), \qquad a \leq x \leq b, \quad y(a) = \alpha, \quad y(b) = \beta:$$

INPUT endpoints a, b; boundary conditions α, β; integer N.
OUTPUT approximations w_i to $y(x_i)$ for each $i = 0, 1, \ldots, N + 1$.

Step 1 Set $h = (b - a)/(N + 1)$;
$$x = a + h;$$
$$a_1 = 2 + h^2 q(x);$$
$$b_1 = -1 + (h/2)p(x);$$
$$d_1 = -h^2 r(x) + (1 + (h/2)p(x))\alpha.$$

Step 2 For $i = 2, \ldots, N - 1$
set $x = a + ih$;
$$a_i = 2 + h^2 q(x);$$
$$b_i = -1 + (h/2)p(x);$$
$$c_i = -1 - (h/2)p(x);$$
$$d_i = -h^2 r(x).$$

Step 3 Set $x = b - h$;
$$a_N = 2 + h^2 q(x);$$
$$c_N = -1 - (h/2)p(x);$$
$$d_N = -h^2 r(x) + (1 - (h/2)p(x))\beta.$$

Step 4 *Set* $l_1 = a_1$; (*Steps 4–10 solve a tridiagonal linear system using Algorithm 6.7.*)
$$u_1 = b_1/a_1.$$

Step 5 For $i = 2, \ldots, N - 1$ set $l_i = a_i - c_i u_{i-1}$;
$$u_i = b_i/l_i.$$

Step 6 Set $l_N = a_N - c_N u_{N-1}$.

Step 7 Set $z_1 = d_1/l_1$.

Step 8 For $i = 2, \ldots, N$ set $z_i = (d_i - c_i z_{i-1})/l_i$.

Step 9 Set $w_0 = \alpha$;
$$w_{N+1} = \beta;$$
$$w_N = z_N.$$

Step 10 For $i = N - 1, \ldots, 1$ set $w_i = z_i - u_i w_{i+1}$.

Step 11 For $i = 0, \ldots, N + 1$ set $x = a + ih$;
$$\text{OUTPUT } (x, w_i).$$

Step 12 STOP.   (*Procedure is complete.*)

EXAMPLE 1   Algorithm 10.3 will be used to approximate the solution to the linear boundary-value problem

$$y'' = -\frac{2}{x}y' + \frac{2}{x^2}y + \frac{\sin(\ln x)}{x^2}, \quad 1 \le x \le 2, \quad y(1) = 1, \quad y(2) = 2,$$

which was also approximated by the shooting method in Example 2 of Section 10.1. For this example, we will use $N = 9$, so that $h = .1$ and we have the same spacing as in Example 2 of Section 10.1. The results are listed in Table 10.3.

TABLE 10.3

x_i	w_i	$y(x_i)$	$\|w_i - y(x_i)\|$
1.0	1.00000000	1.00000000	—
1.1	1.09260052	1.09262930	2.88×10^{-5}
1.2	1.18704313	1.18708484	4.17×10^{-5}
1.3	1.28333687	1.28338236	4.55×10^{-5}
1.4	1.38140205	1.38144595	4.39×10^{-5}
1.5	1.48112026	1.48115942	3.92×10^{-5}
1.6	1.58235990	1.58239246	3.26×10^{-5}
1.7	1.68498902	1.68501396	2.49×10^{-5}
1.8	1.78888175	1.78889853	1.68×10^{-5}
1.9	1.89392110	1.89392951	8.41×10^{-6}
2.0	2.00000000	2.00000000	—

It should be noted that these results are considerably less accurate than those obtained in Example 2 of Section 10.1. This is not surprising since the method used in that example involved a Runge–Kutta technique with truncation error of order $O(h^4)$, while the difference method used here has truncation error of order $O(h^2)$. □

In order to obtain a difference method with greater accuracy, we could proceed in a number of ways. Using fifth-order Taylor series for approximating $y''(x_i)$ and $y'(x_i)$ results in a truncation error term involving h^4. However, this requires using multiples not only of $y(x_{i+1})$ and $y(x_{i-1})$, but also $y(x_{i+2})$ and $y(x_{i-2})$ in the approximation formulas for $y''(x_i)$ and $y'(x_i)$. This leads to difficulty at $i = 0$ and $i = N$. Moreover, the resulting system of equations analogous to (10.27) would not be in tridiagonal form, and the solution to the system would require many more calculations. Additionally, this procedure would require that the solution to the boundary-value problem have a bounded sixth derivative, which, even when true, is difficult to verify for most problems arising from physical situations.

Instead of attempting to obtain a difference method with a higher-order truncation error in this manner, it is generally more satisfactory to consider a reduction in step size. In addition, it can be shown (see, for example, Keller [56], page 81) that the Richardson's extrapolation technique can be utilized for this method, since the error term is expressed in even powers of h with coefficients independent of h, provided y is sufficiently differentiable.

EXAMPLE 2   If we use the Richardson extrapolation method discussed in Sections 4.2, 4.6, and 5.7, to approximate the solution to the boundary-value problem

$$y'' = -\frac{2}{x}y' + \frac{2}{x^2}y + \frac{\sin(\ln x)}{x^2}, \quad 1 \le x \le 2, \quad y(1) = 1, \quad y(2) = 2,$$

with $h = .1, .05,$ and $.025$, we obtain the results listed in Table 10.4. The first extrapolation is

$$\text{Ext}_{1i} = \frac{4w_i(h = .05) - w_i(h = .1)}{3};$$

the second extrapolation is

$$\text{Ext}_{2i} = \frac{4w_i(h = .025) - w_i(h = .05)}{3};$$

and the final extrapolation is

$$\text{Ext}_{3i} = \frac{16\text{Ext}_{2i} - \text{Ext}_{1i}}{15}.$$

All of the results of Ext_{3i} are correct to the decimal places listed. In actuality, this approximation gives results that agree with the exact solution with maximum error of 6.3×10^{-11} at the mesh points. $\square$

TABLE 10.4

x_i	$w_i(h = .1)$	$w_i(h = .05)$	$w_i(h = .025)$	Ext_{1i}	Ext_{2i}	Ext_{3i}
1.0	1.00000000	1.00000000	1.00000000	1.00000000	1.00000000	1.00000000
1.1	1.09260052	1.09262207	1.09262749	1.09262925	1.09262930	1.09262930
1.2	1.18704313	1.18707436	1.18708222	1.18708477	1.18708484	1.18708484
1.3	1.28333687	1.28337094	1.28337950	1.28338230	1.28338236	1.28338236
1.4	1.38140205	1.38143493	1.38144319	1.38144589	1.38144595	1.38144595
1.5	1.48112026	1.48114959	1.48115696	1.48115937	1.48115941	1.48115942
1.6	1.58235990	1.58238429	1.58239042	1.58239242	1.58239246	1.58239246
1.7	1.68498902	1.68500770	1.68501240	1.68501393	1.68501396	1.68501396
1.8	1.78888175	1.78889432	1.78889748	1.78889852	1.78889853	1.78889853
1.9	1.89392110	1.89392740	1.89392898	1.89392950	1.89392951	1.89392951
2.0	2.00000000	2.00000000	2.00000000	2.00000000	2.00000000	2.00000000

For the general nonlinear boundary-value problem

(10.28) $y'' = f(x, y, y'),$ $a \le x \le b,$ $y(a) = \alpha,$ $y(b) = \beta,$

the difference method is similar. Here, however, the system of equations that are derived will not be linear; so an iteration process is required to solve it.

For the development of the procedure, we will assume throughout that f satisfies the following conditions:

i) f and the partial derivatives $f_y \equiv \partial f / \partial y$ and $f_{y'} \equiv \partial f / \partial y'$ are all continuous on

$$D = \{(x, y, y') | a \le x \le b, -\infty < y < \infty, -\infty < y' < \infty\};$$

ii) $f_y(x, y, y') \ge \delta > 0$ on D for some $\delta > 0$;
iii)   constants k and L exist, with

$$k = \max_{(x, y, y') \in D} |f_y(x, y, y')|, \quad L = \max_{(x, y, y') \in D} |f_{y'}(x, y, y')|.$$

This will ensure, by Theorem 10.1, page 463, that a unique solution to Eq. (10.28) exists.

As in the linear case, we divide $[a, b]$ into $(N + 1)$ equal subintervals whose endpoints are at the meshpoints $x_i = a + ih$ for $i = 0, 1, \ldots, N + 1$. Assuming that the exact solution has a bounded fourth derivative allows us to replace $y''(x_i)$ and $y'(x_i)$ in each of the equations

$$(10.29) \qquad y''(x_i) = f(x_i, y(x_i), y'(x_i))$$

by the appropriate centered-difference formula given in equations (10.24) and (10.25), to obtain, for each $i = 1, 2, \ldots, N$,

$$(10.30) \quad \frac{y(x_{i+1}) - 2y(x_i) + y(x_{i-1})}{h^2} = f\left(x_i, y(x_i), \frac{y(x_{i+1}) - y(x_{i-1})}{2h} - \frac{h^2}{6} y'''(\eta_i)\right)$$

$$+ \frac{h^2}{12} y^{(4)}(\xi_i),$$

for some ξ_i, η_i in the interval (x_{i-1}, x_{i+1}).

As in the linear case, the difference method results when the error terms are deleted and the boundary conditions employed:

$$w_0 = \alpha, \qquad w_{N+1} = \beta,$$

and

$$-\frac{w_{i+1} - 2w_i + w_{i-1}}{h^2} + f\left(x_i, w_i, \frac{w_{i+1} - w_{i-1}}{2h}\right) = 0$$

for each $i = 1, 2, \ldots, N$.

The $N \times N$ nonlinear system obtained from this method,

$$2w_1 - w_2 + h^2 f\left(x_1, w_1, \frac{w_2 - \alpha}{2h}\right) - \alpha = 0,$$

$$-w_1 + 2w_2 - w_3 + h^2 f\left(x_2, w_2, \frac{w_3 - w_1}{2h}\right) = 0,$$

$$(10.31) \qquad\qquad\qquad \vdots \qquad\qquad\qquad \vdots$$

$$-w_{N-2} + 2w_{N-1} - w_N + h^2 f\left(x_{N-1}, w_{N-1}, \frac{w_N - w_{N-2}}{2h}\right) = 0,$$

$$-w_{N-1} + 2w_N + h^2 f\left(x_N, w_N, \frac{\beta - w_{N-1}}{2h}\right) - \beta = 0,$$

will have a unique solution provided that $h \le 2/L$, as shown in the previously referenced book by Keller [56], page 86.

To approximate the solution to this system, we will use Newton's method for nonlinear systems discussed in Section 9.2. A sequence of iterates $\{(w_1^{(k)}, w_2^{(k)}, \ldots, w_N^{(k)})\}$ is generated that will converge to the solution of system (10.31), provided that the initial approximation $(w_1^{(0)}, w_2^{(0)}, \ldots, w_N^{(0)})$ is sufficiently close to the solution, $(w_1, w_2, \ldots, w_N)$ and that the Jacobian matrix for the system, defined by Eq. (9.13), is nonsingular. However, for the system (10.31), the Jacobian matrix is tridiagonal, and the assumptions presented at the beginning of this discussion ensure that J is a nonsingular matrix.

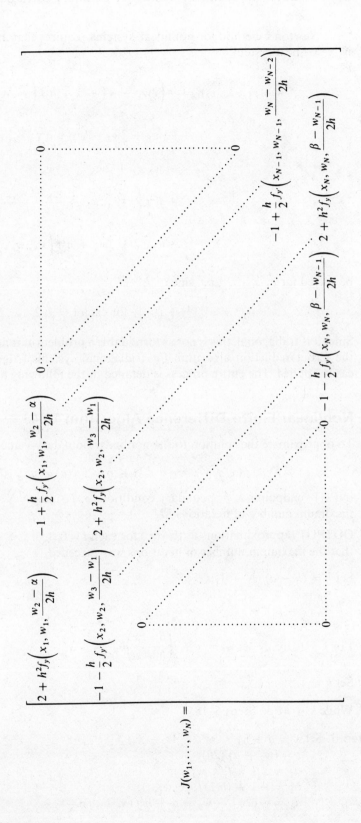

Newton's method for nonlinear systems requires that at each iteration, the $N \times N$ linear system

$$J(w_1, \ldots, w_N)(v_1, \ldots, v_n)^t = -\left(2w_1 - w_2 - \alpha + h^2 f\left(x_1, w_1, \frac{w_2 - \alpha}{2h}\right),\right.$$

$$-w_1 + 2w_2 - w_3 + h^2 f\left(x_2, w_2, \frac{w_3 - w_1}{2h}\right), \ldots,$$

$$-w_{N-2} + 2w_{N-1} - w_N$$

$$+ h^2 f\left(x_{N-1}, w_{N-1}, \frac{w_N - w_{N-2}}{2h}\right),$$

$$\left. -w_{N-1} + 2w_N + h^2 f\left(x_N, w_N, \frac{\beta - w_{N-1}}{2h}\right) - \beta\right)^t$$

be solved for $v_1, v_2, \ldots, v_N$, since

$$w_i^{(k)} = w_i^{(k-1)} + v_i \qquad \text{for each } i = 1, 2, \ldots, N.$$

Since J is tridiagonal, this is not as formidable a problem as it might at first appear; the Crout reduction algorithm for tridiagonal systems (Algorithm 6.7) can be easily applied. The entire process is detailed in the following algorithm.

Nonlinear Finite-Difference Algorithm 10.4

To approximate the solution to the nonlinear boundary-value problem

$$y'' = f(x, y, y'), \qquad a \le x \le b, \qquad y(a) = \alpha, \quad y(b) = \beta:$$

INPUT endpoints a, b; boundary conditions α, β; integer N; tolerance TOL; maximum number of iterations M.

OUTPUT approximations w_i to $y(x_i)$ for each $i = 0, 1, \ldots, N+1$ or a message that the maximum number of iterations was exceeded.

Step 1 Set $h = (b - a)/(N + 1)$;
  $w_0 = \alpha$;
  $w_{N+1} = \beta$.

Step 2 For $i = 1, \ldots, N$ set $w_i = \alpha + i\left(\dfrac{\beta - \alpha}{b - a}\right)h$.

Step 3 Set $k = 1$.

Step 4 While $k \le M$ do Steps 5–18.

  Step 5 Set $x = a + h$;
    $t = (w_2 - \alpha)/(2h)$;
    $a_1 = 2 + h^2 f_y(x, w_1, t)$;
    $b_1 = -1 + (h/2)f_{y'}(x, w_1, t)$;
    $d_1 = -(2w_1 - w_2 - \alpha + h^2 f(x, w_1, t))$.

Step 6 For $i = 2, \ldots, N - 1$
$\qquad$ set $x = a + ih$;
$\qquad\qquad t = (w_{i+1} - w_{i-1})/(2h)$;
$\qquad\qquad a_i = 2 + h^2 f_y(x, w_i, t)$;
$\qquad\qquad b_i = -1 + (h/2) f_{y'}(x, w_i, t)$;
$\qquad\qquad c_i = -1 - (h/2) f_{y'}(x, w_i, t)$;
$\qquad\qquad d_i = -(2w_i - w_{i+1} - w_{i-1} + h^2 f(x, w_i, t))$.

Step 7 Set $x = b - h$;
$\qquad t = (\beta - w_{N-1})/(2h)$;
$\qquad a_N = 2 + h^2 f_y(x, w_N, t)$;
$\qquad c_N = -1 - (h/2) f_{y'}(x, w_N, t)$;
$\qquad d_N = -(2w_N - w_{N-1} - \beta + h^2 f(x, w_N, t))$.

Step 8 Set $l_1 = a_1$; $\quad$ (*Steps 8–14 solve a tridiagonal linear system using Algorithm 6.7.*)
$\qquad u_1 = b_1/a_1$.

Step 9 For $i = 2, \ldots, N - 1$ set $l_i = a_i - c_i u_{i-1}$;
$\qquad\qquad u_i = b_i/l_i$.

Step 10 Set $l_N = a_N - c_N u_{N-1}$.

Step 11 Set $z_1 = d_1/l_1$.

Step 12 For $i = 2, \ldots, N$ set $z_i = (d_i - c_i z_{i-1})/l_i$.

Step 13 Set $v_N = z_N$;
$\qquad\quad w_N = w_N + v_N$.

Step 14 For $i = N - 1, \ldots, 1$ set $v_i = z_i - u_i v_{i+1}$;
$\qquad\qquad w_i = w_i + v_i$.

Step 15 If $\|v\| \le TOL$ then do Steps 16 and 17.

$\qquad$ **Step 16** For $i = 0, \ldots, N + 1$ set $x = a + ih$;
$\qquad\qquad\qquad\qquad\qquad\qquad$ OUTPUT (x, w_i).

$\qquad$ **Step 17** STOP. $\quad$ (*Procedure completed successfully.*)

Step 18 Set $k = k + 1$.

Step 19 OUTPUT ('Maximum number of iterations exceeded');
$\qquad\;$ STOP.

The initial approximations $w_i^{(0)}$ to w_i for each $i = 1, 2, \ldots, N$, are obtained in Step 2 by passing a straight line through (a, α) and (b, β) and evaluating at x_i.

It can be shown (see Isaacson and Keller [52], page 433) that this nonlinear finite-difference method is of order $O(h^2)$ so the stopping criteria in Step 15 could be based on the condition that $|v_j| = O(h^2)$ for each $j = 1, 2, \ldots, N$.

Since a good initial approximation is required when the satisfaction of conditions (i), (ii), and (iii), given at the beginning of this presentation, cannot be verified, an upper bound for k should be specified and, if exceeded, a new initial approximation or a reduction in step size considered.

EXAMPLE 3   Applying Algorithm 10.4, with $h = .1$, to the nonlinear boundary-value problem

$$y'' = \tfrac{1}{8}(32 + 2x^3 - yy'), \qquad 1 \le x \le 3, \quad y(1) = 17, \quad y(3) = \tfrac{43}{3}$$

gives the results in Table 10.5. The stopping procedure used in this example was to iterate until all values of successive iterates differed by less than 10^{-8}. This was accomplished with four iterations. Note that the problem in this example is the same as that considered for the nonlinear shooting method, Example 1 of Section 10.2.   □

TABLE 10.5

x_i	w_i	$y(x_i)$	$|w_i - y(x_i)|$
1.0	17.000000	17.000000	—
1.1	15.754503	15.755455	9.520×10^{-4}
1.2	14.771740	14.773333	1.594×10^{-3}
1.3	13.995677	13.997692	2.015×10^{-3}
1.4	13.386297	13.388571	2.275×10^{-3}
1.5	12.914252	12.916667	2.414×10^{-3}
1.6	12.557538	12.560000	2.462×10^{-3}
1.7	12.299326	12.301765	2.438×10^{-3}
1.8	12.126529	12.128889	2.360×10^{-3}
1.9	12.028814	12.031053	2.239×10^{-3}
2.0	11.997915	12.000000	2.085×10^{-3}
2.1	12.027142	12.029048	1.905×10^{-3}
2.2	12.111020	12.112727	1.707×10^{-3}
2.3	12.245025	12.246522	1.497×10^{-3}
2.4	12.425388	12.426667	1.278×10^{-3}
2.5	12.648944	12.650000	1.056×10^{-3}
2.6	12.913013	12.913846	8.335×10^{-4}
2.7	13.215312	13.215926	6.142×10^{-4}
2.8	13.553885	13.554286	4.006×10^{-4}
2.9	13.927046	13.927241	1.953×10^{-4}
3.0	14.333333	14.333333	—

Richardson's extrapolation procedure can also be used for the nonlinear difference method. If this method is applied to our problem, using $h = .1, .05$, and .025, and four iterations in each case, the results in Table 10.6 are obtained. The notation is the same as in Example 2, and the values of EXT_{3i} are all accurate to the places listed, with an actual maximum error at the mesh points of 3.68×10^{-10}. The values of $w_i(h = .1)$ are omitted from the table since they were listed previously.

TABLE 10.6

x_i	$w_i(h = .05)$	$w_i(h = .025)$	Ext_{1i}	Ext_{2i}	Ext_{3i}
1.0	17.00000000	17.00000000	17.00000000	17.00000000	17.00000000
1.1	15.75521721	15.75539525	15.75545543	15.75545460	15.75545455
1.2	14.77293601	14.77323407	14.77333479	14.77333342	14.77333333
1.3	13.99718996	13.99756680	13.99769413	13.99769242	13.99769231
1.4	13.38800424	13.38842973	13.38857346	13.38857156	13.38857143
1.5	12.91606471	12.91651628	12.91666881	12.91666680	12.91666667
1.6	12.55938618	12.55984665	12.56000217	12.56000014	12.56000000
1.7	12.30115670	12.30161280	12.30176684	12.30176484	12.30176471
1.8	12.12830042	12.12874287	12.12889094	12.12888902	12.12888889
1.9	12.03049438	12.03091316	12.03105457	12.03105275	12.03105263
2.0	11.99948020	11.99987013	12.00000179	12.00000011	12.00000000
2.1	12.02857252	12.02892892	12.02904924	12.02904772	12.02904762
2.2	12.11230149	12.11262089	12.11272872	12.11272736	12.11272727
2.3	12.24614846	12.24642848	12.24652299	12.24652182	12.24652174
2.4	12.42634789	12.42658702	12.42666773	12.42666673	12.42666667
2.5	12.64973666	12.64993420	12.65000086	12.65000005	12.65000000
2.6	12.91363828	12.91379422	12.91384683	12.91384620	12.91384615
2.7	13.21577275	13.21588765	13.21592641	13.21592596	13.21592593
2.8	13.55418579	13.55426075	13.55428603	13.55428573	13.55428571
2.9	13.92719268	13.92722921	13.92724153	13.92724139	13.92724138
3.0	14.33333333	14.33333333	14.33333333	14.33333333	14.33333333

Exercise Set 10.3

1. Use Algorithm 10.3 to approximate the solution to the following boundary-value problems.

   a) $y'' + y = 0$, $0 < x < \pi$, $y(0) = 1$, $y(\pi) = -1$;

   use $h = \pi/3$.

   b) $y'' + 4y = \cos x$, $0 < x < \pi/4$, $y(0) = 0$, $y(\pi/4) = 0$;

   use $h = \pi/12$.

2. Use Algorithm 10.4 to approximate the solution to the following boundary-value problems:

   a) $y'' = -(y')^2 - y + \ln x$, $1 < x < 2$, $y(1) = 0$, $y(2) = \ln 2$;

   use $h = .5$.

   b) $y'' = \dfrac{xy' - y}{x^2}$, $1 < x < 1.6$, $y(1) = -1$, $y(1.6) = -.847994$;

   use $h = .2$.

3. Show that the following linear boundary-value problems satisfy the hypotheses of Theorem 10.4, and use Algorithm 10.3 to approximate their solutions.

   a) $y'' = -\dfrac{4}{x}y' + \dfrac{2}{x^2}y - \dfrac{2\ln x}{x^2}$, $1 < x < 2$, $y(1) = -\frac{1}{2}$, $y(2) = \ln 2$;

   use $h = .05$, and compare results to the actual solution.

b) $y'' = -4y' + 4y$, $0 < x < 5$, $y(0) = 1$, $y(5) = 0$;

use $h = .2$, and compare results to the actual solution.

c) $y'' = -3y' + 2y + 2x + 3$, $0 < x < 1$, $y(0) = 2$, $y(1) = 1$;

use $h = .1$, and compare results to the actual solution.

d) $y'' = 100y$, $0 < x < 1$, $y(0) = 1$, $y(1) = e^{-10}$;

use $h = .1$ and $h = .05$, and compare results to the actual solution $y = e^{-10x}$ and to Exercise 9, Section 10.1.

4.   Repeat Exercise 3 using extrapolation.

5.   Show that the following linear boundary-value problems do not satisfy the hypotheses of Theorem 10.4. Use Algorithm 10.3 to approximate their solutions.

a) $y'' = -\dfrac{4}{x}y' - \dfrac{2}{x^2}y + \dfrac{2}{x^2}\ln x$, $1 < x < 2$, $y(1) = \frac{1}{2}$, $y(2) = \ln 2$;

use $h = .05$.

b) $y'' = -y$, $0 < x < \pi/4$, $y(0) = 1$, $y(\pi/4) = 1$;

use $h = \pi/40$.

6.   Show that if $y \in C^3[x_{i-1}, x_{i+1}]$, then there exists an η_i, $x_{i-1} < \eta_i < x_{i+1}$ with the property that

$$y'(x_i) = \frac{1}{2h}[y(x_{i+1}) - y(x_{i-1})] - \frac{h^2}{6}y'''(\eta_i).$$

7.   Prove Theorem 10.4.

8.   Show that if $y \in C^6[a, b]$ and if $w_0, w_1, \ldots, w_{N+1}$ satisfy Eq. (10.27), then

$$w_i - y(x_i) = Ah^2 + O(h^4)$$

where A is independent of h, provided $q(x) \geq w > 0$ on $[a, b]$ for some w.

9.   Show that the following nonlinear boundary-value problems satisfy the hypotheses

i) $f, f_y, f_{y'}$ are all continuous on

$$D = \{(x, y, y') \mid a \leq x \leq b, -\infty < y, y' < \infty\};$$

ii) $f_y(x, y, y') \geq \delta > 0$ on D for some $\delta > 0$;
iii)   constants k and L exist, with

$$k = \max_{(x, y, y') \in D} |f_y(x, y, y')|, \qquad L = \max_{(x, y, y') \in D} |f_{y'}(x, y, y')|;$$

and use Algorithm 10.4 to approximate their solutions.

a) $y'' = \frac{1}{2}y^3$, $1 < x < 2$, $y(1) = -\frac{2}{3}$, $y(2) = -1$;

use $h = .05$, and compare results to Exercise 3(a), Section 10.2.

b) $y'' = 2y^3$, $1 < x < 5$, $y(1) = \frac{1}{4}$, $y(2) = \frac{1}{8}$;

use $h = .2$, and compare results to Exercise 3(b), Section 10.2.

c) $y'' = y^3 - yy'$, $1 < x < 2$, $y(1) = \frac{1}{2}$, $y(2) = \frac{1}{3}$;

use $h = .1$, and compare results to Exercise 3(c), Section 10.2.

d) $y'' = 2y^3 - 6y - 2x^3$, $1 < x < 2$, $y(1) = 2$, $y(2) = \frac{5}{2}$;

   use $h = .05$, and compare results to Exercise 3(d), Section 10.2.

10. Repeat Exercise 9 using extrapolation.

11. Show that the hypotheses listed in Exercise 9 ensure the nonsingularity of the Jacobian matrix J for $h < 2/L$.

12. The deflection of a uniformly loaded, long rectangular plate under an axial tension force, for small deflections, is governed by a second-order differential equation. Let S represent the axial force and q the intensity of the uniform load. The deflection W along the elemental length is given by:

$$W''(x) - \frac{S}{D}W(x) = \frac{-ql}{2D}x + \frac{q}{2D}x^2, \qquad 0 < x < l, \quad W(0) = W(l) = 0,$$

where l is the length of the plate, and D is the flexural rigidity of the plate. Let $q = 200$ lb/in.^{2}, $S = 100$ lb/in., $D = 8.8 \times 10^7$ lb in. and $l = 50$ in. Approximate the deflection at one inch intervals by the linear finite-difference Algorithm 10.3.

10.4 Rayleigh–Ritz Method

An important linear two-point boundary-value problem in beam-stress analysis is given by the differential equation

(10.32) $$-\frac{d}{dx}\left(p(x)\frac{dy}{dx}\right) + q(x)y = f(x) \qquad \text{for } 0 \le x \le 1,$$

with the boundary conditions

(10.33) $$y(0) = y(1) = 0.$$

This differential equation describes the deflection $y(x)$ on a beam of length one with variable cross section represented by $q(x)$. This deflection is due to the added stresses $p(x)$ and $f(x)$.

   In the discussion that follows we will assume that $p \in C^1[0, 1]$ and $q, f \in C[0, 1]$. Further, we require that there exist a constant $\delta > 0$ such that

$$p(x) \ge \delta > 0 \qquad \text{for } 0 \le x \le 1$$

and that $$q(x) \ge 0 \qquad \text{for } 0 \le x \le 1.$$

These assumptions are sufficient to guarantee that the boundary-value problem (10.32) and (10.33) has a unique solution (see Bailey, Shampine, and Waltman [8]).

   As is the case of many boundary-value problems that describe physical phenomena, the solution to the beam equation satisfies a **variational** property. The variational principle for the beam equation is fundamental to the development of the Rayleigh–Ritz method and characterizes the solution to the beam equation as the function that minimizes a certain integral over all functions in $C_0^2[0, 1]$, the set of those functions u in $C^2[0, 1]$ with the property that $u(0) = u(1) = 0$. The following theorem gives the characterization. The proof of this theorem, while not difficult, is lengthy; it can be found in Schultz [77].

Theorem 10.5   Let $p \in C^1[0, 1]$, q, $f \in C[0, 1]$, and

$$(10.34) \qquad p(x) \geq \delta > 0, \quad q(x) \geq 0 \qquad \text{for } 0 \leq x \leq 1.$$

The function $y \in C_0^2[0, 1]$ is the unique solution to the differential equation

$$(10.35) \qquad -\frac{d}{dx}\left(p(x)\frac{dy}{dx}\right) + q(x)y = f(x), \qquad 0 \leq x \leq 1,$$

if and only if y is the unique function in $C_0^2[0, 1]$ that minimizes the integral

$$(10.36) \qquad I[u] = \int_0^1 \{p(x)[u'(x)]^2 + q(x)[u(x)]^2 - 2f(x)u(x)\} \, dx.$$

In the Rayleigh–Ritz procedure the integral I is not minimized over all the functions in $C_0^2[0, 1]$, but over a smaller set of functions consisting of linear combinations of certain basis functions $\phi_1, \phi_2, \ldots, \phi_n$. The basis functions $\{\phi_i\}_{i=1}^n$ must be linearly independent and satisfy

$$\phi_i(0) = \phi_i(1) = 0 \qquad \text{for each } i = 1, 2, \ldots, n.$$

An approximation $\phi(x) = \sum_{i=1}^n c_i \phi_i(x)$ to the solution $y(x)$ of Eq. (10.35) is obtained by finding constants $c_1, c_2, \ldots, c_n$ to minimize $I[\sum_{i=1}^n c_i \phi_i]$.

From Eq. (10.36),

$$(10.37) \qquad \begin{aligned} I[\phi] &= I\left[\sum_{i=1}^n c_i \phi_i\right] \\ &= \int_0^1 \left\{ p(x)\left[\sum_{i=1}^n c_i \phi_i'(x)\right]^2 + q(x)\left[\sum_{i=1}^n c_i \phi_i(x)\right]^2 - 2f(x)\sum_{i=1}^n c_i \phi_i(x) \right\} dx, \end{aligned}$$

and, in order for a minimum to occur it is necessary, when considering I as a function of $c_1, c_2, \ldots, c_n$, to have

$$(10.38) \qquad \frac{\partial I}{\partial c_j} = 0 \qquad \text{for each } j = 1, 2, \ldots, n.$$

Differentiating (10.37) gives

$$\frac{\partial I}{\partial c_j} = \int_0^1 \left\{ 2p(x)\sum_{i=1}^n c_i \phi_i'(x)\phi_j'(x) + 2q(x)\sum_{i=1}^n c_i \phi_i(x)\phi_j(x) - 2f(x)\phi_j(x) \right\} dx,$$

and substituting into Eq. (10.38) yields

$$(10.39) \qquad 0 = \sum_{i=1}^n \left[\int_0^1 \{p(x)\phi_i'(x)\phi_j'(x) + q(x)\phi_i(x)\phi_j(x)\} \, dx \right] c_i - \int_0^1 f(x)\phi_j(x) \, dx,$$

for each $j = 1, 2, \ldots, n$.

The first choice of basis functions we will discuss involves piecewise linear polynomials. The first step is to form a partition on $[0, 1]$ by choosing points $x_0, x_1, \ldots, x_{n+1}$ with

$$0 = x_0 < x_1 < \cdots < x_n < x_{n+1} = 1.$$

Letting $h_i = x_{i+1} - x_i$ for each $i = 0, 1, \ldots, n$, we define the basis functions $\phi_1(x), \phi_2(x), \ldots, \phi_n(x)$ by

$$
(10.40) \qquad \phi_i(x) = \begin{cases}
0, & 0 \leq x \leq x_{i-1}, \\[2mm]
\dfrac{(x - x_{i-1})}{h_{i-1}}, & x_{i-1} \leq x \leq x_i, \\[2mm]
\dfrac{(x_{i+1} - x)}{h_i}, & x_i \leq x \leq x_{i+1}, \\[2mm]
0, & x_{i+1} \leq x \leq 1,
\end{cases}
$$

for each $i = 1, 2, \ldots, n$. (See Fig. 10.4.)

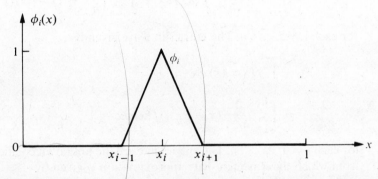

FIGURE 10.4

Since the functions ϕ_i are piecewise linear, the derivatives ϕ_i', while not continuous, are constant on the open subinterval (x_j, x_{j+1}) for each $j = 0, 1, \ldots, n$. Thus, we have

$$
(10.41) \qquad \phi_i'(x) = \begin{cases}
0, & 0 < x < x_{i-1}, \\[2mm]
\dfrac{1}{h_{i-1}}, & x_{i-1} < x < x_i, \\[2mm]
-\dfrac{1}{h_i}, & x_i < x < x_{i+1}, \\[2mm]
0, & x_{i+1} < x < 1,
\end{cases}
$$

for each $i = 1, 2, \ldots, n$.

Because ϕ_i and ϕ_i' are nonzero only on (x_{i-1}, x_{i+1}), the linear system (10.39) reduces to an $n \times n$ tridiagonal linear system:

$$
(10.42) \qquad\qquad\qquad A\mathbf{c} = \mathbf{b}.
$$

The nonzero entries in A are given by

$$
\begin{aligned}
a_{ii} &= \int_0^1 \{ p(x)[\phi_i'(x)]^2 + q(x)[\phi_i(x)]^2 \} \, dx \\[2mm]
&= \int_{x_{i-1}}^{x_i} \left(\frac{1}{h_{i-1}} \right)^2 p(x) \, dx + \int_{x_i}^{x_{i+1}} \left(\frac{-1}{h_i} \right)^2 p(x) \, dx \\[2mm]
&\quad + \int_{x_{i-1}}^{x_i} \left(\frac{1}{h_{i-1}} \right)^2 (x - x_{i-1})^2 q(x) \, dx + \int_{x_i}^{x_{i+1}} \left(\frac{1}{h_i} \right)^2 (x_{i+1} - x)^2 q(x) \, dx
\end{aligned}
$$

for each $i = 1, 2, \ldots, n$;

$$a_{i,i+1} = \int_0^1 \{p(x)\phi_i'(x)\phi_{i+1}'(x) + q(x)\phi_i(x)\phi_{i+1}(x)\}\, dx$$

$$= \int_{x_i}^{x_{i+1}} -\left(\frac{1}{h_i}\right)^2 p(x)\, dx + \int_{x_i}^{x_{i+1}} \left(\frac{1}{h_i}\right)^2 (x_{i+1} - x)(x - x_i)q(x)\, dx$$

for each $i = 1, 2, \ldots, n - 1$; and

$$a_{i,i-1} = \int_0^1 \{p(x)\phi_i'(x)\phi_{i-1}'(x) + q(x)\phi_i(x)\phi_{i-1}(x)\}\, dx$$

$$= \int_{x_{i-1}}^{x_i} -\left(\frac{1}{h_{i-1}}\right)^2 p(x)\, dx + \int_{x_{i-1}}^{x_i} \left(\frac{1}{h_{i-1}}\right)^2 (x_i - x)(x - x_{i-1})q(x)\, dx$$

for each $i = 2, \ldots, n$. The entries in **b** are given by

$$b_i = \int_0^1 f(x)\phi_i(x)\, dx$$

$$= \int_{x_{i-1}}^{x_i} \frac{1}{h_{i-1}}(x - x_{i-1})f(x)\, dx + \int_{x_i}^{x_{i+1}} \frac{1}{h_i}(x_{i+1} - x)f(x)\, dx$$

for each $i = 1, 2, \ldots, n$. The entries in **c** are the unknown coefficients $c_1, c_2, \ldots, c_n$ from which the Rayleigh–Ritz approximation ϕ given by

$$\phi(x) = \sum_{i=1}^n c_i\phi_i(x)$$

is constructed.

   The following algorithm sets up the tridiagonal linear system and incorporates the Tridiagonal Algorithm 6.7 to solve the system. In Steps 3, 4, and 5 integrations must be performed so a quadrature formula must be included. It is recommended that the quadrature be accomplished by interpolating p, q, and f with the piecewise polynomials $\phi_1, \ldots, \phi_n$, unless those integrals can be evaluated easily (see Section 3.6).

Piecewise Linear Rayleigh–Ritz Algorithm 10.5

To approximate the solution to the boundary-value problem

$$-\frac{d}{dx}\left(p(x)\frac{dy}{dx}\right) + q(x)y = f(x), \qquad 0 \le x \le 1, \quad y(0) = y(1) = 0,$$

with the piecewise linear function

$$\phi(x) = \sum_{i=1}^n c_i\phi_i(x):$$

INPUT   integer n; mesh points $x_0 = 0 < x_1 < \cdots < x_n < x_{n+1} = 1$.

OUTPUT   coefficients $c_1, \ldots, c_n$.

Step 1   For $i = 0, \ldots, n$ set $h_i = x_{i+1} - x_i$.

Step 2 For $i = 1, \ldots, n$
define the piecewise linear basis ϕ_i by

$$\phi_i(x) = \begin{cases} 0, & 0 \le x \le x_{i-1}, \\[2mm] \dfrac{x - x_{i-1}}{h_{i-1}}, & x_{i-1} < x \le x_i, \\[3mm] \dfrac{x_{i+1} - x}{h_i}, & x_i < x \le x_{i+1}, \\[3mm] 0, & x_{i+1} < x \le 1. \end{cases}$$

Step 3 Set $IOLD1 = \left(\dfrac{1}{h_0}\right)^2 \displaystyle\int_{x_0}^{x_1} p(x)\, dx;$ (*Note: the integrals in Steps 3–5 can be evaluated using a numerical integration procedure.*)

$$IOLD2 = \left(\frac{1}{h_0}\right)^2 \int_{x_0}^{x_1} (x - x_0)^2 q(x)\, dx.$$

Step 4 For $i = 1, \ldots, n - 1$ set $INEW1 = \left(\dfrac{1}{h_i}\right)^2 \displaystyle\int_{x_i}^{x_{i+1}} p(x)\, dx;$

$$INEW2 = \left(\frac{1}{h_i}\right)^2 \int_{x_i}^{x_{i+1}} (x - x_i)^2 q(x)\, dx;$$

$$I3 = \left(\frac{1}{h_i}\right)^2 \int_{x_i}^{x_{i+1}} (x_{i+1} - x)^2 q(x)\, dx;$$

$$I4 = \left(\frac{1}{h_i}\right)^2 \int_{x_i}^{x_{i+1}} (x_{i+1} - x)(x - x_i) q(x)\, dx;$$

$$I5 = \frac{1}{h_{i-1}} \int_{x_{i-1}}^{x_i} (x - x_{i-1}) f(x)\, dx;$$

$$I6 = \frac{1}{h_i} \int_{x_i}^{x_{i+1}} (x_{i+1} - x) f(x)\, dx;$$

$$\alpha_i = IOLD1 + INEW1 + IOLD2 + I3;$$
$$\beta_i = -INEW1 + I4;$$
$$b_i = I5 + I6;$$
$$IOLD1 = INEW1;$$
$$IOLD2 = INEW2.$$

Step 5 Set $INEW1 = \left(\dfrac{1}{h_n}\right)^2 \displaystyle\int_{x_n}^{x_{n+1}} p(x)\, dx;$

$$I3 = \left(\frac{1}{h_n}\right)^2 \int_{x_n}^{x_{n+1}} (x_{n+1} - x)^2 q(x)\, dx;$$

$$I5 = \frac{1}{h_{n-1}} \int_{x_{n-1}}^{x_n} (x - x_{n-1}) f(x)\, dx;$$

$$I6 = \frac{1}{h_n} \int_{x_n}^{x_{n+1}} (x_{n+1} - x) f(x)\, dx;$$

$$\alpha_n = IOLD1 + INEW1 + IOLD2 + I3;$$
$$b_n = I5 + I6.$$

Step 6 Set $a_1 = \alpha_1$;   (*Steps 6–12 solve a symmetric tridiagonal linear system using Algorithm 6.7.*)

$$\zeta_1 = \beta_1/\alpha_1.$$

Step 7 For $i = 2, \ldots, n - 1$ set $a_i = \alpha_i - \beta_{i-1}\zeta_{i-1}$;

$$\zeta_i = \beta_i/a_i.$$

Step 8 Set $a_n = \alpha_n - \beta_{n-1}\zeta_{n-1}$.

Step 9 Set $z_1 = b_1/a_1$.

Step 10 For $i = 2, \ldots, n - 1$ set $z_i = (b_i - \beta_{i-1}z_{i-1})/a_i$.

Step 11 Set $c_n = z_n$;

OUTPUT (c_n).

Step 12 For $i = n - 1, \ldots, 1$ set $c_i = z_i - \zeta_i c_{i+1}$;

OUTPUT (c_i).

Step 13 STOP.   (*Procedure is complete.*)

The following example uses Algorithm 10.5. Because of the nature of this example, the integrations in Steps 3, 4, and 5 were found directly.

EXAMPLE 1 Consider the boundary-value problem

$$-y'' + \pi^2 y = 2\pi^2 \sin(\pi x), \qquad 0 \le x \le 1, \quad y(0) = y(1) = 0.$$

Let $h_i = h = .1$, so that $x_i = .1i$ for each $i = 0, 1, \ldots, 9$. Following the steps in Algorithm 10.5 gives:

Step 3 $IOLD1 = 100 \displaystyle\int_0^{.1} dx = 10$;

$$IOLD2 = 100 \int_0^{.1} x^2\pi^2 \, dx = \frac{\pi^2}{30}.$$

Step 4 For each $i = 1, \ldots, 8$

$$INEW1 = 100 \int_{.1i}^{.1i+.1} dx = 10;$$

$$INEW2 = 100 \int_{.1i}^{.1i+.1} (x - .1i)^2\pi^2 \, dx = \frac{\pi^2}{30};$$

$$I3 = 100 \int_{.1i}^{.1i+.1} (.1i + .1 - x)^2\pi^2 \, dx = \frac{\pi^2}{30};$$

$$I4 \doteq 100 \int_{.1i}^{.1i+.1} (.1i + .1 - x)(x - .1i)\pi^2 \, dx = \frac{\pi^2}{60};$$

$$I5 = 10 \int_{.1i-.1}^{.1i} (x - .1i + .1)2\pi^2 \sin(\pi x)\, dx$$

$$= -2\pi \cos .1\pi i + 20[\sin(.1\pi i) - \sin((.1i - .1)\pi)];$$

$$I6 = 10 \int_{.1i}^{.1i+.1} (.1i + .1 - x)2\pi^2 \sin(\pi x)\, dx$$

$$= 2\pi \cos .1\pi i - 20[\sin((.1i + .1)\pi) - \sin(.1\pi i)];$$

$$\alpha_i = 20 + \frac{\pi^2}{15};$$

$$\beta_i = -10 + \frac{\pi^2}{60};$$

$$b_i = 40 \sin(.1\pi i)(1 - \cos(.1\pi));$$

$$IOLD1 = 10;$$

$$IOLD2 = \frac{\pi^2}{30}.$$

Step 5 $INEW1 = 100 \int_{.9}^{1} dx = 10;$

$$I3 = 100 \int_{.9}^{1} (1 - x)^2 \pi^2\, dx = \frac{\pi^2}{30};$$

$$I5 = -2\pi \cos(.9\pi) + 20[\sin(.9\pi) - \sin(.8\pi)];$$
$$I6 = 2\pi \cos(.9\pi) + 20 \sin(.9\pi);$$

$$\alpha_9 = 20 + \frac{\pi^2}{15};$$

$$b_9 = 40 \sin(.9\pi) - 20 \sin(.8\pi).$$

Steps 6, 7, and 8 These give the following values of a_i and ζ_i:

$$
\begin{aligned}
a_1 &= 20.65797363, & \zeta_1 &= -.4761118767 \\
a_2 &= 15.97517213, & \zeta_2 &= -.6156745300 \\
a_3 &= 14.60250273, & \zeta_3 &= -.6735493754 \\
a_4 &= 14.03327431, & \zeta_4 &= -.7008704010 \\
a_5 &= 13.76455818, & \zeta_5 &= -.7145530183 \\
a_6 &= 13.6299871, & \zeta_6 &= -.7216081489, \\
a_7 &= 13.56059192, & \zeta_7 &= -.7253006839, \\
a_8 &= 13.52427397, & \zeta_8 &= -.7272483991, \\
a_9 &= 13.50511721.
\end{aligned}
$$

Steps 9 and 10 These generate the following values of z_i:

$$
\begin{aligned}
z_1 &= .0292852890, & z_6 &= .3687722609, \\
z_2 &= .0900626268, & z_7 &= .3842683589, \\
z_3 &= .1691255264, & z_8 &= .3645448400, \\
z_4 &= .2512140726, & z_9 &= .3102866742. \\
z_5 &= .3217362267,
\end{aligned}
$$

Steps 11 and 12 These give the coefficients c_i:

$$
\begin{aligned}
c_9 &= .3102866742, & c_4 &= .9549641893, \\
c_8 &= .5902003271, & c_3 &= .8123410598, \\
c_7 &= .8123410598, & c_2 &= .5902003271, \\
c_6 &= .9549641893, & c_1 &= .3102866742. \\
c_5 &= 1.004108771,
\end{aligned}
$$

The piecewise linear approximation is given by

$$
\phi(x) = \sum_{i=1}^{9} c_i \phi_i(x).
$$

The actual solution to the boundary-value problem is

$$
f(x) = \sin \pi x
$$

and Table 10.7 lists the error in the approximation at x_i for each $i = 1, \ldots, 9$.

$\square$

TABLE 10.7

i	x_i	$\phi(x_i)$	$f(x_i)$	$|\phi(x_i) - f(x_i)|$
1	.1	.3102866742	.3090169943	.00127
2	.2	.5902003271	.5877852522	.00242
3	.3	.8123410598	.8090169943	.00332
4	.4	.9549641896	.9510565162	.00391
5	.5	1.004108771	1.0000000000	.00411
6	.6	.9549641893	.9510565162	.00391
7	.7	.8123410598	.8090169943	.00332
8	.8	.5902003271	.5877852522	.00242
9	.9	.3102866742	.3090169943	.00127

It can be shown that the tridiagonal matrix A given in Eq. (10.42) is symmetric and positive definite (see Exercise 6), so the linear system is stable. Under the hypotheses presented at the beginning of this section we have

$$
|\phi(x) - y(x)| = O(h^2), \qquad 0 \le x \le 1.
$$

A proof of this result can be found in Schultz [77], pages 103 and 104.

Another choice of basis functions involves the cubic spline functions that were studied for interpolating purposes in Section 3.6. The first step in constructing the cubic spline approximation is to form a partition of [0, 1] by choosing a positive integer n, letting $h = 1/(n + 1)$ and defining the mesh points as $x_i = ih$ for each $i = 0, 1, \ldots, n + 1$. A cubic spline function on [0, 1] relative to this partition will belong to $C^2[0, 1]$ and be a cubic polynomial on each of the subintervals $[x_i, x_{i+1}]$ for $i = 0, 1, \ldots, n$.

The functions ϕ_i, $i = 0, 1, \ldots, n + 1$ in this basis are called the **B-splines** or **bell-shaped splines** presented originally by Schoenberg [74]. For each mesh point x_i the associated B-spline S_i is defined by

$$S_i(x) = \begin{cases} 0, & x \leq x_{i-2} \\[2mm] \dfrac{\left[2 - \dfrac{(x - x_i)}{h}\right]^3}{24} - \dfrac{\left[1 - \dfrac{(x - x_i)}{h}\right]^3}{6} - \dfrac{\left[\dfrac{x - x_i}{h}\right]^3}{4} + \dfrac{\left[1 + \dfrac{(x - x_i)}{h}\right]^3}{6}, & \\[2mm] & x_{i-2} < x \leq x_{i-1}, \\[2mm] \dfrac{\left[2 - \dfrac{(x - x_i)}{h}\right]^3}{24} - \dfrac{\left[1 - \dfrac{(x - x_i)}{h}\right]^3}{6} - \dfrac{\left[\dfrac{x - x_i}{h}\right]^3}{4}, & x_{i-1} < x \leq x_i, \\[2mm] \dfrac{\left[2 - \dfrac{(x - x_i)}{h}\right]^3}{24} - \dfrac{\left[1 - \dfrac{(x - x_i)}{h}\right]^3}{6}, & x_i < x \leq x_{i+1}, \\[2mm] \dfrac{\left[2 - \dfrac{(x - x_i)}{h}\right]^3}{24}, & x_{i+1} < x \leq x_{i+2}, \\[2mm] 0, & x_{i+2} < x, \end{cases}$$

where the additional meshpoints needed for S_0, S_1, S_n, and S_{n+1} are defined by $x_{-2} = -2h$, $x_{-1} = -h$, $x_{n+2} = (n + 2)h$, and $x_{n+3} = (n + 3)h$. A typical B-spline is shown in Fig. 10.5.

If we define the cubic spline S by

$$S(x) = \begin{cases} 0, & x \leq -2 \\[2mm] \dfrac{(2 - x)^3}{24} - \dfrac{(1 - x)^3}{6} - \dfrac{x^3}{4} + \dfrac{(1 + x)^3}{6}, & -2 < x \leq -1, \\[2mm] \dfrac{(2 - x)^3}{24} - \dfrac{(1 - x)^3}{6} - \dfrac{x^3}{4}, & -1 < x \leq 0, \\[2mm] \dfrac{(2 - x)^3}{24} - \dfrac{(1 - x)^3}{6}, & 0 < x \leq 1, \\[2mm] \dfrac{(2 - x)^3}{24}, & 1 < x \leq 2, \\[2mm] 0, & x < 2, \end{cases}$$

we can represent S_i by

$$S_i(x) = S\left(\frac{x - x_i}{h}\right) \qquad \text{for each } i = 0, 1, \ldots, n + 1.$$

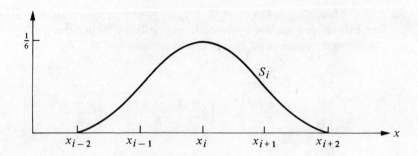

FIGURE 10.5

It is easy to show that $\{S_i\}_{i=0}^{n+1}$ is a linearly independent set of cubic splines (see Exercise 15). In order for the set $\{S_i\}_{i=0}^{n+1}$ to satisfy the boundary conditions $\phi_i(0) = \phi_i(1) = 0$, it is necessary to modify S_0, S_1, S_n, and S_{n+1}. The basis with this modification is defined by

$$\phi_i(x) = \begin{cases} S_0(x) - 4S\left(\dfrac{x+h}{h}\right), & i = 0, \\[2mm] S_1(x) - S\left(\dfrac{x+h}{h}\right), & i = 1, \\[2mm] S_i(x), & 2 \le i \le n-1, \\[2mm] S_n(x) - S\left(\dfrac{x-(n+2)h}{h}\right), & i = n \\[2mm] S_{n+1}(x) - 4S\left(\dfrac{x-(n+2)h}{h}\right), & i = n+1. \end{cases}$$

(See Fig. 10.6.)

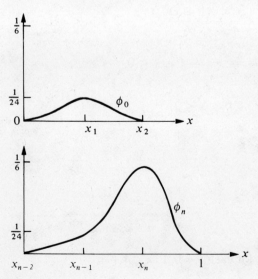

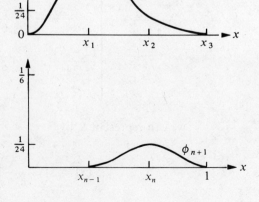

FIGURE 10.6

Since $\phi_i(x)$ and $\phi_i'(x)$ are nonzero only for $x_{i-2} \le x \le x_{i+2}$, the matrix in the Rayleigh–Ritz approximation will be a band matrix with band-width at most seven:

(10.43)

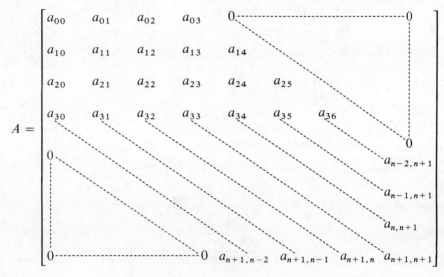

$$A = \begin{bmatrix} a_{00} & a_{01} & a_{02} & a_{03} & 0 & & & & & & 0 \\ a_{10} & a_{11} & a_{12} & a_{13} & a_{14} & & & & & & \\ a_{20} & a_{21} & a_{22} & a_{23} & a_{24} & a_{25} & & & & & \\ a_{30} & a_{31} & a_{32} & a_{33} & a_{34} & a_{35} & a_{36} & & & & \\ 0 & & & & & & & & & & 0 \\ & & & & & & & & & & a_{n-2,n+1} \\ & & & & & & & & & & a_{n-1,n+1} \\ & & & & & & & & & & a_{n,n+1} \\ 0 & & & & & 0 & a_{n+1,n-2} & a_{n+1,n-1} & a_{n+1,n} & a_{n+1,n+1} \end{bmatrix}$$

where

$$a_{ij} = \int_0^1 \{ p(x)\phi_i'(x)\phi_j'(x) + q(x)\phi_i(x)\phi_j(x) \}\, dx,$$

for each $i, j = 0, 1, \ldots, n + 1$. The matrix A is also positive definite (see Exercise 12), so the linear system (10.43) can be easily solved by Choleski's Algorithm 6.6 or by Gaussian elimination. The following algorithm details the construction of the cubic spline approximation $\phi(x)$ by the Rayleigh–Ritz method for the boundary-value problem (10.32) and (10.33) given on page 489.

Cubic Spline Rayleigh–Ritz Algorithm 10.6

To approximate the solution to the boundary-value problem

$$-\frac{d}{dx}\left(p(x)\frac{dy}{dx} \right) + q(x)y = f(x), \qquad 0 \le x \le 1, \quad y(0) = y(1) = 0$$

with the cubic spline

$$\phi(x) = \sum_{i=0}^{n+1} c_i \phi_i(x):$$

INPUT   integer n.

OUTPUT   coefficients $c_0, \ldots, c_{n+1}$.

Step 1   Set $h = 1/(n + 1)$.

Step 2   For $i = 0, \ldots, n + 1$ set $x_i = ih$.

Step 3 Define the function S by

$$S(t) = \begin{cases} 0, & t \le -2, \\ \frac{1}{24}[(2-t)^3 - 4(1-t)^3 - 6t^3 + 4(1+t)^3], & -2 < t \le -1, \\ \frac{1}{24}[(2-t)^3 - 4(1-t)^3 - 6t^3], & -1 < t \le 0, \\ \frac{1}{24}[(2-t)^3 - 4(1-t)^3], & 0 < t \le 1, \\ \frac{1}{24}(2-t)^3, & 1 < t \le 2, \\ 0, & 2 < t. \end{cases}$$

Step 4 Define the cubic spline basis $\{\phi_i\}_{i=0}^{n+1}$ by

$$\phi_0(x) = S\left(\frac{x}{h}\right) - 4S\left(\frac{x+h}{h}\right),$$

$$\phi_1(x) = S\left(\frac{x-x_1}{h}\right) - S\left(\frac{x+h}{h}\right),$$

for $i = 2, \ldots, n-1$, $\phi_i(x) = S\left(\frac{x-x_i}{h}\right),$

$$\phi_n(x) = S\left(\frac{x-x_n}{h}\right) - S\left(\frac{x-(n+2)h}{h}\right),$$

$$\phi_{n+1}(x) = S\left(\frac{x-x_{n+1}}{h}\right) - 4S\left(\frac{x-(n+2)h}{h}\right).$$

Step 5 For $i = 0, \ldots, n+1$ do Steps 6–9.

(Note: the integrals in Steps 6 and 9 can be evaluated using a numerical integration procedure.)

Step 6 For $j = i, i+1, \ldots, min(i+3, n+1)$
set $L = max(x_{j-2}, 0)$;
$U = min(x_{i+2}, 1)$;

$$a_{ij} = \int_L^U [p(x)\phi_i'(x)\phi_j'(x) + q(x)\phi_i(x)\phi_j(x)] \, dx;$$

$a_{ji} = a_{ij}$. *(Since A is symmetric.)*

Step 7 If $i \ge 4$ then for $j = 0, \ldots, i-4$ set $a_{ij} = 0$.

Step 8 If $i \le n-3$ then for $j = i+4, \ldots, n+1$ set $a_{ij} = 0$.

Step 9 Set $L = max(x_{i-2}, 0)$;
$U = min(x_{i+2}, 1)$;

$$b_i = \int_L^U f(x)\phi_i(x) \, dx.$$

Step 10 Solve the linear system $Ac = b$, where $A = (a_{ij})$, $\mathbf{b} = (b_0, \ldots, b_{n+1})^t$ and $\mathbf{c} = (c_0, \ldots, c_{n+1})^t$.

Step 11 For $i = 0, \ldots, n+1$ OUTPUT (c_i).

Step 12 STOP. *(Procedure is complete.)*

In practice it is recommended that the integrations in Steps 6 and 9 be performed by interpolating p, q, and f with the cubic spline interpolating polynomial and then integrating. The hypotheses assumed at the beginning of this section are sufficient to guarantee that

$$|y(x) - \phi(x)| = O(h^4), \qquad 0 \le x \le 1.$$

For a proof of this result, see Schultz [77], pages 107 and 108.

Another basis used often in practice is the piecewise cubic Hermite polynomials. For an excellent presentation of this method see Schultz [77].

Other methods that have received considerable attention recently are Galerkin or "weak form" methods and collocation methods.

For the boundary-value problem we have been considering

$$-\frac{d}{dx}\left(p(x)\frac{dy}{dx}\right) + q(x)y = f(x), \qquad y(0) = y(1) = 0, \quad 0 \le x \le 1,$$

under the assumptions listed on page 489, the Galerkin and Rayleigh–Ritz methods are both determined by Eq. (10.39). This is not the case for an arbitrary boundary-value problem, however. A treatment of the similarities and differences in the two methods and a discussion of the wide application of the Galerkin method can be found in Schultz [77] and in Strang and Fix [85].

The method of collocation begins by selecting a set of basis functions $\{\phi_1, \ldots, \phi_N\}$ and a set of numbers $\{x_1, \ldots, x_n\} \subset [0, 1]$ and requiring that an approximation

$$\sum_{i=1}^{N} c_i \phi_i(x)$$

satisfy the differential equation at each of the numbers x_j for $1 \le j \le n$. If, in addition, we require that $\phi_i(0) = \phi_i(1) = 0$ for $1 \le i \le N$, the boundary conditions are automatically satisfied. Much attention in the literature has been given to the choice of the numbers $\{x_j\}$ and the basis functions $\{\phi_i\}$. One of the more popular choices is to let the ϕ_i be the basis functions for cubic or quintic spline functions relative to a partition of $[0, 1]$, and to let the nodes $\{x_j\}$ be the Gaussian points or roots of certain orthogonal polynomials, transformed to the proper subinterval. A comparison of various collocation methods and finite difference methods is contained in Russell [71]. His conclusion is that the collocation methods using higher degree splines are very competitive with finite difference techniques using extrapolation. Other references for collocation methods are DeBoor and Swartz [29] and Lucas and Reddien [60].

Exercise Set 10.4

1. Use Algorithm 10.5 to approximate the solution to each of the following boundary-value problems.

    a) $y'' + \dfrac{\pi^2}{4} y = \dfrac{\pi^2}{16} \cos \dfrac{\pi}{4} x, \quad 0 < x < 1, \quad y(0) = 0, \quad y(1) = 0;$

    use $x_0 = 0$, $x_1 = .3$, $x_2 = .7$, and $x_3 = 1$.

b) $-\dfrac{d}{dx}(xy') + 4y = 4x^2 - 8x + 1,\quad 0 < x < 1,\quad y(0) = 0,\quad y(1) = 0;$

use $x_0 = 0,\ x_1 = .4,\ x_2 = .8,$ and $x_3 = 1.$

2. Use Algorithm 10.6 with $n = 3$ to approximate the solution to each of the following boundary-value problems.

a) $y'' + \dfrac{\pi^2}{4} y = \dfrac{\pi^2}{16} \cos \dfrac{\pi}{4} x,\quad 0 < x < 1,\quad y(0) = 0,\quad y(1) = 0.$

b) $-\dfrac{d}{dx}(xy') + 4y = 4x^2 - 8x + 1,\quad 0 < x < 1,\quad y(0) = 0,\quad y(1) = 0.$

3. Show that the following boundary-value problems satisfy the hypotheses of Theorem 10.5 and approximate their solutions using Algorithm 10.5.

a) $-y'' + y = x,\quad 0 < x < 1,\quad y(0) = 0,\quad y(1) = 0;$

use $h = .1$, and compare to the actual solution

$$y = x + \left(\frac{e}{e^2 - 1}\right)(e^{-x} - e^x).$$

b) $-x^2 y'' - 2xy' + 2y = -4x^2,\quad 0 < x < 1,\quad y(0) = 0,\quad y(1) = 0;$

use $h = .05$, and compare to the actual solution $y = x^2 - x.$

4. Apply Algorithm 10.5 to the following boundary-value problems, which do not satisfy the hypotheses of Theorem 10.5.

a) $-y'' - y = x,\quad 0 < x < 1,\quad y(0) = 0,\quad y(1) = 0;$

use $h = .1$, and compare to the actual solution.

b) $-y'' = 1,\quad 0 < x < 1,\quad y(0) = 0,\quad y(1) = 0;$

use $h = .1$, and compare to the actual solution.

5. Show that $\{\phi_i\}_{i=1}^n$ is a linearly independent set of functions on $[0, 1]$ for the functions ϕ_i defined in Eq. (10.40).

6. Show that the matrix A in Eq. (10.42) is positive definite.

7. Using $\phi_j(x) = \sin(j\pi x)$ for each $j = 1, 2, \ldots, 5$, find a Rayleigh–Ritz approximation for the boundary-value problem

$$-y'' + 4y = 4x(1 - e^{-2}),\qquad 0 < x < 1,\quad y(0) = 0,\quad y(1) = 0.$$

8. Show that the matrix A given by Eq. (10.42) for the boundary-value problem (10.32) with $p(x) = 1$ and $q(x) = 0$ is the same as the matrix obtained in the finite-difference method.

9. Show that the boundary-value problem

$$-\frac{d}{dx}(p(x)y') + q(x)y = f(x),\qquad 0 < x < 1,\quad y(0) = \alpha,\quad y(1) = \beta$$

can be transformed by the change of variable

$$z = y - \beta x - (1 - x)\alpha$$

into the form

$$-\frac{d}{dx}(p(x)z') + q(x)z = F(x),\qquad 0 < x < 1,\quad z(0) = 0,\quad z(1) = 0.$$

10. Using Exercise 9 and Algorithm 10.5, approximate the solution to the boundary-value problem

$$-y'' + y = x, \qquad 0 < x < 1, \quad y(0) = 1, \quad y(1) = 1 + e^{-1}.$$

11. Show that the boundary-value problem

$$-\frac{d}{dx}(p(x)y') + q(x)y = f(x), \qquad a < x < b, \quad y(a) = \alpha, \quad y(b) = \beta,$$

can be transformed into the form of equations (10.32) and (10.33).

12. Show that the matrix A in (10.43) is positive definite.

13. Solve the problem in Example 1 using Algorithm 10.6. Approximate the integrals using the interpolation as suggested.

14. Repeat Exercise 3 using Algorithm 10.6.

15. Repeat Exercise 5 using the cubic spline basis $\{\phi_i\}_{i=0}^{n+1}$.

16. Repeat Exercise 10 using Algorithm 10.6.

11

Numerical Solutions to Partial-Differential Equations

A body is called isotropic if the thermal conductivity at each point in the body is independent of the direction of heat flow through the point. The temperature, $u \equiv u(x, y, z, t)$ in an isotropic body can be found by solving the partial-differential equation

$$\frac{\partial}{\partial x}\left(k\,\frac{\partial u}{\partial x}\right) + \frac{\partial}{\partial y}\left(k\,\frac{\partial u}{\partial y}\right) + \frac{\partial}{\partial z}\left(k\frac{\partial u}{\partial z}\right) = c\rho\,\frac{\partial u}{\partial t},$$

where k, c, and ρ are functions of (x, y, z) and represent, respectively, the thermal conductivity, specific heat, and density of the body at the point (x, y, z).

When k, c, and ρ are constants, this equation is known as the simple three-dimensional heat equation, and can be expressed as

$$\frac{\partial^2 u}{\partial x^2} + \frac{\partial^2 u}{\partial y^2} + \frac{\partial^2 u}{\partial z^2} = \frac{c\rho}{k}\,\frac{\partial u}{\partial t}.$$

The solution to this equation can be found by using Fourier series if the boundary of the body is relatively simple.

In most situations where k, c, and ρ are not constant, or when the boundary is irregular, the solution to the equation must be obtained by approximation techniques. An introduction to techniques of this type is presented in this chapter.

Physical situations involving more than one variable can often be expressed using equations involving partial derivatives. In this chapter we will present a brief introduction to some of the techniques available for approximating the solution to partial-differential equations involving two variables by showing how these techniques can be applied to certain standard physical problems. We will limit our treatment to problems of this type because most of the more advanced techniques require a background in analysis that would not be expected of a student at the undergraduate level.

11.1 Physical Problems Involving Partial-Differential Equations

The first type of problem we will consider is an example of an **elliptic** partial-differential equation and is known as the **Poisson equation**:

$$(11.1) \qquad \frac{\partial^2 u}{\partial x^2}(x, y) + \frac{\partial^2 u}{\partial y^2}(x, y) = f(x, y).$$

We assume, in this equation, that the function f describes the input to the problem on a plane region R whose boundary we will denote by S. Equations of this type arise naturally in the study of various time-independent physical problems such as the steady-state distribution of heat in a plane region, the potential energy of a point in a plane acted on by gravitational forces in the plane, and two-dimensional steady-state problems involving incompressible fluids.

In order to obtain a unique solution to the Poisson equation additional constraints must be placed on the solution. For example, the study of the steady-state distribution of heat in a plane region requires that $f(x, y) \equiv 0$, resulting in a simplification of Eq. (11.1) to:

$$(11.2) \qquad \frac{\partial^2 u(x, y)}{\partial x^2} + \frac{\partial^2 u(x, y)}{\partial y^2} = 0,$$

which is called **Laplace's equation**. If the temperature within the region is determined by the temperature distribution on the boundary of the region, the constraints are called the **Dirichlet boundary conditions**, given by

$$(11.3) \qquad u(x, y) = g(x, y)$$

for all (x, y) on S, the boundary of the region R. (See Fig. 11.1.)

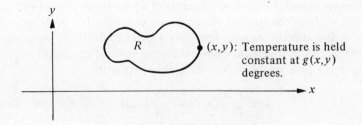

FIGURE 11.1

In Section 11.3 we will consider the numerical solution to a problem involving a **parabolic** partial-differential equation of the form

(11.4)
$$\frac{\partial u(x, t)}{\partial t} - \alpha^2 \frac{\partial^2 u(x, t)}{\partial x^2} = 0.$$

The physical problem considered here concerns the flow of heat along a rod of length l (see Fig. 11.2), which is assumed to have uniform temperature within each cross-sectional element. This condition requires the rod to be perfectly insulated on its lateral surface. The constant α in Eq. (11.4) is determined by the heat-conductive properties of the material of which the rod is composed and is assumed to be independent of the position in the rod.

FIGURE 11.2 0 l

One of the typical sets of constraints for a heat-flow problem of this type is to specify the initial heat distribution in the rod

$$u(x, 0) = f(x),$$

and describe what occurs at the ends of the rod. For example, if we held the ends of the rod at constant temperatures U_1 and U_3, the boundary conditions would have the form

$$u(0, t) = U_1 \quad \text{and} \quad u(l, t) = U_2,$$

and the heat distribution in the rod would approach the limiting temperature distribution

$$\lim_{t \to \infty} u(x, t) = U_1 + \frac{U_2 - U_1}{l} x. $$

If instead we insulated the rod so that no heat would flow through the ends, the boundary conditions would be

$$\frac{\partial u}{\partial x}(0, t) = 0 \quad \text{and} \quad \frac{\partial u}{\partial x}(l, t) = 0,$$

and would result in a constant temperature in the rod as the limiting case.

The parabolic partial-differential equation is also of importance in the study of gas diffusion and, in fact, Eq. (11.4) is known in some circles as the **diffusion equation**.

The problem we will study in Section 11.4 is called the one-dimensional **wave equation** and is an example of a **hyperbolic** partial-differential equation.

Suppose that an elastic string of length l is stretched between two supports at the same horizontal level (see Fig. 11.3). If the string is set in motion so that it vibrates in a vertical plane, then the vertical displacement $u(x, t)$ of a point x at time t satisfies the partial-differential equation

(11.5)
$$\alpha^2 \frac{\partial^2 u}{\partial x^2}(x, t) = \frac{\partial^2 u}{\partial t^2}(x, t), \qquad 0 < x < l, \; 0 < t,$$

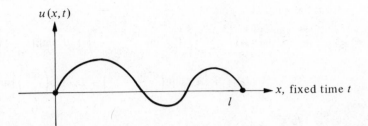

FIGURE 11.3

provided that damping effects are neglected and the amplitude is not too large. To impose constraints on this problem, we will assume that the initial position and velocity of the string are given by

$$u(x, 0) = f(x) \quad \text{and} \quad \frac{\partial u}{\partial t}(x, 0) = g(x), \quad 0 \le x \le l;$$

and we use the fact that the endpoints are fixed. This implies that $u(0, t) = 0$ and $u(l, t) = 0$.

Other physical problems involving the hyperbolic partial-differential equation (11.5) occur in the study of vibrating beams with one or both ends clamped, and in the transmission of electricity in a long transmission line in which there is some leakage of the current to the ground.

Exercise Set 11.1

The following problems summarize the Fourier-series solutions to the partial-differential equations of this section. It will be assumed that each series converges, and that the series obtained by termwise differentiations converge to the appropriate partial derivatives.

1. **Laplace's Equation on a Rectangle**   Assume that the functions $f_1(x), f_2(x), g_1(y)$, and $g_2(y)$ have the Fourier series:

$$f_1(x) = \sum_{n=1}^{\infty} A_n \sin\left(\frac{n\pi x}{a}\right), \quad \text{where } A_n = \frac{2}{a}\int_0^a f_1(x)\sin\left(\frac{n\pi x}{a}\right)dx;$$

$$f_2(x) = \sum_{n=1}^{\infty} B_n \sin\left(\frac{n\pi x}{a}\right), \quad \text{where } B_n = \frac{2}{a}\int_0^a f_2(x)\sin\left(\frac{n\pi x}{a}\right)dx;$$

$$g_1(y) = \sum_{n=1}^{\infty} C_n \sin\left(\frac{n\pi y}{b}\right), \quad \text{where } C_n = \frac{2}{b}\int_0^b g_1(y)\sin\left(\frac{n\pi y}{b}\right)dy;$$

and $$g_2(y) = \sum_{n=1}^{\infty} D_n \sin\left(\frac{n\pi y}{b}\right), \quad \text{where } D_n = \frac{2}{b}\int_0^b g_2(y)\sin\left(\frac{n\pi y}{b}\right)dy.$$

Show that Laplace's equation on a rectangle,

$$\frac{\partial^2 u}{\partial x^2} + \frac{\partial^2 u}{\partial y^2} = 0, \quad 0 < x < a, \ 0 < y < b,$$

with boundary conditions

$$u(x, 0) = f_1(x), \quad u(x, b) = f_2(x), \quad 0 \le x \le a,$$
$$u(0, y) = g_1(y), \quad u(a, y) = g_2(y), \quad 0 \le y \le b$$

has the solution

$$u(x, y) = \sum_{n=1}^{\infty} \left\{ \left[A'_n \sinh\left(\frac{n\pi(b - y)}{a}\right) + B'_n \sinh\left(\frac{n\pi y}{a}\right) \right] \sin\left(\frac{n\pi x}{a}\right) \right.$$
$$\left. + \left[C'_n \sinh\left(\frac{n\pi(a - x)}{b}\right) + D'_n \sinh\left(\frac{n\pi x}{b}\right) \right] \sin\left(\frac{n\pi y}{b}\right) \right\},$$

where

$$A'_n = \frac{A_n}{\sinh\left(\frac{n\pi b}{a}\right)}, \quad B'_n = \frac{B_n}{\sinh\left(\frac{n\pi b}{a}\right)}, \quad C'_n = \frac{C_n}{\sinh\left(\frac{n\pi a}{b}\right)}, \quad \text{and} \quad D'_n = \frac{D_n}{\sinh\left(\frac{n\pi a}{b}\right)}.$$

2. **Heat Equation**   Assume that the function f has the Fourier series

$$f(x) = \sum_{n=1}^{\infty} A_n \sin\frac{n\pi x}{l} \quad \text{where } A_n = \frac{2}{l} \int_0^l f(x)\sin\frac{n\pi x}{l}\, dx.$$

a) Show that the solution to the heat equation

$$\frac{\partial u}{\partial t} - \alpha^2 \frac{\partial^2 u}{\partial x^2} = 0, \quad 0 < x < l, \ 0 < t$$

with boundary conditions

$$u(0, t) = u(l, t) = 0, \quad 0 < t$$

and initial condition

$$u(x, 0) = f(x), \quad 0 \le x \le l$$

is given by

$$u(x, t) = \sum_{n=1}^{\infty} A_n \sin\left(\frac{n\pi x}{l}\right) \exp\left(\frac{-\alpha^2 n^2 \pi^2}{l^2} t\right).$$

b) Let $w(x) = U_1 + x(U_2 - U_1)/l$. Show that if the boundary conditions in part (a) are replaced by

$$u(0, t) = U_1, \quad u(l, t) = U_2,$$

then the solution is given by

$$u(x, t) = w(x) + \sum_{n=1}^{\infty} B_n \sin\left(\frac{n\pi x}{l}\right) \exp\left(\frac{-\alpha^2 n^2 \pi^2}{l^2} t\right)$$

where

$$B_n = \frac{2}{l} \int_0^l [f(x) - w(x)]\sin\left(\frac{n\pi x}{l}\right) dx.$$

3. **Heat Equation**   Assume that the Fourier series for $f(x)$ is given by

$$f(x) = A_0 + \sum_{n=1}^{\infty} A_n \cos\frac{n\pi x}{l},$$

where

$$A_0 = \frac{1}{l} \int_0^l f(x)\, dx \quad \text{and} \quad A_n = \frac{2}{l} \int_0^l f(x)\cos\frac{n\pi x}{l}\, dx$$

for each $n = 1, 2, \ldots$. Show that the solution to the heat equation

$$\frac{\partial u}{\partial t} - \alpha^2 \frac{\partial^2 u}{\partial x^2} = 0, \qquad 0 < x < l, \quad 0 < t,$$

with boundary conditions

$$\frac{\partial u}{\partial x}(0, t) = \frac{\partial u}{\partial x}(l, t) = 0, \qquad 0 < t,$$

and initial condition

$$u(x, 0) = f(x), \qquad 0 \leq x \leq l,$$

is given by

$$u(x, t) = A_0 + \sum_{n=1}^{\infty} A_n \cos\left(\frac{n\pi x}{l}\right) \exp\left(\frac{-\alpha^2 n^2 \pi^2}{l^2} t\right).$$

4. **Wave Equation**   Assume that the functions $f(x)$ and $g(x)$ have Fourier series given by:

$$f(x) = \sum_{n=1}^{\infty} A_n \sin\left(\frac{n\pi x}{l}\right), \qquad \text{where } A_n = \frac{2}{l} \int_0^l f(x) \sin\left(\frac{n\pi x}{l}\right) dx,$$

and

$$g(x) = \sum_{n=1}^{\infty} B_n \sin\left(\frac{n\pi x}{l}\right), \qquad \text{where } B_n = \frac{2}{l} \int_0^l g(x) \sin\left(\frac{n\pi x}{l}\right) dx.$$

Show that the solution to the wave problem,

$$\frac{\partial^2 u}{\partial t^2} - \alpha^2 \frac{\partial^2 u}{\partial x^2} = 0, \qquad 0 < x < l, \quad 0 < t,$$

$$u(0, t) = u(l, t) = 0, \qquad 0 < t,$$

$$u(x, 0) = f(x), \qquad 0 \leq x \leq l,$$

$$\frac{\partial u}{\partial t}(x, 0) = g(x), \qquad 0 \leq x \leq l$$

is given by

$$u(x, t) = \sum_{n=1}^{\infty} \left[A_n \cos\left(\frac{\alpha n\pi}{l} t\right) + B_n' \sin\left(\frac{\alpha n\pi}{l} t\right) \right] \sin\left(\frac{n\pi x}{l}\right)$$

where $B_n' = (l/\alpha n\pi) B_n$.

11.2 Elliptic Partial-Differential Equations

The elliptic partial-differential equation we will consider is the Poisson equation,

(11.6) $$\nabla^2 u(x, y) \equiv \frac{\partial^2 u(x, u)}{\partial x^2} + \frac{\partial^2 u(x, y)}{\partial y^2} = f(x, y)$$

for $(x, y) \in R$ and

$$u(x, y) = g(x, y) \qquad \text{for } (x, y) \in S,$$

where $$R = \{(x, y) \mid a < x < b, c < y < d\},$$

and S denotes the boundary of R. For this discussion we assume that both f and g are continuous on their domains so that a unique solution to Eq. (11.6) is ensured.

The method to be used is an adaptation of the finite-difference method for boundary-value problems, which was discussed in Section 10.3. The first step is to choose integers n and m, and define step sizes h and k by $h = (b - a)/n$ and $k = (d - c)/m$. Partitioning the interval $[a, b]$ into n equal parts of width h and the interval $[c, d]$ into m equal parts of width k (see Fig. 11.4) provides a means of placing a grid on the rectangle R by drawing vertical and horizontal lines through the points with coordinates (x_i, y_j), where

$$x_i = a + ih \qquad \text{for each } i = 0, 1, \ldots, n,$$

and

$$y_j = c + jk \qquad \text{for each } j = 0, 1, \ldots, m.$$

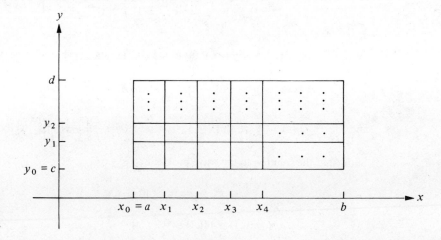

FIGURE 11.4

The lines $x = x_i$ and $y = y_j$ are called the **grid lines** and their intersections are called the **mesh points** of the grid. For each mesh point in the interior of the grid, (x_i, y_j), $i = 1, 2, \ldots, n - 1$, and $j = 1, 2, \ldots, m - 1$, we can use the Taylor series in the variable x about x_i to generate the central-difference formula

(11.7) $$\frac{\partial^2}{\partial x^2} u(x_i, y_j) = \frac{u(x_{i+1}, y_j) - 2u(x_i, y_j) + u(x_{i-1}, y_j)}{h^2} - \frac{h^2}{12} \frac{\partial^4 u}{\partial x^4}(\xi_i, y_j),$$

where $\xi_i \in (x_{i-1}, x_{i+1})$; and the Taylor series in the variable y about y_j to generate the central-difference formula

(11.8) $$\frac{\partial^2}{\partial y^2} u(x_i, y_j) = \frac{u(x_i, y_{j+1}) - 2u(x_i, y_j) + u(x_i, y_{j-1})}{k^2} - \frac{k^2}{12} \frac{\partial^4 u}{\partial y^4}(x_i, \eta_j),$$

where $\eta_j \in (y_{j-1}, y_{j+1})$.

Using these formulas in Eq. (11.6) allows us to express the Poisson equation at the points (x_i, y_j) as:

(11.9)

$$\frac{u(x_{i+1}, y_j) - 2u(x_i, y_j) + u(x_{i-1}, y_j)}{h^2} + \frac{u(x_i, y_{j+1}) - 2u(x_i, y_j) + u(x_i, y_{j-1})}{k^2}$$

$$= f(x_i, y_j) + \frac{h^2}{12} \frac{\partial^4 u}{\partial x^4}(\xi_i, y_j) + \frac{k^2}{12} \frac{\partial^4 u}{\partial y^4}(x_i, \eta_j),$$

for each $i = 1, 2, \ldots, (n - 1)$ and $j = 1, 2, \ldots, (m - 1)$, and the boundary conditions as

$$
\begin{aligned}
u(x_0, y_j) &= g(x_0, y_j) & &\text{for each } j = 0, 1, \ldots, m, \\
u(x_n, y_j) &= g(x_n, y_j) & &\text{for each } j = 0, 1, \ldots, m, \\
u(x_i, y_0) &= g(x_i, y_0) & &\text{for each } i = 1, 2, \ldots, n - 1, \\
u(x_i, y_m) &= g(x_i, y_m) & &\text{for each } i = 1, 2, \ldots, n - 1.
\end{aligned}
$$

(11.10)

In difference-equation form, this results in a method, called the **central-difference** method, with local truncation error of order $O(h^2 + k^2)$, that can be written:

(11.11)

$$
2\left[\left(\frac{h}{k}\right)^2 + 1\right] w_{i,j} - (w_{i+1,j} + w_{i-1,j}) - \left(\frac{h}{k}\right)^2 (w_{i,j+1} + w_{i,j-1}) = -h^2 f(x_i, y_j)
$$

for each $i = 1, 2, \ldots, n - 1$ and $j = 1, 2, \ldots, m - 1$, and

$$
\begin{aligned}
w_{0,j} &= g(x_0, y_j) & &\text{for each } j = 0, 1, \ldots, m, \\
w_{n,j} &= g(x_n, y_j) & &\text{for each } j = 0, 1, \ldots, m, \\
w_{i,0} &= g(x_i, y_0) & &\text{for each } i = 1, 2, \ldots, n - 1, \\
w_{i,m} &= g(x_i, y_m) & &\text{for each } i = 1, 2, \ldots, n - 1,
\end{aligned}
$$

(11.12)

where $w_{i,j}$ approximates $u(x_i, y_j)$.

The typical equation in (11.11) involves approximations to $u(x, y)$ at the points

$$
(x_{i-1}, y_j), \quad (x_i, y_j), \quad (x_{i+1}, y_j), \quad (x_i, y_{j-1}), \quad \text{and} \quad (x_i, y_{j+1});
$$

reproducing the portion of the grid where these points are located (Fig. 11.5) shows that each equation involves approximations in a star-shaped region about (x_i, y_j).

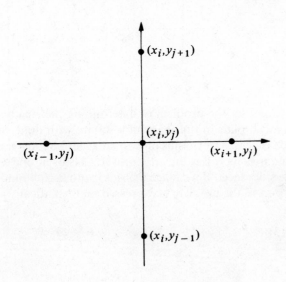

FIGURE 11.5

If we use the information from the boundary conditions (11.12), whenever appropriate, in the system given by (11.11)—that is, at all points (x_i, y_j) that are adjacent to a boundary mesh point—we have an $(n-1) \times (m-1)$ system of linear equations in $(n-1) \times (m-1)$ unknowns, the unknowns being the approximations $w_{i,j}$ to $u(x_i, y_j)$ for the interior mesh points.

The linear system involving these unknowns can be expressed for matrix calculations more efficiently if a relabeling of the interior mesh points is introduced. A recommended labeling of these points (see Varga [87], page 187) is to let

$$P_l = (x_i, y_j) \qquad \text{and} \qquad w_l = w_{i,j},$$

where $l = i + (m - 1 - j)(n - 1)$ for each

$$i = 1, 2, \ldots, n - 1 \qquad \text{and} \qquad j = 1, 2, \ldots, m - 1.$$

This in effect labels the mesh points consecutively from left to right and top to bottom. For example, with $n = 4$ and $m = 5$, the relabeling would result in a grid whose points are shown in Fig. 11.6.

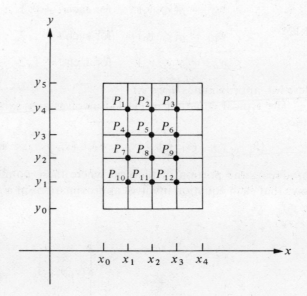

FIGURE 11.6

EXAMPLE 1   Consider the problem of determining the steady-state heat distribution in a thin metal plate in the shape of a square with dimensions .5 meters by .5 meters, which is held at $0°$ Celsius on two adjacent boundaries while the heat on the other boundaries increases linearly from $0°$ Celsius at one corner to $100°$ Celsius where these sides meet. If we place the sides with zero boundary conditions along the x- and y-axes, the problem is expressed mathematically as:

(11.13)
$$\frac{\partial^2 u}{\partial x^2} + \frac{\partial^2 u}{\partial y^2} = 0,$$

for (x, y) in the set $R = \{(x, y)|0 < x < .5, 0 < y < .5\}$, with the boundary conditions

(11.14) $\quad u(0, y) = 0, \quad u(x, 0) = 0, \quad u(x, .5) = 200x, \quad u(.5, y) = 200y.$

If $n = m = 4$, the problem has the grid given in Fig. 11.7, and the difference equation (11.11) is

$$4w_{i, j} - w_{i+1, j} - w_{i-1, j} - w_{i, j-1} - w_{i, j+1} = 0$$

for each $i = 1, 2, 3$, and $j = 1, 2, 3$.

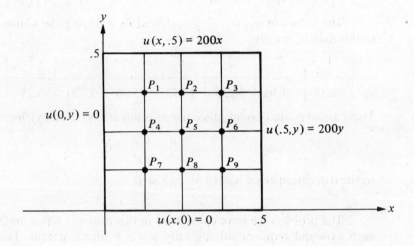

FIGURE 11.7

Expressing this in terms of the relabeled grid points $w_i = u(P_i)$ implies that the equations are:

$$4w_1 - w_2 - w_4 = w_{0, 3} + w_{1, 4},$$

$$4w_2 - w_3 - w_1 - w_5 = w_{2, 4},$$

$$4w_3 - w_2 - w_6 = w_{4, 3} + w_{3, 4},$$

$$4w_4 - w_5 - w_1 - w_7 = w_{0, 2},$$

$$4w_5 - w_6 - w_4 - w_2 - w_8 = 0,$$

$$4w_6 - w_5 - w_3 - w_9 = w_{4, 2},$$

$$4w_7 - w_8 - w_4 = w_{0, 1} + w_{1, 0},$$

$$4w_8 - w_9 - w_7 - w_5 = w_{2, 0},$$

$$4w_9 - w_8 - w_6 = w_{3, 0} + w_{4, 1},$$

where the righthand sides of the equations are obtained from the boundary conditions. In fact, the conditions given in (11.14) imply that

$$w_{1, 0} = w_{2, 0} = w_{3, 0} = w_{0, 1} = w_{0, 2} = w_{0, 3} = 0,$$

$$w_{1, 4} = w_{4, 1} = 25, \quad w_{2, 4} = w_{4, 2} = 50, \quad \text{and} \quad w_{3, 4} = w_{4, 3} = 75.$$

The linear system that is associated with this problem has the form

$$
\begin{bmatrix}
4 & -1 & 0 & -1 & 0 & 0 & 0 & 0 & 0 \\
-1 & 4 & -1 & 0 & -1 & 0 & 0 & 0 & 0 \\
0 & -1 & 4 & 0 & 0 & -1 & 0 & 0 & 0 \\
-1 & 0 & 0 & 4 & -1 & 0 & -1 & 0 & 0 \\
0 & -1 & 0 & -1 & 4 & -1 & 0 & -1 & 0 \\
0 & 0 & -1 & 0 & -1 & 4 & 0 & 0 & -1 \\
0 & 0 & 0 & -1 & 0 & 0 & 4 & -1 & 0 \\
0 & 0 & 0 & 0 & -1 & 0 & -1 & 4 & -1 \\
0 & 0 & 0 & 0 & 0 & -1 & 0 & -1 & 4
\end{bmatrix}
\begin{bmatrix}
w_1 \\ w_2 \\ w_3 \\ w_4 \\ w_5 \\ w_6 \\ w_7 \\ w_8 \\ w_9
\end{bmatrix}
=
\begin{bmatrix}
25 \\ 50 \\ 150 \\ 0 \\ 0 \\ 50 \\ 0 \\ 0 \\ 25
\end{bmatrix}
$$

The values of $w_1, w_2, \ldots, w_9$, found by applying the Gauss–Seidel method to this matrix, are given as follows:

i	1	2	3	4	5	6	7	8	9
w_i	18.75	37.50	56.25	12.50	25.00	37.50	6.25	12.50	18.75

These answers are correct, since the solution $u(x, y) = 400xy$ has

$$
\frac{\partial^4 u}{\partial x^4} = \frac{\partial^4 u}{\partial y^4} \equiv 0;
$$

so the truncation error is zero at each step.   ☐

The problem we have considered in this example has a mesh size of .125 on each axis and requires solving only a 9×9 linear system. This simplifies the situation considerably, and does not introduce the computational problems that are present when the system is much larger. Algorithm 11.1 utilizes the Gauss–Seidel iterative method for solving the linear system that is produced and allows for unequal mesh sizes on the axes.

Poisson Equation Finite-Difference Algorithm 11.1

To approximate the solution to the Poisson equation

$$
\frac{\partial^2 u}{\partial x^2} + \frac{\partial^2 u}{\partial y^2} = f(x, y), \qquad a \le x \le b, \quad c \le y \le d,
$$

subject to the boundary conditions

$$u(x, y) = g(x, y) \qquad \text{if } x = a \text{ or } x = b \text{ and } c \le y \le d$$

and $u(x, y) = g(x, y) \qquad \text{if } y = c \text{ or } y = d \text{ and } a \le x \le b.$

INPUT endpoints a, b, c, d; integers m, n; tolerance TOL; maximum number of iterations $LBOUND$.

OUTPUT approximations $w_{i,j}$ to $u(x_i, y_j)$ for each $i = 1, \ldots, n - 1$ and $j = 1, \ldots, m - 1$ or a message that the maximum number of iterations was exceeded.

Step 1 Set $h = (b - a)/n$;
$k = (d - c)/m.$

Step 2 For $i = 1, \ldots, n - 1$ set $x_i = a + ih$.   (*Steps 2 and 3 construct mesh points.*)

Step 3 For $j = 1, \ldots, m - 1$ set $y_j = c + jk$.

Step 4 For $i = 1, \ldots, n - 1$
   for $j = 1, \ldots, m - 1$ set $w_{i,j} = 0$.

Step 5 Set $\lambda = h^2/k^2$;
   $\mu = 2(1 + \lambda)$;
   $l = 1$.

Step 6 While $l \leq LBOUND$ do Steps 7–20.   (*Steps 7–20 perform Gauss-Seidel iterations.*)

   Step 7 Set $z = (-h^2 f(x_1, y_{m-1}) + g(a, y_{m-1}) + \lambda g(x_1, d) + \lambda w_{1,m-2}$
   $$+ \, w_{2,m-1})/\mu;$$

   $NORM = |z - w_{1,m-1}|$
   $w_{1,m-1} = z$.

   Step 8 For $i = 2, \ldots, n - 2$
   set $z = (-h^2 f(x_i, y_{m-1}) + \lambda g(x_i, d) + w_{i-1,m-1}$
   $\qquad + \, w_{i+1,m-1} + \lambda w_{i,m-2})/\mu$;
   if $|w_{i,m-1} - z| > NORM$ then set $NORM = |w_{i,m-1} - z|$;
   set $w_{i,m-1} = z$.

   Step 9 Set $z = (-h^2 f(x_{n-1}, y_{m-1}) + g(b, y_{m-1}) + \lambda g(x_{n-1}, d)$
   $\qquad + \, w_{n-2,m-1} + \lambda w_{n-1,m-2})/\mu$;
   if $|w_{n-1,m-1} - z| > NORM$ then set $NORM$
   $\qquad = |w_{n-1,m-1} - z|$;
   set $w_{n-1,m-1} = z$.

   Step 10 For $j = m - 2, \ldots, 2$ do Steps 11, 12, and 13.

   Step 11 Set $z = (-h^2 f(x_1, y_j) + g(a, y_j) + \lambda w_{1,j+1} + \lambda w_{1,j-1} + w_{2,j})/\mu$;
   if $|w_{1,j} - z| > NORM$ then set $NORM = |w_{1,j} - z|$;
   set $w_{1,j} = z$.

   Step 12 For $i = 2, \ldots, n - 2$
   set $z = (-h^2 f(x_i, y_j) + w_{i-1,j} + \lambda w_{i,j+1}$
   $\qquad + \, w_{i+1,j} + \lambda w_{i,j-1})/\mu$;
   if $|w_{i,j} - z| > NORM$ then set $NORM = |w_{i,j} - z|$;
   set $w_{i,j} = z$.

   Step 13 Set $z = (-h^2 f(x_{n-1}, y_j) + g(b, y_j) + w_{n-2,j}$
   $\qquad + \, \lambda w_{n-1,j+1} + \lambda w_{n-1,j-1})/\mu$;
   if $|w_{n-1,j} - z| > NORM$ then set $NORM = |w_{n-1,j} - z|$;
   set $w_{n-1,j} = z$.

   Step 14 Set $z = (-h^2 f(x_1, y_1) + g(a, y_1) + \lambda g(x_1, c) + \lambda w_{1,2} + w_{2,1})/\mu$;
   if $|w_{1,1} - z| > NORM$ then set $NORM = |w_{1,1} - z|$;
   set $w_{1,1} = z$.

   Step 15 For $i = 2, \ldots, n - 2$
   set $z = (-h^2 f(x_i, y_1) + \lambda g(x_i, c) + w_{i-1,1} + \lambda w_{i,2} + w_{i+1,1})/\mu$;
   if $|w_{i,1} - z| > NORM$ then set $NORM = |w_{i,1} - z|$;
   set $w_{i,1} = z$.

Step 16 Set $z = (-h^2 f(x_{n-1}, y_1) + g(b, y_1) + \lambda g(x_{n-1}, c) + w_{n-2, 1}$
$+ \lambda w_{n-1, 2})/\mu;$
if $|w_{n-1, 1} - z| > NORM$ then set $NORM = |w_{n-1, 1} - z|;$
set $w_{n-1, 1} = z.$

Step 17 If $NORM \leq TOL$ then do Steps 18 and 19.

   Step 18 For $i = 1, \ldots, n-1$
                  for $j = 1, \ldots, m-1$ OUTPUT $(x_i, y_j, w_{i, j})$.

   Step 19 STOP. (*Procedure completed successfully.*)

Step 20 Set $l = l + 1.$

Step 21 OUTPUT ('Maximum number of iterations exceeded');
          STOP.

Although the Gauss–Seidel iterative procedure is incorporated into Algorithm 11.1 for simplicity, it is generally advisable to use a direct technique such as Gaussian elimination when the system is small, on the order of 100 or less, since the symmetry and positive definiteness will ensure stability with respect to rounding errors. In particular, the generalization of the Crout Reduction Algorithm 6.7, which is discussed in Exercise 13 of Section 6.6 is very efficient for solving this system, since the matrix is in symmetric-block tridiagonal form

$$
\begin{bmatrix}
A_1 & C_1 & 0 & \cdots & & & 0 \\
C_1 & A_2 & C_2 & & & & \\
0 & C_2 & & & & & \\
\vdots & & & & & & 0 \\
& & & & & & C_{m-1} \\
0 & \cdots & & & 0 & C_{m-1} & A_{m-1}
\end{bmatrix},
$$

with square blocks of size $(n-1)$ by $(n-1)$.

For very large systems it is recommended that an iterative method be used, specifically the SOR method discussed in Algorithm 8.3. The choice of ω that is optimal in this situation comes from the fact that when A is decomposed into its diagonal D and upper- and lower-triangular parts U and L,

$$
A = D - L - U,
$$

and B is the Jacobi matrix,

$$
B = D^{-1}(L + U),
$$

then the spectral radius of B is (see Varga [87])

$$
\rho(B) = \frac{1}{2}\left[\cos\left(\frac{\pi}{m}\right) + \cos\left(\frac{\pi}{n}\right)\right].
$$

The value of ω to be used is consequently

$$\omega = \frac{2}{1 + \sqrt{1 - [\rho(B)]^2}} = \frac{4}{2 + \sqrt{4 - \left[\cos\left(\dfrac{\pi}{m}\right) + \cos\left(\dfrac{\pi}{n}\right)\right]^2}}$$

For faster convergence of the SOR procedure, a block technique can be incorporated into the algorithm. For a presentation of the technique involved, see Varga [87], pages 194–199.

EXAMPLE 2   Consider Poisson's equation

$$\frac{\partial^2 u}{\partial x^2} + \frac{\partial^2 u}{\partial y^2} = xe^y, \qquad 0 < x < 2, \quad 0 < y < 1,$$

with the boundary conditions

$$u(0, y) = 0, \qquad u(2, y) = 2e^y, \qquad 0 \le y \le 1,$$

$$u(x, 0) = x, \qquad u(x, 1) = ex, \qquad 0 \le x \le 2.$$

We will use Algorithm 11.1 to approximate the exact solution $u(x, y) = xe^y$ with $n = 6$ and $m = 5$. The stopping criterion for Step 17 required

$$|w_{i,j}^{(l)} - w_{i,j}^{(l-1)}| \le 10^{-10},$$

for each $i = 1, \ldots, 5$, and $j = 1, \ldots, 4$; so the solution to the difference equation was accurately obtained, and the procedure stopped at $l = 61$. The results, along with the correct values, are presented in Table 11.1.   □

TABLE 11.1

i	j	x_i	y_i	$w_{i,j}^{(61)}$	$u(x_i, y_i)$	$|u(x_i, y_j) - w_{i,j}^{(61)}|$
1	1	.3333	.2000	.40726	.40713	1.30×10^{-4}
1	2	.3333	.4000	.49748	.49727	2.08×10^{-4}
1	3	.3333	.6000	.60760	.60737	2.23×10^{-4}
1	4	.3333	.8000	.74201	.74185	1.60×10^{-4}
2	1	.6667	.2000	.81452	.81427	2.55×10^{-4}
2	2	.6667	.4000	.99496	.99455	4.08×10^{-4}
2	3	.6667	.6000	1.2152	1.2147	4.37×10^{-4}
2	4	.6667	.8000	1.4840	1.4837	3.15×10^{-4}
3	1	1.0000	.2000	1.2218	1.2214	3.64×10^{-4}
3	2	1.0000	.4000	1.4924	1.4918	5.80×10^{-4}
3	3	1.0000	.6000	1.8227	1.8221	6.24×10^{-4}
3	4	1.0000	.8000	2.2260	2.2255	4.51×10^{-4}
4	1	1.3333	.2000	1.6290	1.6285	4.27×10^{-4}
4	2	1.3333	.4000	1.9898	1.9891	6.79×10^{-4}
4	3	1.3333	.6000	2.4302	2.4295	7.35×10^{-4}
4	4	1.3333	.8000	2.9679	2.9674	5.40×10^{-4}
5	1	1.667	.2000	2.0360	2.0357	3.71×10^{-4}
5	2	1.667	.4000	2.4870	2.4864	5.84×10^{-4}
5	3	1.667	.6000	3.0375	3.0369	6.41×10^{-4}
5	4	1.667	.8000	3.7097	3.7092	4.89×10^{-4}

Exercise Set 11.2

1. Use Algorithm 11.1 to approximate the solution to the following elliptic partial-differential equation

$$\frac{\partial^2 u}{\partial x^2} + \frac{\partial^2 u}{\partial y^2} = 4, \qquad 0 < x < 1, \quad 0 < y < 2;$$

$$u(x, 0) = x^2, \quad u(x, 2) = (x - 2)^2, \quad 0 \le x \le 1;$$

$$u(0, y) = y^2, \quad u(1, y) = (y - 1)^2, \quad 0 \le y \le 2.$$

Use $h = k = .5$.

2. Use Algorithm 11.1 to approximate the solution to the following elliptic partial-differential equation

$$\frac{\partial^2 u}{\partial x^2} + \frac{\partial^2 u}{\partial y^2} = x, \qquad 0 < x, y < 1;$$

$$u(x, 0) = \tfrac{1}{6}x^3, \quad u(x, 1) = \tfrac{1}{6}x^3, \quad 0 \le x \le 1;$$

$$u(0, y) = 0, \quad u(1, y) = \tfrac{1}{6}, \quad 0 \le y \le 1.$$

Use $h = k = \tfrac{1}{3}$.

3. Approximate the solutions to the following elliptic partial-differential equations, using Algorithm 1.1:

a) $\dfrac{\partial^2 u}{\partial x^2} + \dfrac{\partial^2 u}{\partial y^2} = 0, \qquad 0 < x, y < 1;$

$$u(x, 0) = 0, \quad u(x, 1) = x, \quad 0 \le x \le 1;$$

$$u(0, y) = 0, \quad u(1, y) = y, \quad 0 \le y \le 1.$$

Use $h = k = .1$, and compare results with the solution $u(x, y) = xy$ and with the first five terms of the Fourier-series solution.

b) $\dfrac{\partial^2 u}{\partial x^2} + \dfrac{\partial^2 u}{\partial y^2} = -2, \qquad 0 < x, y < 1;$

$$u(0, y) = 0, \quad u(1, y) = \sinh(\pi)\sin \pi y, \quad 0 \le y \le 1;$$

$$u(x, 0) = u(x, 1) = x(1 - x), \quad 0 \le x \le 1.$$

Use $h = k = .05$, and compare results with the solution

$$u(x, y) = \sinh(\pi x)\sin(\pi y) + x(1 - x).$$

c) $\dfrac{\partial^2 u}{\partial x^2} + \dfrac{\partial^2 u}{\partial y^2} = (x^2 + y^2)e^{xy}, \qquad 0 < x < 2, \quad 0 < y < 1;$

$$u(0, y) = 1, \quad u(2, y) = e^{2y}, \quad 0 \le y \le 1;$$

$$u(x, 0) = 1, \quad u(x, 1) = e^x, \quad 0 \le x \le 2.$$

Use $h = .2$ and $k = .1$, and compare results with the solution $u(x, y) = e^{xy}$.

4. Repeat Exercise 3, parts (a) and (b), using extrapolation with $h_0 = h$, $h_1 = h/2$, and $h_2 = h/4$.

5. Construct an algorithm similar to Algorithm 11.1, except use Choleski's method instead of the Gauss–Seidel iterative method for solving the linear system.

6. Apply the algorithm from Exercise 5 to the problems in Exercise 3.

7. Construct an algorithm similar to Algorithm 11.1, except use the SOR method with optimal ω instead of the Gauss–Seidel method for solving the linear system.

8. Repeat Exercise 3, using the algorithm constructed in Exercise 7.

9. A coaxial cable is made of a 1-inch-square inner conductor and a 5-inch-square outer conductor. Suppose the inner conductor is kept at zero volts while the outer conductor is kept at 110 volts. Find the potential between the two conductors by placing a grid with horizontal mesh spacing $h = 1$ inch and vertical mesh spacing $k = 1$ inch on the region

$$D = \{(x, y) | 0 \leq x, y \leq 5\}.$$

Approximate the solution to Laplace's equation at each grid point, and use the two sets of boundary conditions to derive a linear system to be solved by the Gauss–Seidel method.

10. A 6-cm × 5-cm rectangular silver plate has heat being uniformly generated at each point at the rate $q = 1.5$ cal/cm$^3 \cdot$ sec. Let x represent the distance along the edge of the plate of length 6 cm and y the distance along the edge of the plate of length 5 cm. Suppose the temperature u along the edges is kept at the following temperatures:

$$u(x, 0) = x(6 - x), \quad u(x, 5) = 0, \qquad 0 \leq x \leq 6,$$

$$u(0, y) = y(5 - y), \quad u(6, y) = 0, \qquad 0 \leq y \leq 5,$$

where the origin lies at a corner of the plate with coordinates $(0, 0)$ and the edges lie along the positive x- and y-axes. The steady-state temperature $u = u(x, y)$ satisfies Poisson's equation:

$$\frac{\partial^2 u(x, y)}{\partial x^2} + \frac{\partial^2 u(x, y)}{\partial y^2} = \frac{-q}{K}, \qquad 0 < x < 6, \quad 0 < y < 5,$$

where K, the thermal conductivity, is 1.04 cal/cm $\cdot$ deg $\cdot$ sec. Approximate the temperature $u(x, y)$ using Algorithm 11.1 with $h = .4$ and $k = \frac{1}{3}$.

11.3 Parabolic Partial-Differential Equations

The parabolic partial-differential equation we will study is the heat or diffusion equation

$$(11.15) \qquad \frac{\partial}{\partial t} u(x, t) = \alpha^2 \frac{\partial^2}{\partial x^2} u(x, t), \qquad 0 < x < l, \quad t > 0,$$

subject to the conditions

$$u(0, t) = 0, \quad u(l, t) = 0, \qquad t > 0,$$

and

$$u(x, 0) = f(x), \qquad 0 \leq x \leq l.$$

The approach we will use to approximate the solution to this problem involves finite differences and is similar to the method used in Section 11.2. We first select two mesh constants h and k, with the stipulation that $m = l/h$ is an integer. The grid points for this situation are (x_i, t_j), where $x_i = ih$ for $i = 0, 1, \ldots, m$, and $t_j = jk$, for $j = 0, 1, \ldots$.

We obtain the difference method by using the Taylor series in t to form the difference quotient

$$(11.16) \qquad \frac{\partial u}{\partial t}(x_i, t_j) = \frac{u(x_i, t_j + k) - u(x_i, t_j)}{k} - \frac{k}{2}\frac{\partial^2}{\partial t^2}u(x_i, t_j + \theta_j k),$$

for some $0 < \theta_j < 1$, and the Taylor series in x to form the difference quotient

$$(11.17) \qquad \frac{\partial^2 u}{\partial x^2}(x_i, t_j) = \frac{u(x_i + h, t_j) - 2u(x_i, t_j) + u(x_i - h, t_j)}{h^2}$$

$$- \frac{h^2}{12}\frac{\partial^4 u}{\partial x^4}(x_i + \phi_i h, t_j) \qquad \text{where } -1 < \phi_i < 1.$$

The partial-differential equation (11.15) implies that, at the interior gridpoint (x_i, t_j) for each $i = 1, 2, \ldots, m - 1$ and $j = 1, 2, \ldots$, we have

$$\frac{\partial u}{\partial t}(x_i, t_j) - \alpha^2 \frac{\partial^2 u}{\partial x^2}(x_i, t_j) = 0;$$

so the difference method utilizing the difference quotients (11.16) and (11.17) is

$$(11.18) \qquad \frac{w_{i, j+1} - w_{i, j}}{k} - \alpha^2 \frac{w_{i+1, j} - 2w_{i, j} + w_{i-1, j}}{h^2} = 0,$$

where w_{ij} approximates $u(x_i, t_j)$.

The local truncation error for this difference equation is

$$(11.19) \qquad \tau_{i, j} = \frac{k}{2}\frac{\partial^2}{\partial t^2}u(x_i, t_j + \theta_j k) - \alpha^2 \frac{h^2}{12}\frac{\partial^4 u}{\partial x^4}(x_i + \phi_i h, t_j)$$

for some $0 < \theta_j < 1$ and $-1 < \phi_i < 1$.

If Eq. (11.18) is solved for $w_{i, j+1}$, we have

$$(11.20) \qquad w_{i, j+1} = \left(1 - \frac{2\alpha^2 k}{h^2}\right)w_{i, j} + \alpha^2 \frac{k}{h^2}(w_{i+1, j} + w_{i-1, j})$$

for each $i = 1, 2, \ldots, (m - 1)$ and $j = 1, 2, \ldots$. Since the initial condition $u(x, 0) = f(x)$, for each $0 \le x \le l$, implies that $w_{i, 0} = f(x_i)$, for each $i = 0, 1, \ldots, m$, these values can be used in Eq. (11.20) to find the value of $w_{i, 1}$ for each $i = 1, 2, \ldots, (m - 1)$. The additional condition that $u(0, t) = 0$ and $u(l, t) = 0$ implies that $w_{0, 1} = w_{m, 1} = 0$; so all the entries of the form $w_{i, 1}$ can be determined. If the procedure is reapplied once all the approximations $w_{i, 1}$, are known, the values of $w_{i, 2}, w_{i, 3}, \ldots, w_{i, m-1}$ can be obtained in a similar manner.

The explicit nature of the difference method expressed in Eq. (11.20) implies that the $(m - 1) \times (m - 1)$ matrix associated with this system can be written in the tridiagonal form

$$A = \begin{bmatrix} (1 - 2\lambda) & \lambda & 0 \cdots\cdots\cdots\cdots 0 \\ \lambda \cdots & (1 - 2\lambda) & \lambda & & \\ 0 \cdots & & \ddots & & 0 \\ \vdots & \ddots & & & \lambda \\ 0 \cdots\cdots\cdots\cdots 0 & & \lambda & (1 - 2\lambda) \end{bmatrix},$$

where $\lambda = \alpha^2(k/h^2)$. If we let

$$\mathbf{w}^{(j)} = \begin{bmatrix} w_{1,j} \\ w_{2,j} \\ \vdots \\ w_{m-1,j} \end{bmatrix} \qquad \text{for each } j = 1, 2, \ldots$$

and

$$\mathbf{w}^{(0)} = \begin{bmatrix} f(x_1) \\ f(x_2) \\ \vdots \\ f(x_{m-1}) \end{bmatrix},$$

then the approximate solution is given by

$$\mathbf{w}^{(j)} = A\mathbf{w}^{(j-1)} \qquad \text{for each } j = 1, 2, \ldots .$$

This difference method is known as the **forward-difference** method, and if the solution to the partial-differential equation has four continuous partial derivatives in x and two in t, then Eq. (11.19) implies that the method is of order $O(k + h^2)$.

EXAMPLE 1   Consider the heat equation

$$\frac{\partial u}{\partial t} - \frac{\partial^2 u}{\partial x^2} = 0, \qquad 0 < x < 1, \quad 0 < t,$$

with boundary conditions

$$u(0, t) = u(1, t) = 0, \qquad 0 < t,$$

and initial conditions

$$u(x, 0) = \sin(\pi x), \qquad 0 \le x \le 1.$$

It is easily verified that the solution to this problem is

$$u(x, t) = e^{-\pi^2 t} \sin(\pi x).$$

The solution at $t = .5$ will be approximated using the forward-difference method first with $h = .1$, $k = .0005$, and $\lambda = .05$, and then with $h = .1$, $k = .01$, and $\lambda = 1$. The results are presented in Table 11.2.   □

TABLE 11.2

x_i	$u(x_i, .5)$	$w_{i,1000}$ $k = .005$	$\lvert u(x_i, 5)$ $- w_{i,1000}\rvert$	$w_{i,50}$ $k = .01$	$\lvert u(x_i, .5)$ $- w_{i,50}\rvert$
0	0	0	—	0	—
.1	.00222241	.00228652	6.411×10^{-5}	8.19876×10^7	8.20×10^7
.2	.00422728	.00434922	1.219×10^{-4}	-1.55719×10^8	1.557×10^8
.3	.00581836	.00598619	1.678×10^{-4}	2.13833×10^8	2.138×10^8
.4	.00683989	.00703719	1.973×10^{-4}	-2.50642×10^8	2.506×10^8
.5	.00719188	.00739934	2.075×10^{-4}	2.62685×10^8	2.627×10^8
.6	.00683989	.00703719	1.973×10^{-4}	-2.49015×10^8	2.490×10^8
.7	.00581836	.00598619	1.678×10^{-4}	2.11200×10^8	2.112×10^8
.8	.00422728	.00434922	1.219×10^{-4}	-1.53086×10^8	1.531×10^8
.9	.00222241	.00228652	6.511×10^{-5}	8.03604×10^7	8.036×10^7
1.0	0	0	—	0	—

Although a local truncation error of order $O(k + h^2)$ would be expected, this is certainly not the case in our example when $h = .1$ and $k = .01$. To explain the difficulty we must look at the stability of the forward-difference method. If the error $\mathbf{e}^{(0)} = (e_1^{(0)}, e_2^{(0)}, \ldots, e_{m-1}^{(0)})^t$ was made in representing the initial data $\mathbf{w}^{(0)} = (f(x_1), f(x_2), \ldots, f(x_{m-1}))^t$ or, for that matter, in any particular step (the choice of the initial step is simply for convenience), an error of $A\mathbf{e}^{(0)}$ would be propagated in $\mathbf{w}^{(1)}$ since

$$\mathbf{w}^{(1)} = A(\mathbf{w}^{(0)} + \mathbf{e}^{(0)}) = A\mathbf{w}^{(0)} + A\mathbf{e}^{(0)}.$$

This process would continue, assuming no other errors were introduced. At the nth time step, the error in $\mathbf{w}^{(n)}$ would be $A^n \mathbf{e}^{(0)}$. The method is consequently stable if and only if these errors do not grow as n increases—that is, if and only if $\|A^n \mathbf{e}^{(0)}\| \leq \|\mathbf{e}^{(0)}\|$ for all n. But this implies that $\|A^n\| \leq 1$, a condition which, by Theorem 8.15 (p. 377), implies that the spectral radius $\rho(A^n) = (\rho(A))^n \leq 1$. The forward-difference method will therefore be stable only if $\rho(A) \leq 1$.

To compute the spectral radius of A we first note (see Exercise 7) that the eigenvalues of A are

$$\mu_i = 1 - 4\lambda \sin^2\left(\frac{i\pi}{2m}\right) \qquad \text{for each } i = 1, 2, \ldots, (m-1).$$

The condition for stability reduces to determining whether:

$$\rho(A) = \max_{1 \leq i \leq m-1} \left| 1 - 4\lambda \sin^2\left(\frac{i\pi}{2m}\right) \right| \leq 1,$$

which simplifies to

$$0 \leq \lambda \sin^2\left(\frac{i\pi}{2m}\right) \leq \tfrac{1}{2} \qquad \text{for each } i = 1, 2, \ldots, m-1.$$

Since stability requires that this inequality condition hold as $h \to 0$ or, equivalently, as $m \to +\infty$, the fact that

$$\lim_{m \to +\infty} \sin^2\left(\frac{(m-1)\pi}{2m}\right) = 1$$

means that stability will occur only if $0 \leq \lambda \leq \tfrac{1}{2}$. Since $\lambda = \alpha^2(k/h^2)$ is certainly positive, this inequality essentially requires that h and k be chosen so that

$$\alpha^2 \frac{k}{h^2} \leq \tfrac{1}{2}.$$

This condition was satisfied in our example when $h = .1$ and $k = .0005$; but, when k was increased to .01 with no corresponding increase in h, the ratio became

$$\frac{.01}{(.1)^2} = 1 > \tfrac{1}{2}$$

and stability problems became apparent. Consistent with the terminology of Chapter 5, we call this method **conditionally stable** and remark that the method converges to the solution of Eq. (11.15) with rate of convergence $O(k + h^2)$, provided

$$\alpha^2 \frac{k}{h^2} \leq \tfrac{1}{2}$$

and the required continuity conditions on the solution are met. [For a detailed proof of this fact, see Isaacson and Keller [52], pages 502–505.]

In order to obtain a method that is **unconditionally stable**, we will consider an implicit-difference method that results from using the backward-difference quotient for $(\partial u/\partial t)(x_i, t_j)$ in the form

$$\frac{\partial u}{\partial t}(x_i, t_j) = \frac{u(x_i, t_j) - u(x_i, t_{j-1})}{k} + \frac{k}{2}\frac{\partial^2 u}{\partial t^2}(x_i, t_j + \theta_j k),$$

where $-1 < \theta_j < 0$. Substituting this equation, together with Eq. (11.17) for $\partial^2 u/\partial x^2$, into the partial-differential equation gives

$$\frac{u(x_i, t_j) - u(x_i, t_{j-1})}{k} - \alpha^2 \frac{u(x_{i+1}, t_j) - 2u(x_i, t_j) + u(x_{i-1}, t_j)}{h^2}$$

$$= -\frac{k}{2}\frac{\partial^2 u}{\partial t^2}(x_i, t_j + \theta_j k) - \frac{h^2}{12}\frac{\partial^4 u}{\partial x^4}(x_i + \phi_i h, t_j),$$

for some $-1 < \theta_j < 0$ and $-1 < \phi_i < 1$. The **backward-difference method** that results is

(11.21) $$\frac{w_{i,j} - w_{i,j-1}}{k} - \alpha^2 \frac{w_{i+1,j} - 2w_{i,j} + w_{i-1,j}}{h^2} = 0$$

for each $i = 1, 2, \ldots, m - 1$, and $j = 1, 2, \ldots$.

This method involves, at a typical step, the mesh points

$$(x_i, t_j), \quad (x_i, t_{j-1}), \quad (x_{i-1}, t_j), \quad \text{and} \quad (x_{i+1}, t_j),$$

and, in grid form, involves approximations at the points marked with ×'s in Fig. 11.8.

Since the boundary and initial conditions associated with the problem give information at the circled mesh points, it is clear from the figure that no explicit procedures can be used to solve Eq. (11.21). Recall that, in the forward-difference method (see Fig. 11.9), approximations at

$$(x_{i-1}, t_j), \quad (x_i, t_j), \quad (x_i, t_{j+1}), \quad \text{and} \quad (x_{i+1}, t_j)$$

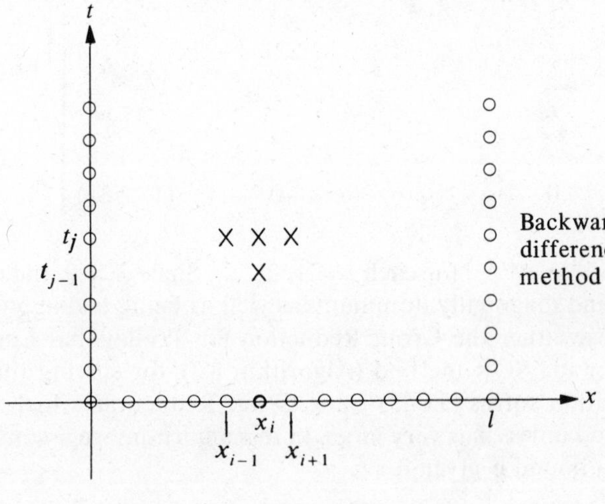

FIGURE 11.8

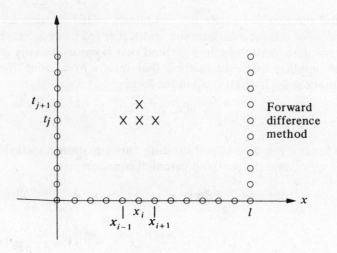

FIGURE 11.9

were used; so an explicit method for finding the approximations, based on the information from the initial and boundary conditions, was available.

If we again let λ denote the quantity $\alpha^2(k/h^2)$, the backward-difference method becomes

(11.22) $$(1 + 2\lambda)w_{i,j} - \lambda w_{i+1,j} - \lambda w_{i-1,j} = w_{i,j-1}$$

for each $i = 1, 2, \ldots, m - 1$, and $j = 1, 2, \ldots$. Using the knowledge that $w_{i,0} = f(x_i)$ for each $i = 1, 2, \ldots, m - 1$, and $w_{m,j} = w_{0,j} = 0$ for each $j = 1, 2, \ldots$, this difference method has the matrix representation:

(11.23)

$$
\begin{bmatrix}
(1 + 2\lambda) & -\lambda & 0 & \cdots & \cdots & 0 \\
-\lambda & (1 + 2\lambda) & -\lambda & & & \vdots \\
0 & & & & & 0 \\
\vdots & & & & & \\
& & & & & -\lambda \\
0 & \cdots & \cdots & 0 & -\lambda & (1 + 2\lambda)
\end{bmatrix}
\begin{bmatrix}
w_{1,j} \\
w_{2,j} \\
\vdots \\
w_{m-1,j}
\end{bmatrix}
=
\begin{bmatrix}
w_{1,j-1} \\
w_{2,j-1} \\
\vdots \\
w_{m-1,j-1}
\end{bmatrix},
$$

or $A\mathbf{w}^{(j)} = \mathbf{w}^{(j-1)}$ for each $j = 1, 2, \ldots$. Since $\lambda > 0$, the matrix A is positive definite and diagonally dominant, as well as being tridiagonal and symmetric; so we can use either the Crout Reduction for Tridiagonal Linear Systems (Algorithm 6.7) or the SOR method (Algorithm 8.3), for solving this system. The following algorithm solves (11.23) using Crout Reduction, which would be an acceptable method unless m is very large. In this algorithm we assume, for stopping purposes, that a bound is given for t.

Heat Equation Backward-Difference Algorithm 11.2

To approximate the solution to the parabolic partial-differential equation

$$\frac{\partial u}{\partial t} - \alpha^2 \frac{\partial^2 u}{\partial x^2} = 0, \qquad 0 < x < l, \ \ 0 < t < T,$$

subject to the boundary conditions

$$u(0, t) = u(l, t) = 0, \qquad 0 < t < T,$$

and the initial conditions

$$u(x, 0) = f(x), \qquad 0 \le x \le l:$$

INPUT endpoint l; maximum time T; constant α; integers m, N.

OUTPUT approximations $w_i(t_j)$ to $u(x_i, t_j)$ for each $i = 1, \ldots, m - 1$ and $j = 1, \ldots, N$.

Step 1 Set $h = l/m$;
$\quad\quad\quad k = T/N$;
$\quad\quad\quad \lambda = \alpha^2 k/h^2$.

Step 2 For $i = 1, \ldots, m - 1$ set $w_i = f(ih)$.   (*Initial values.*)

(*Steps* 3–11 *solve a tridiagonal linear system using Algorithm* 6.7.)

Step 3 Set $l_1 = 1 + 2\lambda$;
$\quad\quad\quad u_1 = -\lambda/l_1$.

Step 4 For $i = 2, \ldots, m - 2$ set $l_i = 1 + 2\lambda + \lambda u_{i-1}$;
$\quad\quad\quad\quad\quad\quad\quad\quad\quad\quad\quad u_i = -\lambda/l_i$.

Step 5 Set $l_{m-1} = 1 + 2\lambda + \lambda u_{m-2}$.

Step 6 For $j = 1, \ldots, N$ do Steps 7–11.

Step 7 Set $t = jk$;   (*Current t_j.*)
$\quad\quad\quad\ z_1 = w_1/l_1$.

Step 8 For $i = 2, \ldots, m - 1$ set $z_i = (w_i + \lambda z_{i-1})/l_i$.

Step 9 Set $w_{m-1} = z_{m-1}$.

Step 10 For $i = m - 2, \ldots, 1$ set $w_i = z_i - u_i w_{i+1}$.

Step 11 OUTPUT (t);
$\quad\quad\quad\ $ For $i = 1, \ldots, m - 1$ set $x = ih$;
$\quad\quad\quad\quad\quad\quad\quad\quad\quad\quad\quad\quad\quad\quad$ OUTPUT (x, w_i).

Step 12 STOP.   (*Procedure is complete.*)

EXAMPLE 2   The backward-difference method (Algorithm 11.2) with $h = .1$ and $k = .01$ will be used to approximate the solution to the heat equation

$$\frac{\partial u}{\partial t} - \frac{\partial^2 u}{\partial x^2} = 0, \qquad 0 < x < 1, \ \ 0 < t,$$

subject to the constraints

$$u(0, t) = u(1, t) = 0, \quad 0 < t, \quad \text{and} \quad u(x, 0) = \sin \pi x, \quad 0 \le x \le 1,$$

which was considered in Example 1. To demonstrate the unconditional stability of the backward-difference method, we will again compare $w_{i, 50}$ to $u(x_i, .5)$ where $i = 0, 1, \ldots, 10$.

The results listed in Table 11.3 should be compared with the fifth and sixth columns of Table 11.2. $\square$

TABLE 11.3

x_i	$w_{i, 50}$	$u(x_i, .5)$	$|w_{i, 50} - u(x_i, .5)|$
0	0	0	—
.1	.00289802	.00222241	6.756×10^{-4}
.2	.00551236	.00422728	1.285×10^{-3}
.3	.00758711	.00581836	1.769×10^{-3}
.4	.00891918	.00683989	2.079×10^{-3}
.5	.00937818	.00719188	2.186×10^{-3}
.6	.00891918	.00683989	2.079×10^{-3}
.7	.00758711	.00581836	1.769×10^{-3}
.8	.00551236	.00422728	1.285×10^{-3}
.9	.00289802	.00222241	6.756×10^{-4}
1.0	0	0	—

The backward-difference method does not have the stability problems of the forward-difference method. The reason for this can again be found by analyzing the eigenvalues of the matrix A. In this case (see Exercise 6), the eigenvalues are of the form

$$\mu_i = 1 + 4\lambda \sin^2\left(\frac{i\pi}{2m}\right) \quad \text{for each } i = 1, 2, \ldots, (m - 1);$$

and since $\lambda > 0$, $\mu_i > 1$ for all $i = 1, 2, \ldots, (m - 1)$. This implies that A^{-1} exists (by Exercise 13 in Section 8.1). An error of $\mathbf{e}^{(0)}$ in the initial data produces an error of $(A^{-1})^n \mathbf{e}^{(0)}$ at the nth step. Since the eigenvalues of A^{-1} are the reciprocals of the eigenvalues of A, the spectral radius of A^{-1} is bounded by 1 and the method is stable, independent of the choice of $\lambda = \alpha^2(k/h^2)$. In the terminology of Chapter 5, we would call the backward-difference method an unconditionally stable method. The local truncation error for the method is of order $O(k + h^2)$, provided the solution of the differential equation satisfies the usual differentiability conditions. In this case the method converges to the solution of the partial-differential equation with this same rate of convergence (see Isaacson and Keller [52], page 508).

The weakness in the backward-difference method results from the fact that the local trucation error has a portion with order $O(k)$, requiring that time intervals be made much smaller than spatial intervals. It would clearly be desirable to devise a procedure with local truncation error of order $O(k^2 + h^2)$. The first step in this direction is to use a difference equation for $u_t(x, t)$, which has $O(k^2)$ error instead of those we have used previously, whose error was $O(k)$. This can be done by using the Taylor series in t for the function $u(x, t)$ at the point (x_i, t_j) and evaluating at (x_i, t_{j+1}) and (x_i, t_{j-1}) to obtain the central-difference formula

$$\frac{\partial u}{\partial t}(x_i, t_j) = \frac{u(x_i, t_{j+1}) - u(x_i, t_{j-1})}{2k} - \frac{k^2}{6} \frac{\partial^3 u(x_i, t_j + \theta_j k)}{\partial t^3}$$

where $-1 < \theta_j < 1$. The difference method that results from substituting this and the usual difference quotient for $(\partial^2 u/\partial x^2)$, Eq. (11.17), into the differential equation is called **Richardson's method**, and is given by

$$\frac{w_{i,j+1} - w_{i,j-1}}{2k} - \alpha^2 \frac{w_{i+1,j} - 2w_{i,j} + w_{i-1,j}}{h^2} = 0.$$

This method does have local truncation error of order $O(k^2 + h^2)$, but unfortunately also has serious stability problems (see Exercise 6).

A more rewarding method can be derived by averaging the forward-difference method at the jth step in t,

$$\frac{w_{i,j+1} - w_{i,j}}{k} - \alpha^2 \frac{w_{i+1,j} - 2w_{i,j} + w_{i-1,j}}{h^2} = 0,$$

which has local truncation error

$$\tau_F = \frac{k}{2} \frac{\partial^2}{\partial t^2} u(x_i, t_j + \theta_j k) + O(h^2),$$

and the backward-difference formula at the $(j + 1)$st step in t,

$$\frac{w_{i,j+1} - w_{i,j}}{k} - \alpha^2 \frac{w_{i+1,j+1} - 2w_{i,j+1} + w_{i-1,j+1}}{h^2} = 0,$$

which has local truncation error

$$\tau_B = -\frac{k}{2} \frac{\partial^2}{\partial t^2} u(x_i, t_{j+1} + \tilde{\theta}_j k) + O(h^2).$$

Since $0 < \theta_j < 1$, while $-1 < \tilde{\theta}_j < 0$, it is reasonable to assume that $t_j + \theta_j k \approx t_{j+1} + \tilde{\theta}_j k$ and, in fact, if these values are equal, then the averaged difference method,

$$\frac{w_{i,j+1} - w_{i,j}}{k}$$

$$-\frac{\alpha^2}{2} \left[\frac{w_{i+1,j} - 2w_{i,j} + w_{i-1,j}}{h^2} + \frac{w_{i+1,j+1} - 2w_{i,j+1} + w_{i-1,j+1}}{h^2} \right] = 0,$$

has local truncation error of order $O(k^2 + h^2)$, provided, of course, that the usual differentiability conditions are satisfied.

This method is known as the **Crank–Nicolson** method, and can be represented in the matrix form $A\mathbf{w}^{(j+1)} = B\mathbf{w}^{(j)}$ for each $j = 0, 1, 2, \ldots$, where

$$\lambda = \alpha^2 \frac{k}{h^2}, \qquad \mathbf{w}^{(j)} = (w_{1,j}, w_{2,j}, \ldots, w_{m-1,j})^t,$$

and the matrices A and B are given by:

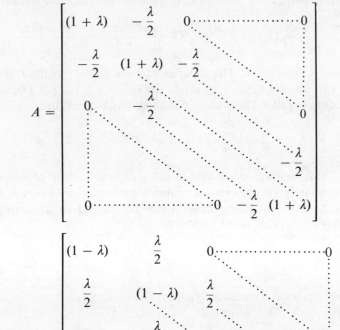

$$A = \begin{bmatrix} (1+\lambda) & -\dfrac{\lambda}{2} & 0 & \cdots\cdots\cdots\cdots\cdots & 0 \\ -\dfrac{\lambda}{2} & (1+\lambda) & -\dfrac{\lambda}{2} & & \\ 0 & -\dfrac{\lambda}{2} & & & 0 \\ & & & & -\dfrac{\lambda}{2} \\ 0 & \cdots\cdots\cdots\cdots\cdots & 0 & -\dfrac{\lambda}{2} & (1+\lambda) \end{bmatrix}$$

and

$$B = \begin{bmatrix} (1-\lambda) & \dfrac{\lambda}{2} & 0 & \cdots\cdots\cdots\cdots\cdots & 0 \\ \dfrac{\lambda}{2} & (1-\lambda) & \dfrac{\lambda}{2} & & \\ 0 & \dfrac{\lambda}{2} & & & 0 \\ & & & & \dfrac{\lambda}{2} \\ 0 & \cdots\cdots\cdots\cdots & 0 & \dfrac{\lambda}{2} & (1-\lambda) \end{bmatrix}$$

Since A is a positive definite, symmetric, strictly diagonally dominant and tridiagonal matrix, it is nonsingular. Either the Crout Reduction for Tridiagonal Linear System (Algorithm 6.7) or the SOR method (Algorithm 8.3) can be used to obtain $\mathbf{w}^{(j+1)}$ from $\mathbf{w}^{(j)}$, for each $j = 0, 1, 2, \ldots$. The following algorithm incorporates Crout Reduction into the Crank–Nicolson technique. As in Algorithm 11.2, a finite length for the time interval must be specified in order to determine a stopping procedure.

Crank–Nicolson Algorithm 11.3

To approximate the solution to the parabolic partial-differential equation

$$\frac{\partial u}{\partial t} - \alpha^2 \frac{\partial^2 u}{\partial x^2} = 0, \qquad 0 < x < l, \quad 0 < t < T,$$

subject to the boundary conditions

$$u(0, t) = u(l, t) = 0, \qquad 0 < t < T,$$

and the initial conditions

$$u(x, 0) = f(x), \qquad 0 \le x \le l:$$

INPUT endpoint l; maximum time T; constant α; integers m, N.

OUTPUT approximations $w_i(t_j)$ to $u(x_i, t_j)$ for each $i = 1, \ldots, m - 1$ and $j = 1, \ldots, N$.

Step 1 Set $h = l/m$;
   $k = T/N$;
   $\lambda = \alpha^2 k/h^2$;
   $w_n = 0$.

Step 2 For $i = 1, \ldots, m - 1$ set $w_i = f(ih)$.   (*Initial values.*)

   (*Steps 3–11 solve a tridiagonal linear system using Algorithm 6.7.*)

Step 3 Set $l_1 = 1 + \lambda$;
   $u_1 = -\lambda/(2l_1)$.

Step 4 For $i = 2, \ldots, m - 2$ set $l_i = 1 + \lambda + \lambda u_{i-1}/2$;
   $u_i = -\lambda/(2l_i)$.

Step 5 Set $l_{m-1} = 1 + \lambda + \lambda u_{m-2}/2$.

Step 6 For $j = 1, \ldots, N$ do Steps 7–11.

   Step 7 Set $t = jk$.   (*Current t_j.*)
      $z_1 = [(1 - \lambda)w_1 + \lambda w_2]/l_1$.

   Step 8 For $i = 2, \ldots, m - 1$ set $z_i = \left[(1 - \lambda)w_i + \dfrac{\lambda}{2}(w_{i+1} + w_{i-1} + z_{i-1})\right]\Big/l_i$.

   Step 9 Set $w_{m-1} = z_{m-1}$.

   Step 10 For $i = m - 2, \ldots, 1$ set $w_i = z_i - u_i w_{i+1}$.

   Step 11 OUTPUT (t);
      For $i = 1, \ldots, m - 1$ set $x = ih$;
         OUTPUT (x, w_i).

Step 12 STOP.   (*Procedure is complete.*)

The Crank–Nicolson method is unconditionally stable and has order of convergence $O(k^2 + h^2)$. The verification of these facts can be found in the previously mentioned book by Isaacson and Keller [52], pages 508–512.

EXAMPLE 3 The Crank–Nicolson method can be used to approximate the solution to the problem in Examples 1 and 2, consisting of the equation

$$\frac{\partial u}{\partial t} - \frac{\partial^2 u}{\partial x^2} = 0, \qquad 0 < x < 1, \quad 0 < t,$$

subject to the conditions

$$u(0, t) = u(1, t) = 0, \qquad 0 < t,$$

and

$$u(x, 0) = \sin(\pi x), \qquad 0 \le x \le 1.$$

The choices $m = 10$, $h = .1$, $N = 50$, $k = .01$, and $\lambda = 1$ are used in Algorithm 11.3, as they were in the previous examples. The results in Table 11.4 indicate the increase in accuracy of the Crank–Nicolson method over the backward-difference method, the best of the two previously discussed techniques. □

TABLE 11.4

x_i	$w_{i, 50}$	$u(x_i, .5)$	$\lvert w_{i, 50} - u(x_i, .5) \rvert$
0	0	0	—
.1	.00230512	.00222241	8.271×10^{-5}
.2	.00438461	.00422728	1.573×10^{-4}
.3	.00603489	.00581836	2.165×10^{-4}
.4	.00709444	.00683989	2.546×10^{-4}
.5	.00745954	.00719188	2.677×10^{-4}
.6	.00709444	.00683989	2.546×10^{-4}
.7	.00603489	.00581836	2.165×10^{-4}
.8	.00438461	.00422728	1.573×10^{-4}
.9	.00230512	.00222241	8.271×10^{-5}
1.0	0	0	—

Exercise Set 11.3

1. Approximate the solution to each of the following parabolic partial-differential equations using Algorithm 11.2.

   a) $\dfrac{\partial u}{\partial t} - \dfrac{\partial^2 u}{\partial x^2} = 0, \quad 0 < x < 2, \quad 0 < t;$

   $u(0, t) = u(2, t) = 0, \quad 0 < t;$

   $u(x, 0) = \sin \dfrac{\pi}{2} x, \quad 0 \le x \le 2.$

   Use $m = 4$, $T = .1$, and $N = 2$.

   b) $\dfrac{\partial u}{\partial t} - \dfrac{1}{16} \dfrac{\partial^2 u}{\partial x^2} = 0, \quad 0 < x < 1, \quad 0 < t,$

   $u(0, t) = u(1, t) = 0, \quad 0 < t;$

   $u(x, 0) = 2 \sin 2\pi x, \quad 0 \le x \le 1.$

   Use $m = 3$, $T = .1$, and $N = 2$.

2. Repeat Exercise 1, using Algorithm 11.3.

3. Use the forward-difference method to approximate the solutions to the following parabolic partial-differential equations.

a) $\dfrac{\partial u}{\partial t} - \dfrac{\partial^2 u}{\partial x^2} = 0, \quad 0 < x < 2, \quad 0 < t;$

$u(0, t) = u(2, t) = 0, \quad 0 < t;$

$u(x, 0) = \sin 2\pi x, \quad 0 \le x \le 2.$

Use $h = .1$ and $k = .01$, and compare your answers at $t = .5$ to the actual solution. Then use $h = .1$ and $k = .005$, and compare answers.

b) $\dfrac{\partial u}{\partial t} - \dfrac{\partial^2 u}{\partial x^2} = 0, \quad 0 < x < 2, \quad 0 < t;$

$u(0, t) = u(2, t) = 0, \quad 0 < t,$

$u(x, 0) = x(2 - x), \quad 0 \le x \le 2.$

Use $h = .1$ and choose an appropriate value for k. Compare your answers to the first five terms of the Fourier-series solution at $t = .5$.

4. Repeat Exercise 3, using the Backward-Difference Algorithm 11.2.

5. Repeat Exercise 3, using the Crank–Nicolson Algorithm 11.3.

6. Repeat Exercise 3, using Richardson's method.

7. Show that the eigenvalues for the $(m - 1) \times (m - 1)$ tridiagonal matrix A given by

$$a_{ij} = \begin{cases} \lambda, & j = i - 1 \quad \text{or} \quad j = i + 1, \\ (1 - 2\lambda), & j = i, \\ 0, & \text{otherwise,} \end{cases}$$

are $\mu_i = 1 - 4\lambda \sin^2\left(\dfrac{i\pi}{2m}\right) \quad$ for each $i = 1, 2, \ldots, m - 1.$

[*Hint*: Consider the form of the eigenvectors.]

8. Show that the $(m - 1) \times (m - 1)$ tridiagonal matrix A given by

$$a_{ij} = \begin{cases} -\lambda, & j = i - 1 \text{ or } j = i + 1, \\ 1 + 2\lambda, & j = i, \\ 0, & \text{otherwise,} \end{cases}$$

where $\lambda > 0$, is positive definite and diagonally dominant and has eigenvalues

$$\mu_i = 1 + 4\lambda \sin^2\left(\dfrac{i\pi}{2m}\right) \quad \text{for each } i = 1, 2, \ldots, m - 1.$$

[*Hint*: Consider the form of the eigenvectors.]

9. Modify Algorithms 11.2 and 11.3 to include the parabolic partial-differential equation

$$\dfrac{\partial u}{\partial t} - \dfrac{\partial^2 u}{\partial x^2} = F(x), \quad 0 < x < l, \quad 0 < t,$$

$$u(0, t) = u(l, t) = 0, \quad 0 < t,$$

$$u(x, 0) = f(x), \quad 0 \le x \le l.$$

10. Use the results of Exercise 9 to approximate the solution to

$$\frac{\partial u}{\partial t} - \frac{\partial^2 u}{\partial x^2} = 2, \qquad 0 < x < 1, \quad 0 < t$$

$$u(0, t) = u(1, t) = 0, \qquad 0 < t,$$

$$u(x, 0) = \sin \pi x + x(1 - x),$$

with $h = .1$ and $k = .01$. Compare your answers to the actual solution $u(x, t) = e^{-\pi^2 t} \sin \pi x + x(1 - x)$ at $t = .25$.

11. Change Algorithms 11.2 and 11.3 to accommodate the partial-differential equation

$$\frac{\partial u}{\partial t} - \alpha^2 \frac{\partial^2 u}{\partial x^2} = 0, \qquad 0 < x < l, \quad 0 < t,$$

$$u(0, t) = \phi(t), \quad u(l, t) = \Psi(t), \quad 0 < t,$$

$$u(x, 0) = f(x), \qquad 0 \leq x \leq l,$$

where $f(0) = \phi(0)$ and $f(l) = \Psi(l)$.

12. The temperature $u(x, t)$ in a long thin rod of constant cross section and homogeneous conducting material is governed by the one-dimensional heat equation. If heat is generated in the material, for example, by resistance to current or nuclear reaction, the heat equation becomes

$$\frac{\partial^2 u(x, t)}{\partial x^2} + \frac{Kr}{\rho C} = K \frac{\partial u(x, t)}{\partial t}, \qquad 0 < x < l, \quad 0 < t,$$

where l is the length, ρ the density, C the specific heat, and K the thermal diffusivity of the rod. The function $r = r(x, t, u)$ represents the heat generated per unit volume. Suppose that

$$l = 1.5 \text{ cm}, \qquad K = 1.04 \text{ cal/cm} \cdot \text{deg} \cdot \text{sec},$$

$$\rho = 10.6 \text{ g/cm}^3, \qquad C = .056 \text{ cal/g} \cdot \text{deg},$$

and
$$r(x, t, u) = 5.0 \text{ cal/sec} \cdot \text{cm}^3.$$

If the ends of the rod are kept at 0°C, then

$$u(0, t) = u(l, t) = 0, \qquad t > 0.$$

Suppose the initial temperature distribution is given by

$$u(x, 0) = \sin \frac{\pi x}{l}, \qquad 0 \leq x \leq l.$$

Use the results of Exercise 9 to approximate the temperature distribution with $h = .15$ and $K = .0225$.

13. V. Sagar and D. J. Payne [73], in analyzing the stress–strain relationships and material properties of a cylinder alternately subjected to heating and cooling, consider the equation

$$\frac{\partial^2 T}{\partial r^2} + \frac{1}{r} \frac{\partial T}{\partial r} = \frac{1}{4K} \frac{\partial T}{\partial t}, \qquad \tfrac{1}{2} < r < 1, \quad 0 < T,$$

where $T = T(r, t)$ is the temperature, r is the radial distance from the center of the cylinder, t is time, and K is a diffusivity coefficient.

a)   Find approximations to $T(r, 10)$ for a cylinder with outside radius one, given the
initial and boundary conditions:

$$T(1, t) = 100 + 40t, \quad 0 \le t \le 10,$$

$$T(\tfrac{1}{2}, t) = t, \quad 0 \le t \le 10,$$

$$T(r, 0) = 200(r - .5), \quad .5 \le r \le 1.$$

Use a modification of the backward-difference method with $K = .1$, $k = .5$, and $h = \Delta r = .1$.

b)   Using the temperature distribution of part (a), calculate the strain by approximating
the integral

$$I = \int_{.5}^{1} \alpha T(r, t) r \, dr$$

where $\alpha = 10.7$ and $t = 10$. Use the Composite Trapezoidal method given in Theorem
4.5 (p. 149) with $m = 5$.

14.   The equation describing one-dimensional, single-phase, slightly compressible flow in a
producing petroleum reservoir is given, for $0 < x < 1000$ and $0 < t$, by

$$\frac{\phi \mu C}{K} \frac{\partial p(x, t)}{\partial t} = \frac{\partial^2 p(x, t)}{\partial x^2} - \begin{cases} 0, & \text{if } x \ne 500 \\ 1000, & \text{if } x = 500, \end{cases}$$

where it has been assumed that the porous medium and the reservoir are homogeneous,
that the liquid is ideal, and that gravitational effects are negligible. The symbols are defined
as x representing distance (in feet), t the time (in days), p the pressure (in pounds per square
inch), ϕ the dimensionless constant porosity of the medium, μ the viscosity (in centipoise),
K the permeability of the medium (in millidarcies), and C the compressibility (in [pounds
per square inch]$^{-1}$). Assuming that $\alpha = \phi \mu C / K = .00004$ days/ft^2, and that the following
conditions hold:

$$p(x, 0) = 2.5 \times 10^7, \quad 0 \le x \le 1000,$$

$$\frac{\partial p(0, t)}{k \, \partial x} = \frac{\partial p(1000, t)}{\partial x} = 0, \quad 0 < t,$$

find the pressure p at $t = 5$, using the Crank–Nicolson method with $k = \Delta t = .5$ and
$h = \Delta x = 100$.

11.4   Hyperbolic Partial-Differential Equations

In this section we shall consider the numerical solution of the wave equation, an
example of a hyperbolic partial-differential equation. The wave equation is given
by the differential equation

(11.24) $$\frac{\partial^2 u}{\partial t^2}(x, t) - \alpha^2 \frac{\partial^2 u}{\partial x^2}(x, t) = 0, \quad 0 < x < l, \quad t > 0,$$

subject to the conditions

$$u(0, t) = u(l, t) = 0, \quad t > 0,$$

$$u(x, 0) = f(x), \quad 0 \le x \le l,$$

and $$\frac{\partial u}{\partial t}(x, 0) = g(x), \quad 0 \le x \le l$$

where α is a constant. To set up the finite-difference method we select an integer $m > 0$ and time-step size $k > 0$. With $h = l/m$, the mesh points (x_i, t_j) are defined by

$$x_i = ih, \quad \text{for each } i = 0, 1, \ldots, m,$$

and $$t_j = jk, \quad \text{for each } j = 0, 1, \ldots .$$

At any interior mesh point (x_i, t_j) the wave equation becomes

(11.25) $$\frac{\partial^2 u}{\partial t^2}(x_i, t_j) - \alpha^2 \frac{\partial^2 u}{\partial x^2}(x_i, t_j) = 0.$$

The difference method is obtained by using the centered-difference quotient for the second partials given by

(11.26) $$\frac{\partial^2 u}{\partial t^2}(x_i, t_j) = \frac{u(x_i, t_{j+1}) - 2u(x_i, t_j) + u(x_i, t_{j-1})}{k^2}$$

$$- \frac{k^2}{12} \frac{\partial^4 u(x_i, t_j + \theta_j k)}{\partial t^4},$$

where $-1 < \theta_j < 1$ and

(11.27) $$\frac{\partial^2 u}{\partial x^2}(x_i, t_j) = \frac{u(x_{i+1}, t_j) - 2u(x_i, t_j) + u(x_{i-1}, t_j)}{h^2}$$

$$- \frac{h^2}{12} \frac{\partial^4 u(x_i + \phi_i h, t_j)}{\partial x^4}$$

where $-1 < \phi_i < 1$. Substituting these into (11.25) gives

(11.28)

$$\frac{u(x_i, t_{j+1}) - 2u(x_i, t_j) + u(x_i, t_{j-1})}{k^2} - \alpha^2 \frac{u(x_{i+1}, t_j) - 2u(x_i, t_j) + u(x_{i-1}, t_j)}{h^2}$$

$$= \frac{1}{12} \left[k^2 \frac{\partial^4 u(x_i, t_j + \theta_j k)}{\partial t^4} - \alpha^2 h^2 \frac{\partial^4 u(x_i + \phi_i h, t_j)}{\partial x^4} \right];$$

and neglecting the truncation error

(11.29) $$\tau_{i,j} = \frac{1}{12} \left[k^2 \frac{\partial^4 u}{\partial t^4}(x_i, t_j + \theta_j k) - \alpha^2 h^2 \frac{\partial^4 u}{\partial x^4}(x_i + \phi_i h, t_j) \right]$$

leads to the difference equation

(11.30) $$\frac{w_{i,j+1} - 2w_{i,j} + w_{i,j-1}}{k^2} - \alpha^2 \frac{w_{i+1,j} - 2w_{i,j} + w_{i-1,j}}{h^2} = 0.$$

If λ is used to denote $\alpha k/h$, we can write the difference equation as

$$w_{i,j+1} - 2w_{i,j} + w_{i,j-1} - \lambda^2 w_{i+1,j} + 2\lambda^2 w_{i,j} - \lambda^2 w_{i-1,j} = 0$$

and solve for $w_{i,j+1}$, the most advanced time-step approximation, to obtain

(11.31) $$w_{i,j+1} = 2(1 - \lambda^2) w_{i,j} + \lambda^2 (w_{i+1,j} + w_{i-1,j}) - w_{i,j-1}.$$

This equation holds for each $i = 1, 2, \ldots, (m - 1)$, and $j = 1, 2, \ldots$. The boundary conditions give

(11.32) $w_{0,j} = w_{m,j} = 0$ for each $j = 1, 2, 3, \ldots,$

and the initial condition implies that

(11.33) $w_{i,0} = f(x_i)$ for each $i = 1, 2, \ldots, m - 1.$

Writing this set of equations in matrix form implies that

(11.34)

$$
\begin{bmatrix} w_{1,j+1} \\ w_{2,j+1} \\ \vdots \\ w_{m-1,j+1} \end{bmatrix} = \begin{bmatrix} 2(1 - \lambda^2) & \lambda^2 & 0 \cdots \cdots 0 \\ \lambda^2 & 2(1 - \lambda^2) & \lambda^2 & \ddots & \vdots \\ 0 & & \ddots & \ddots & 0 \\ \vdots & & \ddots & \ddots & \lambda^2 \\ 0 \cdots \cdots \cdots 0 & & \lambda^2 & 2(1 - \lambda^2) \end{bmatrix} \begin{bmatrix} w_{1,j} \\ w_{2,j} \\ \vdots \\ w_{m-1,j} \end{bmatrix} - \begin{bmatrix} w_{1,j-1} \\ w_{2,j-1} \\ \vdots \\ w_{m-1,j-1} \end{bmatrix}
$$

It is clear from equations (11.31) and (11.32) that the $(j + 1)$st time step requires values from the jth and $(j - 1)$st time steps. This produces a minor starting problem since values for $j = 0$ are given by Eq. (11.33), but values for $j = 1$, which are needed in Eq. (11.31) to compute $w_{i,2}$, must be obtained from the initial-velocity condition

$$\frac{\partial u}{\partial t}(x, 0) = g(x), \qquad 0 \leq x \leq l.$$

The first approach is to replace $(\partial u/\partial t)$ by a forward-difference approximation,

$$\frac{\partial u}{\partial t}(x_i, 0) = \frac{u(x_i, t_1) - u(x_i, 0)}{k} - \frac{k}{2} \frac{\partial^2 u}{\partial t^2}(x_i, \theta_i k), \qquad 0 < \theta_i < 1;$$

then $w_{i,1}$ is given by

(11.35) $w_{i,1} = w_{i,0} + kg(x_i)$ for each $i = 1, \ldots, m - 1.$

Equation (11.35), however, gives only an $O(k)$ approximation to the initial data, while the local truncation error for Eq. (11.31) is $O(k^2 + h^2)$.

An $O(k^2)$ approximation to the derivative can be obtained as follows: since

$$\frac{u(x_i, t_1) - u(x_i, 0)}{k} = \frac{\partial u}{\partial t}(x_i, 0) + \frac{k}{2} \frac{\partial^2 u}{\partial t^2}(x_i, 0) + \frac{k^2}{6} \frac{\partial^3 u}{\partial t^3}(x_i, \theta k)$$

for some $0 < \theta < 1$, suppose the wave equation also holds on the initial line; that is,

$$\frac{\partial^2 u}{\partial t^2}(x_i, 0) - \alpha^2 \frac{\partial^2 u}{\partial x^2}(x_i, 0) = 0 \qquad \text{for each } i = 0, 1, \ldots, m.$$

If f'' exists, then

$$\frac{\partial^2 u}{\partial t^2}(x_i, 0) = \alpha^2 \frac{\partial^2 u}{\partial x^2}(x_i, 0) = \alpha^2 \frac{d^2}{dx^2} f(x_i) = \alpha^2 f''(x_i).$$

But we can write

$$f''(x_i) = \frac{f(x_{i+1}) - 2f(x_i) + f(x_{i-1})}{h^2} - \frac{h^2}{12} f^{(4)}(x_i + \theta_i h),$$

for some $-1 < \theta_i < 1$, if $f \in C^4[0, l]$, which implies, in this case, that the approximation becomes

$$\frac{u(x_i, t_1) - u(x_i, 0)}{k} = g(x_i) + \frac{k\alpha^2}{2h^2}[f(x_{i+1}) - 2f(x_i) + f(x_{i-1})]$$

$$+ O(k^2 + h^2 k)$$

or

$$u(x_i, t_1) = u(x_i, 0) + kg(x_i) + \frac{\lambda^2}{2}[f(x_{i+1}) - 2f(x_i) + f(x_{i-1})] + O(k^3 + h^2 k^2)$$

$$= (1 - \lambda^2)f(x_i) + \frac{\lambda^2}{2}f(x_{i+1})$$

$$+ \frac{\lambda^2}{2}f(x_{i-1}) + kg(x_i) + O(k^3 + h^2 k^2) \qquad \text{where } \lambda = \frac{k\alpha}{h}.$$

Thus, the difference equation

$$(11.36) \qquad w_{i,1} = (1 - \lambda^2)f(x_i) + \frac{\lambda^2}{2}f(x_{i+1}) + \frac{\lambda^2}{2}f(x_{i-1}) + kg(x_i),$$

can be used to find $w_{i,1}$ for each $i = 1, 2, \ldots, m - 1$.

The following algorithm uses Eq. (11.36) to approximate $w_{i,1}$, although Eq. (11.35) could also be used. It is assumed that there is an upper bound for the value of t, to be used in the stopping technique.

Wave Equation Finite-Difference Algorithm 11.4

To approximate the solution to the wave equation

$$\frac{\partial^2 u}{\partial t^2} - \alpha^2 \frac{\partial^2 u}{\partial x^2} = 0, \qquad 0 < x < l, \quad 0 < t < T,$$

subject to the boundary conditions

$$u(0, t) = u(l, t) = 0, \qquad 0 < t < T,$$

and the initial conditions

$$u(x, 0) = f(x), \qquad 0 \le x \le l,$$

$$\frac{\partial u(x, 0)}{\partial t} = g(x), \qquad 0 \le x \le l:$$

INPUT endpoint l; maximum time T; constant α; integers m, N.

OUTPUT approximations $w_{i,j}$ to $u(x_i, t_j)$ for each $i = 0, \ldots, m$ and $j = 0, \ldots, N$.

Step 1 Set $h = l/m$;
$k = T/N$;
$\lambda = k\alpha/h$.

Step 2 For $j = 1, \ldots, N$ set $w_{0,j} = 0$;
$w_{m,j} = 0$.

Step 3 Set $w_{0,0} = f(0)$;
     $w_{m,0} = f(l)$.

Step 4 For $i = 1, \ldots, m - 1$ (*Initialize for* $t = 0$ *and* $t = k$.)
     set $w_{i,0} = f(ih)$;

$$w_{i,1} = (1 - \lambda^2)f(ih) + \frac{\lambda^2}{2}[f((i+1)h) + f((i-1)h)] + kg(ih).$$

Step 5 For $j = 1, \ldots, N - 1$ (*Perform matrix multiplication.*)
     for $i = 1, \ldots, m - 1$
       set $w_{i,j+1} = 2(1 - \lambda^2)w_{i,j} + \lambda^2(w_{i+1,j} + w_{i-1,j}) - w_{i,j-1}$.

Step 6 For $j = 0, \ldots, N$
     set $t = jk$;
     for $i = 0, \ldots, m$
       set $x = ih$;
       OUTPUT $(x, t, w_{i,j})$.

Step 7 STOP.   (*Procedure is complete.*)

EXAMPLE 1 Consider the hyperbolic problem

$$\frac{\partial^2 u}{\partial t^2} - 4\frac{\partial^2 u}{\partial x^2} = 0, \qquad 0 < x < 1, \quad 0 < t,$$

with boundary conditions $u(0, t) = u(1, t) = 0, 0 < t$, and initial conditions

$$u(x, 0) = \sin(\pi x), \quad 0 \le x \le 1, \qquad \text{and} \qquad \frac{\partial u}{\partial t}(x, 0) = 0, \quad 0 \le x \le 1.$$

It is easily verified that the solution to this problem is

$$u(x, t) = \sin(\pi x)\cos(2\pi t).$$

The finite-difference method (Algorithm 11.4) is used in this example with $m = 10$, $T = 1$, and $N = 20$, which implies that $h = .1$, $k = .05$, and $\lambda = 1$. The following table lists the results of the approximation, $w_{i,N}$, for $i = 0, 1, \ldots, 10$. The values listed in Table 11.5 are correct to the places given.   □

TABLE 11.5

x_i	$w_{i,20}$
.0	.0000000000
.1	.3090169944
.2	.5877852523
.3	.8090169944
.4	.9510565163
.5	1.0000000000
.6	.9510565163
.7	.8090169944
.8	.5877852523
.9	.3090169944
1.0	.0000000000

The results of the example were very accurate, more so than the truncation error $O(k^2 + h^2)$ would lead us to believe. The explanation for this phenomenon lies in the fact that the true solution to the equation is infinitely differentiable. When this is the case, it is easy to show, using Taylor series, that:

$$\frac{u(x_{i+1}, t_j) - 2u(x_i, t_j) + u(x_{i-1}, t_j)}{h^2}$$

$$= \frac{\partial^2 u}{\partial x^2}(x_i, t_j) + 2\left[\frac{h^2}{4!}\frac{\partial^4 u(x_i, t_j)}{\partial x^4} + \frac{h^4}{6!}\frac{\partial^6 u(x_i, t_j)}{\partial x^6} + \cdots\right]$$

and

$$\frac{u(x_i, t_{j+1}) - 2u(x_i, t_j) + u(x_i, t_{j-1})}{k^2}$$

$$= \frac{\partial^2 u}{\partial t^2}(x_i, t_j) + 2\left[\frac{k^2}{4!}\frac{\partial^4 u(x_i, t_j)}{\partial t^4} + \frac{k^4}{6!}\frac{\partial^6 u(x_i, t_j)}{\partial t^6} + \cdots\right].$$

Thus,

(11.37)

$$\frac{u(x_i, t_{j+1}) - 2u(x_i, t_j) + u(x_i, t_{j-1})}{k^2} - \alpha^2 \frac{u(x_{i+1}, t_j) - 2u(x_i, t_j) + u(x_{i-1}, t_j)}{h^2}$$

$$= 2\left[\frac{1}{4!}\left(k^2 \frac{\partial^4 u(x_i, t_j)}{\partial t^4} - \alpha^2 h^2 \frac{\partial^4 (x_i, t_j)}{\partial x^4}\right)\right.$$

$$\left. + \frac{1}{6!}\left(k^4 \frac{\partial^6 u(x_i, t_j)}{\partial t^6} - \alpha^2 h^4 \frac{\partial^6 u(x_i, t_j)}{\partial x^6}\right) + \cdots\right].$$

However, by differentiating the wave equation,

$$k^2 \frac{\partial^4 u(x_i, t_j)}{\partial t^4} = k^2 \frac{\partial^2}{\partial t^2}\left[\alpha^2 \frac{\partial^2 u(x_i, t_j)}{\partial x^2}\right] = \alpha^2 k^2 \frac{\partial^2}{\partial x^2}\left[\frac{\partial^2 u(x_i, t_j)}{\partial t^2}\right]$$

$$= \alpha^2 k^2 \frac{\partial^2}{\partial x^2}\left[\alpha^2 \frac{\partial^2}{\partial x^2} u(x_i, t_j)\right] = \alpha^4 k^2 \frac{\partial^4 u(x_i, t_j)}{\partial x^4},$$

we see that

$$\frac{1}{4!}\left[k^2 \frac{\partial^4 u(x_i, t_j)}{\partial t^4} - \alpha^2 h^2 \frac{\partial^4 u(x_i, t_j)}{\partial x^4}\right] = \frac{\alpha^2}{4!}[\alpha^2 k^2 - h^2]\frac{\partial^4 u(x_i, t_j)}{\partial x^4} = 0$$

since we chose

$$\lambda^2 = \frac{\alpha^2 k^2}{h^2} = 1.$$

In this manner, all the terms on the righthand side of (11.37) are zero, implying a zero local truncation error. The only errors in Example 1 are those which are due to the approximation of $w_{i,1}$ and to rounding.

As in the case of the forward-difference method for the heat equation, the explicit finite-difference method for the wave equation has stability problems. In fact, it is necessary that $\lambda = \alpha(k/h) \leq 1$ in order for the method to be stable.

(See Isaacson–Keller [52], page 489.) Although we will not discuss them, there are implicit methods that are unconditionally stable. A discussion of these methods can be found in Ames [5], page 199 or in the books by Mitchell [61] or Smith [83]. The explicit method given in Algorithm 11.4, with $\lambda \leq 1$ is $O(h^2 + k^2)$-convergent if f and g are sufficiently differentiable. For verification of this, see Isaacson–Keller [52], page 491.

Exercise Set 11.4

1. Use Algorithm 11.4 to approximate the solution to the wave equation

$$\frac{\partial^2 u}{\partial t^2} - \frac{\partial^2 u}{\partial x^2} = 0, \qquad 0 < x < 1, \quad 0 < t.$$

$$u(0, t) = u(1, t) = 0, \qquad 0 < t,$$

$$u(x, 0) = \sin 2\pi x, \qquad 0 \leq x \leq 1,$$

$$\frac{\partial u}{\partial t}(x, 0) = \sin \pi x, \qquad 0 \leq x \leq 1,$$

   with $m = 4$, $N = 2$, and $T = .5$.

2. Approximate the solution to the wave equation

$$\frac{\partial^2 u}{\partial t^2} - \frac{1}{4}\frac{\partial^2 u}{\partial x^2} = 0, \qquad 0 < x < .5, \quad 0 < t.$$

$$u(0, t) = u(.5, t) = 0, \qquad 0 < t,$$

$$u(x, 0) = 0, \qquad 0 \leq x \leq .5,$$

$$\frac{\partial u}{\partial t}(x, 0) = \sin 4\pi x, \qquad 0 \leq x \leq .5,$$

   using Algorithm 11.4 with $m = 2$, $N = 2$, and $T = .5$.

3. Approximate the solution $u(x, t) = \sin \pi x \cos \pi t$ to the wave equation

$$\frac{\partial^2 u}{\partial t^2} - \frac{\partial^2 u}{\partial x^2} = 0, \qquad 0 < x < 1, \quad 0 < t.$$

$$u(0, t) = u(1, t) = 0, \qquad 0 < t,$$

$$u(x, 0) = \sin \pi x, \qquad 0 \leq x \leq 1,$$

$$\frac{\partial u}{\partial t}(x, 0) = 0, \qquad 0 \leq x \leq 1,$$

   using Algorithm 11.4 with $h = .1$ and $k = .05$, with $h = .05$ and $k = .1$, and then with $h = .05$ and $k = .05$. Compare your results to the exact solution at $t = .5$.

4. Repeat Exercise 3, using the approximation

$$w_{i, 1} = w_{i, 0} + kg(x_i) \qquad \text{for each } i = 0, 1, \ldots, m.$$

   Can you explain the results?

5. Approximate the solution to the wave equation

$$\frac{\partial^2 u}{\partial t^2} - \frac{\partial^2 u}{\partial x^2} = 0, \qquad 0 < x < 1, \quad 0 < t,$$

$$u(0, t) = u(1, t) = 0, \qquad 0 < t,$$

$$u(x, 0) = \sin 2\pi x, \qquad 0 \le x \le 1,$$

$$\frac{\partial u}{\partial t}(x, 0) = 2\pi \sin 2\pi x, \qquad 0 \le x \le 1,$$

using Algorithm 11.4 with $h = .1$ and $k = .1$. Compare your results to the actual solution, $u(x, t) = \sin 2\pi x[\cos 2\pi t + \sin 2\pi t]$, at $t = .3$.

6. Approximate the solution to the wave equation

$$\frac{\partial^2 u}{\partial t^2} - \frac{\partial^2 u}{\partial x^2} = 0, \qquad 0 < x < 1, \quad 0 < t,$$

$$u(0, t) = u(1, t) = 0, \qquad 0 < t,$$

$$u(x, 0) = \begin{cases} 1, & 0 \le x \le \frac{1}{2}, \\ -1, & \frac{1}{2} < x \le 1, \end{cases}$$

$$\frac{\partial u}{\partial t}(x, 0) = 0, \qquad 0 \le x \le 1.$$

Use $h = .1$ and $k = .1$, and compare your answer to the Fourier series solution at $t = .5$.

7. The air pressure $p(x, t)$ in an organ pipe is governed by the wave equation

$$\frac{\partial^2 p}{\partial x^2} = \frac{1}{c^2}\frac{\partial^2 p}{\partial t^2}, \qquad 0 < x < l, \quad 0 < t$$

where l is the length of the pipe and c is a physical constant. If the pipe is open, the boundary conditions are given by

$$p(0, t) = p_0 \qquad \text{and} \qquad p(l, t) = p_0.$$

If the pipe is closed at the end where $x = l$, the boundary conditions are

$$p(0, t) = p_0 \qquad \text{and} \qquad \frac{\partial p(l, t)}{\partial x} = 0.$$

Assume that $c = 1$, $l = 1$ and the initial conditions are

$$p(x, 0) = p_0 \cos 2\pi x, \qquad 0 \le x \le l,$$

$$p_t(x, 0) = 0, \qquad 0 \le x \le 1.$$

a) Approximate the pressure for an open pipe with $p_0 = .9$ at $x = \frac{1}{2}$ for $t = .5$ and $t = 1$, using Algorithm 11.4 with $h = k = .1$.

b) Modify Algorithm 11.4 for the closed pipe problem with $p_0 = .9$, and approximate $p(.5, .5)$ and $p(.5, 1)$, using $h = k = .1$.

8. In an electric transmission line of length l that carries alternating current of high frequency (called a "lossless" line), the voltage V and current i are described by

$$\frac{\partial^2 V}{\partial x^2} = LC\frac{\partial^2 V}{\partial t^2}, \qquad 0 < x < l, \quad 0 < t,$$

$$\frac{\partial^2 i}{\partial x^2} = LC\frac{\partial^2 i}{\partial t^2}, \qquad 0 < x < l, \quad 0 < t$$

where L is the inductance per unit length and C is the capacitance per unit length. Suppose the line is 200 feet long and the constants C and L are given by

$$C = .1 \text{ farads/ft,}$$

$$L = .3 \text{ henries/ft.}$$

Suppose the voltage and current also satisfy

$$V(0, t) = V(200, t) = 0, \qquad 0 < t$$

$$V(x, 0) = 110 \sin \frac{\pi x}{200}, \qquad 0 \le x \le 200,$$

$$\frac{\partial V}{\partial t}(x, 0) = 0, \qquad 0 \le x \le 200,$$

$$i(0, t) = i(200, t) = 0, \qquad 0 < t,$$

$$i(x, 0) = 5.5 \cos \frac{\pi x}{200}, \qquad 0 \le x \le 200,$$

$$\frac{\partial i}{\partial t}(x, 0) = 0, \qquad 0 \le x \le 200.$$

a) Approximate the voltage and current at $t = .2$ and $t = .5$, using Algorithm 11.4 with $h = 10$ and $k = .1$.

b) Find the first three terms of the Fourier-series solution for V and i, and compare the values $V(50, .2)$, $V(50, .5)$, $i(50, .2)$ and $i(50, .5)$ to the approximations in (a).

11.5 An Introduction to the Finite-Element Method

A method frequently used to solve elliptic partial-differential equations that occur in engineering applications is called the **finite-element method**. This method is similar to the Rayleigh–Ritz procedure discussed in Section 10.4, but generalized to higher dimensions.

One advantage of the finite-element method over finite-difference methods is the relative ease with which the boundary conditions of the problem are handled. Many physical problems have boundary conditions involving derivatives and, in general, the boundary of the region is irregularly shaped. Boundary conditions of this type are very difficult to handle using finite-difference techniques, since each boundary condition involving a derivative must be approximated by a difference quotient at the grid points, and irregular shaping of the boundary makes placing the grid points difficult. The finite-element method includes the boundary conditions as integrals in a functional that is being minimized, so the basis-construction procedure is independent of the particular boundary conditions of the problem.

In our discussion, we will consider the partial differential equation

$$(11.38) \qquad \frac{\partial}{\partial x}\left(p(x, y) \frac{\partial u}{\partial x}\right) + \frac{\partial}{\partial y}\left(q(x, y) \frac{\partial u}{\partial y}\right) + r(x, y)u(x, y) = f(x, y),$$

with $(x, y) \in \mathcal{D}$, where $\mathcal{D}$ is a plane region with boundary $\mathcal{S}$.

Boundary conditions of the form

(11.39) $u(x, y) = g(x, y)$

are imposed on a portion, $\mathscr{S}_1$, of the boundary; and on the remainder of the boundary, $\mathscr{S}_2$, $u(x, y)$ is required to satisfy

(11.40) $p(x, y) \dfrac{\partial u}{\partial x} \cos \theta_1 + q(x, y) \dfrac{\partial u}{\partial y} \cos \theta_2 + g_1(x, y)u(x, y) = g_2(x, y)$

where θ_1 and θ_2 are the direction angles of the outward normal to the boundary at the point (x, y) (see Fig. 11.10).

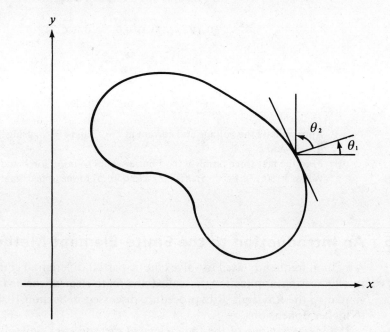

FIGURE 11.10

Physical problems in the areas of solid mechanics and elasticity have associated partial-differential equations similar to Eq. (11.38). The solution to a problem of this type is typically the minimization of a certain functional, involving integrals, over a class of functions determined by the problem. Suppose that p, q, r, and f are all continuous in $\mathscr{D} \cup \mathscr{S}$, p and q have continuous first partials, and g_1 and g_2 are continuous on $\mathscr{S}_2$. Suppose, in addition, that $p(x, y) > 0$, $q(x, y) > 0$, $r(x, y) \leq 0$, and $g_1(x, y) > 0$. Then a solution to Eq. (11.38) uniquely minimizes the functional

(11.41)

$$I[w] = \iint\limits_{\mathscr{D}} \left\{ \frac{1}{2}\left[p(x, y)\left(\frac{\partial w}{\partial x}\right)^2 + q(x, y)\left(\frac{\partial w}{\partial y}\right)^2 - r(x, y)w^2 \right] + f(x, y)w \right\} dx\, dy$$

$$+ \int_{\mathscr{S}_2} \left\{ -g_2(x, y)w + \tfrac{1}{2}g_1(x, y)w^2 \right\} dS$$

over all functions w, satisfying Eq. (11.39) on $\mathscr{S}_1$, which are twice continuously differentiable. The finite-element method approximates this solution by minimizing the functional I over a smaller class of functions.

The first step in the procedure is to divide the region into a finite number of sections, or elements, of a regular shape, either rectangles or, more commonly, triangles (see Fig. 11.11).

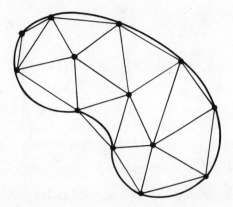

FIGURE 11.11

The set of functions used for approximation is generally a set of piecewise polynomials of fixed degree in x and y, and the approximation requires that the polynomials be pieced together in such a manner that the resulting function will be continuous with an integrable or continuous first or second derivative on the entire region. Polynomials of linear type in x and y

$$\phi(x, y) = a + bx + cy$$

are commonly used with triangular elements while polynomials of bilinear type in x and y,

$$\phi(x, y) = a + bx + cy + dxy,$$

are used with rectangular elements.

For our discussion, suppose that the region $\mathscr{D}$ has been subdivided into triangular elements. The collection of triangles will be denoted by D, and the vertices of these triangles will be called **nodes**. The method seeks an approximation of the form

$$\phi(x, y) = \sum_{i=1}^{m} \gamma_i \phi_i(x, y),$$

where $\phi_1, \phi_2, \ldots, \phi_m$ are linearly independent piecewise polynomials of the first degree and $\gamma_1, \gamma_2, \ldots, \gamma_m$ are constants. Some of these constants, say, $\gamma_{n+1}, \gamma_{n+2}, \ldots, \gamma_m$, are used to ensure that the boundary condition

$$\phi(x, y) = g(x, y)$$

is satisfied on $\mathscr{S}_1$, while the remaining constants $\gamma_1, \gamma_2, \ldots, \gamma_n$ are used to minimize the functional $I[\sum_{i=1}^{m} \gamma_i \phi_i]$.

From Eq. (11.41) the functional is of the form

$$(11.42) \quad I[\phi] = I\left[\sum_{i=1}^{m} \gamma_i \phi_i\right]$$

$$= \iint_{\mathscr{D}} \left(\frac{1}{2} \left\{ p(x, y)\left[\sum_{i=1}^{m} \gamma_i \frac{\partial \phi_i}{\partial x}(x, y)\right]^2 \right.\right.$$

$$+ q(x, y)\left[\sum_{i=1}^{m} \gamma_i \frac{\partial \phi_i}{\partial y}(x, y)\right]^2 - r(x, y)\left[\sum_{i=1}^{m} \gamma_i \phi_i(x, y)\right]^2\right\}$$

$$\left. + f(x, y)\sum_{i=1}^{m} \gamma_i \phi_i(x, y)\right) dy\, dx$$

$$+ \int_{\mathscr{S}_2} \left\{-g_2(x, y)\sum_{i=1}^{m} \gamma_i \phi_i(x, y) + \tfrac{1}{2}g_1(x, y)\left[\sum_{i=1}^{m} \gamma_i \phi_i(x, y)\right]^2\right\} dS.$$

In order for a minimum to occur, considering I as a function of $\gamma_1, \ldots, \gamma_n$, it is necessary to have

$$(11.43) \qquad \frac{\partial I}{\partial \gamma_j} = 0 \qquad \text{for each } j = 1, 2, \ldots, n.$$

Differentiating (11.42) gives:

$$\frac{\partial I}{\partial \gamma_j} = \iint_{\mathscr{D}} \left\{ p(x, y)\sum_{i=1}^{m} \gamma_i \frac{\partial \phi_i}{\partial x}(x, y)\frac{\partial \phi_j}{\partial x}(x, y)\right.$$

$$+ q(x, y)\sum_{i=1}^{m} \gamma_i \frac{\partial \phi_i}{\partial y}(x, y)\frac{\partial \phi_j}{\partial y}(x, y)$$

$$\left. - r(x, y)\sum_{i=1}^{m} \gamma_i \phi_i(x, y)\phi_j(x, y) + f(x, y)\phi_j(x, y)\right\} dx\, dy$$

$$+ \int_{\mathscr{S}_2} \left\{ -g_2(x, y)\phi_j(x, y) + g_1(x, y)\sum_{i=1}^{m} \gamma_i \phi_i(x, y)\phi_j(x, y)\right\} dS;$$

so we must have

$$(11.44) \quad 0 = \sum_{i=1}^{m} \left[\iint_{\mathscr{D}} \left\{ p(x, y)\frac{\partial \phi_i(x, y)}{\partial x}\frac{\partial \phi_j(x, y)}{\partial x} + q(x, y)\frac{\partial \phi_i(x, y)}{\partial y}\frac{\partial \phi_j(x, y)}{\partial y}\right.\right.$$

$$\left. - r(x, y)\phi_i(x, y)\phi_j(x, y)\right\} dx\, dy$$

$$\left. + \int_{\mathscr{S}_2} g_1(x, y)\phi_i(x, y)\phi_j(x, y)\, dS\right]\gamma_i$$

$$+ \iint_{\mathscr{D}} f(x, y)\phi_j(x, y)\, dx\, dy - \int_{\mathscr{S}_2} g_2(x, y)\phi_j(x, y)\, dS$$

for each $j = 1, 2, \ldots, n$. This set of equations can be written as a linear system:

$$(11.45) \qquad\qquad\qquad A\mathbf{c} = \mathbf{b},$$

where $A = (\alpha_{ij})$, $\mathbf{c} = (\gamma_1, \ldots, \gamma_n)^t$ and $\mathbf{b} = (\beta_1, \ldots, \beta_n)^t$ are defined by

$$(11.46) \quad \alpha_{ij} = \iint_{\mathscr{D}} \left[p(x, y) \frac{\partial \phi_i(x, y)}{\partial x} \frac{\partial \phi_j(x, y)}{\partial x} + q(x, y) \frac{\partial \phi_i(x, y)}{\partial y} \frac{\partial \phi_j(x, y)}{\partial y} \right.$$

$$\left. - r(x, y)\phi_i(x, y)\phi_j(x, y) \right] dx\, dy + \int_{\mathscr{S}_2} g_1(x, y)\phi_i(x, y)\phi_j(x, y)\, dS$$

for each $i = 1, 2, \ldots, n, j = 1, 2, \ldots, m$, and

$$(11.47) \quad \beta_i = -\iint_{\mathscr{D}} f(x, y)\phi_i(x, y)\, dx\, dy + \int_{\mathscr{S}_2} g_2(x, y)\phi_i(x, y)\, dS - \sum_{k=n+1}^{m} \alpha_{ik}\gamma_k$$

for each $i = 1, \ldots, n$.

The particular choice of basis functions is very important since the appropriate choice can often make the matrix A positive definite and banded. For our second-order problem, (11.38), we will assume that $\mathscr{D}$ is polygonal and that $\mathscr{S}_1$ is a contiguous set of straight lines so that $\mathscr{D} = D$. To begin the procedure we divide the region D into a collection of triangles $T_1, T_2, \ldots, T_M$ with the ith triangle having three vertices, or nodes, denoted by

$$V_j^{(i)} = (x_j^{(i)}, y_j^{(i)}) \qquad \text{for } j = 1, 2, 3.$$

To simplify the notation, we will write $V_j^{(i)}$ simply as $V_j = (x_j, y_j)$ when working with the fixed triangle T_i. With each vertex V_j we associate a linear polynomial

$$N_j^{(i)} \equiv N_j = a_j + b_j x + c_j y, \qquad \text{where} \quad N_j^{(i)}(x_k, y_k) = \begin{cases} 1, & \text{if } j = k, \\ 0, & \text{if } j \neq k. \end{cases}$$

This produces linear systems of the form.

$$\begin{bmatrix} 1 & x_1 & y_1 \\ 1 & x_2 & y_2 \\ 1 & x_3 & y_3 \end{bmatrix} \begin{bmatrix} a_j \\ b_j \\ c_j \end{bmatrix} = \begin{bmatrix} 0 \\ 1 \\ 0 \end{bmatrix},$$

with the element one occurring in the jth row in the vector on the right.

Let $E_1, \ldots, E_n$ be a labeling of the nodes lying in $D \cup \mathscr{S}$ in a left-to-right, top-to-bottom fashion. With each node E_k, we associate a function ϕ_k which is linear on each triangle, has the value one at E_k, and is zero at each of the other nodes. This choice makes ϕ_k identical to $N_j^{(i)}$ on triangle T_i when the node E_k is the vertex denoted $V_j^{(i)}$.

To generate the entries in the matrix A, consider as an example Fig. 11.12, the upper left portion of the region shown in Fig. 11.11.

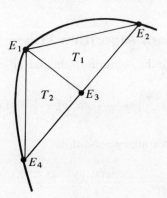

FIGURE 11.12

For simplicity we will assume that E_1 is not one of the nodes on $\mathscr{S}_2$. The relationship between the nodes and the vertices of the triangles for this portion is

$$E_1 = V_3^{(1)} = V_1^{(2)}, \quad E_4 = V_2^{(2)}, \quad E_3 = V_2^{(1)} = V_3^{(2)}, \quad \text{and} \quad E_2 = V_1^{(1)}.$$

Since ϕ_1 and ϕ_3 are both nonzero on T_1 and T_2, the entries $\alpha_{1,3} = \alpha_{3,1}$ are computed by:

$$\alpha_{1,3} = \iint\limits_{D} \left[p \frac{\partial \phi_1}{\partial x} \frac{\partial \phi_3}{\partial x} + q \frac{\partial \phi_1}{\partial y} \frac{\partial \phi_3}{\partial y} - r \phi_1 \phi_3 \right] dx \, dy$$

$$= \iint\limits_{T_1} \left[p \frac{\partial \phi_1}{\partial x} \frac{\partial \phi_3}{\partial x} + q \frac{\partial \phi_1}{\partial y} \frac{\partial \phi_3}{\partial y} - r \phi_1 \phi_3 \right] dx \, dy$$

$$+ \iint\limits_{T_2} \left[p \frac{\partial \phi_1}{\partial x} \frac{\partial \phi_3}{\partial x} + q \frac{\partial \phi_1}{\partial y} \frac{\partial \phi_3}{\partial y} - r \phi_1 \phi_3 \right] dx \, dy$$

$$= b_3^{(1)} b_2^{(1)} \iint\limits_{T_1} p \, dx \, dy + c_3^{(1)} c_2^{(1)} \iint\limits_{T_1} q \, dx \, dy$$

$$- \iint\limits_{T_1} r(a_3^{(1)} + b_3^{(1)} x + c_3^{(1)} y)(a_2^{(1)} + b_2^{(1)} x + c_2^{(1)} y) \, dx \, dy$$

$$+ b_1^{(2)} b_3^{(2)} \iint\limits_{T_2} p \, dx \, dy + c_1^{(2)} c_3^{(2)} \iint\limits_{T_2} q \, dx \, dy$$

$$- \iint\limits_{T_2} r(a_1^{(2)} + b_1^{(2)} x + c_1^{(2)} y)(a_3^{(2)} + b_3^{(2)} x + c_3^{(2)} y) \, dx \, dy.$$

Part of the entry β_1 is contributed by ϕ_1 restricted to T_1 and the remainder by ϕ_1 restricted to T_2. In general, an entry β_k has contributions from ϕ_k restricted to each of the triangles of which E_k is a vertex. In addition, nodes that lie on $\mathscr{S}_2$ will have line integrals adding to their entries in A and $\mathbf{b}$.

The following algorithm performs the finite-element method on a second-order elliptic differential equation. The algorithm sets all values of the matrix A and vector $\mathbf{b}$ initially at zero and, after all the integrations have been performed on all the triangles, adds these values to the appropriate entries in A and $\mathbf{b}$.

Finite-Element Algorithm 11.5

To approximate the solution to the partial-differential equation

$$\frac{\partial}{\partial x} \left(p(x, y) \frac{\partial u}{\partial x} \right) + \frac{\partial}{\partial y} \left(q(x, y) \frac{\partial u}{\partial y} \right) + r(x, y)u = f(x, y), \qquad (x, y) \in D$$

subject to the boundary conditions

$$u(x, y) = g(x, y), \qquad (x, y) \in \mathscr{S}_1$$

and

$$p(x, y) \frac{\partial u}{\partial x} \cos \theta_1 + q(x, y) \frac{\partial u}{\partial y} \cos \theta_2 + g_1(x, y)u(x, y) = g_2(x, y), (x, y) \in \mathscr{S}_2$$

where $\mathscr{S}_1 \cup \mathscr{S}_2$ is the boundary of D, and θ_1 and θ_2 are the direction angles of the normal to the boundary:

Step 0 Divide the region D into triangles $T_1, \ldots, T_M$ such that: $T_1, \ldots, T_K$ are the triangles with all vertices interior to D;
(*Note*: $K = 0$ *implies that no triangle is interior to D*);
$T_{K+1}, \ldots, T_N$ are the triangles with at least one edge on $\mathscr{S}_2$; $T_{N+1}, \ldots, T_M$ are the remaining triangles.
(*Note*: $M = N$ *implies that all triangles have edges on $\mathscr{S}_2$.*)
Label the three vertices of the triangle T_i by
$(x_1^{(i)}, y_1^{(i)})$, $(x_2^{(i)}, y_2^{(i)})$, and $(x_3^{(i)}, y_3^{(i)})$.
Label the nodes (vertices) $E_1, \ldots, E_m$ where
$E_1, \ldots, E_n$ are in $D \cup \mathscr{S}_2$ and $E_{n+1}, \ldots, E_m$ are on $\mathscr{S}_1$.
(*Note*: $n = m$ *implies that $\mathscr{S}_1$ contains no nodes.*)

INPUT integers K, N, M, n, m; vertices $(x_1^{(i)}, y_1^{(i)})$, $(x_2^{(i)}, y_2^{(i)})$, $(x_3^{(i)}, y_3^{(i)})$ for each $i = 1, \ldots, M$; nodes E_j for each $j = 1, \ldots, m$.
(*Note: all that is needed is a means of corresponding a vertex $(x_k^{(i)}, y_k^{(i)})$ to a node $E_j = (x_j, y_j)$.*)

OUTPUT constants $\gamma_1, \ldots, \gamma_m$; $a_j^{(i)}$, $b_j^{(i)}$, $c_j^{(i)}$ for each $j = 1, 2, 3$ and $i = 1, \ldots, M$.

Step 1 For $l = n + 1, \ldots, m$ set $\gamma_l = g(x_l, y_l)$. $(E_l = (x_l, y_l))$.

Step 2 For $i = 1, \ldots, n$
    set $\beta_i = 0$;
    for $j = 1, \ldots, n$ set $\alpha_{i,j} = 0$.

Step 3 For $i = 1, \ldots, M$

$$\text{set } \Delta_i = \det \begin{vmatrix} 1 & x_1^{(i)} & y_1^{(i)} \\ 1 & x_2^{(i)} & y_2^{(i)} \\ 1 & x_3^{(i)} & y_3^{(i)} \end{vmatrix};$$

$$a_1^{(i)} = \frac{x_2^{(i)} y_3^{(i)} - y_2^{(i)} x_3^{(i)}}{\Delta_i}; \quad b_1^{(i)} = \frac{y_2^{(i)} - y_3^{(i)}}{\Delta_i}; \quad c_1^{(i)} = \frac{x_3^{(i)} - x_2^{(i)}}{\Delta_i};$$

$$a_2^{(i)} = \frac{x_3^{(i)} y_1^{(i)} - y_3^{(i)} x_1^{(i)}}{\Delta_i}; \quad b_2^{(i)} = \frac{y_3^{(i)} - y_1^{(i)}}{\Delta_i}; \quad c_2^{(i)} = \frac{x_1^{(i)} - x_3^{(i)}}{\Delta_i};$$

$$a_3^{(i)} = \frac{x_1^{(i)} y_2^{(i)} - y_1^{(i)} x_2^{(i)}}{\Delta_i}; \quad b_3^{(i)} = \frac{y_1^{(i)} - y_2^{(i)}}{\Delta_i}; \quad c_3^{(i)} = \frac{x_2^{(i)} - x_1^{(i)}}{\Delta_i};$$

for $j = 1, 2, 3$
    define $N_j^{(i)}(x, y) = a_j^{(i)} + b_j^{(i)} x + c_j^{(i)} y$.

Step 4 For $i = 1, \ldots, M$ (*The integrals in Steps 4 and 5 can be evaluated using numerical integration procedures.*)

for $j = 1, 2, 3$

for $k = 1, \ldots, j$ (*Compute all double integrals over the triangles.*)

set $z_{j,k}^{(i)} = b_j^{(i)} b_k^{(i)} \iint\limits_{T_i} p(x, y)\, dx\, dy + c_j^{(i)} c_k^{(i)} \iint\limits_{T_i} q(x, y)\, dx\, dy$

$- \iint\limits_{T_i} r(x, y) N_j^{(i)}(x, y) N_k^{(i)}(x, y)\, dx\, dy;$

set $H_j^{(i)} = - \iint\limits_{T_i} f(x, y) N_j^{(i)}(x, y)\, dx\, dy.$

Step 5 For $i = K + 1, \ldots, N$ (*Compute all line integrals.*)

for $j = 1, 2, 3$

for $k = 1, \ldots, j$

set $J_{j,k}^{(i)} = \int_{\mathscr{S}_2} g_1(x, y) N_j^{(i)}(x, y) N_k^{(i)}(x, y)\, dS;$

set $I_j^{(i)} = \int_{\mathscr{S}_2} g_2(x, y) N_j^{(i)}(x, y)\, dS.$

Step 6 For $i = 1, \ldots, M$ do Steps 7–12.   (*Assembling the integrals over each triangle into the linear system.*)

Step 7 For $k = 1, 2, 3$ do Steps 8–12.

Step 8 Find l so that $E_l = (x_k^{(i)}, y_k^{(i)})$.

Step 9 If $k > 1$ then for $j = 1, \ldots, k - 1$ do Steps 10, 11.

Step 10 Find t so that $E_t = (x_j^{(i)}, y_j^{(i)})$.

Step 11 If $l \le n$ then

if $t \le n$ then set $\alpha_{lt} = \alpha_{lt} + z_{k,j}^{(i)};$

$\alpha_{tl} = \alpha_{tl} + z_{k,j}^{(i)};$

else set $\beta_l = \beta_l - \gamma_t z_{k,j}^{(i)};$

else

if $t \le n$ then set $\beta_t = \beta_t - \gamma_l z_{k,j}^{(i)}.$

Step 12 If $l \le n$ then set $a_{ll} = \alpha_{ll} + z_{k,k}^{(i)};$

$\beta_l = \beta_l + H_k^{(i)}.$

Step 13 For $i = K + 1, \ldots, N$ do Steps 14–19.   (*Assembling the line integrals into the linear system.*)

Step 14 For $k = 1, 2, 3$ do Steps 15–19.

Step 15 Find l so that $E_l = (x_k^{(i)}, y_k^{(i)})$.

Step 16 If $k > 1$ then for $j = 1, \ldots, k - 1$ do Steps 17, 18.

Step 17 Find t so that $E_t = (x_j^{(i)}, y_j^{(i)})$.

Step 18   If $l \leq n$ then
if $t \leq n$ then set $\alpha_{lt} = \alpha_{lt} + J_{k,j}^{(i)}$;
$$\alpha_{tl} = \alpha_{tl} + J_{k,j}^{(i)};$$
else set $\beta_l = \beta_l - \gamma_t J_{k,j}^{(i)}$;
else
if $t \leq n$ then set $\beta_t = \beta_t - \gamma_l J_{k,j}^{(i)}$.

Step 19   If $l \leq n$ then set $\alpha_{ll} = \alpha_{ll} + J_{k,k}^{(i)}$;
$$\beta_l = \beta_l + I_k^{(i)}.$$

Step 20   Solve the linear system $A\mathbf{c} = \mathbf{b}$ where $A = (\alpha_{l,t})$, $\mathbf{b} = (\beta_l)$ and $\mathbf{c} = (\gamma_t)$ for $1 \leq l, t \leq n$.

Step 21   OUTPUT $(\gamma_1, \ldots, \gamma_m)$.
(*For each* $k = 1, \ldots, m$ *let* $\phi_k = N_j^{(i)}$ *on* T_i *if* $E_k = (x_j^{(i)}, y_j^{(i)})$.)

Then $\phi(x, y) = \displaystyle\sum_{k=1}^{m} \gamma_k \phi_k(x, y)$ *approximates* $u(x, y)$ *on* $D \cup \mathscr{S}_1 \cup \mathscr{S}_2$.)

Step 22   For $i = 1, \ldots, M$
for $j = 1, 2, 3$ OUTPUT $(a_j^{(i)}, b_j^{(i)}, c_j^{(i)})$.

Step 23   STOP.   (*Procedure is complete.*)

EXAMPLE 1   The temperature, $u(x, y)$, in a two-dimensional region D satisfies Laplace's equation

$$\frac{\partial^2 u}{\partial x^2}(x, y) + \frac{\partial^2 u}{\partial y^2}(x, y) = 0 \qquad \text{on } D.$$

Consider the region D shown in Fig. 11.13 and suppose that the following boundary conditions are given:

$$u(x, y) = 4 \qquad \text{for } (x, y) \in L_6 \text{ and } (x, y) \in L_7,$$

$$\frac{\partial u}{\partial n}(x, y) = x \qquad \text{for } (x, y) \in L_2 \text{ and } (x, y) \in L_4,$$

$$\frac{\partial u}{\partial n}(x, y) = y \qquad \text{for } (x, y) \in L_5,$$

$$\frac{\partial u}{\partial n}(x, y) = \frac{x + y}{\sqrt{2}} \qquad \text{for } (x, y) \in L_1 \text{ and } (x, y) \in L_3,$$

where $\partial u / \partial n$ denotes the normal derivative to the boundary of the region D at the point (x, y).

We will first subdivide D into triangles with the labeling suggested in Step 0 of the algorithm. For this example, $\mathscr{S}_1 = L_6 \cup L_7$ and $\mathscr{S}_2 = L_1 \cup L_2 \cup L_3 \cup L_4 \cup L_5$. The labeling of triangles is shown in Fig. 11.14.

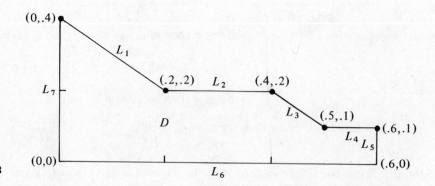

FIGURE 11.13

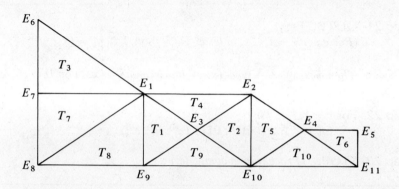

FIGURE 11.14

The boundary condition $u(x, y) = 4$ on L_6 and L_7 implies that $\gamma_l = 4$ when $l = 6, 7, \ldots, 11$. To determine the values of γ_l for $l = 1, 2, \ldots, 5$, we apply the remaining steps of the algorithm and generate the matrix

$$
A = \begin{bmatrix}
2.5 & 0 & -1 & 0 & 0 \\
0 & 1.5 & -1 & -.5 & 0 \\
-1 & -1 & 4 & 0 & 0 \\
0 & -.5 & 0 & 2.5 & -.5 \\
0 & 0 & 0 & -.5 & 1
\end{bmatrix}
$$

and the vector

$$
\mathbf{b} = \begin{bmatrix}
6.0\overline{666} \\
0.06\overline{33} \\
8.0000 \\
6.05\overline{66} \\
2.031\overline{6}
\end{bmatrix}.
$$

The solution to the equation $A\mathbf{c} = \mathbf{b}$ is:

$$
\mathbf{c} = \begin{bmatrix}
\gamma_1 \\
\gamma_2 \\
\gamma_3 \\
\gamma_4 \\
\gamma_5
\end{bmatrix} = \begin{bmatrix}
4.0383 \\
4.0782 \\
4.0291 \\
4.0496 \\
4.0565
\end{bmatrix},
$$

which gives the following approximation to the solution of Laplace's equation and the boundary conditions on the respective triangles:

T_1 : $\phi(x, y) = 4.0383(1 - 5x + 5y) + 4.0291(-2 + 10x) + 4(2 - 5x - 5y)$,

T_2 : $\phi(x, y) = 4.0782(-2 + 5x + 5y) + 4.0291(4 - 10x) + 4(-1 + 5x - 5y)$,

T_3 : $\phi(x, y) = 4(-1 + 5y) + 4(2 - 5x - 5y) + 4.0383(5x)$,

T_4 : $\phi(x, y) = 4.0383(1 - 5x + 5y) + 4.0782(-2 + 5x + 5y) + 4.0291(2 - 10y)$,

T_5 : $\phi(x, y) = 4.0782(2 - 5x + 5y) + 4.0496(-4 + 10x) + 4(3 - 5x - 5y)$,

T_6 : $\phi(x, y) = 4.0496(6 - 10x) + 4.0565(-6 + 10x + 10y) + 4(1 - 10y)$.

T_7 : $\phi(x, y) = 4(-5x + 5y) + 4.0383(5x) + 4(1 - 5y)$,

T_8 : $\phi(x, y) = 4.0383(5y) + 4(1 - 5x) + 4(5x - 5y)$,

T_9 : $\phi(x, y) = 4.0291(10y) + 4(2 - 5x - 5y) + 4(-1 + 5x - 5y)$,

T_{10}: $\phi(x, y) = 4.0496(10y) + 4(3 - 5x - 5y) + 4(-2 + 5x - 5y)$.

The actual solution to the boundary-value problem is $u(x, y) = xy + 4$. Table 11.6 compares the value of u to the value of ϕ at E_i for each $i = 1, \ldots, 5$.   □

TABLE 11.6

x	y	$\phi(x, y)$	$u(x, y)$	$\lvert \phi(x, y) - u(x, y) \rvert$
.2	.2	4.0383	4.04	.0017
.4	.2	4.0782	4.08	.0018
.3	.1	4.0291	4.03	.0009
.5	.1	4.0496	4.05	.0004
.6	.1	4.0565	4.06	.0035

Typically, the error for second-order problems with smooth coefficient functions is $O(h^2)$, where h is the maximum diameter of the triangular elements. Piecewise bilinear basis functions on rectangular elements are also expected to give $O(h^2)$ results, where h is the maximum length of the rectangular elements. Other classes of basis functions can be used to give $O(h^4)$ results, but the construction is more complex.

Efficient error theorems for finite-element methods are difficult to state and apply because the accuracy of the approximation depends on the continuity properties of the solution and the regularity of the boundary.

The finite-element method can also be applied to parabolic and hyperbolic partial-differential equations, but the minimization procedure is more difficult and the finite-difference procedures are more competitive.

A good survey on the advantages and techniques of the finite-element method applied to various physical problems can be found in a paper by Fix [38]. For a more extensive discussion, refer to the book by Strang and Fix [85] or Zienkiewicz [94].

Exercise Set 11.5

1. Use Algorithm 11.5 to approximate the solution to the following partial differential equation (see the following figure).

$$\frac{\partial}{\partial x}\left(y^2 \frac{\partial u}{\partial x}\right) + \frac{\partial}{\partial y}\left(y^2 \frac{\partial u}{\partial y}\right) - yu = -x, \qquad (x, y) \in D,$$

$$u(x, .5) = 2x, \qquad 0 \le x \le .5,$$

$$u(0, y) = 0, \qquad .5 \le y \le 1,$$

$$y^2 \frac{\partial u}{\partial x} \cos\theta_1 + y^2 \frac{\partial u}{\partial y}\cos\theta_2 = \frac{\sqrt{2}}{2}(y - x) \qquad \text{for } (x, y) \in \mathscr{S}_2.$$

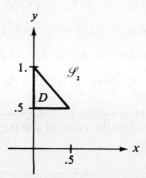

Let $M = 2$; T_1 have vertices $(0, .5)$, $(.25, .75)$, $(0, 1)$; and T_2 have vertices $(0, .5)$, $(.5, .5)$, and $(.25, .75)$.

2. Repeat Exercise 1. using the triangles:

$$T_1: \quad (0, .75), (0, 1), (.25, .75);$$

$$T_2: \quad (.25, .5), (.25, .75), (.5, .5);$$

$$T_3: \quad (0, .5), (0, .75), (.25, .75);$$

$$T_4: \quad (0, .5), (.25, .5), (.25, .75).$$

3. Approximate the solution to the partial-differential equation

$$\frac{\partial^2 u(x, y)}{\partial x^2} + \frac{\partial^2 u(x, y)}{\partial y^2} - 12.5\pi^2 u(x, y) = -25\pi^2 \sin\frac{5\pi}{2}x \sin\frac{5\pi}{2}y, \qquad 0 < x, \ y < .4,$$

subject to the Dirichlet boundary condition

$$u(x, y) = 0,$$

using the Finite-Element Algorithm 11.5 with the elements given in the following figure. Compare the approximate solution to the actual solution

$$u(x, y) = \sin\frac{5\pi}{2}x \sin\frac{5\pi}{2}y$$

at the interior vertices and at the points $(.125, .125)$, $(.125, .25)$, $(.25, .125)$, and $(.25, .25)$.

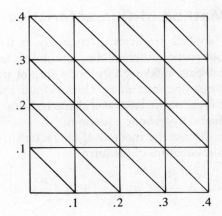

4. Repeat Exercise 3, with $f(x, y) = -25\pi^2 \cos \dfrac{5\pi}{2} x \cos \dfrac{5\pi}{2} y$, using the Neumann boundary condition

$$\frac{\partial u}{\partial n}(x, y) = 0.$$

The actual solution for this problem is

$$u(x, y) = \cos \frac{5\pi}{2} x \cos \frac{5\pi}{2} y.$$

5. Construct an algorithm for the finite element method using rectangular elements and basis functions of the form $a + bx + cy + dxy$.

6. Use the algorithm developed in Exercise 5 to solve the problem discussed in

   a) Exercise 3. b) Exercise 4.

7. A silver plate in the shape of a trapezoid (see the following figure) has heat being uniformly generated at each point at the rate $q = 1.5$ cal/cm^3 · sec. The steady-state temperature $u = u(x, y)$ of the plate satisfies Poisson's equation

$$\frac{\partial^2 u}{\partial x^2} + \frac{\partial^2 u}{\partial y^2} = \frac{-q}{k},$$

where k, the thermal conductivity, is 1.04 cal/cm · deg · sec. Assume that the temperature is held at 15°C on L_2, that heat is lost on the slanted edges L_1 and L_3 according to the boundary condition $\partial u/\partial n = 4$, and that no heat is lost on L_4, that is, $\partial u/\partial n = 0$. Approximate the temperature of the plate at $(1, 0)$, $(4, 0)$ and $(\frac{5}{2}, \sqrt{3}/2)$ by using Algorithm 11.5.

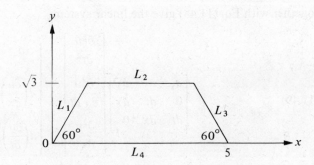

11.6   The Method of Characteristics

Certain applied problems, particularly in the area of solid mechanics, involve hyperbolic partial-differential equations that fail to have continuous initial data. In such problems, difference methods are not used because these methods tend to smooth out discontinuities. One method that takes these discontinuities into account is called the **method of characteristics**. This method can also be applied to more general hyperbolic equations.

To discuss the method of characteristics, we need to consider the general hyperbolic equation of the form

$$(11.48) \qquad a_1 \frac{\partial^2 u}{\partial t^2} + a_2 \frac{\partial^2 u}{\partial x \, \partial t} + a_3 \frac{\partial^2 u}{\partial x^2} = a_4, \qquad 0 < x < l, \quad t > 0,$$

where a_i denotes a function in the variables x, t, u, $(\partial u/\partial x)$ and $(\partial u/\partial t)$, for $i = 1$, 2, 3, 4. A problem of this type is called **quasilinear** since it is linear with respect to the second partial derivatives. The boundary and initial conditions are given as

$$u(x, 0) = f(x), \qquad 0 \le x \le l,$$

$$\frac{\partial u}{\partial t}(x, 0) = g(x), \qquad 0 \le x \le l,$$

and
$$u(0, t) = u(l, t) = 0, \qquad t > 0.$$

We will also assume that $f'(x)$ exists so that $(\partial u/\partial x)(x, 0) = f'(x)$ whenever $0 \le x \le l$.

Since u, $(\partial u/\partial t)$ and $(\partial u/\partial x)$ are known for $0 \le x \le l$ and $t = 0$, we seek conditions under which

$$\frac{\partial^2 u}{\partial x^2}, \qquad \frac{\partial^2 u}{\partial x \, \partial t}, \qquad \text{and} \qquad \frac{\partial^2 u}{\partial t^2}$$

can be found uniquely to satisfy Eq. (11.48). If the first partials are continuous, then the differentials

$$d\left(\frac{\partial u}{\partial x}\right) \qquad \text{and} \qquad d\left(\frac{\partial u}{\partial t}\right)$$

must exist. The equations

$$d\left(\frac{\partial u}{\partial x}\right) = \frac{\partial^2 u}{\partial x^2} \, dx + \frac{\partial^2 u}{\partial x \, \partial t} \, dt, \qquad d\left(\frac{\partial u}{\partial t}\right) = \frac{\partial^2 u}{\partial x \, \partial t} \, dx + \frac{\partial^2 u}{\partial t^2} \, dt,$$

together with Eq. (11.48) give the linear system

$$(11.49) \qquad \begin{bmatrix} a_1 & a_2 & a_3 \\ 0 & dt & dx \\ dt & dx & 0 \end{bmatrix} \begin{bmatrix} \dfrac{\partial^2 u}{\partial t^2} \\[2ex] \dfrac{\partial^2 u}{\partial x \, \partial t} \\[2ex] \dfrac{\partial^2 u}{\partial x^2} \end{bmatrix} = \begin{bmatrix} a_4 \\[2ex] d\left(\dfrac{\partial u}{\partial x}\right) \\[2ex] d\left(\dfrac{\partial u}{\partial t}\right) \end{bmatrix}.$$

This system can be solved uniquely for

$$\frac{\partial^2 u}{\partial t^2}, \qquad \frac{\partial^2 u}{\partial x \, \partial t}, \qquad \text{and} \qquad \frac{\partial^2 u}{\partial x^2}$$

unless

(11.50) $\qquad \det \begin{bmatrix} a_1 & a_2 & a_3 \\ 0 & dt & dx \\ dt & dx & 0 \end{bmatrix} = -a_1(dx)^2 + a_2 \, dt \, dx - a_3(dt)^2 = 0,$

that is, unless

$$a_3 \left(\frac{dt}{dx} \right)^2 - a_2 \frac{dt}{dx} + a_1 = 0.$$

Solving for dt/dx implies that the solution is unique unless

$$\frac{dt}{dx} = \frac{a_2 \pm \sqrt{a_2^2 - 4a_1 a_3}}{2a_3}.$$

This equation for dt/dx depends not only on the point (x, t), but also on the values of u, $(\partial u/\partial x)$, and $(\partial u/\partial t)$ at (x, t), and describes the slope of curves in the x, t-plane called **characteristic curves**. The discriminant of this equation,

$$\Delta = a_2^2 - 4a_1 a_3,$$

indicates whether there are zero, one, or two characteristic curves through each point. As a means of classifying second-order partial-differential equations, elliptic equations have $\Delta < 0$ and no real characteristics, parabolic equations have $\Delta = 0$ and one real characteristic curve through each point, while the hyperbolic equations we are considering in this section have $\Delta > 0$ and consequently have two characteristic curves passing through each point.

For the hyperbolic equations, we denote the characteristic directions at a point P with coordinates (x, t) by:

(11.51) $\qquad m_1(P) = m_1(x, t) = \dfrac{a_2 + \sqrt{a_2^2 - 4a_1 a_3}}{2a_3}$

and

(11.52) $\qquad m_2(P) = m_2(x, t) = \dfrac{a_2 - \sqrt{a_2^2 - 4a_1 a_3}}{2a_3}.$

Even though the notation in equations (11.51) and (11.52) does not express it explicitly, it should be kept in mind that these values also depend on

$$u(x, t), \qquad \frac{\partial u}{\partial x}(x, t), \qquad \text{and} \qquad \frac{\partial u}{\partial t}(x, t).$$

The method of characteristics is derived by examining what happens to the other determinants associated with the linear system (11.49) when these characteristic values are assumed, or, equivalently, when Eq. (11.50) holds. When the determinant in Eq. (11.50) is zero, Cramer's Rule (see Exercise 12 in Section 6.3) implies that a solution to the system cannot occur unless we also have:

$$\det \begin{bmatrix} a_4 & a_2 & a_3 \\ d\left(\dfrac{\partial u}{\partial x}\right) & dt & dx \\ d\left(\dfrac{\partial u}{\partial t}\right) & dx & 0 \end{bmatrix} = a_2\, d\left(\dfrac{\partial u}{\partial t}\right) dx + a_3\left(d\left(\dfrac{\partial u}{\partial x}\right) dx - d\left(\dfrac{\partial u}{\partial t}\right) dt\right) - a_4 (dx)^2 = 0,$$

$$\det \begin{bmatrix} a_1 & a_2 & a_4 \\ 0 & dt & d\left(\dfrac{\partial u}{\partial x}\right) \\ dt & dx & d\left(\dfrac{\partial u}{\partial t}\right) \end{bmatrix} = a_1\left(d\left(\dfrac{\partial u}{\partial t}\right) dt - d\left(\dfrac{\partial u}{\partial x}\right) dx\right) + a_2\, d\left(\dfrac{\partial u}{\partial x}\right) dt - a_4 (dt)^2 = 0,$$

and

$$\det \begin{bmatrix} a_1 & a_4 & a_3 \\ 0 & d\left(\dfrac{\partial u}{\partial x}\right) & dx \\ dt & d\left(\dfrac{\partial u}{\partial t}\right) & 0 \end{bmatrix} = -a_1\, d\left(\dfrac{\partial u}{\partial t}\right) dx - a_3\, d\left(\dfrac{\partial u}{\partial x}\right) dt + a_4\, dx\, dt = 0.$$

In particular, the third equation can be rewritten as

$$(11.53) \qquad a_1 d\left(\frac{\partial u}{\partial t}\right) + a_3 d\left(\frac{\partial u}{\partial x}\right)\frac{dt}{dx} - a_4\, dt = 0.$$

The method of characteristics uses this equation, together with equations (11.51) and (11.52) and the fact that the total differential of u can be written as

$$(11.54) \qquad\qquad du = \frac{\partial u}{\partial t}\, dt + \frac{\partial u}{\partial x}\, dx.$$

To describe the operation of the method, suppose $P = (x_1, t_1)$ and $Q = (x_2, t_2)$, with $t_1 = t_2 = 0$, are two points along the x-axis between 0 and l. From P we draw the straight lines that are tangent to the characteristic curves at P—that is, the lines with slopes $m_1(P)$ and $m_2(P)$; likewise, from Q we draw the lines with slopes $m_1(Q)$ and $m_2(Q)$. We expect that two of these lines, say the lines with slope $m_1(P)$ and $m_2(Q)$, intersect at some point $R_0 = (x^{(0)}, t^{(0)})$ with positive t-coordinate. This point is the initial approximation to R, the intersection of two of the characteristic curves through P and Q (see Fig. 11.15).

In order to better our approximation to R, we first find approximations to

$$u(R_0), \qquad \frac{\partial u}{\partial x}(R_0), \qquad \text{and} \qquad \frac{\partial u}{\partial t}(R_0).$$

The method used is to average the values of m_1, a_1, a_3, and a_4 at P and R_0 to form one part of the approximation, and the values of m_2, a_1, a_3, and a_4 at Q and R_0 to form the remaining portion. For example, approximations for

$$\frac{\partial u}{\partial t}(R_0) \qquad \text{and} \qquad \frac{\partial u}{\partial x}(R_0)$$

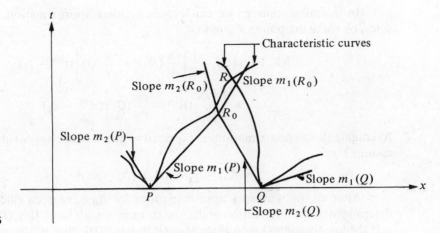

FIGURE 11.15

are obtained by using the averages and differences at the points P and R_0 in Eq. (11.53) to obtain:

$$(11.55) \qquad \tfrac{1}{4}[a_3(R_0) + a_3(P)][m_1(R_0) + m_1(P)]\left[\frac{\partial u}{\partial x}(R_0) - \frac{\partial u}{\partial x}(P)\right]$$

$$+ \tfrac{1}{2}[a_1(R_0) + a_1(P)]\left[\frac{\partial u}{\partial t}(R_0) - \frac{\partial u}{\partial t}(P)\right] = \tfrac{1}{2}[a_4(R_0) + a_4(P)](t^{(0)} - t_1),$$

and the averages and differences at the points Q and R_0 in the same equation to give

$$(11.56) \qquad \tfrac{1}{4}[a_3(R_0) + a_3(Q)][m_2(R_0) + m_2(Q)]\left[\frac{\partial u}{\partial x}(R_0) - \frac{\partial u}{\partial x}(Q)\right]$$

$$+ \tfrac{1}{2}[a_1(R_0) + a_1(Q)]\left[\frac{\partial u}{\partial t}(R_0) - \frac{\partial u}{\partial t}(Q)\right] = \tfrac{1}{2}[a_4(R_0) + a_4(Q)](t^{(0)} - t_2).$$

These equations can be solved simultaneously for the approximations to

$$\frac{\partial u}{\partial x}(R_0) \qquad \text{and} \qquad \frac{\partial u}{\partial t}(R_0).$$

Once these are calculated, we can find one approximation, $u^{(1)}(R_0)$, to $u(R_0)$ by assuming that $du = u^{(1)}(R_0) - u(P)$ and use the averages and Eq. (11.54) to obtain

$$u^{(1)}(R_0) - u(P) = \frac{1}{2}\left[\frac{\partial u}{\partial t}(R_0) + \frac{\partial u}{\partial t}(P)\right](t^{(0)} - t_1)$$

$$+ \frac{1}{2}\left[\frac{\partial u}{\partial x}(R_0) + \frac{\partial u}{\partial x}(P)\right](x^{(0)} - x_1)$$

or $\qquad u^{(1)}(R_0) = u(P) + \frac{1}{2}\left[\frac{\partial u}{\partial t}(R_0) + \frac{\partial u}{\partial t}(P)\right](t^{(0)} - t_1)$

$$+ \frac{1}{2}\left[\frac{\partial u}{\partial x}(R_0) + \frac{\partial u}{\partial x}(P)\right](x^{(0)} - x_1).$$

In a similar manner we can obtain another approximation, $u^{(2)}(R_0)$, for $u(R_0)$ by using the points R_0 and Q,

$$u^{(2)}(R_0) = u(Q) + \frac{1}{2}\left[\frac{\partial u}{\partial t}(R_0) + \frac{\partial u}{\partial t}(Q)\right](t^{(0)} - t_2)$$

$$+ \frac{1}{2}\left[\frac{\partial u}{\partial x}(R_0) + \frac{\partial u}{\partial x}(Q)\right](x^{(0)} - x_2).$$

Averaging these approximations is expected to give a better approximation; so we assume that

$$u(R_0) = \tfrac{1}{2}[u^{(1)}(R_0) + u^{(2)}(R_0)].$$

After all the necessary approximations for R_0 have been calculated R_1 is determined as the intersection of the line through P with slope $\frac{1}{2}[m_1(R_0) + m_1(P)]$ and the line through Q with slope $\frac{1}{2}[m_2(R_0) + m_2(Q)]$; that is, the coordinates of $R_1 = (x^{(1)}, t^{(1)})$ are obtained by solving the equations

$$\tfrac{1}{2}[m_1(R_0) + m_1(P)] = \frac{t^{(1)} - t_1}{x^{(1)} - x_1} \quad \text{and} \quad \tfrac{1}{2}[m_2(R_0) + m_2(Q)] = \frac{t^{(1)} - t_2}{x^{(1)} - x_2}$$

simultaneously for $x^{(1)}$ and $t^{(1)}$. (See Fig. 11.16.)

Approximations to

$$u(R_1), \qquad \frac{\partial u}{\partial x}(R_1), \qquad \text{and} \qquad \frac{\partial u}{\partial t}(R_1)$$

can be obtained by the same methods that were used to find approximations to

$$u(R_0), \qquad \frac{\partial u}{\partial x}(R_0), \qquad \text{and} \qquad \frac{\partial u}{\partial t}(R_0).$$

Continuing in this manner will provide a sequence of points, which we assume will converge to R, the point at the intersection of the characteristic curves, and will also provide approximations to the values of u, $\partial u/\partial x$, and $\partial u/\partial t$ at that point.

The whole procedure is repeated for new values of P and Q, and then repeated again with the values at the new points, in order to move farther up the time

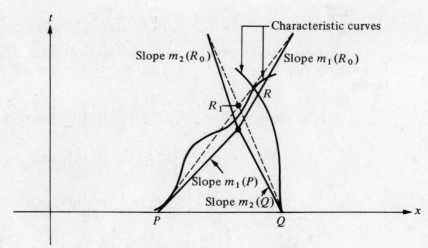

FIGURE 11.16

axis. In this manner a mesh of points is obtained where u and its partial derivatives are approximated, as shown in Fig. 11.17. The following algorithm details the procedure.

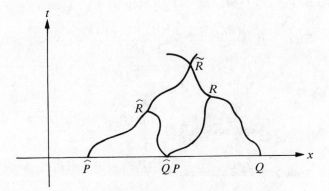

FIGURE 11.17

Method of Characteristics Algorithm 11.6

To approximate the solution to the hyperbolic partial-differential equation

$$a_1(P)\frac{\partial^2 u}{\partial t^2} + a_2(P)\frac{\partial^2 u}{\partial x\,\partial t} + a_3(P)\frac{\partial^2 u}{\partial x^2} = a_4(P), \qquad 0 < x < l, \quad t > 0,$$

subject to the boundary conditions

$$u(0, t) = u(l, t) = 0, \qquad 0 < t,$$

and the initial conditions

$$u(x, 0) = f(x), \qquad 0 \le x \le l,$$

$$\frac{\partial u(x, 0)}{\partial t} = g(x), \qquad 0 \le x \le l,$$

where $P = (x, t, u(x, t), u_t(x, t), u_x(x, t))$:

INPUT endpoint l; integers N, M; tolerance TOL; maximum number of iterations $KBOUND$.

OUTPUT characteristic mesh $(x_{i,j}, t_{i,j})$, approximations $w_{i,j}$ to $u(x_{i,j}, t_{i,j})$, $w'_{i,j}$ to $\dfrac{\partial u}{\partial x}(x_{i,j}, t_{i,j})$, and $\dot{w}_{i,j}$ to $\dfrac{\partial u}{\partial t}(x_{i,j}, t_{i,j})$ for each $i = 0, \ldots, M$ and $j = 0, \ldots, N$ or a message that the maximum number of iterations was exceeded.

Step 1 Set $h = l/M$.

Step 2 For $i = 0, \ldots, M$ set $x_{i,0} = ih$;

$\qquad t_{i,0} = 0$;

$\qquad w_{i,0} = f(x_{i,0})$;

$\qquad \dot{w}_{i,0} = g(x_{i,0})$;

$\qquad w'_{i,0} = f'(x_{i,0})$;

$\qquad P_{i,0} = (x_{i,0}, t_{i,0}, w_{i,0}, \dot{w}_{i,0}, w'_{i,0})$.

Step 3 For $j = 0, \ldots, N - 1$ do Steps 4–15.   (*Time control.*)

Step 4 If j is odd then set $r = 0$
else set $r = 1$.

Step 5 For $i = 0, \ldots, M - 1$ do Steps 6–13.   (*Space control.*)

Step 6 If $\dfrac{a_2(P_{i,j}) + \sqrt{a_2^2(P_{i,j}) - 4a_1(P_{i,j})a_3(P_{i,j})}}{2a_3(P_{i,j})} > 0$

then set $m_1(P_{i,j}) = \dfrac{a_2(P_{i,j}) + \sqrt{a_2^2(P_{i,j}) - 4a_1(P_{i,j})a_3(P_{i,j})}}{2a_3(P_{i,j})}$

else set $m_1(P_{i,j}) = \dfrac{a_2(P_{i,j}) - \sqrt{a_2^2(P_{i,j}) - 4a_1(P_{i,j})a_3(P_{i,j})}}{2a_3(P_{i,j})}$.

Step 7 If $\dfrac{a_2(P_{i+1,j}) - \sqrt{a_2^2(P_{i+1,j}) - 4a_1(P_{i+1,j})a_3(P_{i+1,j})}}{2a_3(P_{i+1,j})} < 0$

then set

$$m_2(P_{i+1,j}) = \dfrac{a_2(P_{i+1,j}) - \sqrt{a_2^2(P_{i+1,j}) - 4a_1(P_{i+1,j})a_3(P_{i+1,j})}}{2a_3(P_{i+1,j})}$$

else set

$$m_2(P_{i+1,j}) = \dfrac{a_2(P_{i+1,j}) + \sqrt{a_2^2(P_{i+1,j}) - 4a_1(P_{i+1,j})a_3(P_{i+1,j})}}{2a_3(P_{i+1,j})}.$$

Step 8 Set $k = 0$;

$$x_{r,j+1}^{(0)} = \dfrac{t_{i+1,j} - t_{i,j} + m_1(P_{i,j})x_{i,j} - m_2(P_{i+1,j})x_{i+1,j}}{m_1(P_{i,j}) - m_2(P_{i+1,j})};$$

$$t_{r,j+1}^{(0)} = m_1(P_{i,j})(x_{r,j+1}^{(0)} - x_{i,j}) + t_{i,j};$$

$$P_{r,j+1} = (x_{r,j+1}^{(0)}, t_{r,j+1}^{(0)}, w_{i,j}, \dot{w}_{i,j}, w'_{i,j});$$

$$FLAG = 1.$$

Step 9 While $FLAG = 1$ do Steps 10–13.

Step 10 Set $A = \frac{1}{2}[a_1(P_{r,j+1}) + a_1(P_{i,j})]$;
$B = \frac{1}{4}[a_3(P_{r,j+1}) + a_3(P_{i,j})][m_1(P_{r,j+1}) + m_1(P_{i,j})]$;
$C = \frac{1}{2}[a_1(P_{r,j+1}) + a_1(P_{i+1,j})]$;
$D = \frac{1}{4}[a_3(P_{r,j+1}) + a_3(P_{i+1,j})]$
$\quad \times [m_2(P_{r,j+1}) + m_2(P_{i+1,j})]$;
$E = \frac{1}{2}[a_4(P_{r,j+1}) + a_4(P_{i,j})][t_{r,j+1}^{(k)} - t_{i,j})]$
$\quad + A\dot{w}_{i,j} + Bw'_{i,j}$;
$F = \frac{1}{2}[a_4(P_{r,j+1}) + a_4(P_{i+1,j})][t_{r,j+1}^{(k)} - t_{i+1,j}]$
$\quad + C\dot{w}_{i+1,j} + Dw'_{i+1,j}$;

$$\dot{w}_{r,j+1} = \dfrac{ED - FB}{AD - BC};$$

$$w'_{r,j+1} = \dfrac{AF - CE}{AD - BC};$$

$$H = \tfrac{1}{2}[\dot{w}_{r,j+1} + \dot{w}_{i,j}][t^{(k)}_{r,j+1} - t_{i,j}]$$
$$+ \tfrac{1}{2}[w'_{r,j+1} + w'_{i,j}][x^{(k)}_{r,j+1} - x_{i,j}];$$
$$I = \tfrac{1}{2}[\dot{w}_{r,j+1} + \dot{w}_{i+1,j}][t^{(k)}_{r,j+1} - t_{i+1,j}]$$
$$+ \tfrac{1}{2}[w'_{r,j+1} + w'_{i+1,j}][x^{(k)}_{r,j+1} - x_{i+1,j}];$$
$$w_{r,j+1} = \tfrac{1}{2}[w_{i,j} + H + w_{i+1,j} + I];$$
$$P_{r,j+1} = (x^{(k)}_{r,j+1}, t^{(k)}_{r,j+1}, w_{r,j+1}, \dot{w}_{r,j+1}, w'_{r,j+1});$$
$$G = \tfrac{1}{2}[m_1(P_{r,j+1}) + m_1(P_{i,j})]x_{i,j} - \tfrac{1}{2}[m_2(P_{r,j+1})$$
$$+ m_2(P_{i+1,j})]x_{i+1,j} + t_{i+1,j} - t_{i,j};$$

$$x^{(k+1)}_{r,j+1} = \cfrac{G}{\tfrac{1}{2}[m_1(P_{i,j}) - m_2(P_{i+1,j}) \\ + m_1(P_{r,j+1}) - m_2(P_{r,j+1})]};$$

$$t^{(k+1)}_{r,j+1} = \tfrac{1}{2}[m_1(P_{r,j+1}) + m_1(P_{i,j})](x^{(k+1)}_{r,j+1} - x_{i,j}) + t_{i,j}.$$

Step 11   Set $k = k + 1$.

Step 12   If $|x^{(k)}_{r,j+1} - x^{(k-1)}_{r,j+1}| < TOL$ and $|t^{(k)}_{r,j+1} - t^{(k-1)}_{r,j+1}| < TOL$
then set $x_{r,j+1} = x^{(k)}_{r,j+1}$;
$$t_{r,j+1} = t^{(k)}_{r,j+1};$$
$$r = r + 1;$$
$$FLAG = 0.$$

Step 13   If $k > KBOUND$ then
OUTPUT ('Maximum number of iterations exceeded');
STOP.

Step 14   If j is odd then set $x_{M,j} = l$;
$$t_{M,j} = t_{M-1,j};$$
$$w_{M,j} = 0;$$
$$\dot{w}_{M,j} = g(l);$$
$$w'_{M,j} = f'(l);$$
$$P_{M,j} = (x_{M,j}, t_{M,j}, w_{M,j}, \dot{w}_{M,j}, w'_{M,j});$$
else set $x_{0,j} = 0$;
$$t_{0,j} = t_{1,j};$$
$$w_{0,j} = 0;$$
$$\dot{w}_{0,j} = g(0);$$
$$w'_{0,j} = f'(0);$$
$$P_{0,j} = (x_{0,j}, t_{0,j}, w_{0,j}, \dot{w}_{0,j}, w'_{0,j}).$$

Step 15   For $i = 0, \ldots, M$
OUTPUT $(j, x_{i,j}, t_{i,j}, w_{i,j}, \dot{w}_{i,j}, w'_{i,j})$.

Step 16   STOP.   (*Procedure is complete.*)

If the values $t_{i,N}$ are not large enough for each $i = 0, 1, \ldots, M$, it is suggested that the procedure be continued until sufficiently large t values are obtained. The reason for using the "switch" variable r in Step 4 is to ensure that the characteristic mesh will remain distributed throughout $\{(x, t) | 0 \leq x \leq l, 0 \leq t \leq t_N\}$ (see Fig. 11.18).

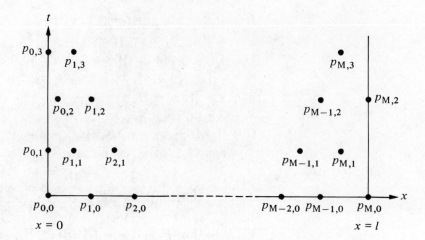

FIGURE 11.18

EXAMPLE 1   The Method of Characteristics (Algorithm 11.6) with $M = 4$ and $N = 2$ will be used to approximate the solution to the hyperbolic partial-differential equation

$$\frac{\partial^2 u}{\partial t^2} - \frac{\partial^2 u}{\partial x^2} - 50\pi^2 t \frac{\partial u}{\partial t} - 50\pi^2 u = 0, \qquad 0 < x < .2, \quad 0 < t,$$

subject to the constraints

$$u(0, t) = u(.2, t) = 0, \qquad 0 < t,$$
$$u(x, 0) = 0, \qquad 0 \le x \le .2,$$
$$u_t(x, 0) = \sin 10\pi x, \qquad 0 \le x \le .2.$$

TABLE 11.7

j	r	$x_{r,j+1}$	$t_{r,j+1}$	$w_{r,j+1}$	$\dot{w}_{r,j+1}$	$w'_{r,j+1}$	$u(x_{r,j+1}, t_{r,j+1})$	$\|w_{r,j+1} - u(x_{r,j+1}, t_{r,j+1})\|$
0	1	.025	.025	.012500	.50000	.50000	.017678	.0051
0	2	.075	.025	.014428	.65421	−.50000	.017678	.0032
0	3	.125	.025	−.012500	−.50000	−.50000	−.017678	.0051
0	4	.175	.025	−.014428	−.65421	.50000	−.017678	.0032
0	0	0	.025	0	0	0	0	0
1	0	.0125	.0375	.0096160	.53855	.53855	.014351	.0047
1	1	.05	.05	.032934	.48047	.09494	.050000	.0171
1	2	.1	.05	.0067491	.38572	−.01249	0	.0067
1	3	.15	.05	−.032934	−.48047	−.09494	−.050000	.0171
1	4	.2	.05	0	0	0	0	0
2	1	.0375	.0625	.036149	.66770	.21686	.057742	.0216
2	2	.075	.075	.039186	.44245	−.71986	.053033	.0138
2	3	.125	.075	−.012718	.65443	−1.3611	−.053033	.0403
2	4	.175	.075	−.029987	−.79387	.36844	−.053033	.0230
2	0	0	.0625	0	0	0	0	0

For this problem,

$$a_1(x, t, u, u_x, u_t) = 1$$

$$a_2(x, t, u, u_x, u_t) = 0$$

$$a_3(x, t, u, u_x, u_t) = -1$$

$$a_4(x, t, u, u_x, u_t) = 50\pi^2(u + tu_t),$$

and it is easy to verify that $u(x, t) = t \sin 10\pi x$ is the solution. The results of the algorithm are listed in Table 11.7.   ◻

Exercise Set 11.6

1. Approximate the solution $u(x, t) = \sin \pi x \cos \pi t$ to the wave equation

$$\frac{\partial^2 u}{\partial t^2} - \frac{\partial^2 u}{\partial x^2} = 0, \quad 0 < x < 1, \quad 0 < t,$$

$$u(0, t) = u(1, t) = 0, \quad 0 < t,$$

$$u(x, 0) = \sin \pi x, \quad 0 \le x \le 1,$$

$$\frac{\partial u}{\partial t}(x, 0) = 0, \quad 0 \le x \le 1.$$

using the method of characteristics with $h = .1$ and also with $h = .05$. Compare your results to the exact solution and to the results obtained in Exercise 3 of Section 11.4.

2. Approximate the solution to the wave equation

$$\frac{\partial^2 u}{\partial t^2} - \frac{\partial^2 u}{\partial x^2} = 0, \quad 0 < x < 1, \quad 0 < t,$$

$$u(0, t) = u(1, t) = 0, \quad 0 < t,$$

$$u(x, 0) = \sin 2\pi x, \quad 0 \le x \le 1,$$

$$\frac{\partial u}{\partial t}(x, 0) = 2\pi \sin 2\pi x, \quad 0 \le x \le 1,$$

using the method of characteristics with $h = .1$. Compare your results to those obtained in Exercise 5 of Section 11.4 and to the exact solution.

3. Approximate the solution to the following hyperbolic partial-differential equation, using the method of characteristics:

$$\frac{\partial^2 u}{\partial t^2} - \frac{\partial^2 u}{\partial x^2} = 0, \quad 0 < x < 1, \quad 0 < t,$$

$$u(0, t) = u(1, t) = 0, \quad 0 < t,$$

$$u(x, 0) = \begin{cases} 1, & 0 \le x \le \frac{1}{2}, \\ -1, & \frac{1}{2} < x \le 1, \end{cases}$$

$$\frac{\partial u}{\partial t}(x, 0) = 0, \quad 0 \le x \le 1.$$

Use $h = .1$, and compare your answer to the Fourier series solution and to the results obtained in Exercise 6 of Section 11.4.

4. Approximate the solution to the following hyperbolic partial differential equation,

$$\frac{\partial^2 u}{\partial t^2} + \frac{\partial^2 u}{\partial x\, \partial t} - \frac{\partial^2 u}{\partial x^2} + 2t - 2x + 1 = 0, \qquad 0 < x < 1, \quad 0 < t.$$

$$u(0, t) = u(1, t) = 0, \qquad 0 < t,$$

$$u(x, 0) = 0, \qquad 0 \le x \le 1,$$

$$\frac{\partial u}{\partial t}(x, 0) = x^2 - x, \qquad 0 \le x \le 1.$$

Use $h = .1$, and compare your answers to the solution $u(x, t) = tx(x - 1)$.

Answers to Selected Exercises

CHAPTER 1

Exercise Set 1.1 (page 8)

1. Consider $f'(0)$ and $f'(1)$.

3. $P_4(x) = 1 + x - \dfrac{x^3}{3} - \dfrac{x^4}{6}$, $P_4\left(\dfrac{\pi}{16}\right) = 1.19357852$, $\left| R_5\left(\dfrac{\pi}{6}\right) \right| \leq 6.0 \times 10^{-6}$

5. $P_3(x) = e^{-1}[1 - (x - 1) + (x - 1)^2/2 - (x - 1)^3/6]$, $e^{-.99} \approx P_3(.99) = .3715766909$, $|R_3(.99)| < 5 \times 10^{-10}$.

7. a) No, $f'(0)$ does not exist.   b) Yes, $c = \frac{64}{27}$.
   c) No, $f'(0)$ does not exist.   d) Yes, $c = \frac{1}{2}$.

Exercise Set 1.2 (page 18)

1. a) $3.14002 < x < 3.14316$

3. a) 14.2, relative error 1.3×10^{-4} c) 20, relative error 4.4×10^{-3}

9. a) -1.827 b) 6.959×10^{-3}

11. 1.0000, .33333, .11111, .037037, .012346, .0041153, .0013718, .00045727, .00015242

13. 86.625, 39.375, 119.50, 71.50

CHAPTER 2

Exercise Set 2.1 (page 25)

1. Consider $f''(x)$ on $[1, 2]$.   3. 2.93

5. 7, 1.320

Exercise Set 2.2 (page 33)

1. .6412, 11 iterations required. With $k = .551$, $p_0 = .5$ Corollary 2.5 gives 15 iterations as a bound.

3. 1.32472   5. 2.92

7. a) $g(x) = \left(\dfrac{e^x}{3}\right)^{1/2}$ on $[0, 1]$; .91001

Exercise Set 2.3 (page 41)

5. b) .91001   d) 1.96887   7. 1.89549

13. a) 1.31907, 15 iterations   c) 1.82938, 8 iterations

15. .1010, 4 iterations, 2,187,946   17. 6.74%, 6.69%

19. 222334, 248476 (in thousands)   21. a) 2.10639, 4.51469, 7.22928, 10.12546

Exercise Set 2.4 (page 53)

1. 1.3247   **3.** 1.732   **5.** 1.89549

7.

	10th iteration	
m	Single precision	Double precision
1	$-.5664420$	$-.5664420$
2	$-.5705472$	$-.5671433$
3	$-.5664667$	$-.5667686$

The correct answer is $-.5671433$.

Exercise Set 2.5 (page 70)

1. 400

7. b) Root in $[\frac{10}{3}, 5]$. Since $P(.2) \approx -.086$, $P(.3) \approx -.0093$, $P(.4) \approx -.009$, and $P(.5) \approx -.079$, there appears to be a root in $[.3, .4]$, which there is at .35.

9. Using the stopping technique of $|e_i^{(k)}| < 10^{-6}$ for some k or when $i = 30$ gave the following results:
a) approximate roots: $-2.49998 \pm 1.32297i$, 4.12307, -4.12310
c) approximate roots: $-.500008 \pm .866030i$, $-7.5 \times 10^{-6} \pm 1.41420i$

CHAPTER 3

Exercise Set 3.1 (page 79)

1. a) $(x - 1)^2 + 2(x - 1) - 2$

3. .04761875. Error bound 3.1×10^{-7}, actual error 3.0×10^{-7}.

5. 8.86×10^{-7} **7.** .0953083   **11.** 2×10^6

Exercise Set 3.2 (page 87)

5. 1.75496. Error bound 1.2×10^{-4}, actual error -9×10^{-5}.

7. 3.05348. Error bound 1.19×10^{-4}, actual error -1.14×10^{-4}.

9. a) 1.32436   b) 2.18350   c) 1.15278, 2.01191

15. The predictions are: for 1910, 31764; for 1965, 193271; for 2000, 55394.

Exercise Set 3.3 (page 92)

1. $P_{0,1,2,3,4} = .4980705$ **3.** .198269   **5.** 1.75496

7. .8095   **9.** $1.708\overline{3}$ **11.** 1.75496

13. .567142   **17.** a) ± 4.12311,   c) No real roots

Exercise Set 3.4 (page 100)

3. 1.05126   **5.** 1.56830.

Exercise Set 3.5 (page 106)

1. .4980704   3. .80903236

5. a) 1.2840246   b) 8.3×10^{-6}

7. 1.169016, 1.169080, 4.43×10^{-4}, 4.82×10^{-5}

Exercise Set 3.6 (page 119)

1. $\int_0^1 S(x)\,dx = .000000$, $S'(.5) = -3.24264$, $S''(.5) = -.000019$

3. $\int_0^1 s(x)\,dx = .000000$, $s'(.5) = -3.134446$, $s''(.5) = -.000021$

5. .492174, .497627 (.498070 using the derivatives from section 3.5.)

7. b) .11070136   c) 1.59×10^{-7} d) 1.04113925, 1.04081053

9. The equation of the spline is

$$S(x) \equiv S_i(x) = f_i + b_i(x - x_i) + c_i(x - x_i)^2 + d_i(x - x_i)^3$$

on the interval $[x_i, x_{i+1}]$, where the results in Table A.1 are obtained.

TABLE A.1

x_i	f_i	b_i	c_i	d_i
.9	.7	3.541214	−23.11823	77.06073
.8	.3	2.605763	32.47269	−185.3031
.7	.2	1.035753	−16.77264	164.1511
.6	0	2.251228	4.617891	−71.3018
.5	−.2	1.959329	−1.698893	21.05594
.4	−.4	1.911449	2.177681	−12.92191
.3	−.5	−.6051273	22.98808	−69.36810
.2	−.3	−2.490925	−4.130146	90.39406
.1	−.1	−1.431164	−6.467466	7.79107
0	0	−.7844177	0	21.55821

19. The equation of the spline is

$$S(x) \equiv S_i(x) = f_i + b_i(x - x_i) + c_i(x - x_i)^2 + d_i(x - x_i)^3$$

on the interval $[x_i, x_{i+1}]$ where the results in Table A.2 are obtained.

TABLE A.2

x_i	Sample 1				Sample 2			
	f_i	b_i	c_i	d_i	f_i	b_i	c_i	d_i
0	6.67	−.44687	0	.06176	6.67	1.6629	0	−.00249
6	17.33	6.2237	1.1118	−.27099	16.11	1.3943	−.04477	−.03251
10	42.67	2.1104	−2.1401	.28109	18.89	−.52442	−.43490	.05916
13	37.33	−3.1406	.38974	−.01411	15.00	−1.5365	.09756	.00226
17	30.10	−.70021	.22036	−.02491	10.56	−.64732	.12473	−.01113
20	29.31	−.05069	−.00386	.00016	9.44	−.19955	.02453	−.00102

21. The free cubic spline predicts the populations to be 88,219; 205,013; and 524,575.

Chapter 4

Exercise Set 4.1 (page 133)

1. Using Eq. (4.12), $f'(1.8) \approx 6.049444$ with error 2.04×10^{-4}.
   Using Eq. (4.13), $f'(1.8) \approx 6.049748$ with error 1.01×10^{-4}.

3. $-3.10457, 3.98 \times 10^{-2}$ 5. $2.27403, 2.27510$

7. a) $f'(1.005) \approx 5.0, f(1.015) \approx 6.0$ b) $f''(1.01) \approx 100$
   c) f' accurate within 1.0, f'' accurate within 200.

9. $f'(x_0) = \dfrac{1}{6h} [-8f(x_0) + 9f(x_0 + h) - f(x_0 + 3h)];$

   $f'(1.8) = 6.049342$ with error 3.05×10^{-4}

13. b) The equation of the spline is $s(x) = S_i(x) \equiv f_i + b_i(x - x_i) + c_i(x - x_i)^2 + d_i(x - x_i)^3$
    on the interval $[x_i, x_{i+1}]$, the results in Table A.3 are obtained.

TABLE A.3

x_i	f_i	b_i	c_i	d_i
.9	.7	4.652147	-3.876282	-226.4519
.8	.3	2.308093	27.31677	-103.9769
.7	.2	1.115502	-15.39093	142.3590
.6	0	2.229899	4.246964	-65.45966
.5	$-.2$	1.964892	-1.596888	19.47949
.4	$-.4$	1.910522	2.140592	-12.45826
.3	$-.5$	$-.6069813$	23.03444	-69.64616
.2	$-.3$	-2.48258	-4.278495	91.04306
.1	$-.1$	-1.462688	-5.920434	5.473132
0	0	$-.6666667$	-2.039782	-12.93550

Exercise Set 4.2 (page 137)

1. $0.0, f(x)$ is symmetric about $x = .5, 5.07 \times 10^{-1}$

3. 6.049680. Error 3.3×10^{-5} 5.   6.0496675. Error 2.0×10^{-5}

7. 6.05000. Error 3.5×10^{-4} due mainly to round-off error

9. 6.0508125. Error 1.165×10^{-3}

Exercise Set 4.3 (page 145)

1. Trapezoidal rule gives .023208, with an error of .01160. Simpson's rule gives .032296, with an error of .00252. The error bounds cannot be calculated because $\lim_{x \to 0^+} f''(x)$ and $\lim_{x \to 0^+} f^{(4)}(x)$ are both infinite.

3. Trapezoidal rule gives .785398, with an error bound of 6.5×10^{-1}, but no actual error. Simpson's rule gives the same result with an error bound of 1.3×10^{-1}.

5. The closed form gives .76692. The open form gives .76658.

7. a) The results in the order requested are: .102459819, .102459821, .102459822, .102466280, .102459824. The actual value is .102459822.
   c) The results in the order requested are: 1.477536117, 1.477528858, 1.477523049, 1.470981472, 1.477511615. The actual value is 1.477523046.

9. The errors in order are $1.6 \times 10^{-6}, 2.7 \times 10^{-8}, -6.7 \times 10^{-7}, 1.6 \times 10^{-6}, -1.8 \times 10^{-6}$.

Exercise Set 4.4 (page 151)

1. a) 1.1167, exact value 1.09861. c) 10.5869, exact value 10.20759.
   e) −5.9568, exact value −6.28319.

3. $m = 25$, $h = .04$. The approximation with these values is 10.95017032, with is correct to the places listed.

5. Using $h = .0002$ gives 10.95017045, although $h < .00039$ will suffice.

7. .907937

9. a) $f^{(3)}$ is discontinuous at $x = .1$. b) .302506. Error bound 1.9×10^{-4}.
   c) .302625. Actual value .302425.

11. a) 2.6209 sec. b) 2.6197 sec. c) 2.6197 sec.

13. $m = 64$ gives 58.47047.   15. b) 2.925304 c) 2.888285

17. 6.945 miles/sec.

Exercise Set 4.5 (page 159)

1. a) 1.0999 c) 10.2013 e) −6.2840   3. 14.025820   5. .7499954

7. Adaptive quadrature: $.25332 \times 10^{-6}$; Simpson's Composite method: $-.45253 \times 10^{-6}$.

Exercise Set 4.6 (page 164)

1. a) 1.0993 c) 10.2046 e) −3.5092   3. $R_{5,5}$ is 10.9501703.

5. $R_{10,10}$ is 14.02585093, which is exact to the places listed.   9. 58.47047

Exercise Set 4.7 (page 171)

1. a) 1.09804 c) 10.21063 e) −5.52485

3. 11.141495, 10.948403, 10.950140

7.

n	Roots	Coefficients
2	.5857864	.8535534
	3.4142136	.1464466
3	.4157746	.7110930
	2.2942804	.2785177
	6.2899451	.0103893

When $n = 2$, the approximation is .43246. When $n = 3$, the approximation is .49603. The exact value is .5.

Exercise Set 4.8 (page 177)

1. .9059444   3. .1479099   5. .6128014
9. 1.469840   11. $\bar{x} = .380644$, $\bar{y} = .382249$

CHAPTER 5

Exercise Set 5.1 (page 185)

1. a) Lipschitz constant $L = 1$; well-posed problem
   c) Lipschitz constant $l = \frac{3}{8}\sqrt{3}$; well-posed problem
3. $y(t) = t^2(e^t - e)$

Exercise Set 5.2 (page 192)

1. a)

i	t_i	w_i
1	1.1	1.2
2	1.2	1.4281

c)

i	t_i	w_i
1	1.5	-1.0
2	2.0	-1.0
3	2.5	-1.0
4	3.0	-1.0

3. a) Representative values:

i	t_i	w_i
2	1.2	.684755
5	1.5	3.18744
7	1.7	6.46640
10	2.0	15.3982

$h \approx .00064$ for absolute error of .1

c) Representative values:

With $h = .2$

i	t_i	w_i
5	1.0	1.32768
15	3.0	3.03518
25	5.0	5.00378

With $h = .1$

i	t_i	w_i
10	1.0	1.34868
30	3.0	3.14055
50	5.0	5.03050

5. Representative values:

i	t_i	w_i
10	1.0	1.20319
20	2.0	2.14982

7. b) $w_{50} = .10430 \approx p(50)$
   c) $p(t) = 1 - .99e^{-.002t}$; $p(50) = .10421$

Exercise Set 5.3 (page 199)

1. a)

i	t_i	w_i
1	.5	.5
2	1.0	1.07686

c)

i	t_i	w_i
1	.5	0
2	1.0	.25
3	1.5	1.0
4	2.0	2.5

3. a) Representative values:

Taylor method of order 2

i	t_i	w_i
10	1.0	-2.0
20	2.0	-3.0

Taylor method of order 4

i	t_i	w_i
10	1.0	-2.0
20	2.0	-3.0

5. $y(5) \approx w_{25} = 5.0134730$

7. a) Representative values:

Taylor method of order 2

i	t_i	w_i
2	.2	5.86595
5	.5	2.82145
7	.7	.84926
10	1.0	-2.09503

Taylor method of order 4

i	t_i	w_i
2	.2	5.86433
5	.5	2.81789
7	.7	.84455
10	1.0	-2.10154

b) .8 seconds

Exercise Set 5.4 (page 207)

1. a)

i	t_i	w_i
1	1.1	1.21405
2	1.2	1.46302

c)

i	t_i	w_i
1	1.5	-1.5
2	2.0	-1.33594
3	2.5	-1.25246
4	3.0	-1.20209

3. Representative values:

i	t_i	Midpoint method	Modified Euler method	Heun's method
10	1.0	1.2530068	1.2536129	1.2532123
20	2.0	2.0963063	2.0958129	2.0961403

5. $y(5) = 5.0134759$, $w_{25} = 5.0134720$

7. a) Representative values:

t	Euler's method	Heun's method	Runge–Kutta 4
1.5	$-.6578870$	$-.6664112$	$-.6666640$
2.0	$-.4855158$	$-.4996025$	$-.4999962$

c) Representative values:

t	Euler's method	Heun's method	Runge–Kutta 4
.5	.6004670	.6190949	.6178911
1.0	1.1176186	1.1365805	1.1353475

9. b) 7.9787 feet

Exercise Set 5.5 (page 214)

1. a)

i	t_i	w_i	h_i
1	1.17783	1.40833	.17783

c)

i	t_i	w_i	h_i
1	1.17783	-1.73767	.17783
2	1.64199	-1.43898	.46416
3	2.14199	-1.30525	.5
4	2.64199	-1.23397	.5

3. a) Representative results:

i	t_i	w_i	h_i
2	.568817	.128956	.239108
4	1.02998	.390269	.231338
6	1.59984	.852058	.316820
8	1.992820	.999971	.101749
11	2.54344	1.34419	.239916
13	2.99257	2.67965	.180949

c) Representative results:

i	t_i	w_i	h_i
5	.0534694	.0258533	.0111953
8	.0969589	.0120053	.0165621
12	.194914	.0379899	.0293748
19	.414326	.171637	.0314950
25	.603228	.363855	.0314687
31	.792007	.627246	.0314658
37	.980881	.962098	.0314338

5. $z(30) \approx 81$, $x(30) \approx 19{,}592$, $y(30) \approx 80{,}327$

Exercise Set 5.6 (page 228)

1. a)

i	t_i	w_i
1	1.05	1.10386
2	1.10	1.21588
3	1.15	1.33684
4	1.20	1.46757

c)

i	t_i	w_i
1	1.5	-1.49541
2	2.0	-1.33056
3	2.5	-1.24804
4	3.0	-1.17988

3. a) Representative results

i	t_i	w_i
2	.2	.1812691
5	.5	.3934668
7	.7	.5034131
10	1.0	.6321199

c) Representative results

i	t_i	w_i
5	1.25	1.206333
10	1.50	3.967653
15	1.75	9.298721
20	2.00	18.683089

5. a) Unique solution for $0 < h < \frac{24}{9}(\max_{0 < t < .2} e^{y(t)})^{-1}$

b) Representative results are:

i	t_i	w_i
10	.1	1.317218
20	.2	1.784511

c) Newton's method will reduce the number of iterations per step from 4 to 3, using the stopping criteria $|w_i^{(k)} - w_i^{(k-1)}| \le 10^{-6}$.

7. Representative results:

i	t_i	w_i	h_i
8	.00921686	.409912	.00115211
18	.0271715	.993578	.00263456
28	.0616987	1.80532	.00427090
38	.141701	3.11764	.0111251
48	.326033	5.11508	.0258907
58	.772227	7.92086	.0642137
68	1.96140	8.80528	.160687

Exercise Set 5.7 (page 236)

1. a)

i	t_i	w_i	h_i
1	1.1	1.21588	.1
2	1.2	1.46756	.1

c)

i	t_i	w_i	h_i
1	1.1	−1.83334	.1
2	1.3	−1.62504	.2
3	1.7	−1.41678	.4
4	2.1	−1.31259	.4
5	2.5	−1.25007	.4
6	2.9	−1.20838	.4

3. 56751

Exercise Set 5.8 (page 242)

1. a) Representative results:

i	t_i	$w_{1,i}$	$w_{2,i}$
2	.2	.632872	1.45160
5	.5	3.85385	4.46038
7	.7	10.8525	11.3490
10	1.0	49.2066	49.5744

c) Representative results:

i	t_i	$w_{1,i}$	$w_{2,i}$
2	.2	−.525560	−.640149
5	.5	−.693561	−.388708
7	.7	−.718129	.229078
10	1.0	−.353308	2.57897

e) Representative results:

i	t_i	w_{1i}	w_{2i}	w_{3i}
5	1	3.73186	4.18087	4.45810
10	2	11.3146	12.5027	13.7536
15	3	34.0455	37.3715	40.7377

g) Representative results:

i	t_i	w_{1i}	w_{2i}
5	.25	2.99512	$-.0572923$
10	.50	2.96354	$-.208374$
15	.75	2.88569	$-.422317$
20	1.0	2.74965	$-.669048$

5. Representative values for $h = .5$:

i	t_i	$w_{1,i}$	$w_{2,i}$
2	1	8716	1435
4	2	7907	2120
6	3	6666	2813

A stable solution is $x_1 = 833$, $x_2 = 1500$.

Exercise Set 5.9 (page 251)

5. b) $w_2 = .18065$ c) $w_{20} = -50.983 \approx y(.2)$,
$\quad\quad w_5 = .35785$ $\quad\quad\quad\quad w_{50} = -1.16 \times 10^{17} \approx y(.5)$,
$\quad\quad w_7 = .15342$ $\quad\quad\quad\quad w_{70} = -4.32 \times 10^{26} \approx y(.7)$
$\quad\quad w_{10} = -9.7414$

7. For $h = .1$: $\quad\quad w_{10} = .367883 \approx y(1) = .3678794$,
$\quad\quad\quad\quad\quad\quad\quad w_{100} = 3.75800 \approx y(10) = .0000454$;

For $h = .01$: $\quad\quad w_{100} = .367879 \approx y(1) = .3678794$,
$\quad\quad\quad\quad\quad\quad\quad w_{1000} = .677194 \approx y(10) = .0000454$

9. a) $4\varepsilon, 13\varepsilon, 40\varepsilon, 121\varepsilon, 364\varepsilon$ b) $w_i = \varepsilon$ for each $i = 2, \ldots, 6$

11. a) $p(\beta) = \left(1 - \dfrac{h\lambda}{3}\right)\beta^2 - \dfrac{4h\lambda}{3}\beta - \left(1 + \dfrac{h\lambda}{3}\right)$

b) If $hL < 0$, there is a root $\beta_i < -1$.

Exercise Set 5.10 (page 257)

3. a) Representative values:

i	t_i	h	Trap.	R–K4
5	.05	.01	.124715	.125129
10	.1	.01	.054810	.055113
20	.2	.075	.046024	.046106
25	.51	.075	.260097	.260669
30	.885	.075	.783220	.783769

b) Representative values:

i	t_i	Trap.	R–K4
5	.05	.4166256	.4178647
10	.1	.2342621	.2351730
20	.2	.2167388	.2169858
25	.51	.4881578	.4881463
30	.885	.7139033	.7737179

5. a) Representative values:

i	t_i	h_i	w_i
5	.05	.01	.136758
10	.1	.01	.064254
20	.2	.075	.049182
25	.51	.075	.263952
30	.885	.075	.786976

b) Representative values:

i	t_i	w_i
5	.05	.451852
10	.1	.261325
20	.2	.224716
25	.51	.487951
30	.885	.772567

CHAPTER 6

Exercise Set 6.1 (page 264)

1. a) $x_1 = -4, x_2 = 2, x_3 = 0, x_4 = 3$ c) $x_1 = 1, x_2 = 2, x_3 = 0$
e) $x_1 = 1, x_2 = 1, x_3 = 1$

3. a) $x_1 = -1, x_2 = 2, x_3 = 0, x_4 = 1$
c) Procedure fails since there are no solutions.

Exercise Set 6.2 (page 273)

1. a) $x_1 = 1.1875, x_2 = 1.8125, x_3 = .875$; row interchange necessary.
c) $x_1 = 1.5, x_2 = 2, x_3 = -1.2, x_4 = 3$; no row interchanges.
e) $x_1 = -.0317460016, x_2 = .5952376656, x_3 = -2.380951197, x_4 = 2.777776938$; no row interchanges.
g) $x_1 = -1, x_2 = 0, x_3 = 1$; no row interchanges.

5. a) $x_1 = x_2 = x_3 = x_4 = 1$ b) $x_1 = x_2 = x_3 = x_4 = 1$

Exercise Set 6.3 (page 285)

1. a) no inverse b) no inverse

c)
$$C^{-1} = \begin{bmatrix} \frac{1}{4} & 0 & 0 & 0 \\ -\frac{3}{14} & \frac{1}{7} & 0 & 0 \\ \frac{3}{28} & -\frac{11}{7} & 1 & 0 \\ -\frac{1}{2} & 1 & -1 & 1 \end{bmatrix}$$

11. c)
$$A^{-1} = \begin{bmatrix} 0 & 2 & 0 \\ 0 & 0 & 3 \\ \frac{1}{6} & 0 & 0 \end{bmatrix},$$
$(A^{-1})_{ij}$ is the number of beetles of age i necessary to produce one of the age j.

Exercise Set 6.4 (page 294)

1. a) $x_1 = .9260$, $x_2 = 14.29$; $x_1 = .8275$, $x_2 = 20.00$; $x_1 = .9260$, $x_2 = 14.29$.
   c) $x_1 = 1.196$, $x_2 = 1.152$; $x_1 = 1.153$, $x_2 = 1.153$; $x_1 = 1.153$, $x_2 = 1.153$.

3. a) $x_1 = 30.0$, $x_2 = .990$; $x_1 = 10.0$, $x_2 = 1.00$; $x_1 = 10.0$, $x_2 = 1.00$
   c) $x_1 = -10.0$, $x_2 = 45.4$, $x_3 = -36.3$; $x_1 = -9.93$, $x_2 = 46.1$, $x_3 = -37.0$;
   $x_1 = -10.3$, $x_2 = 47.7$, $x_3 = -38.5$.

7. $x_1 = .8275$, $x_2 = 20.00$ for 1(a); $x_1 = 1.153$, $x_2 = 1.153$ for 1(c);
   $x_1 = 10.0$, $x_2 = 1.00$ for 3(a); $x_1 = -9.63$, $x_2 = 45.0$, $x_3 = -36.0$ for 3(c).

Exercise Set 6.5 (page 301)

1. (d) is positive definite; (a) and (b) are singular.

11.
$$A = \begin{bmatrix} 1 & 0 & 0 & 0 \\ 2 & 1 & 0 & 0 \\ -1 & -3 & 0 & 1 \\ 3 & 4 & 1 & 0 \end{bmatrix} \begin{bmatrix} 1 & 1 & 0 & 3 \\ 0 & -1 & -1 & -5 \\ 0 & 0 & 3 & 13 \\ 0 & 0 & 0 & -13 \end{bmatrix}$$

$$= \begin{bmatrix} 1 & 0 & 0 & 0 \\ 0 & 1 & 0 & 0 \\ 0 & 0 & 0 & 1 \\ 0 & 0 & 1 & 0 \end{bmatrix} \begin{bmatrix} 1 & 0 & 0 & 0 \\ 2 & 1 & 0 & 0 \\ 3 & 4 & 1 & 0 \\ -1 & -3 & 0 & 1 \end{bmatrix} \begin{bmatrix} 1 & 1 & 0 & 3 \\ 0 & -1 & -1 & -5 \\ 0 & 0 & 3 & 13 \\ 0 & 0 & 0 & -13 \end{bmatrix}$$

Exercise Set 6.6 (page 314)

1. a)
$$L = \begin{bmatrix} 1 & 0 & 0 \\ 1.5 & 1 & 0 \\ 1.5 & 1 & 1 \end{bmatrix}, \quad U = \begin{bmatrix} 2 & -1 & 1 \\ 0 & 4.5 & 7.5 \\ 0 & 0 & -4 \end{bmatrix}$$

   c)
$$L = \begin{bmatrix} 1 & 0 & 0 \\ -.5 & 1 & 0 \\ 2 & 2 & 1 \end{bmatrix}, \quad U = \begin{bmatrix} 2 & -1.5 & 3 \\ 0 & -.75 & 3.5 \\ 0 & 0 & -8 \end{bmatrix}$$

3. a)
$$L = \begin{bmatrix} 1.414213 & 0 & 0 \\ -.7071069 & 1.224743 & 0 \\ 0 & -.8164972 & 1.154699 \end{bmatrix}$$

   c)
$$L = \begin{bmatrix} 2 & 0 & 0 & 0 \\ .5 & 1.658311 & 0 & 0 \\ -.5 & -.4522671 & 2.132006 & 0 \\ 0 & 0 & .9380833 & 1.766351 \end{bmatrix}$$

5. a) $x_1 = .5$, $x_2 = .5$, $x_3 = 1$ c) $x_1 = 1$, $x_2 = -1$, $x_3 = 0$

11. $x_i = 1$ for each $i = 1, 2, \ldots, 10$.   15. $x_i = 1$ for each $i = 1, 2, \ldots, 9$.

17. n square roots, $\frac{1}{6}(n^3 + 9n^2 + 2n)$ multiplications/divisions and $\frac{1}{6}(n^3 + 6n^2 - 7n)$ additions/subtractions.

CHAPTER 7

Exercise Set 7.1 (page 328)

1. $1.70784x + .89968$
3. $.841679x^2 + .876603x + 1$

5. $.22335x - .80283$. For minimal A, 406; for minimal D, 272. The A prediction is not reasonable.

7. $.1795x + 8.2084$
9. $3.87x + 25.70$

Exercise Set 7.2 (page 343)

1. a) $P_1(x) = -x + 2.8333$ c) $P_1(x) = -.2958x + 1.1410$
e) $P_1(x) = -2.4317x + 1.2159$

5. $S_n(x) = \sum_{k=1}^{n} \frac{2}{k\pi}(1 - (-1)^k)\sin kx$
7. Both answers are $-.08002232$.

11. $P_4(x) = \frac{1}{8}(35x^4 - 30x^2 + 3), P_5(x) = \frac{1}{8}(63x^5 - 70x^3 + 15x)$

Exercise Set 7.3 (page 351)

1. a) $\tilde{P}_2(x) = .5320x^2 + 1.1298x + 1$ b) $\tilde{P}_2(x) = -.2320x^2 + 1.3823x - 1.1458$

5. $P_6(x) = x + x^2 + \frac{1}{2}x^3 + \frac{1}{6}x^4 + \frac{1}{24}x^5 + \frac{1}{120}x^6$
$= .00026 + x + .99531x^2 + .5x^3 + .17917x^4 + .04167x^5 + \frac{1}{3840}T_6$

7. 1.04873

Exercise Set 7.4 (page 357)

1. $r(x) = \dfrac{1 + \frac{2}{5}x + \frac{1}{20}x^2}{1 - \frac{3}{5}x + \frac{3}{20}x^2}$
3. $r(x) = \dfrac{x - \frac{7}{60}x^3}{1 + \frac{1}{20}x^2}$

5. a) $r(x) = \dfrac{1}{1 + x + \frac{1}{2}x^2 + \frac{1}{6}x^3 + \frac{1}{24}x^4 + \frac{1}{120}x^5}$

c) $r(x) = \dfrac{1 - \frac{3}{5}x + \frac{3}{20}x^2 - \frac{1}{60}x^3}{1 + \frac{2}{5}x + \frac{1}{20}x^2}$

9. $r_T(x) = \dfrac{.88010T_1(x)}{T_0(x) + .044461T_2(x)}$

Exercise Set 7.5 (page 363)

1. $S_3(x) = -.496899 + .239206\cos(x) + 1.51539\cos(2x) + .239190\cos(3x)$
$- 1.15066\sin(x) + .00001278\sin(2x)$

3. a) $S_4(x) = -4.62637 + 6.67949\cos(x) - 3.701088\cos(2x) + 3.19008\cos(3x)$
$- 1.54212\cos(4x) + 5.95683\sin(x) - 2.46740\sin(2x)$
$+ 1.02203\sin(3x)$

b) $S_4(x) = -4.62637 + 6.67950\cos(x) - 3.70110\cos(2x) + 3.19008\cos(3x)$
$- 1.54212\cos(4x) + 5.95683\sin(x) - 2.46740\sin(2x)$
$+ 1.02204\sin(3x)$

5. a) $.197672$ b) $.223244$

CHAPTER 8

Exercise Set 8.1 (page 379)

5. a) $(0, 0, 0)^t$ c) $(1, -1, 1)^t$ 15. a) 3 c) 7 e) 3

19. $A = \begin{bmatrix} 1 & 1 \\ 0 & 1 \end{bmatrix}, \quad B = \begin{bmatrix} 1 & 0 \\ 1 & 1 \end{bmatrix}$ 21. (f)

27. a) $\lambda = 1$ and $\mathbf{x} = (6, 3, 1)^t$

Exercise Set 8.2 (page 394)

1. `a) $\mathbf{x}^{(2)} = (-23, 17, 35)^t$ c) $\mathbf{x}^{(2)} = (-1.625, 0, -2.1875)^t$

3. a) Jacobi diverges, Gauss–Seidel diverges
   c) Jacobi:

$$\mathbf{x}^{(8)} = (.98497, 1.9575, .98497, 1.96994, .97874, 1.96994)^t$$

   Gauss–Seidel:

$$\mathbf{x}^{(5)} = (.97408, 1.9778, .99054, 1.98434, .98662, 1.99429)^t$$

11. $\mathbf{x}^{(24)} = (-.6394611, -.3360750, -.2553595, -.2153484, -.2447643)^t$

Exercise Set 8.3 (page 402)

1. a) $\tilde{x}_1^{(5)} = 1.09$, $\tilde{x}_2^{(5)} = .488$, $\tilde{x}_3^{(5)} = 1.49$, $K_\infty(A) \approx 5.22$
   b) $\tilde{x}_1^{(6)} = 1.00$, $\tilde{x}_2^{(6)} = 1.01$, $K_\infty(A) \approx 18.7$
   d) $\tilde{x}_1^{(3)} = -227.0781$, $\tilde{x}_2^{(3)} = 47.69253$, $\tilde{x}_3^{(3)} = -17.76931$, $K_\infty(A) \approx 75.5$
5. $\tilde{\mathbf{x}} = (-1.0000, 2.00000)$; yes   7. $1.01(n^3 + 3n^2)$ is closer.

Exercise Set 8.4 (page 419)

1. a) $\mu^{(3)} = 4.999999$, $\mathbf{x}^{(3)} = (-.2599998, 1, -.2399999)^t$
   c) $\mu^{(3)} = 5.236839$, $\mathbf{x}^{(3)} = (1, .5678389, .155779, .487437)^t$
3. a) $\mu^{(3)} = 3.142854$, $\mathbf{x}^{(3)} = (.6767157, -.6767154, .2900211)^t$
   c) $\mu^{(3)} = 5.142565$, $\mathbf{x}^{(3)} = (.8373054, .37017710, .1939021, .3525497)^t$
9. a) $\mu^{(4)} = 4$, $\mathbf{x}^{(4)} = (1, -.40625, .2460937)^t$
   c) $\mu^{(10)} = 2.53703$, $\mathbf{x}^{(10)} = (.748606, .650065, 1)^t$
11. For $q = 7$: $\mu^{(8)} = -1.000028$, $\mathbf{x}^{(8)} = (-1.000028, -.7143087, .2500002)^t$;
       $q = 4$: $\mu^{(17)} = -1.000006$, $\mathbf{x}^{(17)} = (1.000013, .5000097, -1.000006)^t$;
       $q = 0$: $\mu^{(22)} = .4998214$, $\mathbf{x}^{(22)} = (-.00035603, -.00017801, .4998214)^t$
17. a) $\lambda_2 = -3$ c) $\lambda_2 = 1.480148$
21. a) $|\lambda| \le 6$ for all eigenvalues λ.
    b) $\lambda_1 = .6982681$, $\mathbf{x} = (1, .71606, .25638, .04602)^t$
    d) $P(\lambda) = \lambda^4 - \frac{1}{4}\lambda - \frac{1}{16}$;
       $\lambda_1 = .6976684972$,
       $\lambda_2 = -.237313308$,
       $\lambda_3 = -.2301775942 + .56965884i$,
       $\lambda_4 = -.2301775942 - .56965884i$
    e) The beetle population should approach zero since A is convergent.
23. a) $\lambda_1 = -6$, $\lambda_2 = -5$, $\lambda_3 = -2$; yes.
    b) $\lambda_1 = -2$, $\lambda_2 = -1.106711$, $\lambda_3 = -3.94664 + .82970i$, $\lambda_4 = -3.94664 - .82970i$;
       yes.

Exercise Set 8.5 (page 436)

1. a) $\begin{bmatrix} 20.19511 & -.2439012 & 0 \\ -.2439012 & -.1951208 & 6.403123 \\ 0 & 6.403123 & 3.000000 \end{bmatrix}$

3. a) $\begin{bmatrix} 2.199999 & -.3999996 & 0 \\ -.3999996 & 2.799995 & 2.236067 \\ 0 & 2.236067 & -1.000000 \end{bmatrix}$

c) $\begin{bmatrix} 4 & -1.414213 & 0 & 0 \\ -1.414213 & 4 & 0 & 0 \\ 0 & 0 & 4 & -1.414213 \\ 0 & 0 & -1.414213 & 4 \end{bmatrix}$

5. a) 2, 6, 4, 7.46409, .535911 c) 6.84462, 1.084365, 2.268526, -2.197511

CHAPTER 9

Exercise Set 9.1 (page 447)

5. $x_1 = .772, x_2 = .420$ **9.** Yes; $x_1 = 8000, x_2 = 40000$

Exercise Set 9.2 (page 453)

1. a) $x_1 = 1, x_2 = 1$; Newton's method converges in 4 iterations.
  c) $x_1 = .5, x_2 = .8660254$

5. a) $k_1 = 8.77129, k_2 = .259695, k_3 = -1.37228$
  b) 3.19 inches

7. b) $a = 6.588, b = 29.840, c = 1.163$

Exercise Set 9.3 (page 460)

1. b) $x_1 = 0, x_2 = .1, x_3 = 1$ d) $x_1 = 1, x_2 = 2, x_3 = 0$

5.

i	θ_i	i	θ_i
1	.14062	11	.48348
2	.19954	12	.50697
3	.24522	13	.52980
4	.28413	14	.55205
5	.31878	15	.57382
6	.35045	16	.59516
7	.37990	17	.61615
8	.40763	18	.63683
9	.43397	19	.65726
10	.45920	20	.67746

CHAPTER 10

Exercise Set 10.1 (page 468)

1. (a), (d), (e), (g), (h), (i)

3.

i	x_i	w_i
1	.261799	$-.0844105$
2	.523599	$-.0819684$

5. a) Representative values:

i	x_i	w_{1i}
5	1.25	.6431423
10	1.50	.6832421
15	1.75	.6922685

9. Representative values for $h = .05$:

i	x_i	w_{1i}
5	.25	.08224765
10	.50	.006764673
15	.75	.000556349

Exercise Set 10.2 (page 475)

1. $w_1 = .405991 \approx \ln 1.5 = .405465$

3. a) Representative values:

i	x_i	w_{1i}
5	1.25	$-.7272908$
10	1.50	$-.8000371$
15	1.75	$-.8889473$

c) Representative values:

i	x_i	w_{1i}
2	1.2	.4545455
5	1.5	.4000000
7	1.7	.3703704

5. a) Representative values:

x_i	w_{1i}	w_{2i}
1.25	-0.7272728	-0.2644638
1.50	-0.8000001	-0.3200009
1.75	-0.8888891	-0.3950626
2.00	-1.0000000	-0.5000008

c) Representative values:

x_i	w_{1i}	w_{2i}
1.50	0.4000011	-0.1599981
2.00	0.3333333	-0.1111111

7. b) 1a) Representative values:

x_i	w_{1i}	w_{2i}
1.50	-0.8000017	-0.3200122
2.00	-1.000000	-0.5000095

1c) Representative values:

x_i	w_{1i}	w_{2i}
1.50	0.3999999	-0.1599964
2.00	0.3333333	-0.1111077

Exercise Set 10.3 (page 487)

1. a)

i	x_i	w_i
1	1.04720	.525382
2	2.09439	$-.525382$

3. a) Representative values: c) Representative values:

i	x_i	w_i
5	1.25	.16797186
10	1.50	.45842388
15	1.75	.60787334

i	x_i	w_i
2	.2	1.3642
5	.5	.95450
7	.7	.90014

5. a) Representative values:

i	x_i	w_i
5	1.25	.643282
10	1.50	.683328
15	1.75	.692302

9. a) Representative values: c) Representative values:

i	x_i	w_i
5	1.25	$-.7272810$
10	1.50	$-.8000128$
15	1.75	$-.8889005$

i	x_i	w_i
2	1.2	.4545563
5	1.5	.4000130
7	1.7	.3703798

Exercise Set 10.4 (page 501)

1. a) $\phi(x) = -.08057117\phi_1(x) - .07200348\phi_2(x)$

3. a) $c_1 = 1.4777309 \times 10^{-2}$ $c_6 = 5.8304182 \times 10^{-2}$
$c_2 = 2.8700968 \times 10^{-2}$ $c_7 = 5.4549063 \times 10^{-2}$
$c_3 = 4.0908778 \times 10^{-2}$ $c_8 = 4.4328659 \times 10^{-2}$
$c_4 = 5.0521349 \times 10^{-2}$ $c_9 = 2.6538926 \times 10^{-2}$
$c_5 = 5.6633300 \times 10^{-2}$

7. $.045900355 \sin \pi x - .007321101 \sin 2\pi x + .002286058 \sin 3\pi x - .0009829617 \sin 4\pi x$
$+ .0005077925 \sin 5\pi x$

13. $c_0 = 3.05828 \times 10^{-5}$ $c_6 = 3.86732$
$c_1 = 1.256566$ $c_7 = 3.28974$
$c_2 = 2.39014$ $c_8 = 2.39014$
$c_3 = 3.28974$ $c_9 = 1.25656$
$c_4 = 3.86732$ $c_{10} = 4.50001 \times 10^{-5}$
$c_5 = 4.06634$

CHAPTER 11

Exercise Set 11.2 (page 518)

1. $w_3 = 1.00 = u(.5, 1.5)$, $w_6 = .25 = u(.5, 1)$, $w_9 = 0 = u(.5, .5)$

3. a) Representative values: c) Representative values:

i	j	x_i	y_j	$w_{i,j}$
3	3	.3	.3	.089941
3	7	.3	.7	.209928
7	3	.7	.3	.209951
7	7	.7	.7	.489941

i	j	x_i	y_j	$w_{i,j}$
4	3	.8	.3	1.27136
4	7	.8	.7	1.75084
8	3	1.6	.3	1.61675
8	7	1.6	.7	3.06587

9. Representative values:

i	j	x_i	y_j	$w_{i,j}$
1	4	1.0	4.0	88
2	1	2.0	1.0	66
4	2	4.0	2.0	66

Exercise Set 11.3 (page 530)

1. a)

i	j	x_i	t_j	w_{ij}
1	1	.5	.05	.7709764
2	1	1.0	.05	1.090324
3	1	1.5	.05	.7709770
1	2	.5	.1	.8406147
2	2	1.0	.1	1.188807
3	2	1.5	.1	.8406162

3. a) Representative values for $h = .1$ and $k = .01$:

i	j	x_i	t_j	$w_{i,j}$
4	50	.4	.5	-9.3352×10^8
10	50	1.0	.5	-9.1860×10^8
17	50	1.7	.5	2.6047×10^8

5. a) Representative values for $h = .1$ and $k = .01$:

i	j	x_i	t_j	$w_{i,j}$
4	50	.4	.5	2.3541×10^{-9}
10	50	1.0	.5	6.2567×10^{-17}
17	50	1.7	.5	-3.8090×10^{-9}

13. a) Representative values:

i	j	r_i	t_j	$w_{i,j}$
6	50	.6	10	700.23
7	50	.7	10	450.42
8	50	.8	10	325.78
9	50	.9	10	417.30

Exercise Set 11.4 (page 539)

1.

i	j	x_i	t_j	w_{ij}
1	1	.25	.25	.17678
2	1	.50	.25	.25000
3	1	.75	.25	-.17678
1	2	.25	.50	-.75000
2	2	.50	.50	.00000
3	2	.75	.50	1.25000

3. Representative values for $h = .1$ and $k = .05$:

i	j	x_i	t_j	w_{ij}
2	10	.2	.5	.00285202
5	10	.5	.5	.00485216
7	10	.7	.5	.00392548

Representative values for $h = .05$ and $k = .1$.

i	j	x_i	t_j	w_{ij}
4	5	.2	.5	$-.0028789$
10	5	.5	.5	$-.0048977$
14	5	.7	.5	$-.0039623$

Representative values for $h = .05$ and $k = .05$:

i	j	x_i	t_j	w_{ij}
4	10	.2	.5	-1.6792×10^{-15}
10	10	.5	.5	-3.1780×10^{-15}
14	10	.7	.5	-2.8588×10^{-15}

5. Representative values:

i	j	x_i	t_j	w_{ij}
2	3	.2	.3	.6729902
5	3	.5	.3	6.317389×10^{-16}
7	3	.7	.3	$-.6729902$

7. a) $p(.5, .5) \approx .9$ and $p(.5, 1.0) \approx .9$

Exercise Set 11.5 (page 552)

1. With $E_1 = (.25, .75)$, $E_2 = (0, 1)$, $E_3 = (.5, .5)$, and $E_4 = (0, .5)$, basis functions are

$$\phi_1(x) = \begin{cases} 4x & \text{on } T_1 \\ -2 + 4y & \text{on } T_2 \end{cases} \qquad \phi_2(x) = \begin{cases} -1 - 2x + 2y & \text{on } T_1 \\ 0 & \text{on } T_2 \end{cases}$$

$$\phi_3(x) = \begin{cases} 0 & \text{on } T_1 \\ 1 + 2x - 2y & \text{on } T_2 \end{cases} \qquad \phi_4(x) = \begin{cases} 2 - 2x - 2y & \text{on } T_1 \\ 2 - 2x - 2y & \text{on } T_2 \end{cases}$$

and $\gamma_1 = .323825$, $\gamma_2 = 0$, $\gamma_3 = 1.0000$, and $\gamma_4 = 0$.

3. (See diagram on page 584)

$K = 8, N = 8, M = 32, n = 9, m = 25, NL = 0$;

$\gamma_1 = .511023 \qquad \gamma_6 = .720476$
$\gamma_2 = .720476 \qquad \gamma_7 = .507897$
$\gamma_3 = .507898 \qquad \gamma_8 = .720476$
$\gamma_4 = .720475 \qquad \gamma_9 = .511023$
$\gamma_5 = 1.01885 \qquad \gamma_i = 0, \quad 10 \le i \le 25$

$u(.125, .125) \approx .720475$
$u(.125, .25) \approx .690343$
$u(.25, .125) \approx .690343$
$u(.25, .25) \approx .720475$

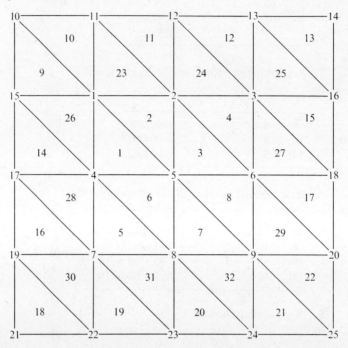

7.   (See diagram)

$K = 0, N = 20, M = 32, n = 20, m = 27, NL = 14$;

$\gamma_1 = 19.25965$	$\gamma_8 = 17.17999$	$\gamma_{15} = 15.30939$	$\gamma_{22} = 15$
$\gamma_2 = 17.12355$	$\gamma_9 = 19.29031$	$\gamma_{16} = 15.54661$	$\gamma_{23} = 15$
$\gamma_3 = 15.94424$	$\gamma_{10} = 21.47760$	$\gamma_{17} = 16.26654$	$\gamma_{24} = 15$
$\gamma_4 = 15.34842$	$\gamma_{11} = 19.28071$	$\gamma_{18} = 17.54011$	$\gamma_{25} = 15$
$\gamma_5 = 15.17034$	$\gamma_{12} = 17.45838$	$\gamma_{19} = 19.41416$	$\gamma_{26} = 15$
$\gamma_6 = 15.36329$	$\gamma_{13} = 16.21978$	$\gamma_{20} = 21.61106$	$\gamma_{27} = 15$
$\gamma_7 = 15.97682$	$\gamma_{14} = 15.52535$	$\gamma_{21} = 15$	

$u(1, 0) \approx 17.45838$

$u(4, 0) \approx 17.54011$

$u\left(\frac{5}{2}, \frac{\sqrt{3}}{2}\right) \approx 15.17034$

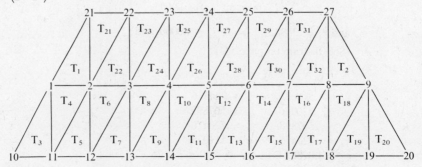

Exercise Set 11.6 (page 563)

1. Representative values for $h = .1$:

i	j	x_{ij}	t_{ij}	w_{ij}
2	10	.24375	.45625	.073624
4	10	.42500	.47499	.081627
6	10	.63749	.53749	$-.11264$
8	10	.84687	.54687	$-.17539$
10	10	1.0000	.54844	0

3. Representative values:

i	j	x_{ij}	t_{ij}	w_{ij}
3	2	.3	.1	1.00
7	2	.7	.1	-1.00
3	4	.3	.2	1.00
7	4	.7	.2	$-.875$

Bibliography

[1.] ABBOTT, M. B. (1966), *The method of characteristics*. Thames and Hudson, London; 243 pp.

[2.] ACTON, F. S. (1970), *Numerical methods that work*. Harper & Row, New York; 541 pp.

[3.] AHLFORS, L. V. (1966), *Complex analysis* (Second edition). McGraw-Hill, New York; 317 pp.

[4.] AHO, A. V., J. E. HOPCROFT, and J. D. ULLMAN (1974), *The design and analysis of computer algorithms*. Addison-Wesley, Reading, Mass.; 470 pp.

[5.] AMES, W. F. (1977), *Numerical methods for partial differential equations* (Second edition). Academic Press, New York; 365 pp.

[6.] BAILEY, N. T. J. (1967), *The mathematical approach to biology and medicine*. John Wiley & Sons, London; 296 pp.

[7.] BAILEY, N. T. J. (1957), *The mathematical theory of epidemics*. C. Griffin, London; 194 pp.

[8.] BAILEY, P. B., L. F. SHAMPINE, and P. E. WALTMAN (1968), *Nonlinear two-point boundary-value problems*. Academic Press, New York; 171 pp.

[9.] BARTLE, R. G. (1976), *The elements of real analysis* (Second edition). John Wiley & Sons, New York; 480 pp.

[10.] BEKKER, M. G. (1969), *Introduction to terrain vehicle systems*. University of Michigan Press, Ann Arbor, Mich.; 846 pp.

[11.] BERNADELLI, H. (1941), "Population Waves." *Journal of the Burma Research Society*, **31**, 1–18.

[12.] BIRKHOFF, G., and C. DE BOOR (1964), "Error bounds for spline interpolation." *Journal of Mathematics and Mechanics*, **13**, 827–836.

[13.] BIRKHOFF, G., and G. ROTA (1962), *Ordinary differential equations*. Blaisdell, Waltman, Mass.; 317 pp.

[14.] BRACEWELL, R. (1965), *The Fourier transform and its application*. McGraw-Hill, New York; 381 pp.

[15.] BRENT, R. (1973), *Algorithms for minimization without derivatives*. Prentice-Hall, Englewood Cliffs, N.J.; 195 pp.

[16.] BRIGHAM, E. O. (1974), *The fast Fourier transform*. Prentice-Hall, Englewood Cliffs, N.J.; 252 pp.

[17.] BROGAN, W. L. (1974), *Modern control theory*. Quantum Publishers, New York; 393 pp.

[18.] BROYDEN, C. G. (1965), "A class of methods for solving nonlinear simultaneous equations." *Mathematics of Computation*, **19**, 577–593.

[19.] BULIRSCH, R., and J. STOER (1966), "Numerical treatment of ordinary differential equations by extrapolation methods." *Numerische Mathematik*, **8**, 1–13.

[20.] BULIRSCH, R., and J. STOER (1968), "Fehlerabschätzungen und extrapolation mit rationalen Funktionen bei Verfahren von Richardson-typus." *Numerische Mathematik*, **6**, 413–427.

[21.] BULIRSCH, R., and J. STOER (1968), "Asymptotic upper and lower bounds for results of extrapolation methods." *Numerische Mathematik*, **8**, 93–104.

[22.] BUNCH, J. R., and D. J. ROSE, editors (1976), *Sparse matrix computations*. Proceedings of a conference held at Argonne National Laboratories, September 9–11, 1975. Academic Press, New York; 453 pp.

[23.]   BUTCHER, J. C. (1965), "On the attainable order of Runge–Kutta methods." *Mathematics of Computation*, **19**, 408–417.

[24.]   CANTONI, A., and P. BUTLER (1976), "Properties of the eigenvectors of persymmetric matrices with applications to communication theory." *IEEE Transactions on Communications*, Vol. Com-24, **8**, 804–809.

[25.]   CHEN, B. H. (1976), "Holdup and axial mixing in bubble columns containing screen cyclinders." *Industrial and Engineering Chemistry, Process Design and Development*, **15**, No. 1, 20–24.

[26.]   CHIARELLA, C., W. CHARLTON, and A. W. ROBERTS (1975), "Optimum chute profiles in gravity flow of granular materials: A discrete segment solution method." *Transactions of the ASME, Journal of Engineering for Industry*, Series B, **97**, 10–13.

[27.]   CHU, Y. (1970), *Introduction to computer organization.* Prentice-Hall, Englewood Cliffs, N.J.; 376 pp.

[28.]   COOLEY, J. W., and J. W. TUKEY (1965), "An algorithm for the machine calculation of complex Fourier series." *Mathematics of Computation*, **19**, No. 90, 297–301.

[29.]   DE BOOR, C., and B. SWARTZ (1973), "Collocation at Gaussian points." *SIAM Journal of Numerical Analysis*, **10**, No. 4, 582–606.

[30.]   DENNIS, J. E. JR., and J. J. MORÉ (1977), "Quasi-Newton methods, motivation and theory." *SIAM Review*, **19**, No. 1, 46–89.

[31.]   DETTMAN, J. W. (1965), *Applied complex variables.* Macmillan, New York; 481 pp.

[32.]   DORN, G. L. and A. B. BURDICK (1962), "On the recombinational structure of complementation relationships in the *m-dy* complex of the *Drosophila melanogaster*." *Genetics*, **47**, 503–518.

[33.]   ENRIGHT, W. H. (1974), "Optimal second derivative methods for stiff systems." *Stiff differential equations*, R. A. Willoughby, editor. Plenum Press, New York; 95–109.

[34.]   ENRIGHT, W. H., T. E. HULL, and B. LINDBERG (1975), "Comparing numerical methods for stiff systems of O.D.E.'s." *BIT*, **15**, 10–48.

[35.]   FEHLBERG, E. (1964), "New high-order Runge–Kutta formulas with step-size control for systems of first- and second-order differential equations." *Zeitschrift für Angewandte Mathematik und Mechanik*, **44**, 17–29.

[36.]   FEHLBERG, E. (1966), "New high-order Runge–Kutta formulas with an arbitrarily small truncation error." *Zeitschrift für Angewandte Mathematik und Mechanik*, **46**, 1–16.

[37.]   FEHLBERG, E. (1970), "Klassische Runge–Kutta Formeln vierter und niedrigerer Ordnung mit Schrittweiten-Kontrolle und ihre Anwendung auf Wärmeleitungs-probleme." *Computing*, **6**, 61–71.

[38.]   FIX, G. (1975), "A survey of numerical methods for selected problems in continuum mechanics." *Proceedings of a Conference on Numerical Methods of Ocean Circulation, National Academy of Sciences*, Durham, N. H., October 17–20, 1972, 268–283.

[39.]   FORSYTHE, G. E., and C. B. MOLER (1967), *Computer solution of linear algebraic systems*, Prentice-Hall, Englewood Cliffs, N.J.; 148 pp.

[40.]   FULKS, W. (1969), *Advanced calculus* (Second edition). John Wiley & Sons, New York; 597 pp.

[41.]   GEAR, C. W. (1971), *Numerical initial-value problems in ordinary differential equations.* Prentice-Hall, Englewood Cliffs, N.J.; 253 pp.

[42.]   GEORGE, J. A. (1973), "Nested dissection of a regular finite-element mesh." *SIAM Journal of Numerical Analysis*, **10**, No. 2, 345–362.

[43.]   GRAGG, W. B. (1965), "On extrapolation algorithms for ordinary initial-value problems," *SIAM Journal of Numerical Analysis*, **2**, 384–403.

[44.] HAMMING, R. W. (1973), *Numerical methods for engineers and scientists* (Second edition). McGraw-Hill, New York; 721 pp.

[45.] HENRICI, P. (1962), *Discrete variable methods in ordinary differential equations*. John Wiley & Sons, New York; 407 pp.

[46.] HENRICI, P. (1964), *Elements of numerical analysis*. John Wiley & Sons, New York; 328 pp.

[47.] HENRICI, P. (1963), *Error propagation for difference methods*. John Wiley & Sons, New York; 73 pp.

[48.] HILDEBRAND, F. B. (1974), *Introduction to numerical analysis* (Second edition). McGraw-Hill, New York; 511 pp.

[49.] HOUSEHOLDER, A. S. (1970), *The numerical treatment of a single nonlinear equation*. McGraw-Hill, New York; 216 pp.

[50.] HULL, T. E., and W. H. ENRIGHT (1976), "Test results on initial-value methods for nonstiff ordinary differential equations." *SIAM Journal of Numerical Analysis*, **13**, No. 6, 944–961.

[51.] HULL, T. E., W. H. ENRIGHT, B. M. FELLEN, and A. E. SEDGEWICK (1972), "Comparing numerical methods for ordinary differential equations." *SIAM Journal of Numerical Analysis*, **9**, No. 4, 603–637.

[52.] ISAACSON, E., and H. B. KELLER (1966), *Analysis of numerical methods*. John Wiley & Sons, New York; 541 pp.

[53.] JACKSON, K. R., W. H. ENRIGHT, and T. E. HULL (1978), "A theoretical criterion for comparing Runge–Kutta formulas." *SIAM Journal of Numerical Analysis*, **15**, No. 3, 618–641.

[54.] JACOBS, D., editor (1977), *The state of the art in numerical analysis*. Academic Press, New York; 978 pp.

[55.] JENKINS, M. A., and J. F. TRAUB (1970), "A three-stage algorithm for real polynomials using quadratic iteration." *SIAM Journal of Numerical Analysis*, **7**, No. 4, 545–566.

[56.] KELLER, H. B. (1968), *Numerical methods for two-point boundary-value problems*. Blaisdell, London; 184 pp.

[57.] KAMMERER, W. J., G. W. REDDIEN, and R. S. VARGA (1974), "Quadratic splines." *Numerische Mathematik*, **22**, 241–259.

[58.] LAMBERT, J. D. (1977), "The initial value problem for ordinary differential equations." *The state of the art in numerical analysis*, D. Jacobs, editor. Academic Press, New York; 451–501.

[59.] LARSON, H. J. (1974), *Introduction to probability theory and statistical inference* (Second edition), John Wiley & Sons, New York; 430 pp.

[60.] LUCAS, T. R., and G. W. REDDIEN, JR. (1972), "Some collocation methods for nonlinear boundary value problems." *SIAM Journal of Numerical Analysis*, **9**, No. 2, 341–356.

[61.] MITCHELL, A. R. (1969), *Computational methods for partial-differential equations*. John Wiley & Sons, London; 255 pp.

[62.] NA, T. Y., and G. M. KURAJIAN (1976), "Initial-curvature and lateral-load effects on thin struts with large elastic displacements." *Transactions of the ASME, Journal of Engineering for Industry*, Series B, **98**, 34–38.

[63.] NOBLE, B., and J. W. DANIEL (1977), *Applied linear algebra* (Second edition). Prentice-Hall, Englewood Cliffs, N.J.; 447 pp.

[64.] ORTEGA, J. M. (1972), *Numerical analysis—A second course*. Academic Press, New York; 193 pp.

[65.] ORTEGA, J. M., and W. C. RHEINBOLDT (1970), *Iterative solution of nonlinear equations in several variables*. Academic Press, New York; 572 pp.

[66.] PANDIT, S. M., T. L. SUBRAMANIAN, and S. M. WU (1975), "Modeling machine-tool chatter by time series." *Transactions of the ASME, Journal of Engineering for Industry*, Series B, **97**, 211–215.

[67.] RALSTON, A., and P. RABINOWITZ (1978), *A first course in numerical analysis* (Second edition). McGraw-Hill, New York; 556 pp.

[68.] RALSTON, A., and H. S. WILF, ed. (1960 and 1967), *Numerical methods for digital computers*. Vols. 1 and 2. John Wiley & Sons, New York; 293 + 287 pp.

[69.] RASHEVSKY, N. (1968), *Looking at history through mathematics*. Massachusetts Institute of Technology Press, Cambridge, Mass.; 199 pp.

[70.] ROSE, D. J., and R. A. WILLOUGHBY, editors (1972), "Sparse matrices and their applications." Proceedings of a conference held at IBM Research Center, New York, September 9–10, 1971. Plenum Press, New York; 215 pp.

[71.] RUSSELL, R. D. (1977), "A comparison of collocation and finite differences for two-point boundary value problems." *SIAM Journal of Numerical Analysis*, **14**, No. 1, 19–39.

[72.] SALE, P. F., and R. DYBDAHL (1975). "Determinants of community structure for coral-reef fishes in experimental habitat," *Ecology*, **56**, 1345–1355.

[73.] SAGAR, V., and D. J. PAYNE (1975), "Incremental collapse of thick-walled circular cyclinders under steady axial tension and torsion loads and cyclic transient heating." *Journal of the Mechanics and Physics of Solids*, **21**, No. 1, 39–54.

[74.] SCHOENBERG, I. J. (1946), "Contributions to the problem of approximation of equidistant data by analytic functions." *Quarterly of Applied Mathematics*, **4**, Part A, 45–99; Part B, 112–141.

[75.] SCHROEDER, L. A. (1973), "Energy budget of the larvae of the moth *Pachysphinx modesta*," *Oikos*, **24**, 278–281.

[76.] SCHROEDER, L. A. (1980), "Thermal tolerances and acclimation of two species of hydras." *Limnology and Oceanography*. To appear.

[77.] SCHULTZ, M. H. (1966), *Spline analysis*. Prentice-Hall, Englewood Cliffs, N.J.; 156 pp.

[78.] SEARLE, S. R. (1966), *Matrix algebra for the biological sciences*. John Wiley & Sons, New York; 296 pp.

[79.] SECRIST, D. A., and R. W. HORNBECK (1976), "An analysis of heat transfer and fade in disk brakes." *Transactions of the ASME, Journal of Engineering for Industry*, Series B, **98**, No. 2, 385–390.

[80.] SHAMPINE, L. F., and C. W. GEAR (1979), "A user's view of solving stiff ordinary differential equations." *SIAM Review*, **21**, No. 1, 1–17.

[81.] SHAMPINE, L. F., H. A. WATTS, and S. M. DAVENPORT (1976), "Solving nonstiff ordinary differential equations—the state of the art." *SIAM Review*, **18**, No. 3, 376–411.

[82.] SINGH, V. P. (1976), "Investigations of attentuation and internal friction of rocks by ultrasonics." *International Journal of Rock Mechanics and Mining Sciences*, 69–72.

[83.] SMITH, G. D. (1965), *Numerical solution of partial-differential equations*. Oxford University Press, Oxford; 179 pp.

[84.] STETTER, H. J. (1973), *Analysis of discretization methods for ordinary differential equations* from Tracts in natural philosophy. Springer-Verlag, Berlin; 388 pp.

[85.] STRANG, W. G., and G. J. FIX (1973), *An analysis of the finite element method*. Prentice-Hall, Englewood Cliffs, N.J.; 306 pp.

[86.] STROUD, A. H., and D. SECRIST (1966), *Gaussian quadrature formulas*. Prentice-Hall, Englewood Cliffs, N.J.; 384 pp.

[87.] VARGA, R. S. (1962), *Matrix iterative analysis*. Prentice-Hall, Englewood Cliffs, N.J.; 322 pp.

[88.]   WANG, H. T. (1975), "Determination of the accuracy of segmented representations of cable shape." *Transactions of the ASME, Journal of Engineering in Industry*, Series B, **97**, No. 2, 472–478.

[89.]   WENDROFF, B. (1966), *Theoretical numerical analysis*. Academic Press, New York; 239 pp.

[90.]   WILKINSON, J. H. (1963), *Rounding errors in algebraic processes*. H.M. Stationery Office, London; 161 pp.

[91.]   WILKINSON, J. H. (1965), *The algebraic eigenvalue problem*. Clarendon Press, Oxford; 662 pp.

[92.]   WILKINSON, J. H., and C. REINSCH (1971), *Handbook for automatic computation*. Volume 2: *linear algebra*. Springer-Verlag, Berlin; 439 pp.

[93.]   YOUNG, D. M. (1971), *Iterative solution of large linear systems*. Academic Press, New York; 570 pp.

[94.]   ZIENKIEWICZ, O. (1971), *The finite-element method in engineering science*. McGraw-Hill, London; 521 pp.

Index